British Dolichopodidae (Diptera)

RES INSECT IDENTIFICATION HANDBOOKS

Series editor: Dr Beulah Garner, Senior Curator of Small Orders, Natural History Museum, UK (b.garner@nhm.ac.uk)

The *RES Insect Identification Handbooks* series is published jointly by CABI and the Royal Entomological Society. It provides a vital resource on insect taxonomy for researchers, industry specialists and amateurs, providing identification keys and other biological information including on distribution and morphology.

This series builds upon Royal Entomological Society Collection Handbooks for the identification of British Insects, which was first published in 1949, and now includes handbooks on insect groups from outside the UK.

If you are interested in contributing to this prestigious series and would like to know more, please contact Ward Cooper, Senior Commissioning Editor, CABI (w.cooper@cabi.org) or Emilie Aimé, Head of Publishing, Royal Entomological Society (emilie@royensoc.co.uk) or visit www.royensoc.co.uk/handbook-proposal-form

RES INSECT IDENTIFICATION HANDBOOKS

British Dolichopodidae (Diptera)

C. Martin Drake

Photographs by Stuart G. Ball, illustrations by Martin Drake and Dawn Painter

Royal Entomological Society

CABI

CABI is a trading name of CAB International

CABI
Nosworthy Way
Wallingford
Oxfordshire OX10 8DE
UK

Tel: +44 (0)1491 832111
E-mail: info@cabi.org
Website: www.cabi.org

CABI
200 Portland Street
Boston
MA 02114
USA

Tel: +1 (617)682-9015
E-mail: cabi-nao@cabi.org

A catalogue record for this book is available from the British Library, London, UK.

ISBN-13: 9781800628335 (hardback)
9781800628342 (ePDF)
9781800628359 (ePub)

DOI: 10.1079/9781800628359.0000

Commissioning Editor: Ward Cooper
Editorial Assistant: Emma McCann
Production Editor: Rosie Hayden

Typeset by Straive, Pondicherry, India
Printed and bound in the UK by Halstan & Co. Ltd, Amersham

Contents

Colour plate section between pp. viii and 1.

Abstract

This Handbook covers British species of the family Dolichopodidae, popularly called long-legged flies. These often have an attractive appearance and striking behaviour so the larger species will be familiar to many naturalists. The sexes are often strongly dimorphic, males having their legs and wings modified for signalling during complex courtship displays. Both the larvae and adults are predators, with the exception of one genus whose larvae mine the leaves of monocotyledons, and may impact some pest insects. They have no medical importance.

Most are associated with wetlands from the sea-shore to montane pools, a small number have more xerophilic tendencies, and others are associated with rot-holes or bark beetles in trees.

Dolichopodids can play an important role in nature conservation assessment of sites as their habitat affinities, rarity status and distributions are fairly well known.

This Handbook covers the British fauna of 326 species in 47 genera placed in 12 subfamilies. One species so far found only in Ireland is also included. The limits of several subfamilies are poorly defined and, as phylogenetic relationships based on morphological and molecular data are also not always in agreement, the placement in this book of some genera differs from previous British checklists. In particular, some genera are assigned to the Peloropeodinae which has not been used previously in the British literature.

Owing to the male secondary sexual characteristics, males are easier to identify than females so separate keys are provided for nearly all genera. Some males can be separated reliably only by using genitalia characters which are therefore often included in the keys and are illustrated for most genera. In a few cases no reliable differences have been found for females.

Line drawings illustrate many of the characters used in the couplets and photographs show representative species of each genus. Most keys are derived *de novo* rather than updating those already existing. Species accounts briefly describe the most important characters of each genus and species to confirm the identification arrived at using the key, distinctions from similar species that may yet be found in Britain, the distribution in Britain and habitat affinities.

Acknowledgements

I am particularly pleased to have the skill of Dawn Painter for the Dolichopodinae genitalia drawings and Stuart Ball for his photography. I am most grateful to the Royal Entomological Society Publications Committee for their support in seeing this book through to publication, in particular Rebecca Farley-Brown in early years and Emilie Aimé in its final stages, and CABI editors Rosie Hayden and Ray Loughlin for making such a good job in producing the final version. I thank David Gibbs and Iain MacGowan for reviewing the draft. Many colleagues and friends have contributed in different ways. I would especially like to mention Andrew Cunningham, David Gibbs, Andrew Godfrey, Martin Greenland, Richard Lane, Colin Le Boutillier, Iain MacGowan, Ivan Perry, Donald Smith and Robert Wolton. Marc Pollet has been most helpful over the years in carefully reviewing my papers on dolichopodids. Dan Bickel, Zbyněk Kejval and Miroslav Barták kindly helped with difficult genera. At museums, Duncan Sivell and Zoe Adams at the Natural History Museum, London, Zoe Simmons and Robert Douglas at Oxford University Museum and Tony Hunter at the World Museum, Liverpool, welcomed me during my many visits and helped in loans of specimens. The species accounts would have been far less detailed without the contributions of numerous recorders to the national Empidid and Dolichopodid Recording Scheme. I would like to acknowledge the late Jon Cole's guidance in my early years with the study of the family.

I am grateful to the following organisations and publishers for permission to reproduce figures. E. Schweizerbart'sche Verlagsbuchhandlung OHG (www.schweizerbart.de/132062904), Negrobov, O.P. and Stackelberg, A.A. 1972. 29. Dolichopodidae. In: Lindner, E. (ed.) *Die Fliegen der palaearktischen Region*, Lieferung 302 Figs 486, 487, 498, 511–514, 518, 579, 588, 590, 592; Lieferung 303 Figs 644, 691, 692, 699, 749, 815–817. Negrobov, O.P. 1979. 29. Dolichopodidae. In: Lindner, E. (ed.) *Die Fliegen der palaearktischen Region* **4** (5), Lieferung 322 Fig. 1859. Royal Entomological Society for d'Assis-Fonseca, E.C.M. 1978. Diptera Orthorrhapha Brachycera Dolichopodidae. *Handbooks for the Identification of British Insects* **9** (5), Figs 92, 146; and Pollet, M. 1996. Systematic revision and phylogeny of the Palaearctic species of the genus *Achalcus* Loew (Diptera: Dolichopodidae) with the description of four new species. *Systematic Entomology* **21**, 353–386, Figs 11d and 11h. Dipterists Forum for numerous figures in *Dipterists Digest* in publications by C.M. Drake, Iain MacGowan and David Gibbs. British Entomological and Natural History Society for Cole, J.H. 1989. Two species of *Medetera* Fischer (Diptera, Dolichopodidae) new to Britain. *British Journal of Entomology and Natural History* **2**, 115–118, Fig. 2. *Irish Naturalists' Journal* for Chandler, P. 1989. *Campsicnemus dasycnemus* (Loew) (Diptera: Dolichopodidae): first record for the British Isles from Ireland. *Irish Naturalists' Journal* **23**, 17–19, Fig. 1.2.

Notes for Plates

Representative species of each genus. Males unless stated otherwise.

Scale bar = 1mm

Plate 1

Achalcus flavicollis

Acropsilus niger female *

Anepsiomyia flaviventris

Aphrosylus celtiber

Argyra leucocephala

Argyra leucocephala

Argyra perplexa

Asyndetus latifrons

Australachalcus melanotrichus *

* Specimens stored in alcohol resulting in faded colours and overall appearance not like a fresh specimen.

Plate 2

Campsicnemus compeditus

Campsicnemus picticornis

Campsicnemus scambus

Chrysotimus molliculus male

Chrysotimus molliculus female

Chrysotus blepharosceles

Chrysotus gramineus

Corindia viridis female

Cyrturella albosetosa

Plate 3

Diaphorus nigricans

Diaphorus oculatus

Dolichophorus kerteszi

Dolichopus atratus

Dolichopus brevipennis

Dolichopus diadema

Dolichopus discifer

Dolichopus picipes

Dolichopus urbanus

Plate 4

Ethiromyia chalybea

Gymnopternus assimilis

Gymnopternus cupreus

Hercostomus chetifer

Hercostomus gracilis

Hercostomus nigripennis

Hercostomus nigriplantis

Hydrophorus balticus

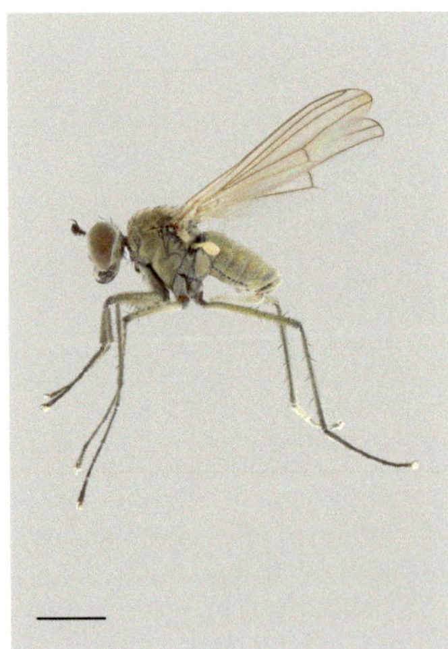
Hydrophorus oceanus

Plate 5

Lamprochromus semiflavus

Liancalus virens

Machaerium maritimae

Medetera impigra

Medetera jacula

Melanostolus melancholicus

Micromorphus Sp A

Microphor anomalus

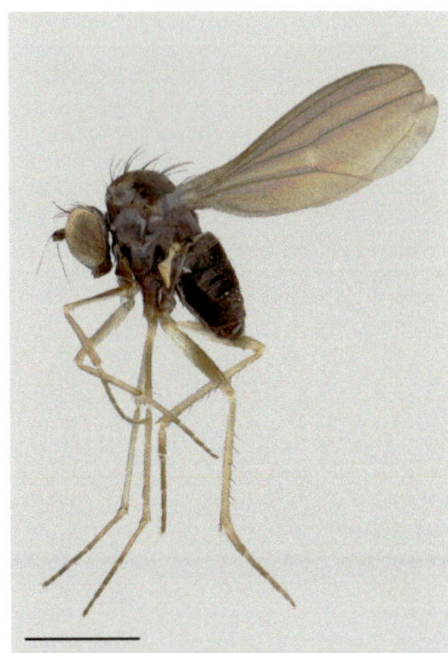

Micropygus vagans

Plate 6

Muscidideicus praetextatus

Nematoproctus distendens

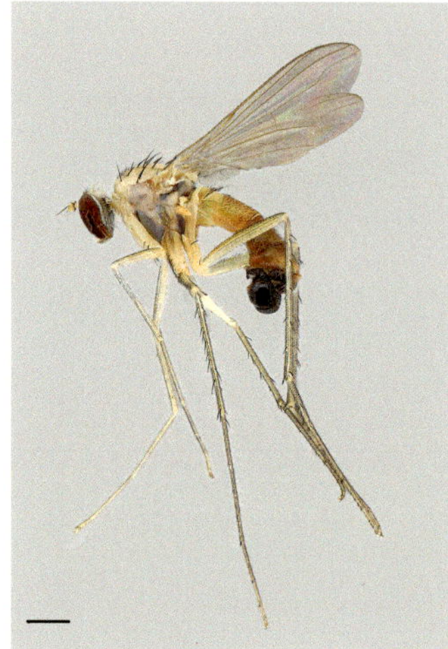

Neurigona pallida

Neurigona suturalis

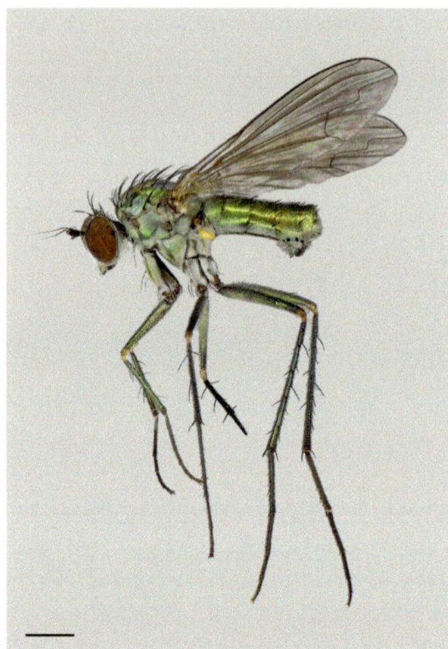

Orthoceratium sabulosum

Ortochile nigrocoerulea

Poecilobothrus chrysozygos

Poecilobothrus nobilitatus

Rhaphium appendiculatum

Plate 7

Rhaphium laticorne

Rhaphium longicorne

Scellus notatus

Schoenophilus versutus

Sciapus laetus

Sciapus platypterus

Sybistroma obscurellum

Sympycnus pulicarius

Syntormon aulicum

Plate 8

Syntormon denticulatum

Syntormon pallipes

Systenus bipartitus

Tachytrechus consobrinus

Telmaturgus tumidulus

Teuchophorus calcaratus

Thinophilus flavipalpis

Thrypticus tarsalis

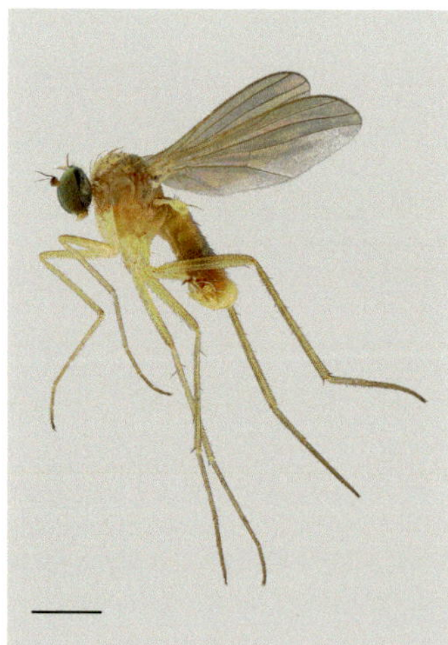

Xanthochlorus galbanus

Introduction

Dolichopodidae (long-legged flies) is a large family of attractive flies ranging in size from nearly 10 mm down to just over a millimetre. The ever-increasing British fauna stands at 326 species at the time of writing although with some doubt over a few species of *Medetera*. Many are readily recognised in the field by their bottle-green metallic colour, moderately slender shape and perky stance. Green is an uncommon colour among flies, so quickly helps to narrow the choice, although Latreille's (1809) family name 'long-legged' is unhelpful as there are numerous long-legged flies.

Males often have obvious modifications of the legs and sometimes their face, antennae, palps and wings, which have been shown to be used in courtship (Lunau 1996). Indeed, their conspicuous and varied courtship behaviour is one of their most endearing characteristics. Many naturalist will be familiar with the wing-waving and hopping dance that has given the common name of semaphore fly to *Poecilobothrus nobilitatus* as it congregates in huge numbers on puddles and ponds in midsummer. This extravagant example has attracted detailed analysis (e.g. Land 1993a) and many others have simpler courtships, although the furious bobbing up and down of male *Chrysotus* over a solitary female must involve considerable energy expenditure and neural complexity. Pheromone-secreting organs are suggested for males of *Dolichopus* that have a ciliolarium on the hind tibia (Negrobov *et al.* 2018), other dolichopodines whose males have a pair of eversible glands between the fourth and fifth sternites (Couturier 1974), and for *Scellus* whose males have a 'flag' projecting from between the fourth and fifth tergites. Mate-guarding by the common *Hydrophorus oceanus* can be seen throughout summer on saltmarsh pools (Dyte 1988a).

As far as is known, all adults are predators of small invertebrates (Ulrich 2004). Only a tiny proportion of dolichopodids feed at flowers, in the British fauna seen regularly only in *Hercostomus nigripennis* and *H. germanus*, and a larger number feed on honey-dew (aphid excretion) or other deposits on leaves. Some species of *Microphor* feed on insects caught in spiders' webs (Downes and Smith 1969, Laurence 1948). The larvae are also predators, mostly of a generalist nature although many species of *Medetera* feed on the immature stages of bark beetles (Scolytinae in Curculionidae), and *Aphrosylus celtiber* are predators of barnacles (Poulding 1998, 2010). The exception to a predatory lifestyle is found in *Thrypticus* whose larvae mine the leaves of monocotyledons growing at water margins (Dyte 1993). It has been suggested that adult dolichopodids may exert similar predation pressure to that of spiders (Cicero *et al.* 2017) but the affinity of most dolichopodids to wetlands makes it unlikely that they would alleviate the impact on British crops which are grown in well-drained conditions. *Medetera* has been well-studied with a view to the potential in controlling wood-boring beetles on which their larvae feed (Kenis *et al.* 2004).

Dolichopodids are recognised as a useful group when making conservation evaluations of wetland habitats, both in Britain and on mainland Europe (e.g. Bellstedt 2007, Crossley 1996, Drake *et al.* 2007, Gelbič and Olejníček 2011, Pollet 2001, 2002). Pollet and Grootaert (1999) and Pollet (2009) argue that they meet the criteria for a reliable bioindicator, being relatively easy to identify and collect using repeatable methods, sufficiently species-rich to cover a range of habitat types, having known and fairly narrow habitat affinities, being sensitive to environmental changes, and with known rarity and threat statuses. The rarity status of the British fauna was assessed by Falk and Crossley (2005) based on distribution and later by Drake (2018a) using the IUCN criteria. The British Empidid and Dolichopodid Recording Scheme has been amassing records since 1989 and is still active at the time of writing.

As with any large group of organisms, identification ranges from easy to difficult but the majority of genera present few problems other than sometimes having to trudge through rather long keys. The ornamentation and often easily visible genitalia of the males of most species are huge aids in their identification although the less flamboyant females present more of a challenge. Dolichopodids cannot compete for popularity with the large and colourful hoverflies and soldierflies but are one of the families that can be more easily investigated once those flies have been mastered. It is hoped that this book will provide a spur to reinvigorating the study of this family.

Recognising the family

The family Dolichopodidae falls within the superfamily Empidoidea, a well-defined group that in Britain also includes the Empididae, Hybotidae, Brachystomatidae and Atelestidae. The Empidoidea is usually regarded as the sister group of Cyclorrhapha, the 'higher' flies, and together these form the clade Eremoneura. The Empidoidea may thus be regarded as intermediate between the more recent Cyclorrhapha and more ancient lower Brachycera that includes such familiar families as Stratiomyidae and Asilidae. At least 7400 species of dolichopodid have been described globally including about 100 known as fossils, representing almost 5% of described Diptera (Grichanov *et al.* 2011a, b, Pape *et al.* 2009). In Europe about 60 genera include at least 790 species (Pollet and Hulcr 2007); the British fauna is about 40% of this and represents one of the most speciose dipteran families in Britain.

The family may be distinguished using the key to families of European Diptera (Oosterbroek 2006). Most species of dolichopodids are readily recognised as moderately slender flies with metallic green bodies and simple wing venation. Few other flies have a metallic green cuticle, and the only narrow-bodied ones among these are stratiomyids, all of which have a more complicated venation. The few green calyptrates in the families Calliphoridae, Muscidae and Tachinidae have broad and robust bodies. Recognising dolichopodids that are not obviously green but yellow, black or drab brown or grey, often dulled by dusting, requires closer examination. Their size varies from tiny (wing length 1.5 mm) to moderately large (9.5 mm).

The following characters will distinguish Dolichopodidae *sensu stricto*, leaving out the subfamily Microphorinae which is dealt with later:

- Three long veins reaching well over halfway to the wing tip (R_{2+3}, R_{4+5}, M_1). All are unforked except in *Sciapus* where M forks into M_1 and M_2 and a few *Dolichopus* where M_2 forms a weak spur on M_1. Vein R_1 is short and the subcostal (Sc) weak and either fused at its tip with R_1 or fades and does not join any vein.
- Only one obvious crossvein, dm-m, spanning M_1 and M_4 at about the middle of the wing (varying between one- and two-thirds from the base); the large discal cell dm thus emits only two veins. Less obvious but important in the family's definition is the absence of the small crossvein (bm-m) that normally joins these two long veins near their bases resulting in the discal medial (dm) and basal medial (bm) cells forming a single cell. Other crossveins (h, r-m) are obscure, lying in the basal quarter of the wing or even closer to the base. Some small acalyptrates have a similar arrangement (e.g. Aulacigastridae, Camillidae, Ephydridae) but do not have the simple face described next.
- Face between antennae and mouth edge either clearly a single unit or partially to completely divided horizontally into two sections (epistoma above, clypeus below); in acalyptrate and calyptrate flies the area between the eyes is divided vertically by the ptinal suture into lateral parafacials and a central undivided face. These facial features cannot be seen in males of a few dolichopodids whose huge eyes touch in front.
- Antennae composed of three segments, the third usually much larger than the basal segments, and carrying a fine arista-like style that is bare to pubescent but without long hairs in British species.
- Palps flat and leaf-shaped, rather than cylindrical as in most other flies.
- Males with genitalia asymmetrical, rotated around the long axis and swung forward beneath the preceding abdominal segments, with the pregenital segments also partially twisted or rotated. This is obvious in species with conspicuously exposed genitalia but not in about half the genera where the genitalia are encapsulated within the preceding segments. The resulting upside-down orientation can be deduced from the position of the cerci, which are hairy lamellae located on the apparent underside, whereas they are clearly dorsal in other flies. Hybotidae are an exception where the cerci may be 'lateral' as the genitalia are rotated.

Species in the subfamily Microphorinae superficially resemble some Hybotidae rather than other dolichopodids. They are small black or dull grey to brown flies characterised by possessing all of the following characters (the same condition may be found in some other empidoids but not all at the same time):

- Male genitalia and pregenital segments asymmetrical, rotated and flexed forward as in Dolichopodidae *sensu stricto*.
- Costa continuing around the entire wing margin, still visible on the hind edge (stops at or before M_1 in Dolichopodidae *sensu stricto*).
- Subcostal vein Sc ending in costa.

In addition, but shared with other empidoids:

- Crossvein bm-m present and distinct, although close to the base of the long medial veins so that the basal cells are very short, less than half length of cell dm.
- Cell dm emitting three veins (M_1, M_2, M_4).
- All veins unforked.

Keys and species accounts

Keys

Most keys were developed with no reference to previous keys with exceptions where there are good recent keys, e.g. for *Achalcus* and *Gymnopternus* (Pollet 1990, 1996) and *Sciapus* (Meuffels and Grootaert 1990). The other major exception is *Medetera* for which existing keys were modified since this Handbook is not the appropriate place for the huge task of revising this difficult genus. If the new keys fail there remains the opportunity to use previous keys to approach an answer from a different direction. The keys were derived from species occurring in Britain, and take no account of others in the Western Palaearctic. In particular, the keys to genera should not be expected to be accurate outside the British Isles.

The keys are based primarily on dry specimens but in most cases will also work for wet-preserved material. Both methods of preservation have their advantages: dry material retains colour and dusting well and is easier to hold steady when taking measurements; wet material is usually better for examining genitalia. The keys are intended for use with a microscope as many features cannot be seen using a hand lens, except on some larger species. Some species are most quickly identified by looking at the hypopygium and in some cases this is the only reliable method of determination. Pulling the genitalia away from the abdomen will often give an adequate view of the cerci and some appendages. However, for species in the genera *Achalcus*, *Chrysotus*, *Medetera*, *Microphor* and *Thrypticus*, the genitalia need to be mounted or partially dismembered in alcohol and viewed at high magnification.

Males are almost invariably easier to identify than females as, apart from often conspicuous genitalia, they usually possess ornamentations called male secondary sexual characters, mostly on their legs but also their wings, face, antennae and palps. Taking male secondary sexual characters and the genitalia together, most species can be identified using illustrations alone. Apart from species of Sympycninae, whose genitalia are small, encapsulated and of less use than male secondary sexual characters, the genitalia of nearly all others are illustrated. In contrast, females are rather dull and, had both sexes looked like this, the family would never have attracted much of a following. In this Handbook separate keys are provided for each sex, following the example set by Kowarz (1874). Genera for which keys to females are incomplete are *Chrysotus* and *Microphor*, and occasional inseparable pairs or trios of species occur in *Argyra*, *Dolichopus* and *Rhaphium*. No key to female *Medetera* is provided as it would be far too incomplete to be useable.

Characters are listed in order of their perceived reliability and ease of use. If the first characters appear to fit well, there is no need to go through every other character, but more are provided should confirmation be needed and for specimens in poor condition where the most useful parts are damaged. Inevitably not all variation can be taken into account, and I may have resorted more frequently than users would like to qualifying characters with 'usually'. In the key to genera the qualifying terms 'often', 'usually', 'rarely' refer to the representation of species in a genus or group with a particular character, not variation within that character.

The terms height and depth are used interchangeably, meaning the vertical distance. Width is obvious except for antennae and femora where, because their cross-section is often strongly oval, the term is reserved for the front-to-back distance as viewed from above, as this is not the same as depth, which is the top-to-bottom distance viewed from the side. This distinction is important when a comparing the length of a seta with one of these leg dimensions, or the length-to-depth ratio of the postpedicel.

All couplet numbers are followed by the number they came from, e.g. 7(2) means that couplet 7 was derived from couplet 2. Figures within the keys are labelled with lower case letters, with one letter, e.g. (a), referring to the first part of the couplet and two letters, e.g. (aa), referring to its counterpart in the second half. Sometimes there is no counterpart figure.

Species accounts

For both genera and species, the accounts start with a summary of the characters to confirm the identity arrived at using the key. Species accounts treat males first (without stating 'males'), then females. Distribution, habitat and flight times are derived from the British Empidid and Dolichopodid Recording Scheme (https://dipterists.org.uk/empid-allies-scheme/home). Distributions are described with a broad brush as finer detail will quickly become out of date as recording effort increases. Habitat data have been condensed into one or two sentences; contrasting information based on mainland Europe is sometimes added. Flight times are derived from the national recording scheme. Where there are sufficient data, the peak or main period is given as well as the earliest and latest months. As most records derive from casual recording, it is rarely possible to discern separate cohorts.

Plants are called by their English name: alder *Alnus glutinosa*; apple *Malus*; ash *Fraxinus excelsior*; aspen *Populus tremula*; beech *Fagus sylvatica*; birch *Betula*; club-rush, common *Schoenoplectus lacustris*; dog's mercury *Mercurialis perennis*; elm *Ulmus*; English elm *Ulmus procera*; gorse *Ulex*; grey poplar *Populus canescens*; horse chestnut *Aesculus hippocastanum*; larch *Larix decidua*; Lombardy poplar *Populus nigra* var. *italica*; maple *Acer*; oak *Quercus*; pine *Pinus*; poplar *Populus*; purple moor-grass *Molinia caerulea*; reed *Phragmites australis*; reed sweet-grass *Glyceria maxima*; rush *Juncus*; sallow *Salix cinerea*; Scots Pine *Pinus silvestris*; sea club-rush *Bolboschoenus maritimus*; sedge *Carex*; spike-rush, common *Eleocharis palustris*; spruce *Picea*; sycamore *Acer pseudoplatanus*; water hyacinth *Pontederia crassipes*; willow *Salix*.

Measurements

The preferred measure of size is wing length rather than body length which varies with shrinkage in dry material. Wing length is usually similar to body length when averaged over many specimens except in a few genera with particularly long wings, such as *Hydrophorus*. Length is measured from the wing tip to the base of the basicosta which is the short swollen section of the costa basal to the slightly thicker section carrying a long seta. The error in this measurement for the purpose of identification is trivial so locating the basicosta is not strictly necessary, and the complex of basal sclerites is an adequate approximation. Measurements are the mean and range of ten individuals of each sex where available, from a wide geographic area in Britain and rarely from non-British material; individuals larger and smaller than sizes quoted will inevitably be found. Figure 1 gives the range of mean wing lengths of all the species in each genus. This provides a guide to the expected size although it is less helpful for large genera with a wide size range, such as *Rhaphium* and *Hercostomus*.

Ratios of leg segments and wing veins are often used in the keys and these need to be measured fairly accurately with rather less than 10% error and cannot be guessed by eye. For this, an eye-piece graticule is needed. It can either be kept permanently in the eye-piece of one's working microscope, where it quickly stops becoming an irritation, or in a spare eye-piece that is inserted when needed.

Figures and photographs

Most figures were drawn by me using a camera lucida (drawing tube) from dry, wet or mounted parts. Wings were usually drawn in glycerol under a cover-slip so they are very slightly wider than in life owing to the natural corrugations along the veins being flattened. Many of the figures are the original pencil drawings which are not as crisp as inked figures. The genitalia of the subfamily Dolichopodinae were drawn by Dawn Painter from stacked photographs of dry specimens. Wing photographs were taken by Stuart Ball and are presented as monochrome images. Figures are shown with the same orientation; in particular the leg segments are shown with the dorsal or anterior side upwards, except when indicated. Most genera have a distinct jizz that cannot be captured in a key but comes across well in photographs. At the level of genus, these provide both a first port-of-call and a useful check once the key has been used. The photographs are stacked images taken by Stuart Ball (Plates 1–16).

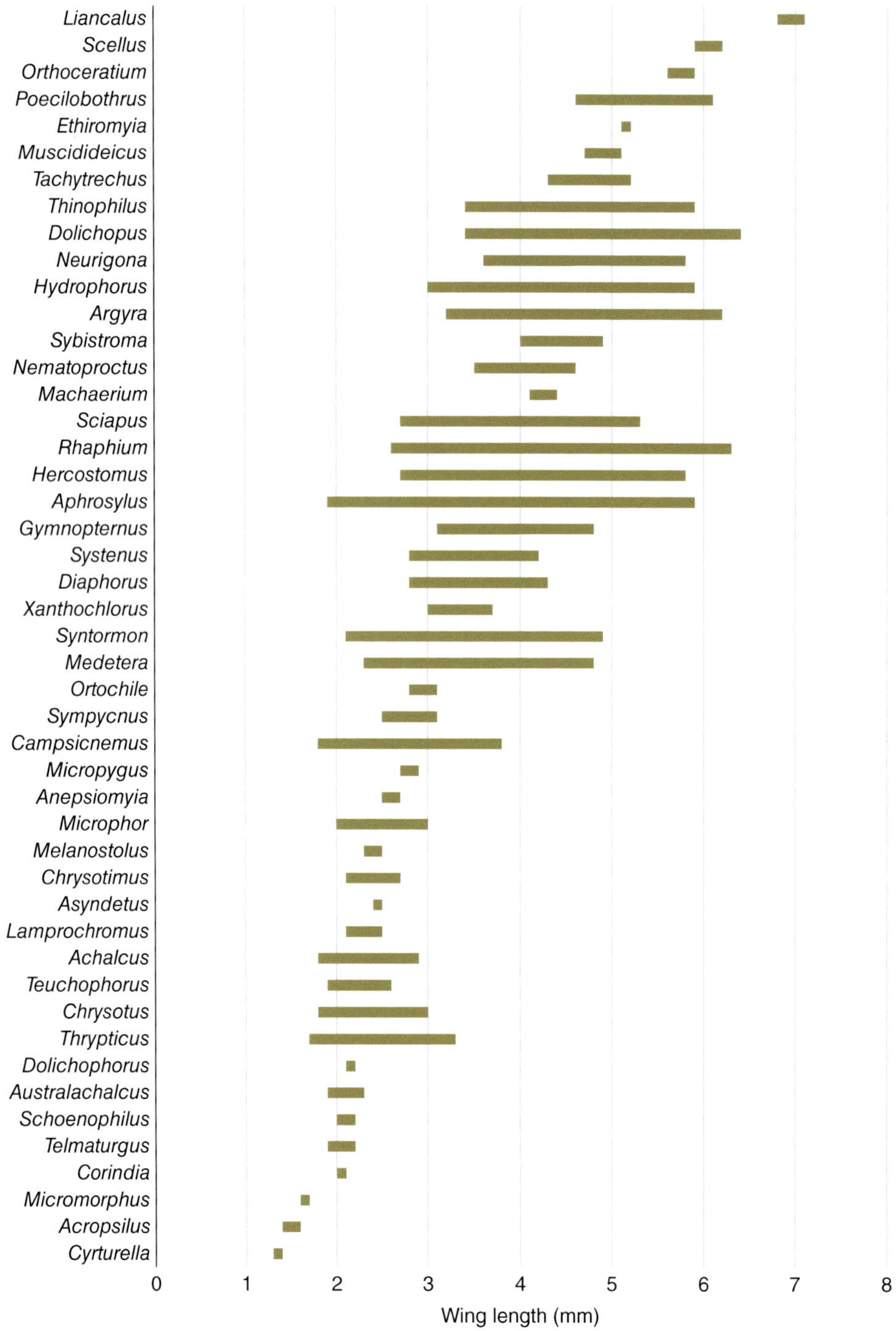

Fig. 1. Wing lengths of species within each genus.

Other identification resources

The first RES Handbook on dolichopodids (d'Assis-Fonseca 1978) is still useful although a few couplets in keys to females do not work, and over 50 species have been added to the fauna since then. The excellent monograph by Parent (1938) and *Die Fliegen der palaearktischen Region,* particularly the later issues (Stackelberg 1930–1941; Negrobov and Stackelberg 1971–1974, Negrobov 1977–1979) are usually the first works to consult if a species appears to be new to the British fauna. Both works give full descriptions and have adequate to good figures. Keys to European genera and species are provided by Grichanov (2006, 2007) based largely on Parent's work but lack figures. Grichanov *et al.* (2011a) cover the entire Palaearctic fauna and include a useful compilation of illustrations and photographs of wings, antennae and hypopygia for each genus, and Grichanov *et al.* (2011b) provides additional information on the important characters of each genus.

Collecting and preservation

Few recorders will specifically target dolichopodids, for which the methods used for most Diptera are appropriate (Irwin 2010a, b). Most species are associated with damp habitats but some need to be sought in other places. Specialists of tree trunks, such as *Medetera* and *Neurigona*, can be collected by sweeping up and down the trunk, and the higher foliage of trees and higher shrubs is where *Chrysotimus*, *Microphor* and *Xanthochlorus* are most often found. *Hydrophorus* adults sit on water and fly just above it, so require dexterity to net them. Vacuum sampling is particularly effective for species with secretive behaviour, in particular *Achalcus*, and where vegetation makes sweep-netting difficult, e.g. in reedbeds. Malaise traps and white water (pan) traps are productive methods (e.g. Pollet and Grootaert 1994), and blue or black water traps are most effective for arboreal species such as *Medetera* and *Systenus* (Pollet and Grootaert 1987, Diestelhorst and Lunau 2001). Rearing is productive for obtaining adults of *Systenus* and *Australachalcus* whose larvae develop in tree rot-holes and sap-runs, and *Medetera* whose larvae are often associated with wood-boring beetles in dead wood.

The heads of dolichopodids often collapse when killed using ethyl acetate, thus obscuring characters on the face and base of the antennae. Killing by freezing or carbon dioxide results in specimens in better condition. Carbon dioxide dispensers can be made by adapting equipment used for soda siphons and inflating bicycle tyres.

Dry specimens are best pinned using a micropin through the side of the thorax rather than dorsally, which can damage characters useful in identification, or glued to card rectangles or triangular points. Identification of males is usually helped if the hypopygium is drawn away from the abdomen. A few genera almost routinely require examination of the genitalia at high magnification. Alcohol (70–80%) is the commonly used wet preservative. Concentrations greater than 80% or the use of formalin or glycerol are less satisfactory as the flies become stiff and brittle.

Genitalia preparations require the specimen to be cleared using dilute potassium or sodium hydroxide, either hot for a few minutes or cold for a few hours, or lactic acid for a few hours. An easy-to-use and reliable mounting medium is dimethyl hydantoin formaldehyde (DMHF) which is water-soluble and prepared using around 70% alcohol. As it is water-soluble, the specimen can be quickly retrieved and re-mounted if it is poorly orientated. As the genitalia are best inspected under high magnification with transmitted light, a neat permanent mount can be made from a strip of card (e.g. 15 × 6 mm) in which a small hole (2.5–3 mm) is made using a leather hole-punch. Small cover-slips can be made by cutting a large cover-slip into squares of about 5 mm using a carborundum stone or diamond glass-cutter to score lines along which it easily breaks into little squares. The cover-slip is placed under the card, a tiny dab of DMHF added and the specimen put into this drop. It needs inspecting for a few minutes to re-orientate the genitalia using a fine needle as they slowly gyrate until enough alcohol has evaporated for the medium to stiffen. There is no need for a top cover-slip. If the medium wrinkles, it can be instantly smoothed by gently breathing on it. The card is staged with the pinned insect, taking up no more space than the staging and labels. An alternative method of storing genitalia is in glycerol (glycerine) in microvials attached to the pin.

Morphology of adults

Terminology follows Cumming and Wood (2017) except for the genitalia for which Brooks (2005) is used.

Surface features and their orientation

Bristles are called **setae** if they are large, **setulae** if tiny or **hairs** if thin. Fine short hair covering much of the surface of the legs is called **vestiture**. All these structures are the same morphological feature, and they arise from a socket, whereas minute 'hairs' without a socket is called **dusting** (pruinescence, tomentum) which dulls the cuticle. **Pubescence** describes minute and often dense hairing visible only at high magnification. The position of setae on the leg segments are described with reference to the orientation of the surfaces (dorsal, ventral, anterior, posterior) and these are most easily envisaged if a leg is imagined extended sideways from the body. Many setae are in the intermediate positions, e.g. anterodorsal, and these positions have sometimes been abbreviated in the text here as **ad** for anterodorsal, **pd** for posterodorsal, and **av** and **pv** for the equivalent ventral positions. These terms are awkward to apply to positions on the front coxa where 'inner' and 'outer' are used instead. Setae at the tip of the tibia are termed **apicals** to distinguish them from those along the shaft. Important setae on the femora are the anterior and posterior **preapicals** which are near the femur's tip, usually at a distance of 2–3 times the diameter of the shaft. A row of short stout setulae on the dorsal or anterodorsal face of the front tibia is the **comb** (pecten). The relative position of setae along the shaft of a leg segment is referred to as basal if closer to the base or apical if nearer the apex, but there are occasions when these terms become ambiguous and the more correct but less familiar terms **distal** and **proximal** are used for positions that are respectively further from or closer to the core of the body. The arrangement of setae is called chaetotaxy.

Head (Fig. 2)

Large eyes dominate the head and are separated on top above the antennae by the **frons** (except in males of *Diaphorus* and *Microphor*) and below the antennae by the **face**. The face is divided into the upper **epistoma** and lower **clypeus** which are sometimes clearly distinct or the clypeus merely projects slightly; the recommendation that the term epistoma is dropped in favour of face is not followed here as the distinction is well embedded in dolichopodid literature. A small protrusion at the top of the frons, the **ocellar triangle**, bears three ocelli and a pair of ocellar setae. The back of the head is divided into the upper **occiput** and lower **postgena**. In nearly all species the eyes in profile obscure the head above the proboscis but in some *Hydrophorus* the **gena** (cheek) is visible as a narrow strip below the eye. The mouthparts consist of the **proboscis** flanked by a pair of flat single-segmented **palps**. The antennae consist of four parts: the basal **scape** (first segment), the **pedicel** (second segment), the **postpedicel** (third segment, first flagellomere) and the **arista** which is the hair-like segment projecting from the postpedicel. Technically the arista should be referred to as the arista-like stylus in Dolichopodidae and some related Brachycera as it is not composed of the last three flagellomeres, but for brevity it is referred to as the arista in this Handbook. The scape often projects anteroventrally as a small point, the **conus**. The chaetotaxy consists of paired **outer vertical** setae close to the top corner of the eye and which are usually large and curve forward, paired **inner vertical** setae (regarded as postvertical or postocellar by some authors) that are usually small and often appear to be the uppermost setae of the tidy line of **postocular** setae that run around the hind margin of the eye, paired **ocellar** setae arising from the ocellar triangle and which point backwards, diverge and are usually as large as the outer verticals, and **postocellars** behind the ocellar triangle which are small or absent. The postgena behind the postoculars bears a few to numerous hairs which, when conspicuously long and dense hairs, are referred to as **whiskers**.

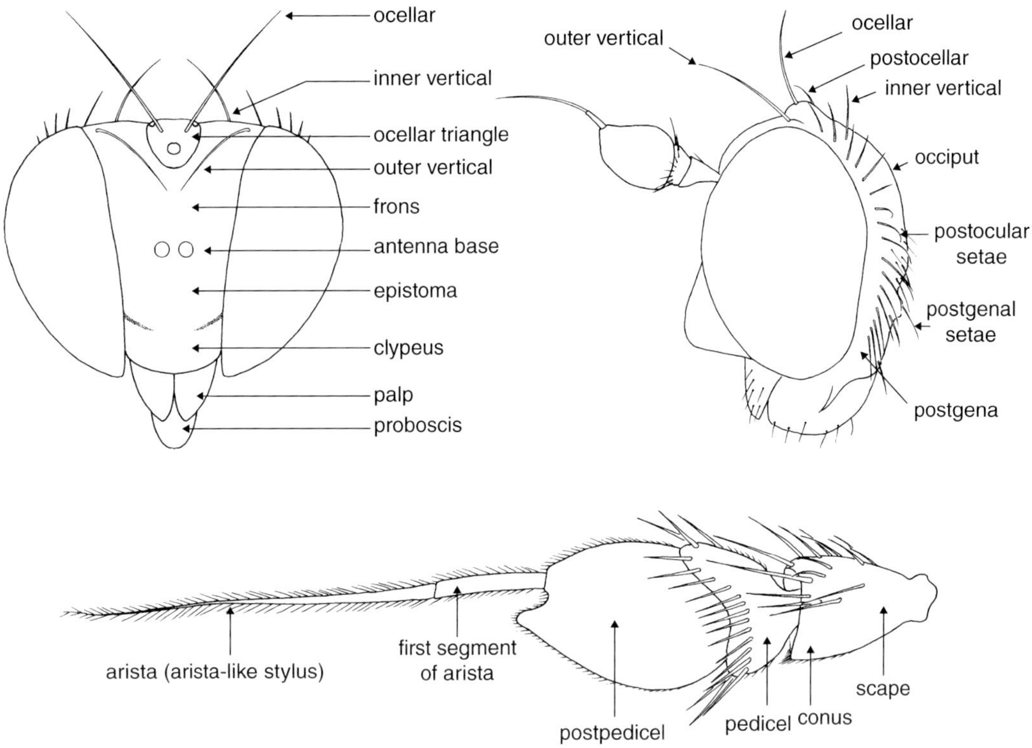

Fig. 2. Head parts and setae.

Thorax (Fig. 3)

The top of the thorax is the **scutum**, often called the mesonotum but technically this term also includes the scutellum and postnotum. Behind the scutum is the small semicircular to rectangular **scutellum**. The scutum shows indications of a weak division or furrow at its sides, the **transverse suture**, which provides a useful landmark for the position of setae. At its front corner is a small slightly protruding lobe, the **postpronotal lobe** (humerus). The **notopleuron** is the triangular sclerite lying in front of the wing base and fused with the scutum. In some genera, the posterior part of the scutum, between the last few dorsocentral setae, is flat and clearly distinct from the domed profile of rest of the scutum and is described here as the **prescutellar depression** although it is not concave and not all authors use this term. Below the scutellum is the **postnotum**, a domed structure connected to the first tergite of the abdomen.

The sides of the thorax are the **pleura** (singular pleuron) composed of several sclerites. The sclerites of the pleura reflect their origins from three thoracic segments. From front to back, these are the **proepisternum** (propleuron) above the front coxa and just in front of the anterior spiracle; the large square **anepisternum** (mesopleuron) between the proepisternum and the wing base; the large triangular **katepisternum** (sternopleuron) that is fused under the thorax to its equivalent on the other side; the **epimeron** (pteropleuron) behind the anepisternum and below the wing base; the **laterotergite** in front of the halter and posterior spiracle and is poorly delineated from the postnotum; below this and above the mid coxa is the **meron** (hypopleuron); and finally the **metepimeron** which joins the thorax to the abdomen and in dolichopodids is far more conspicuous than in many Diptera as it flares out to partly envelop the first abdominal segment.

The thoracic setae (Fig. 4) referred to in this key are, from the midline of the scutum outwards, the **acrostichals** which are usually small, in one or two rows (up to four rows in *Microphor*) or absent; the **dorsocentrals** which are usually large and vary in number from three to seven pairs, sometimes

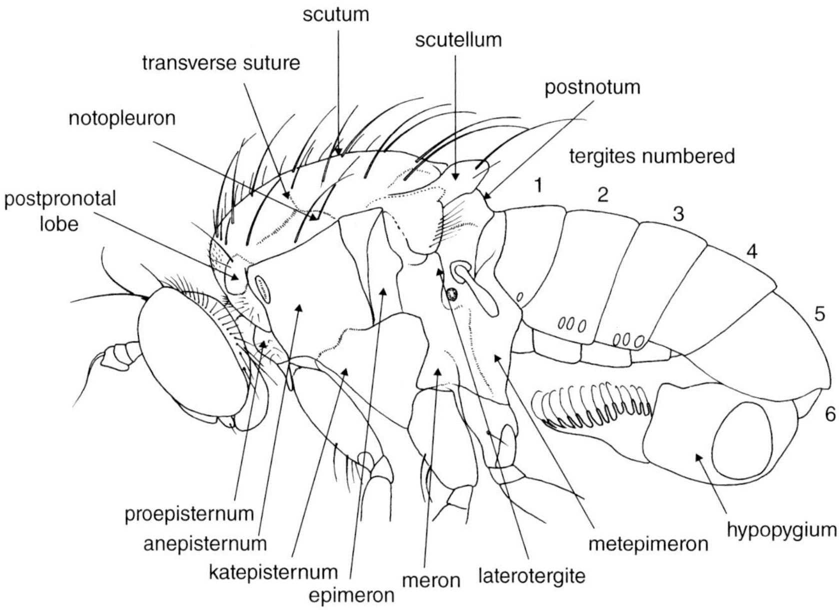

Fig. 3. Parts of the thorax and abdomen.

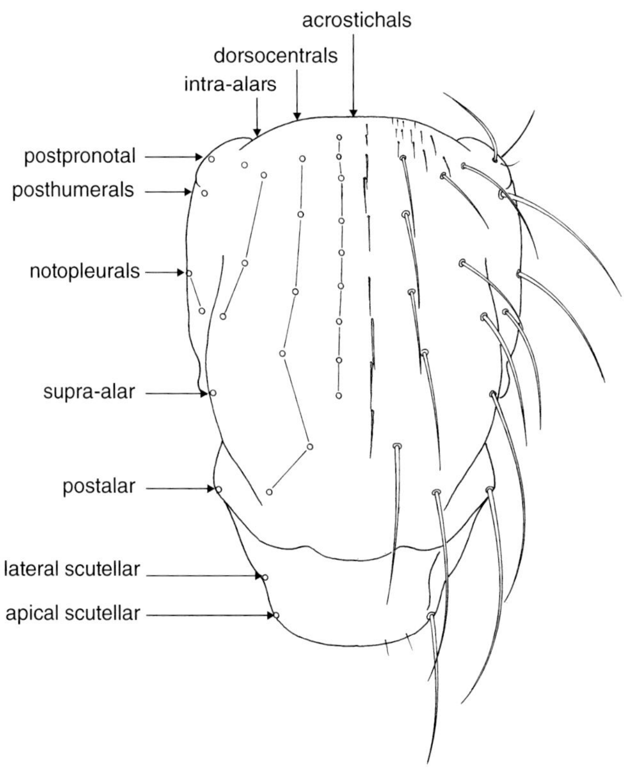

Fig. 4. Thoracic setae.

more in some Hydrophorinae, and are in a single row in most subfamilies but multiserial in *Microphor*; an untidy series outside of the dorsocentrals of four setae consisting of three **intra-alars** (presutural, sutural and postsutural) and below these one **supra-alar** behind the suture, although it is sometimes unclear which name to apply to these setae. Close to the lower margin of the scutum is the **postpronotal** seta on the lobe of that name (usually one, sometimes with accompanying hairs), two **posthumeral** setae, **notopleural** setae (usually two, rarely one) and one **postalar** seta on the posterior corner of the scutum. In some genera, some of these setae may be reduced in number or lost entirely and in others there are additional setae that are indistinguishable in structure from these named ones. The scutellum usually has a pair of long **apical** (median) setae on its outer 'corners' and varying numbers of often smaller **lateral** setae in front of the apicals and sometimes some more between the apicals.

Legs (Fig. 5)

These have the same five elements as in other Diptera: coxa, trochanter, femur, tibia, tarsus. The trochanter is inconspicuous, short and partly fused to the femur but its colour is an important character in some genera. The individual segments of the tarsi are **tarsomeres**, the first most basal of which is the **basitarsus** (metatarsus), and the following termed second to fifth tarsomeres. There are usually paired claws on the last tarsomere and at their bases are two feathery or pubescent pads called **pulvilli**. The hind tibia of *Dolichopus* often has a **ciliolarium** (tibial organ, osmeterium) consisting of a modified patch of vestiture on the posterior side, often extending towards the tip of the tibia as a strip of pubescence (Steyskal 1973).

Wings (Fig. 6)

Vein names suggested by Wootton and Ennos (1989) were recommended by Cumming and Wood (2017), resulting in some changes to names used in previous Diptera Manuals (Cumming and Wood 2009, McAlpine 1981); the earlier abbreviations are given in parentheses. Abbreviations of longitudinal veins are given in capitals, those for crossveins and cells in lower case; cells take the letter and

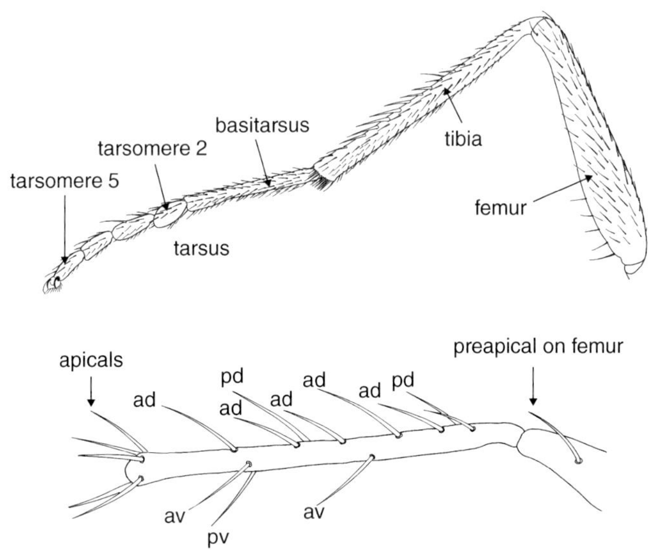

Fig. 5. Front leg of *Syntormon* and setae of mid tibia of *Dolichopus*, anterior face: ad – anterodorsal, av – anteroventral, pd – posterodorsal, pv – posteroventral.

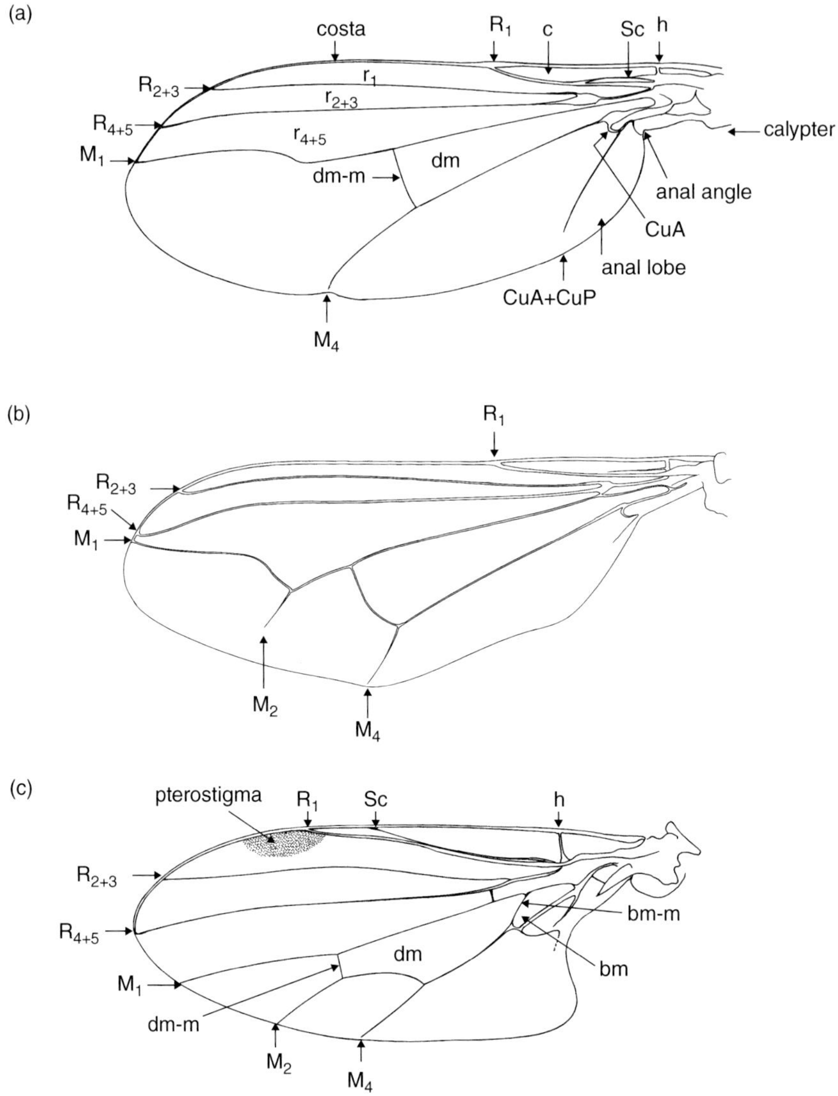

Fig. 6. Wings. a. *Dolichopus*; b. *Sciapus*; c. *Microphor*.

number of the vein on its front margin. The **costa** is the thickened leading edge of the wing, usually extending to vein M_1 but continuing around the whole wing in *Microphor*. The short **subcostal** vein (**Sc**) is the first longitudinal vein, and fuses with R_1 except in *Hydrophorus* in which it fades out in the middle of cell c, and in *Microphor* in which it fuses with the costa. In most subfamilies the **radial** veins (**R**) are unbranched; $\mathbf{R_1}$ is short except in *Microphor*, $\mathbf{R_{2+3}}$ and $\mathbf{R_{4+5}}$ extend to near the wing tip. The **medial** veins (**M**) are the longitudinal veins in the rear half of the wing. In most genera $\mathbf{M_1}$ is single and the longest of the veins, but in *Sciapus* it forks into M_1 and M_2, in a few *Dolichopus* M_1 has a vestigial stub of M_2, and in *Microphor* there are three veins (M_1, M_2, M_4) arising from the discal cell (dm). $\mathbf{M_4}$ (previously the anterior cubital, CuA_1) ends at about the middle of the hind margin of the wing and, unlike those in front of it, has no connection with the costa except in *Microphor*. The **cubital** vein **CuA + CuP** (previously the anal vein, A_1) is short, weak or absent. The anterior cubital vein, **CuA**, forms a characteristic tiny loop enclosing the small **cubital cell**, **cua**, and is sometimes used here as a landmark from which to measure the length of M_4. Between the cubital vein and the hind margin of the wings is the **anal lobe**. The angle that the margin of the anal lobe makes with the most basal part of the wing is the **anal angle** which is a convenient measure of the development of the anal lobe. In a few genera there is an exceptionally weak fold posterior to the cubital vein which is presumably the newly renamed anal vein, A_1. The discal medial crossvein **dm-m** (dm-cu) is at about the middle of the wing and humeral crossvein **h** lies at the wing base near the origin of Sc. Two useful ratios are the basal to apical sections of M_4 either side of the crossvein, and the sections of the costa either side of R_{4+5}. In some genera there is a slight depression in the membrane partway along the apical section of M_1 which Parent (1938) called the 'bosse alaire' and which is referred to here in English (**wing boss**). *Microphor* have a dark patch on the membrane, the **pterostigma**, at the end of R_1. In some *Dolichopus* the costa is swollen at its junction with R_1, here termed the **costal stigma** (the stigma of d'Assis-Fonseca 1978). The **halter** (pleural halteres) is the small club-shaped vestigial hind wing located just behind the base of the wing. A small approximately circular flap of membrane at the hind base of the wing is the **calypter** (squama) fringed with hairs.

Abdomen (Fig. 7)

Each segment comprises an upper **tergite** and small ventral **sternite**. The last segments are the terminalia which are the structures associated with copulation and oviposition. Male terminalia are the **hypopygium** (genitalia) which is curled forward (reflexed) under the abdomen so that, when in its normal position, the morphological ventral side is uppermost but appears to be dorsal. This should be born in mind when the dorsal or ventral surfaces are referred to in the text. As well as being reflexed, the hypopygium is also rotated so the connection to the abdomen has become lateral on the left-hand side. When the hypopygium is dissected, the position of the connection between the preceding segment and the capsule is seen as a round or oval hole, the foramen. The segment forming the connection is **sternite 8** which appears to be a tergite because it is quite large and in an almost dorsal position, but tergite 8 is vestigial or has been completely lost. Sternite 8 sometimes bears setae useful in identification of Diaphorinae. The main part of the hypopygium is composed of a large dorsal **epandrium** (tergite 9) and a smaller ventral **hypandrium** (sternite 9)

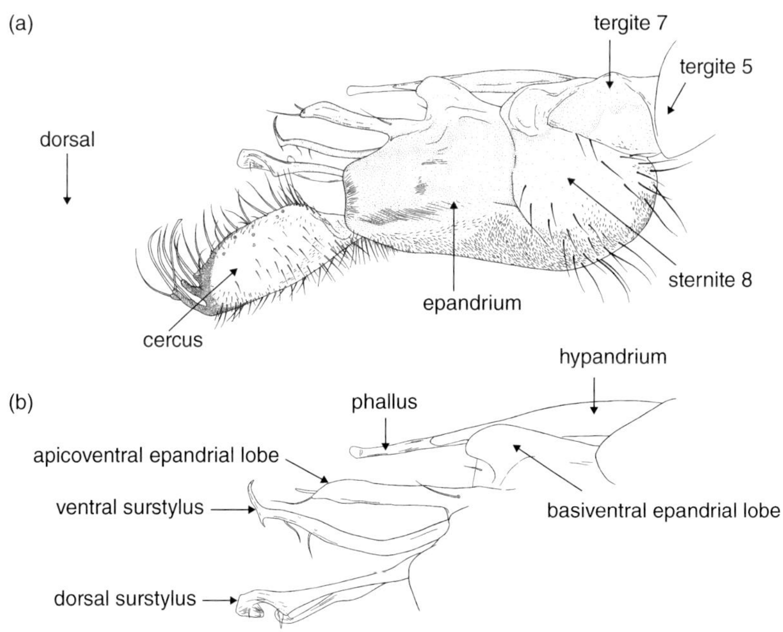

Fig. 7. Male terminalia of *Dolichopus acuticornis*. a. Entire hypopygium with adjacent tergites and sternite; b. internal appendages.

which often projects from the epandrium to provide support or protection for the phallus. Projecting from the ventral margin of the epandrium in many genera are two pairs of lobes, the more basal being the **basiventral epandrial lobes** and the more apical the **apicoventral epandrial lobes**. These are distinct and useful for confirmation of identification in Dolichopodinae, and tiny but nevertheless useful in Medeterinae. Elongate **surstyli** arise dorsal to these lobes and may be fused to or articulated with the epandrium. Further appendages are found in some genera, forming a confusing array and whose precise homology is not clear in *Achalcus*, *Micromorphus* and *Xanthochlorus*. In *Sciapus*, immediately ventral and sometimes fused to the cercus is an appendage which Becker (1918) called **organ X** and the name has stuck. In *Xanthochlorus* the term **paramere** is used for one pair of well-developed appendages. The **phallus** is the fused aedeagus (the intromittent organ) and its supporting parameral sheath which become indistinguishable and form a single long median tube that curls around the inside of the epandrium. At the posterior dorsal opening of the capsule is the pair of **cerci** (singular cercus) which are hairy and often leaf-shaped or elongate.

Female terminalia are comparatively simple. The terminal structure, tergite 10, is partly or completely divided medially into two sclerites called **acanthophorites**. In most genera, the tip of each acanthophorite bears a transverse row of 1–6 pairs of short **acanthophorite spines** which are sometimes reduced to slender setae. Below these is a pair of stylets and small lateral cerci. In *Thrypticus*, segments 8 and 10 are fused into a blade-shaped **ovipositor**.

Sexing

In females, the last large tergite has a clear counterpart sternite of approximately the same length; seen from behind, the opening between these sclerites is lens-shaped or oval, and where the transverse row of acanthophorite spines and stylets are usually visible. The tip of the abdomen is often slightly flattened and wider than deep, and in side view is a truncated cone. In males whose sex is not obvious from externally visible genitalia or conspicuous male secondary sexual characters, the last long tergite (sometimes followed by a shorter one) has no obvious matching sternite and forms a deep sheath that encapsulates the genitalia so that the tip of the abdomen is not conical. Seen from behind the opening is higher than wide and in which the globular end of the epandrium is visible.

Phylogeny

The monophyly of the Dolichopodidae *sensu lato* (see later) within the Empidoidea is well-established (e.g. Chvála 1983, Moulton and Wiegmann 2007). Below this taxonomic level there is greater uncertainty about subfamilies, in both their composition and relationship to one another.

Several classifications of subfamilies have been proposed. Most relevant to the British fauna are those of Robinson (1970) whose system was adopted by Negrobov (1991) in his Palaearctic catalogue, and Ulrich (1981) whose slightly different arrangement was followed with minor changes by Chandler (1998) in the British checklist. One difference between these two systems was Robinson's erection of Peloropeodinae to include genera previously in Sympycninae and which are distinguished by their more robust build, a prescutellar depression and relatively large and complex male genitalia. Subfamilies included later are Achalcinae, erected by Grootaert and Meuffels (1997), and Microphorinae which is most likely the sister group of Dolichopodidae *sensu stricto* plus the non-British Parathalassiinae (Sinclair and Cumming 2006), although this conclusion may not be strongly supported (Lim *et al.* 2010). Chandler (1998 and updates in *Dipterists Digest*) gives other details of genera moved between subfamilies in the British list.

Germann *et al.* (2011a) and Lim *et al.* (2010) analysed subfamilies using molecular data. When the subfamilies occurring in Britain are compared to phylogenies based on morphology, Achalcinae, Dolichopodinae, Sciapodinae and Sympycninae are monophyletic with strong support in one or both of these analyses. Diaphorinae, Medeterinae and Rhaphiinae are paraphyletic (not all included taxa have a common ancestor) although British genera are placed unambiguously within the major lineages. An exception is *Lamprochromus* sharing a clade with *Rhaphium* and *Nematoproctus*, so should be regarded as belonging to Rhaphiinae and not Sympycninae. The Hydrophorinae and Peloropeodinae are polyphyletic (having more than one common ancestor resulting in groups separated ancestrally by other monophyletic groups), which affects the placing of British genera. The Hydrophorinae is divided into three tribes by Negrobov (1986, 1991) and these correspond well to widely separated clades recognised by Germann *et al.* (2011a) and to two distinct clades recognised by Lim *et al.* (2010). It is likely that this subfamily will in future be split into three, with *Aphrosylus* in Aphrosylinae, *Thinophilus* and *Schoenophilus* in another as yet unnamed subfamily equivalent to Negrobov's tribe Thinophilini, and the other genera that occur in Britain remaining in the well-established Hydrophorinae. In Robinson's Peloropeodinae, Negrobov (1991) included the British genera *Acropsilus*, *Anepsiomyia*, *Micromorphus* and *Chrysotimus*. This subfamily was not accepted by Chandler (1998) but some genera assigned to it are clearly distinct from Sympycninae based on molecular data. It is, however, polyphyletic, with one lineage including *Micromorphus* and *Chrysotimus*, and the other including *Acrospilus* along with the non-British type genus *Peloropeodes*. However, *Acropsilus* has particularly uncertain affinities (Ulrich 1981, Bickel 1998). The genus *Anepsiomyia* does not fall neatly in either Sympyncninae where it traditionally lay, or Peloropeodinae where Negrobov (1991) placed it, and may represent an isolated lineage. Given these uncertainties, some placements in the following checklist may be incorrect from a phylogenetic viewpoint but for convenience the Hydrophorinae is treated in the traditional sense of a single subfamily, *Micromorphus*, *Chrysotimus* and *Acrospilus* are included in Peloropeodinae, *Lamprochromus* is moved to the Rhaphiinae, and *Anepsiomyia* is retained in the Sympycninae.

Checklist

The list follows the same format as the checklist of British Isles Diptera (Chandler 1998) with references to changes that have occurred since the appearance of that list. Since that checklist, 42 more species of dolichopodids have been recorded in Britain, two have been removed or synonymised, and nomenclatural changes made to seven others. As in the 1998 checklist, the original genus is given in parentheses for species that have changed their generic assignment. Valid names are in bold type while synonyms and misidentified names are inset.

The gender of the name *Syntormon* is, at the time of writing, the subject of a pending Opinion to the ICZN arguing that the gender is neuter (Drake and Welter-Schultes, 2025). Drake (2021a) explains the confusing endings *monile* and *submonilis*.

ACHALCINAE
ACHALCUS Haliday in Loew, 1857
bimaculatus Pollet, 1997
britannicus Pollet, 1997
cinereus (Haliday in Walker, 1851 – *Rhaphium*)
flavicollis (Meigen, 1824 – *Porphyrops*)
nigropunctatus Pollet and Brunhes, 1997 Added by Drake (2008)
phragmitidis Pollet, 1997 Added here
thalhammeri Lichtwardt, 1913
vaillanti Brunhes, 1987

AUSTRALACHALCUS Pollet, 2005 Split from *Achalcus* by Pollet (2005)
melanotrichus (Mik, 1879 – *Achalcus*) Transferred from *Achalcus* by Pollet (2005)

DIAPHORINAE
ARGYRA Macquart, 1834 Usage resolved by ICZN (2014)
argentina (Meigen, 1824 – *Porphyrops*)
argyria (Meigen, 1824 – *Porphyrops*)
atriceps Loew, 1857
auricollis (Meigen, 1824 – *Porphyrops*)
diaphana (Fabricius, 1775 – *Musca*)
elongata (Zetterstedt, 1843 – *Dolichopus*)
grata Loew, 1857
ilonae Gosseries, 1988
leucocephala (Meigen, 1824 – *Porphyrops*)
perplexa Becker, 1918
vestita (Wiedemann, 1817 – *Dolichopus*)

ASYNDETUS Loew, 1869
latifrons (Loew, 1859 – *Diaphorus*) Added by Godfrey and Drake (2012)

CHRYSOTUS Meigen, 1824
angulicornis Kowarz, 1874 Deleted from British list by Negrobov and Chandler (2006), reinstated by Drake (2020a)
blepharosceles Kowarz, 1874
cilipes Meigen, 1824
collini Parent, 1923
cupreus (Macquart, 1827 – *Rhaphium*)
femoratus Zetterstedt, 1843
gramineus (Fallén, 1823 – *Dolichopus*)
laesus (Wiedemann, 1817 – *Dolichopus*)
longipalpus Aldrich, 1896 Introduced, only in hot-houses
  pallidipalpus van Duzee, 1933 Syn. by Evenhuis (1996)
melampodius Loew, 1857
monochaetus Kowarz, 1874
neglectus (Wiedemann, 1817 – *Dolichopus*)
obscuripes Zetterstedt, 1838
palustris Verrall, 1876
pulchellus Kowarz, 1874
suavis Loew, 1857
tricaudatus Negrobov, Barkalov and Selivanova, 2014 Added by Drake (2020a)
verralli Parent, 1923
Species A Undescribed

DIAPHORUS Meigen, 1824
hoffmannseggii Meigen, 1830
nigricans Meigen, 1824
oculatus (Fallén, 1823 – *Dolichopus*)
[***winthemi*** Meigen, 1824 Deleted from British list by Drake (2018b)]

MELANOSTOLUS Kowarz, 1884
melancholicus (Loew, 1869 – *Diaphorus*)

DOLICHOPODINAE
DOLICHOPUS Latreille, 1796
acuticornis Wiedemann, 1817
agilis Meigen, 1824
andalusiacus Strobl, 1899
arbustorum Stannius, 1831
argyrotarsis Wahlberg, 1850
atratus Meigen, 1824
atripes Meigen, 1824

brevipennis Meigen, 1824
caligatus Wahlberg, 1850
calinotus Loew, 1871 Added by Drake and Pollet (2017)
campestris Meigen, 1824
cilifemoratus Macquart, 1827
claviger Stannius, 1831
clavipes Haliday, 1832
diadema Haliday, 1832 Ex subgenus *Macrodolichopus* (Brooks 2005)
discifer Stannius, 1831
excisus Loew, 1859 Added by Gibbs (2006)
festivus Haliday, 1832
griseipennis Stannius, 1831
laticola Verrall, 1904
latilimbatus Macquart, 1827
latipennis Fallén, 1823 Ex subgenus *Hygroceleuthus* (Pollet *et al.* 2004)
lepidus Staeger, 1842
linearis Meigen, 1824
lineatocornis Zetterstedt, 1843
longicornis Stannius, 1831
longitarsis Stannius, 1831
maculipennis Zetterstedt, 1843
mediicornis Verrall, 1875
melanopus Meigen, 1824
migrans Zetterstedt, 1843
nigripes Fallén, 1823
nitidus Fallén, 1823
notatus Staeger, 1842
nubilus Meigen, 1824
pennatus Meigen, 1824
phaeopus Haliday in Walker, 1851
picipes Meigen, 1824
planitarsis Fallén, 1823
plumipes (Scopoli, 1763 – *Musca*)
plumitarsis Fallén, 1823
popularis Wiedemann, 1817
rupestris Haliday, 1833
sabinus Haliday, 1838
signatus Meigen, 1824
signifer Haliday, 1838
simplex Meigen, 1824
strigipes Verrall, 1875
subpennatus d'Assis-Fonseca, 1976
trivialis Haliday, 1832
ungulatus (Linnaeus, 1758 – *Musca*)
urbanus Meigen, 1824
virgultorum Haliday in Walker, 1851
vitripennis Meigen, 1824
wahlbergi Zetterstedt, 1843

ETHIROMYIA Brooks in Brooks and Wheeler, 2005
chalybea (Wiedemann, 1817 – *Dolichopus*) Transferred here by Brooks and Wheeler (2005)

GYMNOPTERNUS Loew, 1857 Generic rank (Pollet 2003)
aerosus (Fallén, 1823 – *Dolichopus*)
angustifrons (Staeger, 1842 – *Dolichopus*)
assimilis (Staeger, 1842 – *Dolichopus*)
blankaartensis (Pollet, 1990 – *Hercostomus*)
brevicornis (Staeger, 1842 – *Dolichopus*)
celer (Meigen, 1824 – *Dolichopus*)
cupreus (Fallén, 1823 – *Dolichopus*)
metallicus (Stannius, 1831 – *Dolichopus*)
silvestris (Pollet, 1990 – *Hercostomus*)

HERCOSTOMUS Loew, 1857 Stat. rev. by Pollet (2003)
argentifrons Oldenberg, 1916 Added by Godfrey and Drake (2024)
chetifer (Walker, 1849 – *Porphyrops*)
fulvicaudis (Haliday in Walker, 1851 – *Sybistroma*)
germanus (Wiedemann, 1817 – *Dolichopus*)
gracilis (Stannius, 1831 – *Dolichopus*)
nanus (Macquart, 1827 – *Dolichopus*)
nigrilamellatus (Macquart, 1827 – *Dolichopus*)
nigripennis (Fallén, 1823 – *Dolichopus*)
nigriplantis (Stannius, 1831 – *Dolichopus*)
parvilamellatus (Macquart, 1827 – *Dolichopus*)
plagiatus (Loew, 1857 – *Gymnopternus*)
rothi (Zetterstedt, 1859 – *Dolichopus*) Added by Drake *et al.* (2013)
rusticus (Meigen, 1824 – *Dolichopus*) Added by Drake and Painter (2023)
sahlbergi (Zetterstedt, 1838 – *Dolichopus*)
verbekei Pollet, 1993 Added by Pollet (1993)

MUSCIDIDEICUS Becker, 1917
praetextatus (Haliday, 1855 – *Dolichopus*)

ORTOCHILE Latreille, 1809
nigrocoerulea Latreille, 1809

POECILOBOTHRUS Mik, 1878
chrysozygos (Wiedemann, 1817 – *Dolichopus*) Transferred from *Hercostomus* by Brooks (2005)
ducalis (Loew, 1857 – *Gymnopternus*)
majesticus d'Assis-Fonseca, 1976
nobilitatus (Linnaeus, 1767 – *Musca*)
principalis (Loew, 1861 – *Gymnopternus*)

SYBISTROMA Meigen, 1824
crinipes Staeger, 1842
discipes (Germar, 1821 – *Dolichopus*)
obscurellum (Fallén, 1823 – *Dolichopus*)

TACHYTRECHUS Haliday in Walker, 1851
consobrinus (Haliday in Walker, 1851 – *Dolichopus*)
insignis (Stannius, 1831 – *Dolichopus*)
notatus (Stannius, 1831 – *Dolichopus*)
ripicola Loew, 1857

HYDROPHORINAE
APHROSYLUS Haliday in Walker, 1851
celtiber Haliday, 1855
ferox Haliday in Walker, 1851
mitis Verrall, 1912
raptor Haliday in Walker, 1851

HYDROPHORUS Fallén, 1823
albiceps Frey, 1915
balticus (Meigen, 1824 – *Medeterus*)
bipunctatus (Lehmann, 1822 – *Dolichopus*)
litoreus Fallén, 1823
nebulosus Fallén, 1823
oceanus (Macquart, 1838 – *Medeterus*)
pilipes Frey, 1915
rufibarbis authors, misident., not Gerstäcker, 1864 Corrected by MacGowan and Drake (2023)
praecox (Lehmann, 1822 – *Dolichopus*)
viridis (Meigen, 1824 – *Medeterus*)

LIANCALUS Loew, 1857
virens (Scopoli, 1763 – *Musca*)

MACHAERIUM Haliday, 1832
maritimae Haliday, 1832

ORTHOCERATIUM Schrank, 1803
sabulosum (Becker, 1907 – *Alloeoneurus*)
lacustre: authors, misident., not (Scopoli, 1763) Corrected by Pollet and Stark (2018)

SCELLUS Loew, 1857
notatus (Fabricius, 1781 – *Musca*)

SCHOENOPHILUS Mik, 1878
versutus (Haliday in Walker, 1851 – *Thinophilus*)

THINOPHILUS Wahlberg, 1844
flavipalpis (Zetterstedt, 1843 – *Rhaphium*)
ruficornis (Haliday, 1838 – *Medeterus*)

MEDETERINAE
CORINDIA Bickel, 1986
viridis (Parent, 1932) Transferred from *Thrypticus* and added by Drake *et al.* (2023)

CYRTURELLA Collin, 1952
albosetosa (Strobl in Czerny and Strobl, 1909 – *Micromorphus*)

DOLICHOPHORUS Lichtwardt, 1902
kerteszi Lichtwardt, 1902 Added by Drake (2005)

MEDETERA Fischer von Waldheim, 1819
abstrusa Thuneberg, 1955
acanthura Negrobov and Thuneberg, 1970 Added by Drake (2024)
ambigua (Zetterstedt, 1843 – *Hydrophorus*)
belgica Parent, 1936 Added by Drake (2024)
bilineata Frey, 1915 – possibly synonymous with *veles* Loew
bispinosa Negrobov, 1967
borealis Thuneberg, 1955
cuspidata Collin, 1941 (*Medeterus*)
dendrobaena Kowarz, 1877 (*Medeterus*)
diadema (Linnaeus, 1767 – *Musca*)
excellens Frey, 1909
fasciata Frey, 1915 Added in Chandler (1998), formally by MacGowan (2001)
feminina Negrobov, 1967
oscillans Allen, 1976 Syn. by Drake *et al.* (2024)
flavipes Meigen, 1824 (*Medeterus*)
freyi Thuneberg, 1955 Added by MacGowan (2001)
grisescens de Meijere, 1916 Added by Halstead (2003); introduced glasshouse species
impigra Collin, 1941 (*Medeterus*)
infumata Loew, 1857
insignis Girschner, 1888 Added by Gibbs (2007)
inspissata Collin, 1952 (*Medeterus*)
jacula (Fallén, 1823 – *Hydrophorus*)
jugalis Collin, 1941 (*Medeterus*)
melancholica Lundbeck, 1912 (*Medeterus*)
micacea Loew, 1857
mixta Negrobov, 1967 Added by Drake (2024)
muralis Meigen, 1824 (*Medeterus*)
nigrohalteralis Pollet and Drake, 2025 Added by Pollet *et al.* (2025)
nitida (Macquart, 1834 – *Hydrophorus*) Requires confirmation as British (Dyte 1996)
obscura (Zetterstedt, 1838 – *Hydrophorus*)
pallipes (Zetterstedt, 1843 – *Hydrophorus*)

parenti Stackelberg, 1925
peloria Negrobov, 1967 Added by Drake (2024)
petrophila Kowarz, 1877 (*Medeterus*)
petrophiloides Parent, 1925
pinicola Kowarz, 1877 (*Medeterus*)
pseudoapicalis Thuneberg, 1955 Added here
saxatilis Collin, 1941 (*Medeterus*)
setiventris Thuneberg, 1955 Added in Chandler (1998), formally by MacGowan (2001)
striata Parent, 1927 Requires confirmation as British (MacGowan 2001)
tristis (Zetterstedt, 1838 – *Hydrophorus*)
truncorum Meigen, 1824 (*Medeterus*)
unisetosa Collin, 1941 (*Medeterus*)
veles Loew, 1861

SYSTENUS Loew, 1857
bipartitus (Loew, 1850 – *Rhaphium*)
leucurus Loew, 1859
mallochi MacGowan, 1997
alpinus: Speight *et al.*, 1992, misident., not Vaillant, 1978 Corrected by Alexander *et al.* (2008)
pallipes (von Roser, 1840 – *Rhaphium*)
pallidus Vaillant, 1978 Added by Speight and Meuffels (1989) but syn. by Kassebeer (1998)
scholtzii (Loew, 1850 – *Rhaphium*)
tener Loew, 1859 Syn. with *S. bipartitus* in Negrobov (1991) but restored by Kassebeer (1998)

THRYPTICUS Gerstäcker, 1864
atomus Frey, 1915 Added by Drake *et al.* (2023)
bellus Loew, 1869
cuneatus (Becker, 1917 – *Submedeterus*)
cynicus Drake and Godfrey, 2023 Added by Drake *et al.* (2023)
divisus (Strobl, 1880 – *Chrysotus*)
intercedens Negrobov, 1967 Added by Drake (2016), formally by Drake *et al.* (2023)
laetus Verrall, 1912
lichanus Drake, 2023 Added by Drake *et al.* (2023)
nigricauda Wood, 1913
pollinosus Verrall, 1912
smaragdinus Gerstäcker, 1864
tarsalis Parent, 1932

MICROPHORINAE Ex Microphoridae (Sinclair and Cumming 2006)
MICROPHOR Macquart, 1827
anomalus (Meigen, 1824 – *Platypeza*)
crassipes Macquart, 1827
holosericeus (Meigen, 1804 – *Empis*)
intermedius Collin, 1961 Added here
strobli Chvála, 1986 Added by Plant and Cole (2005)

NEURIGONINAE
NEURIGONA Rondani, 1856
abdominalis (Fallén, 1823 – *Dolichopus*)
biflexa Strobl in Czerny and Strobl, 1909
erichsoni (Zetterstedt, 1843 – *Dolichopus*) Added by Drake and Hunnisett (2014)
pallida (Fallén, 1823 – *Dolichopus*)
quadrifasciata (Fabricius, 1781 – *Musca*)
suturalis (Fallén, 1823 – *Dolichopus*)

PELOROPEODINAE
ACROPSILUS Mik, 1878 Possibly *insertae sedis* Bickel (1998)
niger (Loew, 1869 – *Chrysotus*)

CHRYSOTIMUS Haliday in Loew, 1857
flaviventris (von Roser, 1840 – *Chrysotus*)
molliculus (Fallén, 1823 – *Dolichopus*)

MICROMORPHUS Mik, 1878
Species A
Species B
Species C
albipes (Zetterstedt, 1843 – *Hydrophorus*) Identity uncertain

RHAPHIINAE
LAMPROCHROMUS Mik, 1878
bifasciatus (Macquart, 1827 – *Medeterus*)
kowarzi Negrobov and Chalaja, 1988 Added by Drake (2017)
semiflavus (Strobl, 1880 – *Diaphorus*)
strobli Parent, 1925 syn. by Grichanov and Ahmadi (2017)

NEMATOPROCTUS Loew, 1857
distendens (Meigen, 1824 – *Chrysotus*)
praesectus Loew, 1857 Added by Drake (2019)

RHAPHIUM Meigen, 1803
albomaculatum (Becker, 1891 – *Xiphandrium*)
antennatum (Carlier, 1835 – *Anglearia*)
appendiculatum Zetterstedt, 1849
auctum Loew, 1857
brevicorne Curtis, 1835

caliginosum Meigen, 1824
commune (Meigen, 1824 – *Porphyrops*)
consobrinum Zetterstedt, 1843
crassipes (Meigen, 1824 – *Porphyrops*)
elegantulum (Meigen, 1824 – *Porphyrops*)
fasciatum Meigen, 1824
fascipes (Meigen, 1824 – *Porphyrops*)
fractum Loew, 1850
gravipes Haliday in Walker, 1851
lanceolatum Loew, 1850
laticorne (Fallén, 1823 – *Hydrophorus*)
longicorne (Fallén, 1823 – *Hydrochus*)
micans (Meigen, 1824 – *Porphyrops*)
monotrichum Loew, 1850
nasutum (Fallén, 1823 – *Hydrochus*)
patulum (Raddatz, 1873 – *Porphyrops*)
pectinatum (Loew, 1859 – *Porphyrops*)
penicillatum Loew, 1850
riparium (Meigen, 1824 – *Porphyrops*)
rivale (Loew, 1869 – *Porphyrops*)
suave (Loew, 1859) Added by Drake (2007b)

SCIAPODINAE
SCIAPUS Zeller, 1842
basilicus Meuffels and Grootaert, 1990 Added in Howe (1997; J.H. Cole)
contristans (Wiedemann, 1817 – *Dolichopus*)
heteropygus Parent, 1926 (*Sciopus*)
laetus (Meigen, 1838 – *Psilopus*)
longulus (Fallén, 1823 – *Leptopus*)
maritimus Becker, 1918
pallens (Wiedemann, 1830 – *Psilopus*) Added by Sivell (2019)
platypterus (Fabricius, 1805 – *Dolichopus*)
wiedemanni (Fallén, 1823 – *Leptopus*)
zonatulus (Zetterstedt, 1843 – *Psilopus*)

SYMPYCNINAE
ANEPSIOMYIA Bezzi, 1902
flaviventris (Meigen, 1824 – *Porphyrops*)

CAMPSICNEMUS Haliday in Walker, 1851
alpinus (Haliday, 1833 – *Medeterus*)
armatus (Zetterstedt, 1849 – *Dolichopus*)
compeditus Loew, 1857
curvipes (Fallén, 1823 – *Dolichopus*)
dasycnemus Loew, 1857 Irish only; added by Chandler (1989)
loripes (Haliday, 1832 – *Medeterus*)
magius (Loew, 1845 – *Medeterus*)
marginatus Loew, 1857
picticornis (Zetterstedt, 1843 – *Dolichopus*)
pumilio (Zetterstedt, 1843 – *Dolichopus*)
pusillus (Meigen, 1824 – *Medeterus*)
scambus (Fallén, 1823 – *Dolichopus*)
umbripennis Loew, 1856
subspecies ***hispanicus*** Strobl, 1899 Added by Perry (1999)

MICROPYGUS Bickel and Dyte, 1989
vagans Parent, 1933

SYMPYCNUS Loew, 1857
aeneicoxa (Meigen, 1824 – *Porphyrops*)
cirripes (Haliday in Walker, 1851 – *Porphyrops*)
pulicarius (Fallén, 1823 – *Dolichopus*)
desoutteri Parent, 1925 Syn. by Pollet *et al.* (2015)
annulipes Meigen, 1824 Syn. by Pollet *et al.* (2015)
septentrionalis Pollet, Persson, Bøggild and Crossley, 2015 Added by Pollet *et al.* (2015)
spiculatus Gerstäcker, 1864

SYNTORMON Loew, 1857
aulicum (Meigen, 1824 – Porphyrops)
bicolorellum (Zetterstedt, 1843 – *Dolichopus*)
denticulatum (Zetterstedt, 1843 – *Rhaphium*)
filiger Verrall, 1912
fuscipes (von Roser, 1840 – *Porphyrops*)
macula Parent, 1927
metathesis (Loew, 1850) Added by Drake (2023)
mikii Strobl, 1899
setosum Parent, 1938 Added to Irish list by Speight and Meuffels (1989), regarded as synonym in Britain (Drake 2021a)
monile (Haliday in Walker, 1851 – *Rhaphium*)
pallipes (Fabricius, 1794 – Musca)
pseudospicatum Strobl, 1899 New to List (Dyte, pers. comm.), confirmed by Drake (2020b)
pumilum (Meigen, 1824 – *Rhaphium*)
submonilis Negrobov, 1975 Added by Drake (2021a)
silvianum Pârvu, 1989 Synonymised by Drake (2021a)
sulcipes (Meigen, 1824 – *Rhaphium*)
tarsatum (Fallén, 1823 – *Hydrochus*)
zelleri (Loew, 1850 – *Rhaphium*)

TELMATURGUS Mik, 1874
tumidulus (Raddatz, 1873 – *Sympycnus*)

TEUCHOPHORUS Loew, 1857
calcaratus (Macquart, 1827 – *Medeterus*)
monacanthus Loew, 1859
nigricosta (von Roser, 1840 – *Chrysotus*)
simplex Mik, 1881
spinigerellus (Zetterstedt, 1843 – *Dolichopus*)

XANTHOCHLORINAE
XANTHOCHLORUS Loew, 1857
galbanus Chandler and Negrobov, 2008
luridus: Dyte, 1987, misident. not Negrobov, 1978 Added by Dyte (1987), revised by Chandler and Negrobov (2008)
ornatus (Haliday, 1832 – *Medeterus*)
tenellus (Wiedemann, 1817 – *Dolichopus*)
silaceus Chandler and Negrobov, 2008
Added by Chandler and Negrobov (2008)

Key to genera and some species of Dolichopodidae (males)

1 Wing with discal medial cell (dm) having 3 veins apically, and a dark lens-shaped pterostigma beyond end of R_1 (a). Tiny to small dark species with no hint of metallic colouration, wing length 2–3.5 mm........ ***Microphor*** (5 spp.; pp. 250)

– Wing with discal medial cell having 2 veins at its tip and no pterostigma........2

(a)

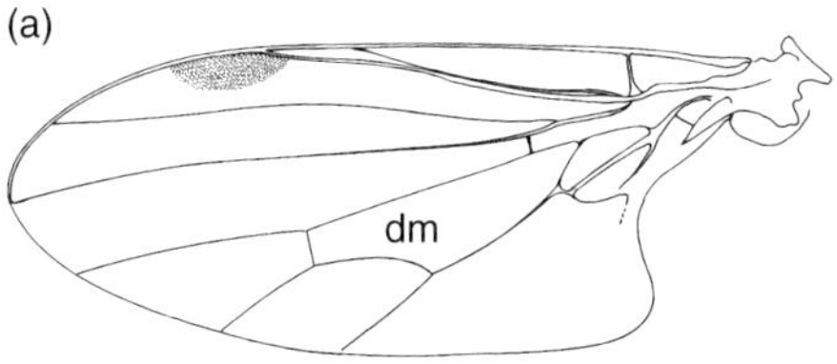

2(1) Vein M forked beyond crossvein dm-m, anterior branch (M_1) reaching wing margin close to R_{4+5}, posterior branch (M_2) long and reaching at least halfway to wing margin (a)........ ***Sciapus*** (10 spp.; p. 300)

– Vein M not forked, or when apparently so (some *Dolichopus*) then posterior branch very short (aa).........3

(a)

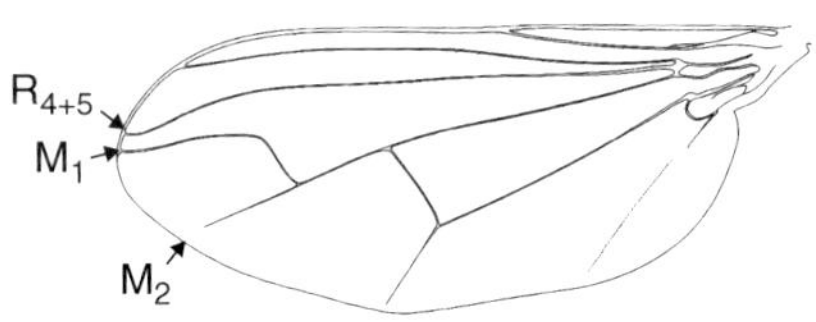

(aa)

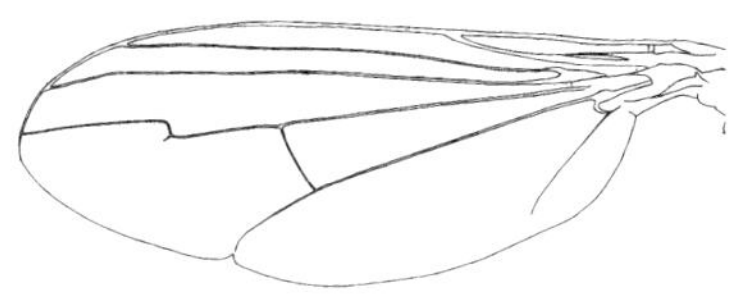

3(2) Setae of thorax black or dark brown.........4

– Setae of thorax pale, at least in anterior view. Mainly tiny to small species, wing length 1.5–2.5 mm and a few up to 3.5 mm.........44

4(3) Large species with extensive dark wing markings, wing length about 6 mm (fig. a of female couplet 13). Front tibia with a stout anteroventral forked peg at basal third and a broad extension at tip (a). Tergite 5 with a long yellow flag protruding from each side (b). Dorsocentral setae in front of wing base tiny, far shorter than distance between rows, and numerous (8–12 pairs)......... ***Scellus*** (1 sp. ***notatus***; p. 299)

– Large to tiny species, wing membrane marked or clear. Front tibia without an anteroventral peg and apical extension. Tergite 5 without lateral flag. Dorsocentral setae various.........5

(a)

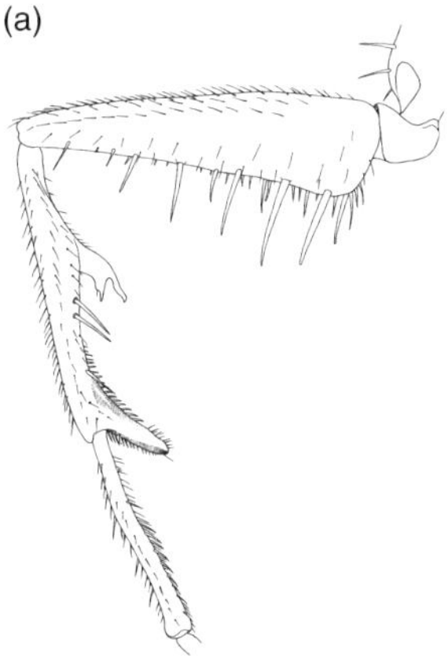

(b)

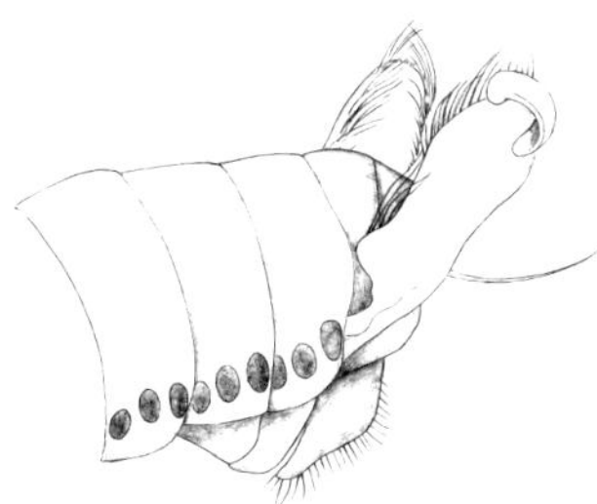

5(4) Tergite 6 without vestiture hairs (a–c) or tergite not visible. (Ignore fine pubescence. Do not confuse with sternite 8 attached to hypopygium. Tiny species in which tergite 6 is almost entirely hidden and any hairs are almost impossible to discern are included in this section. If in doubt, following the first half of couplet works in most cases.) 6

– Tergite 6 with vestiture hairs (aa) even if only on hind and lateral margins (bb, cc), tergite always visible, 0.25–1.0 times as long as tergite 5. 29

(a) *Dolichopus*

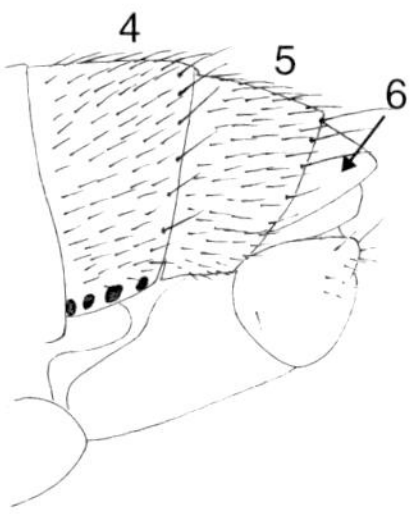

(b) *Hydrophorus*

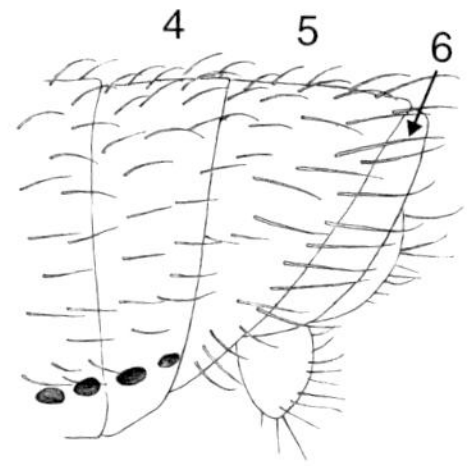

(c) *Diaphorus*

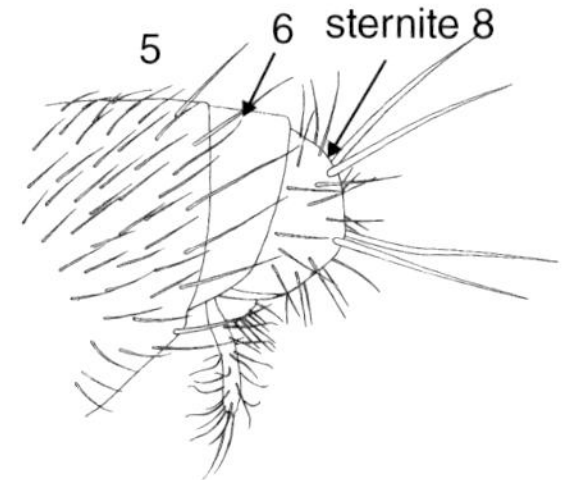

(aa) *Thinophilus*

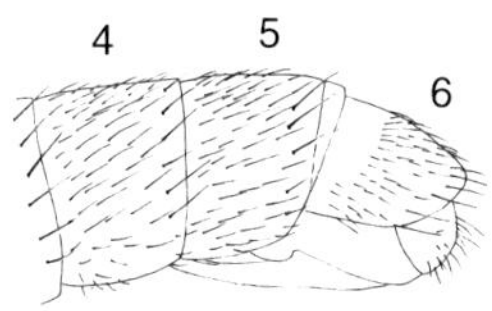

(bb) *Nematoproctus*

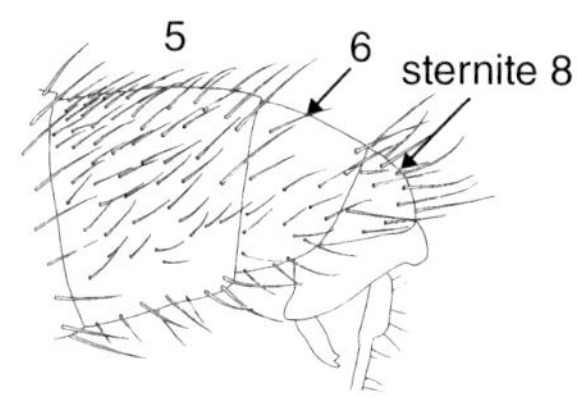

(cc) *Rhaphium*

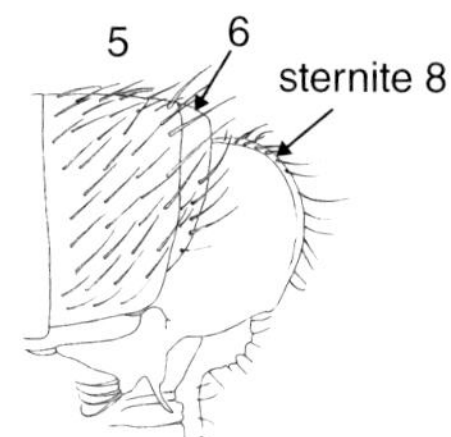

6(5) Sternite 8 at tip of abdomen without outstanding setae, any long setae finer than arista. Hind coxa either with lateral seta absent or placed at more than one-fifth distance from base. Front tarsus with large or small pulvilli. 7

– Sternite 8 at tip of abdomen with 4–8 conspicuously long stout setae at least as thick as base of arista (fig. c of couplet 5). Hind coxa with lateral seta placed close to base, at about one-fifth distance from base or about 2–3 diameters of seta socket from base (aa). Front tarsus with large pulvilli about as long as last tarsomere (bb) (small in *Argyra elongata*). 26

(aa)

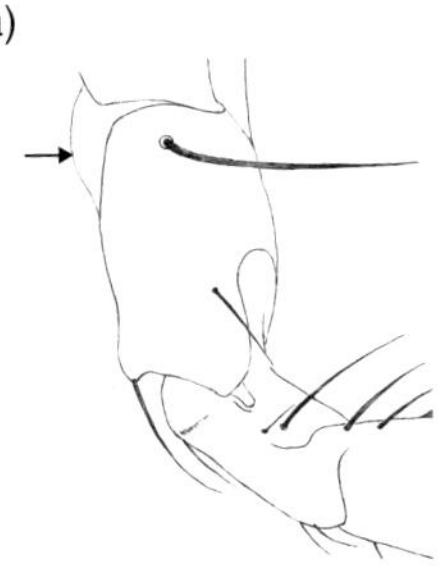

(bb)

7(6) Acrostichal setae biserial between first and penultimate dorsocentral setae, the 2 rows of setae distinctly separated and not irregular except sometimes beyond penultimate dorsocentral seta 8

– Acrostichal setae either absent or mainly uniserial, sometimes partly irregular and forming 2 closely spaced rows. If irregular and mouthparts as long as depth of head (fig. d of next couplet), follow first half of couplet (*Ortochile*) 20

8(7) Arista subapical or dorsal, usually well removed from tip of postpedicel (a) or only doubtful when postpedicel is almost round. Palps short, usually much less than one-quarter height of eye (b) except *Hercostomus nigripennis* (c) and *Ortochile* (d) where about half or more height of eye. Tergite 6 visible, at least one-third length of tergite 5 (fig. a of couplet 5), or if unclear then mid femur without an anterior preapical seta. 9

– Arista apical or almost so on elongate (aa) or pointed-triangular postpedicel (bb). Palp large, at least a third height of eye (aa, bb). Tergite 6 not visible in dry material but may be so in wet material (ee). 19

(a)

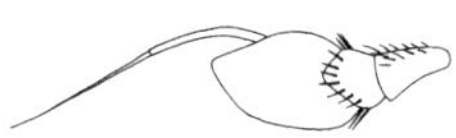

(b)

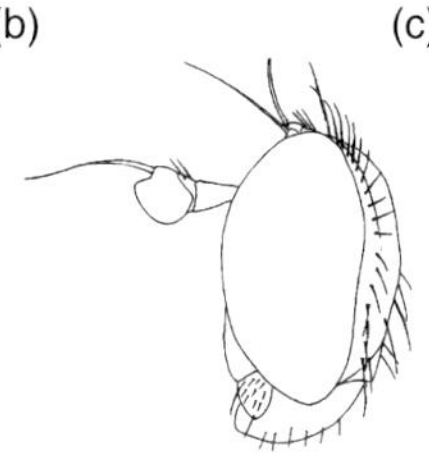

(c)

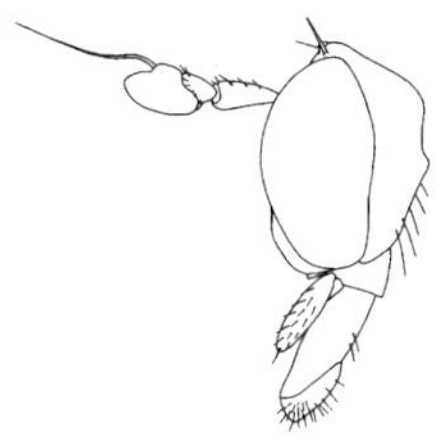

(d)

(aa)

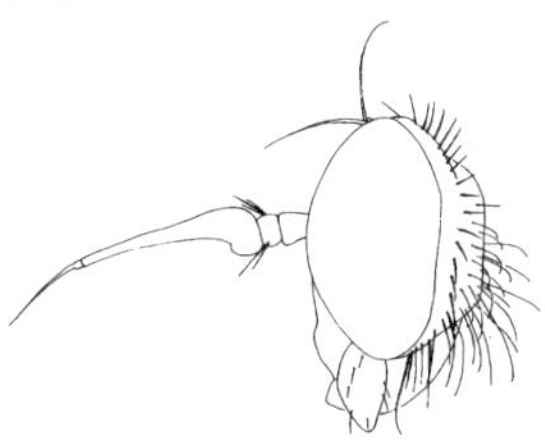

(bb)

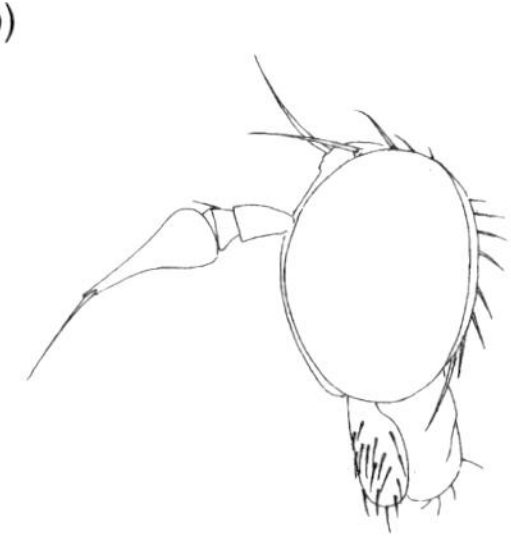

(ee)

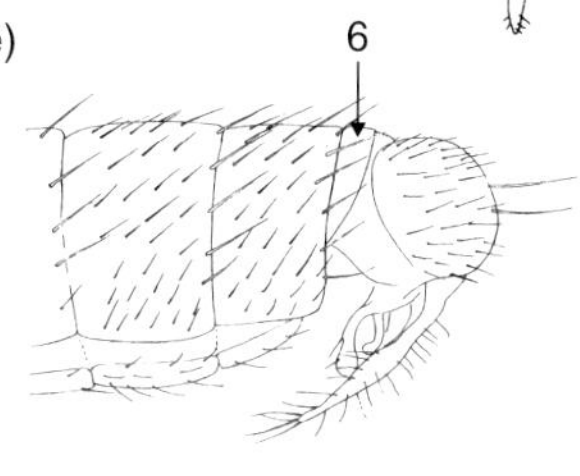

9(8) Abdomen partly or entirely yellow. Mid and hind femora lacking anterior or anterodorsal preapical seta. Hind tibia with weak dorsal setae (ad, pd) scarcely longer than width of shaft. Scape bare dorsally. Scutum in front of scutellum flat or depressed between last 2 pairs of dorsocentral setae .... ***Neurigona*** (6 spp.; p. 258)

– Abdomen dark except in *Hercostomus fulvicaudis* and *H. rothi*. Mid femur and usually hind femur with 1 or more anterior or anterodorsal preapical seta (aa) (lost setae are indicated by the socket and irregularly spaced vestiture nearby). Hind tibia with strong dorsal setae (bb). Scape with tiny hairs dorsally (fig. a of couplet 8). Scutum in front of scutellum rounded, undifferentiated from rest of scutum. 10

(aa)

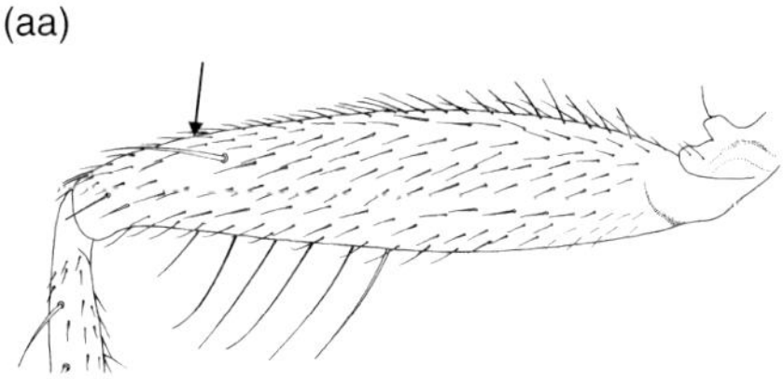

(bb)

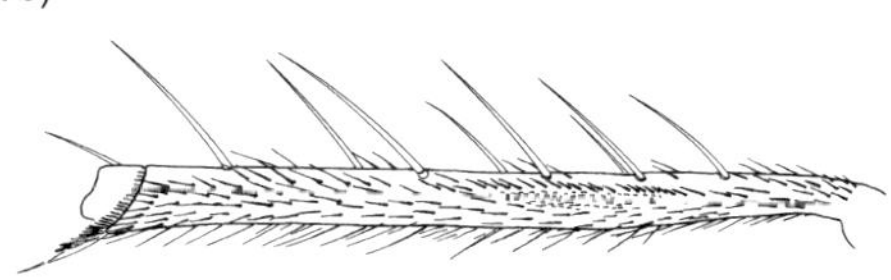

10(9) Hind basitarsus with at least 1 strong dorsal seta (a). Vein M_1 nearly always with a pronounced double inflexion midway between dm-m and wing tip (b) or sometimes less obvious (c). .... ***Dolichopus*** (55 spp.; p. 115)

– Hind basitarsus without strong dorsal setae. Vein M_1 either almost straight or gently sinuous, rarely as fig. c. 11

(a)

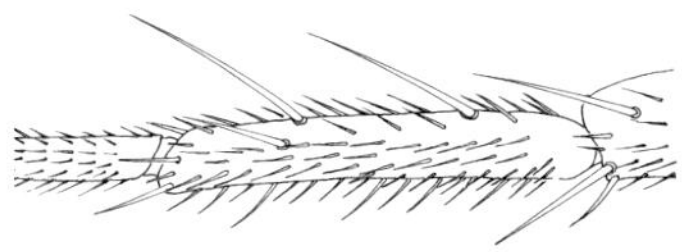

(b)

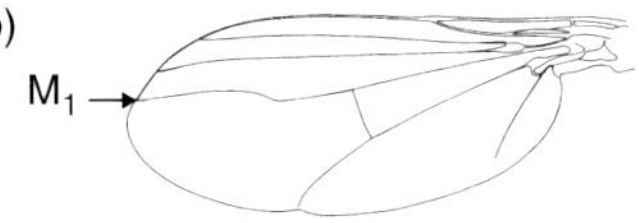

(c)

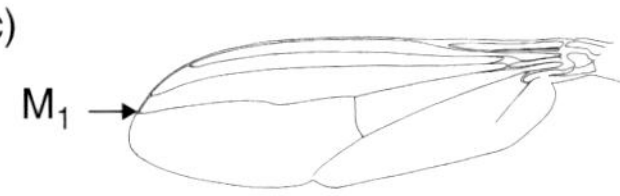

11(10) Hind femur with several strong anterodorsal (almost dorsal) setae on apical half in addition to 1 or 2 anterior preapical setae (a). Epistoma narrowest just below antennae, widening towards clypeus (b). Clypeus free at sides and extending to or below bottom of eye (b). .... ***Tachytrechus*** (4 spp.; p. 340)

– Hind femur without a row of nearly dorsal setae in addition to 1 (rarely 2) anterior preapical seta. Epistoma tapering towards clypeus or parallel-sided. Clypeus fused with eye margin and not reaching bottom of eye. 12

(a)

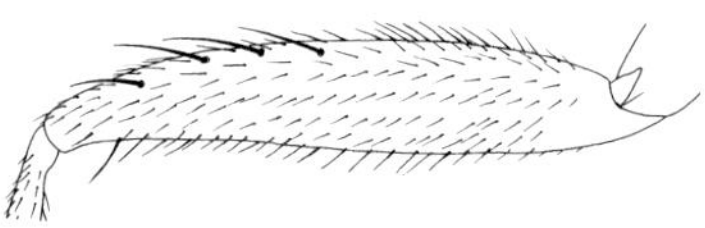

(b)

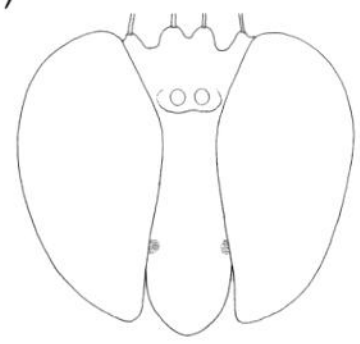

12(11) Epandrial lobe and its long untidy curled hairs extending well beyond cercus and its hairs (a). Tergite 5 slightly longer than tergite 4 and its lateral setae far longer than those on tergite 4 (b, dorsal view); whole abdomen slender and at least twice as long as thorax. Hind tibia with 4–7 setae on ventral face (v or pv), all fine and scarcely longer than width of shaft (c). .... ***Sybistroma*** (3 spp.; p. 311)

– Epandrial lobe and its hairs inconspicuous, not extending as far as cercus which is the dominant haired appendage (figs in next couplet) except in some *Hercostomus* (aa). Tergite 5 not longer than tergite 4, its lateral setae not markedly longer than those on tergite 4; whole abdomen more robust and not more than 1.5 times as long as thorax. Hind tibia as above or more often with at least 1 seta on ventral face (av, v, pv) clearly distinct from ventral hairs and longer than width of shaft 13

(a)

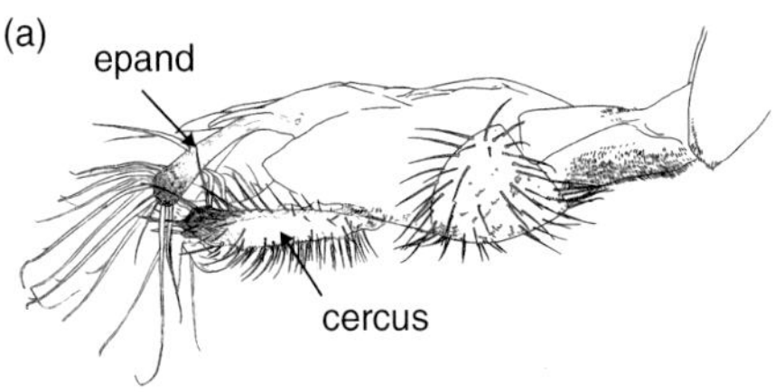

(aa)

(b)

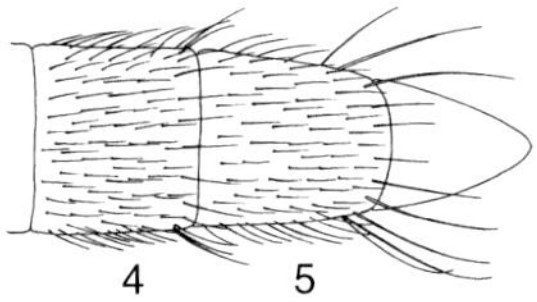

(c)

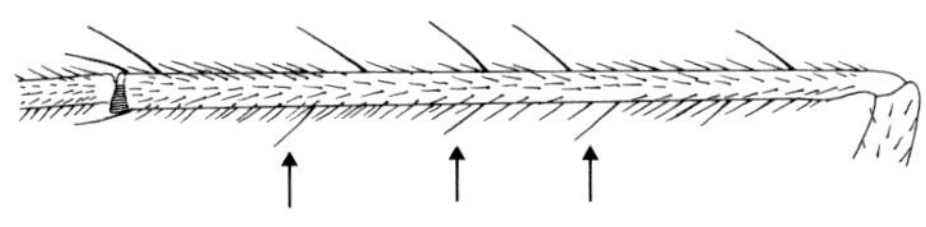

13(12) Cercus with long marginal hairs placed on thin tubercles several times longer than wide; cercus large, its longest dimension about half or more length of epandrium (a). Larger species, wing length 4–6.5 mm. 14

– Cercus with marginal hairs not placed on tubercles or on short tubercles up to 2 times longer than wide; cercus usually smaller (aa). Mainly smaller species, wing length 3–5 mm. 15

(a)

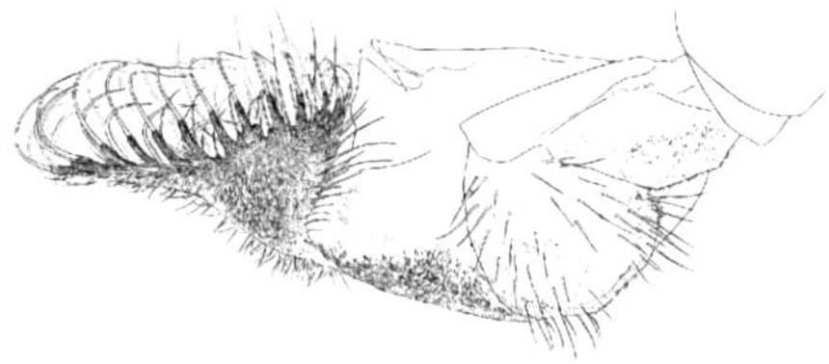

(aa)

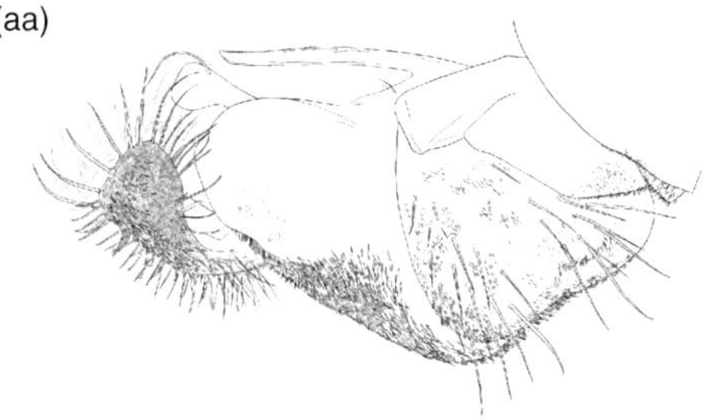

14(13) Front tibia without a long posteroventral apical seta (a). Scutum with a small matt blackish purple patch above notopleuron and extending from transverse suture to postsutural intra-alar seta, contrasting with surrounding shining green cuticle. .... ***Poecilobothrus*** (5 spp.; p. 264)

– Front tibia with a long posteroventral apical seta (aa). Scutum without an obvious purple patch .... 18 (not 15)

(a)

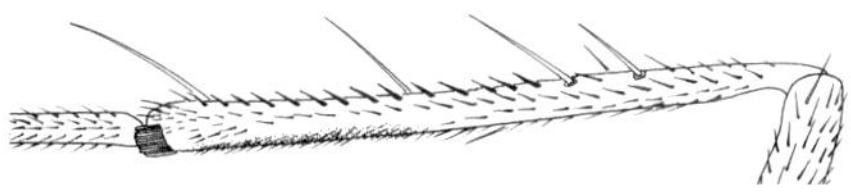

(aa)

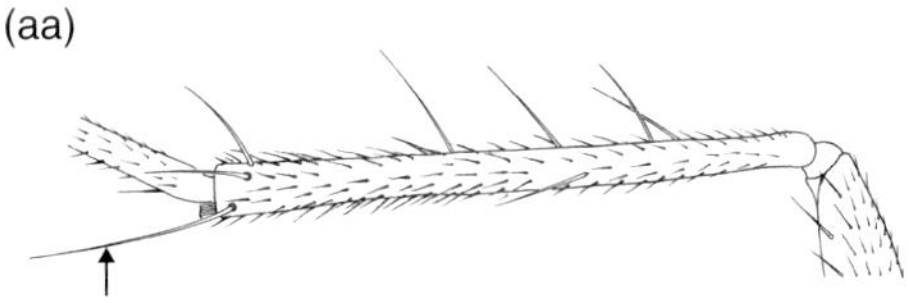

15(13) Abdomen broad and dorsoventrally flattened, tergites curling strongly inwards at sides, the downward-facing surface with long white hairs. Wing tip narrowly dark. R_{4+5} and M_1 both strongly bowed but remaining almost parallel beyond dm-m (a). Front coxa short, ending well above lower margin of katepisternum (b). Front of scutum with conspicuously pale tiny hairs. 7 pairs of dorsocentral setae. .... ***Muscidideicus*** (1 sp. ***praetextatus***; p. 256)

– Abdomen normal, deeper than wide with no demarcated underside to tergites, any pale vestiture on lower margin short and inconspicuous. Wing tip without dark apical mark. R_{4+5} and M_1 either almost parallel but straight or clearly converging towards wing tip. Front coxa long, extending to or beyond lower margin of katepisternum. Front of scutum with tiny black hairs. 6 pairs of dorsocentral setae. 16

(a)

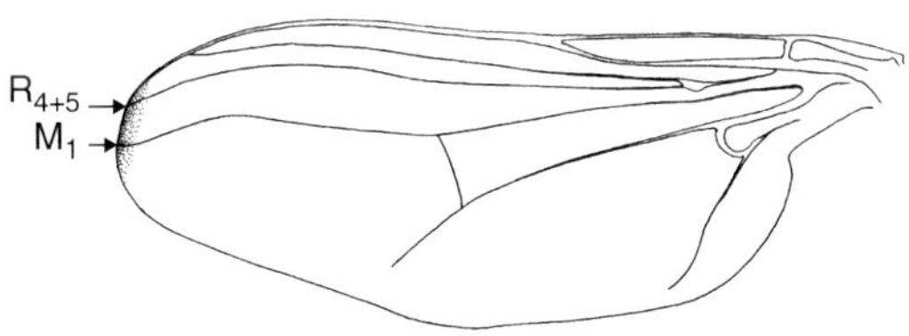

(b)

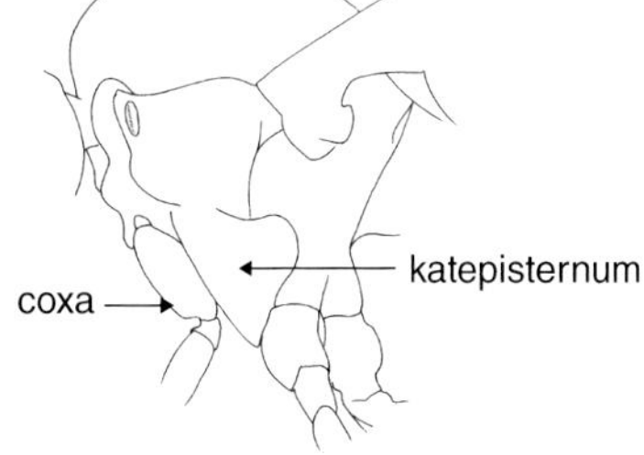

16(15) Palp and proboscis remarkably long and thin, pointing downwards; palp nearly 5 times longer than wide and clearly longer than antenna, dusted pale grey (a); proboscis narrow, its length 1–1.5 times height of head (a) (apparent length varies with degree of extension). Mid femur black on basal two-thirds, yellow on apical third (c)............................***Ortochile*** (1 sp. ***nigrocoerulea***; p. 263)

– Palp small, at most twice as long as wide, much shorter than antenna (aa); proboscis orientated horizontally (aa) except in *Hercostomus nigripennis* and *H. germanus* but then clearly shorter than height of head (bb). Mid femur without this colour combination, either black or yellow with only tips contrasting in colour. 17

(a) (aa) (bb) (c)

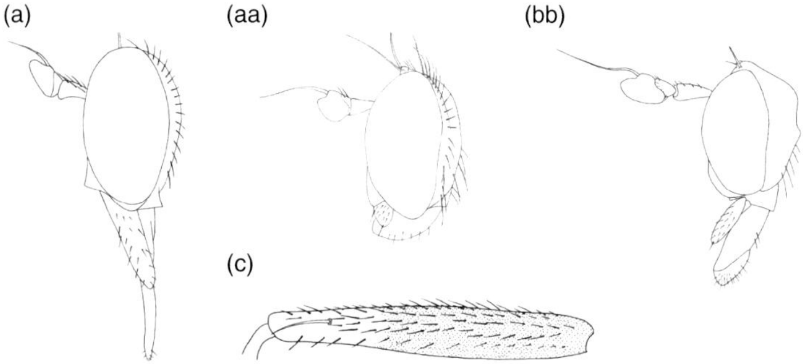

17(16) Vein M_1 beyond dm-m straight or almost so, and parallel to R_{4+5}, greatest distance between R_{4+5} and M_1 less than 1.5 times their separation at wing tip (a). Acrostichal row ending at or behind last (sixth) dorsocentral seta, usually forming a cluster of unpaired irregularly spaced setae in front of scutellum (c). Mid tibia at tip without a long dorsal seta between antero- and posterodorsal apical setae (a dorsal seta may be present but far shorter than other setae) (d). ....***Gymnopternus*** (9 spp.; p. 182)

– Vein M_1 beyond dm-m slightly sinuous or bowed, converging with R_{4+5} towards wing tip so their greatest separation is at least 1.5 times their separation at wing tip (aa), except in *H. nigripennis* with a long downward-pointing proboscis and *Ethiromyia* with violet scutum (bb). Acrostichal row ending before last dorsocentral seta and remaining strictly biserial (cc). Mid tibia with dorsal apical seta as stout as antero- and posterordorsal setae, even if shorter (dd).................................18

(a) (c) (d) (aa) (cc) (dd) (bb)

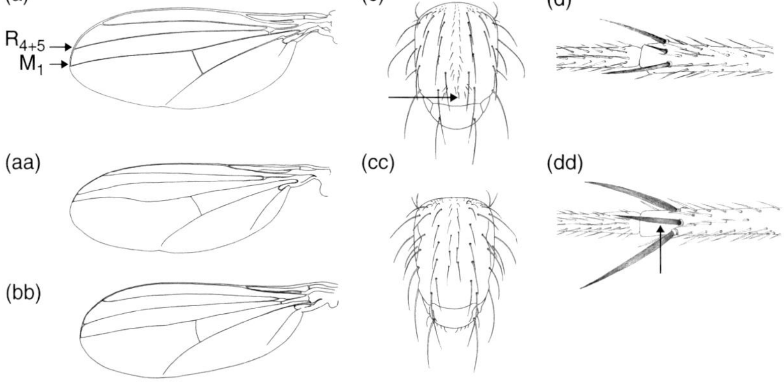

18(14,17) Scutum and frons shining blue- or violet-black with no pattern, when viewed from above. Wing strongly darkened (a) AND femora yellow. Scutellum on hind margin with a fringe of black hairs about one-fifth as long as scutellum (b). Pleuron in front of posterior spiracle with a cluster of tiny hairs (c). Palps as large as postpedicel. Front tibia with a long posteroventral apical seta. ....***Ethiromyia*** (1 sp. ***chalybea***; p. 182)

– Scutum green to greenish black, rarely with a shifting pale blue tinge. Wing strongly darkened only if femora black. Scutellum on hind margin without a fringe or with an inconspicuous fringe of tiny pale hairs. Pleuron in front of posterior spiracle without tiny hairs. Palps smaller than postpedicel. Front tibia with or without a long posteroventral apical seta. ....***Hercostomus*** (15 spp.; p. 192)

(a)

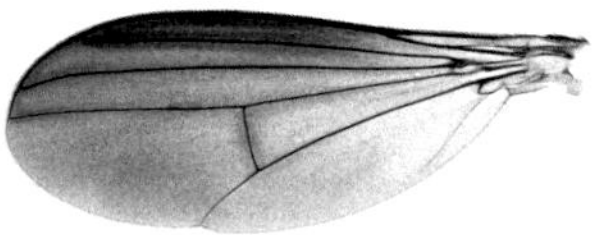

(b)

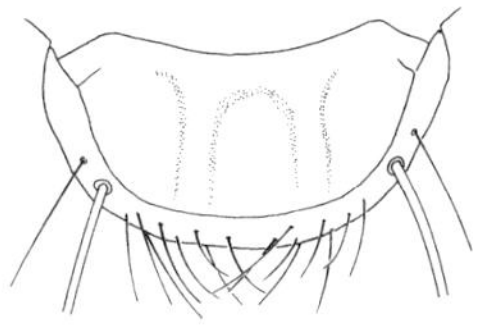

(c)

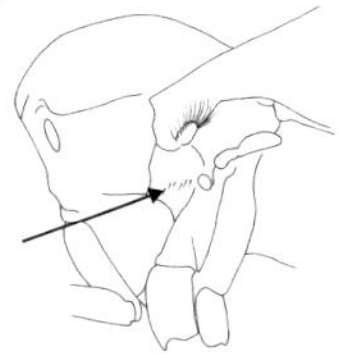

19(8) Medium-sized green fly, wing length 3.5–4.5 mm. Postpedicel long and narrow with bulbous base (a). Postgena with dense whiskers on lower part. Face widening from below antennae to mouth.......... ***Machaerium*** (1 sp. ***maritimae***; p. 218)

– Tiny to small yellow or dark brown species, wing length 1.7–3.0 mm. Postpedicel triangular, 2–2.5 times longer than deep (aa). Postgena with sparse setae on lower part. Face tapering from antennae to mouth.......... ***Achalcus*** (part, 8 spp.; p. 54)

(a)

(aa)

20(7) Small to large species, wing length at least 3.0 mm. Acrostichal setae present. Crossvein dm-m longer than apical section of M_4 (a) except in *Hydrophorus nebulosus* with mottled wings. Postgena with whiskers usually dense. Face margins parallel (b) or diverging from antennae to mouth (c).......... 21

– Tiny species, wing length up to 2.5 mm. Acrostichal setae absent. Crossvein dm-m shorter than apical section of M_4 (aa). Postgena with sparse hairs. Face margins either parallel, or tapering from below antennae to mouth (bb), or narrowest in middle.......... 23

(a)

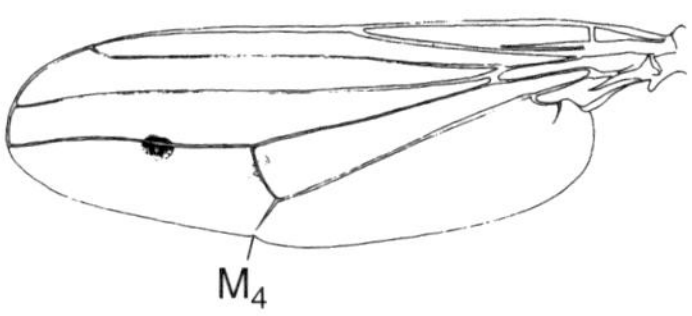

(aa)

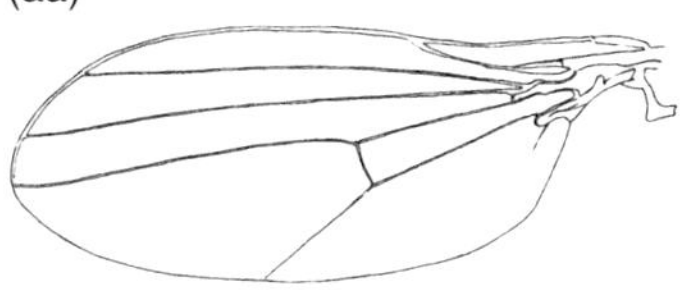

(b)

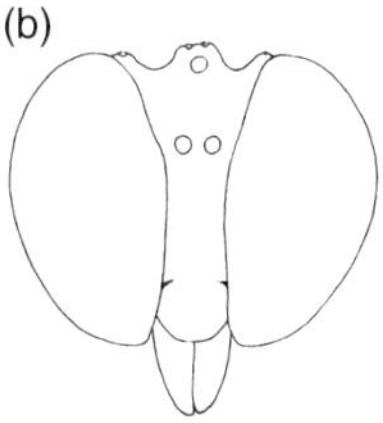

(c)

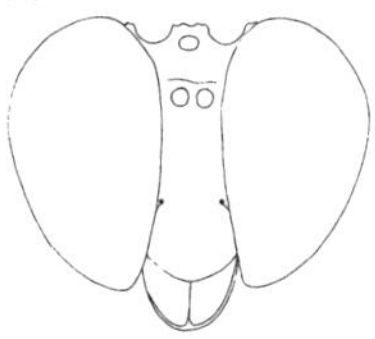

(bb)

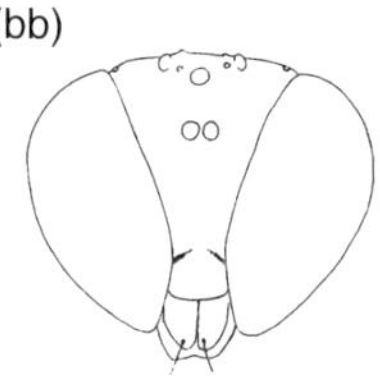

21(20) Front femur on basal half thicker than mid and hind femora, armed ventrally with spines or stout setae (a). Tergite 1 (and other tergites) with marginal setae scarcely distinguishable from vestiture. Hind and mid coxae without a lateral seta, at most with some tiny hairs. Postpedicel below tip with a small pimple (b).......... ***Hydrophorus*** (9 spp.; p. 207)

– Front femur with basal half not thicker than mid and hind femora, unarmed. Tergite 1 and often other tergites with distinct black marginal setae. Hind and mid coxae with a lateral seta. Postpedicel tip simple.......... 22

(a)

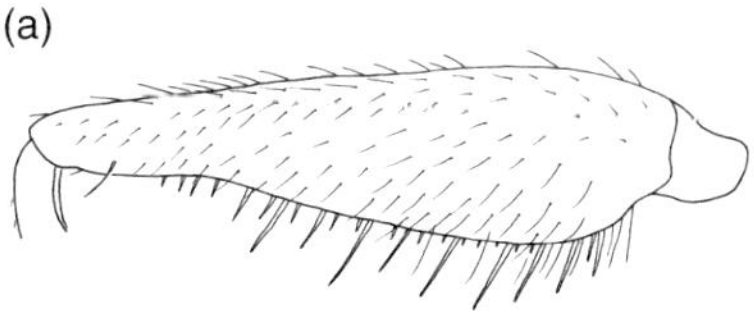

(b)

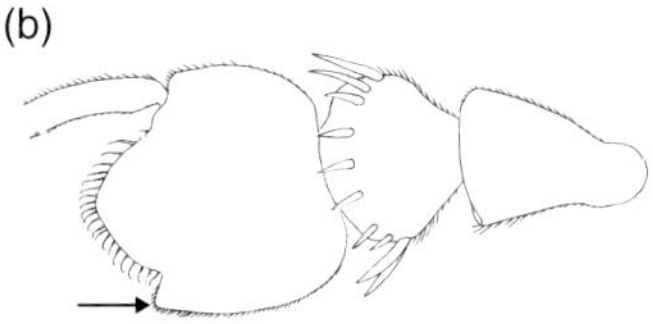

22(21) Wing with dark smudges on apical half and a small clear spot at extreme tip (a). Vein M_1 beyond dm-m slightly sinuous, converging with R_{4+5} towards wing margin (a). Scutellum with 3 pairs of long setae. Front coxa without an isolated long seta on anterior face. Face broad and widening towards mouth (fig. c of couplet 20). .... ***Liancalus*** (1 sp. ***virens***; p. 217)

– Wing clear. Vein M_1 with strong bends beyond dm-m, its last section parallel with R_{4+5} (aa). Scutellum with 1 long and 1 shorter pairs of setae. Front coxa with a single long seta on anterior face (bb). Face narrower and parallel-sided (fig. b of couplet 20). .... ***Orthoceratium*** (1 sp. ***sabulosum***; p. 262)

(a)

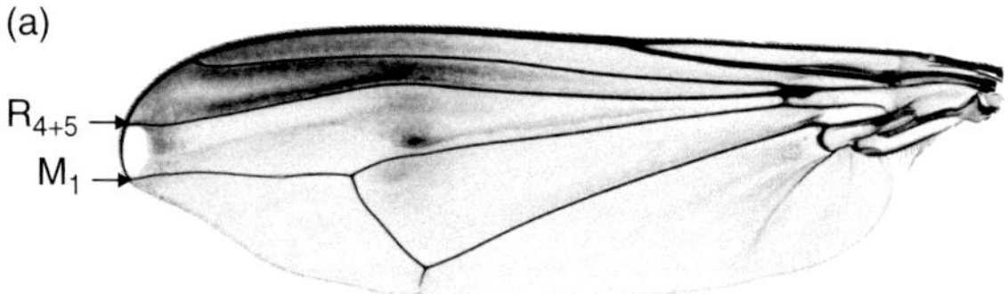

(bb)

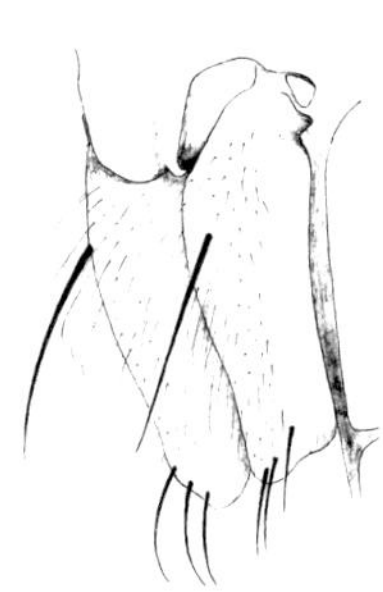

(aa)

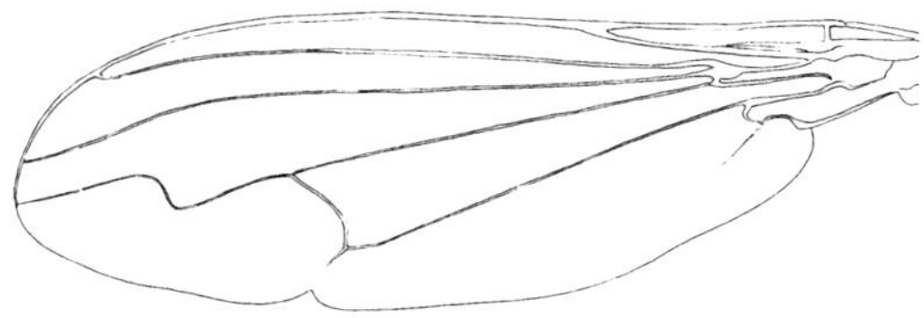

23(20) Pleura and part of abdomen bright yellow. Front femur and tibia armed ventrally with long setae (a). Postpedicel elongate, almost 3 times longer than deep, arista placed close to its base (b). .... ***Anepsiomyia*** (1 sp. ***flaviventris***; p. 62)

– Pleura and abdomen dark, at most dusky yellow above coxae. Front femur and tibia unarmed. Postpedicel scarcely longer than deep with arista placed at or beyond middle of dorsal margin. 24

(a)

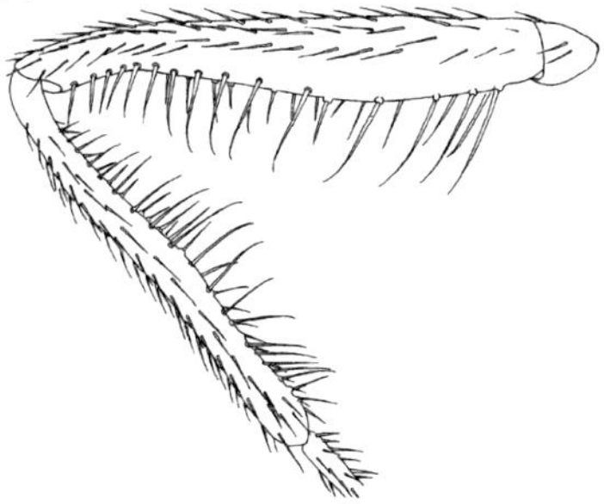

(b)

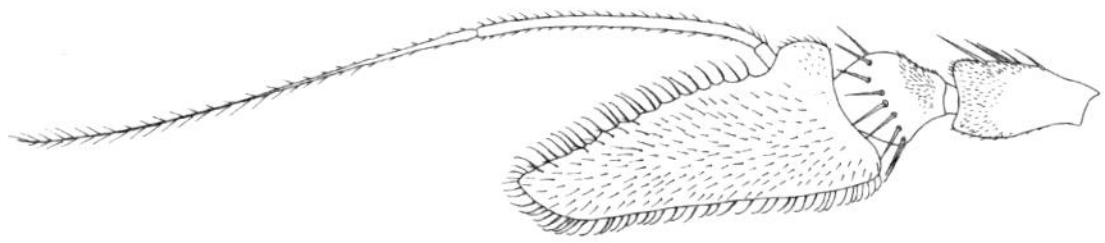

24(23) Scutum shining black. Hind basitarsus less than half length of tarsomere 2 (a). Cercus white and prominent with long hairs (b). .... ***Acropsilus*** (1 sp. ***niger***; p. 61)

– Scutum brown or grey, dull or slightly shining. Hind basitarsus and tarsomere 2 similar in length. Cercus small and inconspicuous, not white with long hairs. 25

(a)

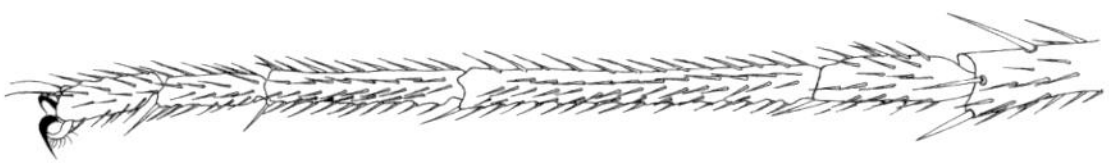

(b)

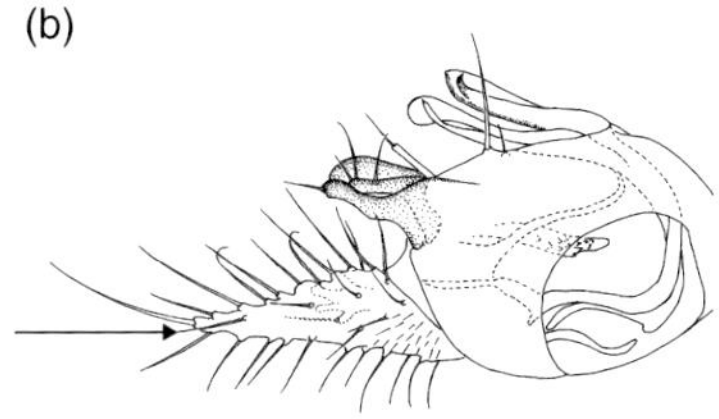

25(24) Palp as large as postpedicel; proboscis prominent and deeper than palp (a). Apical section of M_4 about 1.2–1.5 times length of dm-m (b). Apical section of M_1 with a faint small dark spot (b). Face brown-dusted. 4 dorsocentral and 1 notopleural setae. Tergite 6 not or scarcely visible. .... ***Schoenophilus*** (1 sp. ***versutus***; p. 299)

– Palp smaller than postpedicel; proboscis small, protruding below eyes no further than palp (aa). Apical section of M_4 about 4 times length of dm-m (bb). Apical section of M_1 without a mark. Face matt black. 5 dorsocentral and 2 notopleural setae. Tergite 6 visible, about half length of tergite 5. .... ***Micromorphus*** (3 spp.; p. 247)

(a)

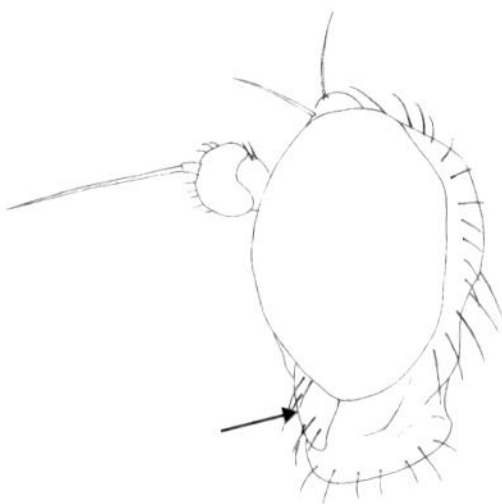

(aa)

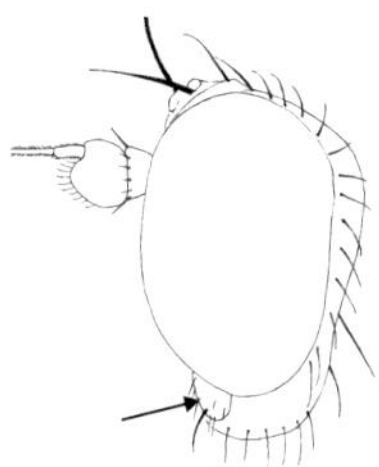

(b)

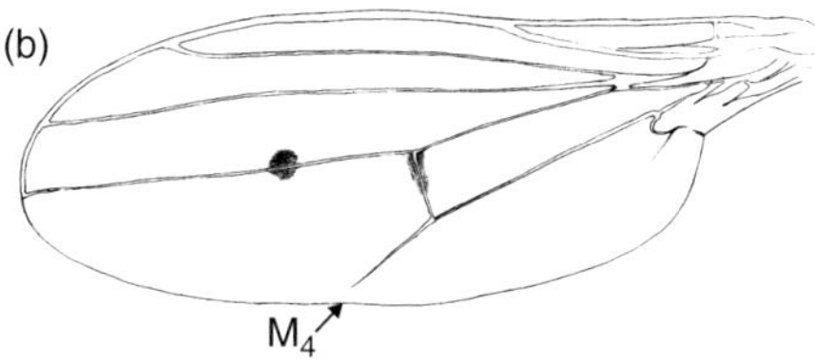

(bb)

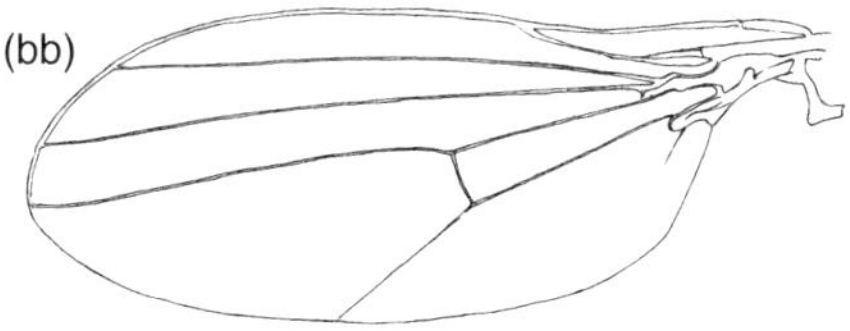

26(6) Eyes touching above antennae, obscuring nearly all of frons (a). Head with only 1 pair of long setae (ocellars) (a). .... ***Diaphorus*** (3 spp.; p. 111)

– Eyes well-separated by wide frons. Head with 2 pairs of long setae (ocellars and outer verticals). 27

(a)

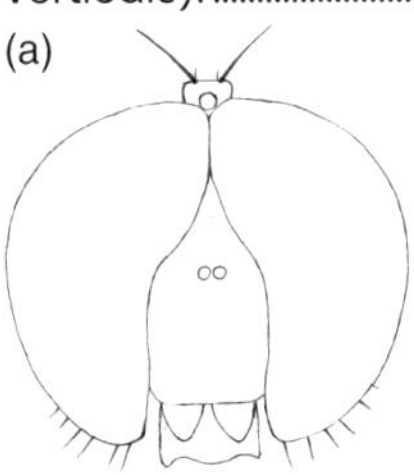

27(26) Larger species, wing length more than 2.5 mm. Legs yellow. Antenna longer than head, postpedicel twice as long as deep, scape with hairs (a) .... ***Argyra*** (part: ***elongata***; p. 65)

– Small species, wing length up to 2.5 mm. Legs black. Antenna much shorter than head, postpedicel not longer than deep, scape bare (aa) 28

(a)

(aa)

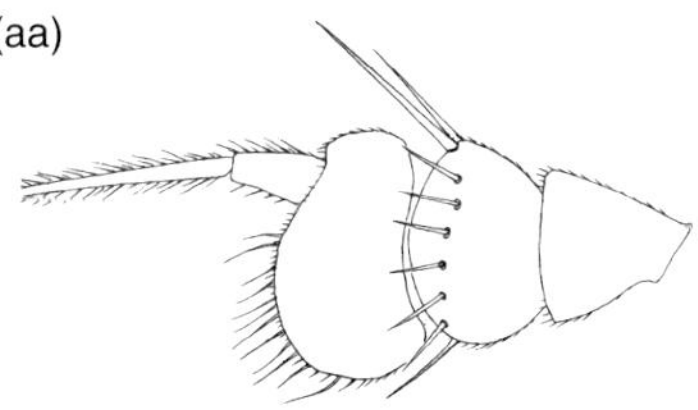

28(27) Halter black. Costa ending at M_1 (a). Veins R_{4+5} and M_1 almost parallel beyond dm-m (a); R_{4+5} not much further from R_{2+3} than from M_1. Mid tibia with 1 strong anterodorsal seta.......................... .... ***Melanostolus*** (1 sp. ***melancholicus***; p. 245)

– Halter pale. Costa ending at R_{4+5} (aa). Veins R_{4+5} and M_1 diverging widely and almost continuously towards wing margin (aa); R_{4+5} far closer to R_{2+3} than to M_1. Mid tibia with 2 strong anterodorsal setae... ***Asyndetus*** (1 sp. ***latifrons***; p. 77)

(a)

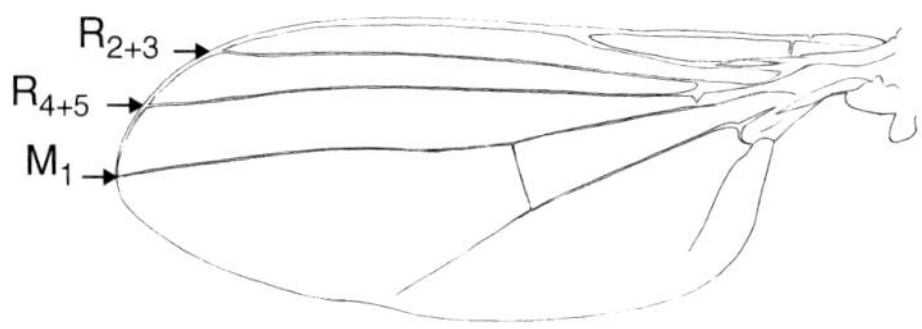

(aa)

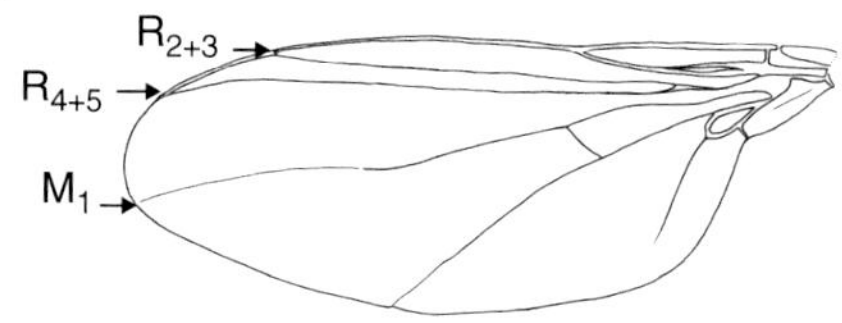

29(5) Hind tibia almost devoid of setae, at most a tiny dorsal seta on basal and apical quarters, vestiture rarely longer than half width of shaft. Epandrium large, most of it visible in its resting position, held on stalk of tergites 6 and 7 visible on left-hand side (a). R_{2+3} and M_1 almost parallel beyond dm-m, both straight or gently curved but not sinuous, R_{4+5} cutting diagonally between them (b)... ***Medetera*** (43 spp.; p. 218)

– Hind tibia usually with several distinct setae along shaft, rarely only a single dorsal seta near base and tip. Epandrium either completely encapsulated in preceding tergites, or small and only partly visible, or large but not on a stalk of tergites 6 and 7 except in *Systenus*. Venation not like this; if R_{4+5} runs diagonally then R_{2+3} or M_1 often curved or sinuous without being parallel to one another beyond dm-m..30

(a)

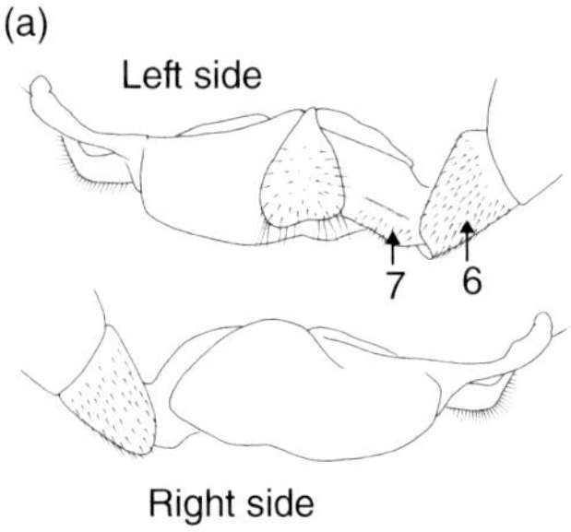

(b)

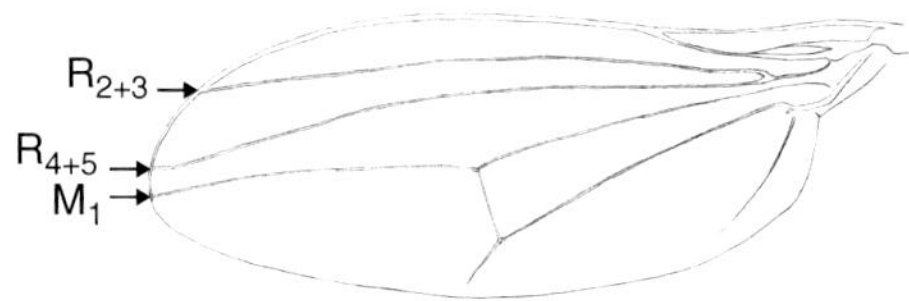

30(29) Hind coxa with 2 or more lateral setae in a vertical row, uppermost longest and others becoming progressively shorter, uppermost well above middle (a, b), without white fine hairs but black hairs may be present (ignore tiny single hair on apical quarter). Scape with tiny dorsal hairs (c) (absent in *vestita*). Tergites or scutum in many species with silver dusting. Abdomen often at least twice length of thorax.................................... ***Argyra*** (most of 11 spp.; p. 65)

– Hind coxa with 0 or 1 lateral seta, or if 2 clearly visible setae then accompanied by fine white hairs. Scape with or without hairs dorsally. Tergites and scutum never with silver dusting. Abdomen at most twice as long as thorax, usually clearly less...31

(a)

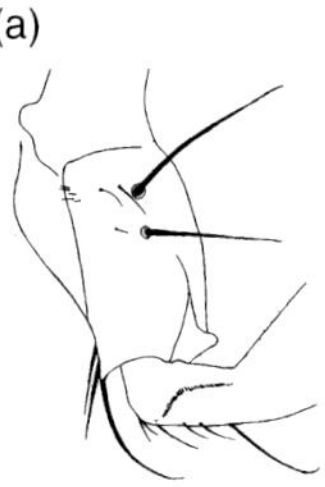

(b)

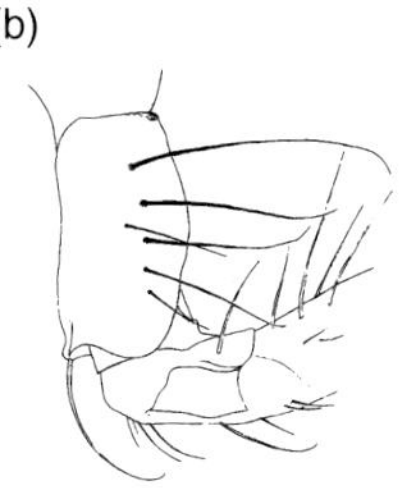

(c)

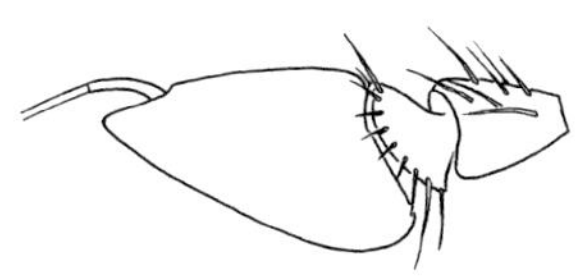

31(30) Hind coxa with lateral seta very close to base, at about one-sixth of distance from base to tip and only 2–3 times diameter of socket from base (a). Postpedicel shorter or scarcely longer than deep, arista inserted at or just above middle of anterior face of postpedicel (b). Hind femur near tip of anterior face with a row of a few (2–5) long hairs almost at ventral edge with no vestiture row below them, any true anterior preapical above not easy to distinguish and sometimes apparently absent (c). (Proepisternal seta present.) Small species, wing length 1.7–2.8 mm.............................. ***Chrysotus*** (19 spp.; p. 91)

– Hind coxa with lateral seta, when present, further from base, at least one-fifth of distance from base to tip and more than 4 times diameter of socket from base; may be tiny or absent. Postpedicel short or long; when only as long as deep, arista either strictly apical on a triangular postpedicel, or obviously dorsal or clearly above middle of anterior face on rounded or D-shaped postpedicel (bb), or if still uncertain then no proepisternal seta (only hairs) above front coxa. Hind femur either clearly without or with an anterior preapical seta, and then placed clearly above ventral edge (about one-third up) and usually with 1 or more rows of vestiture hairs below it; may additionally have preceding longish hairs or a solitary long anteroventral seta.............................. 32

(a)

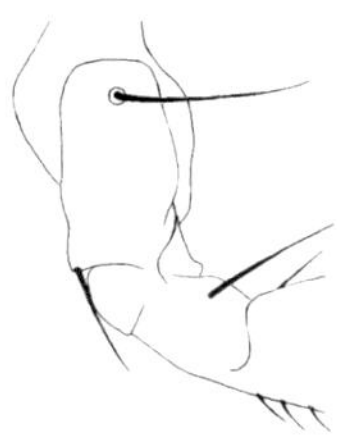

(b)

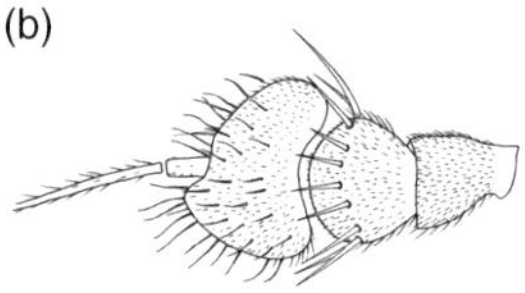

(bb)

(c)

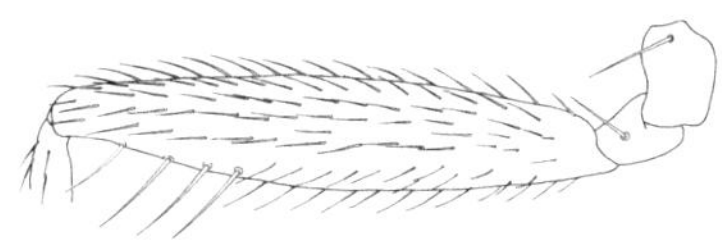

32(31) Arista apical (a, b) or subapical and close to end of pointed (c) or square-tipped postpedicel (d).............................. 33

– Arista basal (aa) or dorsal and well removed from tip of postpedicel (bb), or arising above middle of anterior face on rounded or D-shaped postpedicel (cc).............................. 37

(a)

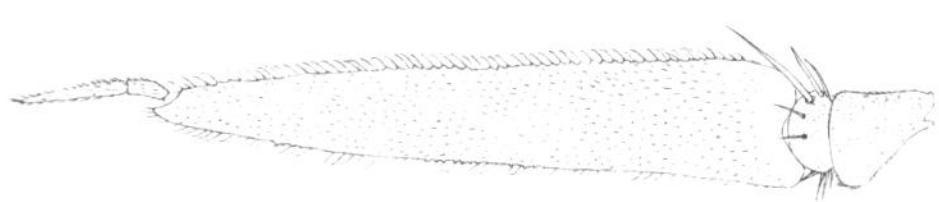

(aa)

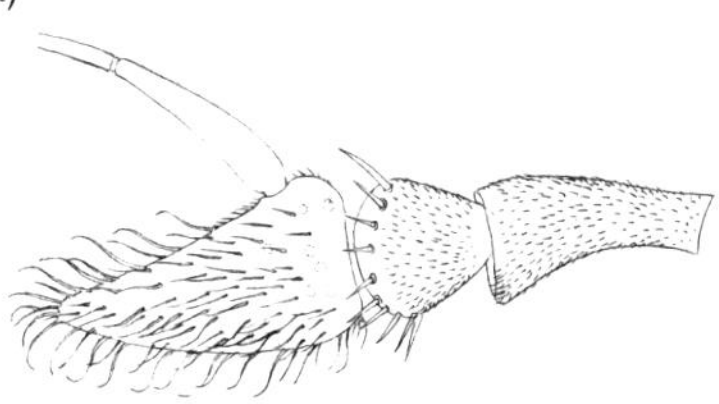

(b)

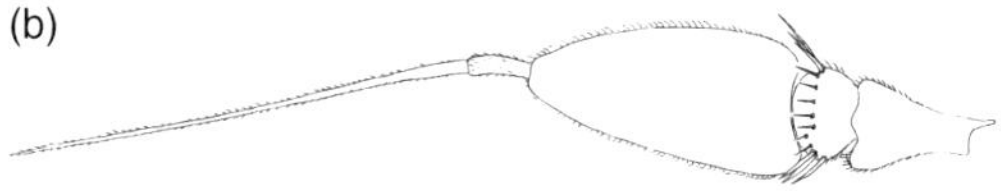

(bb)

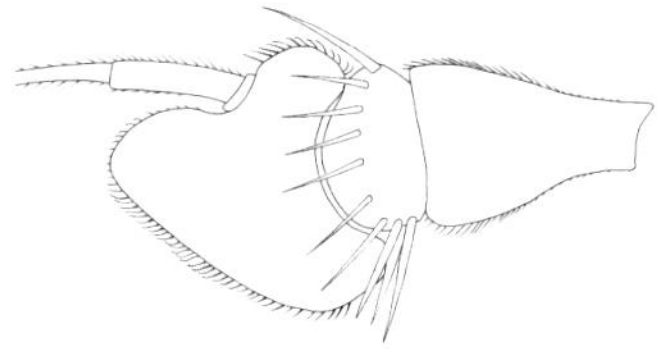

(c)

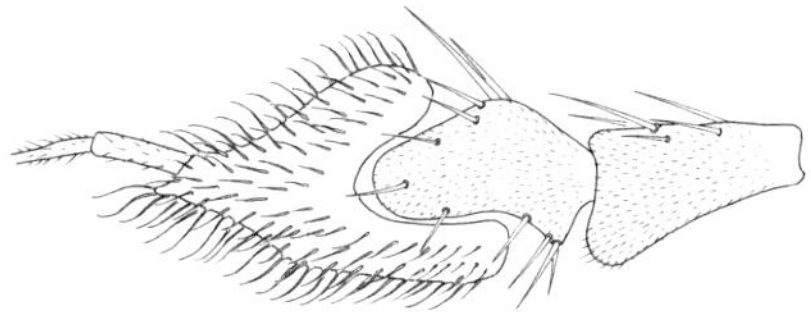

(d)

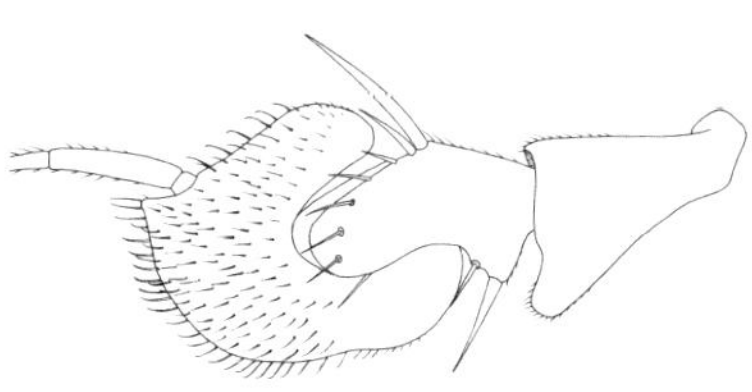

(cc)

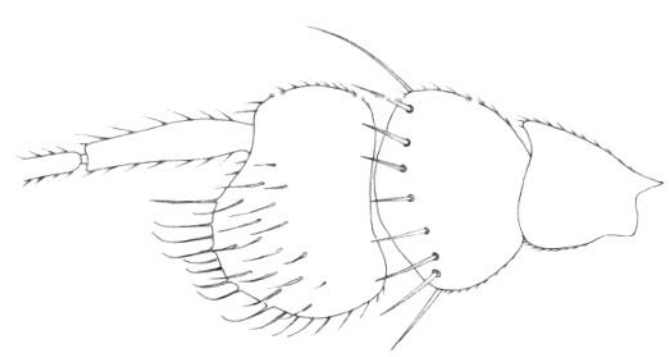

33(32) Palp large and broadly oval, as long as half or more height of an eye, covered with glistening silver dusting (a). Proboscis long, pointing downwards, with a small backwardly pointing hook at its tip (a). Front coxa with numerous stubby spikes over much of surface (b). Vein dm-m longer than apical section of M_4..***Aphrosylus*** (4 spp.; p. 62)

– Palp small, not more than one-third height of eye, with or without silver dusting. Proboscis with its long axis horizontal, no hook at tip. Front coxa with normal setae and hairs. Vein dm-m shorter than or almost equal to apical section of M_4 34

(a)

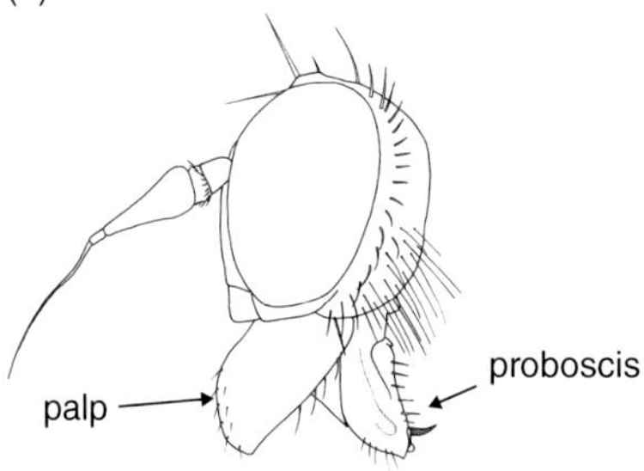

(b)

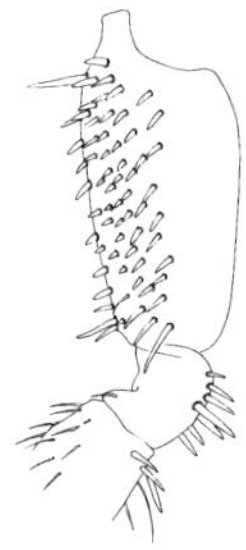

34(33) Mid femur with both anterior and posterior preapical setae (a). Front tibia with setae on shaft, at least 1 small anterodorsal on basal half or dorsal series of close-set short setae on apical half (b) (setae often entirely absent in *Syntormon aulicum*)..35

– Mid femur without a posterior preapical seta, with or without an anterior preapical seta. Front tibia lacking setae on shaft (a small apical may be present)..36

(a)

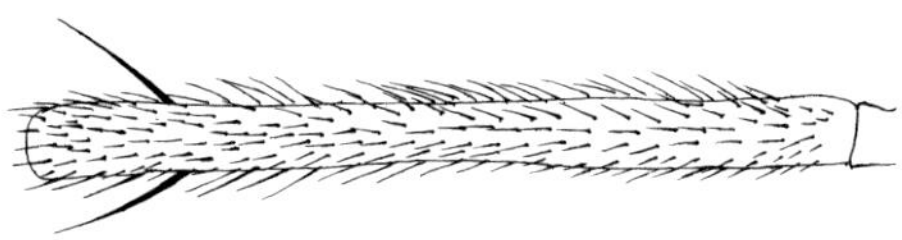

(b)

35(34) Pedicel with a thumb-like peg protruding into cavity on inner face of postpedicel (a), less pronounced in *bicolorellum* with yellow antennae (fig. c of couplet 32). Hypopygium almost entirely encapsulated within preceding tergites, cercus oval and shorter than surstylus. Hind coxa with a black lateral seta. Hind basitarsus often with modified setae or processes. Postgena behind postocular row with sparse setae, about 10 on each side. Mainly small species, wing length 2.0–4.0 mm, exceptionally 5.0 mm. ....***Syntormon*** (part, 16 spp.; p. 318)

– Pedicel without a peg on its inner face, the articulating ends of pedicel and postpedicel more-or-less straight or gently curved (figs a and b of couplet 32). Hypopygium often with long cercus visible externally. Hind basitarsus always simple cylindrical with short even vestiture. Postgena behind postocular row with dense or multiserial hairs far too numerous to count (sparse in *fasciatum*). Small to large species, wing length 2.0–5.5 mm..***Rhaphium*** (26 spp.; p. 268)

(a)

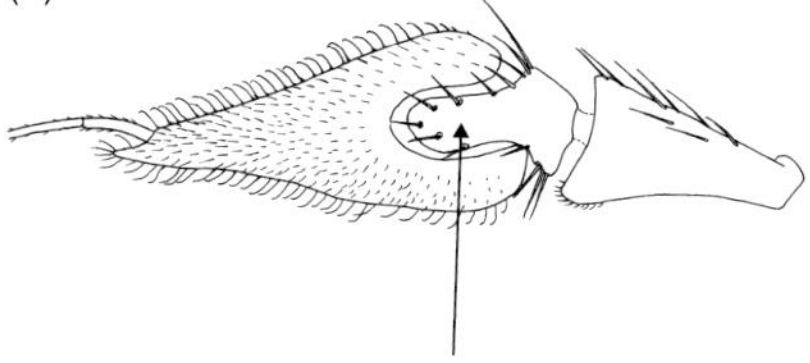

36(34) Hypopygium large and stalked on elongated tergites 6 and 7 (a). Tergite 6 as long as tergite 5. Front coxa with white hairs and apical setae. M_1 beyond dm-m gently or strongly sinuous, cell r_{4+5} obviously narrower near the wing margin than halfway back to dm-m (b). Hind basitarsus less than half length of tarsomere 2 (c). (*Xanthochlorus* with dark setae may fall here: pleura yellow on at least lower half, hypopygium bulbous; see couplet 46.) .... ***Systenus*** (6 spp.; p. 334)

– Hypopygium small and largely encapsulated within preceding tergites. Tergite 6 half length of tergite 5 (aa). Front coxa with black hairs and apical setae. M_1 beyond dm-m straight, cell r_{4+5} as wide at wing margin as halfway back to dm-m (bb). Hind basitarsus slightly shorter than tarsomere 2 (cc)................................ ***Australachalcus*** (1 sp. ***melanotrichus***; p. 78)

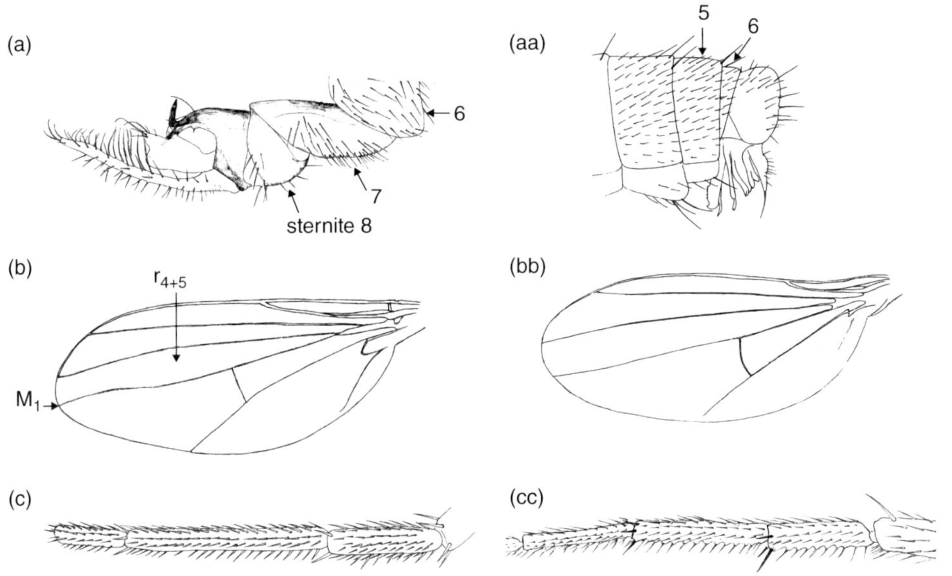

37(32) Postgena with dense whiskers (a); if in doubt, hairs pale yellow and palps large and yellow (b). Front tibia on dorsal side with at least 2 setae longer than width of shaft, and often with a ventral row of vestiture almost as long as width of shaft. Inner vertical seta (iv) well-separated from and longer than adjacent postocular seta (po, 1.5 times) (a). Medium-sized to large species, wing length 3.0–6.5 mm. 38

– Postgena without whiskers; if in doubt, hairs white and palps inconspicuous (aa). Except in *Campsicnemus*, front tibia on dorsal side without setae or with 1 weak seta about as long as width of shaft within basal half; ventral vestiture much shorter than width of shaft. Inner vertical seta scarcely longer than adjacent postocular seta (at most 1.5 times) (aa), except in *Lamprochromus* which has a partly yellow abdomen (bb). Medium-sized to small species, wing length less than 4.1 mm..39

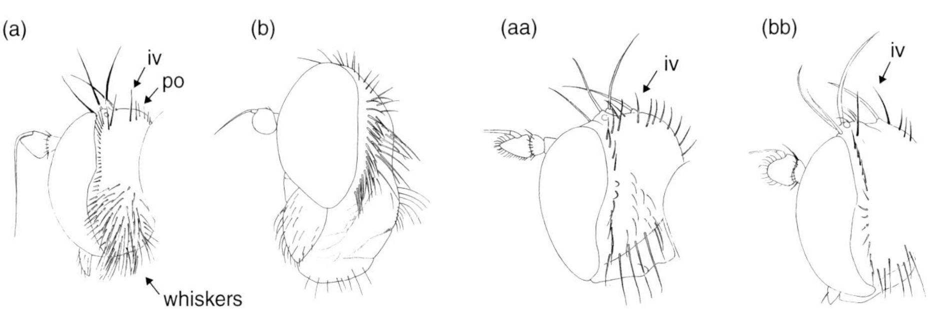

38(37) Acrostichal setae absent. Palp yellow, larger than postpedicel; proboscis stout and bulbous (fig. b of couplet 37). Face constricted in middle and broad, about 2.5 times higher than wide, bronze or yellow. .... ***Thinophilus*** (2 spp.; p. 350)

– Acrostichal setae biserial and long. Palp black, small, not larger than postpedicel. Proboscis small. Face parallel-sided and narrow, 6 times higher than wide, silver. .... ***Nematoproctus*** (2 spp.; p. 256)

39(37) Face narrow in middle of epistoma and much wider at clypeus which is at least twice as wide as epistoma under antennae (a). Proepisternum above front coxa with a distinct white or black seta that is only slightly weaker than lateral seta on hind coxa (b). Hind femur extending beyond abdomen by about half its length, not or scarcely longer than thorax (c). .... ***Campsicnemus*** (12 spp.; p. 78)

– Face either tapering to mouth or constricted in middle but clypeus at most two-thirds width of epistoma under antennae (aa). Proepisternum lacking a seta, with only tiny pale hairs. Hind femur not or scarcely extending beyond abdomen which is at least as long as thorax and usually clearly longer. 40

(a) (aa) (b) (c)

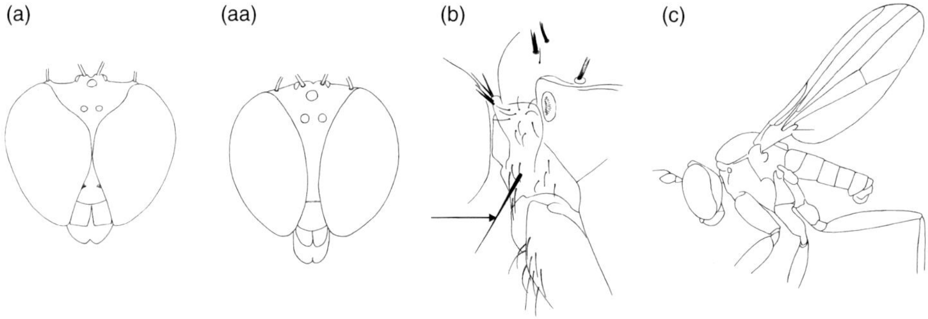

40(39) 5–6 pairs of dorsocentral setae, arranged 2 presutural, 3–4 postsutural. 41

– 3–4 pairs of dorsocentral setae, arranged 1 presutural, 2–3 postsutural. 43

41(40) Vein dm-m and surrounding membrane pale, appearing as a small white spot (a). No acrostichal setae. 5 pairs of dorsocentral setae. Hind and front leg simple without modified shape or arrangement of setae. .... ***Micropygus*** (1 sp. ***vagans***; p. 255)

– Vein dm-m same colour as adjacent veins, not enclosed in a pale spot. Acrostichal setae present. 6 pairs of dorsocentral setae. Hind tibia or tarsus or front tarsus usually with modifications in shape or arrangement of setae. 42

(a)

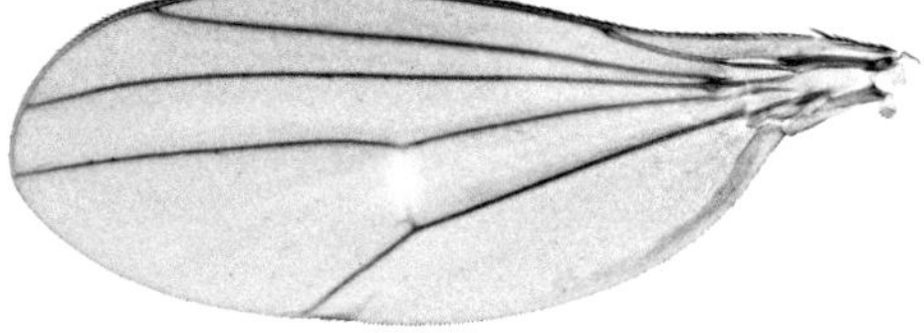

42(41) Costa of wing fused with R_1 forming a thick black sliver (a). Vein dm-m oblique, forming an obvious to slight acute angle with basal section of M_4; M_1 just beyond dm-m bowed anteriorly for a short distance (a). Palp dark. Hind tibia usually with modified shape or setae. .... ***Teuchophorus*** (5 spp.; p. 345)

– Costa of wing not thickened, R_1 distinct (aa). Vein dm-m at right angles to M_1 and not forming an acute angle with M_4, or at most a few degrees from a right angle; M_1 just beyond dm-m straight and aligned with section just before dm-m (aa). Palp pale. Hind tibia simple. .... ***Sympycnus*** (5 spp.; p. 315)

(a)

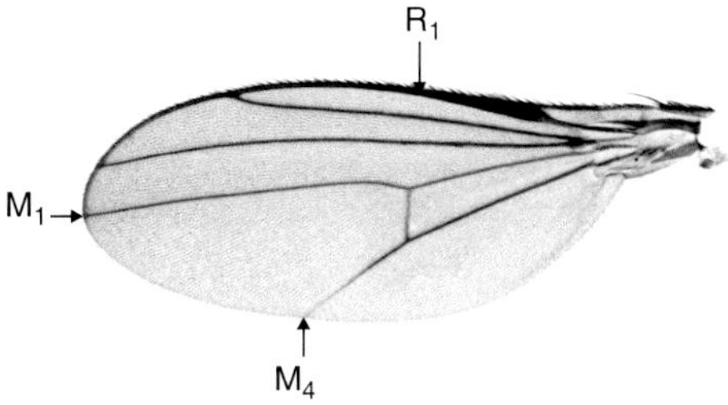

(aa)

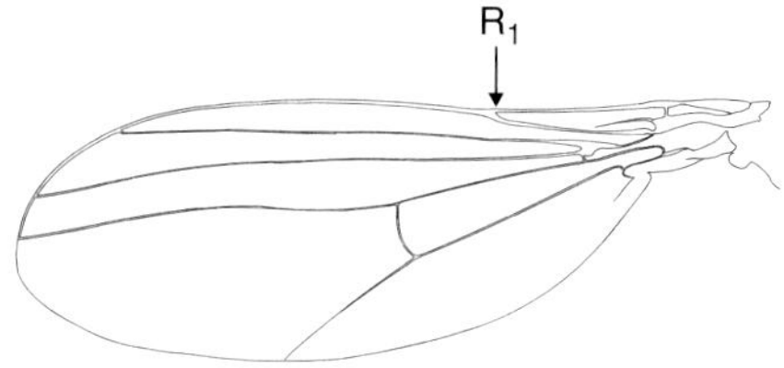

43(40) Tergites 2 and 3, metepimeron and coxae yellow. Scutum with a large matt black lateral mark above wing base between transverse suture and postalar seta. Arista tip tapered. Front basitarsus simple. Acrostichal setae biserial. .... ***Lamprochromus*** (3 spp.; p. 215)

– Tergites, metepimeron and coxae dark. Scutum without a black lateral mark. Arista tip thickened by long pubescence (aa). Front basitarsus with a strong ventral swelling (bb). Acrostichal setae uniserial. .... ***Telmaturgus*** (1 sp. ***tumidulus***; p. 345)

(aa)

(bb)

44(3) Thorax black. Postpedicel pointed-triangular, arista almost apical and about as long as antenna (a). .... ***Achalcus*** (part, ***cinereus***; p. 54)

– Thorax green or yellow, entirely or partly. Postpedicel rounded or square, arista far longer than antenna. 45

(a)

45(44) Mid tibia with 2–3 anterodorsal setae and 1–2 posterodorsal setae (a). Thorax green or partly to entirely yellow. Epandrium either yellow and globular to ovoid, or dark and almost entirely encapsulated within preceding tergites. Epandrium apparently attached to tergite 5 or 6, tergite 7 hidden in resting position. 46

– Mid tibia with only 1 or no anterodorsal seta, no posterodorsal seta (aa). Thorax always green. Epandrium dark, either elongate and tapering and with apical appendages held fairly close together (bb, cc), or rarely short with appendages widely spaced (dd). Epandrium base partly enveloped by tergite 7, together with tergite 6 forming a short stem. 47

(a)

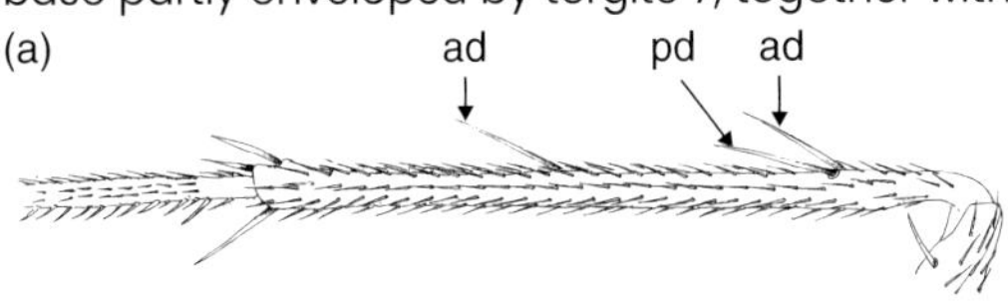

(aa)

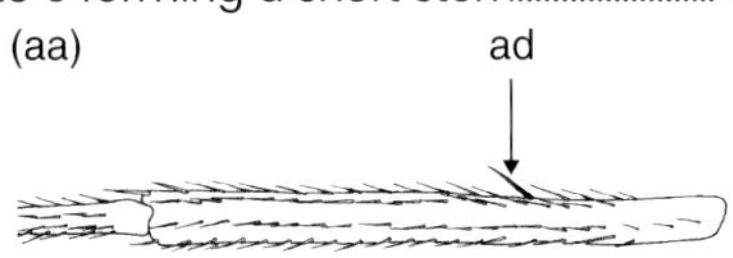

(bb)

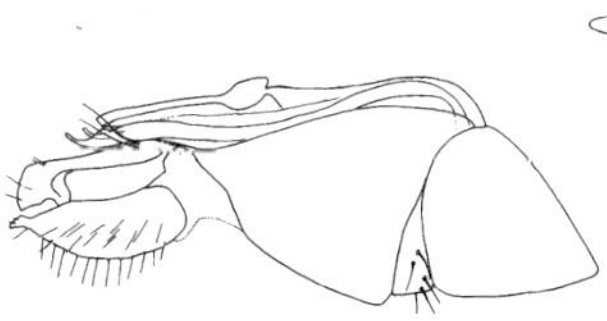

(cc)

(dd)

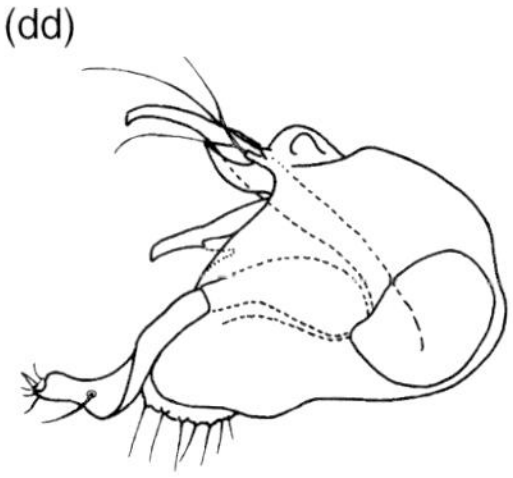

46(45) Body entirely or partly yellow, at least margins of tergites 2 and 3, epandrium, and usually lower part of pleura. R_{4+5} and M_1 converge in apical quarter, distinctly wider apart shortly beyond dm-m than near wing tip (a). Tergite 6 as long as tergite 5, with hairs. Face narrow and tapering to mouth (b). Small species, wing length 2.8–3.7 mm.. ***Xanthochlorus*** (4 spp.; p. 364)

– Body almost entirely metallic green. R_{4+5} and M_1 parallel beyond dm-m, their separation at dm-m the same as near wing tip (aa). Tergite 6 short or not visible, bare. Face wide and almost parallel-sided (bb). Tiny species, wing length 2.1–2.8 mm .... ***Chrysotimus*** (2 spp.; p. 89)

(a)

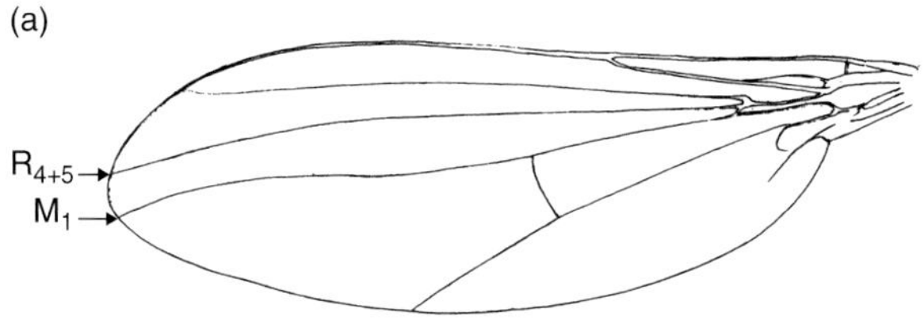

(b)

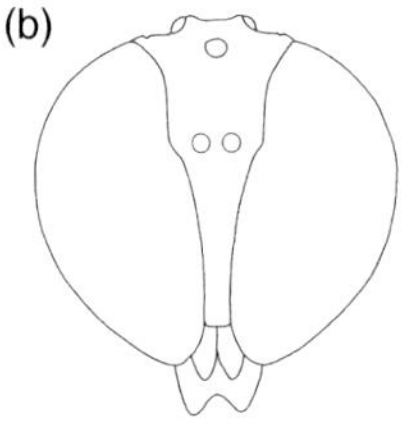

(aa)

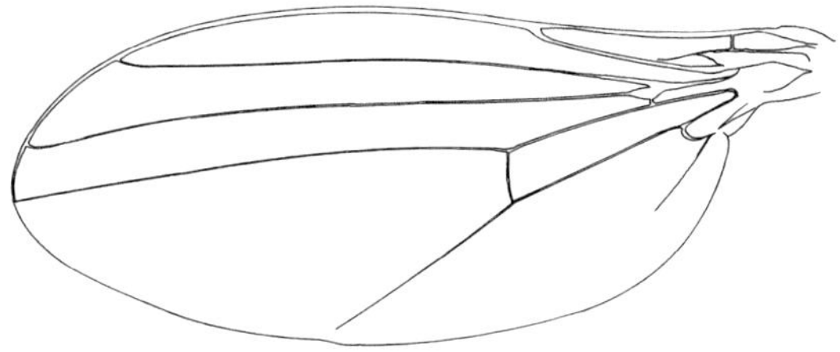

(bb)

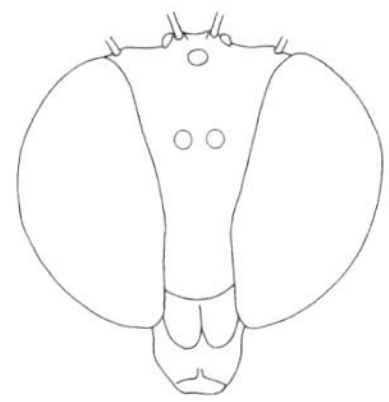

47(45) Hind coxa with 2 subequal lateral setae (a). R_{4+5} and M_1 parallel beyond dm-m, their separation at dm-m scarcely wider than near wing tip (1.25 times at most) (b)...................... 48

– Hind coxa with 1 lateral seta. R_{4+5} and M_1 converging beyond dm-m, their separation at dm-m almost twice that near wing tip. 49

(a)

(b)

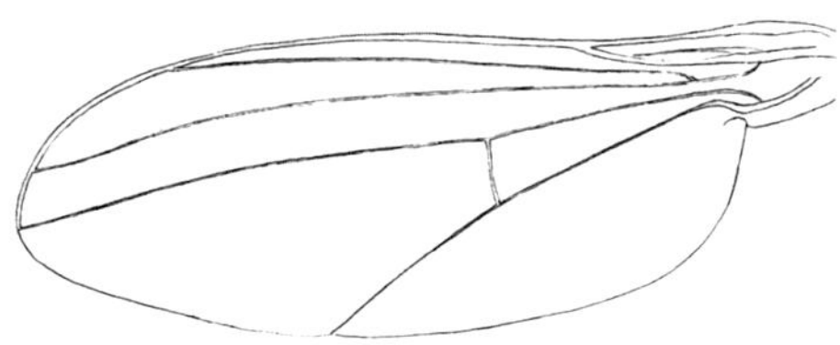

48(47) Scape and pedicel dark. Mid tibia with a weak anterodorsal seta on basal third (a). Cercus without a long apical extension but may have a short triangular extension (b, c). Surstylus with a clear articulation with epandrium (c)...................... ***Thrypticus*** (12 spp.; p. 352)

– Scape and pedicel yellow. Mid tibia with a strong anterodorsal seta on basal third (aa). Cercus with a long apical extension (bb). Surstylus fused to epandrium (cc) .... ***Corindia*** (1 sp. ***viridis***; p. 111)

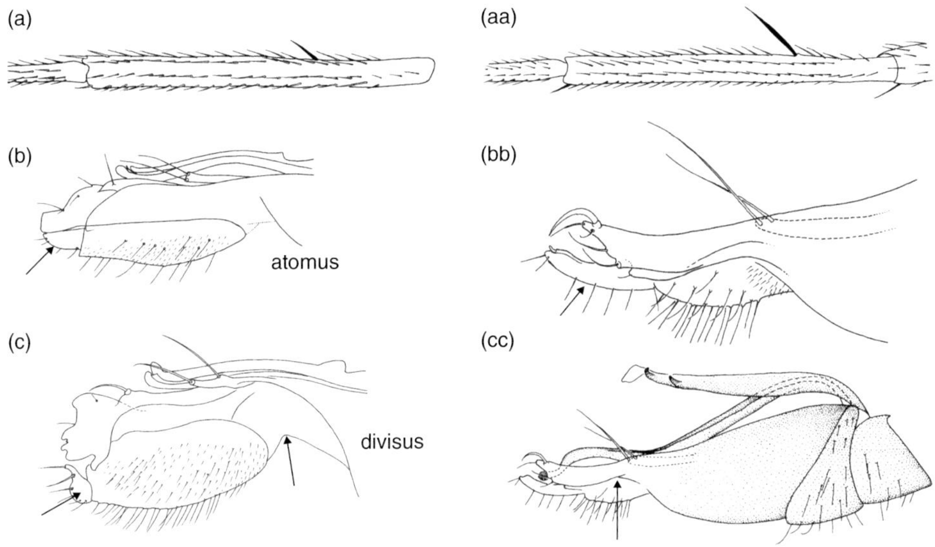

49(47) Cell r_1 at leading edge of wing dark (a). Front coxa at tip with a long spur, almost as long as length of coxa (b). Front basitarsus almost as long as tarsomeres 2–5 combined, with a ventral projection at tip (c). Coxae bright yellow. Scape and pedicel yellow. R_{4+5} and M_1 at wing tip parallel for a short distance of about twice their separation here (a). Tiny species, wing length 2.1 mm..***Dolichophorus*** (1 sp. ***kerteszi***; p. 114)

– Cell r_1 as pale as rest of wing (aa). Front coxa without a long spur. Front basitarsus shorter than tarsomeres 2–5 combined, with a simple tip. Coxae mainly dark. Scape and pedicel black. R_{4+5} and M_1 at wing tip parallel for a distance of about 4 times their separation here (aa). Minute species, wing length 1.3 mm..***Cyrturella*** (1 sp. ***albosetosa***; p. 111)

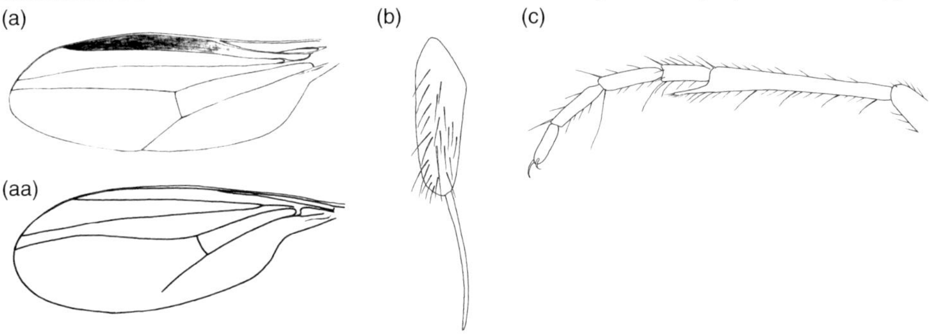

Key to genera and some species of Dolichopodidae (females)

1 Wing with discal medial cell (dm) having 3 veins apically, and a dark lens-shaped pterostigma beyond end of R_1 (a). Tiny to small dark species with no hint of metallic colouration, wing length 2–3.5 mm.......... ***Microphor*** (5 spp.; p. 253)

– Wing with discal medial cell having 2 veins at its tip and no pterostigma.......... 2

(a)

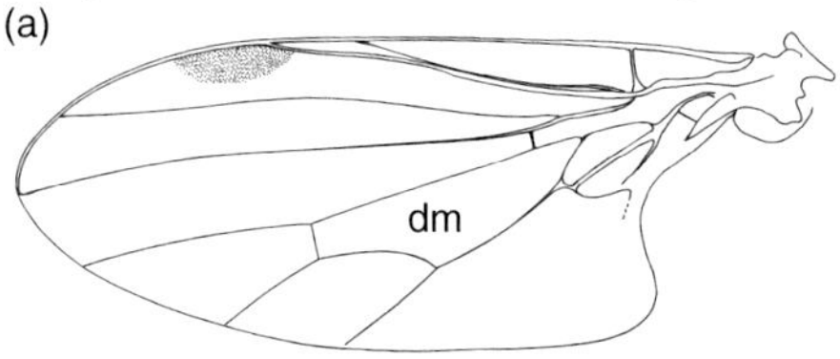

2(1) Vein M forked beyond crossvein dm-m, anterior branch (M_1) reaching wing margin close to R_{4+5}, posterior branch (M_2) long and reaching at least halfway to wing margin (a).......... ***Sciapus*** (10 spp.; p. 304)

– Vein M not forked, or when apparently so (some *Dolichopus*) then posterior branch very short (aa).......... 3

(a)

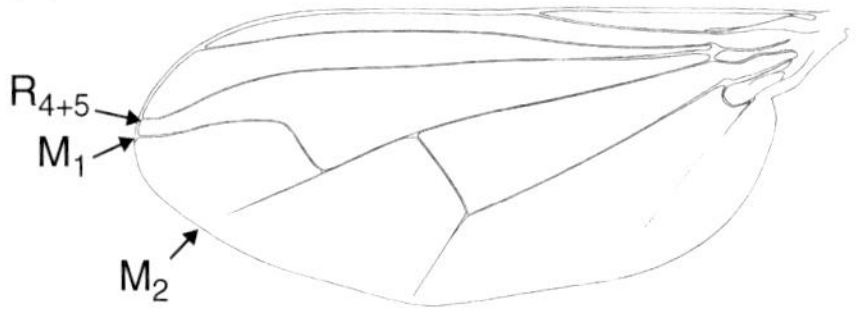

(aa)

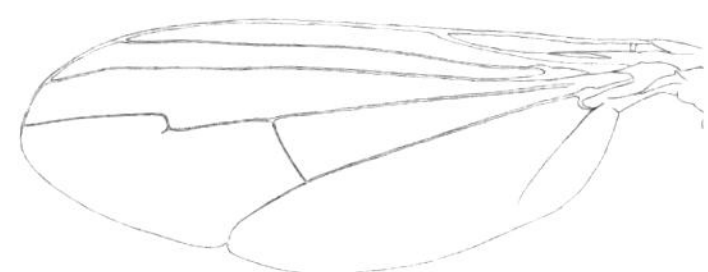

3(2) Setae of body black or dark brown.......... 4

– Setae of body pale, at least in anterior view. Mainly tiny to small flies, wing length 1.5–2.5 mm and some up to 4 mm.......... 49

4(3) Antenna excluding arista longer than head, postpedicel several times longer than deep.......... 5

– Antenna excluding arista shorter than head, postpedicel not longer than twice its depth.......... 6

5(4) Postpedicel cylindrical at base (a). Scutum blackish green. Legs mainly black. .... ***Rhaphium*** (part: ***longicorne***; p. 280)

– Postpedicel with swelling at base (aa). Scutum bright green. Legs mainly yellow. .... ***Machaerium*** (1 sp. ***maritimae***; p. 218)

(a)

(aa)

6(4) Hind coxa with 3 or more lateral setae in a vertical row, approximately equally spaced, the uppermost longest and the others becoming progressively shorter, thinner and hair-like, the uppermost usually at or above top third of coxa well above the middle (a, b). Scape with tiny dorsal hairs (c).......... ***Argyra*** (most of 11 spp.; p. 70)

– Hind coxa with 0, 1 or 2 strong lateral setae; ignore small to tiny isolated hairs. Scape with or without dorsal hairs.......... 7

(a) (b) (c)

7(6) Hind coxa with lateral seta placed close to base, not further than a fifth distance from base or only 2–3 times diameter of its socket from base (a). Postpedicel shorter than deep or rarely slightly longer than deep, anterior profile rounded or straight with no or scarcely any point; arista placed on front margin, at middle (b) or nearer the top (c).. 8

– Hind coxa with lateral seta absent, tiny or strong and placed further than one-fifth from base or more than 4 times diameter of its socket from base; in the few exceptions with lateral seta close to base, postpedicel not as described above...11

(a) (b) (c)

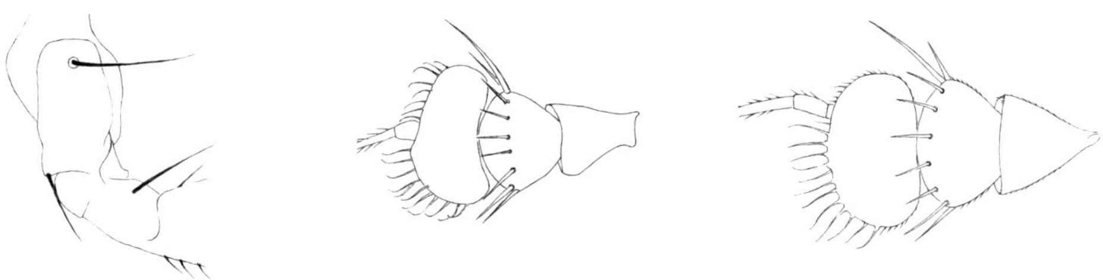

8(7) Costa ending at R_{4+5}. Veins R_{4+5} and M_1 diverging widely and almost continuously towards wing margin.. ***Asyndetus*** (1 sp. ***latifrons***; p. 77)

– Costa ending at M_1. Veins R_{4+5} and M_1 approximately parallel or bowed away from each other beyond dm-m. 9

9(8) Arista arising on front face of postpedicel at midpoint or lower (fig. b of couplet 7). Face narrower, clypeus almost square to 1.5 times wider than deep (a) (except *laesus* with pale hairs on front coxa). Frons wider and usually shorter than face (a); in cases of doubt with wide-faced species, front coxa with white hairs and setae................... ***Chrysotus*** (19 spp.; p. 97)

– Arista arising at upper quarter to third of postpedicel (fig. c of couplet 7). Face broader, clypeus 1.5–2 times wider than deep. Frons as wide and long as face, or nearly so (aa). Front coxa with hairs and setae entirely black. 10

(a) (aa)

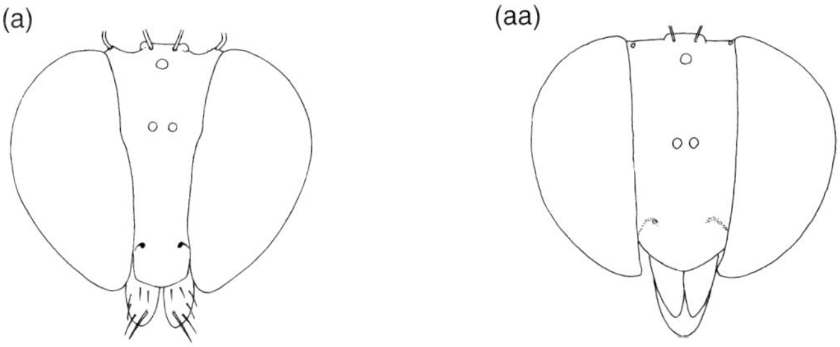

10(9) Halter yellow. .... ***Diaphorus*** (3 spp.; p. 113)

– Halter black. .... ***Melanostolus*** (1 sp. ***melancholicus***; p. 245)

11(7) Front femur on basal third thicker than mid and hind femora, often armed ventrally with small or long spine-like setae (a), sometimes only at base (b) or absent when wing has dark spots on M_1 and dm-m (c). Vein dm-m longer than apical section of M_4 (c) except in *Hydrophorus nebulosus* which has mottled wings. 12

– Front femur not thicker than mid and hind femora, not strongly armed ventrally. Vein dm-m variable, shorter to longer than apical section of M_4. 14

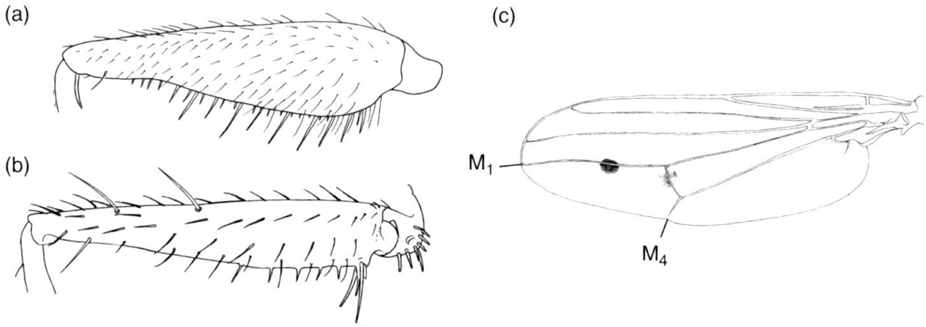

12(11) Front coxa with numerous stubby spikes over much of surface (a). Arista apical. Acrostichal setae absent; 4–6 pairs of long dorsocentral setae. R_{2+3} as long as M_1 and straight at tip where it meets costa. Hind coxa with 1 lateral seta (rarely 2 in aberrant specimens) .... ***Aphrosylus*** (4 spp.; p. 62)

– Front coxa with dense fine hairs, sometimes with a single row of black setae on lateral edge. Arista dorsal. Acrostichal setae present; at least 7 pairs of dorsocentral setae, most of them short. R_{2+3} shorter than M_1 and upturned at tip where it meets costa. Hind coxa without a lateral seta. 13

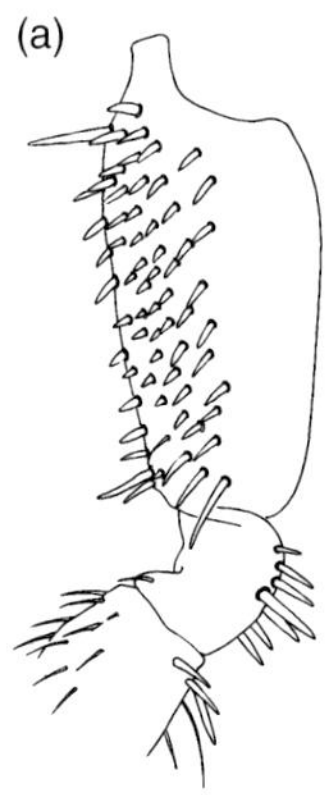

13(12) Wing membrane in front of R_{2+3} yellow on basal half and dark on apical half, wing dark-marked more extensively behind (a). R_{4+5} and M_1 converge more strongly, maximum separation 3 times their separation at tip (a). Front tibia with several long posteroventral setae about twice width of shaft (b). 1 notopleural seta. Postpedicel smoothly rounded below arista. .... ***Scellus*** (1 sp. ***notatus***; p. 299)

– Wing membrane colourless or with dark spots or blotches all the same colour. R_{4+5} and M_1 converge weakly, maximum separation less than twice their separation at tip. Front tibia with setae absent or not longer than width of shaft. 2 notopleural setae. Postpedicel below tip with a small pimple (cc). .... ***Hydrophorus*** (9 spp.; p. 207) (*Campsicnemus umbripennis* may fall here with face constricted in middle, not widening from antennae to mouth.)

(a)

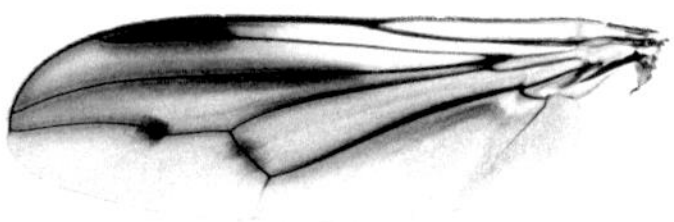

(b)

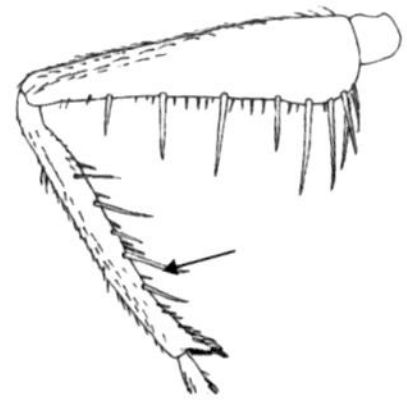

(cc)

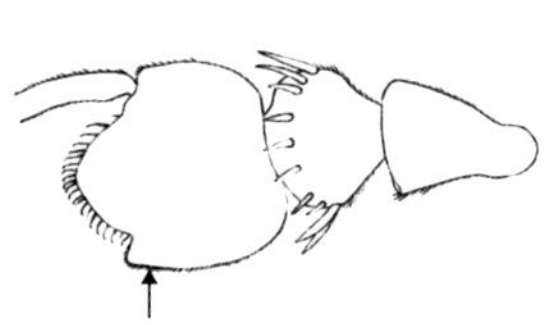

14(11) Mid or hind femur (usually both) with 1 or more distinct anterior preapical setae; ignore longer hairs at end of lowermost row of vestiture (If seta is broken off, then socket is visible and vestiture immediately apical to socket missing 1 or a few hairs, forming a bare area. Species where arrangement of seta is ambiguous are keyed in both halves of couplet)......15

– Mid and hind femora lacking anterior preapical seta, no area bare of vestiture near femur tip anteriorly; an anterior or anterodorsal seta may be present at one-quarter length of femur from its tip, well removed from the usual position of preapical seta close to tip..........43

15(14) Pedicel with a thumb-like extension protruding into a cavity on inner face of postpedicel (a, b) AND no proepisternal seta but only tiny hairs above front coxa. Face broad, at least two-thirds height of face and about equal to width of an eye (c) except in *Syntormon bicolorellum* (d). Clypeus strongly protruding (e) except in *S. bicolorellum* (f).. ***Syntormon*** (16 spp.; p. 324)

– Pedicel on inner face with square or gently rounded tip, or if protruding into postpedicel then 1–3 black or white proepisternal setae present above front coxa. Face and clypeus variable. 16

(a)

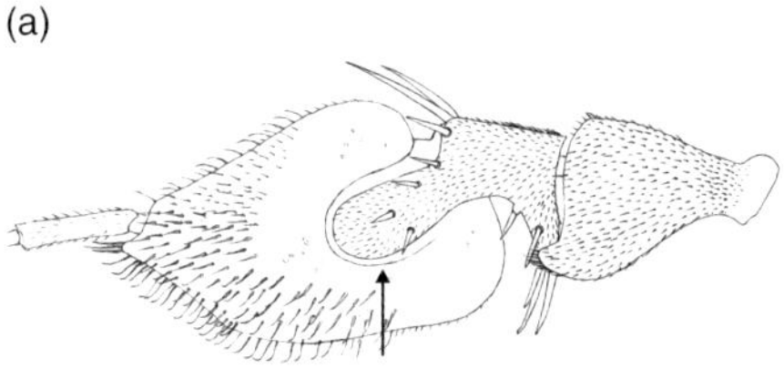

(c)

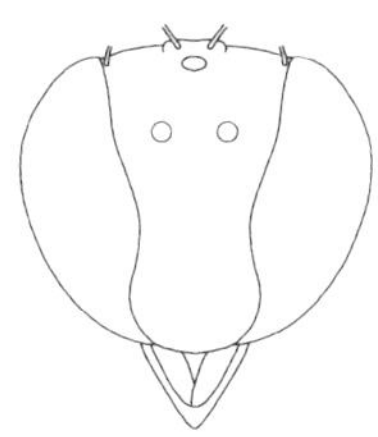

(d)

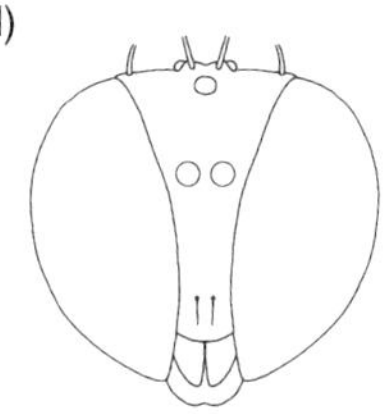

(b)

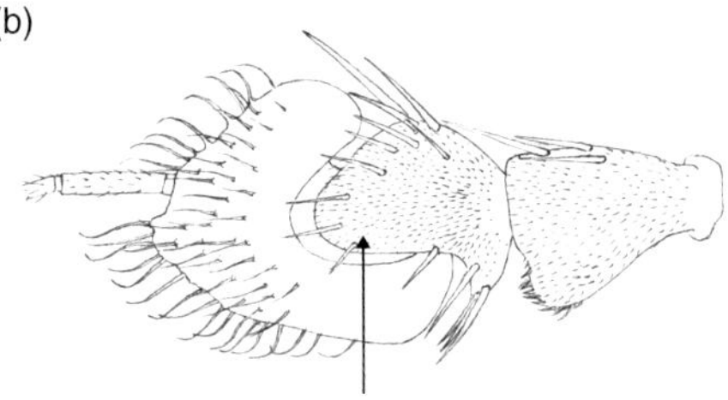

(e)

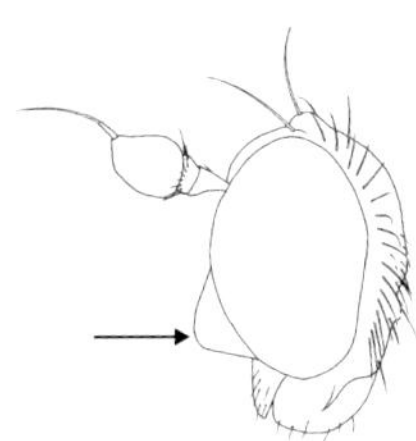

(f)

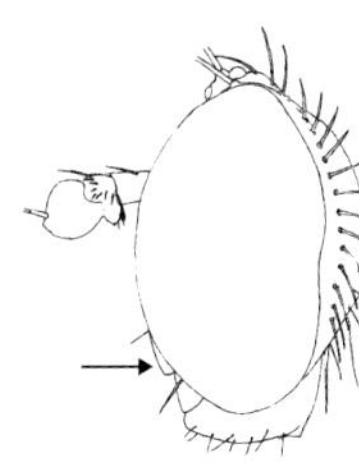

16(15) Scape with small dorsal hairs (a, b) (see notes below). Arista dorsal, well removed from tip of postpedicel or at least clearly above middle of front margin when postpedicel is nearly round. Acrostichal setae clearly biserial except in *Ortochile* with very long proboscis and palps. Small to large species. 17

– Scape bare (see notes below). Arista position variable: basal, dorsal or apical. Acrostichal rows variable. 26

(a) (b)

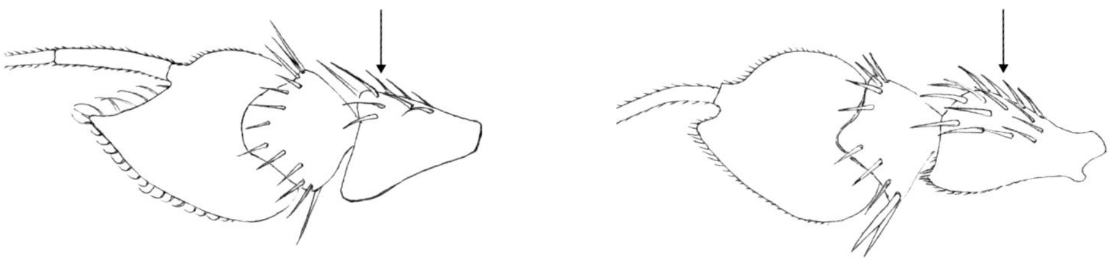

Notes:

1. If arista is dorsal, acrostichal setae biserial AND wings at least 2.5 mm long, check the scape. If arista is apical (or nearly so), acrostichal rows not biserial, OR species is tiny with wings less than 2.5 mm, the scape will be bare, so no need to search for hairs (go to 26).
2. *Argyra grata* may have only 2 lateral setae on hind coxa and will fall here: scape with dorsal hairs, postpedicel shorter than deep, acrostichal setae very irregular, tergites 1–4 yellow laterally.

17(16) Hind basitarsus with at least 1 strong dorsal seta (a). Vein M_1 nearly always with a slight to pronounced double inflexion midway between dm-m and wing tip (b). .... ***Dolichopus*** (55 spp.; p. 134)

– Hind basitarsus without strong dorsal setae. Vein M_1 either almost straight or gently sinuous. 18

(a) (b)

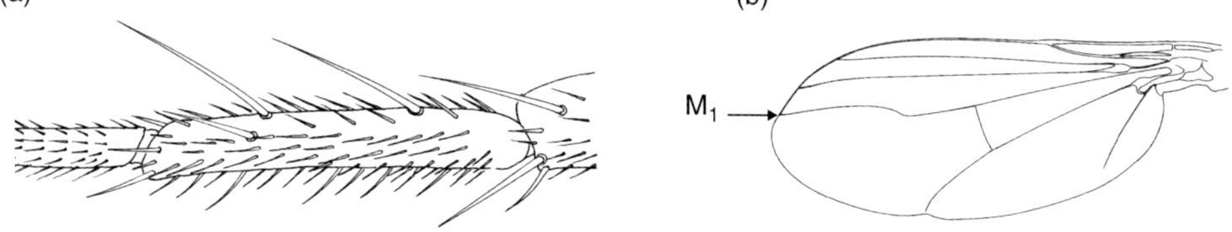

18(17) Front tibia with 1–4 posteroventral setae at least as long as width of shaft, not markedly weaker than smaller setae on dorsal face of shaft (a); with 2–4 apical setae, one of which is longer than width of shaft near tip (a) (except in *Muscidideicus* with front coxa not reaching tip of katepisternum). 19

– Front tibia with no posteroventral setae or with only 1–2 weak setae shorter than width of shaft; no or weak apical setae rarely as long as 1.5 times width of shaft near tip (aa). 23

(a) (aa)

19(18) Hind femur with at least 2 strong anterodorsal (almost dorsal) setae on apical half in addition to 1 or 2 anterior preapical setae (a). Epistoma narrowest just below antennae, widening towards clypeus (b). Clypeus free at sides and extending to or below bottom of eye (b).. ***Tachytrechus*** (4 spp.; p. 342)

– Hind femur with only 1 anterior preapical seta. Epistoma and clypeus parallel-sided. Clypeus fused with eye margin and not reaching bottom of eye. 20

(a)

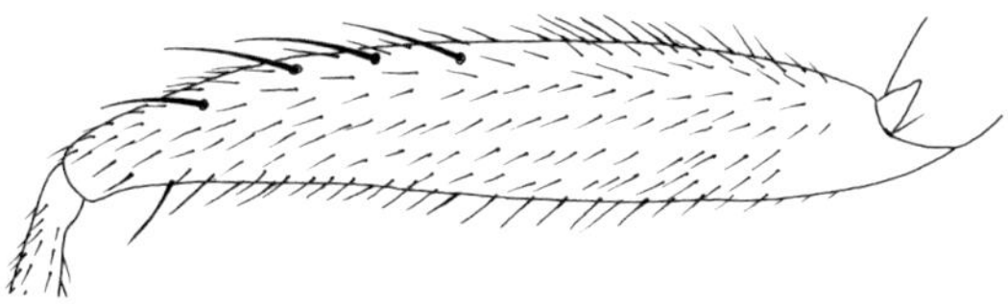

(b)

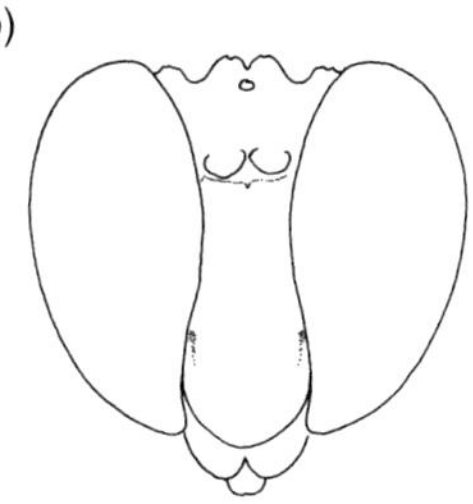

20(19) Scutum and frons shining blue- or violet-black with no pattern when viewed from above. Face broad, its width about three-fifths its height (a). Clypeus strongly protruding in profile (b). Proboscis stout, palp large (b). R_{4+5} and M_1 almost straight and almost parallel until close to wing margin (c).. ***Ethiromyia*** (1 sp. ***chalybea***; p. 182)

– Scutum green, bronze, copper, brown or brownish black, never bluish. Face width half or less its height. Clypeus only slightly bulging in profile. Proboscis and palp small. R_{4+5} and M_1 gently sinuous and converging towards wing tip...21

(a)

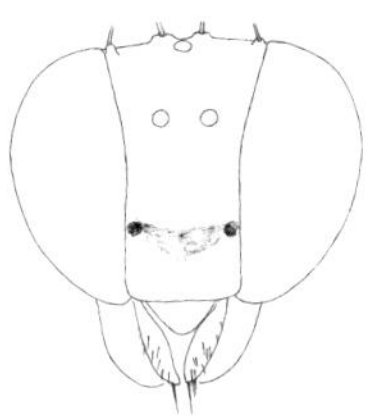

(b)

(c)

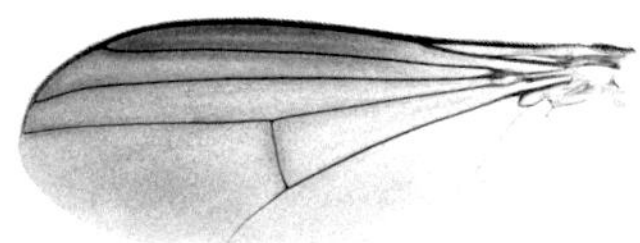

21(20) Front coxa short, ending well above lower margin of katepisternum (a). Leg vestiture longer and less dense, hind femur on apical quarter with dorsal hairs about one-third depth of shaft (b). 7 pairs of dorsocentral setae. Proepisternum with distinct long white hairs up to half length of proepisternal seta (c)....................................... ***Muscidideicus*** (1 sp. ***praetextatus***; p. 256)

– Front coxa long, extending to or beyond lower margin of katepisternum (aa). Leg vestiture fine and short, hind femur on apical quarter with dorsal hairs tiny, one-fifth or less depth of shaft. 6 pairs of dorsocentral setae. Proepisternum with indistinct hairs about one-quarter length of seta...22

(a)

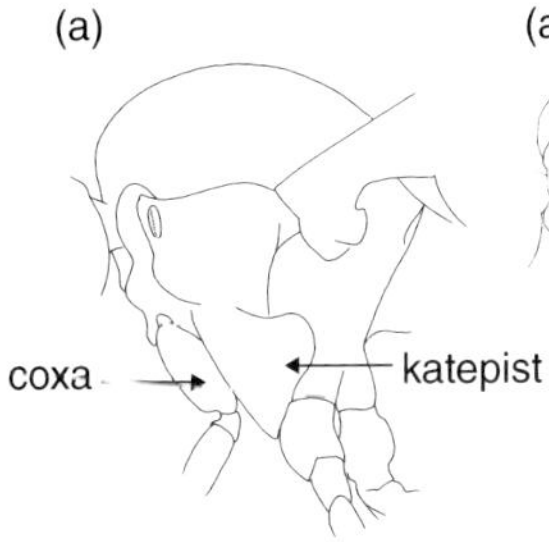

(aa)

(b)

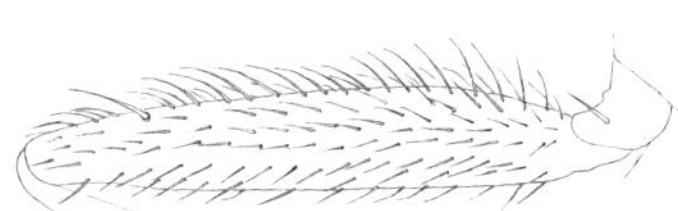

(c)

22(21) Scutum with a small matt blackish purple patch above notopleuron extending from transverse suture to postsutural intra-alar seta, contrasting with surrounding shining green cuticle. Mid tibia with 2–3 anteroventral and 1–3 posteroventral setae (a) (*H. germanus* may fall here but lacks purple patch on scutum). Arista pubescence either inconspicuous or as long as basal diameter of arista (b)..***Poecilobothrus*** (5 spp.; p. 265)

– Scutum with any patch of different colour above notopleuron ill-defined or any purplish marks less extensive. Mid tibia with 1–2 anteroventral and no posteroventral setae (aa). Arista pubescence inconspicuous..***Hercostomus*** (part: 4 spp.; p. 198)

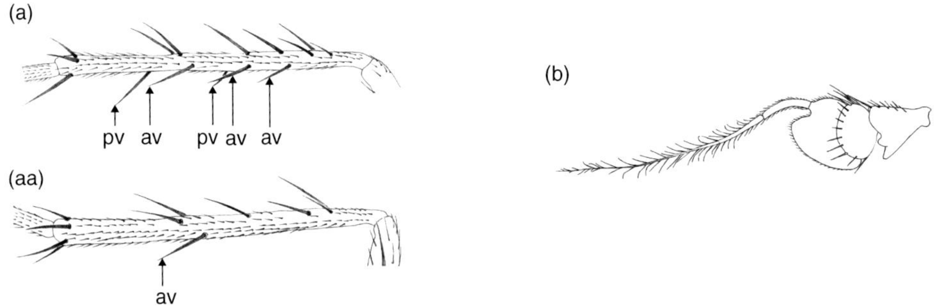

23(18) Vein M_1 straight or almost so, and parallel to R_{4+5} beyond dm-m (a). Acrostichal row ending at or behind last (sixth) dorsocentral seta, usually forming a cluster of unpaired irregularly spaced setae in front of scutellum (b). Pleuron in front of halter with a cluster of tiny hairs. ....***Gymnopternus*** (9 spp.; p. 185)

– Vein M_1 slightly sinuous or bowed beyond dm-m, usually converging with R_{4+5} towards wing tip so their greatest separation is at least 1.5 times their separation at wing tip (aa), except in *H. nigripennis* (cc) with long downward-pointing proboscis. Acrostichal row ending before or at last dorsocentral seta and remaining strictly biserial, except in *Ortochile* with very long narrow proboscis. Pleuron in front of halter entirely bare...24

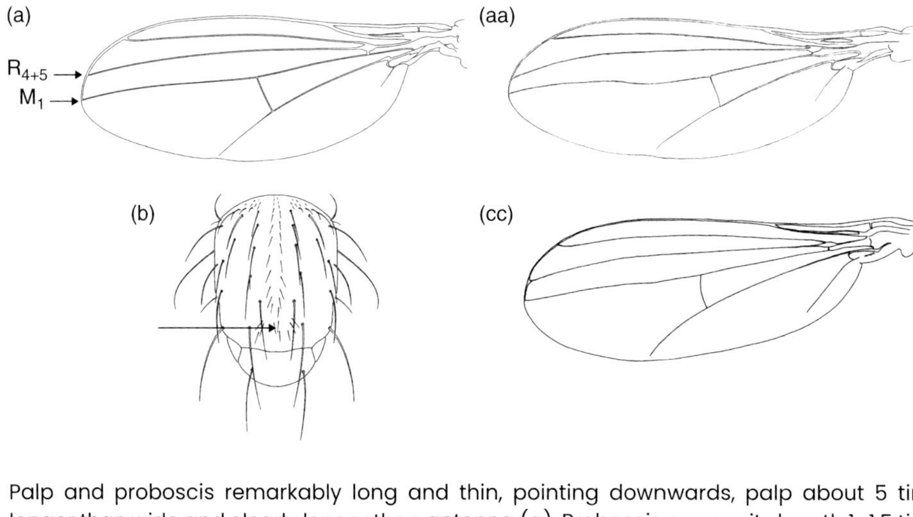

24(23) Palp and proboscis remarkably long and thin, pointing downwards, palp about 5 times longer than wide and clearly longer than antenna (a). Proboscis narrow, its length 1–1.5 times depth of head (a) (apparent length varies with degree of extension). Mid femur black on basal two-thirds, yellow on apical third (b)..........................***Ortochile*** (1 sp. ***nigrocoerulea***; p. 263)

– Palp small, at most twice as long as wide, much shorter than antenna (aa). Proboscis orientated horizontally (aa) except in *Hercostomus nigripennis* (cc) and *H. germanus* but then clearly shorter than depth of head. Mid femur without this colour combination, either black or yellow with only tips contrasting in colour...25

(a)

(aa)

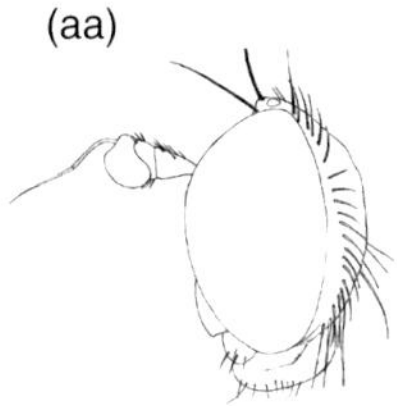

(cc)

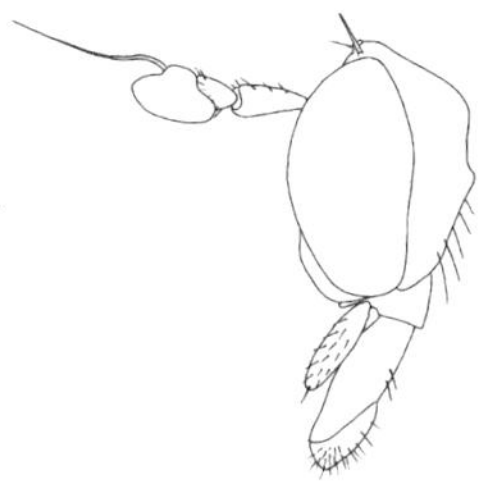

(b)

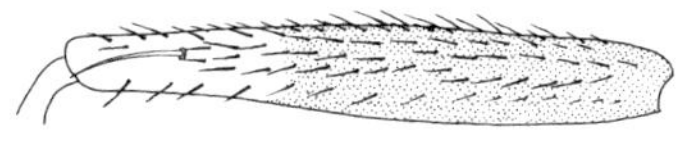

25(24) Front tarsus considerably longer than tibia (1.3–1.7 times – if in doubt then metepimeron yellow) AND palp dark or brown. .... ***Sybistroma*** (3 spp.; p. 312)

– Front tarsus about 1.1 times as long as tibia or if much longer (1.4 times) then palp yellow .... ***Hercostomus*** (part: most of 15 spp.; p. 198)

26(16) Face strongly constricted in middle where half or less as wide as clypeus (a, b). Hind femur extending well beyond tip of abdomen. Proepisternum above front coxa with a distinct black or white seta similar in size to lateral seta on hind coxa (c) .... ***Campsicnemus*** (12 spp.; p. 82)

– Face margins converging towards mouth, or parallel or weakly constricted below middle of epistoma so epistoma is at least two-thirds as wide as clypeus. Hind femur shorter to longer than abdomen. Proepisternal seta either distinct, weak or absent. 27

(a)

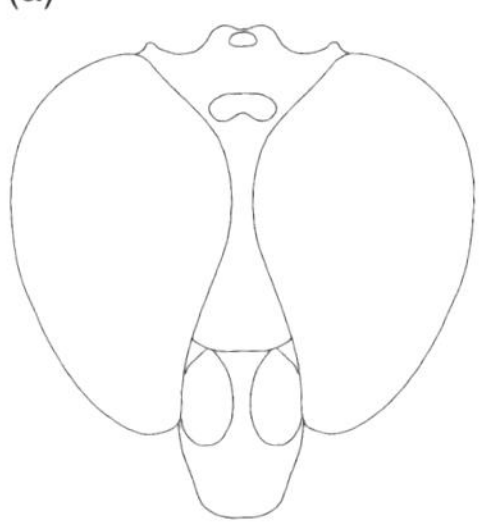

(b)

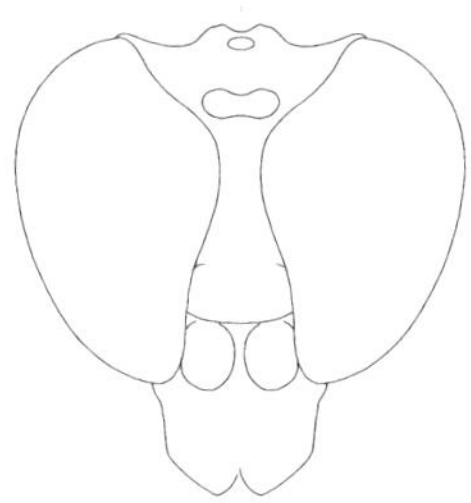

(c)

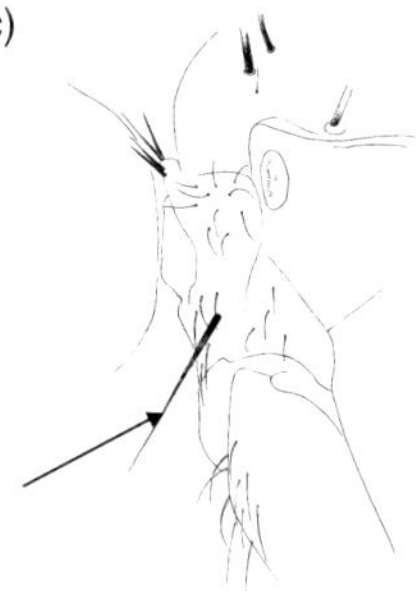

27(26) Postgena with whiskers of dense long hairs behind postocular row and continuous with it, nearly as closely spaced as postocular setae and almost evenly spaced from mouth edge to neck, too numerous to count (a) (indeterminate species key out in both halves of couplet). Small to large species, wing length 3.0–7.5 mm. 28

– Postgena with sparse hairs, widely spaced and often short, not continuous with postocular setae, at most 20 hairs can be counted on each side. If hind coxa with 2 lateral setae, follow second half of couplet. Small to tiny species, wing length 1.5–3.6 mm. 31

(a)

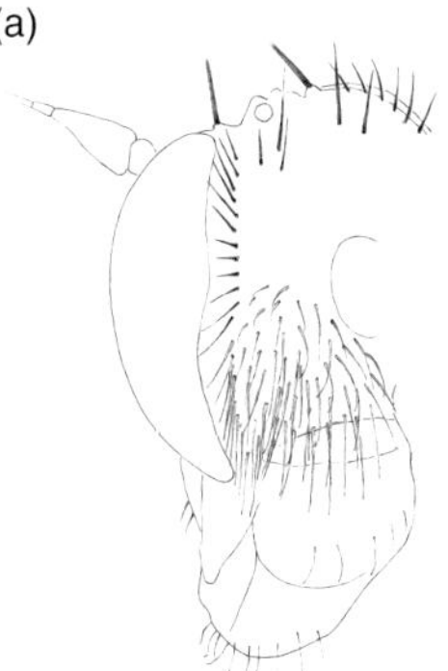

28(27) Crossvein dm-m as long as or shorter than apical section of M_4 (a). Acrostichal setae either absent or biserial and sometimes partly irregular. Hind basitarsus as long as or shorter than tarsomere 2. Scutum in front of scutellum gently domed, not flat............29

– Crossvein dm-m longer than apical section of M_4 (figs in couplet 30). Acrostichal setae uniserial. Hind basitarsus longer than tarsomere 2. Scutum in front of scutellum flat............30

(a)

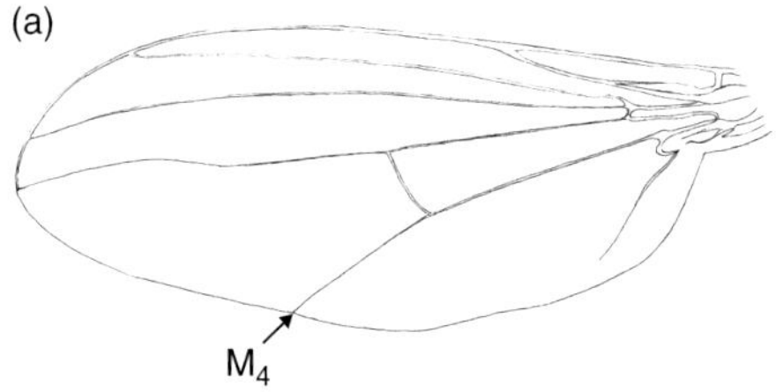

29(28) Antenna partly yellow. Palp entirely bright yellow, wider than face in middle. Acrostichal setae absent. 1 notopleural seta. Front tibia with 2–3 posteroventral seta............ ***Thinophilus*** (2 spp.; p. 350)

– Antenna black. Palp mainly or entirely dark, narrower than face in middle. Acrostichal setae present. 2 notopleural setae. Front tibia without a posteroventral seta or with 1 only as long as width of shaft............***Rhaphium*** (part: most of 26 spp.; p. 280)

30(28) Vein M_1 beyond dm-m smoothly sinuous (a). Wing with a small dark patch in cell r_{4+5} near dm-m. Scutellum with 3 pairs of long setae. Front coxa without an isolated long seta on anterior face............***Liancalus*** (1 sp. ***virens***; p. 217)

– Vein M_1 with strong bends beyond dm-m, its last section parallel with R_{4+5} (aa). Wing clear. Scutellum with 1 long and 1 shorter pairs of setae. Front coxa with a single long setae on anterior face (bb)............***Orthoceratium*** (1 sp. ***sabulosum***; p. 262)

(a)

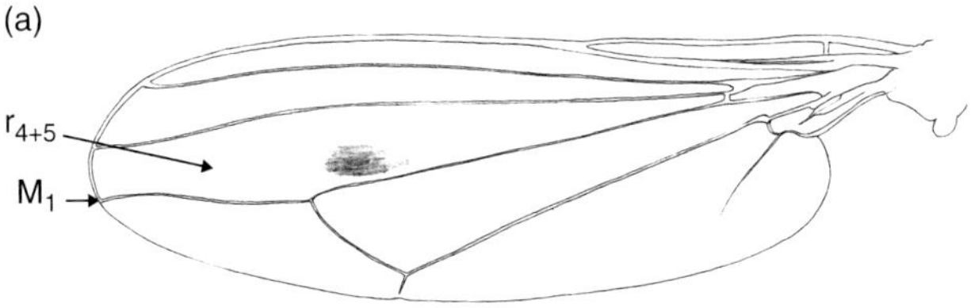

(bb)

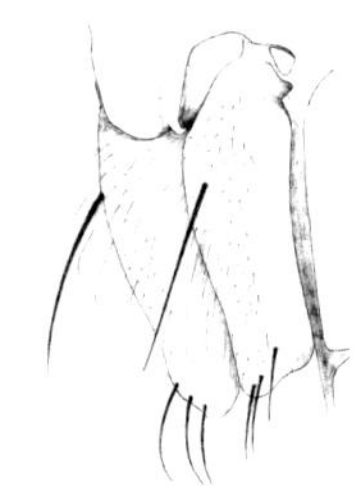

(aa)

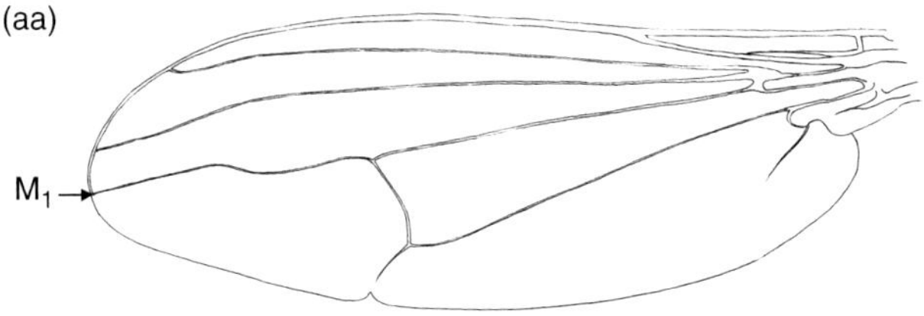

31(27) Tibiae almost devoid of setae, mid tibia with 0 or 1 pair at basal quarter and hind tibia with 1 anterodorsal seta at basal quarter, small apical setae on mid and hind tibiae. R_{2+3} and M_1 almost parallel beyond dm-m, both straight or gently curved but not sinuous, R_{4+5} cutting diagonally between them (a). Posterior part of scutum in profile from almost parallel to about 20° to front margin of katepisternum (fig. b of couplet 48). Postpedicel almost round with arista apical or nearly so. Occiput concave. Prescutellar depression clearly marked. Tiny to small species. (See also couplet 48.)............***Medetera*** (no key to females; p. 218)

– Hind tibia and often mid tibiae with several setae. Venation not like this; if R_{4+5} runs diagonally then R_{2+3} or M_1 often curved or sinuous without being parallel to one another beyond dm-m. Posterior part of scutum in profile at an angle of 20° or more to front margin of katepisternum. Other characters not like this at the same time............32

(a)

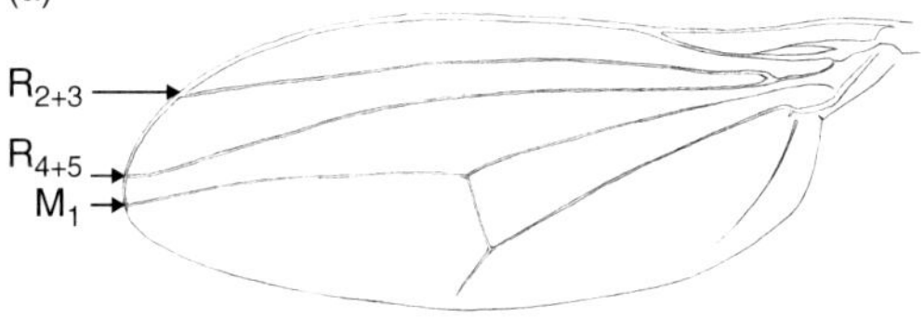

32(31) Arista apical or nearly apical on a triangular to round postpedicel. 33
– Arista dorsal; in species with a round postpedicel, arista placed clearly above middle of anterior face. 36

33(32) Only 5 tergites visible. Face broad, at least as wide as depth of postpedicel (a). M_1 ending at wing tip, slightly longer than R_{4+5} (b). 34
– Tergite 6 visible, about three-quarters as long as tergite 5. Face at its narrowest less than depth of postpedicel (aa). M_1 ending behind wing tip, as long as or slightly shorter than R_{4+5} (bb). 35

(a)

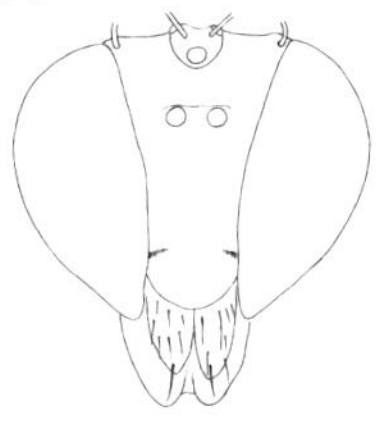

(aa)

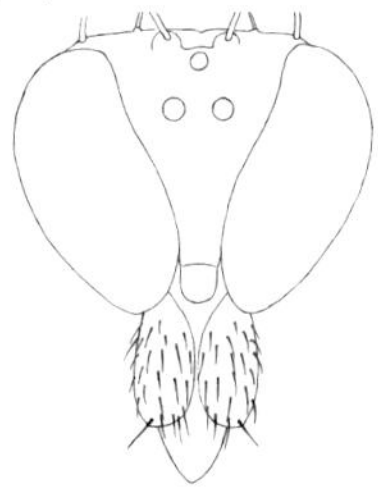

(b)

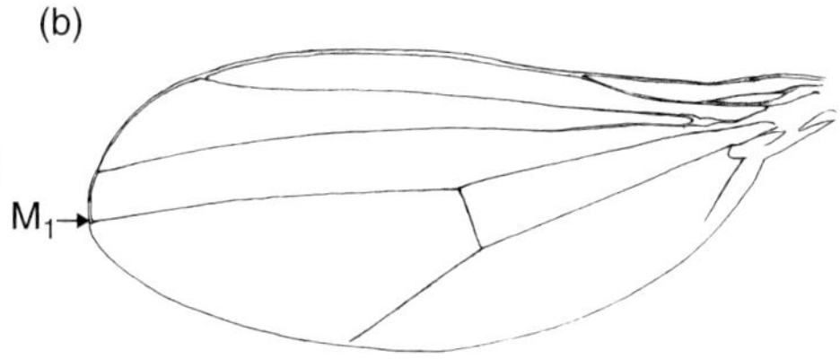

(bb)

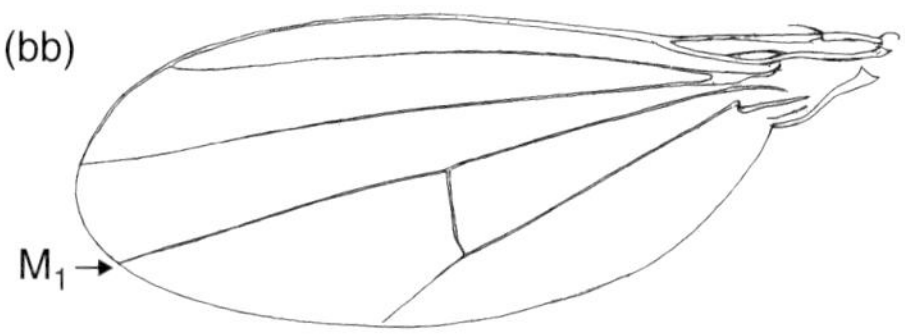

34(33) Postpedicel short triangular. Acrostichal setae biserial. Scutum and tergites greenish black, shining through thin dust. Wing without spots on apical section of M_1 or dm-m. .... ***Rhaphium*** (part of 26 spp.; p. 280)
– Postpedicel almost round (aa). Acrostichal setae absent. Scutum and tergites grey or brown, scarcely shining through dense dust. Wing with a faint small dark spot on apical section of M_1 and on dm-m (bb). .... ***Schoenophilus*** (1 sp. ***versutus***; p. 299)

(aa)

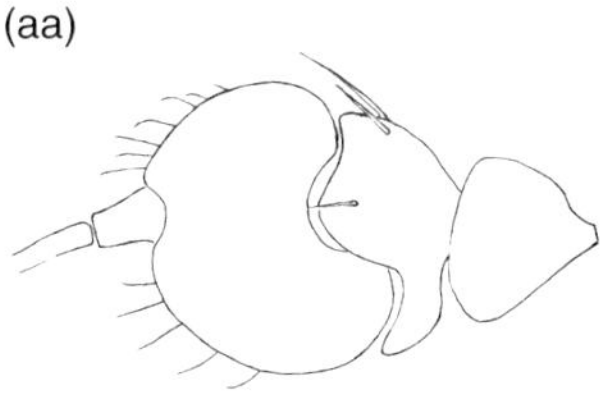

(bb)

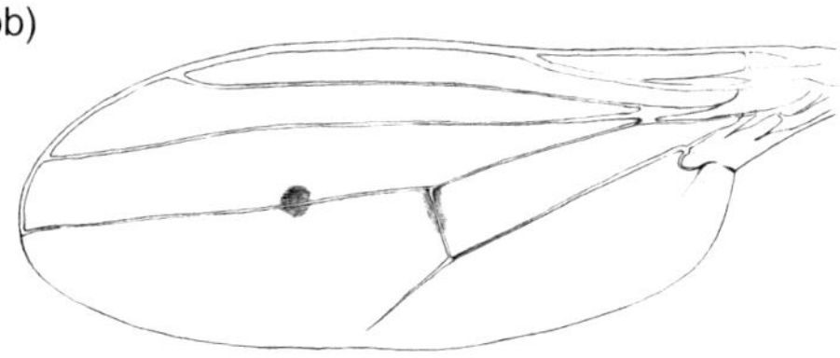

35(33) 5 pairs of dorsocentral setae. Acrostichal setae short, about one-quarter length of the adjacent dorsocentral seta. Upper postocular setae pale. Thorax and abdomen yellow or dark. .... ***Achalcus*** (8 spp.; p. 54)
– 6 pairs of dorsocentral setae. Acrostichal setae long, about half length of the adjacent dorsocentral seta. Upper postocular setae dark. Thorax and abdomen dark. .... ***Australachalcus*** (1 sp. ***melanotrichus***; p. 78)

36(32) Acrostichal setae present. 37
– Acrostichal setae absent. 41

37(36) Hind coxa with 2 lateral setae, the lower one may be fine (a). M_1 sinuous and clearly further from R_{4+5} just beyond dm-m than near tip (b). Proepisternal seta small but distinct. Wing length 3.0–3.6 mm. .... ***Argyra*** (part: ***vestita***; p. 70)
– Hind coxa with 1 lateral seta. M_1 almost parallel with R_{4+5} beyond dm-m (bb). Proepisternal seta absent. wing length 1.5–3.3 mm. 38

(a)

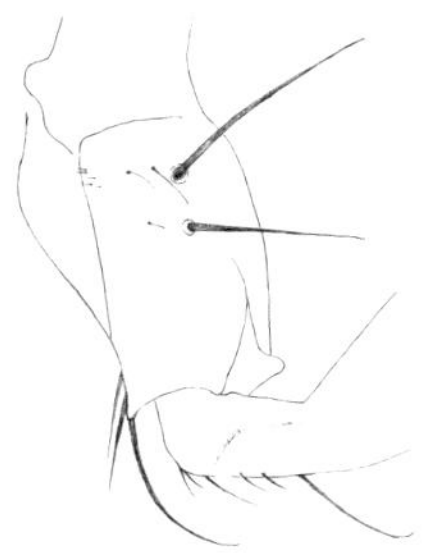

(b)

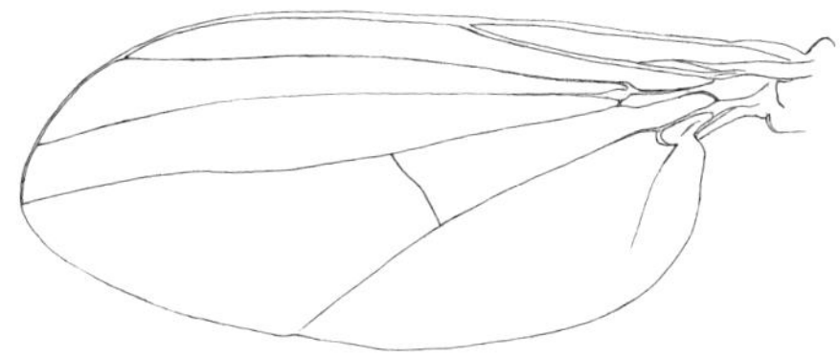

(bb)

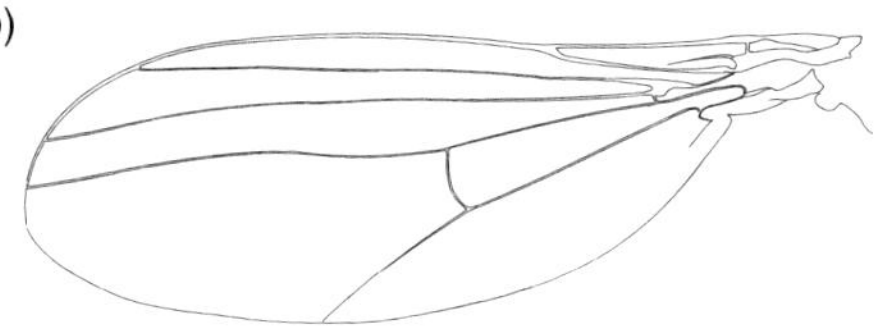

38(37) Face shining violet. Scutum blackish violet. Face almost as wide as high (a), clypeus strongly protruding (b). 3 pairs of dorsocentral setae .... ***Telmaturgus*** (1 sp. ***tumidulus***; p. 345)
– Face dusted white or grey. Scutum green or brown. Face narrower, width about one-quarter to half its height (aa), clypeus almost flat (bb). 4–6 pairs of dorsocentral setae 39

(a)

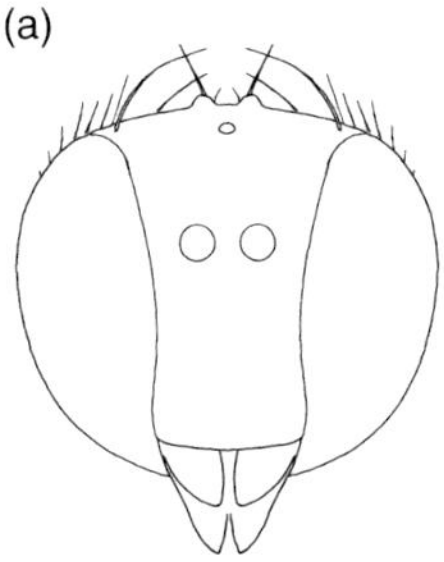

(b)

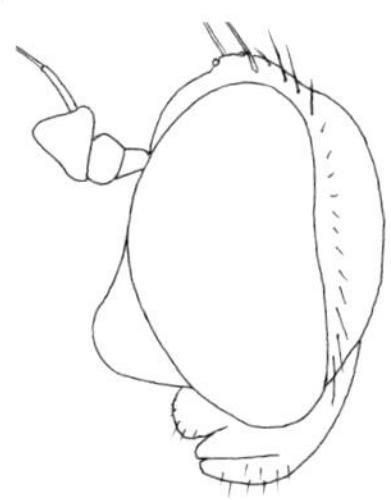

(aa)

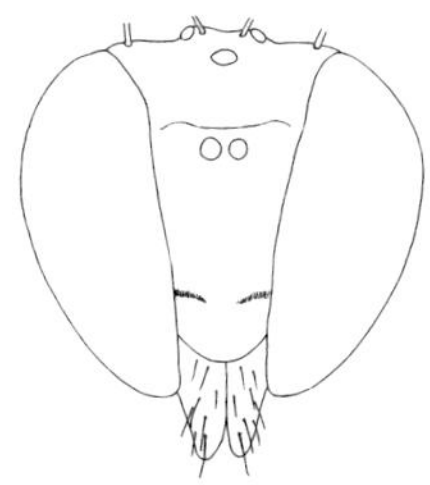

(bb)

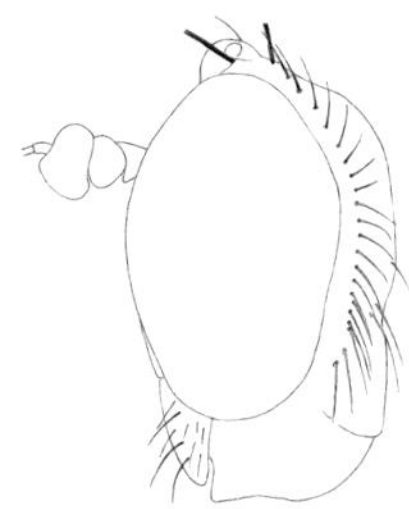

39(38) Scutum with a large matt black lateral mark above wing base between transverse suture and postalar seta. 4–5 pairs of dorsocentral setae. All coxae and metepimeron yellow. Frons shining violet. .... ***Lamprochromus*** (3 spp.; p. 215)
– Scutum without black lateral mark. 6 pairs of dorsocentral setae. Coxae, metepimeron and frons various. 40

40(39, 41) Vein dm-m oblique, forming an obvious to slight acute angle with basal section of M_4; M_1 just beyond dm-m bowed anteriorly for a short distance (a). Frons partly shining green or violet. Mid tibia with 2 anterodorsal and 1 posterodorsal setae (b) .... ***Teuchophorus*** (5 spp.; p. 347)

– Vein dm-m at right angles to M_1, not forming acute angle with M_4, or at most a few degrees from a right angle; M_1 just beyond dm-m straight and aligned with section just before dm-m (aa). Frons olive or brown and entirely dulled by dust. Mid tibia with 3 anterodorsal and 2 posterodorsal setae (bb)........ ***Sympycnus*** (5 spp.; p. 316)

(a)

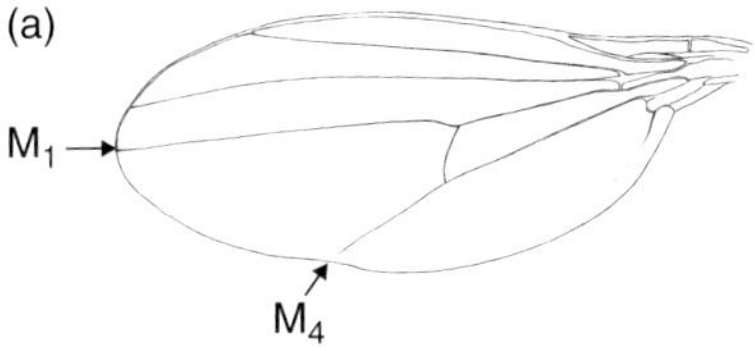

(aa)

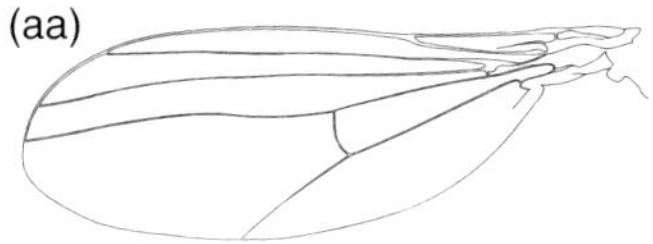

(b)

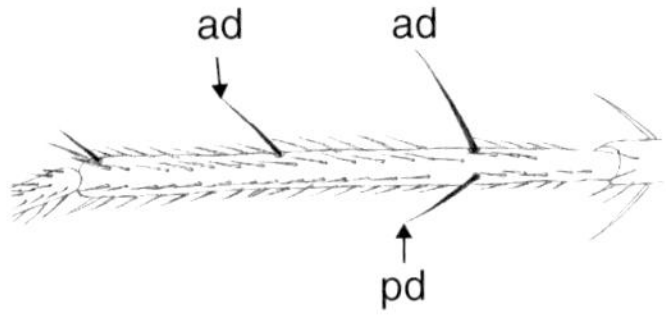

dorsal view

(bb)

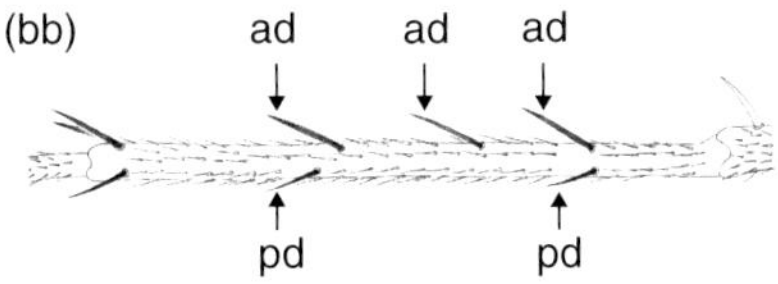

dorsal view

41(36) Scutum dark green. (If antennae partly yellow and palps large and yellow, see *Thinophilus*, couplet 29.)........ return to 40

– Scutum brown or grey with no hint of green........ 42

42(41) Vein dm-m and surrounding membrane pale, appearing as a small white spot (a). Apical section of M_4 about twice as long as dm-m (a). Face pale on lower two-thirds and narrower than postpedicel. Mid tibia with 2 antero- and 1 posterodorsal setae. No proepisternal seta above front coxa........ ***Micropygus*** (1 sp. ***vagans***; p. 255)

– Vein dm-m and surrounding membrane same shade as rest of wing. Apical section of M_4 about 3-4 times longer than dm-m (aa). Face black and broader than postpedicel. Mid tibia with 1 antero- and 1 posterodorsal setae. Proepisternal seta present above front coxa........ ***Micromorphus*** (no key to females; p. 247)

(a)

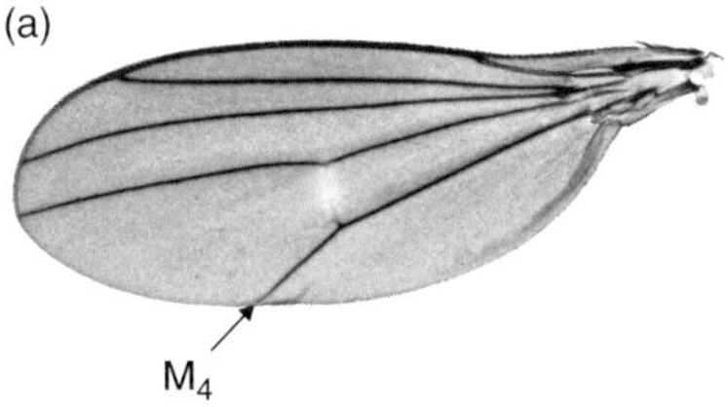

(aa)

43(14) Abdomen or pleura partly to entirely yellow. (*Argyra* with apparently only 1 seta on hind coxa may run to here, with strongly concave occiput and no prescutellar depression, see couplet 6)........ 44

– Abdomen and pleura entirely dark........ 45

44(43) Medium-sized to large species, wing length more than 4 mm. Abdomen extensively or entirely yellow, scutum yellow or dark, dusted and not shining. Proboscis and most of antenna yellow. Acrostichal setae present........ ***Neurigona*** (6 spp.; p. 259)

– Tiny species, wing length less than 3 mm. Abdomen entirely black, scutum shining black. Proboscis and most of antenna black. Acrostichal setae absent........ ***Anepsiomyia*** (1 sp. ***flaviventris***; p. 62)

45(43) Postgena with dense long whiskers behind postocular row and continuous with it, nearly as closely spaced as postocular setae and almost evenly spaced from mouth edge to neck, too numerous to count (a). Hind basitarsus about as long as tarsomere 2. Arista clearly dorsal, arising on basal two-thirds of postpedicel..46

– Postgena with few sparse widely spaced setae (about 5 or fewer). Hind basitarsus much shorter than tarsomere 2. Arista apical or subapical, or if apparently dorsal then far closer to tip of postpedicel than base. 47

(a)

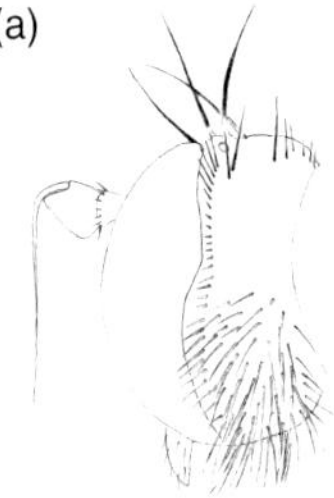

46(45) Acrostichal setae absent. Antenna extensively yellow. Face dusting yellow, gold or brown, its lower margin well above bottom of eye (a). Palp yellow and large, about as wide as middle of face (a, b). 1 notopleural seta.................... ***Thinophilus*** (2 spp.; p. 350)

– Acrostichal setae present. Antenna black. Face white-dusted, almost reaching bottom of eye. Palp partly yellow or entirely black, smaller, about three-quarters as wide as middle of face. 2 notopleural setae....................***Nematoproctus*** (2 spp.; p. 257)

(a) (b)

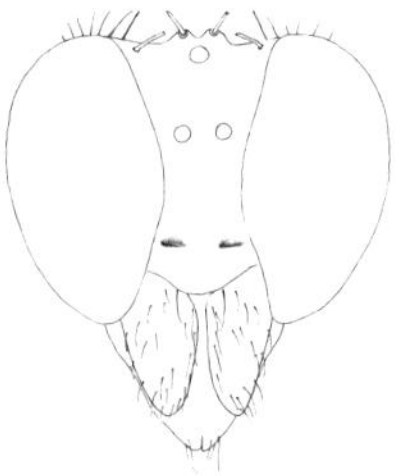

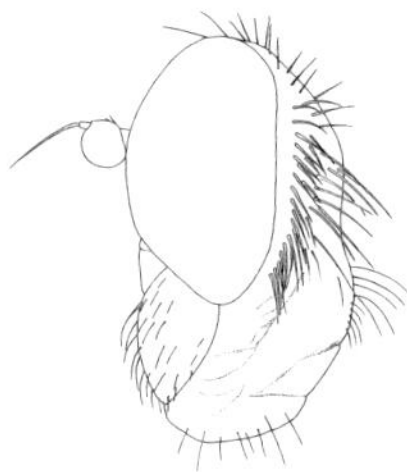

47(45) Tiny species with a shining black scutum without dusting or pattern. Clypeus projecting and bearing 2–4 long setae (a). Acrostichal setae absent. ....***Acropsilus*** (1 sp. ***niger***; p. 61)

– Tiny to medium-sized species, scutum never shining black, dusted and usually patterned with dust or colour. Clypeus not or slightly projecting, lacking long setae. Acrostichal setae present (may be minute). 48

(a)

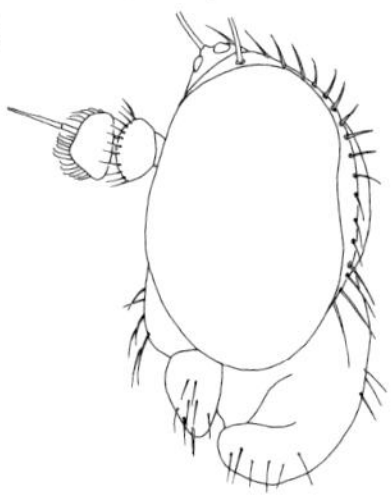

48(47) Tibiae almost devoid of setae on shaft, at most 1 pair at basal quarter of mid tibia (a) and 1 anterodorsal seta on basal quarter of hind tibia. Postpedicel usually almost round with apical or nearly apical arista, rarely 1.5 times as long as deep. Posterior part of scutum in profile making an angle with katepisternum from almost parallel to about 20° (b). 5 visible tergites. Medium-sized to tiny species, wing length 2.0–5.0 mm.. ***Medetera*** (no key to females; p. 218)

– Mid and hind tibiae with several distinct setae which may be short. Postpedicel long-triangular or elongate-oval with apical arista (cc). Posterior part of scutum making an angle of about 30–45° to katepisternum. 6 visible tergites, sixth almost as long as fifth. Medium-sized species, wing length 3.0–4.7 mm..***Systenus*** (6 spp.; p. 337)

(a)

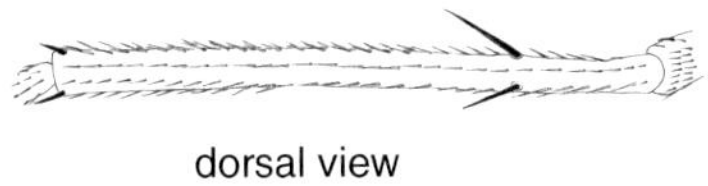

(b)

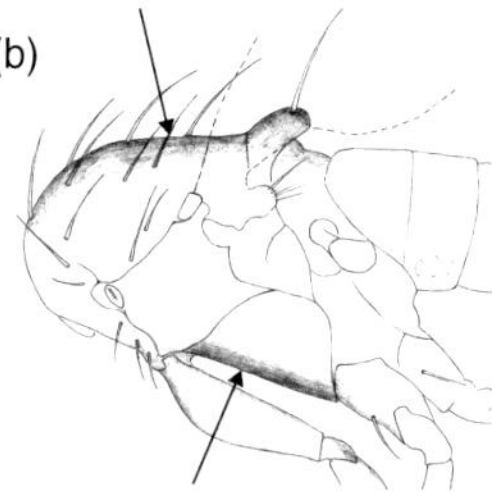

(cc)

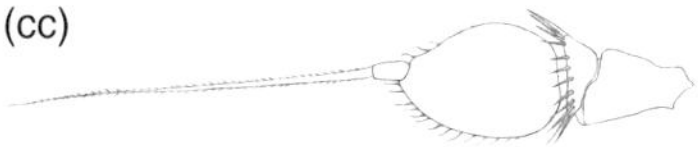

49(3) Abdomen mainly or partly yellow. 50

– Abdomen dark. 51

50(49) Pleura partly or entirely yellow. 6 obvious tergites, sixth as long as fifth, and seventh often also visible. Last few tergites same shade of yellow or discoloured brown as preceding tergites. Mid and hind femora without a preapical seta. Larger species, wing length 3.1–4.0 mm. .... ***Xanthochlorus*** (4 spp.; p. 364)

– Pleura entirely dark. 5 obvious tergites, sixth small or not visible. Last tergite dark green in strong contrast to preceding bright yellow tergites. Mid and hind femora each with a pale preapical seta. Small species, wing length 2.4–2.9 mm. .... ***Chrysotimus*** (2 spp.; p. 89)

51(49) Vein M_1 reaching costa behind wing tip (a). Vein dm-m at about middle of wing, lengths of M_1 either side of it about equal (a). Postpedicel oval with a bluntly pointed tip (b). Hind tibia with several long setae twice as long as width of shaft. Mid and hind femora each with a strong preapical seta. .... ***Achalcus*** (part: ***cinereus***; p. 57)

– Vein M_1 reaching costa at or before wing tip (aa). Vein dm-m at about basal third of wing, basal section of M_1 about half as long as apical section (aa). Postpedicel short and approximately round (bb). Hind tibia without setae, at most some stouter pale hairs equal to width of shaft. Mid and hind femora without a preapical seta. 52

(a)

(b)

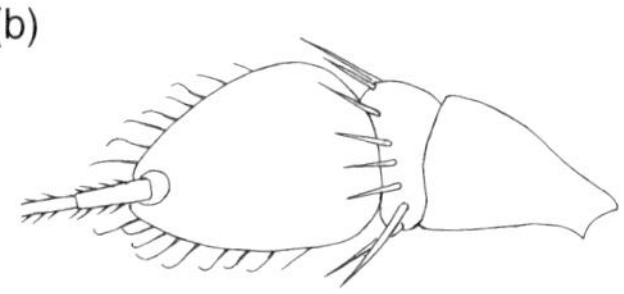

(aa)

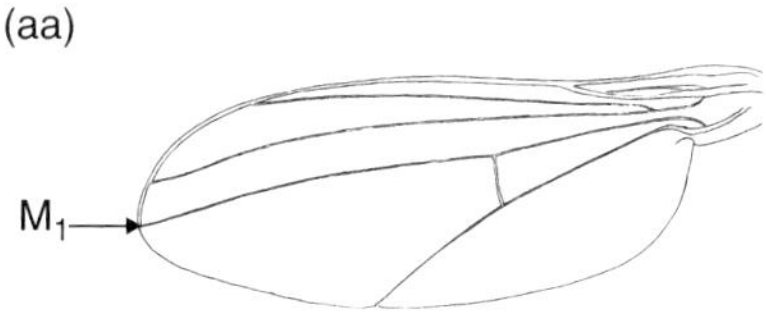

(bb)

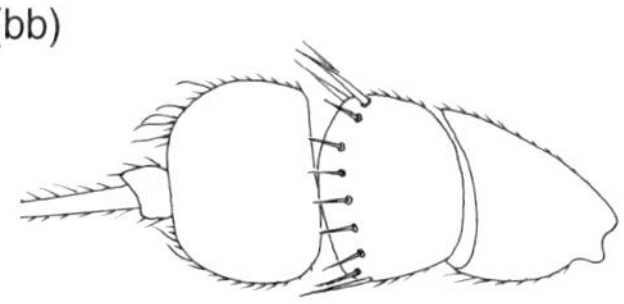

52(51) Hind coxa with 2 lateral setae, lower one may be fine (a). R_{4+5} and M_1 parallel beyond dm-m, their separation at dm-m scarcely wider than near wing tip (1.25 times at most) (fig. aa of couplet 51). 53

– Hind coxa with 1 lateral seta. R_{4+5} and M_1 converging weakly beyond dm-m, their separation at dm-m about 1.5 times that near wing tip. 54

(a)

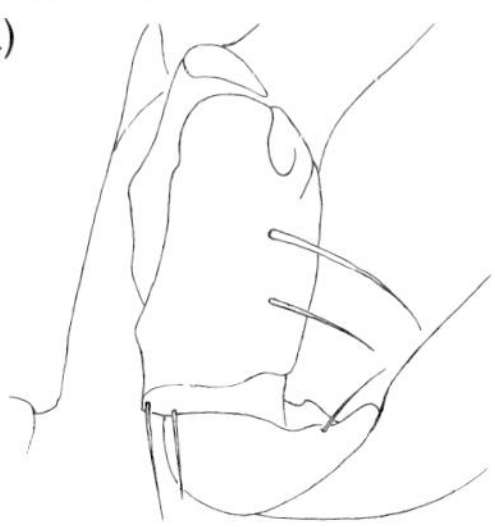

53(52) Ovipositor blade-shaped in profile (a). Scape and pedicel dark. Mid tibia with a weak anterodorsal seta on basal third, at most 1.5 times width of shaft (b). .... ***Thrypticus*** (12 spp.; p. 358)

– Ovipositor in profile stocky with an almost circular toothed structure above paired basal arms (aa, lateral and dorsal views). Scape and pedicel yellow. Mid tibia with a strong anterodorsal seta on basal third, twice width of shaft (bb)............***Corindia*** (1 sp. ***viridis***; p. 111)

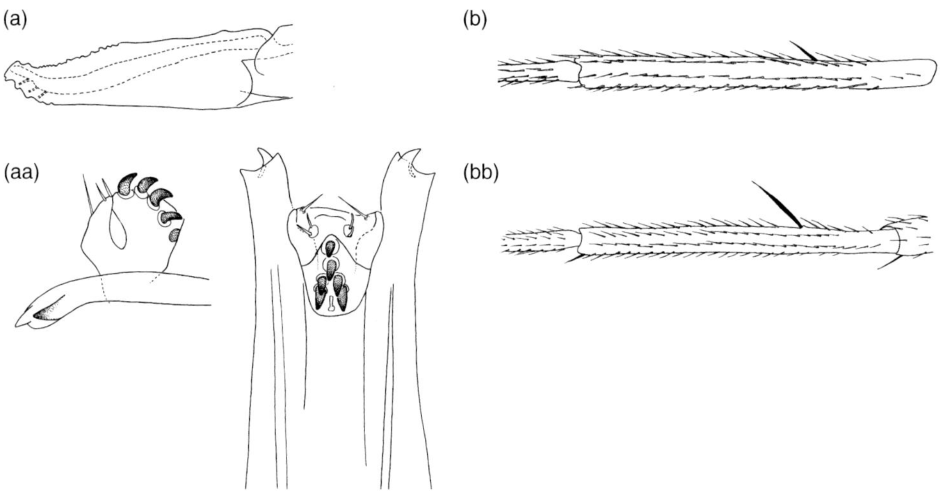

54(52) Front coxa with a spur on inner apical margin. Coxae bright yellow. Scape and pedicel yellow. 3 pairs of dorsocentral setae. Tiny species, wing length 2.1 mm.. .... ***Dolichophorus*** (1 sp. ***kerteszi***; p. 115)

– Front coxa without an apical spur. Coxae mainly dark. Scape and pedicel black. 5 pairs of dorsocentral setae. Minute species, wing length 1.3 mm.. ....***Cyrturella*** (1 sp. ***albosetosa***; p. 111)

ACHALCUS Haliday in Loew, 1857 (Achalcinae)

Achalcus are small to tiny yellow, dark brown or black flies with no trace of green or violet colouring, rather stout-bodied and with slender pale legs (Plate 1). The genus can be recognised in both sexes by the combination of the arista being apical or subapical, five pairs of dorsocentral setae, one or two anterior preapical setae on the mid and hind femora, and veins R_{4+5} and M_1 being practically straight and usually diverging from one another, with M_1 reaching the margin behind the wing tip. Additionally, males have a ventral seta close to the base of the front femur, an elongate-triangular postpedicel which usually has a straight lower margin and a rounded upper one thus giving the appearance of being upside-down, and five almost equally long tergites (six in females). Both sexes have biserial acrostichal setae and sparse leg setae without any ventral (av and pv) setae on any tibia. Differences from the similar *Australachalcus* are given in its account.

The keys are based on Pollet's (1996) revision the European fauna consisting of eight species, all occurring in Britain. *Australachalcus melanotrichus* is included in this key as it may be mistaken for an *Achalcus*. The subfamily Achalcinae was erected by Grootaert and Meuffels (1997), the genus *Achalcus* having previously been placed in a number of subfamilies (Germann *et al.* 2011a).

Specimens stored in alcohol for a long time can fade, especially the abdomens; as a result, some colour characters in the keys become less reliable although the thoracic spots and antennal shading usually remain discernible. Females are sometimes not identifiable using this key.

Species of *Achalcus* are reluctant to fly and, compared to all other genera, a far larger proportion of records come from vacuum sampling and trapping than from sweep-netting. They can be collected using a pooter in dense tussocks. All are associated with wetlands, mainly in open habitats and only occasionally within woodlands. The flight period of several species is long; *cinereus*, *britannicus* and *flavicollis* have been recorded in midwinter, and *thalhammeri* may also be active in winter (Cunningham 2015).

Key to species of *Achalcus* (males)

1 Dark species, with at least thorax and tergites brown to black. 2
– Pale species, with at least thorax yellow 5

2(1) Front femur with a fine ventral seta at basal quarter (a). 5 pairs of dorsocentral setae. 5 large tergites; sixth not visible in dry material, but narrow and lacking hairs in wet-preserved material (b). Vein dm-m placed at about middle of M_1, basal section of M_4 at least twice length of apical section (c) .... ***Achalcus*** 3
– Front femur without a fine ventral seta at basal quarter. 6 pairs of dorsocentral setae. 6 large visible tergites, sixth with hairs as on previous tergites (bb). Vein dm-m placed before middle of M_1, basal section of M_4 about 1.7 times apical section (cc). .... ***Australachalcus***

(a)

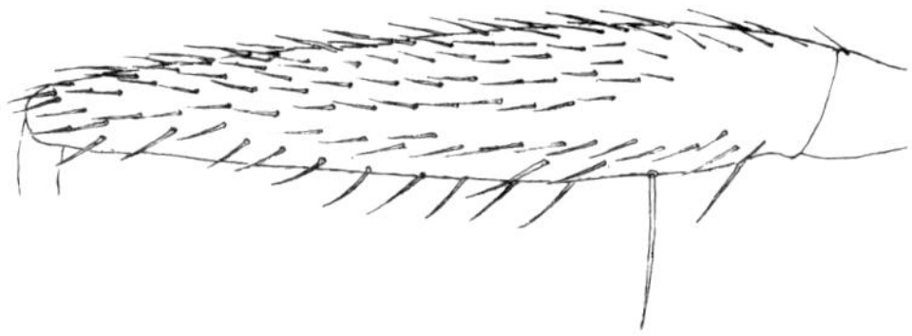

(b)

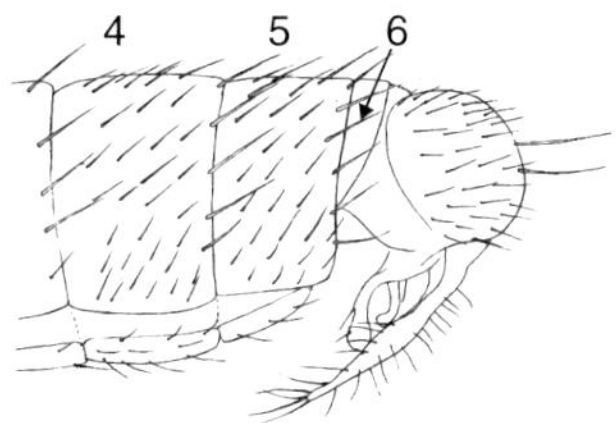

(c)

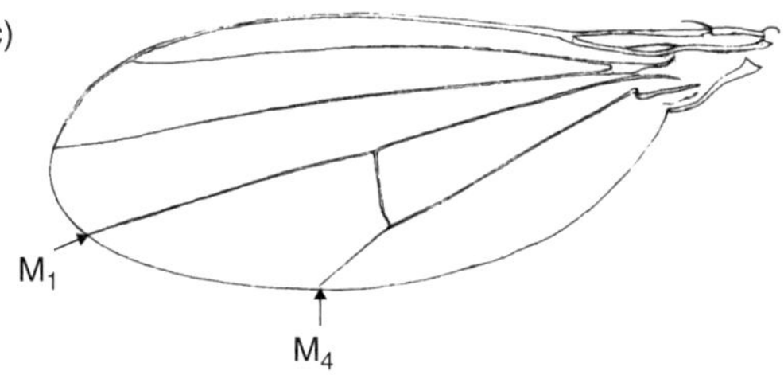

(bb)

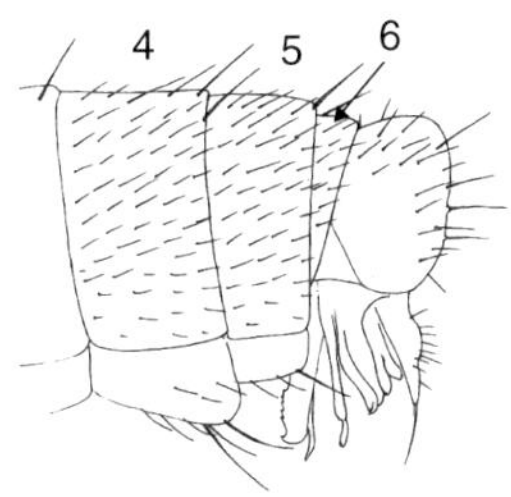

(cc)

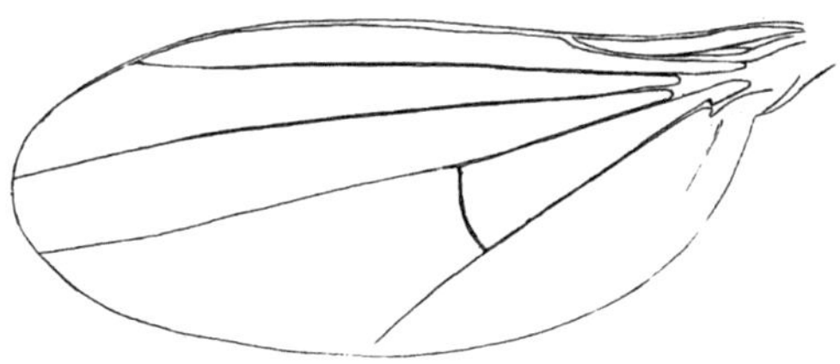

3(2) Sternites and tergites equally dark. Setae on thorax and head yellow to pale brown. Antenna mainly yellow, postpedicel short triangular, about 1.5 times longer than deep (a) .... ***cinereus***
– Sternites pale, clearly contrasting with darker tergites. Setae on thorax and head dark brown. Antenna mainly dark, postpedicel tapered, twice as long as deep (aa) 4

(a)

(aa)

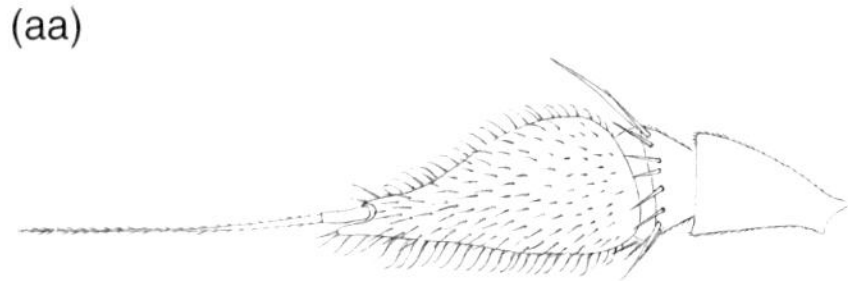

4(3) Cercus long and tapered, with long curved hairs at apex (a). Thorax usually uniformly dark brown.. ***thalhammeri***

– Cercus shorter, as long as other appendages, ovoid with acute apex, without curved apical hairs (aa[1]). Thorax usually dark brown with reddish brown areas along dorsocentral setae .... ***phragmitidis***

(a)

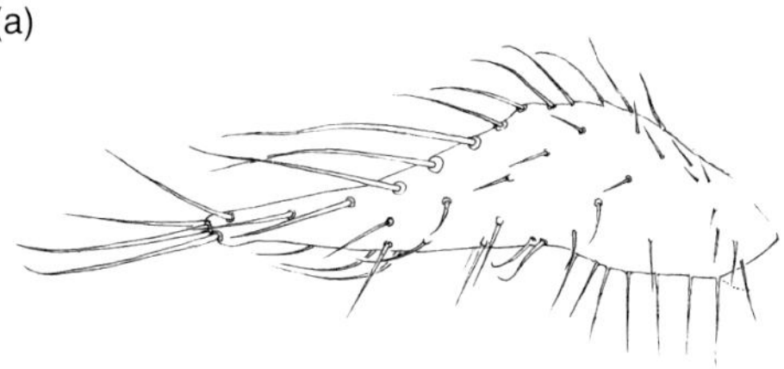

(aa)

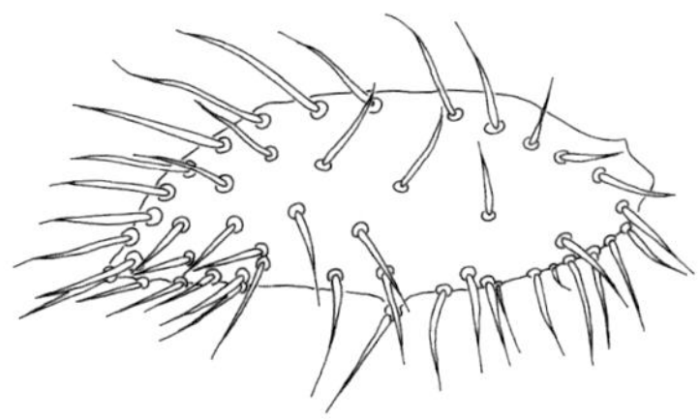

5(1) Scutum with a dark elongate spot on prescutellar depression (a). Hind tibia with 2 anterodorsal setae. Palp long, nearly half height of eye (b)...................................***nigropunctatus***

– Scutum without central dark spot. Hind tibia always with 3 anterodorsal setae. Palp small, a quarter to third height of eye.. 6

(a)

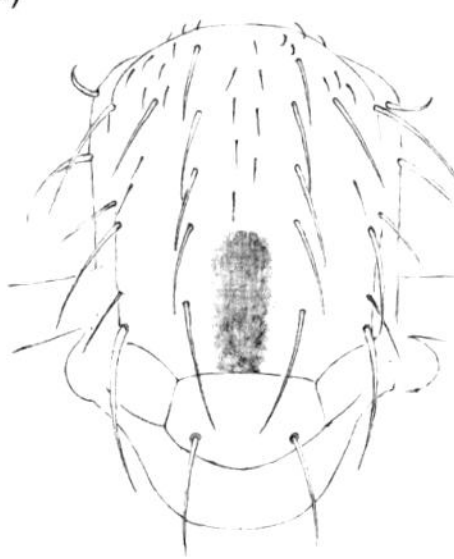

(b)

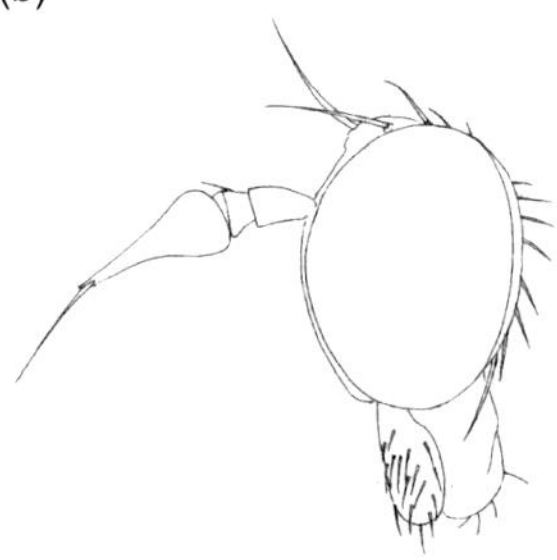

6(5) Entire body pale yellow, head contrastingly grey. Laterotergite (between wing-base and halter) with an isolated almost round dark brown to black spot (a)............................***bimaculatus***

– At least some abdominal segments brownish. Laterotergite with any contrasting dark brown spot concave behind or continuous with other dark marks on upper part of pleura..................7

(a)

7(6) Hind femur with only 1 (anteroventral) preapical seta (a). Mid tibia along its whole length with a row of ventral to posteroventral short setulae, their length about two-thirds depth of tibia (b). Postpedicel tapered (c). First and last tergites usually paler than intermediate tergites. Genital capsule largely yellow..***vaillanti***

[1] Pollet 1996. *Systematic Entomology* **21**, 353-386.

– Hind femur with 2 (anteroventral and anterodorsal) preapical setae (aa). Mid tibia without ventral setulae, all vestiture hairs very short (bb). Postpedicel triangular (cc). All tergites similarly brown, or at most the first slightly paler anteriorly and the last slightly paler posteriorly. Genital capsule usually entirely dark.. 8

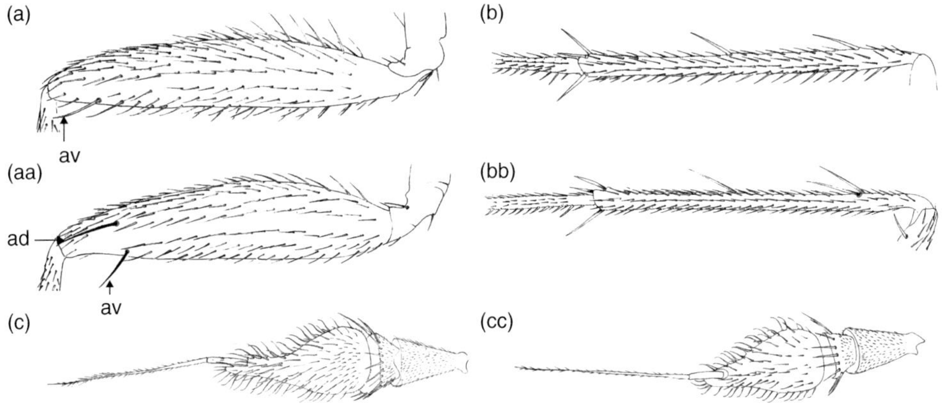

8(7) Cercus long and tapered, longer than epandrium and extending beyond other appendages, with curved setae at apex (a). Antenna usually mainly dark brown, pedicel sometimes yellow. Clypeus with only sparse pubescence..***flavicollis***

– Cercus shorter, about as long as other appendages, without differentiated setae at apex (aa). Antenna mainly yellow. Clypeus with dense pale pubescence ....***britannicus***

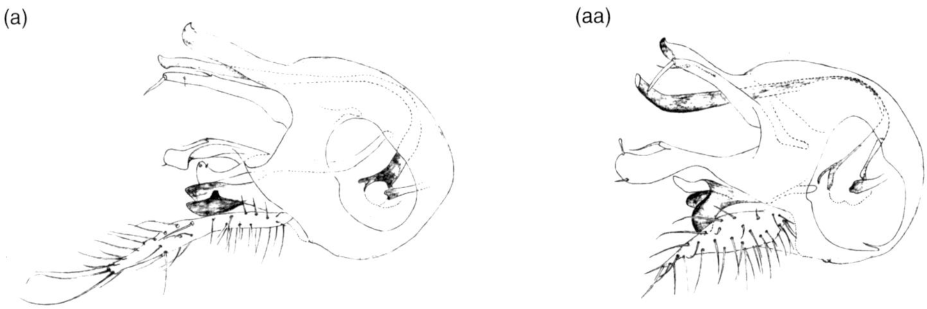

Key to species of *Achalcus* (females)

1 Dark species, at least thorax and tergites concolorous brown to black. 2

– Pale species, at least thorax reddish yellow (*flavicollis* sometimes has a brownish thorax but then showing pale palps).. 5

2(1) 5 pairs of dorsocentral setae. Acrostichal setae short, about a quarter length of adjacent dorsocentral setae...***Achalcus*** 3

– 6 pairs of dorsocentral setae. Acrostichal setae long, about half length of adjacent dorsocentral setae...***Australachalcus***

3(2) Sternites and tergites equally dark. Setae on thorax and head yellow to pale brown. Pedicel entirely black...***cinereus***

– Sternites pale, clearly contrasting with dark tergites. Setae on thorax and head dark brown. Pedicel yellow on inner face. 4

4(3) Thorax usually uniformly dark brown. Lower postnotal area with a broad dark front edge, paler behind. Sutural seta (small intra-alar lying in transverse suture) usually absent .... ***thalhammeri***

– Thorax usually dark brown with reddish brown areas along dorsocentral setae. Lower postnotal area entirely dark brown. Sutural seta usually present .... ***phragmitidis***

5(1) Scutum with a dark elongate spot in front of scutellum (fig. a of male couplet 5). Hind tibia with 2 anterodorsal setae. Tergites 7 and 8 shining dark brown, strongly sclerotised, contrasting with white sternites. .... ***nigropunctatus***

– Scutum without a dark spot. Hind tibia with 3 anterodorsal setae. Tergites 7 and 8 shining pale brown or yellow, usually showing little contrast with pale sternites 6

6(5) Entire body pale yellow, tergites and sternites the same colour. Laterotergite (between wing-base and halter) with a conspicuous almost round dark brown to black spot (fig. a of male couplet 6). Last abdominal segment with 6 acanthophorite spines........ ***bimaculatus***

– Tergites entirely brown or at least slightly darker than sternites. Laterotergite with any contrasting dark brown spot concave behind or continuous with other dark marks on upper part of pleura. Last abdominal segment with 8 acanthophorite spines........ 7

The following couplets from Pollet (1996) do not work well for British material; it may be safer to disregard females of these three species if the palp colour, sutural seta (small intra-alar in transverse suture) and ventral setulae on the hind tibia do not agree. Other colour characters are unreliable.

7(6) Palp yellow to whitish yellow, rarely slightly darkened. Antenna mainly dark brown. Sutural seta usually absent. (Tarsomere 5 distinctly infuscated. Slender species.) .... ***flavicollis***

– Palp brown to dark brown. Scape and pedicel mainly pale, postpedicel darker. Sutural seta usually present. 8

8(7) Sternites entirely whitish yellow. Tergites usually mainly yellowish white with central dark zone from tergite 2 onwards, tergites 4–6 usually with front half dark. Palp dark brown with basal third paler. Scape yellow, slightly darkened dorsally. Postpedicel triangular with short tapered apex. Hind tibia with about 5 short ventral setae. Rather slender species .... ***vaillanti***

– Sternites whitish yellow with broad dark lateral edges. Tergites usually pale yellow, sometimes entirely brownish but always with dark front and lateral edges. Palp brown with darker lower edge to entirely dark brown. Scape mainly dark, yellow only ventrally. Postpedicel triangular. Hind tibia with about 10 short ventral setae. Rather stout species. .... ***britannicus***

Achalcus bimaculatus Pollet, 1996

In both sexes the only dark areas on this tiny yellow species are the head and the round black spot in front of the halter; all other yellow species have at least partly dark tergites and either a crescent-shaped pleural spot or a dark band on the upper part of the pleura. Wing length ♂: 2.0 (1.9–2.1) mm, ♀: 2.2 (2.2–2.3) mm.

Achalcus bimaculatus is widely but sparsely distributed in England and Wales. It has been found in swamps and fens, in tall emergent vegetation at the margins of ponds and ditches, and occasionally in lightly shaded wet woodlands and in damp tussocky grasslands. Flight: vi–x.

Achalcus britannicus Pollet, 1996

This species most resembles *flavicollis* in colour with all tergites being similarly dark, although in *britannicus* the first and last tergites have slightly paler anterior or posterior margins, respectively,

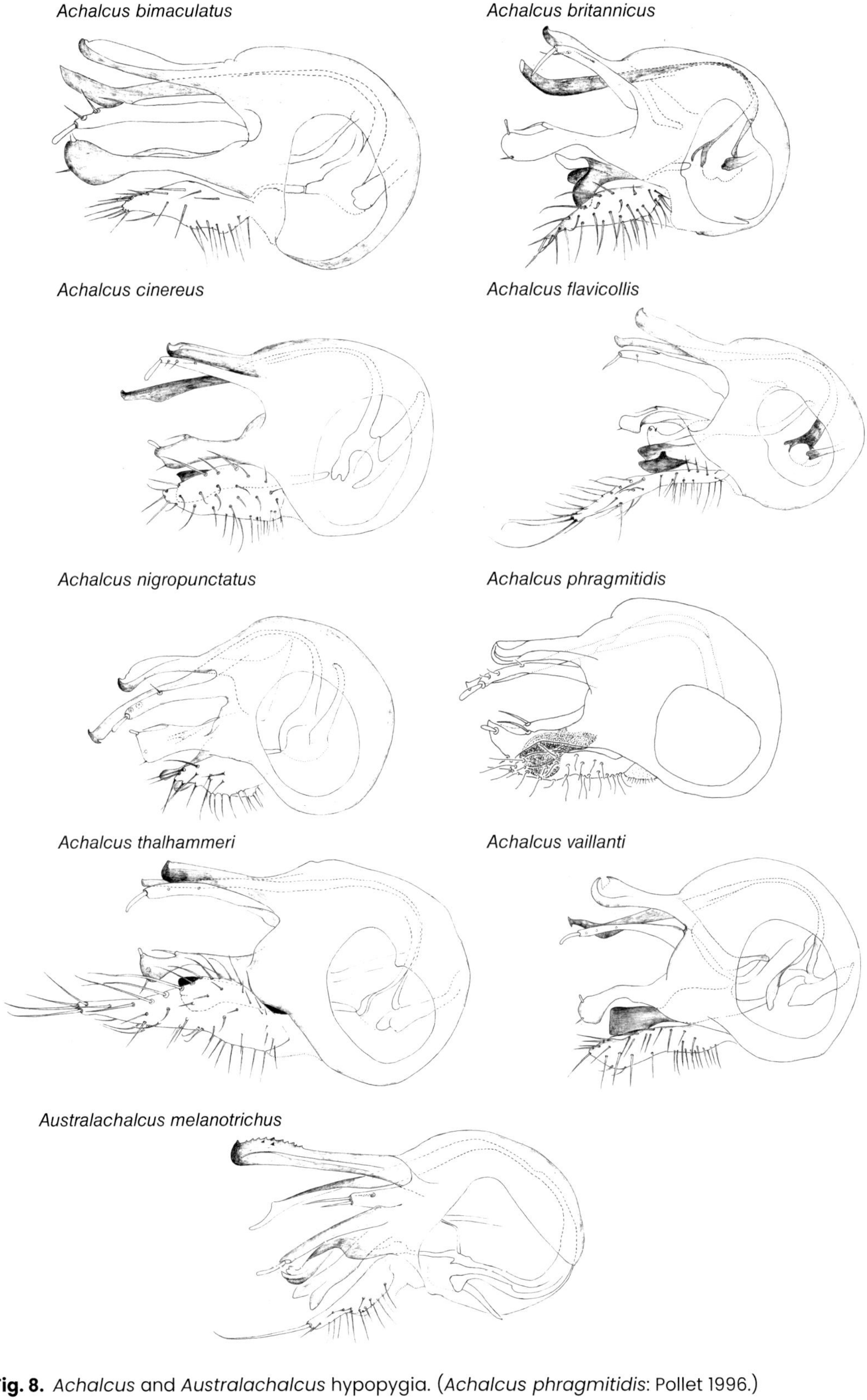

Fig. 8. *Achalcus* and *Australachalcus* hypopygia. (*Achalcus phragmitidis*: Pollet 1996.)

rather than being entirely dark. Pale pubescence on the clypeus, with individual hairs longer than the diameter of the arista, is found only in this species. The phallus resembles that of *vaillanti* in being a narrow black rod with a kink shortly before the tip and thus differs from that of *flavicollis* which is completely straight. Females can be difficult to distinguish from *vaillanti* and *flavicollis* but the larger number of ventral setulae on the hind tibia appears to differ from the smaller number in these two other species. The presence of a sutural setula may separate it from most *flavicollis* although it is tiny. No colour character works convincingly. Wing length ♂: 2.4 (2.2–2.6) mm, ♀: 2.5 (2.5–2.5) mm.

Achalcus britannicus has been recorded sparsely but widely in England north to Yorkshire. It has been found most often in floodplain and valley fens and swamps, at reed-lined water margins of ditches and slow rivers, and rarely in carr woodland. Flight: i, iii, vi–xi; probably adult all year.

Achalcus cinereus **(Haliday in Walker, 1851)**

This species has a slightly shining black body with yellow coxae and legs, and contrastingly pale setae on the body and head; this combination distinguishes both sexes from other dark *Achalcus* and *Australachalcus* and from all other British dolichopodids. Wing length ♂: 2.4 (2.2–2.5) mm, ♀: 2.5 (2.1–2.8) mm.

This is the most frequently recorded *Achalcus*, widely distributed in England but becoming scarce in the north and with very few Scottish records; it is known from the south of Wales. It is found in wetland dominated by tall monocotyledons (e.g. reed, reed sweet-grass, sedges) of fens, swamps, floodplain grasslands and the margins of ponds, ditches and rivers. It may have a slight preference for flushes within fens. Flight: i–xii; adults present all year, peak in vi.

Achalcus flavicollis **(Meigen, 1824) (Plate 1)**

Of the 'yellow' *Achalcus*, this species has all tergites entirely dark, along with the metepimeron and the genital capsule; *britannicus* is similar but usually has a slightly pale discontinuity at the first and last tergites. The male cercus is distinct from those of other pale species in having long curled apical hairs. The female's pale palp is nearly always clearly distinct from the usually darker palp of *vaillanti* and *britannicus*, and the scutum lacks the tiny sutural seta of these two species. Other colour characters are too variable to be reliable. Wing length ♂: 2.3 (2.2–2.4) mm, ♀: 2.2 (2.1–2.4) mm.

This is the most widespread *Achalcus*, found throughout Britain north to the Outer Hebrides and Orkney with no obvious gradient in its distribution. It occurs in many wetlands including fens, swamps, wet grasslands with ditches, acid mires, seepages, including those in woodland, and the margins of ponds, streams and rivers. It is often found within tall vegetation but also in vegetation shorter than occupied by other species of *Achalcus*. While old records will include other yellow species, confirmed records suggest that the distribution and habitat preferences are accurate. Flight: v–x, xii, peak in vi–vii.

Achalcus nigropunctatus **Pollet and Brunhes, 1996**

Unique among *Achalcus* in both sexes is the elongate spot on the scutum and only two anterodorsal setae on the hind tibia. The male antenna resembles that of *vaillanti* in being long, tapered and mainly pale, unlike the dark short triangular antennae of other yellow *Achalcus*. The pleura are distinctively marked with a dark more-or-less continuous stripe from the dark tergites to the neck; other species may have a similar but far less clear-cut stripe. Wing length ♂: 1.8 (1.7–2.0) mm, ♀: 1.8 (1.6–2.2) mm.

The Norfolk fens are the only known sites for this species where it is found in both tall and short fen vegetation, and in carr woodland over fen vegetation (Drake 2008). Flight: vi.

Achalcus phragmitidis **Pollet 1996**

This large dark species resembles *thalhammeri* but both sexes have reddish lines along the dorsocentral rows. Males have a rather shorter cercus.

The single record from Britain is from the Vale of York. It appears to be restricted to reedbeds (Pollet 1996) which is also the habitat where it was recorded in England. Flight: vii.

Achalcus thalhammeri **Lichtwardt, 1913**

This is the largest British *Achalcus* and has relatively long legs. It closely resembles *phragmitidis* both in size and dark colour but *thalhammeri* may be distinguished by its long cercus with long apical hairs. The pale sternites clearly distinguish both sexes of *thalhammeri* from *cinereus* and *Australachalcus* but care is needed to distinguish females from *phragmitidis* as the colour of the laterotergite of *thalhammeri* is variable in British specimens. Wing length ♂: 2.8 (2.6–2.9) mm, ♀: 2.9 (2.7–3.1) mm.

Achalcus thalhammeri is found mainly on the southeast English coast from Sussex to Yorkshire, inland in Cambridgeshire and at a brine pit in Worcestershire. The habitat is often ditch margins in fen and coastal grazing marsh. The close association with ditches within fens is striking. Flight: iii–ix.

Achalcus vaillanti **Brunhes, 1987**

Males are easily distinguished as this is the only British *Achalcus* with only one preapical seta on the hind femur and a ventral row of slightly longer hairs on the mid tibia. The antennae are particularly large and entirely yellow. Females are difficult to separate convincingly from those of *britannicus* and *flavicollis*; *vaillanti* has fewer setulae on the hind tibia than *britannicus* and has a small sutural seta that is usually absent in *flavicollis*. The palps may sometimes be pale so resemble those of *flavicollis*. Wing length ♂: 2.5 (2.2–2.8) mm, ♀: 2.5 (2.3–2.7) mm.

Achalcus vaillanti has a sparse distribution in lowland England from Devon to Yorkshire, and there are a few records from southern Scotland. It has a strong affinity to dense reedbeds and reed-lined ditches. It sometimes also occurs among other tall monocotyledons, usually within fens and sometimes swamps, and it is occasionally found on coastal reedbeds with a slightly brackish influence. Flight: vi–vii.

ACROPSILUS Mik, 1878 (Peloropeodinae)

The single European species is a tiny stocky species with a short deep abdomen scarcely longer than the thorax, the scutum entirely shining black with no trace of green but a slight violet tinge in life (Plate 1). In both sexes, the genus can be recognised by the combination of no acrostichal setae, no preapical setae on the mid and hind femora, a dark or black halter and dark legs with a very short hind basitarsus. The downwardly directed white cercus of the male is distinctive (fig. b of couplet 24, key to ♂ genera). Females have a cluster of four conspicuous hairs on the clypeus, each as long as the postpedicel (fig. a of couplet 47, key to ♀ genera); some *Syntormon* have two facial setae but otherwise such setae are not found in British species.

Acropsilus niger **(Loew, 1869) (Plate 1)**

This tiny species can be recognised using the generic diagnosis. Wing length (non-British specimens) ♂: 1.3 (1.2–1.5) mm, ♀: 1.6 mm.

The only British records are from two sites in Cornwall (1902, 1905) and one in Hampshire (1990, 2000). One of these sites included seepages by a pond. In France it has been found on sunlit stones by streams and rivers, and in Portugal and Czech Republic at sandy or gravelly lake margins (Pollet *et al.* 2019, Kejval 2022, Drake unpublished). Flight: vi–vii.

ANEPSIOMYIA Bezzi, 1902 (Sympycninae)

There is only one species worldwide. No other British genus has the combination of tiny hairs on the antennal scape and no acrostichal setae. There are no preapical setae on the mid and hind femora, unlike other species in the Sympycninae. Males have strongly armed front and mid femora reminiscent of *Hydrophorus* and *Aphrosylus*, and an elongate-triangular postpedicel resembling those of some *Syntormon* and *Rhaphium*, but with the arista inserted very close to its base (figs in couplet 23 of ♂ genus key; Plate 1). In both sexes the basal section of the arista is about half as long as the apical section.

The subfamily placement of *Anepsiomyia* remains uncertain: Peloropeodinae (Negrobov 1991), Diaphorinae (Grichanov 2017) or Sympycninae where it has been traditionally placed and is followed here.

Anepsiomyia flaviventris (Meigen, 1824) (Plate 1)

The shiny black upper-side with no trace of green, contrasting strongly with yellow pleura and coxae, distinguishes this species from all other British dolichopodids. Males also have a yellow-ringed abdomen with part of the second and all the third and fourth tergites yellow, which females lack. Wing length ♂: 2.5 (2.4–2.7) mm, ♀: 2.7 (2.4–2.9) mm.

Anepsiomyia flaviventris is common and widespread in England and Wales but scarcer in Scotland and not yet recorded in the northern half of the country. It is found in woodlands, or at least in shaded places, which are at least damp and more usually wet, particularly alder, sallow and birch woodland. Females often outnumber males or occur in abundance in the absence of males. Flight: v–viii, peak in vi to early vii.

APHROSYLUS Haliday, 1851 (Hydrophorinae)

The genus is immediately distinguished from all others by having short dense spikes rather than hairs on the front coxae (Plate 1). British species are either moderately large or tiny, thus adding confusion to recognising the genus. They are dull brown and grey flies with little if any reflective surface. The palps are particularly large and rounded, as otherwise found only in *Thinophilus* and *Schoenophilus*, bright silver in males and greyish in females. The proboscis is stout, narrow and directed ventrally, terminating in a short curved spine directed backwards. Additionally, the postgena has dense whiskers, the postpedicel is short- to long-triangular with an apical arista, the mid and hind femora have one to two preapical setae and often additional anterodorsal setae, and the marginal setae of tergites are not differentiated from hairs on the disc except at the lateral margin.

Four of the 16 European species of *Aphrosylus* occur in Britain. Of the remaining species, only *venator* Loew, 1857 is likely to be found here as it occurs on the west coast of France. It is a small species intermediate in size between the large and tiny British species and will be recognised by having only four dorsocentral setae, like *ferox* and *mitis*, but with the front femur having a complete ventral row of regularly spaced spines, unlike the sparse and irregular ventral setae in these two tiny species.

The hypopygium is clearly visible in three of the species but small and hidden in *mitis* so the sex of this species is not immediately apparent. Males have six equally long tergites but females have only five, and the male palps have a yellow ground colour below bright silver pubescence compared to black or dark yellow with brown or pale grey pubescence in females.

Leg colour and thoracic chaetotaxy are rather variable and make unreliable characters for separating species. The two larger British *Aphrosylus* often have a second seta placed close to the first dorsocentral, which is interpreted as the anterior intra-alar.

Species of *Aphrosylus* are confined to saline habitats and are found on rocky shores and sheltered estuaries. Larvae of *celtiber* are predators of the barnacle *Chthamalus montagui* Southward (Poulding 1998). The scarcity of this barnacle on the British east coast where *celtiber* is found suggests that *Semibalanus balanoides* (Linnaeus), which occurs extensively on the east coast, may be the principal prey here. Adults of *raptor* also inhabit barnacle-covered shores and so their larvae probably also feed on barnacles, but larvae of *ferox* and *mitis* may have different prey. Poulding (1998) suggested that this may also sometimes be the case for *celtiber*. Dyte and Poulding (1992) and Poulding and Tompsett (2007) mention the co-occurrence of *mitis* with the Australian barnacle *Austrominius modestus* (Darwin) in Cornwall but point out that no direct association has been shown. Crossley (2014) suggested that larvae of *ferox* may be associated with decaying seaweed on the strandline. When submerged, larvae respire by means of an air bubble over the posterior spiracles acting as a plastron that supplements cuticular respiration and allows the larva to remain active for longer (Poulding 2010). Males wave their shining silver palps and, in sunshine, the rapidly flashing signal of the larger species can be seen several metres away. This is assumed to alert females whose palps are less reflective.

Key to species of *Aphrosylus* (males and females)

1 Large species, wing length at least 4.0 mm, usually about 5.0 mm. Wings narrower, on average nearly 3 times as long as wide (a). 5–7 dorsocentral setae. 2

– Tiny species, wing length less than 2.5 mm. Wings broader, on average 2.5 times as long as wide (aa). 4 dorsocentral setae. 3

(a)

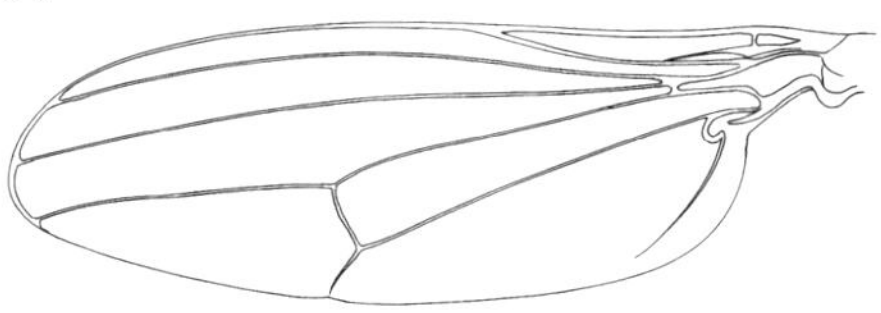

(aa)

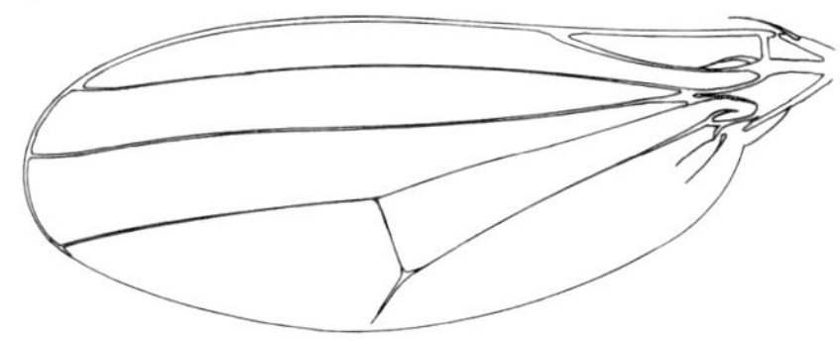

2 Tergite hairs stouter and less dense, about 3-4 rows deep on each tergite, those on side of tergites as strong and long as posteroventral and anteroventral setae of hind femur. Scutum with shifting patches of almost black subshining ground colour showing through dull pale dusting when tilted back-and-forth and viewed from above. Hind femur with anterodorsal setae forming a partly interrupted row from base to tip, the basal setae sometimes rather short but longer than dorsal vestiture between basal and apical parts of the row (a). Male front tarsomere 2 dilated on basal half to two-thirds (b). .... ***celtiber***

– Tergite hairs finer and denser, about 4–5 rows deep on each tergite, those on side of tergites clearly finer than posteroventral and anteroventral setae of hind femur. Scutum almost uniformly dusted pale brown when viewed as above, any pattern not becoming sharply demarcated and even in anterior view not showing almost black patches. Hind femur with anterodorsal setae becoming smaller and almost indistinguishable from vestiture on basal quarter (aa). Male front tarsomere 2 swollen at base only (bb). .... ***raptor***

(a)

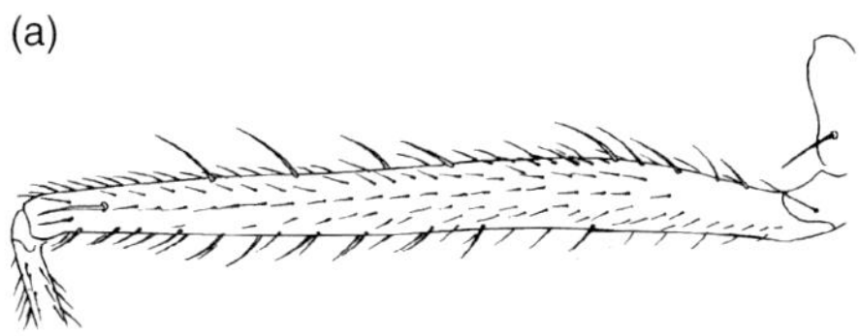

(b)

(aa)

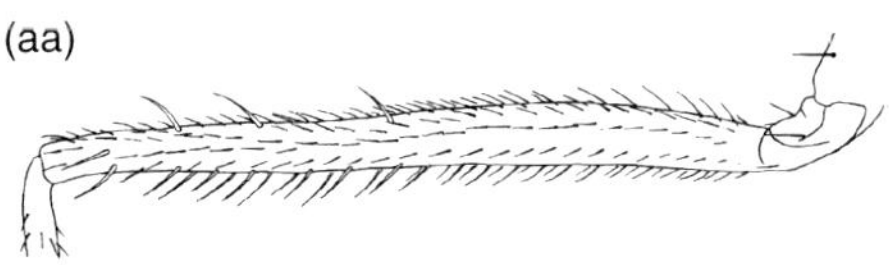

(bb)

3 Wing costa between R_1 and R_{2+3} with two sizes of setae, a long upright seta between every 3–6 short setae on both upper and lower margins (a). Front tibia at apex with a ventral extension bearing a spur at its tip (b). Front femur with several long posteroventral setae but no posterodorsals (c). Postpedicel conical, not tapered into an apical extension (d). *Male*: Hypopygium large, deeper than depth of abdomen. Mid femur with irregularly spaced posteroventral setae, with clusters on basal quarter and midway, and 2–3 setae on apical quarter (e). .... ***ferox***

– Wing costa between R_1 and R_{2+3} with all setae the same size (aa). Front tibia simple at apex (bb). Front femur with no long posteroventral setae apart from a preapical but with 2 posterodorsal setae (cc). Postpedicel bulbous with narrow drawn-out apical extension not clearly distinct from arista (dd). *Male*: Hypopygium tiny and hidden. Mid femur with posteroventral setae regularly spaced. .... ***mitis***

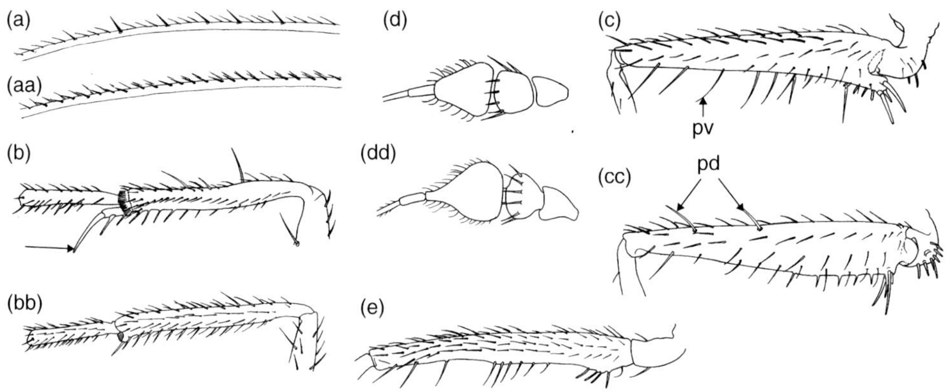

Aphrosylus celtiber Haliday, 1855 (Plate 1)

This species and *raptor* form a pair of similar large species whose separation needs care. No pronounced male secondary sexual characters clearly distinguish them; the shape of the front basitarsus is not always clear, but *celtiber* males have two to three minute curved spines beneath the tip of the mid tibia. Both sexes of *celtiber* are more spiny than those of *raptor* and have slightly longer setae over the whole body. Both sexes have distinctly uneven dusting on the scutum such that, in slightly anterior view, the front half has a central brown-dusted patch and the rear half is almost black and undusted, and this pattern reverses as the specimen is tilted back so it is seen in slightly posterior view. In contrast *raptor* is far more evenly dusted at all angles of view, and even in anterior view only two shades of brown become evident. The front femur of both sexes is usually yellow on slightly less than the apical half whereas in *raptor* the yellow usually extends to slightly more than the apical half. Wing length ♂: 4.9 (4.4–5.4) mm, ♀: 5.1 (4.4–5.5) mm.

Aphrosylus celtiber is found around the British coast from Cornwall to Sutherland and the Outer Hebrides although with few records on southeast English coasts from Lincolnshire to Dorset where soft Cenozoic rocks and the Cretaceous chalk rarely form rocky sea shores which *celtiber* requires. Flight: iv–xi, with indications of continuous breeding over this period.

Aphrosylus ferox Haliday in Walker, 1851

Aphrosylus ferox and *mitis* are similar in size and far smaller than *celtiber* and *raptor* but, unlike that pair, they are amply distinct from one another. As well as the modified mid femur mentioned in the key, some other male secondary sexual characters are the ventrally swollen front basitarsus and second tarsomere, markedly flattened fourth and fifth front tarsomeres, and the hind basitarsus having several fine dorsal hairs. The deep and robust hypopygium is very obvious. Wing length ♂: 2.1 (2.0–2.3) mm, ♀: 2.6 (2.5–2.7) mm.

This species is found round the British coast northward to Orkney and the Outer Hebrides with relatively few records on the coast of lowland England from Lincolnshire to Dorset. It occupies a

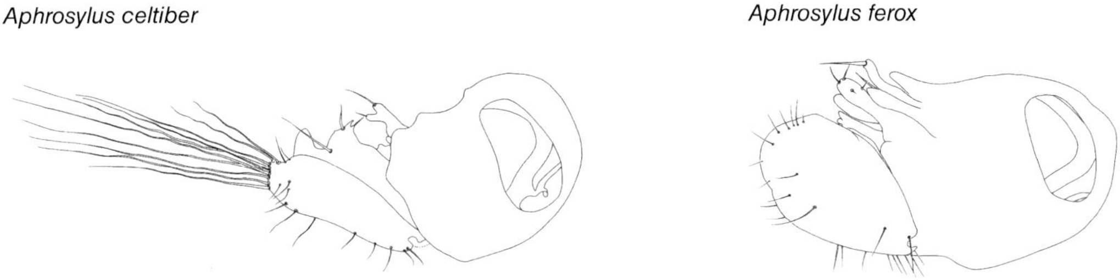

Fig. 9. *Aphrosylus* hypopygia.

greater range of sea-shore habitats than other *Aphrosylus* and is found on rocky shores, bouldery and stony beaches, and more sheltered upper shore habitats with mud and brown or green seaweeds such as *Enteromorpha* (Dyte and Poulding 1992). It can sometimes be numerous. Flight: v–xi, with indications of continuous breeding over this period.

Aphrosylus mitis **Verrall, 1912**

This is a tiny species, like *ferox*. Males have no obvious secondary sexual characters and the hypopygium is so small and hidden that the sex is not immediately obvious (see introduction for sexing). Both sexes have an onion-shaped postpedicel whose terminal extension appears to resemble the base of the arista. Leg colour is particularly variable, from entirely yellow (except the mid and hind coxae) to entirely dark grey. Wing length ♂: 1.9 (1.8–2.0) mm, ♀: 2.3 (2.2–2.5) mm.

Aphrosylus mitis is an uncommon species found in southern England and south Wales, from Essex to Pembrokeshire. It is usually found in sheltered places such as estuaries where brown seaweeds on rocks or boulders are frequent, muddy creeks and saltmarsh, although it can sometimes be found on more exposed rocky shores. As with *ferox*, it has a wide tolerance of salinity variation (Dyte and Poulding 1992). Flight: v–x, peak in vii.

Aphrosylus raptor **Haliday in Walker, 1851**

Aphrosylus raptor is very similar to *A. celtiber*, and the principal distinction is described under the account for that species. Wing length ♂: 5.2 (5.0–5.4) mm, ♀: 5.9 (5.6–6.3) mm.

The distribution is mainly from Dorset to Anglesey, with isolated records scattered on the coast northwards to Orkney although these outliers may need revision, and is markedly less frequent than *celtiber*. It is found on exposed rocky shores, usually on barnacle-covered rocks. Flight: v–ix.

ARGYRA Macquart, 1834 (Diaphorinae)

These are medium-sized to large elongate flies, metallic green and sometimes with yellow markings on the tergites (Plate 1). Males of many species have striking silver-white dusting on the thorax and abdomen not found in other genera. A defining character of the genus is a vertical row of two to six erect setae on the hind coxa, decreasing in size from the large one placed close to the base (Germann *et al.* 2011b) except in *grata* which sometimes has only one seta and fine hairs below it. In both sexes the vertex is deeply excavated so that the ocellar triangle is strongly raised, and all species except *vestita* have tiny hairs on the scape, as in Dolichopodinae, some *Syntormon* and *Anepsiomyia*. The acrostichals are usually biserial, at least in front, but often become irregular behind and are quadriserial in *diaphana*. Males have a short but high head when viewed from the side, erect antennae placed well above the middle of the eye, dense whiskers on the postgena, two to five distinct setae on sternite 8 and often quite hairy legs with fringes that obscure any preapical setae on the femora. Females usually have conspicuous pale dust patches on the tergites, as in *Rhaphium* and a few other genera.

Eleven species of *Argyra* occur in Britain out of the 26 known from the Western Palaearctic (Pollet 2011). Other species that may possibly occur in Britain are *hoffmeisteri* (Loew, 1850) which is widespread in central Europe, *loewi* Kowarz, 1879 and *setimana* Loew, 1859, both with limited distributions in central Europe, and *magnicornis* (Zetterstedt, 1838) and *spoliata* Kowarz, 1879, both found in the northern half of Europe to Scandinavia. All of these except *spoliata* have fine hairs on the scutellum so resemble *diaphana* and are discussed under the account for that species. *Argrya spoliata* resembles the *argyria* group but has thinner dusting on the scutum and tergites and longer setae on sternite 8.

The following is a guide to males based on colour and dusting, for dry specimens in good condition:

Scutum		white-dusted	green	green
Tergites		white-dusted	white	green
Face	white-dusted	*argentina* *argyria* *perplexa*	*auricollis* *vestita*	*elongata* *grata*
	black	*leucocephala*	*diaphana* *ilonae*	*atriceps*

Many common species of *Argyra* are found in woodlands or shaded places. More specific habitats are used by some species: fens and bogs (*vestita*, *elongata*), saltmarsh (*vestita*) and the margins of stony rivers and streams (*atriceps*, *auricollis*). Males of some species patrol sun-dappled damp areas in woodlands where their silver dusting flashes in the sunlight, often returning to a favoured perch. This behaviour presumably attracts females. As summer progresses, males become scarce compared to often numerous females.

Key to species of *Argyra* (males)

1 Pedicel extending over postpedicel so that dorsal margins of all 3 segments are about the same length (a). Arista inserted at midpoint of dorsal margin of postpedicel, its length slightly less than whole antenna, with thick pubescence (a). Hind femur with a strong anterior and posterior preapical seta placed just above lowest row of long anteroventral hairs (b). Sternite 8 with stout setae about as long as tergite 5 (c). Tergite 6 without hairs. .... ***elongata***

– Pedicel far shorter than postpedicel. Arista inserted closer to tip of postpedicel within apical third, its length shorter to longer than postpedicel, nearly bare. Hind femur without a true anterior or posterior preapical seta but only longer and stouter anteroventral hairs in lowermost row near tip. Sternite 8 with long or short setae. Tergite 6 with hairs. 2

(a)

(b)

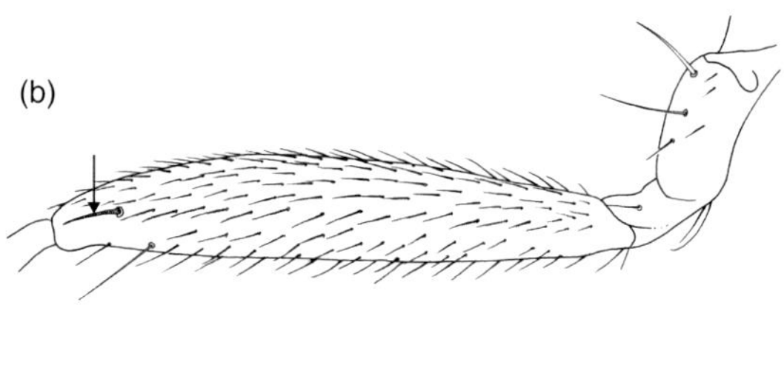

(c)

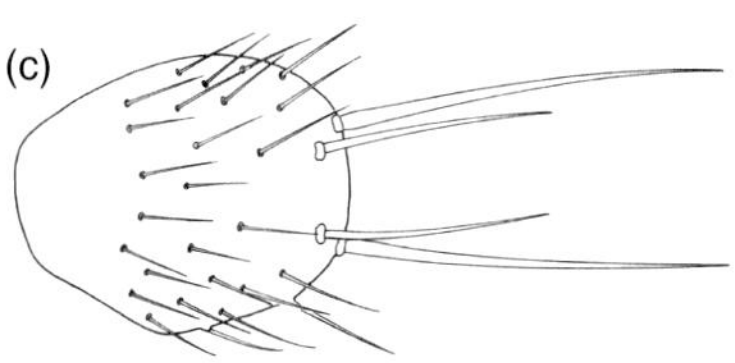

2(1) Tergite 2 with large yellow patches. 3
– Tergites without yellow patches. 10

3(2) Scutellum disc with numerous short fine hairs (a). Mid femur with an anteroventral (nearly ventral) row of very long stout hairs more than twice depth of femur, as well as finer long hairs (b). Vein M_1 on apical half with a pronounced inflexion. .... ***diaphana***
– Scutellum disc bare. Mid femur with postero- and anteroventral hairs less than twice depth of femur, and all similarly fine. Vein M_1 on apical half with a gentle inflexion. 4

(a)

(b)

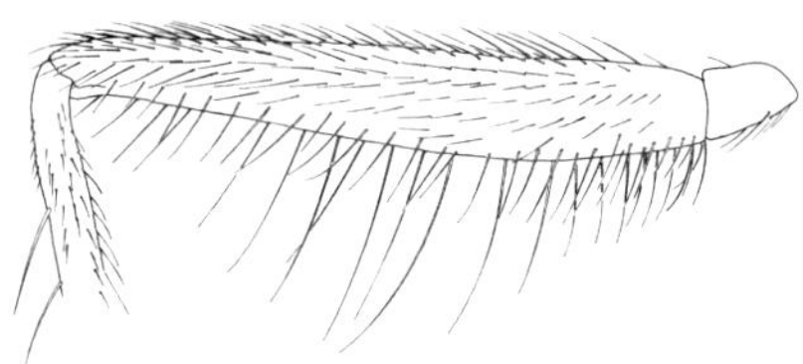

4(3) Scutum with silver dusting. Hind femur on anterior face with vestiture more-or-less even with no longer hairs at middle of femur. Hind tibia with vestiture on anterior and ventral faces clearly shorter than on dorsal face. Front basitarsus with ventral vestiture shorter than width of shaft. 5
– Scutum green, without silver dusting. Hind femur on anterior face with vestiture uneven, lowermost row of hairs at middle of femur at least as long as its depth (aa). Hind tibia with vestiture on anterior or ventral faces about as long as or longer than on dorsal face, forming a moderately dense ciliation (bb). Front basitarsus with ventral setae as long as width of shaft (cc). 9

(bb)

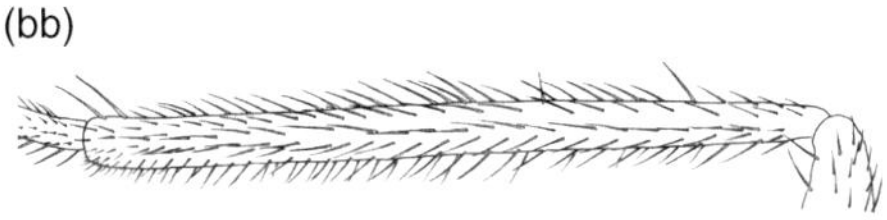

(aa)

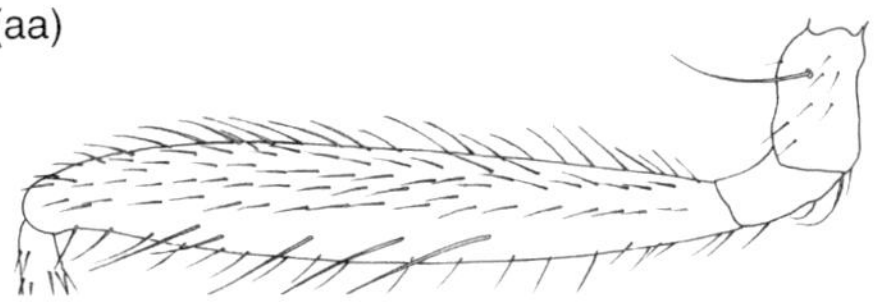

(cc)

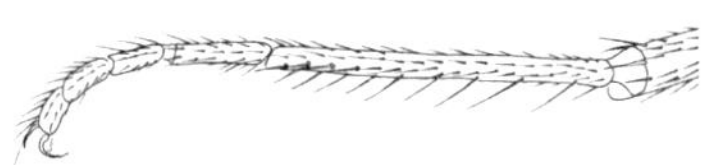

5(4) Mid femur with posteroventral hairs at least as long as greatest depth of femur (a). Femora usually dark but can be yellow. 6
– Mid femur with posteroventral shorter than greatest depth of femur, or absent. Femora entirely or mainly yellow, hind femur dark at most on apical quarter. 7

(a)

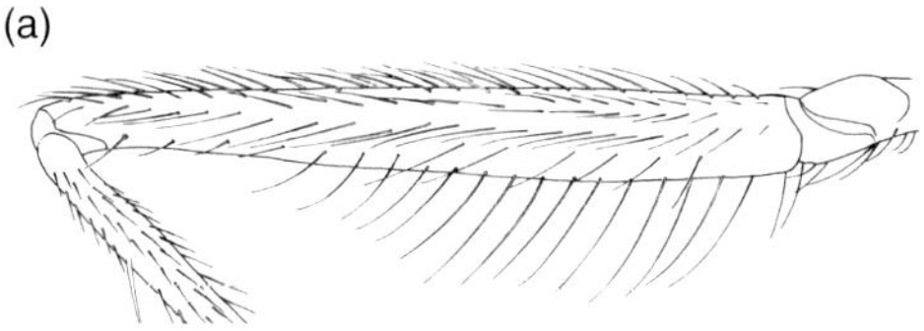

6(5) Face black from all angles of view. Postpedicel far shorter than head length, only slightly longer than deep (a, b). Arista at least 3.5 times longer than postpedicel (b). Lateral hair tuft on first tergite black or with only a few yellow hairs. Tergite 1 in side view yellow on hairless lower corner behind marginal setae and continuous with yellow patch of tergite 2 (c). Sternites 2 and 3 mainly as pale as tergites patches.. ***leucocephala***

– Face silver-dusted when viewed from slightly above. Postpedicel large, about two-thirds head length and twice as long as deep (aa). Arista clearly less than twice postpedicel length (1.25–1.75 times) (bb). Lateral hair tuft on first tergite yellow in front of black marginal setae. Tergite 1 in side view entirely dark, including hairless area behind marginal setae (cc). Sternites usually darker than yellow tergite patches. (See also couplet 7.)... ***argyria*** (dark form)

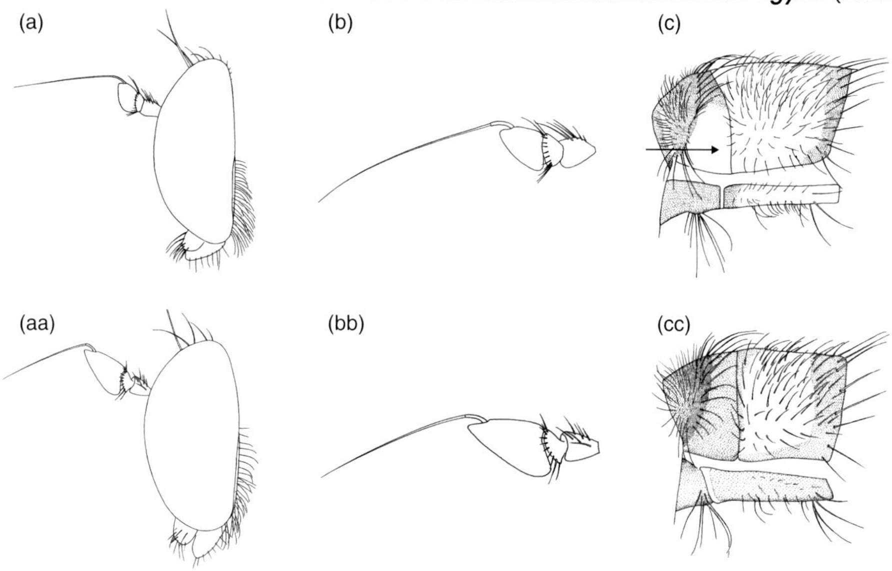

7(5) Arista longer than 1.75 times length of postpedicel (fig. bb of couplet 6). Postpedicel short, about two-thirds length of head (fig. aa of couplet 6). Ventral epandrial lobe narrow and straight in lateral view (b) but both lobes similar in width and shape with rounded or triangular tips when viewed more anteriorly (c). Cercus short-triangular, usually clearly shorter than epandrial lobes (d). (See also couplet 6.)... ***argyria*** (pale form)

– Arista shorter than 1.5 times length of postpedicel which is long, at least three-quarters length of head and usually greater than this (aa). Dorsal and ventral epandrial lobes differ in shape and width when viewed more anteriorly (figs c and cc of next couplet). Cercus usually long-triangular to parallel-sided, usually as long as epandrial lobes (figs d and dd of next couplet)................................8

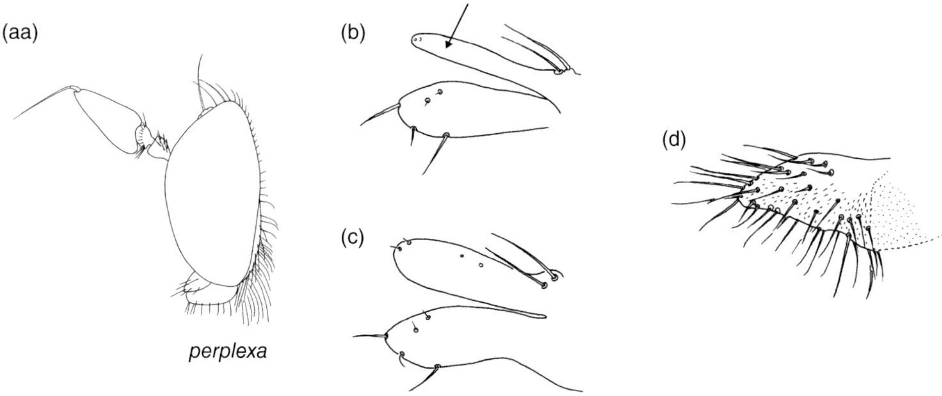

8(5) Arista as long as or shorter than postpedicel, arising close to its tip and usually no further from tip than 1.5 times diameter of arista base (a). Ventral epandrial lobe in side view broad on basal half and abruptly narrowed to blunt hook at the tip (b); in more anterior view with both lobes similar in width (c). Cercus elongate, nearly parallel-sided (d). First tergite with lateral hairs (not setae) mainly dark, and posterolateral corner dark behind marginal setae.............................***perplexa***

– Arista more than 1.25 times as long as postpedicel, arising further from postpedicel tip by 2–3 times diameter of arista base (aa). Ventral epandrial lobe in side view uniformly narrow and gently curved with a triangular projection on the inner side (not always easy to see) (bb); in more anterior view, this lobe narrower than dorsal lobe (cc). Cercus elongate-triangular (dd). First tergite with lateral hairs (not setae) mainly yellow and posterolateral corner pale behind marginal setae. .... ***argentina***

(a)

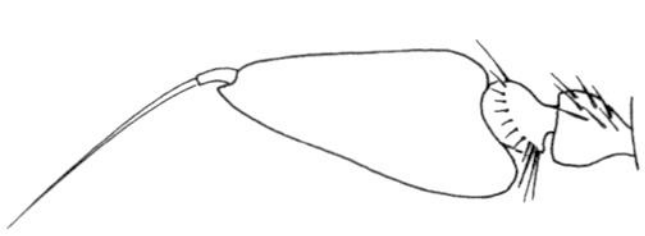

(b)

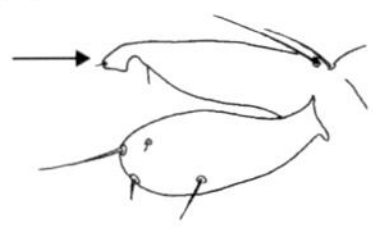

(bb)

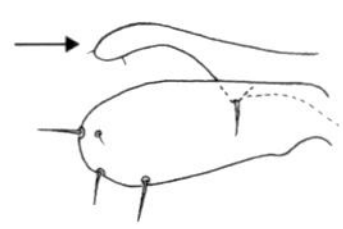

(d)

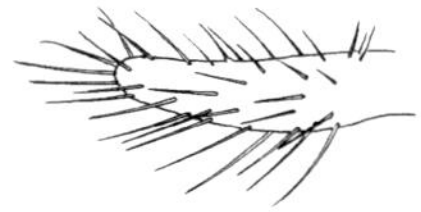

(aa)

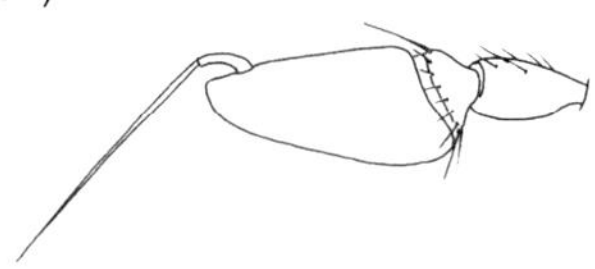

(c)

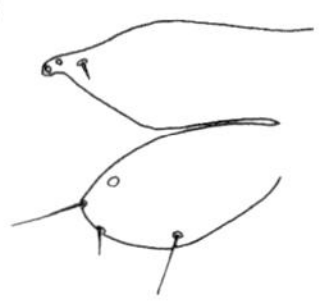

(cc)

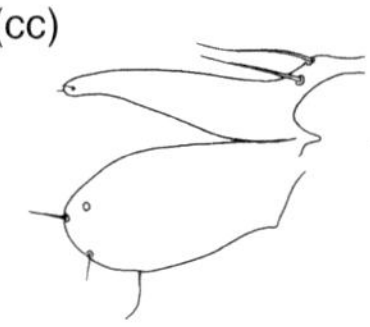

(dd)

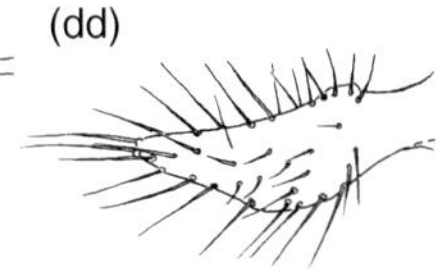

9(4) Coxae and metepimeron dark. Tergites silver-dusted. Face black. Hind basitarsus 1.75 times longer than tarsomere 2, with a long fine ventral hair at base directed towards tibia and an untidy mix of long and short plus stout and fine hairs, some as long as width of shaft (a). .... ***ilonae***

– Coxae and most of metepimeron yellow. Tergites undusted. Face white. Hind basitarsus scarcely longer than tarsomere 2 with all hairs similar in size and orientation......................***grata***

(a)

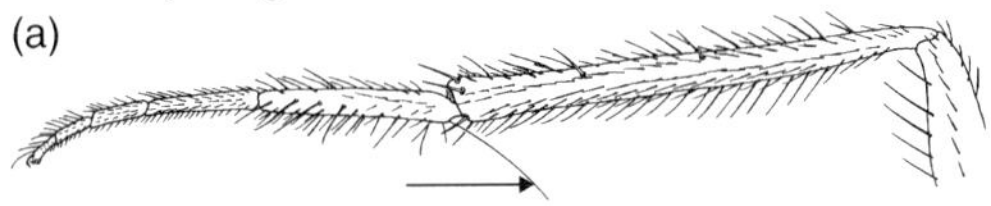

10(2) Postpedicel large, almost as long as head and its depth about a quarter head height (a). Arista short, extending beyond antenna by less than postpedicel length (a). Mid and hind femora with only a few longer hairs on apical quarter, none longer than width of shaft. Scape without dorsal hairs. Face wider than greatest depth of front femur, silver-dusted. Front basitarsus about as long as tarsomeres 2–5 combined, vestiture on ventral face much shorter than width of shaft. Small species, wing length about 3.0–3.5 mm..........................***vestita***

– Postpedicel smaller, shorter than head and its depth less than a fifth head height (aa, bb). Arista longer, extending beyond antenna by at least 1.25 times length of postpedicel (aa, bb). Mid and hind femora with many hairs longer than width of shaft, extending well below ventral face. Scape with dorsal hairs. Face equal to or narrower than greatest depth of front femur, silver or black. Front basitarsus clearly longer than tarsomeres 2–5 combined, on ventral face with hairs as long as shaft width. Larger species, wing length about 4–5 mm. 11

(a)

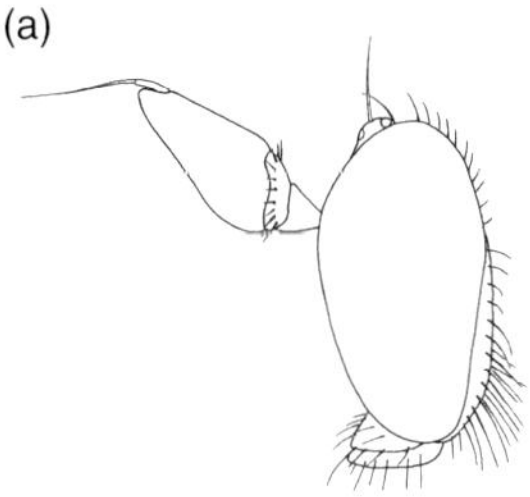

(aa)

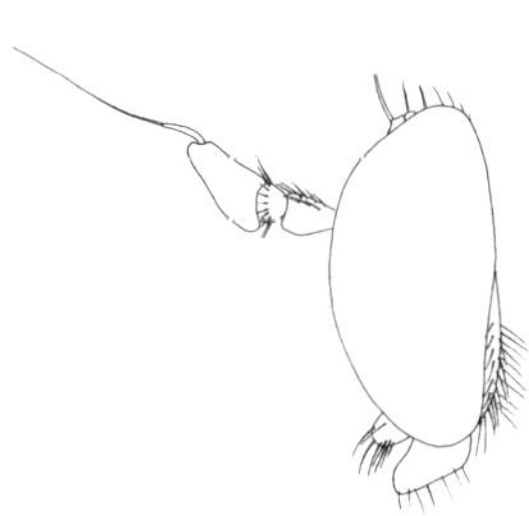

(bb)

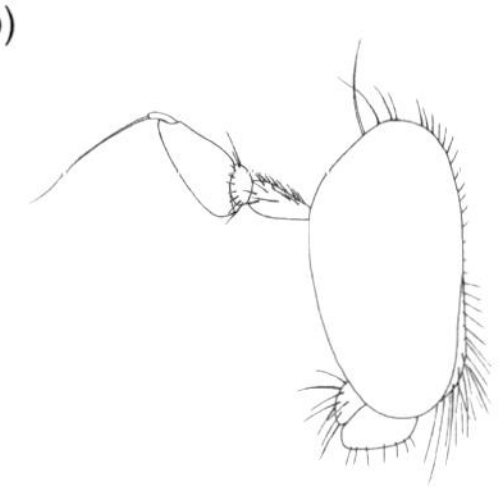

11(10) Tergites without dense pale dust, dark green ground colour visible in front view. Face and frons black in front view. Hind femur and tibia yellow on the basal half or more. Hind tibia with ventral hairs about as long as shaft width and lying at about 45° to shaft axis (a). Arista at least twice length of postpedicel. Front tibia with 2 fine posterodorsal setae......... ***atriceps***

– Tergites with dense silver dust. Face and frons white-dusted. Hind femur and tibia entirely black. Hind tibia with ventral hairs about half width of shaft and standing more upright, at about 70° to shaft axis (aa). Arista about 1.5 times length of postpedicel. Front tibia with row of about 4–6 posterodorsal setae. (If lower postocular and postgenal setae white, see *argyria*.) .... ***auricollis***

(a)

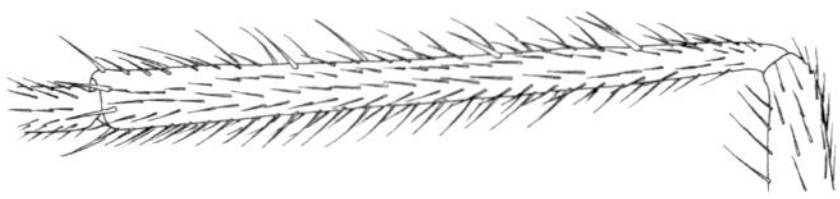

(aa)

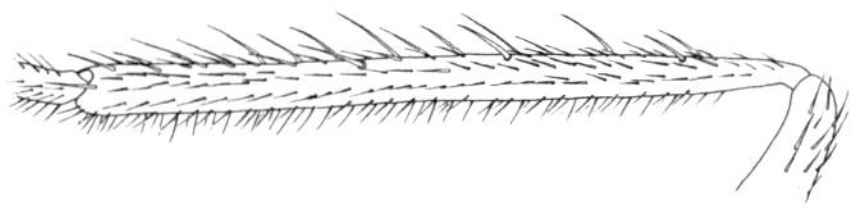

Key to species of *Argyra* (females)

1 Scutellum disc with numerous short fine hairs (a). Vein M_1 on apical section with pronounced inflexion (b). Lower postocular seta black. Large robust species, wing length about 5–6.5 mm. .... ***diaphana***

– Scutellum disc bare. Vein M_1 on apical section with gentle inflexion (bb). Lower postocular seta white. Small to moderately large species, wing length less than 6 mm. 2

(a)

(b)

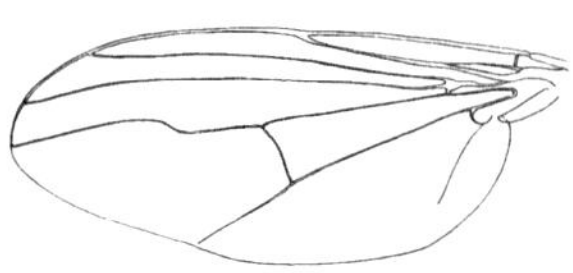

(bb)

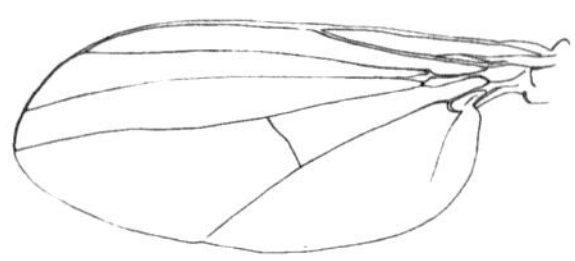

2(1) All coxae and trochanters yellow. Tergites 1–4 with yellow lateral patches, extensive on first and progressively smaller on remaining tergites. .... ***grata***

– Mid and often hind coxae dark. Tergites less extensively yellow, tergite 4 always dark.............. 3

3(2) Hind basitarsus scarcely longer than tarsomere 2 (about 1.2 times maximum) and usually shorter (a). 4

– Hind basitarsus obviously longer than tarsomere 2, usually at least 1.3 times but sometimes only 1.2 times (aa). 7

(a)

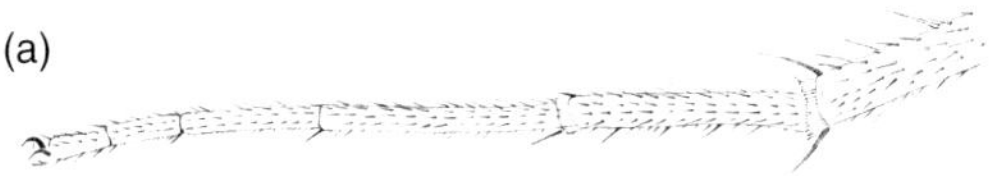

(aa)

4(3) Front tibia on dorsal face with strong setae twice as long as width of shaft. Pedicel elongate, extending over base of postpedicel, its upper margin about as long as postpedicel's lower margin (a). Arista as long as whole antenna, with thick pubescence. Hind femur with a stout anterior preapical seta placed above lowest row of anteroventral hairs. .... ***elongata***

– Front tibia on dorsal face with weak setae scarcely longer than width of shaft. Pedicel clearly shorter than lower margin of postpedicel. Arista at least 1.5 times as long as whole antenna and almost bare. Hind femur without a preapical seta but with only some longer anteroventral hairs. 5

(a)

5(4) Scape bare on dorsal margin. Postpedicel large, about a quarter height of eye (b). Tergites entirely dark or at most vaguely yellow at extreme anterolateral corner. Hind coxa with only 2 setae on outer face, rarely with tiny hairs below them. .... ***vestita***

– Scape with tiny hairs on dorsal margin (aa). Postpedicel smaller, about a fifth or less height of eye (figs b and bb of next couplet). Tergite 2 usually with yellow patches laterally. Hind coxa usually with more than 2 setae on outer face, lower ones becoming finer and shorter. 6

(b)

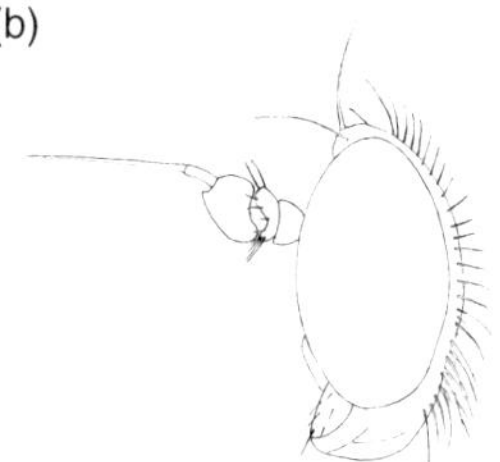

(aa)

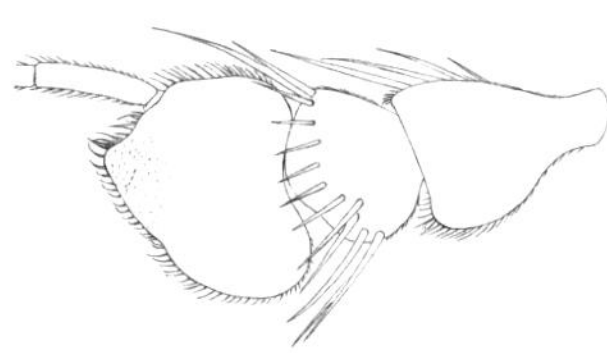

6(5) Calypter hairs black. Face slightly narrower, its height about twice its width, clypeus almost square (a). Antennae slightly smaller relative to eyes (b, arista curtailed). Front femur with posterior hairs longer and denser, forming a continuous long fringe from tip to base (c, dorsal view). Mid femur with the fine posteroventral hairs on basal half longer, usually some almost as long as apical diameter of tibia (not femur) (d). Large species, wing length at least 4.5 mm. .... ***leucocephala***

– Calypter hairs pale, rarely mixed with a few black hairs. Face slightly broader, its height less than twice its width (c. 1.85 times), clypeus wider than deep (aa). Antennae slightly larger relative to eyes (bb). Front and mid femora with posterior hairs either as *leucocephala* or sparser and shorter on front femur (cc, dorsal view) or almost absent on mid femur (dd). Large to small species .... '***argyria***' group (***perplexa***, ***argentina***, ***argyria***)

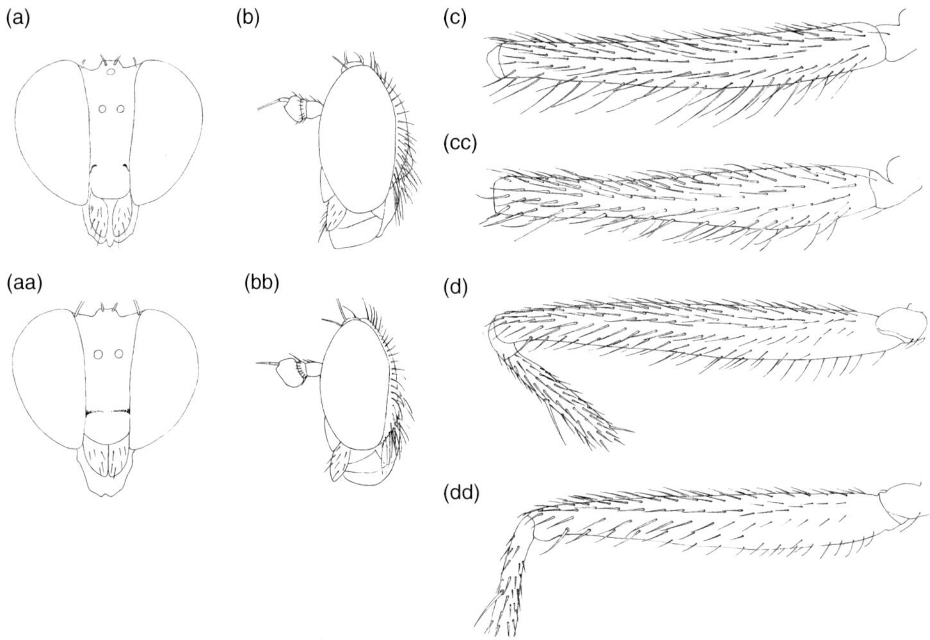

7(3) All coxae darkened, front one yellowish at apex. Some setulae mixed with and very near to the line of strong dorsocentral setae. .... ***auricollis***

– At least hind coxa partly and front coxa largely yellow. Only strong setae in dorsocentral row. 8

8(7) Calypter hairs pale, sometimes with occasional black hairs. Metepimeron entirely yellow. Tergite 1 usually entirely yellow, tergite 2 with large lateral yellow patches. Face narrower, its height measured from bottom of antennal insertion 2.7–3.0 times its width at midpoint (a). Postpedicel often longer than deep, triangular with lower margin usually straight (b). .... ***ilonae***

– Calypter hairs black. Metepimeron largely or entirely dark, only rarely entirely yellow. Tergites 1 and 2 usually almost entirely dark but may have large yellow patches. Face broader, its height 2.0–2.3 times its width (aa). Postpedicel not longer than deep, more rounded especially on lower margin (but may be more triangular) (bb). (*elongata* may come here but has different antenna, couplet 4.) .... ***atriceps***

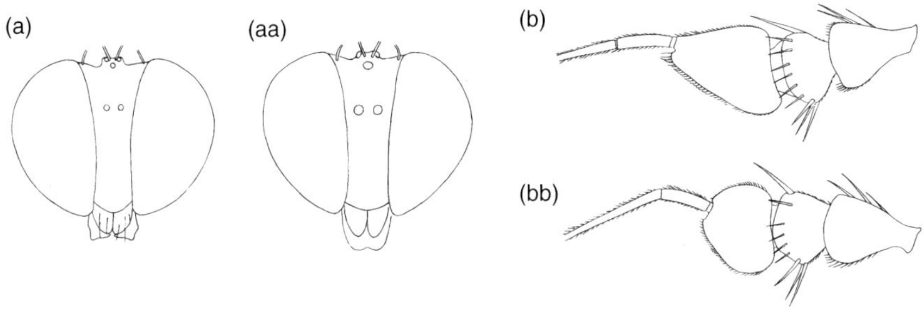

Argyra argentina (Meigen, 1824)

This is a medium-sized species with silver-dusted scutum, tergites and face, as in *argyria* and *perplexa*. Its arista is set back a little from the tip of the postpedicel, as in *argyria* from which it is reliably distinguished by the shape of the ventral epandrial lobe. It has large yellow lateral patches on the tergites, those on the second segment three-quarters the tergite's length and only narrowly separated on the midline, and they merge with the yellow hind lower corner of the first tergite and yellow sternites, making the abdomen particularly pale. See the account for *argyria* for information on females. Wing length ♂: 4.4 (4.1–4.8) mm.

Argyra argentina is widespread in England, less frequent in Wales and Scotland but found on Orkney and the Outer Hebrides. It is a woodland species associated with streams and seepages. Flight: v–x, peak in vii.

Argyra argyria (Meigen, 1824)

Argyra argyria belongs to the group with silver-dusted scutum, tergites and face, and among these is most reliably separated using the shape of the epandrial lobe. If the hypopygium is not exposed, the short stocky cercus and relatively long arista set back a little from the tip of the postpedicel provide fairly reliable characters. The length of the mid femur fringe, used by d'Assis-Fonseca (1978), is an unreliable character. *Argyra argyria* is a variable species, as pointed out by Loew (1859) who listed several combinations of leg colour and calypter hair colour, and concluded that the small pale form that Zetterstedt (1843) described as *argentella* was part of this variation; this synonymy was later adopted (e.g. Negrobov 1991). In Britain two fairly distinct forms may be distinguished, a large dark one which is presumably the true *argyria* and smaller pale agreeing with Zetterstedt's *argentella*. Collin (1943) claimed that there was a consistent difference in the cercus and length of eye hairs. Cole (1990) found no differences in the hypopygia of the largest and smallest specimens but also did not find indeterminate specimens that were intermediate between the forms, although rare examples have since been found. It is felt prudent to record them separately as data can be lumped but not disentangled later. The dark form is about 10% larger (wing length, body length) than the pale form, has dark to black femora with strongly hairy fringes and smaller yellow patches on the tergites, including an entirely dark first tergite; it appears earlier in the year when it is most frequent in May and June. The pale form has almost entirely yellow femora and poorly developed femoral fringes, apart from the posterior one on the front leg and anterior one on the mid leg, and larger tergite marks, including the lower hind corner of the first tergite, giving the fly the same pale appearance as *perplexa* and *argentina*, from which it is most reliably separated by the epandrial lobes. This pale form appears later in the year, from mid June and July. Females of *argyria*, *argentina* and *perplexa* form a continuous series from large and hairy-legged specimens to small far less hairy-legged ones. No clear separation into discrete taxa can be discerned apart from some large darker-legged specimens that possibly represent the dark form of *argyria*. They have therefore been lumped into an '*argyria*' group. Some *leucocephala* may also be inseparable from large '*argyria*' and occasional specimens of the '*argyria*' group have black calypter hairs so resemble *leucocephala*. Wing length ♂ dark form: 5.1 (4.4–5.7) mm, ♂ pale form: 4.6 (4.2–5.0) mm, ♀ dark form: 4.9 (4.6–5.2) mm.

Argyra argyria is widespread and common in England and lowland Wales but infrequent in Scotland where it reaches the Sutherland coast. Older records of the pale form under the name *argentella* show a similar distribution. It is found in a wide range of damp to wet habitats, including

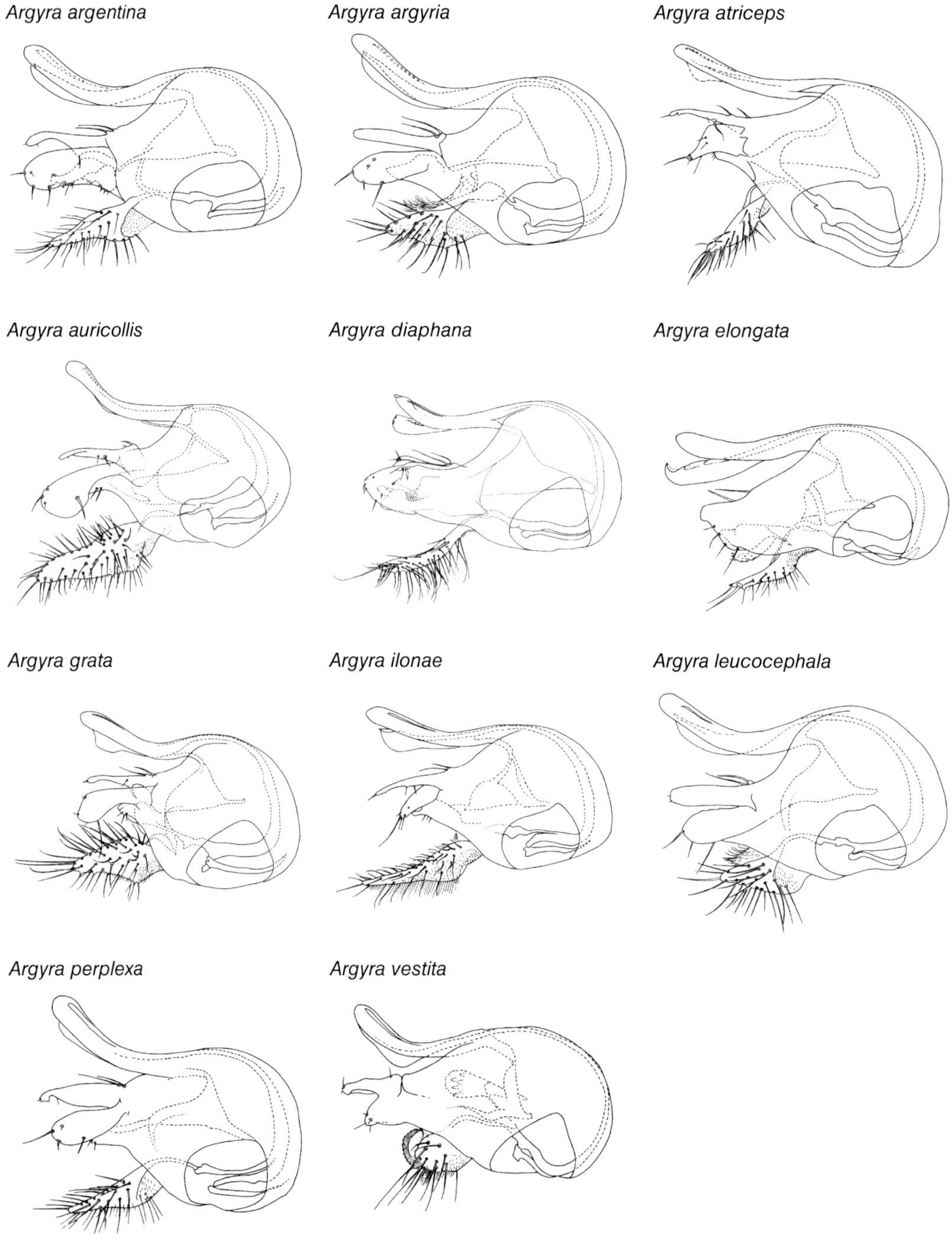

Fig. 10. *Argyra* hypopygia.

the margins of rivers, stream and pond, marshes, fens and ditches. These are more often unshaded than within woodland. Flight: iv–x, peak in vii.

Argyra atriceps Loew, 1857

This is a small, narrow and rather dark green species distinguished from all other species in lacking silver dusting on both the scutum and tergites and having a black face. The dorsal epandrial lobe of the hypopygium is pointed-triangular and makes a characteristic 'V' shape with the narrow ventral lobe, similar to that in *ilonae*. Females need comparing only with *ilonae*; *atriceps* is darker with the first and second tergites dark on the disc and usually laterally too (nearly always pale in *ilonae*), the metepimeron is usually at least partly dark but may be mostly yellow as in *ilonae*, and the postpedicel is about as deep as long, rarely triangular in shape. Some specimens fall between these conditions and cannot be named. Wing length ♂: 4.4 (4.1–4.9) mm, ♀: 4.0 (3.8–4.4) mm.

Argyra atriceps is recorded sparsely in England but not in the East Midlands or East Anglia, and rarely in south Wales and south Scotland. It is found by small streams and rivers in shade and has a moderate association with sediment-depositing stretches of sandy and stony rivers. It is only rarely found by still water bodies. Flight: v–viii.

Argyra auricollis (Meigen, 1824)

Both sexes of this moderately large dark species are distinct from others except *diaphana* as all the coxae are predominantly dark. The setulae among and outside of the dorsocentrals are mostly shorter than the acrostichals so may be overlooked; they are also found in *elongata*. Males have silver tergites and a shining green scutum, and particularly long black hairs and setae, notably those of the dense whiskers behind the head and on the front coxa. The very short fairly upstanding ventral hairs towards the tip of the hind tibia resemble those in *grata*, and are not seen in other *Argyra*. Wing length ♂: 4.3 (4.1–4.7) mm; ♀: 4.6 (4.1–5.3) mm.

This rare species is found in scattered sites in England, Wales and more frequently in upland Scotland. It appears to be associated with stony rivers and fast streams, either sunlit or shaded. Flight: vi–viii.

Argyra diaphana (Fabricius, 1775)

The largest British *Argyra* is a robust species with a green scutum, silver tergites with yellow side-patches and a black face. It is particularly hairy, not only with very long femoral fringes but with a covering of numerous short hairs on the scutellum and scutum, with the quadriserial acrostichal setae reduced to similarly small hairs. Females also have the unique hairy scutellum and stout black proepisternal setae. The tergites have a completely dark ground colour. All coxae and trochanters and often parts of the femora are black. Wing length ♂: 6.2 (5.3–6.6) mm, ♀: 6.2 (5.2–6.6) mm.

Four species similar to *diaphana* in having hairs on the scutellum may occur in Britain but can be recognised as different from *diaphana*: *hoffmeisteri* (Loew) is most similar to *diaphana* in having a black face and mainly black femora but lacks yellow patches on the tergites; *loewi* Kowarz, *magnicornis* (Zetterstedt) and *setimana* Loew all have white-dusted faces and predominantly yellow femora. Parent (1938) and Grichanov (2006) provide keys to these species.

Argyra diaphana is a common species through most of its range, becoming scarcer in northern Scotland but reaching Orkney, being found in woodlands or partial shade over wet ground. Flight: iii–ix, peak in vi.

Argyra elongata (Zetterstedt, 1843)

This is small species with no silver dusting on the body but with a white face, superficially resembling *grata* but amply distinct from other *Argyra* in many respects. Males have a bare sixth tergite (hairy in other species) and the four setae on the eighth sternite are exceptionally long and thick compared to those in other species, more resembling those of *Diaphorus*. Both sexes share the following features. The yellow tergite patches are variable in size from complete broad rings on the second and third tergites to a narrow anterolateral patch on the second only (or absent in females). The large pedicel extending as a hood over the postpedicel and the short thick pubescent arista are not found in any other species which all have a simple pedicel and bare arista. All femora have both a true anterior and weaker posterior preapical seta, in addition to the usual anteroventral long preapical hairs; *diaphana* is the only other species with a weak true anterior preapical seta. All tibiae have particularly strong setae that show up clearly against the yellow legs, and the proepisternal seta is black (white in all but *diaphana*). Wing length ♂: 3.5 (3.1–4.1) mm, ♀: 3.8 (3.4–4.2) mm.

Argyra elongata is sparsely but widely distributed in England and Wales but more frequent in highland Scotland north to Sutherland. It is a fen species in England and probably also in acid mire in Scotland, and may have a strong association with *Carex*-dominated mires. Flight: v–ix, peak in vi.

Argyra grata Loew, 1857

This species is fairly small and lacks silver dusting on the body but has a white face. It superficially resembles *elongata* in colour and in the absence of fringes on the femora, apart from a few very long anteroventrals on the hind femur, the longest being at least 1.5 times the greatest depth of the femur. Both sexes are unique among British *Argyra* in having all the coxae entirely yellow, the metepimeron usually entirely or mainly yellow, and yellow side-patches reaching to the fourth tergite (tergite 1 dark in males). The male hind tibia has very short nearly upright apical hairs resembling those of *auricollis*. As the hind coxa often has only one distinct seta and only one or two (or no) short hairs below this, *grata* may be mistaken for a Dolichopodinae but differs in having no anterior preapical seta on the hind femur. Wing length ♂: 3.8 (3.5–4.2) mm, ♀: 4.2 (3.9–4.5) mm.

Argyra grata is a rare species found mainly in three disjunct populations: Kent where it is possibly most frequent, English counties bordering Wales, and Yorkshire. There is a single record from south Wales. It is associated with deciduous woodland in Kent, and elsewhere with wet woodlands bordering small to large rivers. Flight: v–viii.

Argyra ilonae Gosseries, 1988

This is a moderately large species although rather variable in size, with silver tergites, green scutum, black face and hairy legs. Its legs are characteristically coloured with a strongly contrasting black apical third or quarter on both the femora and tibiae. The relatively long hind basitarsus with untidy hairs and a long basal hair are unique. Females need comparing only with *atriceps* from which they differ in their pale calypter hairs, much more extensively yellow tergites and metepimeron, and usually longer and more triangularly pointed postpedicel. A few specimens have conditions that fall between these characters and cannot be named with certainty. Wing length ♂: 4.8 (4.0–5.4) mm, ♀: 4.2 (3.9–4.6) mm.

This species is widespread in England north to Northumberland but very rare in southern Scotland and apparently scarce in Wales. It is found in deciduous woodland, nearly always close to ponds and streams and less frequently at seepages. Flight vi–x, peak in vii.

Argyra leucocephala (Meigen, 1824) (Plate 1)

This large *Argyra* is the only British species with silver dusting on the thorax and abdomen and a black face. It is also unique in having small antennae and a conspicuously long arista. It looks fairly pale as the yellow patches on the tergites are usually large, extending to the front and lower margins and running into the yellow sternites. Females should be relatively easily identified by their black calypter hairs, longer hairs on the rear face of the front and mid femora and, compared to the '*argyria*' group, also by their small antennae and slightly narrower face. Small females usually have shorter femoral hairs so look more like the '*argyria*' group, and their identification has to rely mainly on the black calypter hairs. Wing length ♂: 6.2 (5.6–6.8) mm, ♀: 5.4 (4.8–5.7) mm, sometimes smaller.

This is a common and widespread species through much of England, south Wales and south Scotland and, although found on Orkney, it is uncommon in northwest Scotland. Its only requirements are woodlands or partial shade of tall herbage and some dampness. Flight: iv–xi, peak in vi–vii.

Argyra perplexa Becker, 1918 (Plate 1)

This species is marginally smaller than the others in the '*argyria*' group with a silver scutum, tergites and face, and slightly darker than *argentina* and the pale form of *argyria* as the yellow tergite patches are smaller. The short nearly apical arista, slightly larger antenna and hooked tip of the ventral epandrial lobe distinguish *perplexa* from these related species. See the account for *argyria* for information on females. Wing length ♂: 3.9 (3.7–4.1) mm.

Argyra perplexa is widespread in Britain, occurring north to Shetland and the Outer Hebrides. It is found at seepages, streams and sometimes rivers, usually but not invariably in shade, and only rarely by standing water. Flight: iv–x, peak in vii.

Argyra vestita (Wiedemann, 1817)

This is the smallest British *Argyra*, with silver tergites (only 2–5 rather than 2–6), undusted shining green scutum, a broad white face and antennae relatively larger than in other species. Among British *Argyra*, it is unique in both sexes in having no hairs on the antennal scape and only two subequal setae on the hind coxa (some *grata* may have only two setae of very unequal size). Wing length ♂: 3.2 (2.9–3.5) mm, ♀: 3.2 (3.0–3.6) mm.

Argyra vestita is widespread in England and Wales, becoming scarcer northwards and is rare in Scotland although present in the Outer Hebrides. It is associated mainly with saltmarsh and brackish marshes and less frequently with high-quality, often calcareous inland fens and marshes where it is often found close to ponds or ditches. Flight: iv–xi, peak in vi–vii.

ASYNDETUS Loew, 1869 (Diaphorinae)

These are small dark species characterised in both sexes by unusual venation: the costa ends at R_{4+5} rather than at M_1, the apical section M_1 has a tiny weak break midway and is slightly disjointed here, R_{4+5} runs far closer to R_{2+3} than to M_1, and dm-m is close to the wing-base at the basal third so that the apical section of M_4 is far longer than crossvein dm-m (fig. aa of couplet 28, ♂ genera key; Plate 1). Males have conspicuously large front pulvilli, as long as the fifth tarsomere, and all tarsi lack claws, as in *Diaphorus* and *Melanostolus*. The male frons and face are about equally wide (Fig. 11a), as in *Melanostolus* but unlike *Diaphorus* (eyes touch at the frons) or *Chrysotus* (eyes nearly touch on the face, except in *laesus*), and the palps are narrow and project forward obscuring the proboscis, and have several conspicuous downward-pointing setae at the tip (Fig. 11b). Sternite 8 carries two pairs of long conspicuous setae, as in some other diaphorines. Females have a flat triangular last sternite that protrudes beyond the acanthophorites. In addition, both sexes have a short postpedicel bearing a dorsal arista, biserial acrostichal setae, four to five dorsocentral setae, and no true anterior preapical setae on the mid and hind femora.

Among the nine European species of *Asyndetus*, only *latifrons* is widespread and no others are expected in Britain. *Cryptophlebs kerteszi* Lichtwardt, 1898, although unlikely, could occur in Britain; its venation is similar to that of *Asyndetus* but it has yellow legs, no crossvein dm-m and no long setae on sternite 8 in the males. In life, *Cryptophlebs* has a striking horizontal eye-band (Roháček *et al.* 2021).

Asyndetus latifrons (Loew, 1859) (Plate 1)

Asyndetus latifrons is a small dark green or dark bronze fly, including the legs apart from the knees which are narrowly yellow in males and more broadly so in females. Wing length ♂: 2.4 (2.3–2.6) mm, ♀: 2.4 (2.3–2.6) mm.

Asyndetus latifrons is known only from two large rivers in northern England, the R. Eden in Lancashire and R. Coquet in Northumberland. The habitat was wide expanses of coarse exposed riverine sediment with very little vegetation (Godfrey and Drake 2012). Flight: vi–viii.

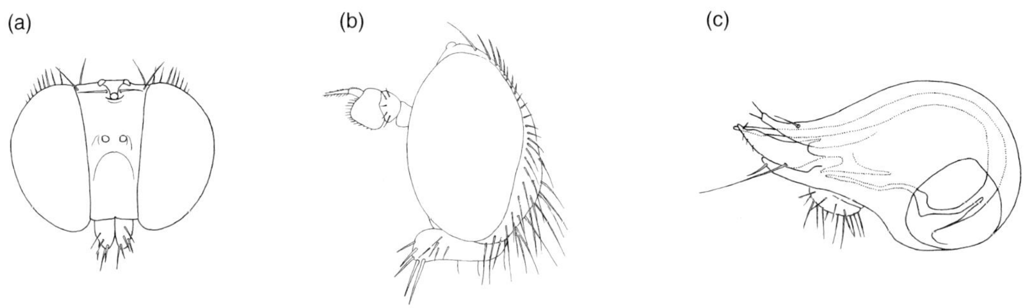

Fig. 11. *Asyndetus latifrons* head (a, b) and hypopygium (c).

AUSTRALACHALCUS Pollet, 2005 (Achalcinae)

Australachalcus closely resembles dark species of *Achalcus* (Plate 1). Characters distinguishing both sexes of *Australachalcus*, with the condition in *Achalcus* in parentheses, are all postoculars black (pale above), six black dorsocentral setae (five), the front tibia without a dorsal seta near the base (present), the hind tibia with two anterodorsal setae (usually three) and vein R_{4+5} slightly curved (straight). Males differ in having six clearly visible pubescent segments (five), and the front femur lacks an erect ventral seta near the base but has a row of short anteroventral setae.

Australachalcus develops in rot-holes of a wide range of deciduous trees, summarised by Pollet (1996) and Alexander (2020). No other species of *Australachalcus* are known from the Western Palaearctic.

Australachalcus melanotrichus **(Mik, 1879)**

Separation from similar dark *Achalcus* is covered above. The male genitalia differ from those of dark *Achalcus* in the cercus having a single strong apical hair (several in *Achalcus thalhammeri*, none in the other dark species) and fine teeth on the ventral edge of the hypandrium (Fig. 8). Wing length ♂: 2.2 (2.0–2.5) mm, ♀: 2.5 (2.2–2.7) mm.

The distribution is mainly lowland England from Devon to Yorkshire and south Wales, and a few records from eastern Scotland. *Australachalcus* has been found mainly in woodland and wood-pasture. Many records are of specimens reared from rot-holes and decaying wood. Flight: iii, v–x.

CAMPSICNEMUS Haliday in Walker, 1851 (Sympycninae)

The genus can be distinguished from most others by the combination of an hour-glass-shaped face which is narrowest shortly below the antennae and widens markedly to the mouth, the position of the arista which is dorsal or close to the base of the postpedicel, and an anterior preapical seta on the hind femur (Plate 2). These characters are also found in *Micropygus*, *Tachytrechus* and a few Hydrophorinae genera; *Micropygus* is most likely to be confused with *Campsicnemus* but it has a pale spot over vein dm-m which is lacking in *Campsicnemus*, while the others are mainly larger flies with wings longer than 3.5 mm. Species of *Campsicnemus* are small dark metallic green flies with no pale colour on the body. They are distinctive as the body appears disproportionately small in relation to the long thin wings and legs as the abdomen is only about as long as the thorax (longer in females). The wing boss is often distinct, appearing as a slightly dark spot. Males of nearly all species can be recognised by their legs alone, whose modifications include some of the most bizarre in the British fauna.

Of the 25 species in Europe, 13 occur in the British Isles. *Campsicnemus dasycnemus* is known from Ireland and has been included in the key. Males of species not yet recorded in Britain can be recognised from their leg shape and setae. *Campsicnemus armoricanus* Parent, 1926 is very similar to *pusillus* and may occur at coastal sites; its recognition is described in the account for *pusillus*. *Campsicnemus lumbatus* Loew, 1857 has its front femur strongly swollen at the base and dense ciliation on the tarsus but otherwise simple legs. *Campsicnemus simplicissimus* Strobl, 1906 and *varipes* Loew, 1859 have simple shafts to the legs, silver faces and no wing marks; *simplicissimus* is a tiny species with no unusual chaetotaxy and shares with the far larger *compeditus* the unusual characters of having only three dorsocentral setae; *varipes* has strongly ciliated front and mid tarsi.

As the hypopygium is completely encapsulated, determining the sex can be difficult in species with weak leg modifications. Males have a small short sixth tergite, whereas in females the last visible tergite is the fifth which is usually relatively flat and square-ended. The colour of the legs and halter can vary hugely in the females of some species, so that the darkest specimens may be mistaken for a new species when compared with typical pale-legged forms. The colour of setae, hairs and face pubescence and the extent of dusting on the frons appear to be more stable.

Species of *Campsicnemus* live in wetlands including saltmarsh and freshwater, and predominantly in unshaded sites. About half of the species have restricted habitat requirements. Several species congregate on bare wet mud, sand or peat, often with a strong preponderance of males. Some species appear to overwinter as adults and can be found through much of the year.

Key to species of *Campsicnemus* (males)

1 Femur, tibia or basitarsus of front or mid leg with modified shape, either bent, strongly thickened, or with protrusions or indentations (ignore all setae). Mainly larger species, wing length 2.3–3.9 mm. 2

– Legs simple, at most tibia very gently bowed and slightly thicker in the apical two-thirds (all femora are broader towards base; ignore all setae). Mainly smaller species, wing length 1.6–3.1 mm. 9

2(1) Front tarsus simple without remarkable projections. 2 notopleural setae. 3

– Front tarsus with remarkable projections. 1 notopleural seta. 8

3(2) Mid basitarsus shorter and slightly stouter than tarsomere 2. 4

– Mid basitarsus longer but not wider than tarsomere 2. 7

4(3) Mid tibia expanded into a flattened segment-shape, in dorsal view far wider than femur, with long dorsal setae on front and rear edges (a). Front femur, tibia, basitarsus and tarsomere 2 with differentiated setae and hairs (b). .... ***scambus***

– Mid tibia not remarkably swollen but may be bent or uneven in width along its length. Front leg with even vestiture, no differentiated hairs. 5

(a) (b)

5(4) Wing without anal lobe (a), membrane dark apically and anteriorly, cell r_1 obviously brown. R_{4+5} and M_1 converge near wing tip, their greatest separation at apical two-thirds about 1.3 times that at tip (a). Clypeus and epistoma silver. Hind basitarsus and tarsomere 2 nearly as wide as tibia (b). Front coxa with apical setae white. Legs entirely black except for knees. [Mid leg (c).] .... ***umbripennis***

– Wing anal lobe distinct (aa), membrane almost uniformly pale, cell r_1 only slightly darker. R_{4+5} and M_1 almost parallel near wing tip. Clypeus yellow or gold, epistoma greyish. Hind basitarsus and tarsomere 2 narrower than tibia. Front coxa with apical setae black. Legs mainly yellow, rarely mostly black. 6

(a)

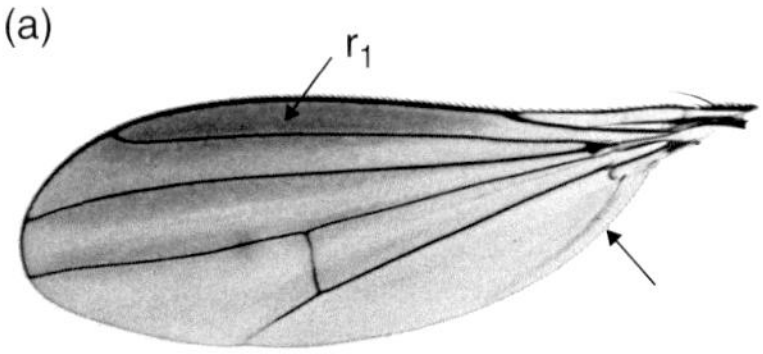

(b)

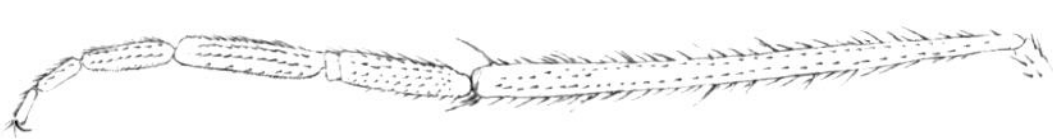

(aa)

(c)

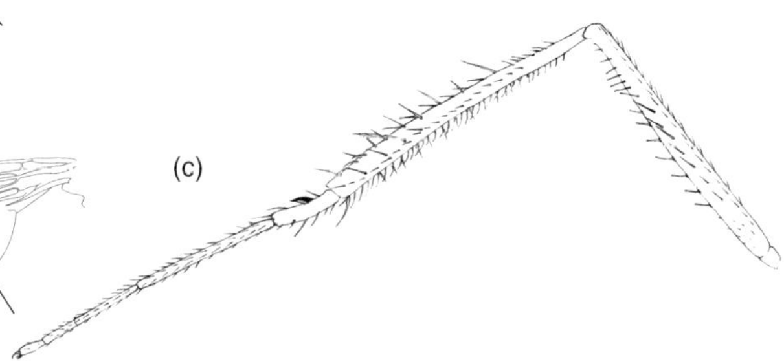

6(5) Hind femur with a row of ventral setae as long as width of shaft (a). Mid tibia twice as broad on apical two-third than on basal third, with no bend halfway (b). Both mid tibia and femur with uninterrupted but untidy ventral rows of long setae (b). .... ***curvipes***

– Hind femur without ventral setae. Mid tibia about same width along whole length and abruptly bent in middle (bb). Both mid femur and tibia with neater rows of setae with a gap in middle of the rows (bb). .... ***loripes***

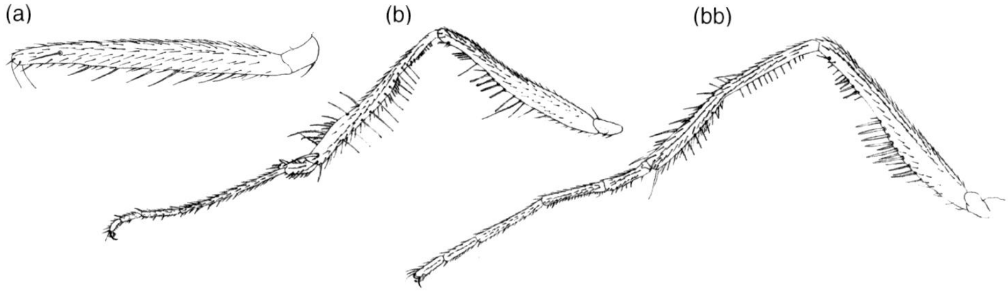

7(3) Mid tibia flattened, twice as wide as deep, appearing curved in anterior view due to being thicker near base and at tip (a). Mid femur with a short posteroventral row of fine hairs confined to apical half, hairs shorter than about 1.5 times greatest depth of shaft (a). Front tarsus with tarsomeres 4 and 5 much broader than basal tarsomeres and with long lateral fringes (b). Face entirely matt black. Front coxal hairs black. .... ***pusillus***

– Mid tibia cylindrical with a small flange at base ventrally (aa). Mid femur with a ventral row of hairs along most of shaft, several setae at least 1.5 times longer than depth of shaft (aa[2]). Front tarsus with all tarsomeres similar. Face dusting yellow. Front coxal hairs pale but apical setae black. (If mid tibia has an even posteroventral row of blunt setae, see *armatus*, couplet 12.) .... ***dasycnemus***

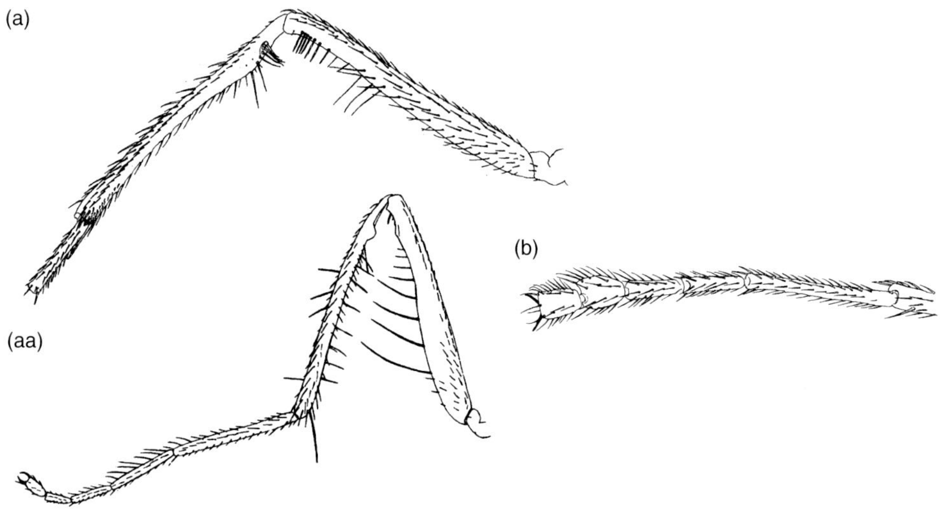

8(2) Front tibia strongly swollen (a). Front tarsus with 3 long complex projections (a). Front coxa with white apical setae. Antenna entirely black. Femora black on basal half, yellow apically. Scutum with 1 pre- and 3 postsutural dorsocentral setae. .... ***magius***

[2] Chandler 1989. *Irish Naturalists' Journal* 23, 17–19.

– Front tibia stout but approximately cylindrical (aa). Front tarsus with 1 long curved projection (aa). Front coxa with black apical setae. Scape and pedicel yellow. Femora mainly yellow, dark dorsally. Scutum with 1 pre- and 2 postsutural dorsocentral setae. .... ***compeditus***

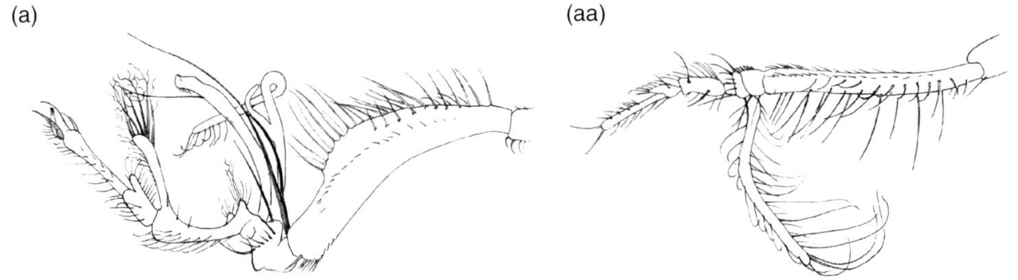

9(1) Wing membrane darker towards tip and over dm-m. Face silver. 10

– Wing membrane with a uniform shade. Face yellow or black. 11

10(9) Wing wedge-shaped, much broader on apical half than basal half by about 1.7 times, anal lobe absent, hind margin nearly parallel to vein M_4 (a). Membrane mainly dark with a pale patch beyond dm-m almost completely enclosed by dark membrane. Hind femur without long anteroventral hairs. Mid femur with a few short stout antero- and posteroventral setae on apical half (b). Scutellum and frons shining metallic blue or green. .... ***alpinus***

– Wing shape normal, with distinct anal lobe and broad membrane between M_4 and wing margin (aa). 2 front wing cells (c, r_1) and wing tip brown but dark shade not extending across whole width on apical third. Hind femur with a row of long fine anteroventral hairs, often facing forward so more obvious in dorsal view (cc). Mid femur with continuous antero- and posteroventral rows of small hairs becoming gradually longer towards tip. Scutellum and frons dull brown-dusted. .... ***marginatus***

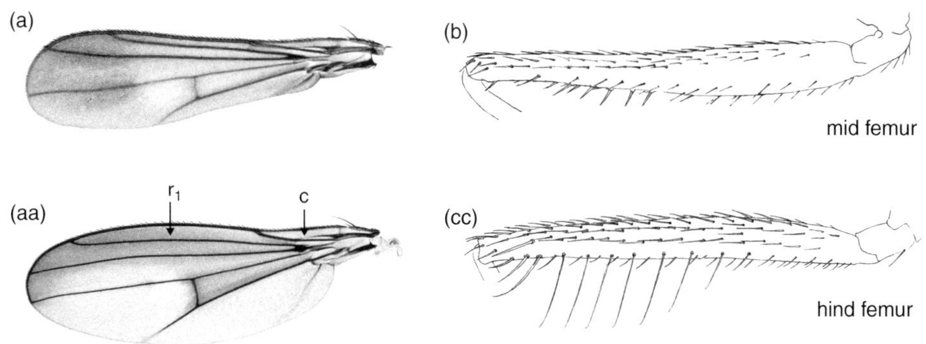

11(9) Scape and pedicel yellow. Mid leg with no unusual setae. Mid trochanter pale. Face entirely golden. Lower proepisternal seta white. .... ***picticornis***

– Antenna entirely black. Mid femora and tibia with rows of setae. Mid trochanter dark. Face bicoloured, clypeus yellow or brownish, epistoma pale grey or black (rarely entirely brownish gold). Lower proepisternal seta black. 12

12(11) Mid tibia along most of its length with a posteroventral row of evenly spaced blunt setae 1.5 times as long as width of tibia (a). Mid femur ventrally with 2–3 rows of setae, some of which more than two-thirds as long as greatest depth of femur (b). Postpedicel long-triangular, at least twice as long as deep (c). Clypeus yellow, epistoma pale grey, rarely face entirely brownish gold. .... ***armatus***

– Mid tibia on only the basal half with a posteroventral row of blunt setae as long as width of tibia (aa). Mid femur ventrally with very short sparse setae (aa). Postpedicel rounded and short, 1.5 times longer than deep (cc). Clypeus brown or dark yellow, epistoma matt black. .... ***pumilio***

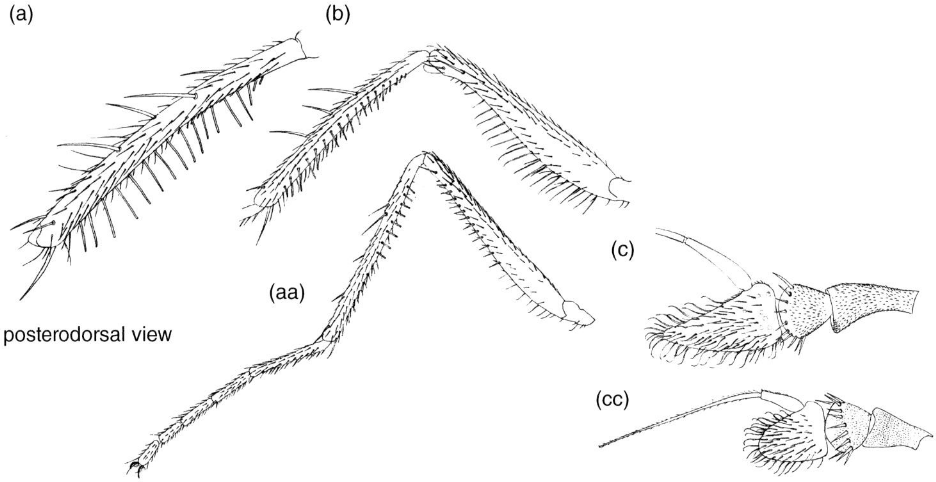

Key to species of *Campsicnemus* (females)

C. dasycnemus is not included and probably falls out at couplet 9.

1 Notopleuron with 2 setae. Mid and front tibiae each with 1 long posterodorsal seta. Hind basitarsus equal to or shorter than tarsomere 2. 2

– Notopleuron with 1 seta. Mid tibia with 3 long posterodorsal setae (aa), front tibia with 2 long posterodorsal setae (bb). Hind basitarsus longer than tarsomere 2. 11

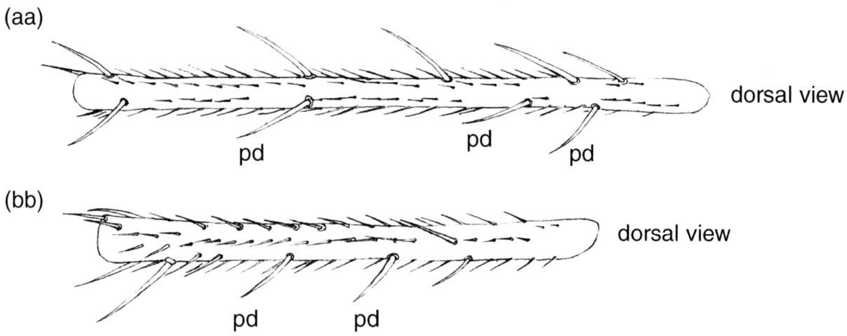

2(1) Front coxa with at least some apical setae black, hairs white or black. Proepisternal seta black. Clypeus various colour, including silver. 3

– Front coxa with apical setae white, hairs white. Proepisternal seta pale. Clypeus silver. 10

3(2) Wing costa between R_1 and R_{2+3} with all hairs the same size or with only slightly differentiated spinules near R_{2+3}. 4

– Wing costa between R_1 and R_{2+3} with well-differentiated stouter and longer spinules between every 3–5 finer and shorter hairs along whole of this section on anterior or anteroventral edge (aa). 9

(aa)

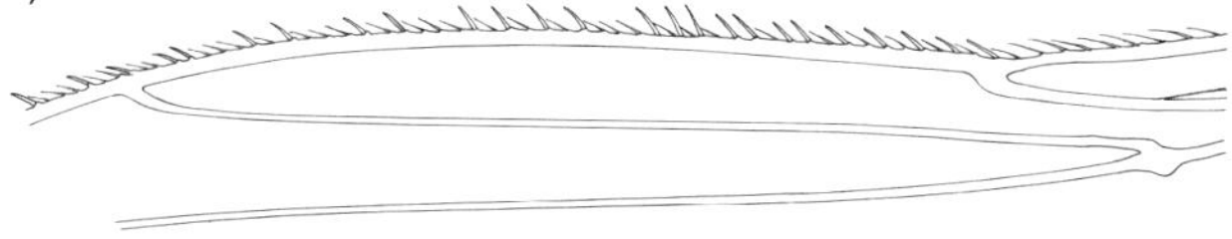

4(3) Front coxa with hairs black on inner face, but may be pale on outer face (setae are black). Legs and coxae mainly greenish black, usually only tips of femora yellow. Clypeus black or dark brown.......... ***pusillus***

– Front coxa with hairs pale, sometimes with occasional black hairs (setae are black). Legs usually mainly yellow but if entirely dark then tips of femora not clearly different in colour. Clypeus silver, yellow or brownish.......... 5

5(4) Wing mainly dark with the pale patch beyond dm-m almost completely enclosed by dark membrane (a). Clypeus and lower half of epistoma silver. Face very narrow, eyes almost touching, minimum width less than diameter of front ocellus (b). Small species, wing length about 1.9–2.2 mm.......... ***alpinus***

– Wing slightly darker anteriorly but no distinct pale patch beyond dm-m. Clypeus yellow to brownish, epistoma pale grey or brownish grey. Face broader, minimum width slightly to obviously wider than front ocellus (bb). Larger species, wing length 2.4–3.7 mm.......... 6

(a)

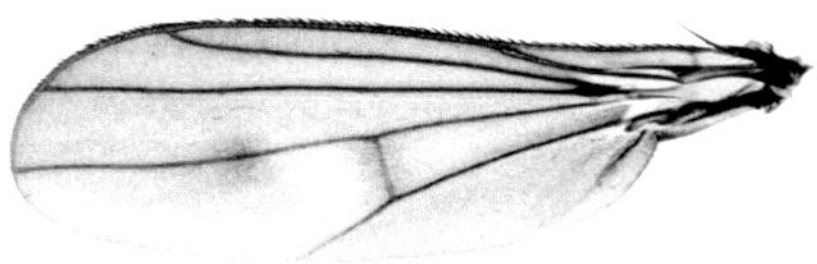

(b)

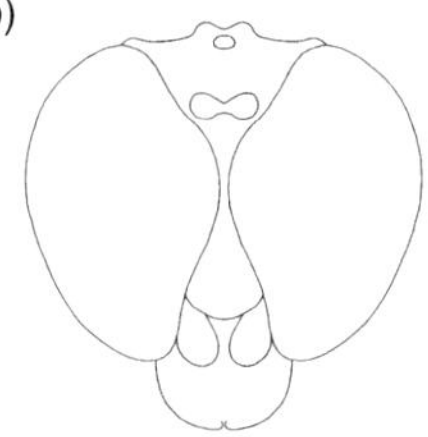

(bb)

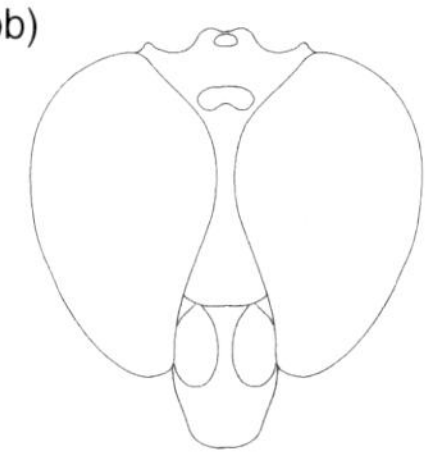

6(5) Lower postocular setae black (but adjacent postgenal setae white; black postocular setae may have yellow reflections so view from different angles). Vein CuA + CuP short and wide to its tip, fuzzy-edged, reaching no further than halfway to wing margin if imaginarily extended, no faint fold continuing its projection near margin (a). Outline of hind margin of wing to where CuA + CuP points slightly indented so anal lobe is broader (a). Costal hairs between R_1 and R_{2+3} always identical. Clypeus brown, not or only slightly paler than brown epistoma. Front coxa usually mostly yellow, at least on inner face in dark-legged individuals.......... ***scambus***

– Lower postoculars white. CuA + CuP tapered from base to tip, extending more than halfway to wing margin and continued by a curved fine fold running just posterior to vein, at low magnification appearing as the vein itself reaching margin (aa). Outline of wing margin smoothly curved near end of CuA + CuP (aa). Costa near R_{2+3} with a few spinules slightly longer and stouter between every 2–5 fine hairs (bb). Clypeus usually a shade of yellow (sometimes almost pale grey), contrasting with paler grey epistoma. Front coxa yellow or dark. Occasional intermediate specimens of the following three species cannot be identified; see Table 1 at end of key.......... 7

(a)

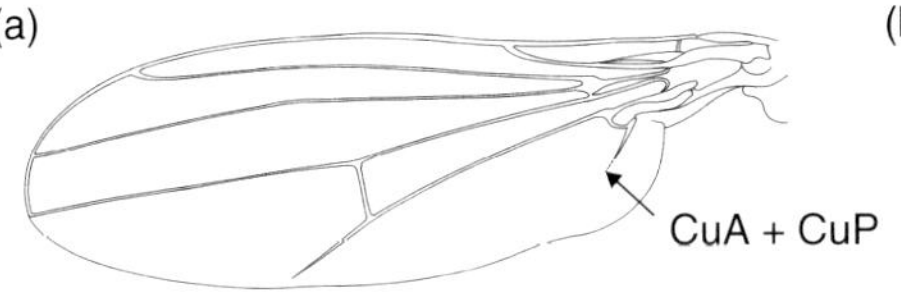

(bb)

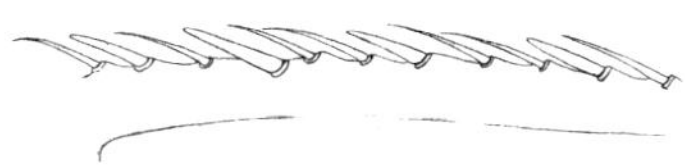

(aa)

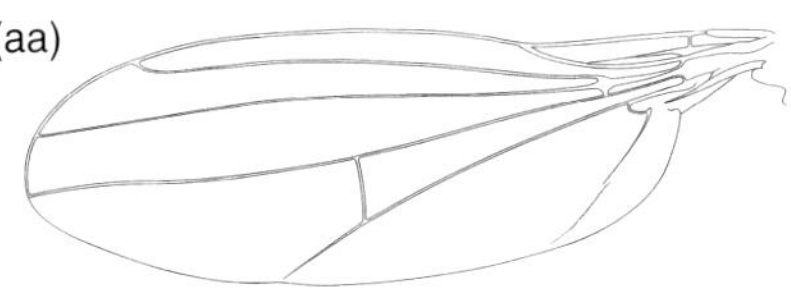

7(6) Frons pale-dusted on front corners above antennae, completely obscuring the ground colour, continuous with same colour and density of dusting on epistoma. Front coxa black with yellow tip, rarely entirely yellow. ***curvipes***

– Frons shining or glistening on front corners above antennae, ground colour showing through, not dusted like epistoma. Front coxa yellow; in dark-legged individuals, yellow at least on inner face. 8

8(7) Hind basitarsus slightly shorter than tarsomere 2 when viewed on ventral or posterior faces (a). Clypeus pale yellow, contrasting with pale grey epistoma. Epistoma narrower, about 0.3–0.4 times width of tip of clypeus, or clearly narrower than distance between inside edges of ocellar setae. Apical section of M_4 about twice or less length of dm-m. Apical section of M_1 beyond dm-m less than 1.5 times section from dm-m to node at fork of R_{2+3} and R_{4+5}. ***loripes***

– Hind basitarsus and tarsomere 2 equal in length when viewed on ventral or posterior faces (aa). Clypeus very pale yellow-grey, cream or sometimes pale grey, similar in colour and shade to epistoma, or epistoma darker. Epistoma wider, about 0.4–0.5 times width of tip of clypeus, or about equal to distance between inside edges of ocellar setae (cc). Apical section of M_4 slightly more than twice length of dm-m (bb). Apical section of M_1 beyond dm-m more than 1.5 times section from dm-m to node at fork of R_{2+3} and R_{4+5} (bb). ***armatus***

(a)

(aa)

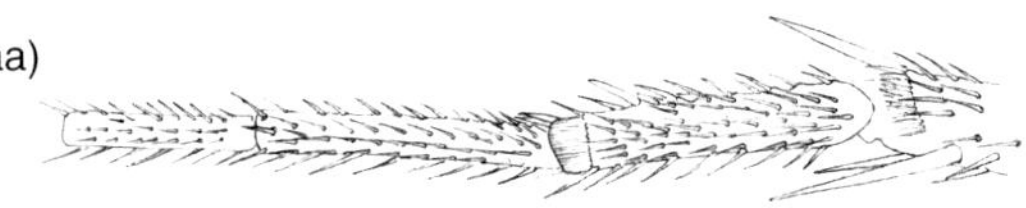

(bb)

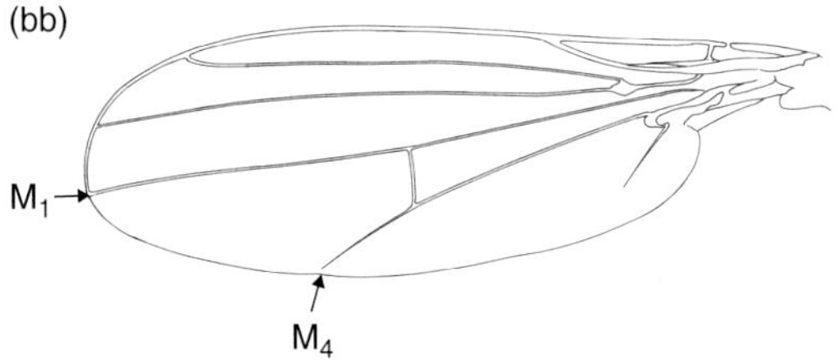

(cc)

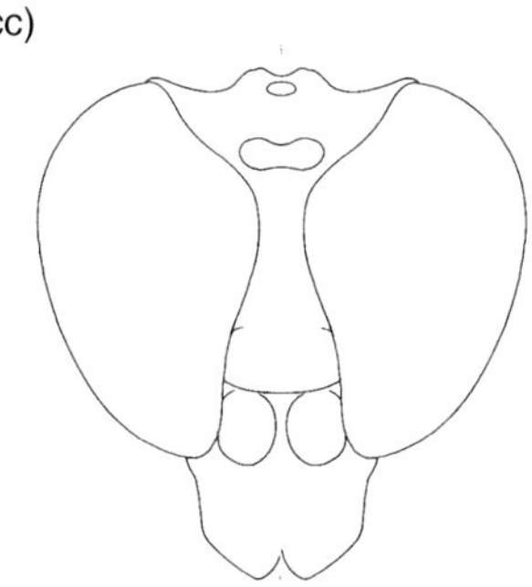

9(3) Frons pale-dusted on front half above antennae. Epistoma and clypeus with a similar shade but clypeus very pale yellow and epistoma pale grey. Epistoma narrower at middle, about half width of clypeus (a). Pedicel and scape yellow below (usually difficult to see). ***picticornis***

Note: ***dasycnemus*** probably falls here; it has yellowish grey epistoma and yellow clypeus.

– Frons almost undusted, shining violet above antennae. Epistoma pale grey to almost black, clypeus dark yellow to brown. Epistoma broader in middle, about 3 times width of clypeus (aa). Antenna entirely black. ***pumilio***

(a)

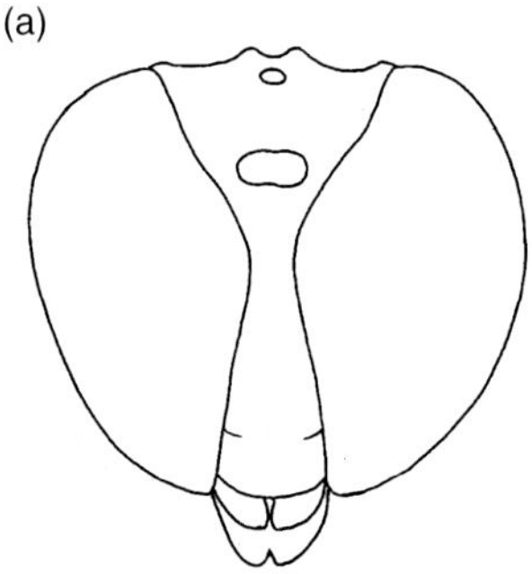

(aa)

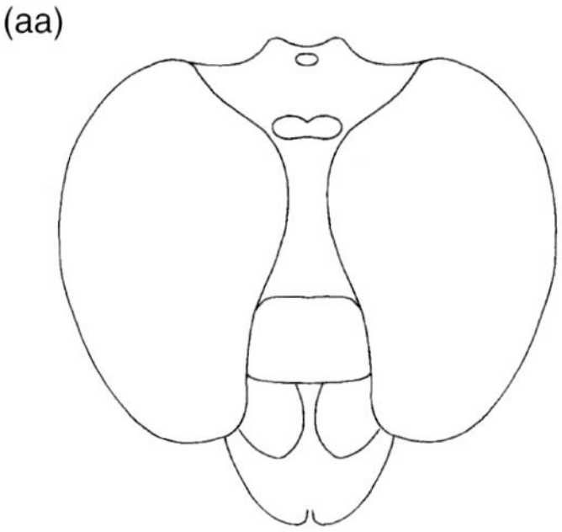

10(2) Femora and tibiae almost completely yellow. Frons and scutum dull-dusted brown without shine. R_{4+5} and M_1 almost parallel beyond dm-m. Halter knob pale. Hind coxa with lateral seta black. .... ***marginatus***

– Legs completely black except for yellow tips of femora. Frons shining blue-black, scutum subshining blackish. R_{4+5} and M_1 converge near wing tip, their greatest separation on apical two-thirds about 1.3 times than at tip (aa). Halter knob black. Hind coxa with lateral seta white. .... ***umbripennis***

(aa)

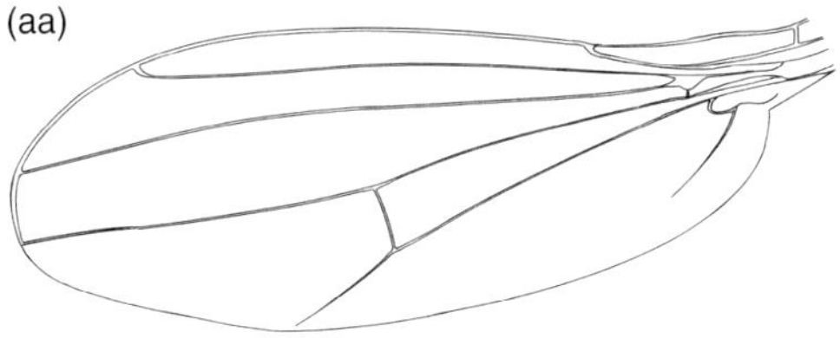

11(1) Hind femur mainly yellow, front and hind femora dark dorsally. 2 postsutural dorsocentral setae. Front coxa with apical setae black. Proepisternal seta black. Scape and pedicel yellow ventrally but sometimes entirely black. Face narrower, epistoma at narrowest about 0.28 times width of clypeus (a). .... ***compeditus***

– Hind femur black basally, yellow on apical quarter to half. 3 postsutural dorsocentral setae. Front coxa with apical setae pale. Proepisternal seta white. Antenna entirely black. Face broader, epistoma at narrowest about 0.35 times width of clypeus. .... ***magius***

(a)

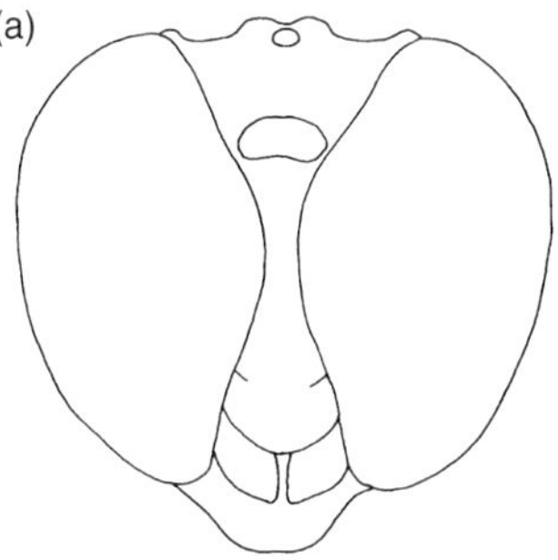

Campsicnemus alpinus (Haliday, 1833)

In both sexes, the wing marking with a pale patch beyond the dm-m crossvein and, in the male, splayed long veins make this tiny species unmistakable. There are no acrostichal setae (present but tiny in other species) and often only one intra-alar seta (usually three). Wing length ♂: 1.9 (1.8–2.1) mm, ♀: 2.0 (1.9–2.2) mm.

Campsicnemus alpinus is widely distributed in upland Britain from southwest England to Orkney but is absent from lowland England except at a few southern heathlands and bogs. It is found in bogs, moorlands, acid valley mires and wet heathlands, where it is usually recorded in sparsely vegetated or bare peat by pools and seepages. Flight: iv–x.

Campsicnemus armatus (Zetterstedt, 1849)

Male *armatus* are most easily recognised by the long row of evenly spaced upright blunt setae on the mid tibia. The mid tibia shaft may be construed as being modified in structure as it is very weakly bowed and as thick as the hind tibia, leading to *dasycnemus* in the key. The triangular postpedicel is the longest in the genus with its length twice its height compared to 1.5 times or less in other species. The legs, including the front coxa, are usually yellow but occasional very dark specimens have almost entirely dark legs and halteres. Females may closely resemble those of *loripes*, both having yellow front coxae, but the variable nature of *loripes* does not allow complete separation on characters

Table 1. Some ratios to help separate female *Campsicnemus* in couplets 7 and 8 (range for ten individuals).

	M_4 apical section/dm-m	M_1 apical/M_1 basal section	Clypeus/epistoma
curvipes	1.7–2.4	1.3–1.5	0.3–0.4
loripes	1.6–2.0	1.2–1.4	0.3–0.4
armatus	2.1–2.5	1.5–1.7	0.4–0.5

other than those used in the key; additionally, although overlapping with some specimens of *loripes*, the apical section of vein M_4 is consistently more than twice the length of the dm-m crossvein, and the apical section of M_1 beyond dm-m is more than 1.5 times the distance from dm-m to the node where R forks. Northern populations may comprise specimens with entirely dark legs, dark epistoma and grey clypeus, superficially resembling *pusillus* (which has dark hairs on the front coxa), and quite unlike the typical pale-legged southern populations of *armatus*. Wing length ♂: 2.4 (2.1–2.8) mm, ♀: 2.5 (2.4–2.6) mm.

This species occurs around much of the British coast including Orkney and the Outer Hebrides. It is found on saltmarshes, estuarine mudflats, brackish ditches and lagoons, and wet dune slacks. Flight: ii–x, xii, mainly vi–vii.

Campsicnemus compeditus Loew, 1857 (Plate 2)

This is one of the three species whose males have bizarrely developed legs, in this species having the femur and tibia of the front leg far stouter than on the other legs and, on the very short front tarsomere 2, a single projection that is longer than the entire tarsus and fringed either side with long hairs. Its basal antennal segments are usually yellow, a character shared only by the tiny *picticornis*, but in females this can sometimes be difficult to see or they may be entirely black. Females are distinct from others in the genus with their more setose legs, having four setae (rarely fewer) on the dorsal side (antero- and posterodorsal) of the front tibia (two in most other species) and 6–7 on the dorsal side (ad, pd) of the mid tibia (four in most other species), only one notopleural setae (as in *magius*) and two postsutural dorsocentral setae, unlike any other species. Wing length ♂: 2.6 (2.4–2.9) mm, ♀: 2.8 (2.4–2.9) mm.

The species is found in mainly upland areas from southwest England to northern Scotland, including Wales and northern England, with outliers in heathland bogs in lowland England. It is a species of especially wet bogs, and found particularly at pools and sometimes at seepages. The inaccessibility of its habitat may account for its apparent scarcity. Flight: v–viii, x, peak in vi.

Campsicnemus curvipes (Fallén, 1823)

Campsicnemus curvipes males are readily recognised by the row of strong ventral setae on the hind femur; *pusillus* and *dasycnemus* also have a row in this position but the setae are weaker, and these species lack the short dumpy mid basitarsus of *curvipes*. Females have a black front coxa with only its tip yellow and only rarely with the inner face slightly dingy yellow. The front coxa in the similar species *armatus*, *loripes* and *scambus* is usually yellow or, in dark specimens, at least pale on the inner face. Leg colour is particularly variable in both sexes and, although in most specimens the legs are mainly yellow, rare specimens may be almost entirely black-legged. Females also differ from most specimens of these similar species in having the apical section of vein M_4 obviously less than twice as long as crossvein dm-m. The face shape is the same as that of *loripes* but narrower than in *armatus*. Wing length ♂: 3.1 (2.9–3.4) mm, ♀: 3.3 (3.0–3.7) mm.

This is one of the most common and widespread British dolichopodids with records extending to Orkney and the Outer Hebrides. It is found in many types of wetland but uncommon on saltmarsh; it may require accessible water or saturated surfaces, and may show a slight preference for unshaded sites. Flight: i–xii (all year), mainly mid vi–vii.

Campsicnemus dasycnemus Loew, 1857

Unique characters for males are the row of very long ventral setae on the mid femur and small ventral semicircular flange near the base of the mid tibia. It has mainly yellow legs and yellow facial dusting. The hind femur resembles that of *curvipes* in having a series of ventral setae but their mid legs are amply distinct. No female specimens have been seen but according to Parent (1938) it appears to have costal spinules similar to those of *pumilio* and *picticornis* but is differentiated from *picticornis* in having entirely black antennae and from *pumilio* in having a yellowish grey epistoma and yellow clypeus. Wing length about 2 mm.

Campsicnemus dasycnemus has been recorded in Ireland (Chandler, 1989) but not in Britain although it is likely to be found here. The Irish record came from freshwater marsh and wet grassland.

Campsicnemus loripes (Haliday, 1832)

The bent mid tibia with its two combs of short setae separated by a short gap is unique to males of this species. Its legs are usually mainly yellow. Females are sometimes difficult to identify as they are particularly variable in leg colour and in ratios of wing veins, although the front coxae are nearly always obviously yellow. Differences from *curvipes* and *armatus*, which are sometimes confusingly similar, are mentioned in those accounts. Some specimens will probably be unidentifiable. Wing length ♂: 2.8 (2.5–3.0) mm, ♀: 3.0 (2.6–3.3) mm.

This species is widespread in Britain, although scarce in lowland southeast England, and is the commonest *Campsicnemus* in Scotland, recorded even from the remote St Kilda archipelago. It is found in many types of wetland and appears to prefer shaded and acid sites but not exclusively so. Flight: i–xi, mainly vi–vii.

Campsicnemus magius (Loew, 1845)

Males exhibit the most extreme development of secondary sexual characters among British dolichopodids. It superficially resembles *compeditus* but has more complex front legs, including particularly long ventral setae on the front femur. Females are unusual in the genus in having bicoloured femora, black basally and yellow apically. Wing length ♂: 3.1 (2.9–3.3) mm, ♀: 3.5 (3.2–3.7) mm.

This is an obligate brackish marsh species recorded on the coast from East Sussex to Norfolk, with a single western population in Cheshire. The habitat is brackish marsh, not true saltmarsh. It has occurred rarely at inland brackish sites, both natural and artificially created, in Cambridgeshire, South Yorkshire and Cheshire (Drake *et al.* 2019, Skidmore 1977). Flight: iv–ix, mainly vii–ix.

Campsicnemus marginatus Loew, 1857

This is one of the three British species with dark wing marks but compared to the blackish shade of the wings of *alpinus* and *umbripennis* those of *marginatus* are brownish on the front and tip and narrowly along most veins. It also shares with these two species, but no others, a silver face in both sexes. Like *pusillus*, the hind femur has an anteroventral row of fine hairs. Among British species it is unusual in that, in both sexes, the scutum and scutellum are strongly dusted over a pale brown or dull olive-green ground and so is far less shining than other species. The frons is also densely dusted so unlike the shiny metallic blue found in nearly all other species. Both sexes also have two sizes of spinules along the whole of the costa between R_1 and R_{2+3}, as in *picticornis*. Wing length ♂: 2.8 (2.6–3.1) mm, ♀: 3.1 (2.8–3.2) mm.

Campsicnemus marginatus has a western and northern distribution reaching the Scottish Highlands but is almost absent from lowland England. It is strongly associated with the bare margins of stony and sandy rivers, reflected in its national distribution, and more rarely with smaller streams and soft-rock cliff seepages. Flight: iv–x, peak in vii.

Campsicnemus picticornis **(Zetterstedt, 1843) (Plate 2)**

Campsicnemus picticornis is a tiny species uniquely lacking any modifications in the structure or setae of its legs. The identity of males is confirmed by the prominently yellow basal antennal segments but in females only the underside of these is yellow and usually difficult to see. However, females need be compared only with the similarly small *pumilio* as they both have clearly differentiated spinules along the whole length of the wing costa between R_1 and R_{2+3}, otherwise found only in *marginatus* (and *dasycnemus*?), and not just on the apical quarter of this costal section as in some other species. The leg colour of female *picticornis* is particularly variable, ranging from mainly yellow to mainly black, whereas the legs of *pumilio* tend to be mainly yellow. Wing length ♂: 2.2 (2.1–2.3) mm, ♀: 3.0 (2.7–3.2) mm.

The distribution is concentrated in East Anglia and the Thames estuary, and sparse in other lowland floodplains in eastern and southern England. It is found on the south Welsh coast but not in Scotland or northernmost England. Its habitats are grazing marsh, washlands, swamps, reedbeds and sedgebeds; saturated ground and perhaps regularly flooding seem to be common features of its sites. Flight: iii–x, peak in vii.

Campsicnemus pumilio **(Zetterstedt, 1843)**

The setae on the male's mid leg are short and inconspicuous, no longer than the shaft width. The matt black epistoma in combination with yellow legs and simple shafts is diagnostic (*pusillus* has an entirely black face and mainly dark legs). Females resemble *picticornis* in being small and having obvious differentiated spinules along the whole length of the wing costa between R_1 and R_{2+3} but their femora are usually entirely yellow rather than often partly dark as in *picticornis*, and lack the pale underside of the first two antennal segments. Wing length ♂: 1.8 (1.6–2.1) mm, ♀: 2.4 (2.2–2.6) mm.

Campsicnemus pumilio is an uncommon species found in southern England and coastal Wales, with isolated records in Yorkshire and the Moray Firth that need confirming. Its habitat requirements are uncertain owing to possible errors in identification of females. A large proportion of records suggest an association with rivers and large ponds on sandy soils, and dunes. Reliable records suggest that bare mud or sand at water margins is important. Flight: iv–x, mainly vii–viii.

Campsicnemus pusillus **(Meigen, 1824)**

This is a dark shiny species with greenish black legs with only the femora yellow-tipped, although occasional pale-legged specimens occur. Being so dark provides an initial clue to identification as few other *Campsicnemus* are quite like this. The shape of the front tarsus and the entirely matt black face in males are unique but *pumilio* also has a matt black epistoma. The hind femur has a posteroventral row of fine hairs, similar to that of *marginatus* and *pumilio*. Females are similarly dark so resemble only *umbripennis*, and the entirely dark postocular setae are shared only with *scambus*. Wing length ♂: 2.5 (2.3–2.8) mm, ♀: 2.7 (2.4–2.9) mm.

Care should be taken to check for the very similar non-British *armoricanus* which in mainland Europe occurs in similar habitats to those of *pusillus*. It differs in having the last two segments of the front tarsus far less obviously widened, a pronounced anteroventral apical projection on the mid tibia (barely protruding in *pusillus*), the row of short ventral setulae at the tip of the mid femur occupying a very short length equal to the width of the shaft here (twice this distance in *pusillus*), and the mid basitarsus being almost as long as the second and third tarsomeres together (these segments all similarly long in *pusillus*); Parent (1938) provides an illustration. Pollet (1991) distinguishes females by their brownish black halteres and two ventral setae on the hind tibia but all British *pusillus* have two ventrals on the hind tibia (and pale halteres).

Campsicnemus pusillus has a predominantly southwest distribution with most records from Wales and the southern counties of England. There are isolated records, particularly in Norfolk and a few scattered in Scotland. Its habitats include acid mires with seepages and the stony margins of rivers, as well as base-rich fens although it is possible that *pusillus* occupies small slightly acidic pockets within these fens. Flight: iii, v–xi, mainly vi–vii.

Campsicnemus scambus (Fallén, 1823) (Plate 2)
This is the most common of the large species whose males have strongly expanded tibiae and a complex arrangement of setae. Both front and mid leg have strongly modified femora, tibiae and basitarsi. Females are easily distinguished from other species by their entirely black postocular setae, found only in the smaller black *pusillus*, and the short dumpy CuA + CuP vein which is more obvious at low magnification; nearly all other *Campsicnemus* have a longer tapered CuA + CuP (shorter only in *alpinus* and *compeditus*). The frons is entirely shining right to the front corners above the antennae; between the antennal bases a bright shining reddish triangle can often be seen when viewed from above which is absent in *loripes* but sometimes present in *armatus*. Wing length ♂: 3.6 (3.4–3.9) mm, ♀: 3.3 (2.9–3.7) mm.

Campsicnemus scambus is widespread throughout Britain and is everywhere frequent, showing no geographic bias. The northernmost record is from Shetland. It is found in a wide range of wetlands with no common feature except perhaps a requirement for water margins or saturated soil. Flight: i–xii (all year), mainly mid vi–vii.

Campsicnemus umbripennis Loew, 1856
This is a large black species whose males have very long darkish wings, long thin legs, ventral rows of short setae in the apical half of the mid femur, and a rather spiky mid tibia with rows of small setae on all faces. The wings of females are less obviously dark and not narrowed basally but R_{4+5} and M_1 bow gently away from each other, quite unlike the parallel veins of *pusillus* which has similarly coloured legs. The only known British specimen agrees with the subspecies *hispanicus* Strobl, 1899, which differs from the nominative species *umbripennis* in small differences in the arrangement of setae of the mid leg described by Parent (1938) and Perry (1999). Wing length of non-British specimens of the nominative subspecies ♂: 3.8 (3.6–4.1) mm, ♀: 3.8 (3.4–4.1) mm.

There is a single record of a male from Dorset on grassland by soft-rock cliffs (Perry 1999). On mainland Europe it occupies a range of habitats including river banks and pine woods (Naglis and Barták 2015), streams, seepages and marsh (Pârvu 2000, Drake unpublished). Flight: vii.

CHRYSOTIMUS Haliday in Loew, 1857 (Peloropeodinae)

These are tiny stocky species with a bright shining almost lime-green thorax and almost entirely yellow legs (Plate 2, male and female). The genus can be recognised by the combination of all hairs and setae, including all postoculars, being pale, the arista placed dorsally or above the middle of the short postpedicel, and an anterior preapical seta on the mid and hind femora. Females are conspicuous in having an almost entirely bright yellow abdomen not found in any other British genus (Plate 1); *Xanthochlorus*, which also has pale setae, does not have the combination of entirely green thorax and yellow abdomen. Males may be mistaken for those of *Thrypticus* but these differ most obviously (in the British fauna) in having partly or completely dark coxae and femora, and a conspicuous large hypopygium rather than the small and encapsulated one of *Chrysotimus*. There are almost no male secondary sexual characters; *flaviventris* has a cluster of short black setae ventrally at the base of its hind basitarsus.

No other Western Palaearctic species are likely to be found in Britain. *Chrysotimus meridionalis* Naglis and Barták, 2015 resembles *C. molliculus* but the first and sixth tergites of the males are yellow, not green as in *molliculus* and *flaviventris*.

Chrysotimus are nearly always collected from shrub and tree foliage.

Key to species of *Chrysotimus* (males and females)

– Abdomen green .... males
– Abdomen bright yellow .... females

1 Antenna entirely black. Crossvein dm-m clearly beyond tip of R_1, section of M_4 beyond dm-m equal to or shorter than section from dm-m to looped vein CuA (a). Last tarsomere of all legs mainly and obviously dark. Acrostichal setae small but distinct, about a quarter length of dorsocentral setae. Male hind basitarsus with only yellow setae ventrally at base. Female first tergite usually dark .... ***molliculus***

– Antenna with 2 basal segments bright yellow, postpedicel partly dark. Crossvein dm-m almost level with tip of R_1, section of M_4 beyond dm-m slightly longer (about 1.25 times) than section from dm-m to CuA (aa). Last tarsomere of all legs mainly pale, dark in apical half or less. Acrostichal setae absent or minute and scarcely visible. Male hind basitarsus with a cluster of dark setae ventrally at base (bb). Female first tergite yellow. .... ***flaviventris***

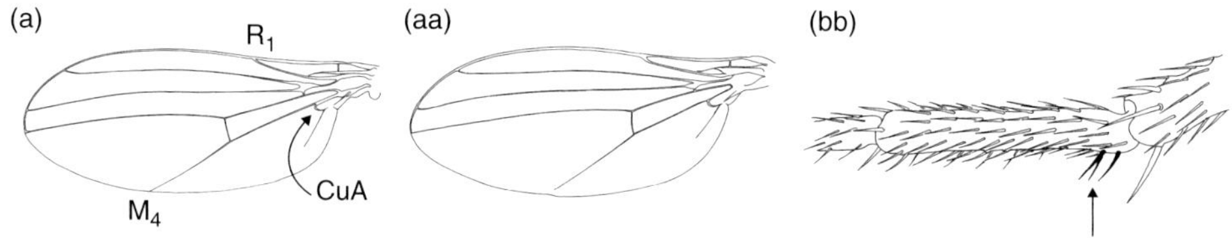

Chrysotimus flaviventris (von Roser, 1840)

There should be no problem distinguishing *flaviventris* from *molliculus* in either sex. Wing length ♂: 2.5 (2.4–2.7) mm, ♀: 2.7 (2.4–2.9) mm.

The species has a wide distribution in Britain but is most frequent in the south of England and becomes scarce north of the Midlands. There are a few widely scattered Scottish records and very few for Wales. Despite the wide distribution, it is rather uncommon. In England it is nearly always found within deciduous woodland which may be damp or dry. In Scotland it appears to be found more often in open sites that probably have shrubs rather than closed canopy. Flight: vi–x, peak in vii.

Chrysotimus molliculus (Fallén, 1823) (Plate 2)

This species is clearly distinct from *flaviventris* in both sexes. The female is remarkable in having a small dark tubercle on the hind margin of the swollen second tergite, obvious in wet-preserved material, and sometimes with the discal hairs of the second and third tergites, which are always short, reduced almost to pubescence. Wing length ♂: 2.6 (2.4–2.8) mm, ♀: 2.7 (2.5–2.8) mm.

Chrysotimus molliculus is common in the southern half of England and Wales but scarce north of Yorkshire. There is a single verified Scottish record from Perthshire. The species is found in a wide range of habitats and has no obvious habitat affinity, but is nearly always recorded on tree and shrub foliage, or on other tall plants well above the ground-layer vegetation. Flight: vi–ix, peak in vii.

CHRYSOTUS Meigen, 1824 (Diaphorinae)

Chrysotus are small compact flies, with metallic green to almost black bodies without yellow patches on the tergites (Plate 2). Their small postpedicel is round, D-shaped or kidney-shaped with the arista arising at about the middle of the anterior face or just dorsal of any slight projecting tip. Their wings have almost straight veins with R_{4+5} and M_1 being almost parallel beyond dm-m, and the long veins appear crowded together in the front of the wing. There is no true preapical seta on the mid or hind femora but a cluster of 3–4 longer hair-like anteroventral setae near the tip may be mistaken for this seta. The combination of the seta on the hind coxa arising close to its base and approximately parallel veins R_{4+5} and M_1 is found in the related *Diaphorus* and *Melanostolus melancholicus*. *Chrysotus* males are distinguished from *Diaphorus* males by their eyes being clearly separated on the frons (eyes touch above the antennae in *Diaphorus*) and from *Melanostolus* males by their eyes nearly touching on the face (widely separated in *Melanostolus*); *C. laesus* is somewhat intermediate. The pulvilli of *Chrysotus* are small compared to those of male *Diaphorus* and *Melanostolus*, and the claws are small but absent in male *Diaphorus*. Female *Chrysotus* differ from these two other genera in the position of the arista, as in males, and the frons being obviously wider than the face; in particular, *obscuripes* and some *palustris* with almost entirely black legs may be confused with the small all-black *Diaphorus nigricans* and *Melanostolus*.

The fringe of hairs on the tibiae is drawn from an anterodorsal view which shows its maximum extent, and the figures show the tibia's anterodorsal side at the top and the anterior face towards the bottom, contrary to the convention in the rest of the book. The fringe's width is the perpendicular distance from the shaft to a line level with the hair tips, not the length of individual hairs which is longer, unless made clear in the text. Identification of males using external characters can sometimes be uncertain but the hypopygia are distinctive and are illustrated and described for most species. Sufficient of the tip of the phallus can often be seen if the capsule is drawn away from the abdomen. In fresh or wet-preserved material, the tip can usually be coaxed out by gently squeezing the capsule. While most female specimens can be identified with moderate confidence, some pairs of species run into one another at the extremes of their variation: *blepharosceles* (common) and *collini* (rare), *gramineus* (common) and *angulicornis* (rare), *neglectus* (common) and *femoratus* (uncommon), *suavis* (frequent) and *palustris* (uncommon).

Chrysotus is a huge genus of several hundred species worldwide and about 28 species in the Western Palaearctic (Pollet 2011, Yang *et al.* 2006). Britain currently has 19 species, including one hot-house species not apparently established in the wild. Many more have been described from the Palaearctic, some of which appear to be very similar or even identical to known British species. Keys to the Palaearctic fauna are given by Maslova *et al.* (2011), Negrobov and Maslova (1996) and Negrobov *et al.* (2000).

Chrysotus includes several species with no obvious habitat affinity, including *blepharosceles*, *cupreus*, *gramineus* and *neglectus*, and others that are found in places drier than normally associated with dolichopodids, including *femoratus* and *laesus*. The remainder are found in a variety of wetlands of which exposed river sediments, wet dune slacks and acid mires support a few species almost confined to these habitats. Apart from the ubiquitous *gramineus* and to some extent *neglectus*, they tend to be found in unshaded places.

Key to species of *Chrysotus* (males)

1	Front and mid femora entirely or mainly black.	2
–	Front and mid femora entirely or mainly yellow.	17
2(1)	Front coxa with apical setae black.	3
–	Front coxa with apical setae pale	13
3(2)	Front and mid tibiae black or dark brown.	4
–	Front and mid tibiae yellow, or rarely at most brownish	7

4(3) Postpedicel large, its depth nearly equal to frons length from insertion of antenna to ocellar seta seen in profile (a), upper margin extending backwards as a hood over pedicel (b). Vein M_4 almost straight, not curved beyond dm-m (c). .... ***obscuripes***

– Postpedicel smaller, its depth about half frons length measured as above (aa). Vein M_4 gently bowed beyond dm-m (cc). 5

(a)

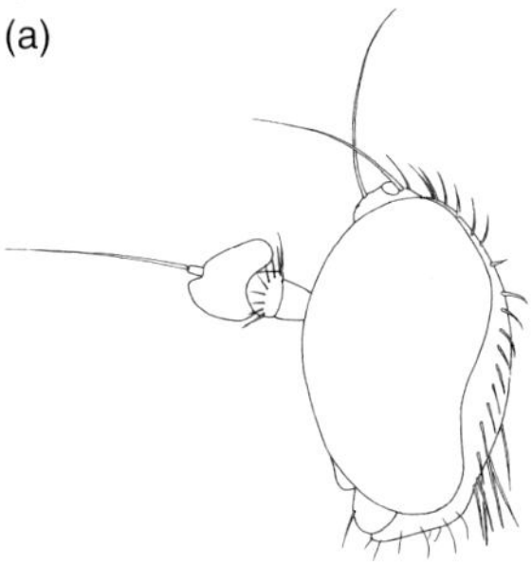

(b)

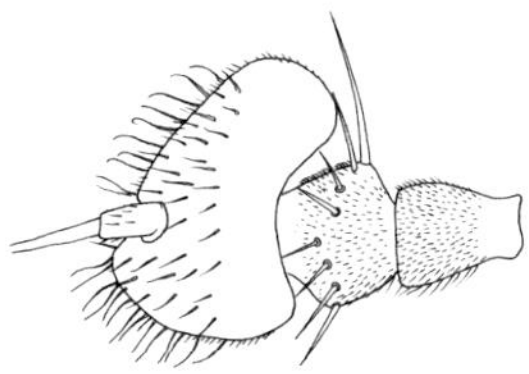

(c)

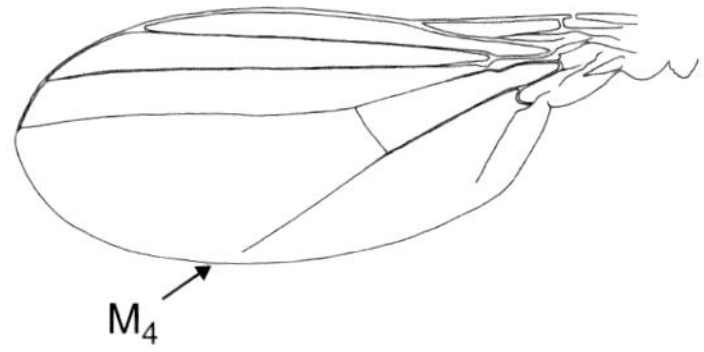

(aa)

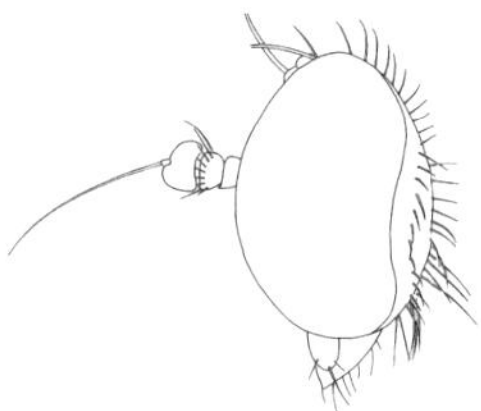

(cc)

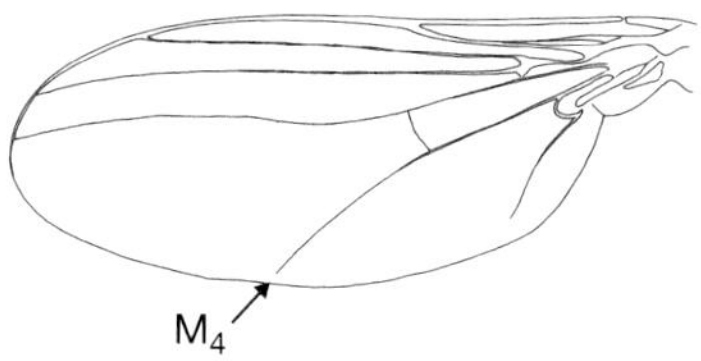

5(4) Front coxa and trochanter pale or almost white, coxa darker at base. Calypter hairs yellow or brown. Palp yellow. Hind tibia with narrow anterodorsal and posterodorsal fringes, at most as wide as shaft (a). .... ***cupreus***

– Front coxa black; trochanter may be paler than femur. Calypter hairs black. Palp entirely black or with a pale tip. Hind tibia with anterior fringe either at least as wide as shaft or poorly developed. 6

(a)

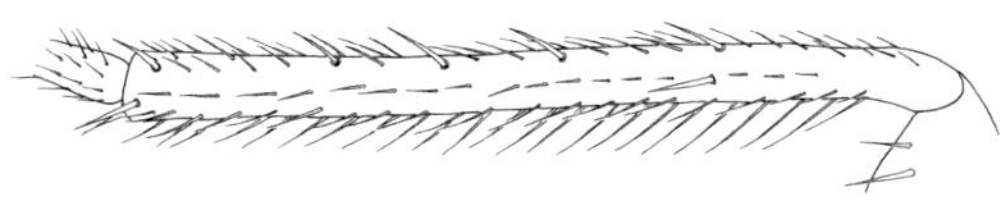

6(5) Hind tibia with a conspicuous anterior fringe wider than shaft in anterodorsal view (a). Front tibia with fringe on posterior to ventral faces as wide as shaft. Front trochanter as dark as mid and hind trochanters. Palp entirely black. [Sternite 8 with narrow tip with short stout setae (b).] .... ***blepharosceles***

– Hind tibia with a short anterior fringe not wider than shaft. Front tibia with fringes poorly developed. Front trochanter yellowish brown, paler than mid and hind trochanters. Palp with pale tip. .... ***melampodius***

(a)

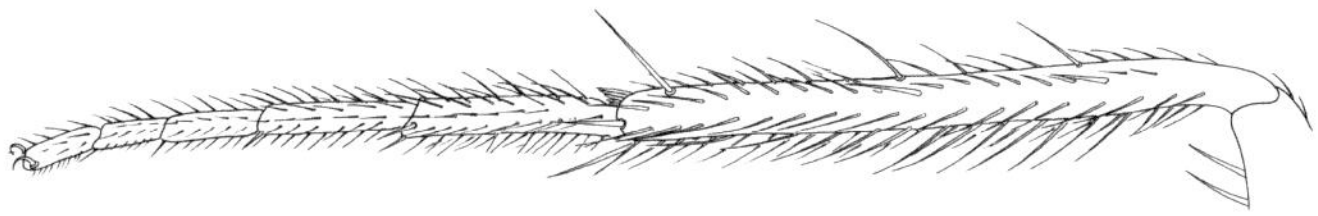

(b)

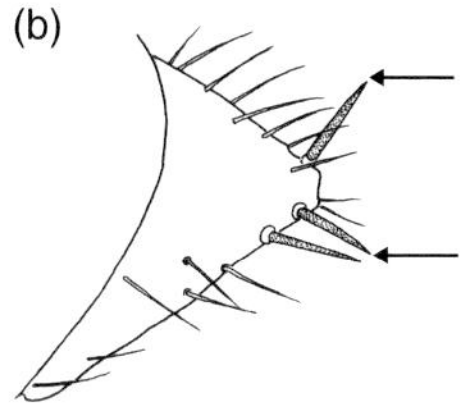

7(3) Mid and hind trochanters black or brown, scarcely different in colour to dark base of femora. 8

– Mid and hind trochanters yellow, or at least clearly paler than base of femora. 12

8(7) Front coxa mainly reddish yellow, black basally on lateral face. Hind tibia with a wide anterior fringe, similar to that in *blepharosceles* (fig. a of couplet 5) .... ***verralli***

– Front coxa mainly or entirely black. Hind tibia with a narrower fringe. 9

9(8) Postpedicel large, about as long as frons from antennal insertion to ocellar seta, and nearly twice as deep as pedicel (a). Mid tibia with 1 long anterodorsal seta at basal quarter, about 3 times as long as width of shaft (if with 2 setae and large postpedicel, check *pulchellus*). .... ***monochaetus***

– Postpedicel small, its depth at most two-thirds length of frons, and scarcely 1.5 times as deep as pedicel. Mid tibia with 1 or 2 strong anterodorsal setae 10

(a)

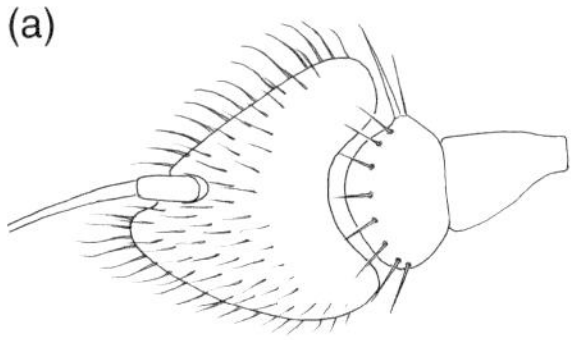

10(9) Sternite 8 with short thick apical setae with forked tips (a), at most half length of marginal setae on preceding tergite (rarely absent or reduced to only 1 forked seta). Ventral surstylus strongly curved and tapered to a pointed tip (b). Front tibia with hairs almost as long as width of shaft on at least one of dorsal, ventral or posterior faces (c). .... ***gramineus***

– Sternite 8 with several long apical setae with simple pointed tips (aa, two examples), the longest subequal to marginal setae on preceding tergite. Ventral surstylus different, as in next couplet. Front tibia hairs clearly not as long as width of shaft on any face. 11

(a)

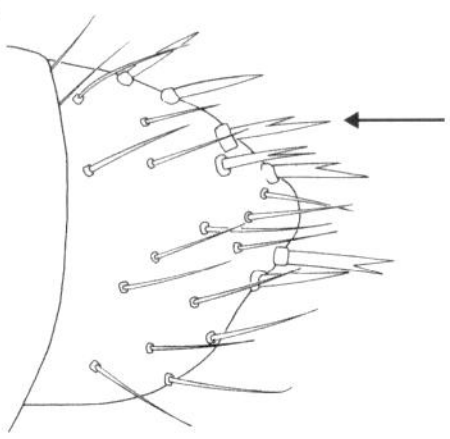

(aa)

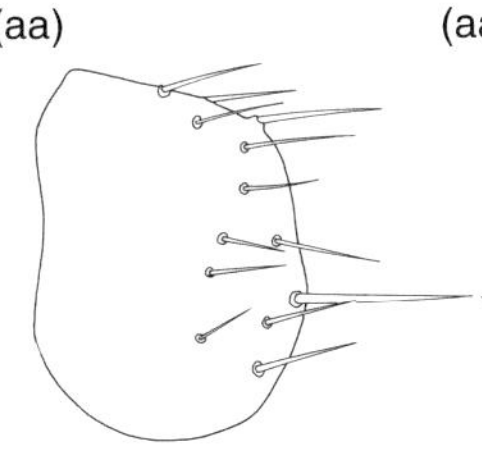

(aa)

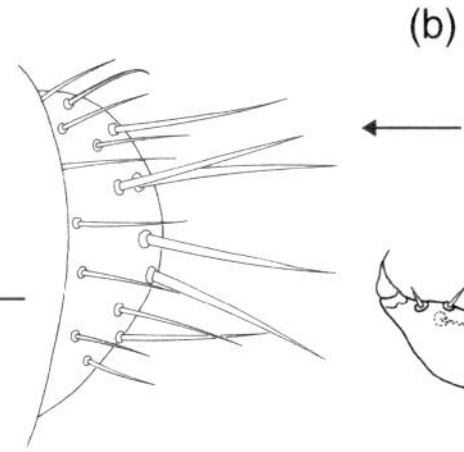

(b)

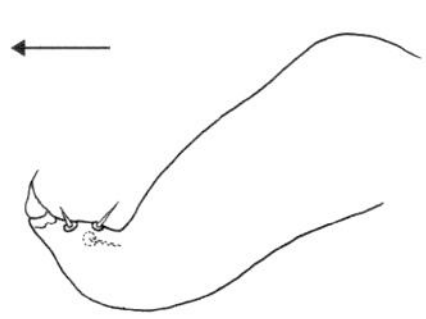

(c)

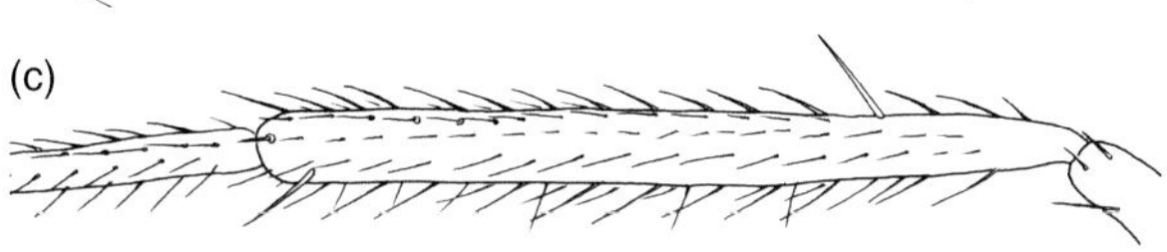

11(10) Postpedicel triangular, as long as deep (a), its depth about half length of frons between ocellar seta and antennal insertion (b). Arista just subapical, tip of postpedicel produced slightly beyond arista insertion (a). Hind tibia with hairs of anteroventral fringe slightly but clearly longer on basal half than on apical half (c). Palp white on apical half or more. Ventral surstylus gently curved and tapered to a pointed tip (d). .... ***angulicornis***

– Postpedicel slightly kidney-shaped to rounded-triangular, usually 1.7 times deeper than long (sometimes less) (aa); its depth two-thirds length of frons measured as above (bb). Arista approximately apical, arising about midway between top and bottom edges of postpedicel (aa). Hind tibia with anteroventral hairs equally long on whole length of tibia (cc). Palp black. Ventral surstylus strongly curved and almost parallel-sided to blunt tip (dd)................ .... ***collini***

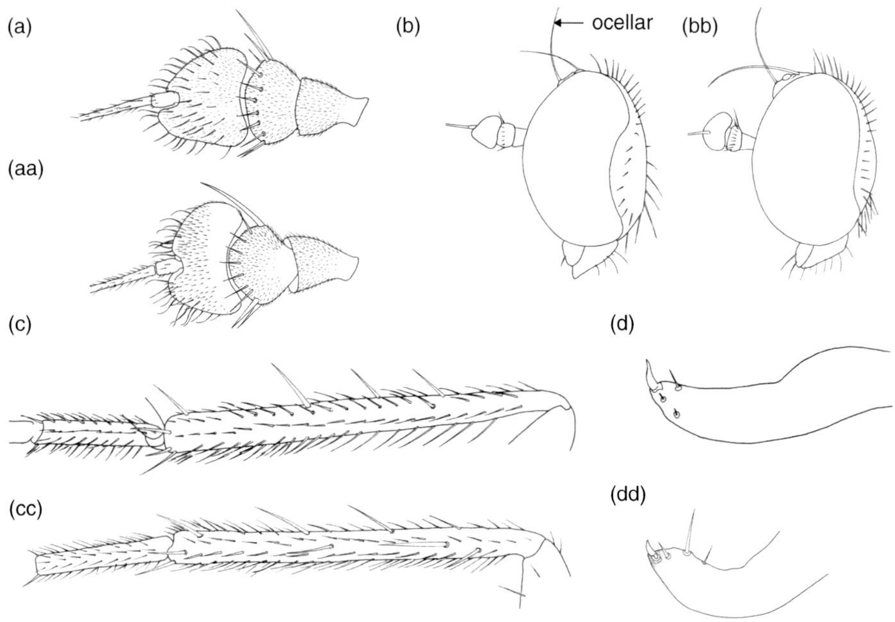

12(7) Postpedicel large, far larger than ocellar triangle, as deep as length of frons from antennal insertion to ocellar seta (a) and about twice as deep as pedicel (b). Hind tibia with anterior fringe not wider than shaft, dorsal fringe scarcely differentiated on hind tarsus (c). Front tibial hairs all obviously shorter than shaft width. Front and mid basitarsi usually yellow. Smaller species, wing length 1.8–2.0 mm. *Note*: If only 1 ad on mid tibia then check *monochaetus*....... .... ***pulchellus***

– Postpedicel small, about the size of ocellar triangle, its depth about two-thirds length of frons measured as above (aa). Hind tibia with dense wide anterior fringe 1.5 times wider than shaft (cc), continuing conspicuously on dorsal (or antero- or posterodorsal) surface of hind basitarsus and tarsomere 2 (dd). Front tibia with ventral hairs almost as long as width of shaft. Larger species, wing length 2.0–2.4 mm. .... ***femoratus***

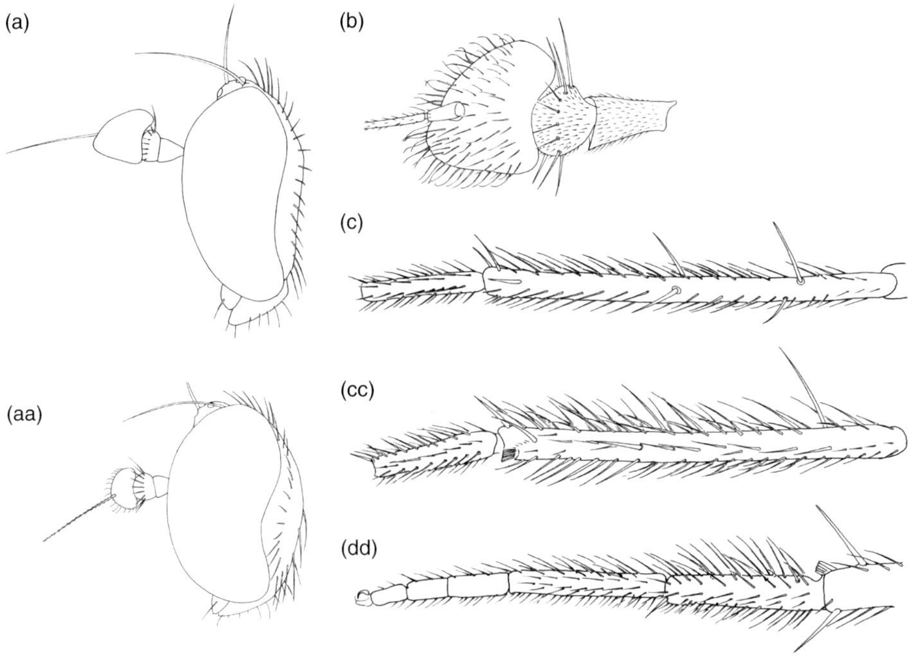

13(2) Face broad, at mouth edge half as wide as frons at level of front ocellus (a). Mid tibia usually without setae on shaft (1 anterodorsal may be present) (b). Postpedicel much deeper than long at its middle (by 2.5 times) (c). Legs except knees dark. .... ***laesus***

– Face narrow, eyes touching or separated on lower half of face by at most one-sixth of frons width at front ocellus. Mid tibia always with distinct setae, even if small, on both antero- and posterodorsal faces. Postpedicel may be large but its depth less than about 1.5 times its length in middle (cc). Front and mid tibiae usually yellow. 14

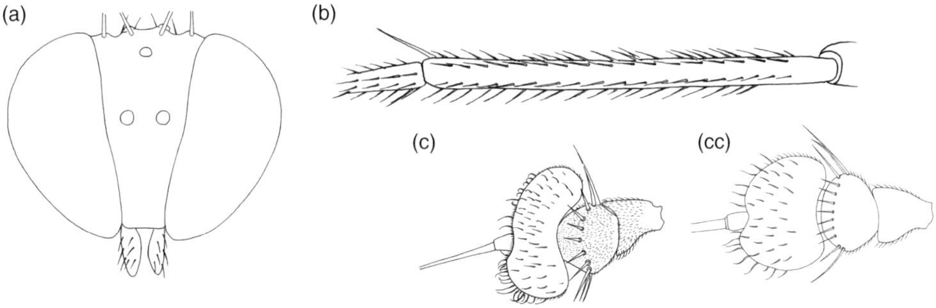

14(13) Mid tibia with only 1 long anterodorsal seta at about basal third (may be 1 short anterodorsal near base). Whiskers of postgena long and dense (a). Hind trochanter dark in contrast to yellow mid and front trochanters. 15

– Mid tibia with 2 long anterodorsal setae. Whiskers sparse and shorter (aa). Hind trochanter clear or dark yellow but as pale as mid and front trochanters. 16

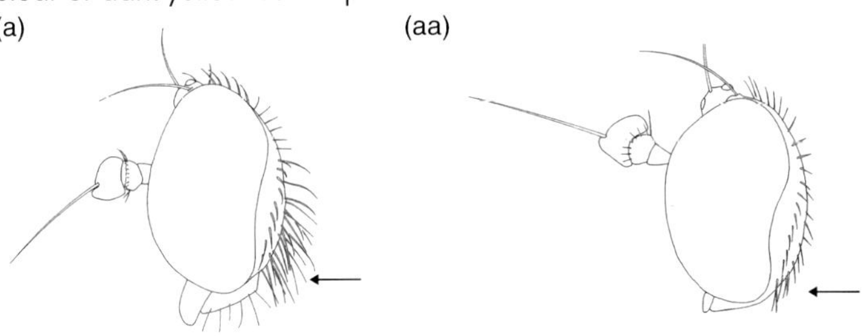

15(14) Scutum violet or blue seen from above. Frons white-dusted on lateral thirds, green and undusted on central stripe (a). Eyes touch on face (b). Front tibia always clear yellow. Hind femur in dorsal view with anterior fringe slightly longer than width of shaft. Phallus 'Y'-shaped, often protruding (Fig. 12). .... ***suavis***

– Scutum dark green. Frons evenly grey-dusted on green ground. Eyes separated on lower half of face by width of 3 adjacent eye facets (bb). Front tibia often dark. Hind femur in dorsal view with anterior fringe shorter than width of shaft. Phallus usually hidden. .... ***palustris***

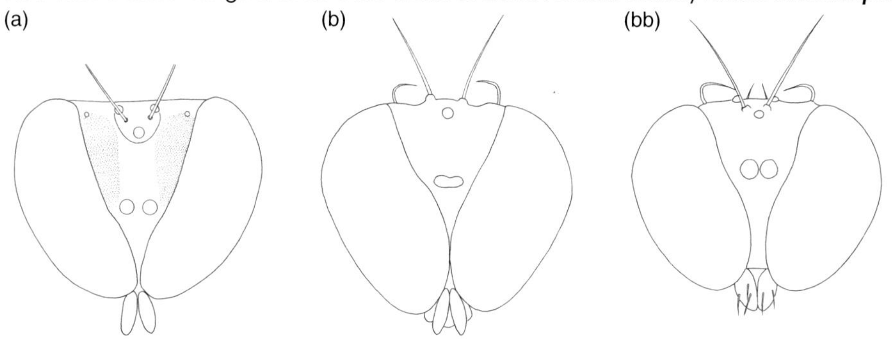

16(14) Front tibia with dorsal fringe only half as wide as shaft. All femora with poorly developed fringes. Hind tibia markedly stouter than mid tibia (c) and with dense black fringes. All trochanters and basal fifth to sixth of hind femur bright yellow. .... **Species A**

– Front tibia with dorsal and ventral fringes almost or fully equal to width of shaft (aa). All femora with ventral (antero- or posteroventral) fringes as wide as shaft (bb). Hind tibia not much stouter than mid tibia and with sparser pale fringes. All trochanters dark yellow, not strongly contrasting with femora. .... ***tricaudatus***

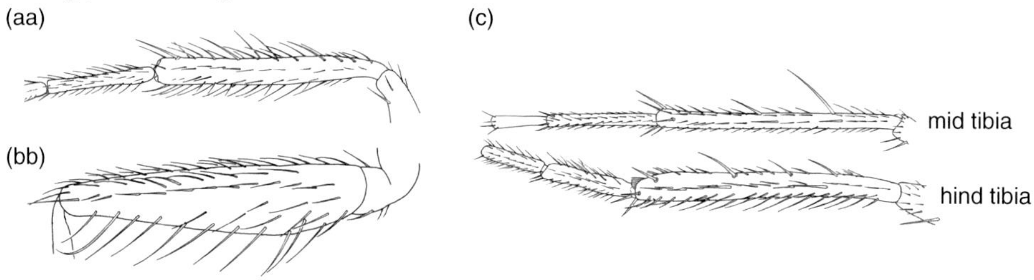

17(1) Antenna entirely yellow. Palp exceptionally long, more than two-thirds as long as eye height. Hind tibia without wide anterior fringe but with posterior fringe as wide as shaft. .... ***longipalpus***

– Antenna black. Palp inconspicuous. Hind tibia with wide anterior fringe. 18

18(17) Hind femur and tibia black on apical two-fifths to one-third (a). Front coxal hairs and setae pale. Hind tibia and basitarsus ventrally on black section near their tips with hairs slightly longer and denser than proximally (b). Mid tibia with posterodorsal setae inconspicuous, one-third or less as long as anterodorsal setae. Small species, wing length 1.0–2.0 mm. .... ***cilipes***

– Hind femur and tibia entirely yellow, at most with a dark smudge on femur tip. Front coxal hairs and setae black or hairs rarely brown. Hind tibia and basitarsus near their tips without a differentiated cluster of longer hairs but hairs becoming gradually denser (bb). Mid tibia with posterodorsal setae at least half as long as anterodorsal setae and 1.5–2 times longer than width of shaft. Larger species, wing length 2.3–2.8 mm. [Sternite 8 with short stout apical setae with ragged tips (cc).] .... ***neglectus***

(a, tibia)

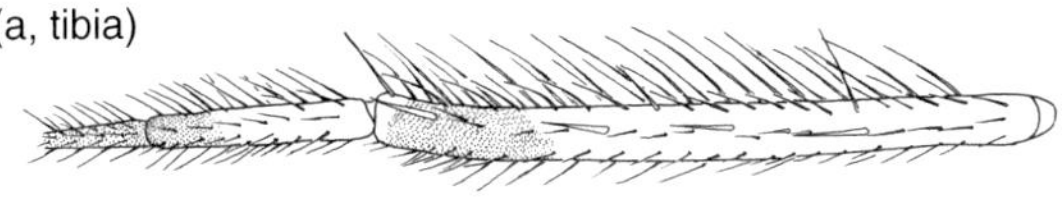

(b)

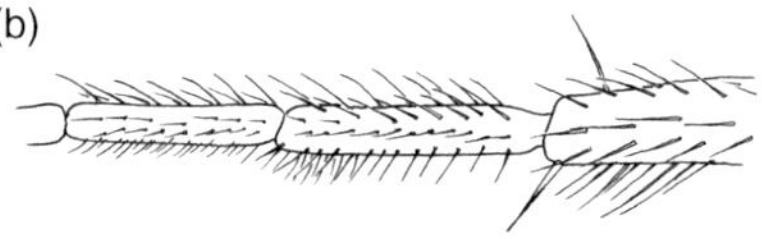

(bb)

(cc)

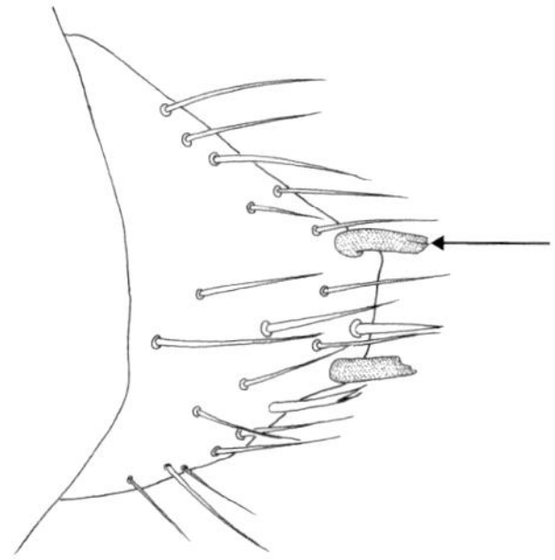

Key to species of *Chrysotus* (females)

This key does not include *melampodius* or *verralli* whose females are unknown. Two easily confused species in related genera are included here.

1 Arista arising near anterodorsal corner of postpedicel. Lower postocular and postgenal setae black. Face wide, clypeus 1.5–2 times wider than high 2

– Arista arising near middle of anterior face of postpedicel. Lower postocular and postgenal setae pale. Face narrow, clypeus 1–1.5 times wider than high 3

2(1) Halter black. .... ***Melanostolus melancholicus***

– Halter yellow. .... ***Diaphorus nigricans***

3(1) Front coxa with black apical setae and usually hairs also black. 4

– Front coxa with pale apical setae and hairs 12

4(3) Hind trochanter darkened. 5

– Hind trochanter clear yellow 9

5(4) Hind tibia entirely yellow or darkened at either end; in cases of doubt when tibia is particularly dark, front and mid femora with conspicuously bright yellow tips, the yellow extending back from tip by at least 2 rows of vestiture hairs .... ***gramineus***, ***angulicornis***

– Hind tibia black or dark brown. Knees very narrowly dark yellow or brown, pale colour not extending further from tip than 2 rows of vestiture hairs. 6

6(5) All tibiae brownish black. Postpedicel large and kidney-shaped with lower margin rounded, at least twice as deep as pedicel (a) and about same width as face, height more than a quarter eye height (b) .... ***obscuripes***

– Front and mid tibiae yellow. Postpedicel either large but triangular, or small, about 1.5 times as deep as pedicel (best seen on inner face) (aa) and not as deep as width of face, its height about a fifth eye height (bb). 7

(a)

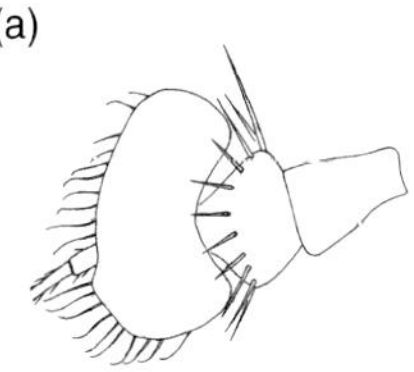

(aa)

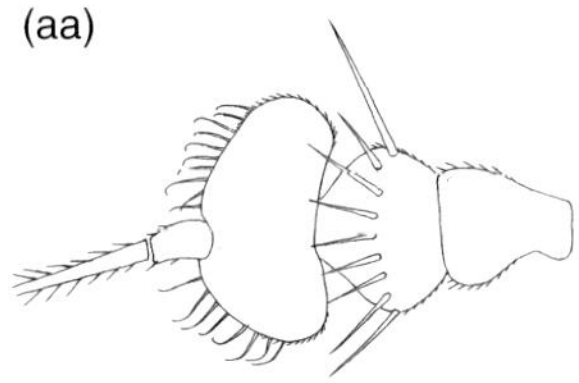

(b)

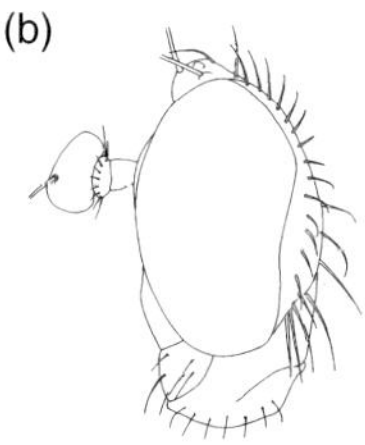

(bb)

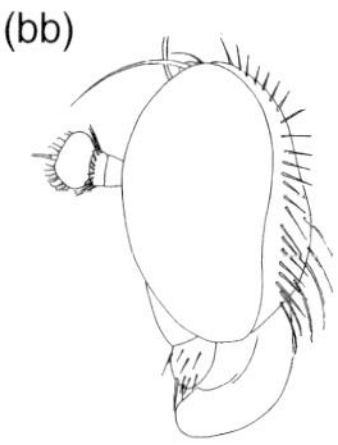

7(6) Front trochanter pale yellow, often also apex of front coxa. Crossvein dm-m further from base of wing so apical section of M_4 usually the same as or shorter than basal section measured from tiny looped vein CuA (a)..***cupreus***

– Front coxa entirely and trochanter largely dark or brownish yellow but not pale. Crossvein dm-m closer to base of wing so apical section of M_4 usually clearly longer than basal section measured as above (aa). 8

(a)

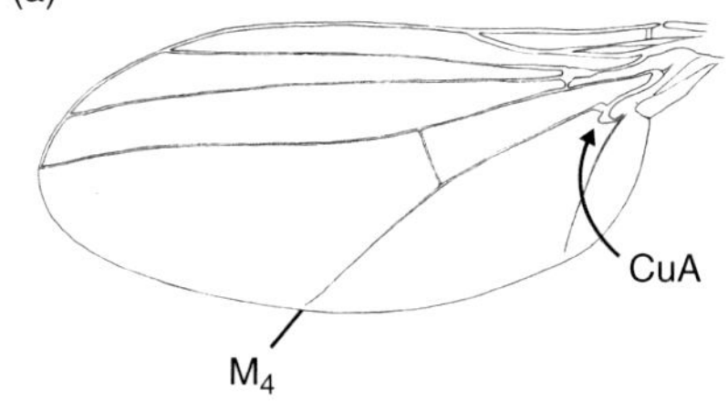

(aa)

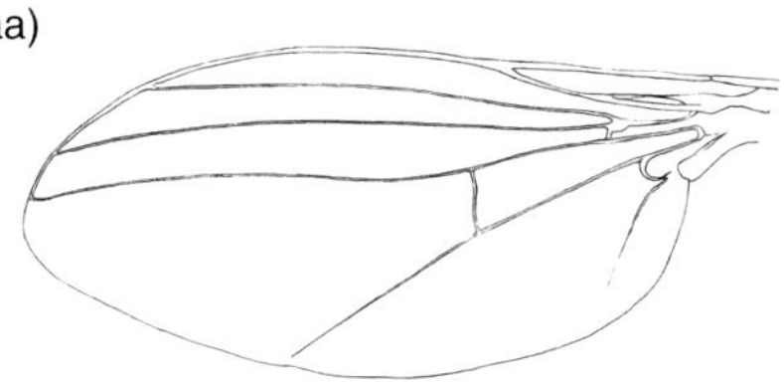

8(7) Mid tibia with 2 anterodorsal setae at a quarter and three-quarters along shaft, sometimes markedly different in size. Postpedicel small and roughly kidney-shaped (a).................................. .
.... ***blepharosceles***, ***collini***

– Mid tibia with only 1 anterodorsal seta at basal quarter. Postpedicel large and more triangular, lower margin almost straight (aa)... ***monochaetus***

(a)

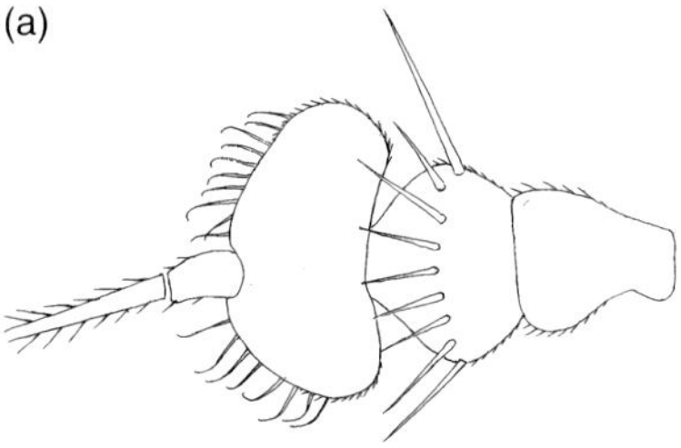

(aa)

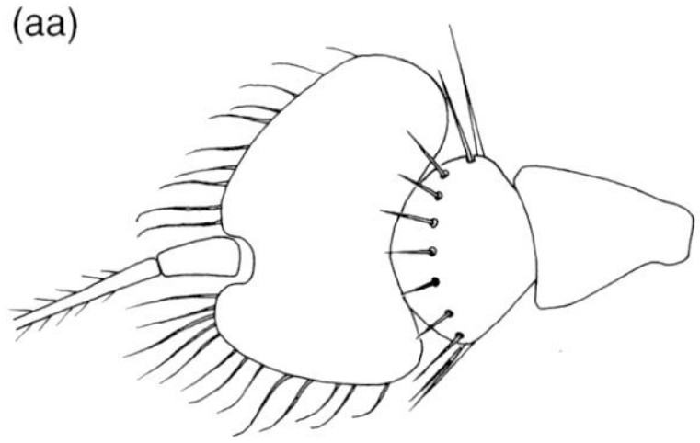

9(4) All femora mainly yellow, hind femur usually with a conspicuous black ring on apical third to half, front femur usually with a narrow dark dorsal streak, anterior and posterior faces yellow. Front coxa yellow on about apical half or more, black colour usually not reaching apical setae, partly pale-haired. 2–2.25 mm..***cilipes***

– Femora entirely black or with a wide black ring occupying most of shaft, yellow at base and tip; if paler, front femur with a broad smudge at about midpoint on lateral as well as on dorsal surface. Front coxa entirely or mainly black as far as apical setae, black-haired.. 10

10(9) Postpedicel large, rounded-triangular, about as long as deep but sometimes shorter (a), depth a quarter height of eye (b). Face broad, about two-thirds width of an eye in front view (c). ....***pulchellus*, Species A**

– Postpedicel small, 1.5–2 times deeper than long (aa), its depth one-fifth height of eye (bb). Face narrow, about half width of an eye in front view (cc).. 11

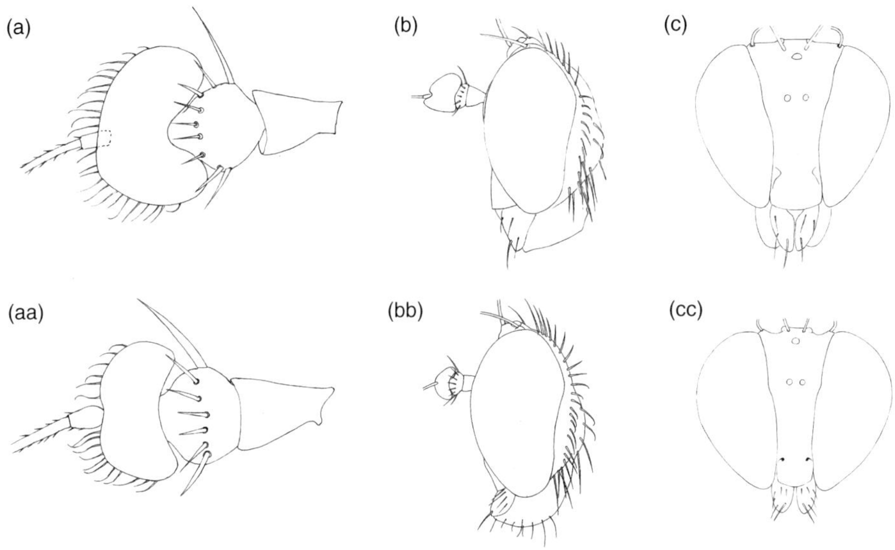

Note: The following two species are sometimes indistinguishable so three options are given:

11(10) Femora narrowly to widely yellow at base, colour continuous with yellow of trochanters. Hind tibia usually yellow at tip, colour continuous with yellow of basitarsus. Wing profile just proximal to M_4 at margin with a very slight bulge (a). .... ***neglectus***

– Femora black to junction with trochanter, usually contrasting strongly with yellow trochanters. Hind tibia usually with a slightly dusky tip, especially posteriorly; hind basitarsus either entirely dark or yellow basally. Wing profile where M_4 meets margin smoothly rounded (as figs in couplet 13) .... ***femoratus***

– Neither suite of characters fits well .... ***neglectus***, ***femoratus***

(a)

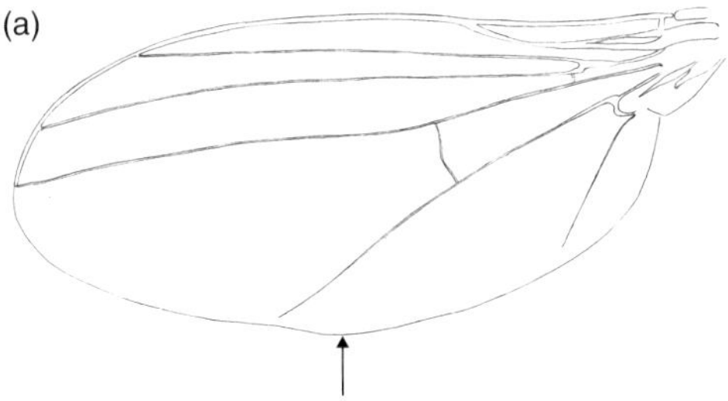

12(3) All coxae and femora black. 13

– Front coxa yellow, at least on apical half. Femora entirely or mainly yellow. 16

13(14) Apical section of M_4 shorter than basal section (about 0.8 times) measured from tiny looped CuA, and 2.5 times longer than dm-m (a). Postpedicel large and kidney-shaped, depth at least twice its length, upper hind margin reaching back to overhang pedicel (b). All trochanters dark or at least not bright yellow, viewed from below. Legs either entirely dark or front and mid tibiae yellow, although rarely bright yellow; overall appearance is of dark legs. .... ***laesus***

– Apical section of M_4 longer than basal section, about 1.1 times, rarely equal in length, and about 3–4.5 times longer than dm-m (aa). Postpedicel fairly large, usually as long as deep although may be shorter, rounded-triangular to more evenly rounded (bb). Front and mid trochanters pale or dark yellow. Legs variable with all tibiae entirely yellow to mainly dark. 14

(a)

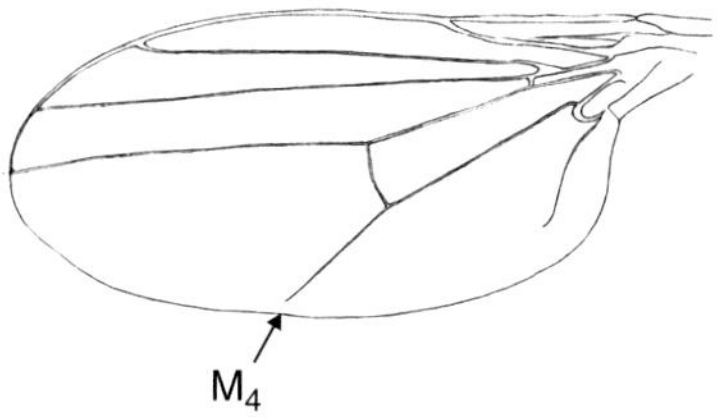

(b)

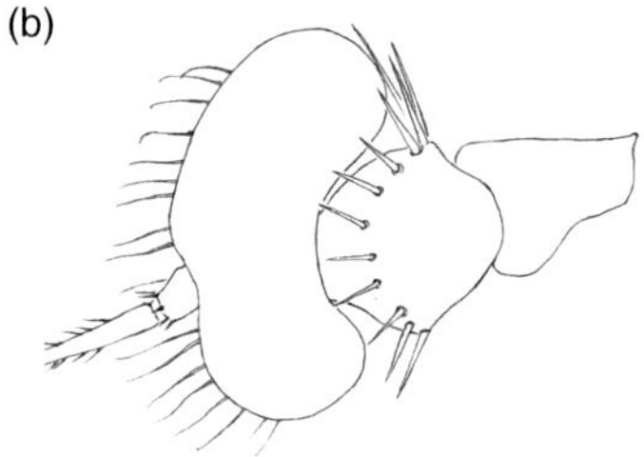

(aa)

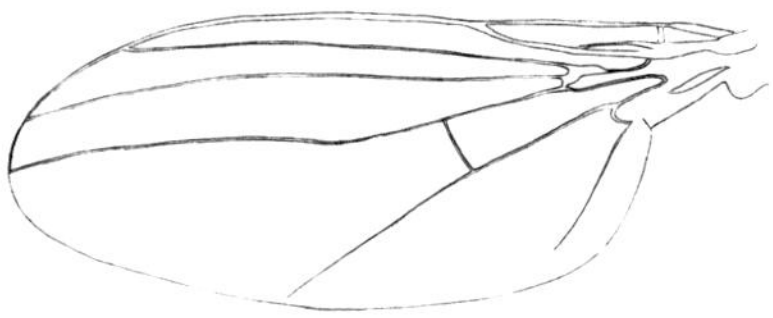

(bb)

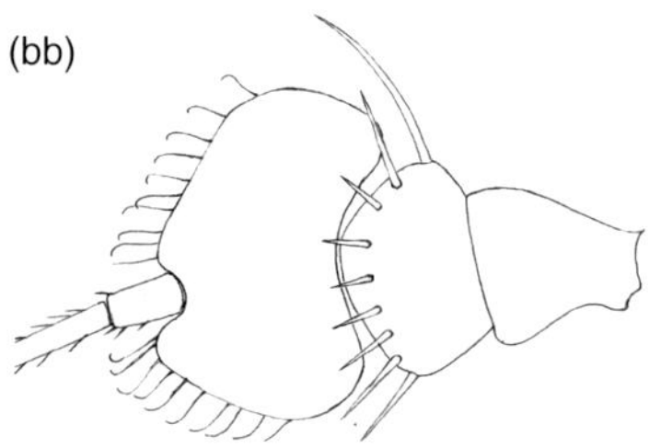

14(13) Mid tibia with 2 strong anterodorsal and 1–2 strong posterodorsal setae in approximate pairs (a, dorsal view). Front tibia with 1 fine anterodorsal and 2 fine posterodorsal setae. Wing membrane slightly brownish. Hind trochanter paler than base of femur. .... ***tricaudatus***

– Mid tibia usually with 1 strong anterodorsal, with or without additional short seta (ad or pd) (aa). Front tibia with only 1 short anterodorsal seta about as long as width of shaft; very fine posterodorsals may be present. Wing membrane pale. Hind trochanter as dark as base of femur. 15

(a)

(aa)

Note: The following two species are not always clearly separable so three options are given:

15(14) All tibiae clear yellow or sometimes hind tibia slightly darkened towards tip. Hind basitarsus yellow on basal half, dark apically. Mid tibia with at most 2 posterodorsal fine setae (as well as usual strong anterodorsal seta). .... ***suavis***

– Tibiae dark yellow to brown, entirely or in part. Hind basitarsus entirely dark. Mid tibia usually with additional posterodorsal fine setae in addition to 1 on basal quarter and 1 at about midpoint; sometimes additional small setae near strong anterodorsal seta. .... ***palustris***

– Intermediate. .... dark ***suavis***, pale ***palustris***

16(12) Hind femur with a more-or-less broad black ring near apex. Face entirely covered with dense dusting. Hind tibia with some longish black setae above. Small species, wing length 2.0–2.5 mm. .... ***cilipes***

– Femora entirely yellow. Face very thinly dusted so that metallic green colour is largely visible. Hind tibia with only short pale setae. Tiny species, wing length 1.0–1.25 mm. .... ***longipalpus***

Chrysotus angulicornis Kowarz, 1874

This species closely resembles *gramineus* in body and leg colour although the hind tibia can vary from mostly yellow to almost entirely dark. The knees of the front and mid legs are clearly yellow, but not those of the hind leg. The mid tibia has a strong anterodorsal seta at the basal quarter and another small one at about two-thirds, which is sometimes absent so resembling *monochaetus* which has a much larger postpedicel. Compared with *gramineus*, the tibial fringes are poorly developed, particularly those of the front tibia. The white palp is distinctive. The triangular postpedicel is distinct from the shorter more rounded profile of *collini* which it closely resembles. The phallus is particularly slender in lateral view, with two small almost symmetrical lateral flanges near the tip, superficially resembling that of *cilipes*. Probable females are known from association with males but it is not possible to be sure of their identity. They closely resemble *gramineus* from which they may perhaps be separated by the following characters although these remain untested and may not be unique to *angulicornis*: the frons, face and scutum are rather dull without a shine; the hind tibia has a slightly dusky tip and base (as many *gramineus*) and its colour is slightly more orange than clear yellow; the front coxa is narrowly orange at its tip below the apical setae (not nearly or completely black at tip); the palps are sometimes yellow (not white) at the tip; the hairs of the hind basitarsus are clearly shorter than the shaft's width. Wing length ♂: 2.0 (1.8–2.1) mm, ♀: 2.3 (2.2–2.5) mm.

Confirmed records of *angulicornis* as defined by Negrobov and Chandler (2006) are from Dorset, where they were recorded from soft-rock cliff seepages, and East Sussex from a disused clay pit with ponds (Drake 2020a). Flight: vi–vii.

Chrysotus blepharosceles Kowarz, 1874 (Plate 2)

This is a large dark species with small antennae. Its legs are particularly hairy; all femoral fringes are at least as wide as the width of the femora seen from above, the front tibia is fringed on all faces except anteriorly, and the hind tibia's anterior fringe is at least as wide as the shaft and extends onto the basitarsus. At first glance the legs appear entirely dark but the front trochanter and tibia, and sometimes the mid tibia, are often brown rather than black so may resemble dark specimens of several other species, such as *femoratus*, *palustris* and *suavis*. No other male has such a pronounced narrow tip on sternite 8. The phallus tip has an almost rectangular ventral plate, a right lateral arm protruding at a marked angle and which can be seen when the phallus tip is visible, and a short stout tip to the ejaculatory duct. Females stand out as a large species with small antennae and entirely black hind legs contrasting with yellow front and mid tibiae. This is identical to *collini* from which it cannot be separated. In occasional specimens the hind tibia is vaguely brown or very dark yellow, usually the same shade along the entire length; very dark *gramineus* may overlap with this condition and are indistinguishable. Wing length ♂: 2.3 (2.1–2.5) mm, ♀: 2.7 (2.5–2.9) mm.

Chrysotus blepharosceles is widely distributed in England but common only in the south, scarce in Wales and probably absent from Scotland. It occupies a wide range of habitats which are usually wet to damp but do occasionally include dry heathland and chalk grassland. There is no common factor apart from the vegetation being moderately tall, from rank grassland and scrub to woodland. The species is scarce in exposed wetlands where tall cover is absent. Flight: iv-x.

Chrysotus cilipes Meigen, 1824

This small species is the only British *Chrysotus* with pale legs and strongly bicoloured hind femur and tibia, each with a broadly black tip. The front and mid legs have short hairs and setae and no fringes, but the hind tibia has a dense anterior fringe almost twice the shaft's width. The hind basitarsus has unique hairing, denser at either end which gives a dumbbell appearance. Some individuals have dark dorsal streaks on front and mid femora, and these will run to *logvinovskii* Negrobov and Tsurikov, 2000 in the key by Negrobov *et al.* (2000). The phallus tip has almost symmetrical lateral arms that appear to form a ring at the base of the short slender tip of the ejaculatory duct. Females are conspicuous by their mainly yellow legs with a wide black tip on the hind femur. Sometimes specimens in which more than half of the mid and hind femora are dark are not clearly separable from paler *neglectus* although the smaller size of *cilipes* usually helps to distinguish it. Wing length ♂: 1.8 (1.7–2.0) mm, ♀: 2.3 (2.1–2.5) mm.

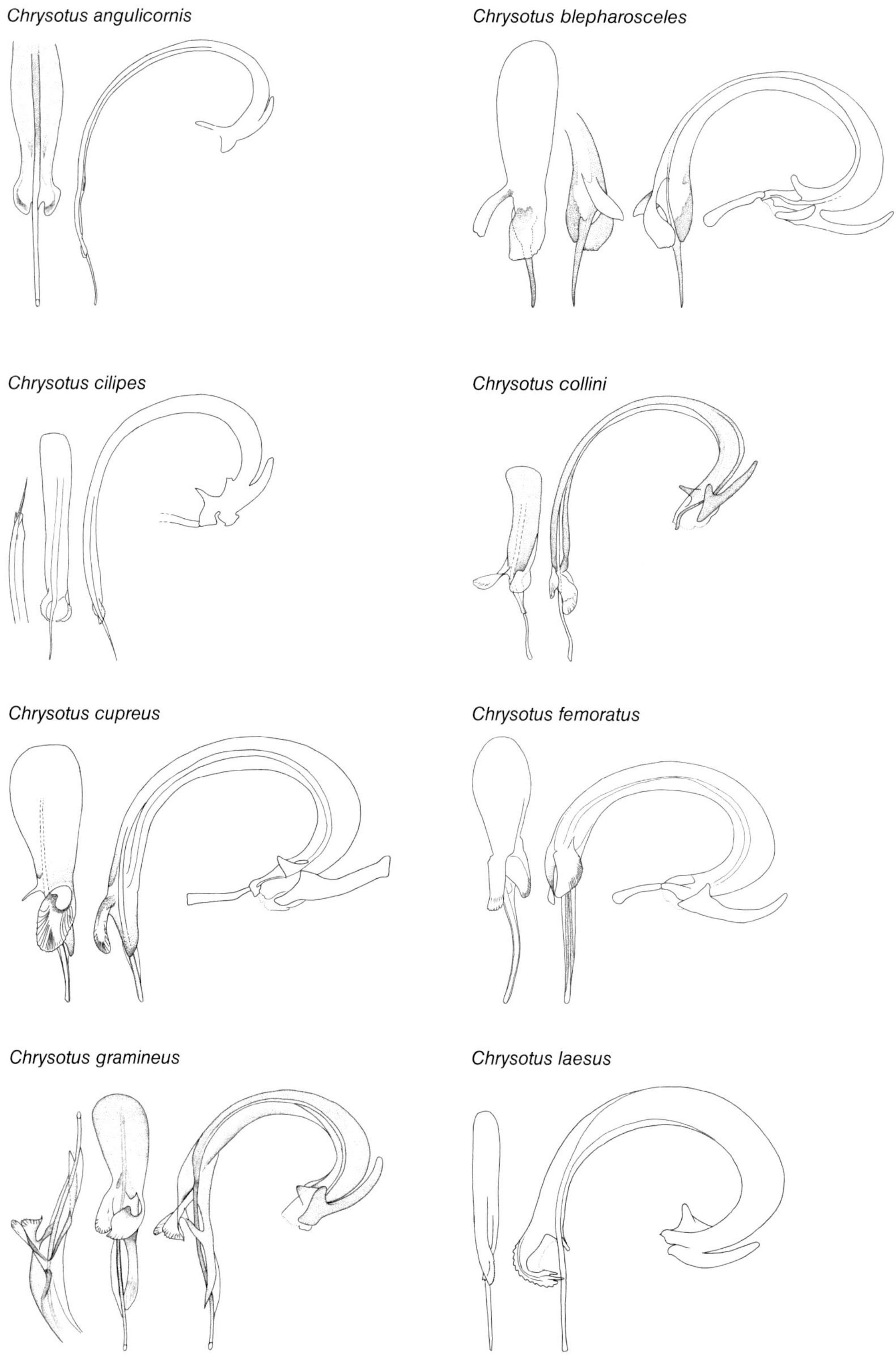

Fig. 12. *Chrysotus* phalli, excluding *melampodius*, *verralli* and *longipalpus*.

Chrysotus monochaetus

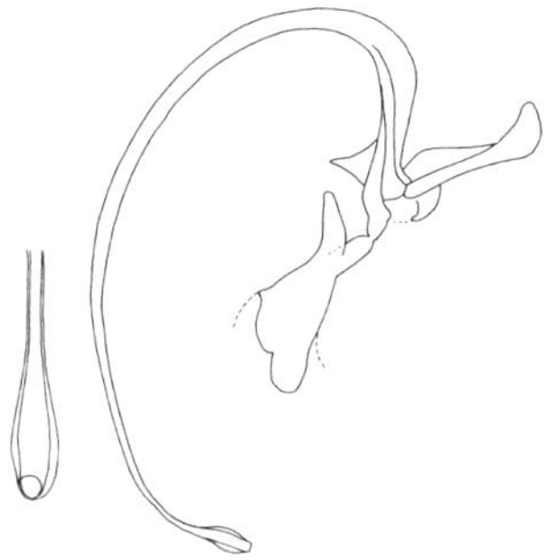

Chrysotus neglectus

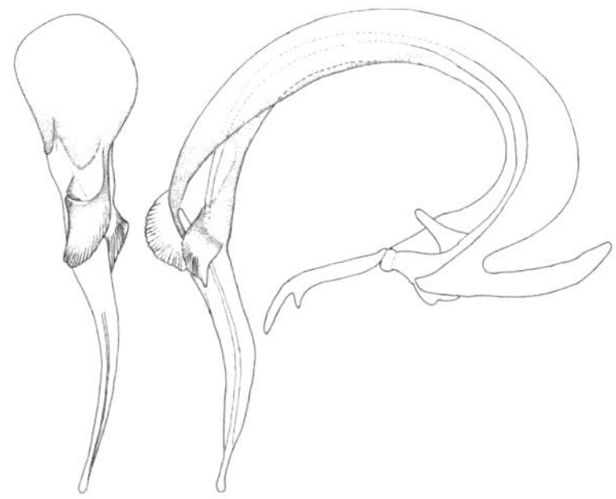

Chrysotus obscuripes

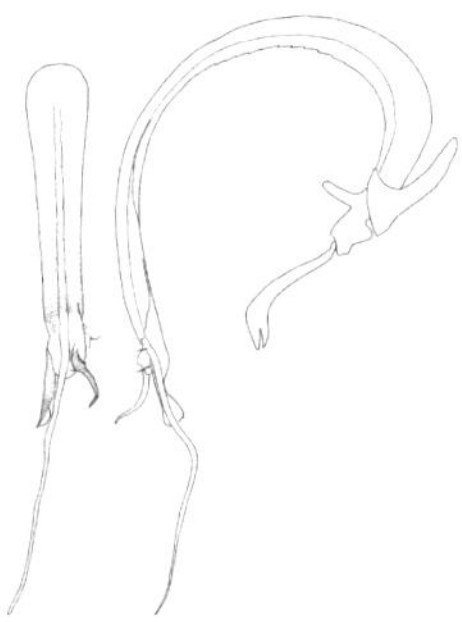

Chrysotus palustris

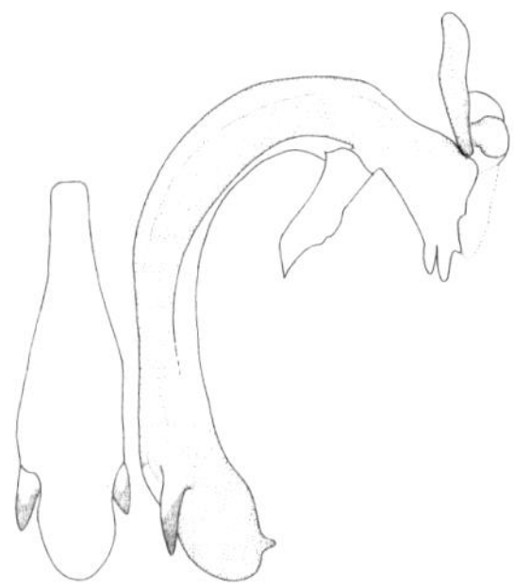

Chrysotus pulchellus

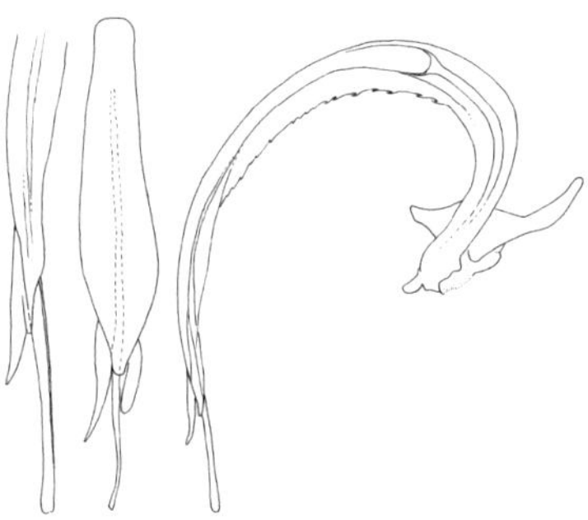

Chrysotus suavis

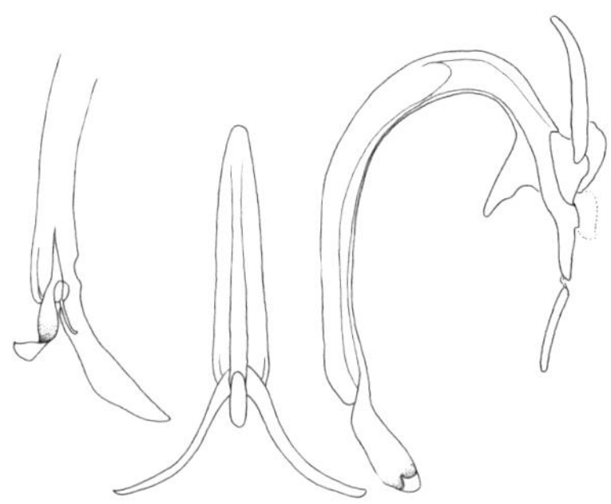

Chrysotus tricaudatus

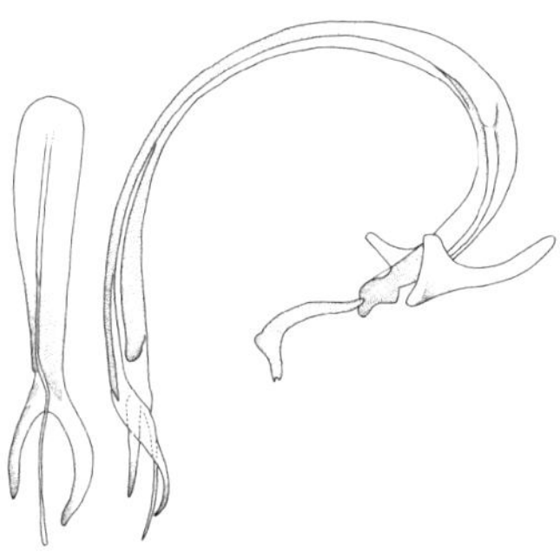

Chrysotus Species A

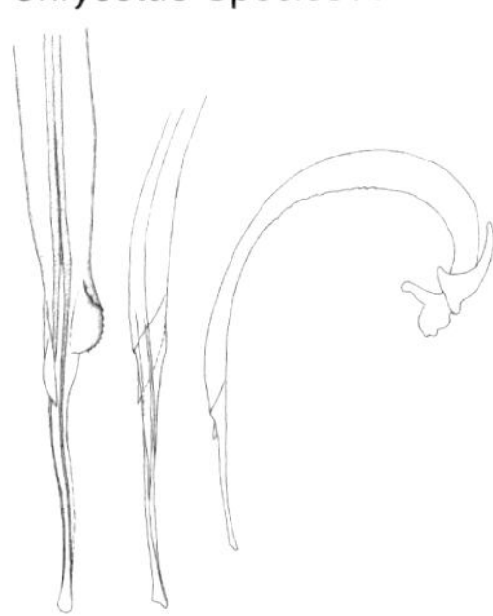

Fig. 12. (*Continued*)

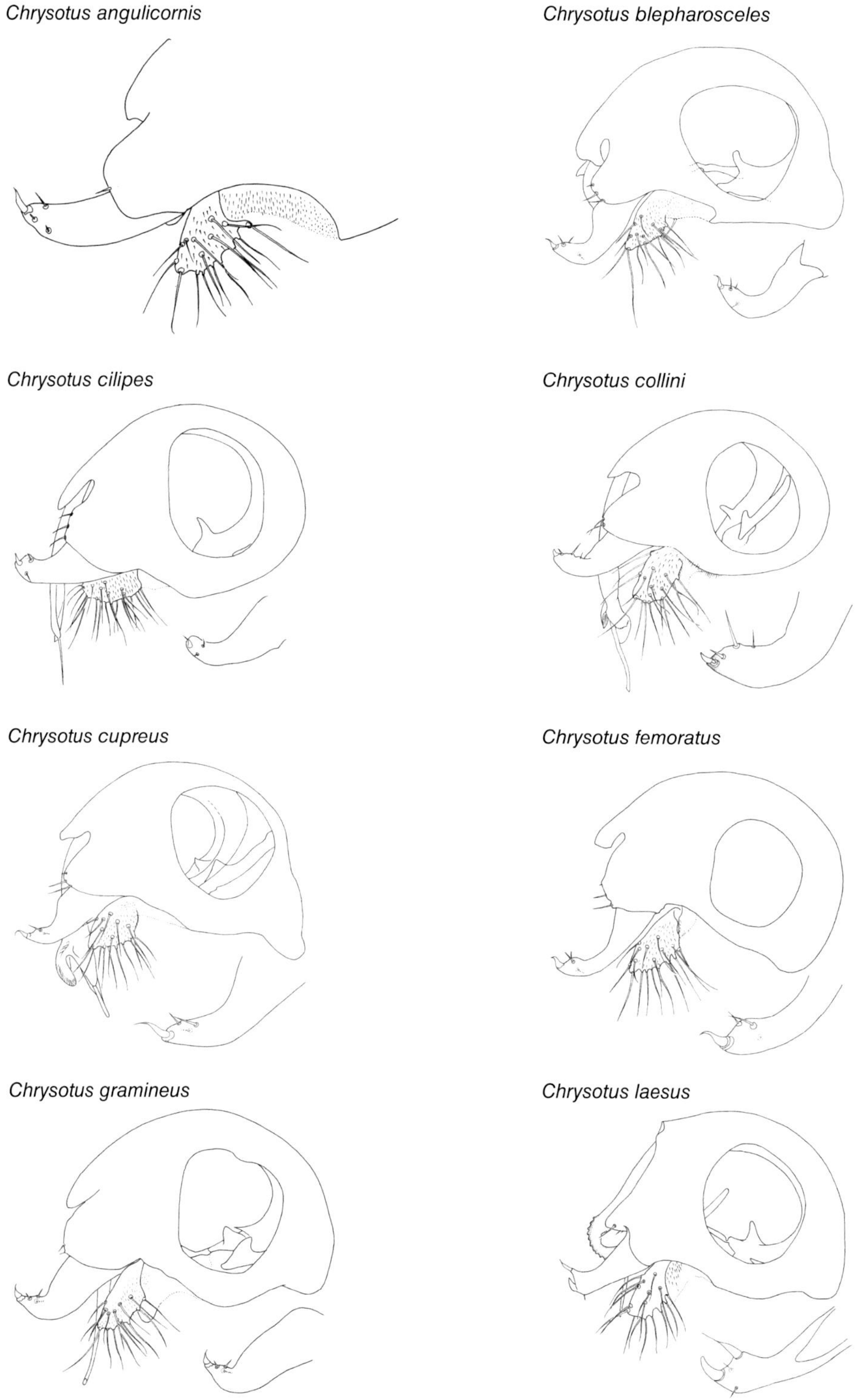

Fig. 13. *Chrysotus* hypopygia, excluding *melampodius*, *verralli* and *longipalpus*.

Chrysotus monochaetus

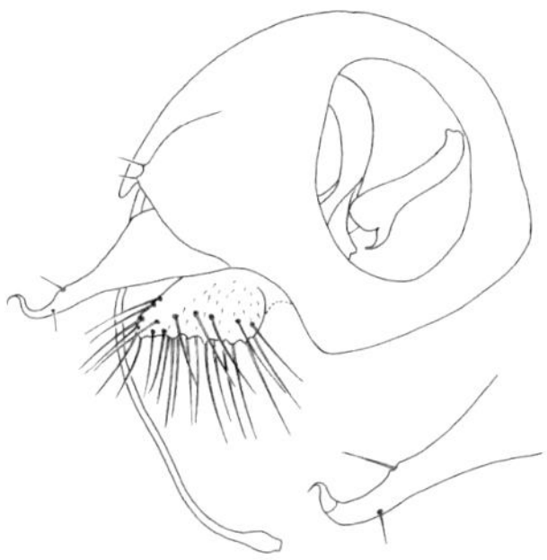

Chrysotus neglectus

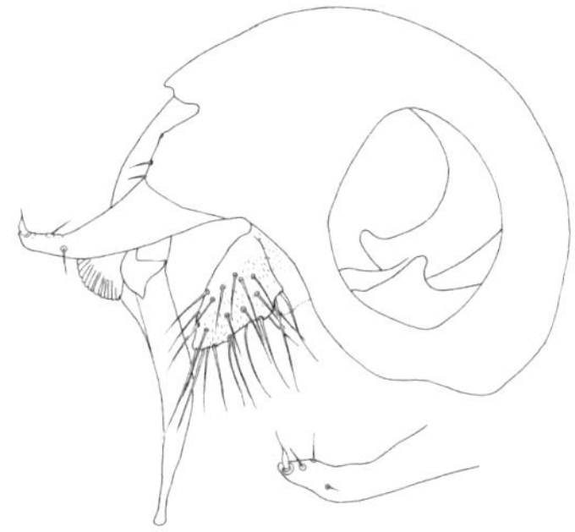

Chrysotus obscuripes

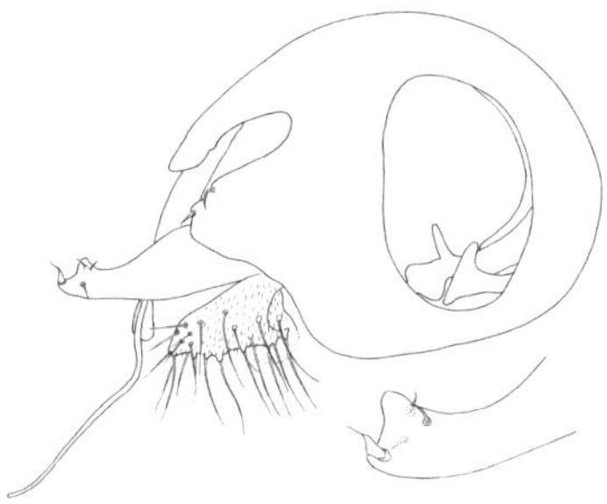

Chrysotus palustris

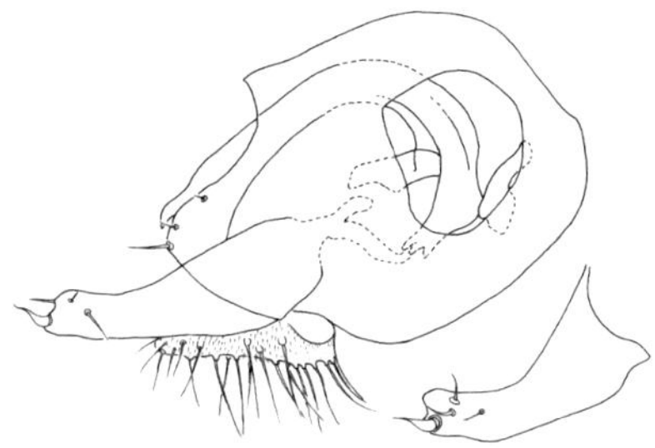

Chrysotus pulchellus

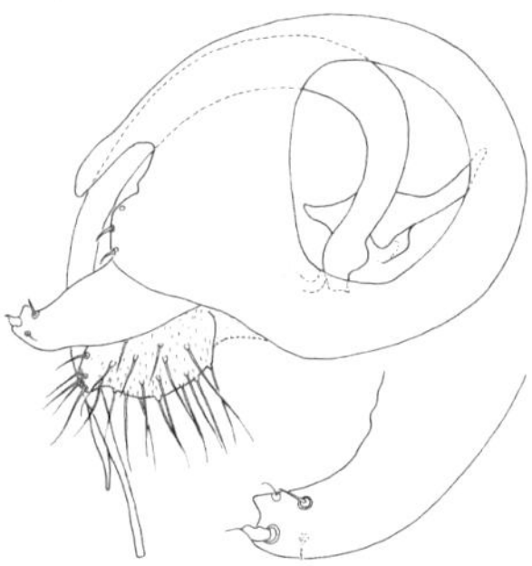

Chrysotus suavis

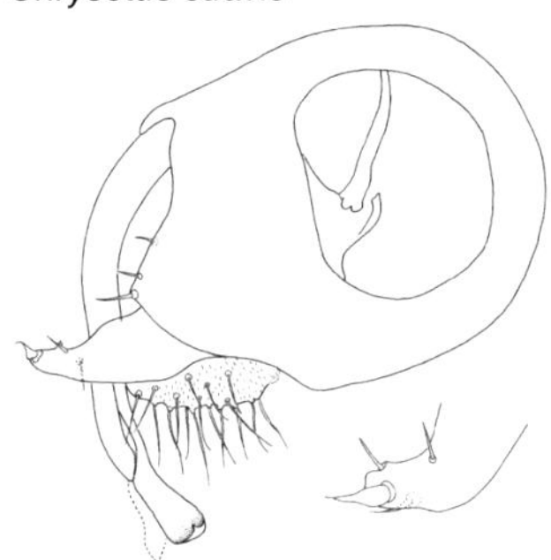

Chrysotus tricaudatus

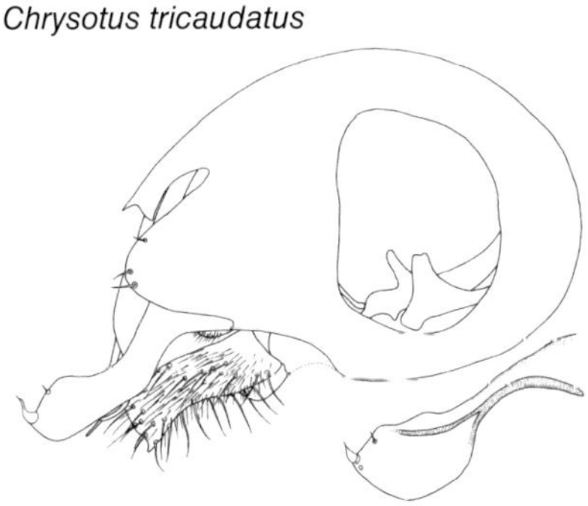

Chrysotus Species A

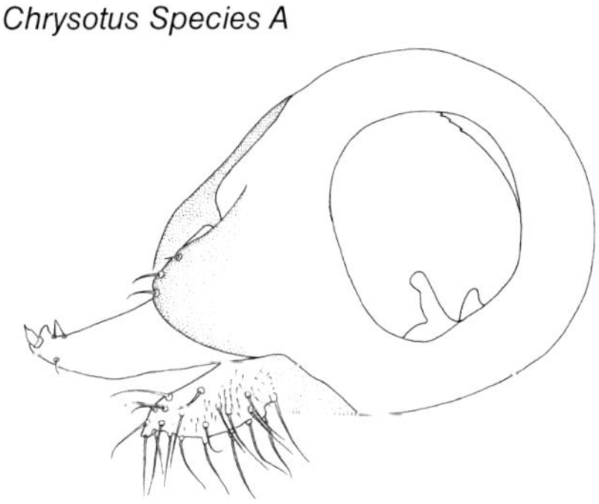

Fig. 13. (*Continued*)

Chrysotus cilipes is widespread in Britain, occurring as far north as Fair Isle and Orkney. It is most often associated with margins of rivers, streams and sometimes seepages, and is a consistent member of the assemblage of exposed riverine sediments. It is also found at margins of still water such as ponds, dunes slacks and ditches in wet grasslands, particularly those on peat, and sometimes in fens. It is frequently found on base-rich sites and is scarce on acid sites except when the water is flowing. Sites are usually unshaded. Flight: v–x, peak in vii.

Chrysotus collini Parent, 1923

This medium-sized species is very similar to *gramineus* and *angulicornis*. Leg colour and development of femoral and tibial fringes can overlap with some specimens of *gramineus*, so it is usually essential to check the small but distinct difference in the setae on sternite 8. As well as the characters used in the key, *C. collini* differs from *angulicornis* by its slightly more widely separated eyes, leaving a strictly parallel pale grey strip on the lower half of the face. With minimal manipulation, the ventral surstylus can be seen to be almost equally thick to its blunt tip in contrast to the tapered tip of *gramineus* and *angulicornis*. The phallus tip has a prominent left arm, superficially resembling that of *gramineus* but is distinguished by the tip of the ejaculatory duct projecting relatively further from the enfolding sheaths, and this is often distinct without having to prepare a mount. d'Assis-Fonseca (1978) had female *collini* in the group with yellow hind tibiae, along with *gramineus* and *angulicornis*, but this is incorrect. They appear to be inseparable from *blepharosceles*, based on large populations where *blepharosceles* and *gramineus* were scarce; they have a dark hind tibia and relatively small antenna that is about a fifth the height of the eye. Wing length ♂: 2.2 (2.0–2.4) mm, ♀: 2.6 (2.2–2.8).

Chrysotus collini is sparsely recorded in coastal areas from East Anglia to North Wales, and occasionally inland in lowland England. Reliable determinations are from coastal sites on clay, including brackish grazing marsh and sea walls, inland clay pits of which one is brackish, and herb-rich grassland. For these confirmed records, the common factor appears to be summer-parched grasslands on clay. Flight: vi–vii.

***Chrysotus cupreus* (Macquart, 1827)**

As a large dark species with small antennae, *cupreus* can be confused with only *blepharosceles* and *verralli*, but only *verralli* has similarly pale front coxae and trochanters. Most fringes on the femora and tibia are only moderately developed, less so than in *blepharosceles*, and the tibial setae are rather short, especially on the hind leg. The tip of the ejaculatory duct is short and stout, about as long as the leaf-shaped right arm. Females are dark-legged with rather dark yellow to dark brown front and mid tibiae but with conspicuously contrasting nearly white front trochanter and tip of the coxa. Wing length ♂: 2.5 (2.3–2.7) mm, ♀: 2.9 (2.3–3.3) mm.

Chrysotus cupreus is recorded widely but sparsely in England and Wales, with Scottish records for the island of Arran. Reliable records are from woodland rides and glades, hedgerows in pasture, ponds, gravel pits and river shores. Flight: v–viii.

***Chrysotus femoratus* Zetterstedt, 1843**

This is a medium-sized species with small antennae and bicoloured legs in which all trochanters, front and mid tibiae and basal half of the basitarsi are usually clear yellow, giving the front leg a characteristic black-yellow-black-yellow alternating pattern. Sometimes the legs can be quite dark and do not show this pattern so may be confused with *gramineus* as both these species have a conspicuous anterior fringe on the hind tibia that runs onto the basitarsus, making the hind tarsus look superficially thick. The front tibia usually has moderately dense short hairs on all faces except anteriorly. There are usually only 6–7 acrostichal setae in each row, distinguishing it from *neglectus* which has 8–11 pairs. The phallus tip is stout and projects well beyond the two broad basal arms. Females can be confused with those of dark *neglectus*. Wing length ♂: 2.1 (2.0–2.4) mm, ♀: 2.7 (2.4–2.9) mm.

Owing to the confusion in females with those of the common *neglectus*, the distribution is uncertain, particularly as *femoratus* is clearly far less common than *neglectus*. The few confirmed records of males are from England, Wales and the Scottish Highlands. Habitats where males have been recorded include grassland areas within woodlands, scrubby dunes systems, soft-rock cliffs, gravelly river shores and saltmarsh. Flight: v–xi, peak in vii.

Chrysotus gramineus (Fallén, 1823) (Plate 2)

This is a variable species but typical *gramineus* have clear yellow front and mid tibiae, black hind tibiae, brown or black hind trochanters, and front tibiae with black ventral vestiture which is scarcely longer than on other faces. A character unique to *gramineus* is the short stout apical setae with forked tips on sternite 8 which can be seen at ×40 magnification without special preparation. Short stout setae on sternite 8 are found in *blepharosceles*, but with simple tips, and *neglectus*, with blunt ragged-ended tips. Pârvu (2001) illustrates *Diaphorus unguiculatus* Parent, 1925 with forked setae on sternite 8 identical to those of *gramineus* but this may be based on a misidentification as the hypopygium resembles that of *Chrysotus blepharosceles* and the wing and antenna are also that of a *Chrysotus* rather than a *Diaphorus*. The phallus of *gramineus* is unusual in having a long dorsal extension as well as the usual two shorter projecting lateral arms beyond which it extends some distance, and consequently the tip of the ejaculatory duct appears to project only a short way beyond these supporting structures. This can be seen if the tip is visible in an unprepared specimen. The highly variable leg colour, fringe development and antennal shape have led to several species being split from *gramineus* only to be synonymised later as their genitalia are indistinguishable. The typical form can be confused with pale-legged *blepharosceles* but differs in having yellow palps, marginally narrower anal lobe about the width of either cell r_1 or r_{2+3} compared to about 1.5 times in *blepharosceles*, and the calypter hairs not so completely black. A second consistent form, possibly included within Kowarz's '*varians*', has the same shade of yellow on the hind tibia and basal half of the basitarsus as on the front and mid tibiae; no British species has all the femora dark and all tibiae yellow so this form cannot be confused with other species. It also has a slightly narrower fringe on the hind tibia than the typical form, and the ventral hairs of the front tibia are usually brownish rather than dark. A third variant has longer ventral hairs on the front tibia, usually large front pulvilli, and long antennal pubescence. Females in which all the tibiae are yellow, the hind trochanters dark and the postpedicel is small are nearly always *gramineus*; only the rare *angulicornis* is similar as described in the account for that species. Wing length ♂: 2.3 (1.9–2.6) mm, ♀: 2.6 (2.3–2.9) mm.

Chrysotus gramineus is a common species through much of England and Wales but relatively scarce in Scotland where records extend to the Outer Hebrides. There seems to be no habitat between saltmarsh and upland bogs where it cannot be found, whether wet or dry, shaded or open, although it is most abundant in unshaded grasslands. Flight: iv–x, peak in vi–vii, males dominant earlier, females dominant to exclusive later.

Chrysotus laesus (Wiedemann, 1817)

This is the most easily distinguished of British *Chrysotus*, being a small to medium-sized species with entirely dark legs (the front tibiae sometimes pale brown), a shining violet or blue scutum, a wide face, conspicuously large kidney-shaped antennae, and long forward-pointing palps. A violet scutum is found otherwise only in *suavis*, a wide face together with kidney-shaped antenna in *Melanostolus*. The phallus tip is unusual in having a strongly curled thick tip, serrated on its outer side, which can be seen in profile with the minimum of preparation. The surstylus has two stout apical setae that, at low magnification, give it the appearance of a spanner which is easy to pick out in wet-preserved material. Females are unique among British *Chrysotus* in having the apical section of M_4 obviously shorter than the basal section; the small dark species *Melanostolus melancholicus* and *Diaphorus nigricans* are similar but have a different postpedicel and arista position. Wing length ♂: 2.0 (1.2–2.1) mm, ♀: 2.3 (1.7–2.5) mm.

The distribution is predominantly southern England and on the west side of Britain to Cumbria, with an isolated population in the Breckland of East Anglia and scattered records in the northeast. A single old Scottish record has not been confirmed. This is a species of unshaded rough grassland and grassy heathland, from acid to base-rich, and often but not invariably dry. Grass tussocks may be a key feature. Flight: v–viii, peak in vi.

Chrysotus longipalpus Aldrich, 1896

This tiny non-native originating probably from Hawaii has almost completely yellow legs, coxae and antenna and exceptionally long tapered palps. It has been found in heated greenhouses at the botanic gardens at Cambridge and Kew (London) and at a nearby site, and in Manchester (identified as *pallidipalpus* van Duzee) (Parmenter 1942). Wing length ♂: 1.4 (1.2–1.5) mm, ♀: 1.7 (1.5–1.8) mm.

Chrysotus melampodius Loew, 1857

Based on the descriptions by Kowarz (1874) and Parent (1938), *melampodius* has a yellowish brown front trochanter, dark mid and hind trochanters, dark tibiae with pale knees on the first and second legs, short tibial fringes, small antennae and dark palps with pale tips. It thus resembles a small *blepharosceles* but with far shorter tibial fringes and small differences in colour. Wing length ♂: about 2 mm.

The records, if correct, suggest a very sparse but wide distribution in England and Wales, and a single record in central Scotland. Few records post-date 1980. Flight: probably vi–vii.

Chrysotus monochaetus Kowarz, 1874

There is a close resemblance to *pulchellus* in the large postpedicel, rather short fringes on the tibiae and general leg colouration but *monochaetus* differs in having only one anterodorsal seta on the mid tibia (two in *pulchellus*), dark hind trochanters (not yellow) and the postpedicel being more elongate and pointed below the arista. The ejaculatory duct is unmistakable in being exceptionally long and thin with a small bulbous tip but without any lateral support structures; the surstylus is finely tapered. Females similarly resemble *pulchellus* but are distinguished by the same set of characters as for males. Some other species have only one anterodorsal seta on the mid tibia; the specific name could mislead. Denninger (1950) treated *monochaetus* as a junior synonym of *viridifemoratus* von Roser, 1984, but Dyte (1993a) suggested that this may be unreliable because the type material examined by Denninger consisted of two females. Dyte's opinion is followed here. Wing length ♂: 1.5 (1.4–1.7) mm, ♀: 2.1 mm.

The few records, all pre-1990, are strung out across southern England. No habitat affinities can be deduced from the records but Pollet *et al.* (2009) give banks of rivers and pools, and reed marshes. Flight: probably vii–viii.

Chrysotus neglectus (Wiedemann, 1817)

Among native species, this is unmistakable as the only large one with almost entirely yellow legs and small antennae. The tiny non-native *longipalpus* has yellow legs but otherwise differs markedly. There are more acrostichal setae (usually 10 pairs, range 8–11) than in most species (6–8 pairs). The apical setae of sternite 8 are short, stout and have ragged ends that appear broken and are visible at ×40 magnification without preparation. Colour variation is modest, and includes the front coxa being mainly dark with brownish hairs, and the hind femur having an apical dark mark although never as extensive as in *cilipes*. Rarely the femora are so dark as to resemble *femoratus* from which *neglectus* can be distinguished by its shorter fringe on the hind basitarsus and wide anal lobe with a sharp anal angle (narrow and with shallow anal angle in *femoratus*). The tip of the phallus is thick and strongly tapered, and with a slight curve in both ventral and lateral view. It is similar to but stouter than that of *femoratus*. Females are usually distinct, being fairly large with a small postpedicel, having extensively yellow bases to the legs, and all tibiae being yellow. Occasional pale specimens resemble *cilipes* which is much smaller, and dark specimens may be inseparable from *femoratus*. The slight bulge on the wing's hind margin is discernible in most specimens when the wing is flat but sometimes it is scarcely evident in specimens that otherwise appear to be *neglectus*. Wing length ♂: 2.5 (2.3–2.8) mm, ♀: 3.0 (2.4–3.2) mm.

Chrysotus neglectus is a fairly common species in most of England and lowland Wales, becoming scarcer northwards and probably genuinely rare in Scotland. It is mostly found in shaded damp or wet places, such as at woodland streams and glades, well-vegetated shaded river margins, and scrubby fen and grasslands. It is not confined to such places and can sometimes be found in open habitats and dry woods. Flight: v–xi, peak in vi–vii.

Chrysotus obscuripes Zetterstedt, 1838

This is a dark, medium-sized, black-legged species with narrowly yellow knees on the front and mid legs, and large antennae in which the postpedicel forms a hood over the pedicel. The hind tibia has a moderately developed anterior fringe but this does not continue onto the basitarsus. All wing veins are noticeably straight; in particular M_4 is absolutely straight either side of the crossvein and M_1 lacks the usual slight but obvious inflexion beyond the crossvein. The face is similar to that of *collini* with the eyes narrowly but clearly separated on the lower half. The tip of the ejaculatory duct is particularly hair-like and wavy, and can be seen with the minimum of preparation. Females also have a large postpedicel and similarly dark legs with the front and mid tibiae often brown but never yellow. It superficially resembles *laesus* which has pale hairs and setae on the front coxa, and *blepharosceles* which has a small postpedicel and narrower face. Wing length ♂: 2.1 (1.9–2.3) mm, ♀: 2.6 (2.4–2.8) mm.

Chrysotus obscuripes occurs on strongly acidic areas in Wales, western England with outlying populations in Yorkshire, and probably in much of Scotland. It is confined to acid mire, wet heath and wet moorland. Two records given by d'Assis-Fonseca (1978) are sites on the chalk and seem highly improbable. Flight: v–ix.

Chrysotus palustris Verrall, 1876

This is a small dull greenish grey species with the abdomen rather more evenly cylindrical than in most *Chrysotus*. The antennae are moderately large but not outstanding. The front and mid tibiae are dark yellow to almost black, and the front and mid trochanters dark yellow but clearly paler than the black hind trochanter. All hairs are short and the fringes are scarcely developed. Like *obscuripes*, the eyes are clearly separated and the palps protrude conspicuously although are mainly white or yellow rather than black. The phallus is unusually thick throughout its length, with its tip broadly rounded and flanked by two small lateral arms. Females are very similar to those of *suavis*, with confusion arising with pale-legged specimens that are indistinguishable from dark-legged *suavis*. It is very likely that Verrall's *palustris* is the same as *albibarbis* Loew, 1857 which has identical genitalia but a bluer body. Wing length ♂: 1.9 (1.8–2.1) mm, ♀: 2.3 (2.1–2.5) mm.

The poor distinction between the females of this species and *C. suavis* has probably resulted in many wrong records so the distribution is not known although, based on males only, it extends from the southern English coast to the Scottish Grampians, including Wales. Habitats recorded for males include coastal dune slacks, the backwater of a stony river, and pools on disused inland clay pits which are mildly brackish. Flight: v–ix, peak in iv.

Chrysotus pulchellus Kowarz, 1874

This small fly has brightly marked legs and conspicuously large antennae. Contrasting with the dark green-black of the femora is the bright yellow of all trochanters, front and mid tibiae and their basitarsi bases, tips of the femora and often up to the basal third of the hind femora. The hind tibia may sometimes be paler than its usual black colour. Most leg hairs are short and only the fringe on the hind tibia is as wide as the shaft's width, reaching weakly onto the basitarsus. This species resembles Species A and differs externally in having dark coxal hairs (pale in Sp A) and little difference in the thickness of the mid and hind tibia (hind tibia markedly thicker in Sp A). The phallus is also rather similar but in *pulchellus* the right-hand arm near the tip splays away from the ejaculatory duct, whereas it lies close to the duct in Sp A. Possible confusion with *monochaetus* is discussed under that species. Females of *pulchellus* and Sp A appear to be indistinguishable but both are uniquely characterised by black setae on the front coxa, yellow trochanters and a large postpedicel. The colour of the apical setae of the front coxa needs to be checked as the hairs may be pale. Wing length ♂: 1.9 (1.8–2.0) mm, ♀: 2.2 (2.0–2.4) mm.

As *pulchellus* has been confused with Species A, the distribution is uncertain but, as there is a marked discontinuity between records in Wales and England as far north as Yorkshire and those in the northern half of Scotland, it is assumed that the English and Welsh records are *pulchellus* and the Scottish records are Sp A. From Hampshire to north Wales, nearly all the sites are coastal but confirmed sparse inland records occur within 50 km of the coast. The coastal habitat is wet slacks on dunes and inland habitats are acid mire and wet heath, raised bog and moorland streams and ponds. Flight: v–viii, peak in vii.

Chrysotus suavis Loew, 1857

Chrysotus suavis is an attractive small species with a violet scutum, similar to that of *laesus* except that it is pale-dusted. Unique to *suavis* are the pale hairs on almost the entire body and legs, as well as the setae on all coxae, and the frons having a central green area flanked by white-dusted bands of the same width. No leg fringes are well developed apart from the anterior fringe of the hind tibia. The unusual Y-shaped tip of the phallus can usually be seen without special preparation. The mid and hind tibiae vary in colour. Females are usually distinct in having pale hairs on the front coxa, all femora black and all tibiae pale yellow. Problems arise with darker-legged specimens that are inseparable from paler-legged *palustris*. Wing length ♂: 1.9 (1.8–2.1) mm, ♀: 2.4 (2.0–2.6) mm.

The species is distributed widely but sparsely in England and Wales, but is very rare in Scotland where the most northern record borders the Moray Firth. The key feature of the habitat is bare or only sparsely or shortly vegetated well-drained damp to wet ground. *Chrysotus suavis* is consequently found on soft-rock cliff seepages, the shores of sandy and gravelly rivers, dune slacks, gravel pits, and occasionally saltmarsh and heavily grazed wet pastures. It is a consistent member of the assemblages of soft-rock cliffs and exposed riverine sediments. Flight: iv–ix, peak in vi–vii.

Chrysotus tricaudatus Negrobov, Barkalov and Selivanova, 2014

This is a medium-sized species with moderately large antennae. Its unique feature is the fringes of long pale fine hairs on the front femur and tibia, and mid femur. Among species with white setae on the front coxa, *tricaudatus* most resembles *suavis* in leg colour (front and mid tibiae yellow, hind tibia dark but variable). The clubbed ventral surstyli and tridentate tip of the phallus are unmistakable. Females are distinct among species with pale hairs on the front coxa in having 3–4 strong setae on the mid tibia rather than only one, and more than a single fine seta on the front tibia. Wing length ♂: 2.1 (1.8–2.2) mm, ♀: 2.7 (2.6–3.1) mm.

The species is known from two nearby areas on saltmarshes of Morecambe Bay, Cumbria (Drake 2020a). Elsewhere it is known only from the type specimens collected in northern Siberia north of the Arctic Circle (Negrobov *et al.* 2014). Flight: vii.

Chrysotus verralli Parent, 1923

This species resembles *cupreus* in its pale front coxa, which is reddish yellow rather than dirty white, whereas all other *Chrysotus* have dark front coxae. However, in specimens collected on the same occasion as the holotype (Isle of Wight, 1885) this paler colour in *verralli* is far more restricted than described by Parent. The hind tibia has strong setae and a fringe about 1.5 times the width of the shaft, similar to the condition in *blepharosceles*. The female remains undescribed.

There are very few records which are widely scattered from Cornwall to Cambridgeshire, and Cumbria. There is no information on the habitat. Flight: vii.

Chrysotus Species A

This small species closely resembles *pulchellus*, particularly in the large antennae and strongly contrasting alternating black-and-yellow leg pattern in which all trochanters and the basal fifth of the hind femora, front and mid tibiae and most of their basitarsi are bright yellow. It differs in having pale hairs and setae on the front coxa (black in *pulchellus*) and in the hind tibia being markedly thicker than the mid tibia. The phallus is very similar to that of *pulchellus*, particularly in the unusual serrated inner curve of the shaft within the capsule, but the ejaculatory duct projects much further beyond the lateral arms, the right arm lies close to the duct (splays out in *pulchellus*) and the left arm is shorter and more leaf-shaped (longer and narrower in *pulchellus*). Females taken in association with males appear to be inseparable from *pulchellus* although the hind tibia of Species A does appear to be marginally wider than the mid tibia, as in the male. Wing length ♂: 1.9 (1.8–2.1) mm, ♀: 2.4 (2.1–2.5) mm.

The species has been recorded in mid Wales and Northumberland in England but mainly in the northern half of Scotland where it is probably moderately frequent. The known sites are exposed sediments of stony rivers. Flight: vi.

CORINDIA Bickel, 1986 (Medeterinae)

Corindia are tiny green moderately dusted flies with yellow legs (Plate 2). The concave occiput, round postpedicel with apical arista, near absence of setae on the tibiae, and neatly aligned vestiture of the tibiae place them within the Medeterinae. They closely resemble *Thrypticus* in having two lateral setae on the hind coxa, pale thoracic setae (at least when viewed from in front) and an exceptionally short ocellar triangle on which the ocelli are nearly in a straight line, but differ in having a yellow scape and pedicel (as in a few *Medetera*), a single strong anterodorsal seta on the mid tibia and, in the female, an unusual ovipositor (see ♀ genus key couplet 53). Males differ from *Thrypticus* in the surstyli being fused with the epandrium (a clear articulation is visible in *Thrypticus*) and the cercus having an apical extension that is as long as the basal part of the cercus (absent or short in *Thrypticus*) (see ♂ genus key couplet 48). Bickel (1986) provides a detailed description of the genus.

Corindia was recognised in Europe when Parent's *Thrypticus viridis* was first discovered in Britain and moved to *Corindia* (Drake *et al.* 2023). No other species of *Corindia* has been recorded in the Western Palaearctic.

Corindia viridis (Parent, 1932) (Plate 2)

This species has probably been overlooked as records come from Dorset, Surrey and Norfolk, which suggest a wide southern distribution. It has been recorded from soft-rock cliffs with seepages and freshwater wetlands. Wing length ♂: 2.0 mm, ♀: 2.1 (1.9–2.3) mm. Flight: vi–viii.

CYRTURELLA Collin, 1952 (Medeterinae)

The single species known in Europe is a tiny shining green species with yellow legs, including parts of the coxae, and white setae and hairs with the exception of some darker head setae (Plate 2). The legs are devoid of setae apart from tiny apicals on the tibiae but the front coxa has strong white apicals. Vein M_1 is sinuous beyond crossvein dm-m, compared to straight or gently curved in other medeterines and, after converging with R_{4+5} beyond dm-m, these two veins run almost parallel to each other in the apical fifth of the wing (see ♂ genus key couplet 49). The crossvein is well within the basal half of the wing so that M_1 beyond dm-m is twice as long as its basal section. The hind basitarsus is as long as the second tarsomere. Long tergite 7 forms a stem for the genitalia whose epandrium is almost triangular rather than elongate as in other medeterines (see ♂ genus key couplet 45, fig. dd).

Cyrturella albosetosa (Strobl in Czerny and Strobl, 1909) (Plate 2)

The minute size, comparable only with *Micromorphus* which are grey-bodied species, and the wing venation make this species straight forward to identify. Wing length ♂: 1.3 (1.0–1.4) mm, ♀: 1.3 (1.2–1.5) mm.

The species is known only from base-rich fen and fen-meadows in Cambridgeshire and Norfolk (Drake 2007a). Flight: vi–vii.

DIAPHORUS Meigen, 1824 (Diaphorinae)

Diaphorus are small to medium-sized metallic green or black flies (Plate 3). In both sexes the genus may be distinguished from most others by the combination of the seta on the hind coxa being placed close to its base, as in other Diaphorinae, and the arista arising clearly above the middle of the very short postpedicel (as in *Melanostolus*). Males are readily recognised by their eyes touching, or nearly so, on the frons so that the frons is absent in front of the protruding ocellar triangle; this has resulted in the loss of the outer vertical seta. Only the males of *Microphor* are similar but are amply distinguished by their venation. The front pulvilli of *Diaphorus* are as long as the fifth tarsomere, making their 'feet' conspicuous, and sternite 8 has outstanding setae. The wing of males is noticeably triangular (Fig. 14a) but less so in females (b). Female *Melanostolus* closely resemble *D. nigricans* but their halteres are black, not yellow.

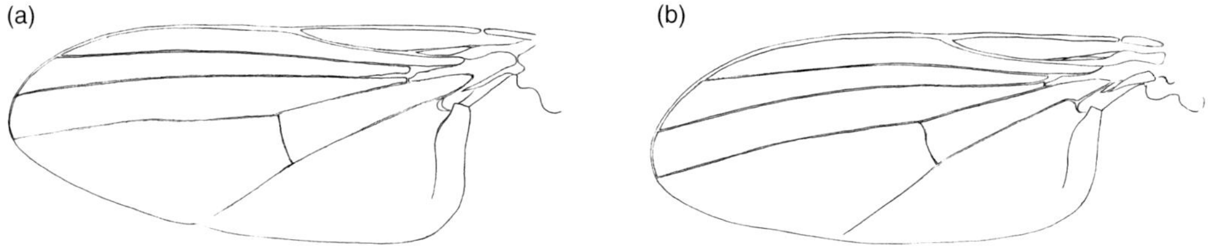

Fig. 14. *Diaphorus nigricans* wing of male (a) and female (b).

Three species of *Diaphorus* are known from Britain out of about 21 European species (Pollet 2011). Some are moderately widespread in mainland Europe so may be found in Britain: *halteralis* Loew, 1869 (discussed under *nigricans*), *disjunctus* Loew, 1857 (listed as British in error in Negrobov 1991), *vitripennis* Loew, 1859 and *winthemi* Meigen, 1824, although *winthemi* was removed from the British list (Drake 2018b). Both *disjunctus* and *vitripennis* might be mistaken for a *Chrysotus*, having a dark abdomen and pale tibiae and halter, but the very narrow frons and large front pulvilli will place them in *Diaphorus*; *disjunctus* has eight long setae on sternite 8, and *vitripennis* has white calypter hairs (black in British species). Parent (1938) provides a key to nearly all European species and descriptions of these widespread species, and Negrobov *et al.* (2007) give a key to 37 Palaearctic species.

British *Diaphorus* are wetland species of rivers margins, seepages and mires.

Key to species of *Diaphorus* (males)

1 Tergite 2 at least partly yellow on disc, sternites 1 and 2 entirely yellow or white. Halter yellow. Legs partly yellow. Larger species, wing length about 3.5–4.2 mm. 2

– Abdomen, halter and legs apart from knees entirely dark. Smaller species, wing length about 2.8–3.2 mm. 3

2 Hind femur entirely black; mid femur black on basal half. Hind femur on apical fifth with a ventral cluster of long hairs (a). Front tibia with posteroventral hairs only slightly longer than width of shaft (b). Lower postocular and postgenal setae and hairs pale. .... ***oculatus***

– Hind femur yellow on basal half, black apically; mid femur entirely yellow. Hind femur without a ventral cluster of long hairs near tip but with a posterior row of hairs along its whole length, visible in dorsal view but scarcely in anterior view (aa). Front tibia with a conspicuous fringe of posteroventral hairs, hairs twice as long as width of shaft (bb), with 1–3 outstanding long fine anteroventral hairs at apical quarter (cc). Lower postocular and postgenal setae and hairs black. .... ***hoffmannseggii***

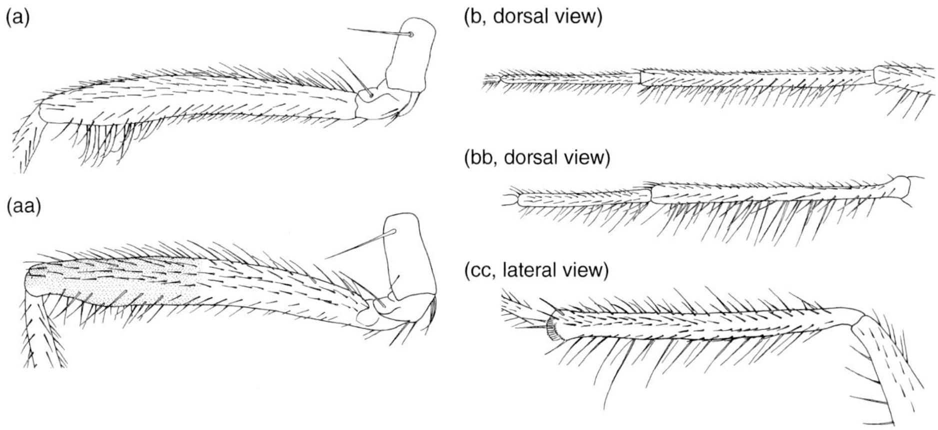

3 Scutum black with deep blue hue and a matt black triangle in front of scutellum when viewed from in front. Dusting on front margin of scutum sparse, usually confined to 2 narrow rows. Sternite 8 with about 10–12 short setae in addition to 2 pairs of long setae. Marginally larger flies. .... ***nigricans*** typical 'blue' form

– Scutum dark green under grey dusting with no differentiated pattern in front of scutellum from any angle of view. Dusting on front margin of scutum dense, extensive and usually not differentiated into rows. Sternite 8 with only a few weak or no additional short setae. Marginally smaller flies. .... ***nigricans*** 'green' form

Key to species of *Diaphorus* (females)

1 All tibiae mainly or entirely black, yellow only on apical quarter to third of front and mid legs (in rare cases tibiae entirely yellow). Mid tibia without ventral setae. Smaller flies, wing length 2.6–3.3 mm. .... ***nigricans***

– All tibiae at least half yellow. Mid tibia with 1–2 strong ventral setae. Larger flies, wing length 3.5–4.5 mm. 2

2 Femora mainly black. Lower postocular and postgenal setae and hairs white. .... ***oculatus***

– Femora mainly yellow, hind one black on apical third. Lower postocular and postgenal setae and hairs black. .... ***hoffmannseggii***

Diaphorus hoffmannseggii **Meigen, 1830**

Males have a markedly hairy front tibia whose posterior multiserial fringe of long hairs continues onto the basitarsus, and the abdomen is less obviously ringed with yellow at the base compared to *oculatus*. Females have entirely dark tergites and bright yellow front and mid femora. Wing length ♂: 4.1 (4.0–4.3) mm, ♀: 4.3 (4.0–4.5) mm.

Diaphorus hoffmannseggii is a rare species with reliable records from the border of England and Wales in Gwent and Herefordshire, Cheshire, Dorset, Hampshire and Surrey. It is strongly associated with exposed sediments of sandy and gravelly rivers, usually with sparse vegetation and in light shade. Flight: v–viii, peak in vii.

Diaphorus nigricans **Meigen, 1824 (Plate 3)**

This small and entirely black *Diaphorus* closely resembles *Melanostolus melancholicus* in size and having black halteres in the males; the wide frons distinguishes male *Melanostolus*. The male genitalia of these two species in different genera are identical. Similar all-black *Chrysotus* (*blepharosceles*, *laesus*, *obscuripes*) have pale halteres. Female *nigricans* have yellow halteres, separating them from female *Melanostolus* in which they are black; further differences are given in the account for *Melanostolus*. Rare female specimens of *nigricans* have yellow tibiae and basitarsi but can be separated from similarly coloured dark *Chrysotus* by the arista arising well above the midpoint of the postpedicel and in having entirely black postocular and postgenal setae and hairs. Wing length ♂: 3.0 (2.8–3.2) mm, ♀: 3.0 (2.6–3.3) mm.

Two colour forms are distinguished in the key but it is not clear whether they represent different species. As well as the characters used in the key, the blue form has slightly but consistently longer vestiture on the tergites and sternites, although not easy to quantify. The colour of the blue form agrees approximately with Parent's (1938) redescription. The hypopygia of the two forms are identical (Fig. 15) and similar to those illustrated Negrobov *et al.* (2007) for *graecus* Parent, 1932 and *nigrotibia* Strobl (and *M. melancholicus*, Fig. 24). *Diaphorus halteralis* is a similarly all-black species that may occur in Britain and is distinguished by having eight (not four) long setae on sternite 8 and its eyes very narrowly separated on the frons.

Diaphorus nigricans is widespread in England and south Wales but rare in Scotland where it has been found sparsely mainly near the west coast. It occupies two contrasting habitats corresponding to the two colour forms: the blue form occurs in unshaded acid mire and heath with seepages or

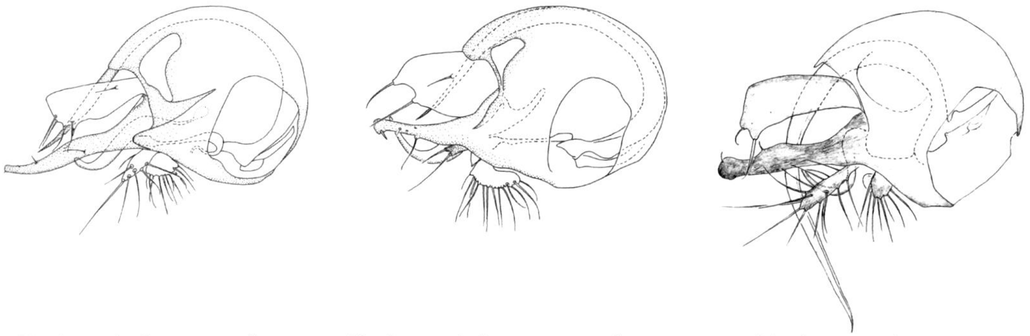

Diaphorus hoffmannseggii *Diaphorus nigricans* – green form *Diaphorus oculatus*

Fig. 15. *Diaphorus* hypopygia.

pools, and more rarely in reedbed or fen, and the green form lives in seepages on coastal soft-rock cliffs which are often base-rich, and where it can be abundant, often in company with *Melanostolus*. Flight: v–viii, peak in vi.

Diaphorus oculatus **(Fallén, 1823) (Plate 3)**

Males are distinguished from *hoffmannseggii* by the leg colour and chaetotaxy and also by the more marked pale anterior ring on the abdomen. Females have a dark abdomen and mainly black front and mid femora. Wing length ♂: 4.0 (3.6–4.2) mm, ♀: 4.0 (3.5–4.4) mm.

Diaphorus oculatus is widespread in England and Wales and rare in Scotland, reaching north to Ayrshire. Its habitat requirements are unclear but features in common to many sites are a degree of shade, either from trees or tall herbage, and usually an adjacent wetland feature including streams, seepages, ponds and drainage ditches, but the species does not appear to be dependent on extensive wetlands. Flight: v–viii, peak in vi–vii.

DOLICHOPHORUS Lichtwardt, 1902 (Medeterinae)

The single European species is a tiny shining green fly with bright yellow coxae and legs, and pale setae and hairs (Plate 3). Males have a conspicuously darkened cell r_1, a very long pale apical spur composed of fused setae on the front coxa, a modified front basitarsus with a ventral extension at its tip, a very long fine anterior hair at the tip of the second and third front tarsomeres, and a conspicuous row of five anterior setae on the hind femur. The genitalia are typically medeterine, held on a pedicel formed from tergite 7 (see ♂ genus key couplet 45, fig. cc). Females have venation similar to that of *Medetera*, and, as in males, only three dorsocentral setae and no preapical seta on the mid and hind femora. The few other genera with these characteristics lack the coxal spur which is shorter than in the males but still about half the length of the coxa. Grichanov (2009) provides a key to the world fauna.

Dolichophorus kerteszi **Lichtwardt, 1902 (Plate 3)**

This species could be mistaken for a *Thrypticus* but has only one seta on the hind coxa, not two as in that genus. The scape and pedicel are yellow contrasting with the black postpedicel, resembling only a few species in other genera of similarly small flies (e.g. *Corindia*, some *Medetera*). Wing length: ♂ 2.1 (2.1–2.2) mm.

The species has been recorded only twice, in the Fens area of eastern England at the Ouse Washes, Cambridgeshire (Drake 2005) and a site near King's Lynn, Norfolk. These sites were winter-flooded swampy pasture and a drainage ditch in arable land. Pollet *et al.* (2005) suggest that it may be associated with trees. Flight: vii.

DOLICHOPUS Latreille, 1796 (Dolichopodinae)

These are medium-sized to large robust flies with metallic green, brown or nearly black bodies (Plate 3). *Dolichopus* are members of the subfamily Dolichopodinae characterised by the combination of the scape having tiny dorsal hairs, the mid and hind femora each with one or more strong anterior preapical setae, the male sixth and seventh tergites bare and the hypopygium folded under the abdomen and not concealed within the previous tergites (Brooks 2005). They also have strictly biserial acrostichal setae. In the British fauna, they are recognised by uniquely possessing 1 to many, but most often 2–3, dorsal setae on the hind basitarsus. Vein M_1 has a short obtuse to angular double inflexion beyond crossvein dm-m; this inflexion is weak in only a few species and occasionally has a rudimentary stub of M_2. Males have conspicuous genitalia with large hair-fringed cerci and the legs of almost half the British species are ornamented in some way.

There are at least 131 species of *Dolichopus* in the Western Palaearctic. In Britain it is the largest genus of dolichopodids with 55 species although *melanopus* is probably extinct here. A number of species not yet found in Britain are widespread but mainly uncommon in the countries closest to Britain so are less likely to occur here. An exception is *tanythrix* Loew, 1869 which would fall out with *phaeopus* in the current key and is discussed under that species.

Recent keys to the Western Palaearctic *Dolichopus* may be found in Grichanov (2006, 2007). The older works of Stackelberg (1930) and Parent (1938) remain useful. Brooks (2005) considers the subgenera *Hygroceleuthus* and *Macrodolichopus* to have no phylogenetic value since they render the remaining large subgenus *Dolichopus s.str.* paraphyletic.

Many male *Dolichopus* can be recognised from their hypopygia, even if the differences between species is small. A few appear to be inseparable using the easily visible characters of the cercus and epandrial lobes, e.g. most of the species with a modified mid tarsus along with *simplex* have essentially identical hypopygia. The figures show the surstyli but these are usually obscured by the cerci, distorted in dry material and have a complex shape so vary with the angle of view. When checking a specimen against the figures, examine the following:

- Cercus overall shape, any outstandingly long apical projections or protruding 'corners'.
- Epandrium overall shape, particularly whether it is somewhat pear-shaped with a slightly concave dorsal surface.
- Shape and degree of separation of the two epandrial lobes and the complex that the basiventral lobe makes with the hypandrium. These lobes sometimes distort in dry material.

Most species of *Dolichopus* are found in wetlands including many types of freshwater, including the margins of ponds and ditches, seepages, wet woodland, fens and bogs; a few are confined to saltmarsh. A small number are found in slightly drier habitats such as pasture farmland and heathland. Some species form conspicuous aggregations on wet mud or shallow water where feeding and courtship occur. The courtship of some species has been described but for most it remains unknown.

Key to species of *Dolichopus* (males)

To help navigate this long key once some familiarity has been gained, major divisions are given as artificial groups that have no taxonomic significance:

Black femora...couplet 2
Yellow femora, tarsi modified: enlarged, fringed with long or flattened hairs, or with silver pubescence on at least anterior face of last 1–3 tarsomeres, as figured in couplets 19 to 30................ couplet 18
Yellow femora, unmodified tarsi, hind femur with long ventral fringe..couplet 32
Yellow femora, unmodified tarsi, no long fringe on hind femur, antenna black..................................... couplet 41
Yellow femora, unmodified tarsi, no long fringe on hind femur, antenna partly yellow. ....couplet 49

1 Mid and hind femora entirely or mainly black, or at least one of these completely encircled with black on at least part of their length......2
– Mid and hind femora entirely or mainly yellow, neither completely encircled with black. 17

Black femora

2(1) Front tibia black. 3
– Front tibia entirely or mainly yellow......9

3(2) Wing dark on apical half of the 3 anterior cells (r_1, r_{2+3}, r_{4+5}) and clearly darker apically than basally (a). 4
– Wing uniformly dark along whole length of 2 anterior cells (r_1, r_{2+3})......5

(a)

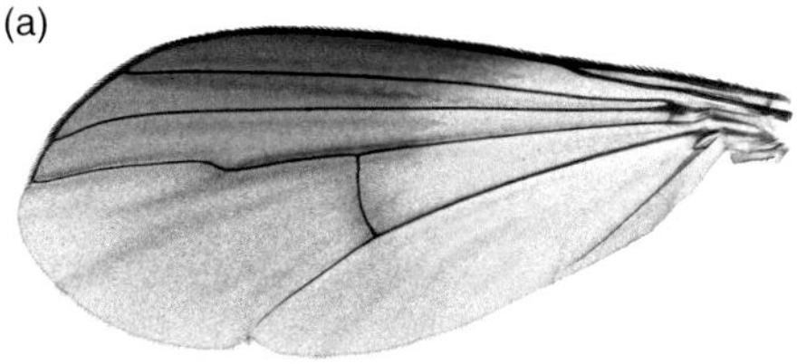

4(3) Hind basitarsus with numerous (c. 20) setae (a). Hind femur with only short hairs ventrally. Mid and hind femora with 1 anterodorsal preapical seta. Face dusted greyish white. Postpedicel nearly twice as long as deep (b). ......***atratus***
– Hind basitarsus with only 3 dorsal setae. Hind femur with a long ventral fringe. Mid and hind femora with at least 2 anterodorsal preapical setae. Face dusted brownish. Postpedicel about as long as deep......***maculipennis***

(a)

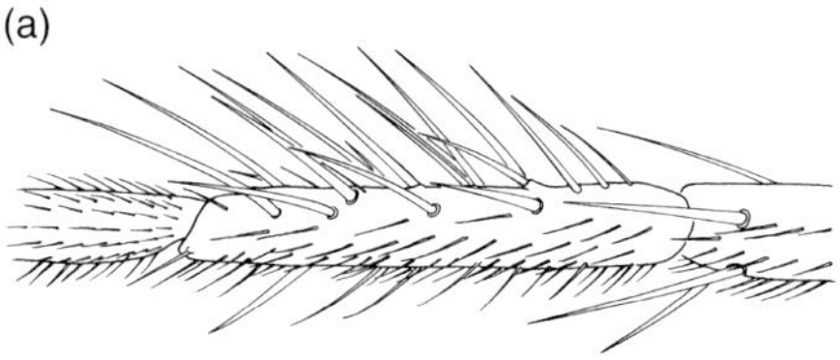

(b)

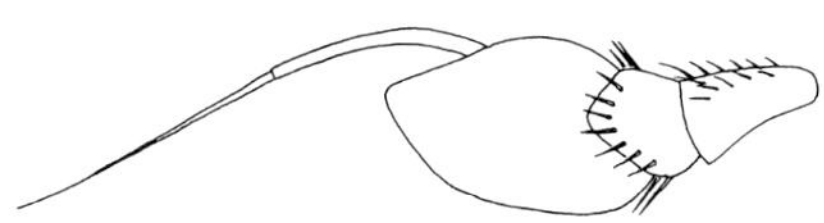

5(3) Front tarsomere 5 flattened dorsoventrally or laterally. 6
– Front tarsus simple. 7

6(5) Front tarsomere 5 flattened dorsoventrally, black, tarsomeres 2–4 pale (a). Hind femur with a ventral fringe of short fine pale hairs at base. Mid femur ventrally at base with a cluster of short fine pale hairs similar to those on hind femur. Hind tibia only slightly thicker than mid tibia. Postpedicel 1.5 times longer than deep. Hind basitarsus with 4–5 dorsal setae. Cercus rectangular, about twice as long as wide; apicoventral epandrial lobe not visible in profile (b). ......***nigripes***
– Front tarsomere 5 flattened laterally, whole tarsus entirely black (aa). Hind femur with a ventral fringe of long dark hairs. Mid femur with only normal black vestiture. Hind tibia thick, twice as wide as mid tibia. Postpedicel about as long as deep. Hind basitarsus with 3 dorsal setae. Cercus almost round; apicoventral epandrial lobe rounded and projecting from epandrial surface (bb). ......***melanopus***

(a, dorsal view)

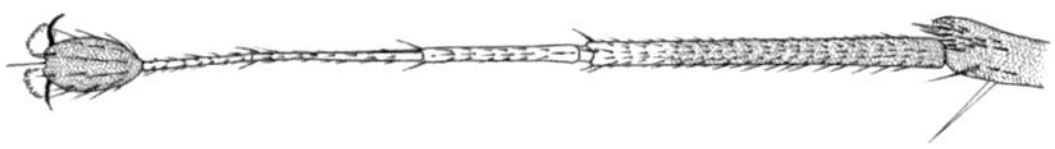

(b)

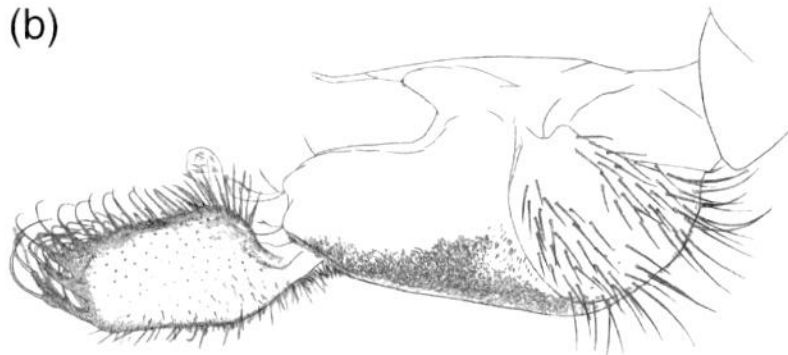

(aa, lateral view)

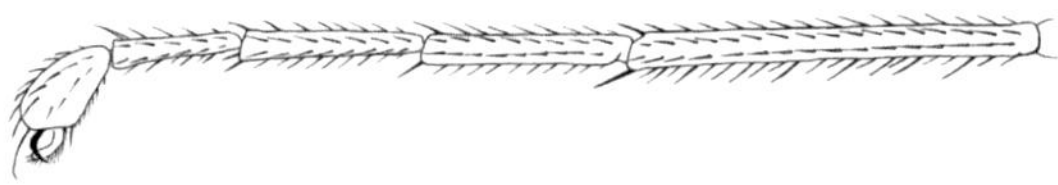

(bb)

7(5) Smaller species, wing length less than about 4.0 mm. Wing with a small costal stigma, particularly obvious in front view (a). Mid basitarsus entirely black. Hind basitarsus with 2–3 dorsal setae. Hind femur with a posteroventral row of fine pale hairs as long as about one-third depth of femur, often pointing backwards so not always visible in front view. (See also couplet 16.) .... ***phaeopus***

– Larger species, wing length at least 4.5 mm. Wing without a costal stigma. Mid basitarsus ground colour pale on basal third to half below dense black vestiture. Hind basitarsus with 3 dorsal setae. Hind femur either with no differentiated row of posteroventral hairs or these hairs not longer than dorsal hairs. 8

(a)

8(7) Face white-dusted. Lower postocular setae black. Knees of front and mid legs narrowly but obviously yellow. Front tarsus 1.2 times longer than tibia. Hind tibia scarcely thicker than mid tibia, with about 15 dorsal (ad, pd) setae. Cercus rectangular with an approximately straight dorsal margin, its apex with long projections bearing many hairs longer than half its width (a). .... ***picipes***

– Face dusted yellow or brown. Lower postocular setae pale or rarely 1–2 lowest setae black. Knees of front and mid legs scarcely differentiated in colour. Front tarsus as long as tibia. Hind tibia about 1.25 times thicker than mid tibia, with about 11 dorsal (ad, pd) setae. Cercus comma-shaped with rounded outer margin, apical hairs one-fifth its width and not born on long projections (aa). .... ***atripes***

(a)

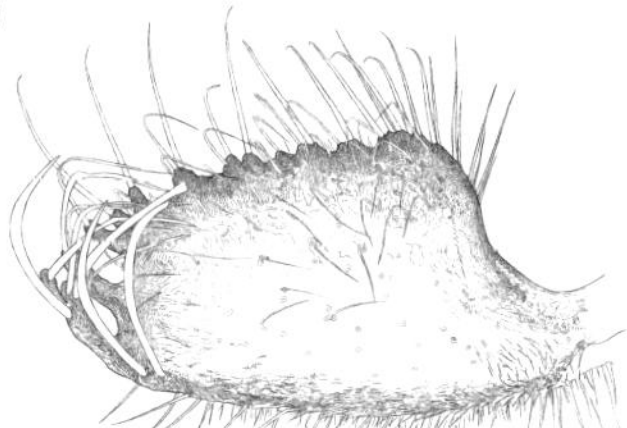

(aa)

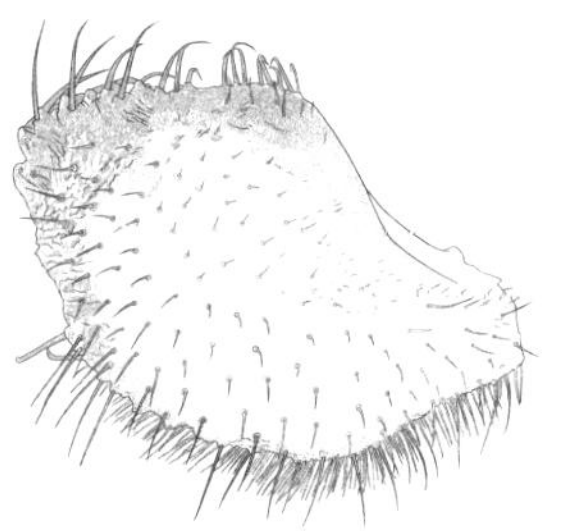

9(2) Lower postocular setae black. Hind basitarsus with 3 true dorsal setae in a line. Hind tibia more bristly (11 or more ad and pd setae on dorsal face). Face dusted white, yellow or golden. 10

– Lower postocular setae pale. Hind basitarsus with 1–2 true dorsal setae. Hind tibia with fewer setae (11 or fewer ad and pd setae). Face white-dusted.. 13

10(9) Mid leg with tarsomere 5 flattened (a). Postpedicel twice as long as deep, tip pointed (b). Arista thick, obviously much wider than ocellar or dorsocentral setae, and clearly pubescent. .... ***planitarsis***

– Mid tarsus unmodified. Postpedicel scarcely longer than deep, at most 1.25 times, tip square or rounded. Arista barely wider than ocellar or dorsocentral setae with scarcely discernible pubescence.. 11

(a)

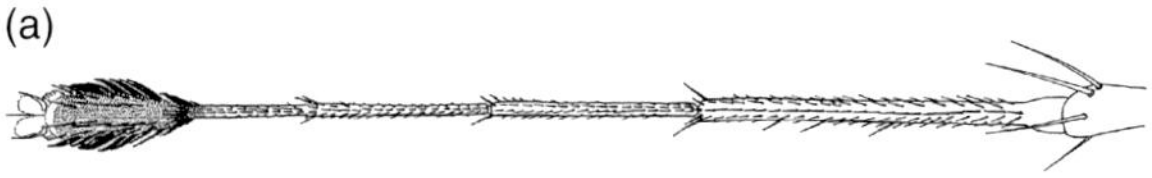

(b)

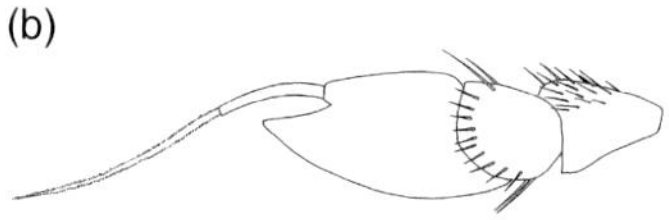

11(10) Hind tibia mainly yellow. Mid femur with 2 or more anterior preapical setae. Hind femur with a ventral fringe of black hairs about half depth of femur (a). Face dusted pale yellow. Hind tibia without ciliolarium. Cercus often completely dark but sometimes pale. .... ***campestris***

– Hind tibia black. Mid femur with only 1 anterior preapical seta. Hind femur fringe either much longer or absent. Face dusted white or gold. Hind tibia with ciliolarium. Cercus pale with narrow black margin.. 12

(a)

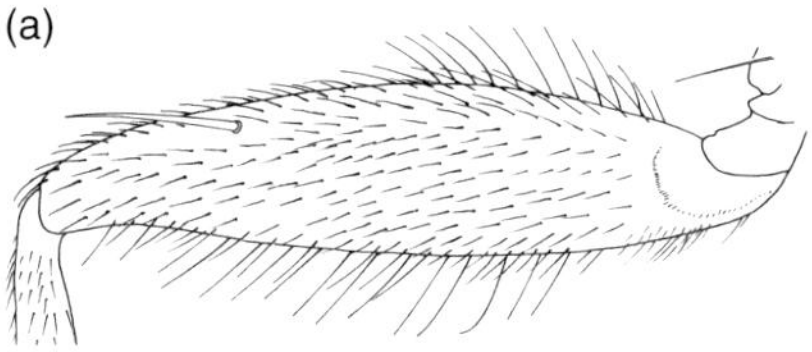

12(11) Hind femur with ventral fringe of long black hairs on apical half and pale shorter hairs basally (a). Face gold-dusted. Hind tibia stout, about twice as thick as mid tibia; ciliolarium on its posterior face with very short yellow or dark pubescence on basal half (b). Mid tibia entirely yellow. Cercus with only short pale hairs near base (c).. ***lepidus***

– Hind femur without ventral fringe. Face white-dusted. Hind tibia 1.5 times wider than mid tibia; ciliolarium small and oval on slight swelling on basal third (bb). Mid tibia with a narrow sub-basal darker ring. Cercus with long black hairs near base on ventral edge (cc).............. ***laticola***

(b)

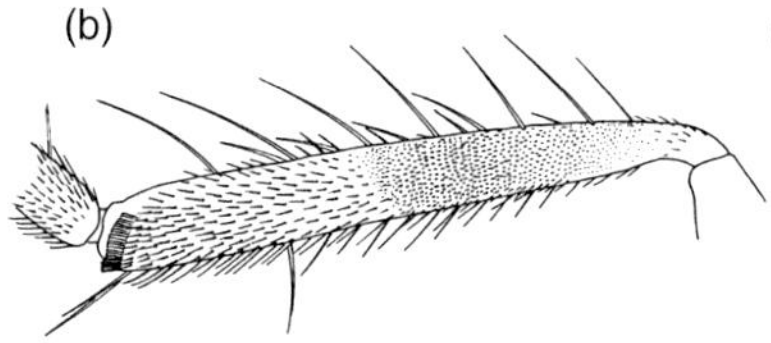

(c)

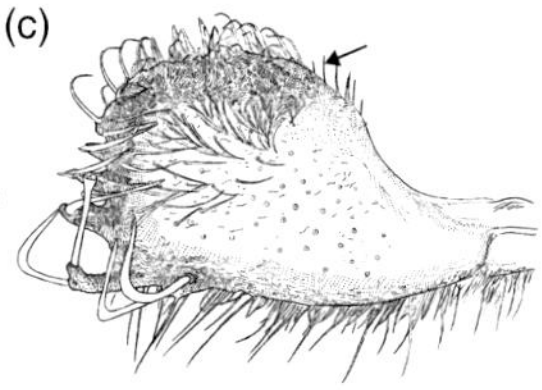

(a)

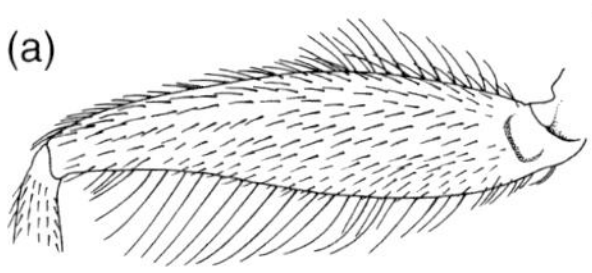

(bb)

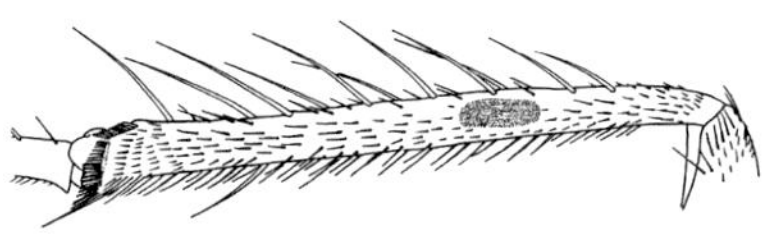

(cc)

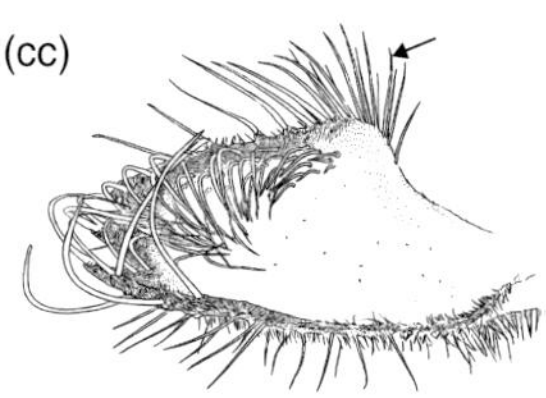

13(9) Hind femur with a ventral fringe of long hairs, as long as depth of femur. Hind basitarsus with 1 dorsal seta. Hind tibia on posterior face with a ciliolarium of fine yellow or black pubescence or points extending from one- to three-fifths from base. Front and mid femora usually black but sometimes largely yellow.......14

– Hind femur without a ventral fringe of long hairs. Hind basitarsus with 2 dorsal setae. Hind tibia without a ciliolarium. Front and mid femora black.......15

14(13) Hind femur with ventral fringe pale. Wing tip clear. Hind tibia very stout and spindle-shaped (a). Epandrial lobes separated (b).......***clavipes***

– Hind femur with ventral fringe dark. Wing tip narrowly darkened. Hind tibia widening gradually to tip. Epandrial lobes touching at their bases (bb). (See also couplet 35.).......***signifer***

(a)

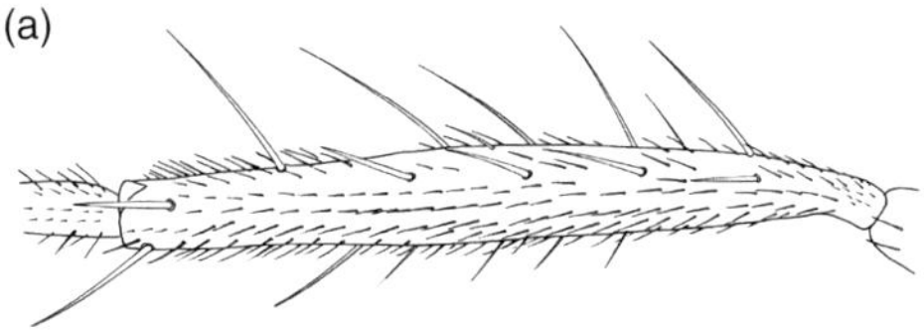

(b)

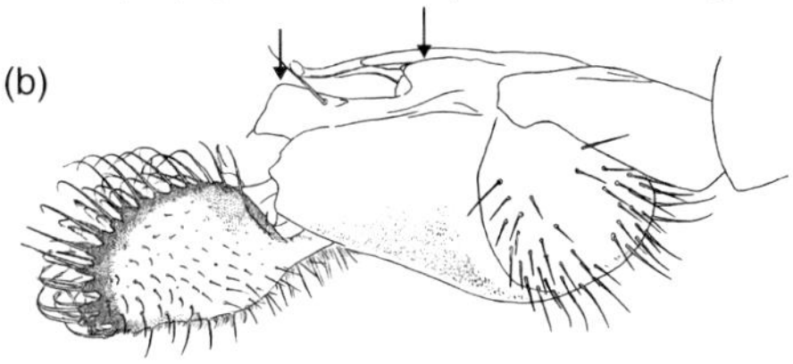

(bb)

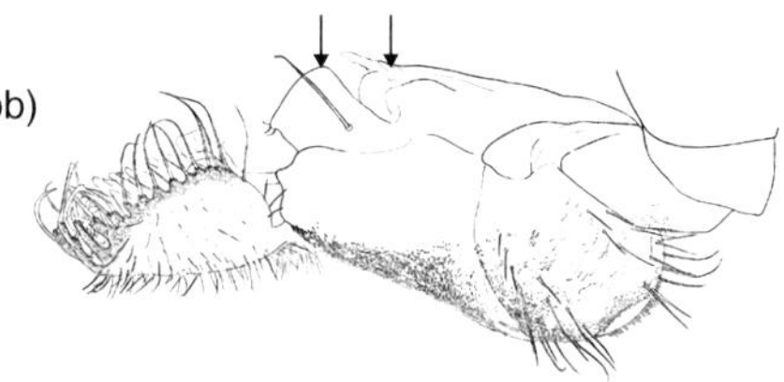

15(13) Vein M_1 with a stub-vein on apical section or at least with sharp right-angled bends. (See also couplet 46 for characters not present in next two species.).......***diadema***

– Vein M_1 with smooth slight bends at middle of apical section.......16

16(15) Mid and hind tibiae black, front and mid basitarsi mainly black. Hind tibia markedly broader than mid tibia, gradually widening towards tip. Hind femur with a posteroventral row of fine short pale hairs as long as about one-third depth of femur, often pointing backwards so not always visible in anterior view. Cercus rectangular without a ventral protrusion near base (a). (See also couplet 7.).......***phaeopus***

– Mid and hind tibiae yellow, front and mid basitarsi mainly yellow. Hind tibia scarcely broader than mid tibia. Hind femur without a ventral fringe. Cercus triangular with a small dark protrusion near base bearing a cluster of short black hairs (aa).......***vitripennis***

(a)

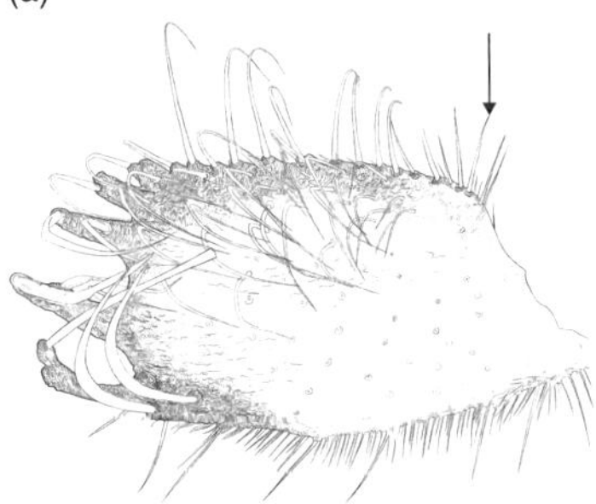

(aa)

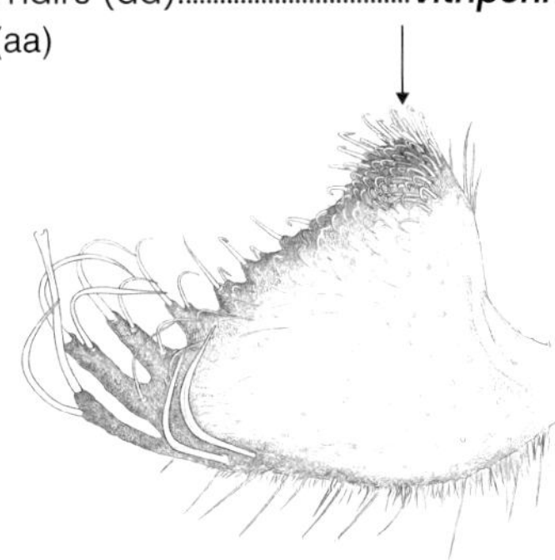

17(1) Front or mid tarsus clearly modified in some way, either tarsomeres enlarged, or fringed with long or flattened hairs, or with silver pubescence on at least anterior face of 1 or more of tarsomeres 3–5 (this is easily overlooked)....18

– Front and mid tarsi unmodified, at most with an anterior fringe of fine short hairs or tarsomeres slightly curved....31

Yellow femora, modified tarsi

18(17) Front tarsus modified, tarsomere 5 deeper than wide. Wing without expanded anal lobe....19

– Mid tarsus modified, either fringed with wide hairs or silver on front face of at least 1 of last 3 tarsomeres. Wing with an expanded anal lobe (aa) except in *signatus*....24

(aa)

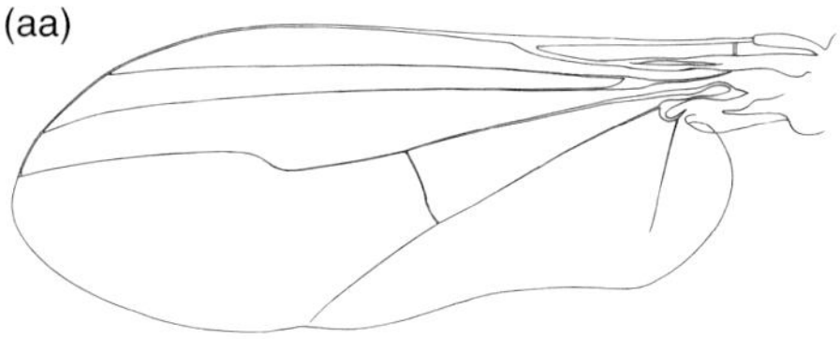

19(18) Front tarsomere 5 scarcely enlarged compared to preceding segments (a); tarsomeres 2–5 black. Hind femur with a ventral fringe of long black hairs (b). Ciliolarium on hind tibia consisting of a wide band of tiny yellow points across whole width of tibia on yellow section (c). Smaller species, wing length less than 4.5 mm. (See also couplet 35.)....***longitarsis***

– Front tarsomere 5 clearly expanded and black; tarsomeres 1–3 or 1–4 pale. Hind femur with fringe pale or absent. Ciliolarium less extensive or absent. Large species, wing length at least 5.5 mm....20

(a) (b) (c)

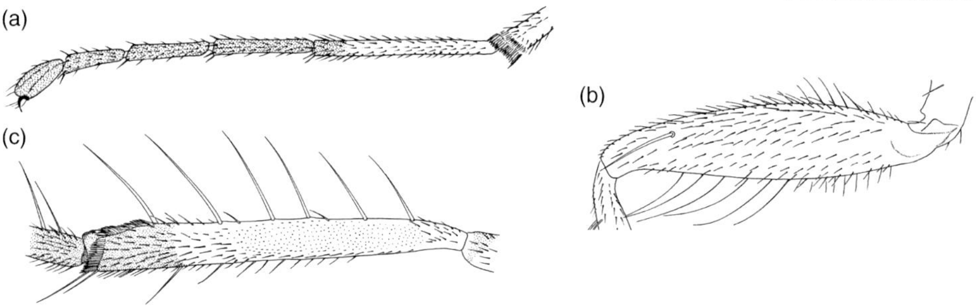

20(19) Front tarsomeres 4 and 5 black and enlarged, tarsomeres 1–3 cylindrical and yellow. Face dusting white or gold....21

– Front tarsus with only tarsomere 5 black and enlarged, tarsomeres 1–4 cylindrical and yellow. Face dusting white....22

21(20) Front tarsomere 5 far larger than 4 (a). Face dusting gold. Hind femur with a fringe of long yellow posteroventral hairs (b). Lower postocular setae black. Antennae entirely black. Cercus large, about two-thirds length of epandrium, with margin broadly rounded and widely black (c)....***brevipennis***

– Front tarsomeres 4 and 5 similar in size, with a dorsal fringe of flattened hairs almost as long as depth of segment (aa). Face dusting white. Hind femur with only a few pale hairs in apical third, none as long as depth of femur. Lower postocular setae pale, only a few upper ones black. Antennal scape yellow ventrally. Cercus smaller, about half length of epandrium, slightly narrowed near tip, margin narrowly black (cc)....***plumitarsis***

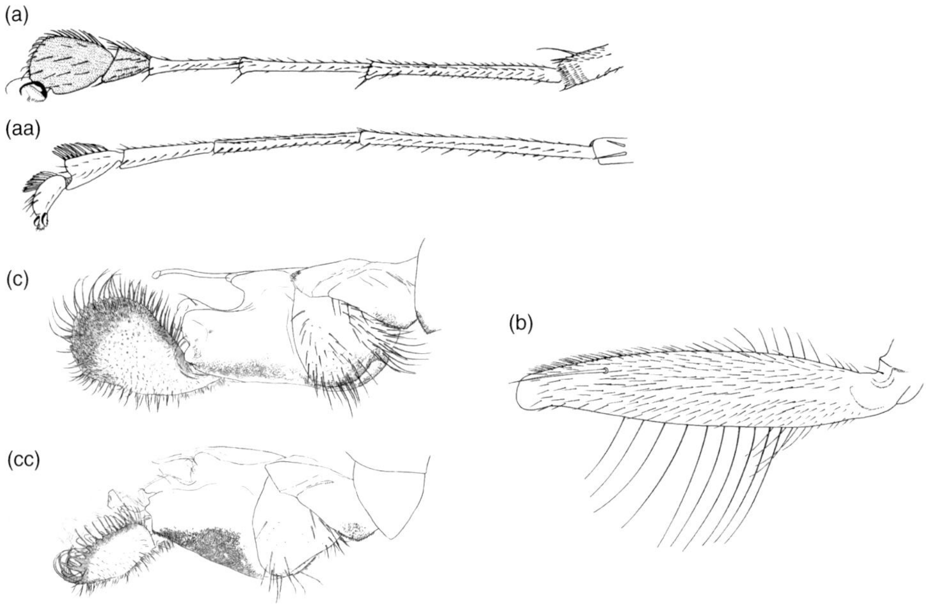

22(20) Mid and hind femora with 2–3 anterodorsal preapical setae (a). Front and mid femora with a posteroventral fringe of long white and black hairs (a). Calypter with a fringe of pale fine wavy hairs that partly coalesce. Mid tibia with about 7 long setae on ventral surfaces. Cercus large, twice as long as wide, expanding apically (b). Front tarsus (c)................................... ***claviger***

– Mid and hind femora with only 1 anterodorsal preapical seta. Front and mid femora without a fringe. Calypter with a fringe of simple black hairs. Mid tibia with 1–3 long setae ventrally. Cercus short, at most 1.5 times as long as wide (bb). 23

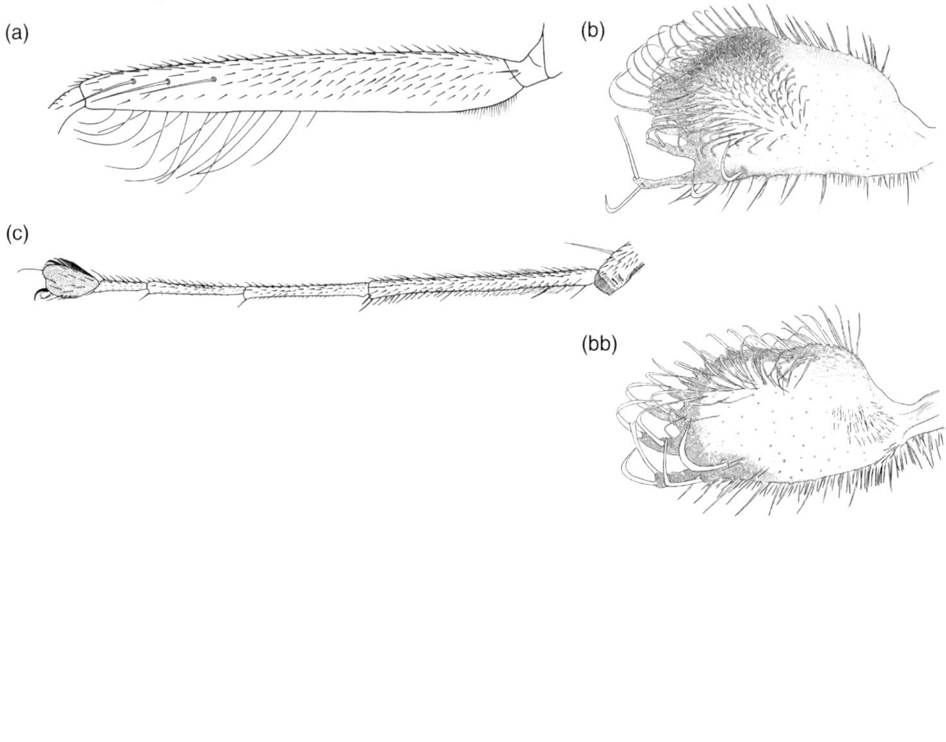

23(22) Front tarsomere 4 twice as long as tarsomere 5 (a). Wing long and narrow, 3 times longer than wide measured from anal lobe, with only slight inflexions on apical section of M_1 (b). Postpedicel twice as long as deep, tip pointed (c). Front coxa and its hairs entirely yellow (but apical setae black). Hind tibia with 1–2 long ventral setae. .... ***discifer***

– Front tarsomere 4 as long as tarsomere 5 (aa). Wing short and broad, about 2.25 times longer than wide with pronounced inflexions in apical section of M_1 (bb). Postpedicel 1.5 times longer than deep, tip more obtuse (cc). Front coxa yellow with a black mark at base on outer face, hairs black. Hind tibia with 4 or more long ventral setae. .... ***migrans***

(a)

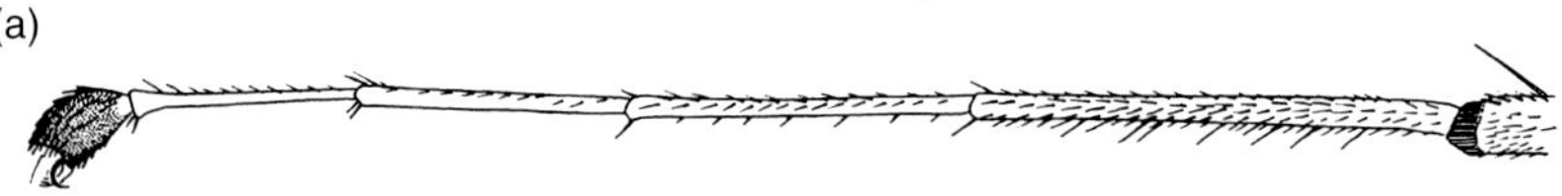

(aa)

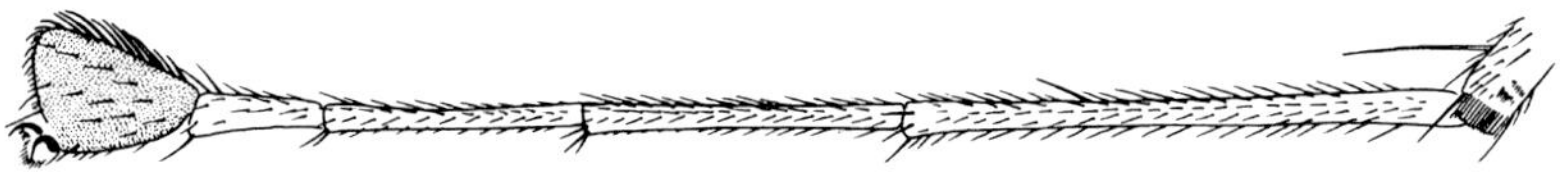

(b)

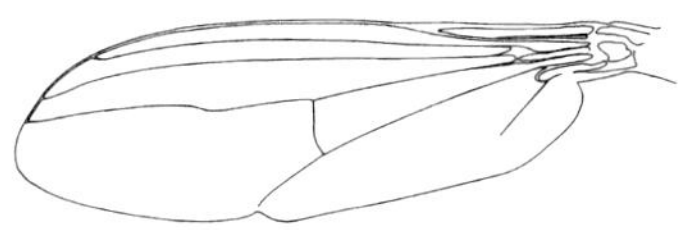

(c)

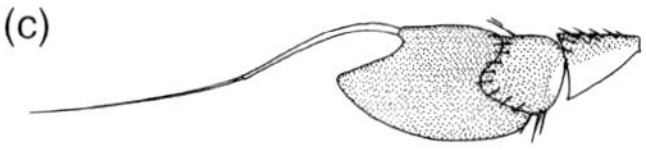

(bb)

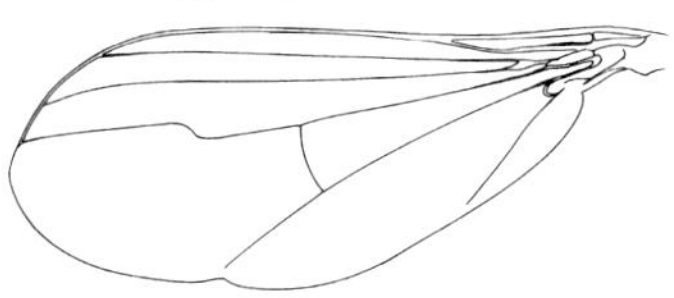

(cc)

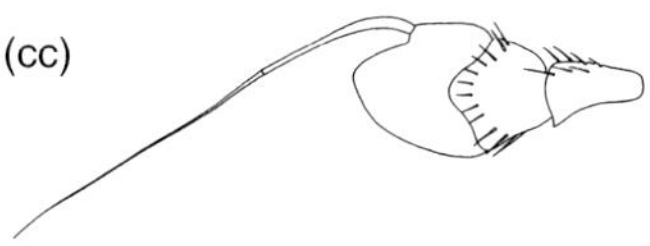

24(18) Mid tarsus with a conspicuous fringe of flat hairs, either dorsal on tarsomeres 3 and 4 or lateral on basitarsus; no tarsomere with a white anterior face. Scape and pedicel entirely yellow. 25

– Mid tarsus without a conspicuous fringe, at most some middle tarsomeres appearing thicker due to dense short hairs; at least 1 tarsomere white on anterior face. At least pedicel mainly black. 27

25(24) Mid tarsomeres 3 and 4 black, flattened, with a black dorsal fringe contrasting with the long thin yellow tarsomeres 1 and 2 and white tarsomere 5 (a). Hind femur with 3 anterodorsal preapical setae (b). Hind basitarsus at least half yellow. .... ***popularis***

– Mid tarsomeres 3 and 4 unmodified, basitarsus with a lateral fringes of flat hairs, all segments black. Hind femur with 1 anterodorsal preapical seta. Hind basitarsus at least half black. 26

(a)

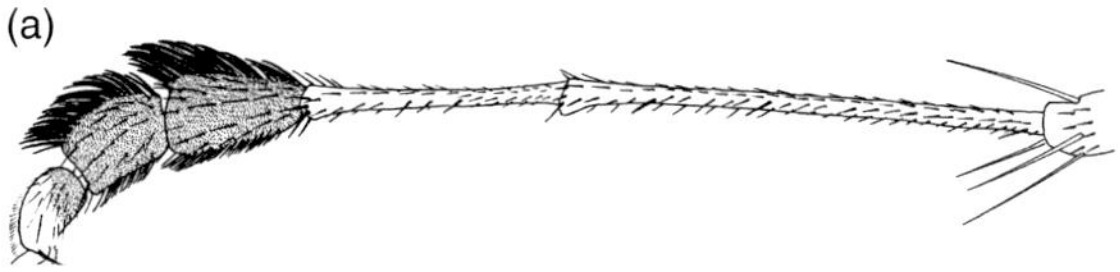

(b)

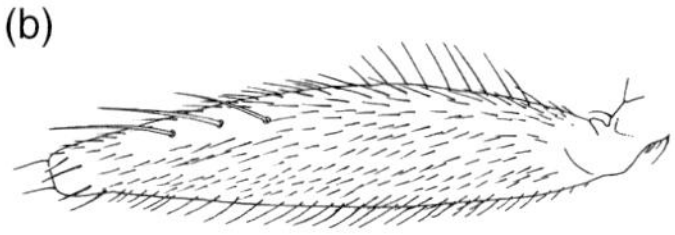

26(25) Mid tibia anterodorsally with a dark shiny streak along its length, ending just before tip at 2 preapical setae, tip contrastingly white; shaft exceptionally flattened (a). Postpedicel about 1.5 times longer than deep (b). Hind basitarsus almost entirely black. .... ***plumipes***

– Mid tibia without a dorsal dark streak and no preapical setae, tip yellow like rest of shaft; shaft cylindrical (aa). Postpedicel about twice as long as deep (bb). Hind basitarsus pale in basal third. .... ***wahlbergi***

(a)

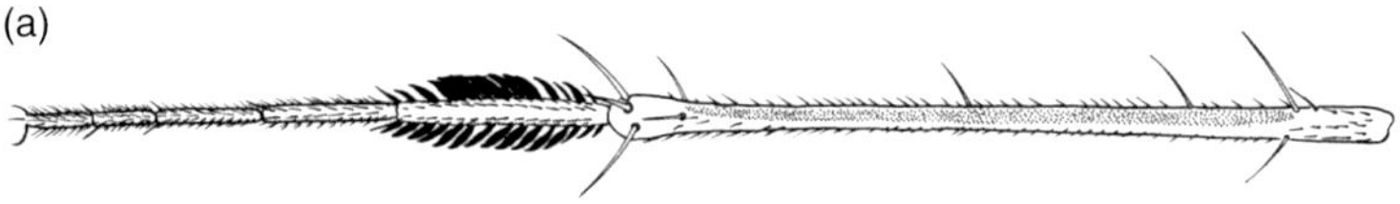

(aa)

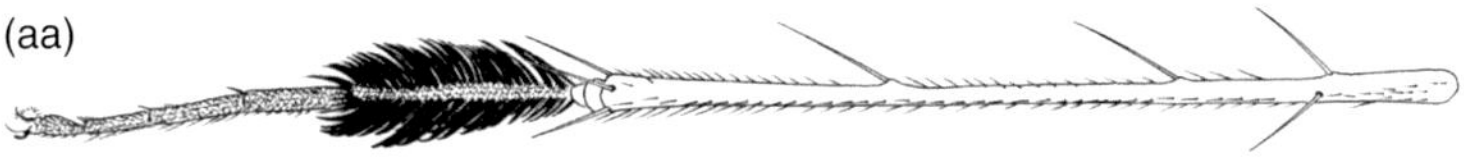

(b)

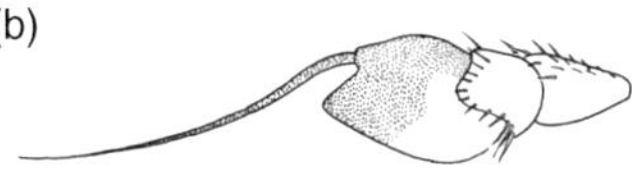

(bb)

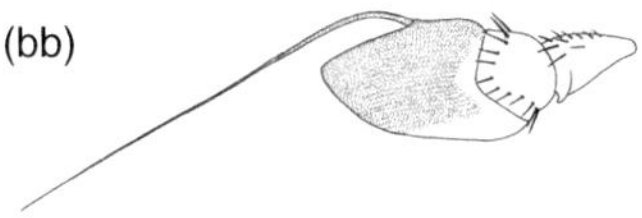

27(24) Hind tibia black. Mid tarsus with only tarsomere 5 white, on all faces (a). Face white. Wings strongly darkened, particularly anteriorly (cell r_1).. ***urbanus***

– Hind tibia mainly yellow. Mid tarsus with at least tarsomeres 4 and 5 white on anterior face, darker posteriorly. Face pale yellow to gold. Wings not markedly darker anteriorly..................28

(a)

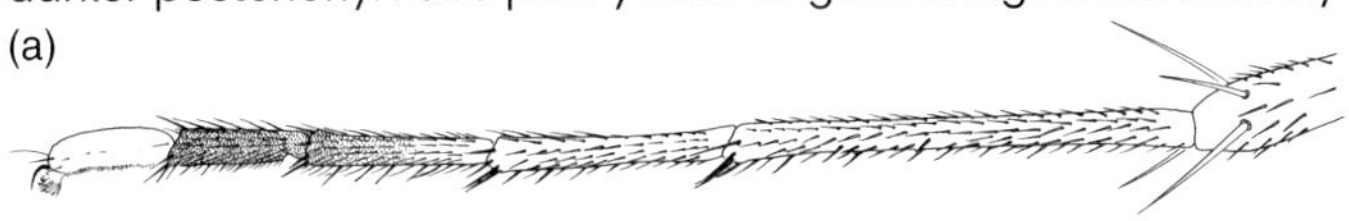

28(27) Mid tarsomeres 2 and 3 not deeper than tarsomeres 4 and 5 (a). Ciliolarium on hind tibia consisting of a short dark oval tubercle (b) clearly protruding in dorsal view (c). Wing costal stigma with ventral swelling when viewed anteriorly (d), anal lobe with smoothly rounded profile (e)... ***signatus***

– Mid tarsomeres 2 and 3 slightly or obviously deeper than tarsomeres 4 and 5. Ciliolarium without dark tubercle but hind tibia may be swollen near base. Wing costal stigma without ventral swelling, anal lobe expanded giving hind margin a sinuous profile (fig. aa of couplet 18)..29

(a)

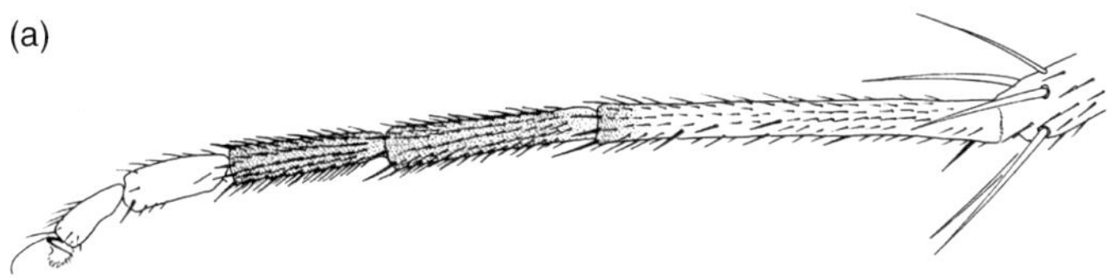

(b)

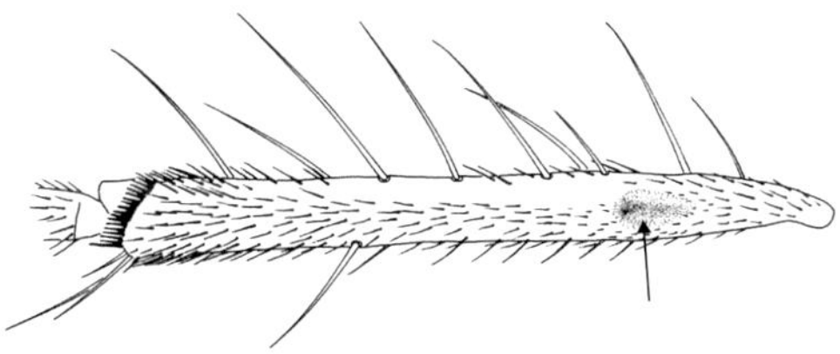

(d)

(c)

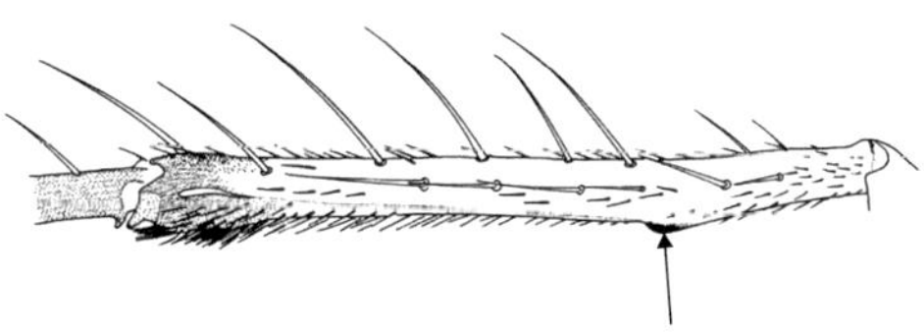

(e)

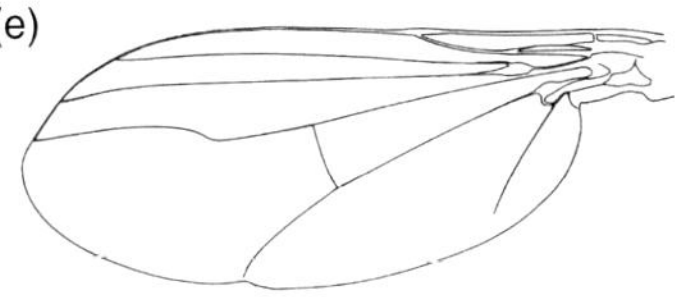

29(28) Mid tarsomeres 3, 4 and 5 white on anterior face; difference in depth between tarsomeres 2 + 3 and 4 + 5 not pronounced (a); basitarsus about three-quarters length of 2–5 combined (a). Hind tibia about 1.5 times wider than mid tibia, black on apical sixth or at least at extreme tip posteriorly. Ciliolarium on hind tibia consisting of a weak narrow strip of yellow pubescence, shaft not swollen here (c).. ***argyrotarsis***

– Only mid tarsomeres 4 and 5 white; difference in depth between tarsomeres 2 + 3 and 4 + 5 pronounced; basitarsus as long as 2–5 combined (aa, bb). Hind tibia stout, twice as wide as mid tibia, its tip entirely yellow. Ciliolarium more obvious and set on swollen area near tibia base.. 30

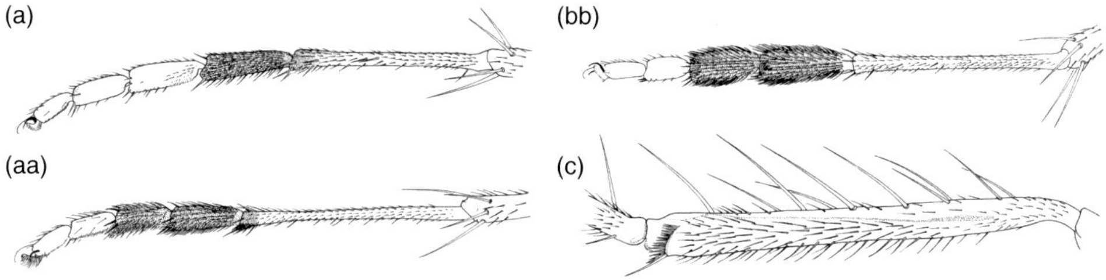

30(29) Ciliolarium on hind tibia consisting of an oval patch of yellow pile and a narrow bare matt strip running from this patch to dorsal edge at tip of tibia; rows of black vestiture hairs clearly diverging around oval patch (a). Hind femur with posteroventral fringe of fine pale hairs twice as long as anterior vestiture. Scutellum fringe of fine hairs in more than 1 row. Postpedicel with a more acute tip (b).. ***pennatus***

– Ciliolarium consisting of a narrow patch of tiny dark spinules directed inwardly and a very narrow bare matt strip, as in *pennatus*; rows of black vestiture hairs scarcely diverging around patch (aa). Hind femur with posteroventral fringe of fine pale hairs scarcely longer than anterior vestiture. Scutellum fringe in a single sparse row. Postpedicel with tip more obtuse (bb).. ***subpennatus***

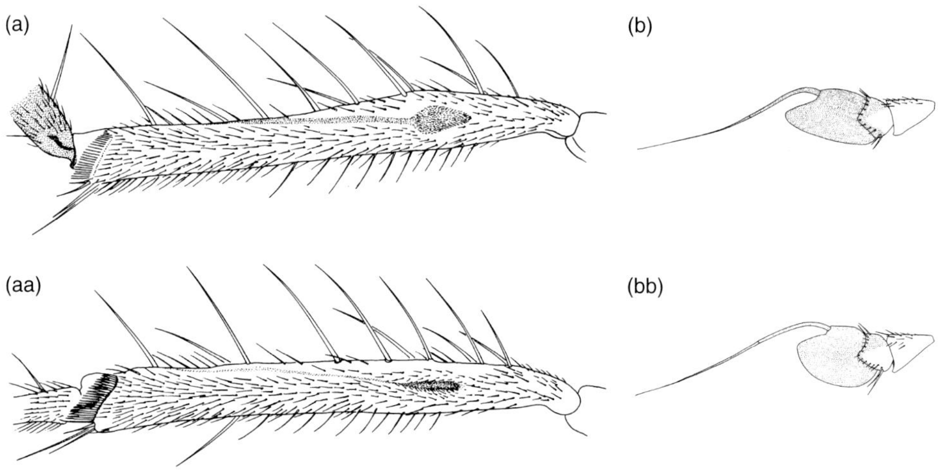

31(17) Hind femur with an obvious ventral fringe with hairs usually as long as depth of femur (a). Hind tibia with a ciliolarium.. 32

– Hind femur either without an obvious ventral fringe or one with hairs at most half depth of femur. Ciliolarium present or absent.. 40

(a)

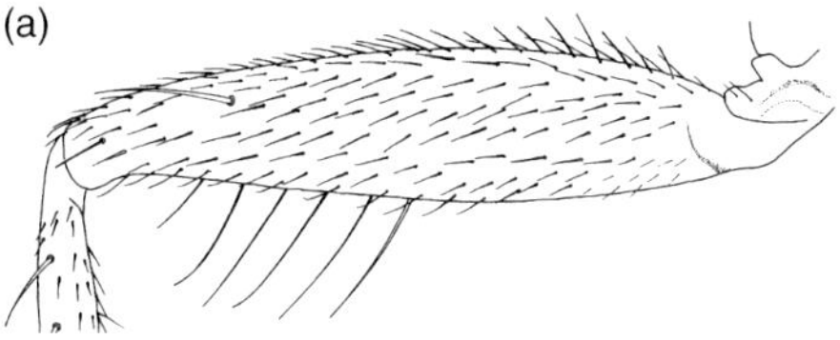

Yellow femora, unmodified tarsi, hind femur with ventral fringe

32(31) Hind femur with fringe dark on at least apical half (may be pale basally). Front coxa dark to black on at least the basal third..33

– Hind femur with fringe obviously pale yellow. Front coxa almost entirely yellow..........................36

33(32) Mid and hind femora with 3 or more anterior preapical setae (a). Mid femur with a posteroventral fringe of long hairs (b, dorsal view). Lower postocular setae black. Cercus large, twice as long as wide, expanding apically, apical third dark (c)....................***ungulatus***

– Mid and hind femora with only 1 anterior preapical seta. Mid femur without a fringe. Lower postocular setae pale (except in *longitarsis*). Cercus small, rectangular or round, scarcely longer than wide, white with narrow black margin (cc)...34

(a)

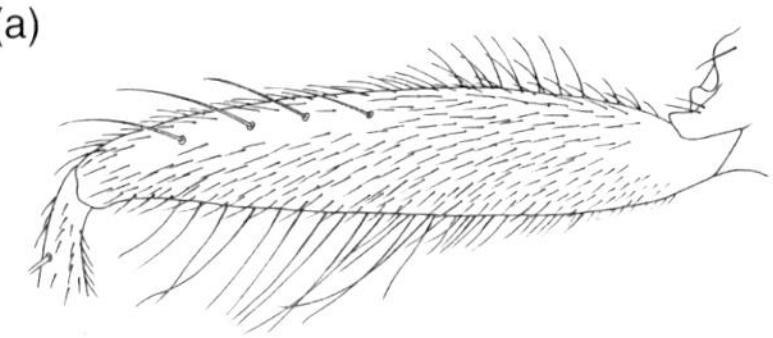

(c)

(b)

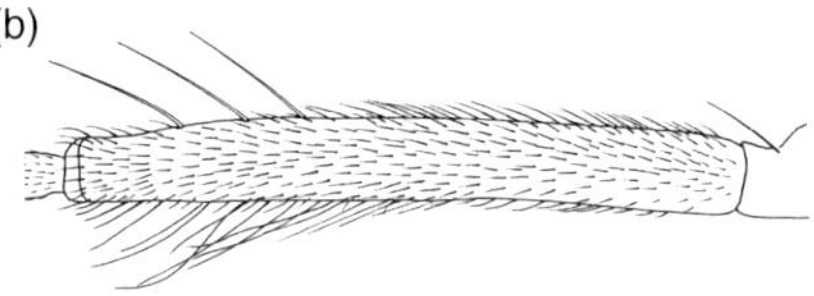

(cc)

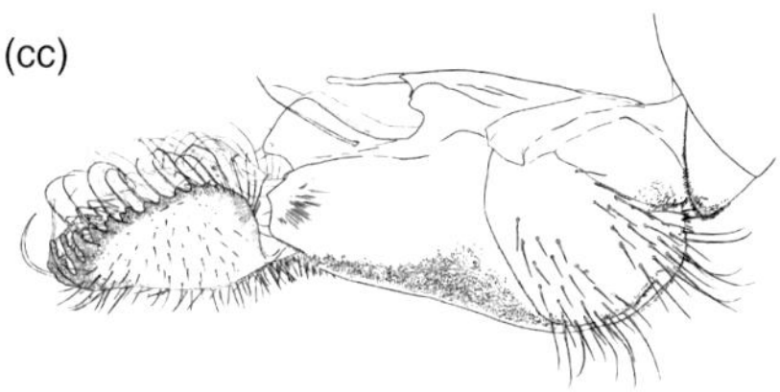

34(33) Vein M_1 with a sharp right-angled bend beyond dm-m and nearly always with a stub-vein (a). Wing membrane more-or-less uniform in colour. Hind tibia usually with only tip darker dorsally and anteriorly. Hind basitarsus with 1 dorsal seta (b). Larger species, wing length usually 5 mm or more...***griseipennis***

– Vein M_1 without a stub-vein. Wing membrane darker anteriorly or apically. Hind tibia with at least apical third black. Hind basitarsus with 1 or 3 dorsal setae. Smaller species, wing length not more than 5 mm...35

(a)

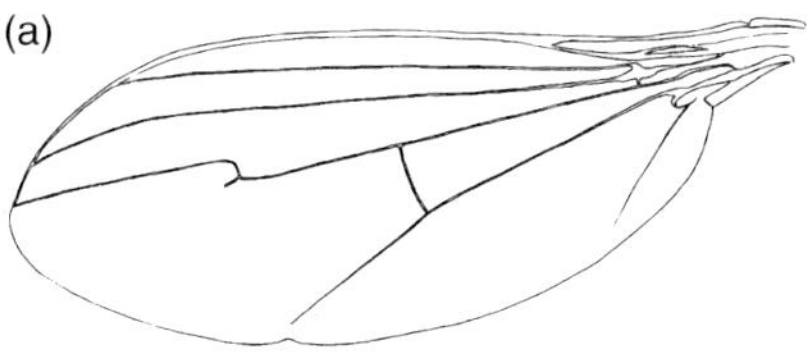

(b)

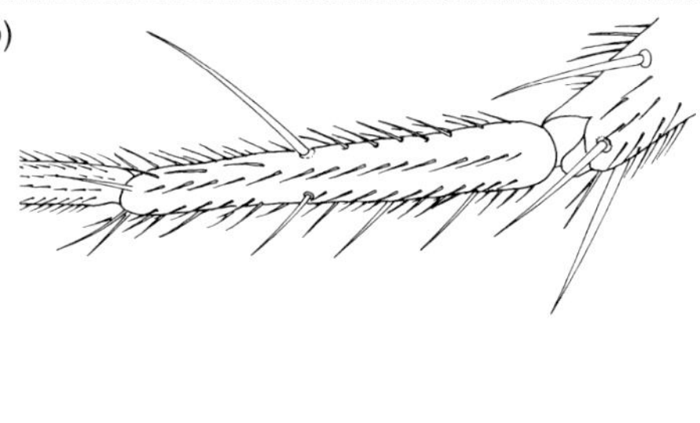

35(34) Wing tip narrowly darkened (a). Hind basitarsus with 1 dorsal seta. Front and hind coxae almost entirely black. Front tibia with a long fine apicoventral seta (b). Lower postocular setae pale. Postpedicel 1.5 times longer than deep (c). (See also couplet 14.) .... ***signifer***

– Wing tip not darkened but anterior cells darker than those behind. Hind basitarsus with 3 dorsal setae. Front coxa darker on basal half, yellow on apical half; hind coxa with a yellow tip. Front tibia without a long fine apicoventral seta. Lower postocular setae black. Postpedicel as long as deep (cc). (See also couplet 19.) .... ***longitarsis***

(a)

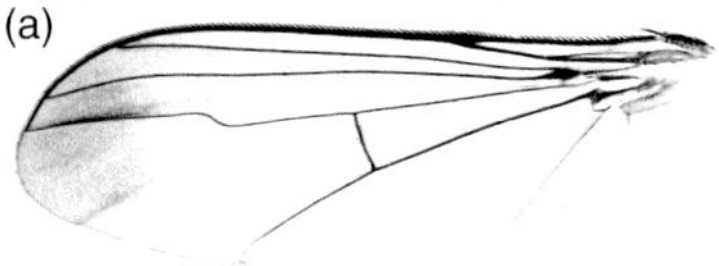

(b)

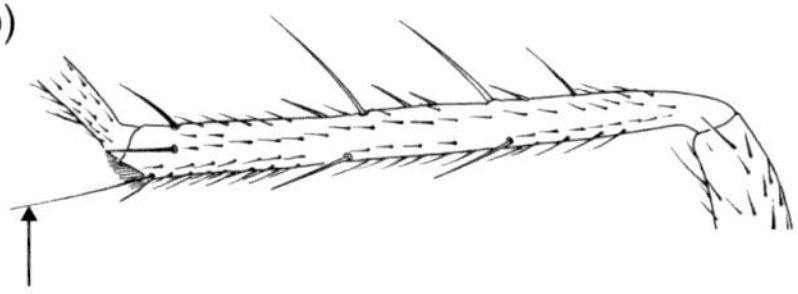

(c)

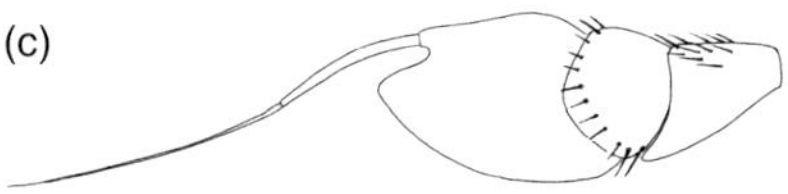

(cc)

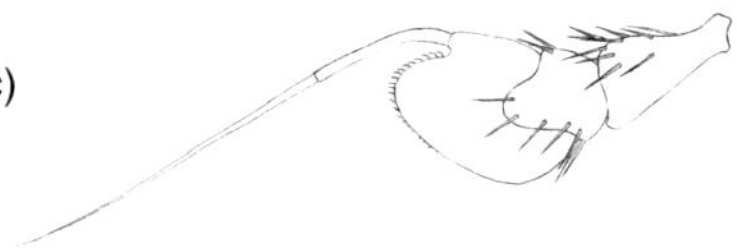

36(32) Front basitarsus with an obvious anteroventral apical seta (fig. b of next couplet). Small costal stigma present (a). Hind coxa with a black wedge from base to lateral seta (b). 37

– Front basitarsus with only a tiny anterior apical seta. No costal stigma. Hind coxa mainly yellow, sometimes with a dark smudge that rarely extends to lateral seta (bb). 39

(a)

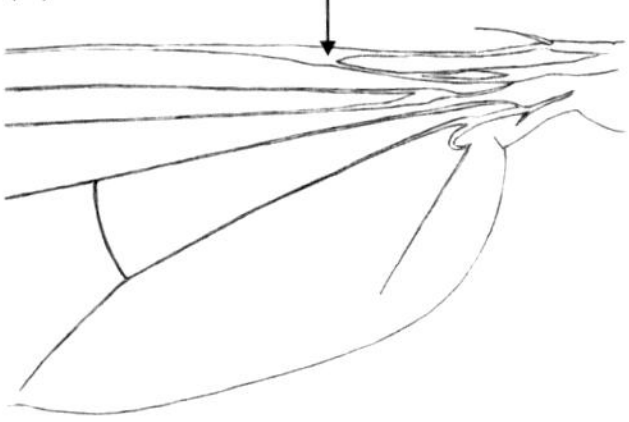

(b)

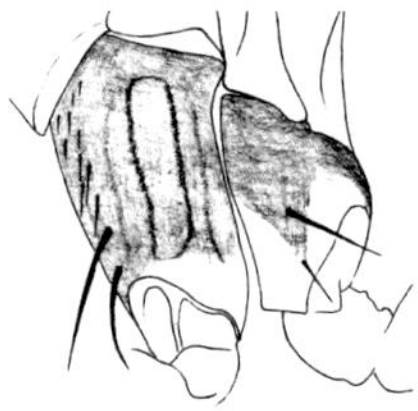

(bb)

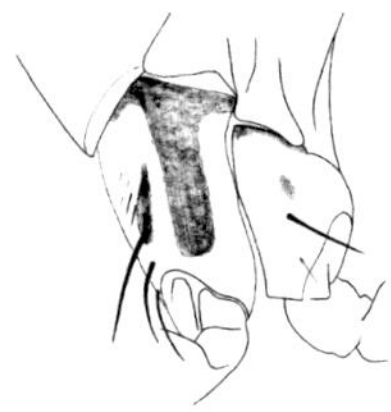

37(36) Antenna with scarcely any yellow visible from above, postpedicel almost entirely black and twice as long as deep (a). Front tarsus with an obvious apical seta on basitarsus, succeeding tarsomeres with only tiny apical setae (b, dorsal view). Face tapered, broad (c). Apicoventral epandrial lobe narrow and pointing backwards (d). Smaller species, wing length 4.1–4.6 mm........................ ***trivialis***

– Antenna with obvious yellow areas seen from above, postpedicel usually largely yellow at base, at most 1.5 times as long as deep (aa). Front tarsus with obvious apical setae on tarsomeres 1 and 2. Face parallel-sided, narrow (cc). Apicoventral epandrial lobe broad and domed, or not pointing backwards. Larger species, wing length 6 mm. 38

(a)

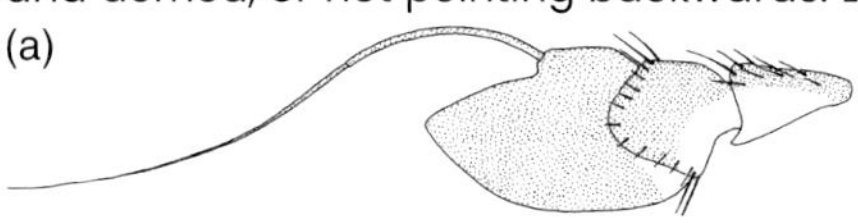

(aa)

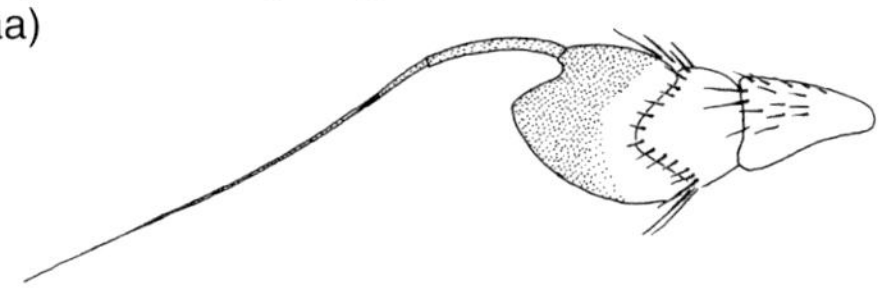

(b)

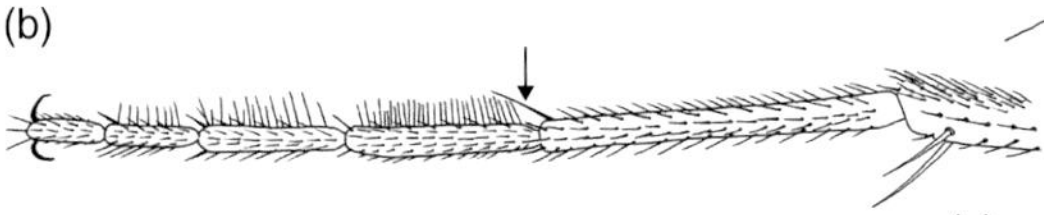

(d)

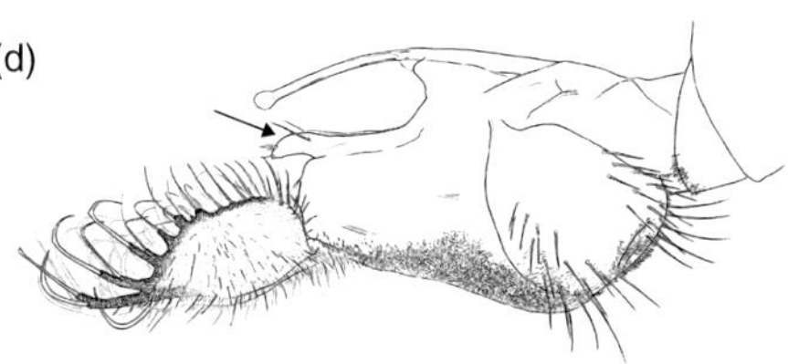

(c)

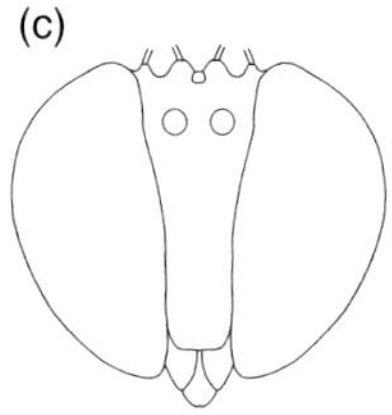

(cc)

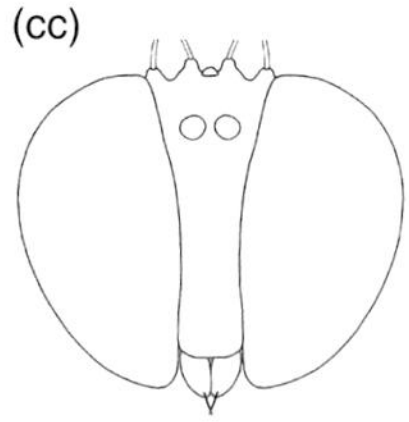

38(37) Front tarsomeres 2 to 4 almost straight in both dorsal and anterior views; tarsomere 3 without a curved apical seta (a). Tip of hind tibia black on all faces. Apicoventral epandrial lobe broad and rounded (b).. ***festivus***

– Front tarsomeres 2 to 4 curved in anterior view with strongly concave ventral faces; tarsomere 3 with a curved apical seta (aa). Tip of hind tibia yellow, sometimes black anteriorly. Apicoventral epandrial lobe narrow, more angular at tip (bb)..................................... ***cilifemoratus***

(a)

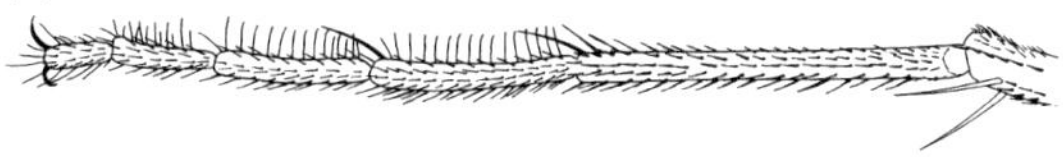

(b)

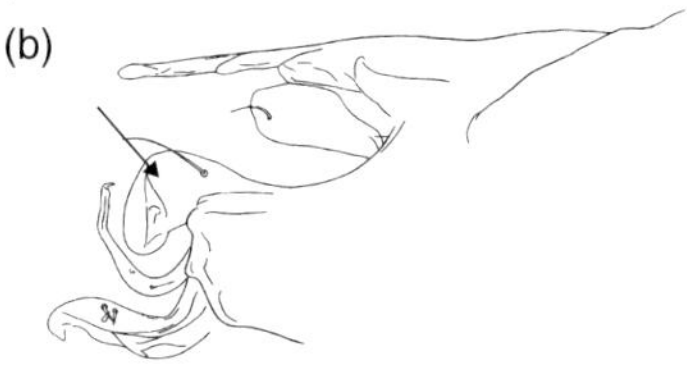

(aa)

dorsal view

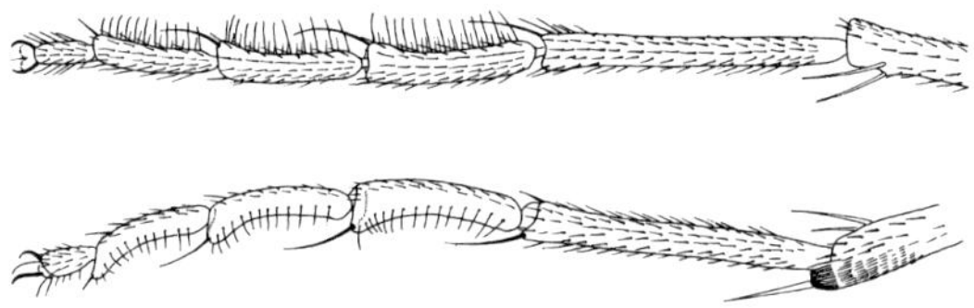

anterior view

(bb)

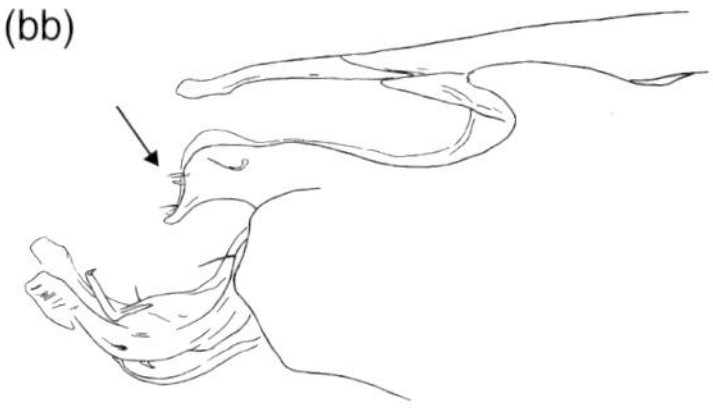

39(36) Hind tibia entirely yellow except at extreme tip, basitarsus yellow except sometimes at tip. Front tarsomeres 2 to 4 with sparse fine upright hairs on anterior face of their apical halves (a, dorsal view). Front basitarsus only slightly shorter (by about 10%) than the remaining segments combined (a). Front tarsus usually entirely pale......................... ***virgultorum***

– Hind tibia usually brownish at tip, basitarsus dusky yellow to black. Front tarsomeres 2 to 4 with normal black vestiture on all faces (aa). Front basitarsus clearly shorter (by about 25%) than the remaining segments combined (aa). Front tarsomeres 2–5 dark. .... ***arbustorum***

(a)

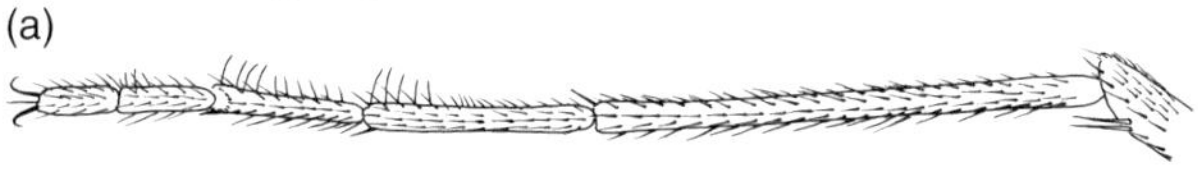

(aa)

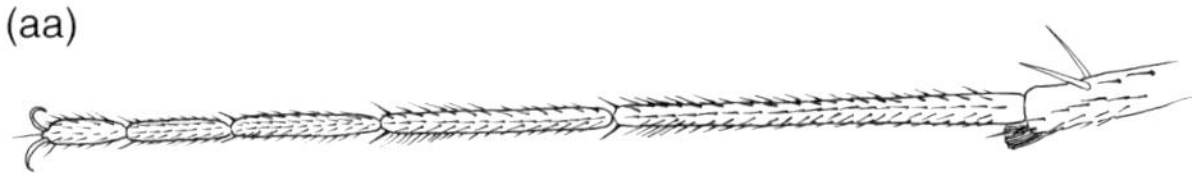

40(31) Antenna entirely black, not yellow basally on scape and pedicel, or rarely at tip of scape ventrally; if postpedicel very long, follow alternative option. *caligatus* and *notatus* may go either way.. 41

– Antenna clearly yellow ventrally on scape and often also on pedicel.. 49

Yellow femora, unmodified tarsi, no long fringe on hind femur, antenna black

41(40) Scape and pedicel exceptionally elongated, both longer than postpedicel, scape with dense long hairs (a). Mid and hind femora with 2–5 anterior setae on apical half. Wing remarkably short and broad with a short thick costal stigma (b). Face descending to bottom of eye....................... ***latipennis***

– Scape and pedicel unremarkable, both obviously shorter than postpedicel. Mid and hind femora with only 1 preapical anterior seta. Wing not broad, costal stigma when present not so swollen. Face ending above bottom of eye. 42

(a)

(b)

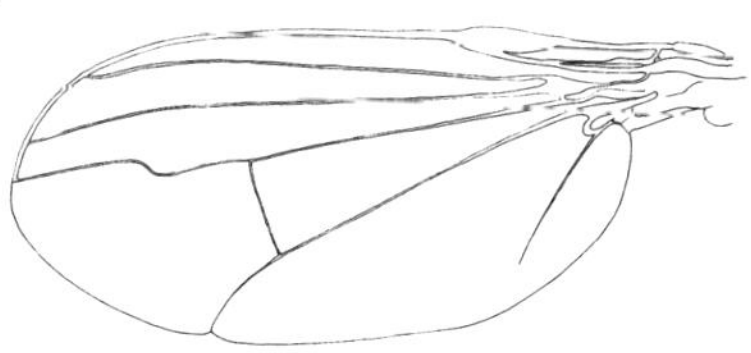

42(41) Rows of acrostichal setae splaying out into an unordered cluster in front of scutellum (a). Face covered with tiny white hairs (b). Wing without a costal stigma (c)................................ 43

– Acrostichal setae strictly biserial, ending near or before the penultimate (fifth) dorsocentral seta. Face bare. Wing with a small costal stigma (cc).. 46

(a)

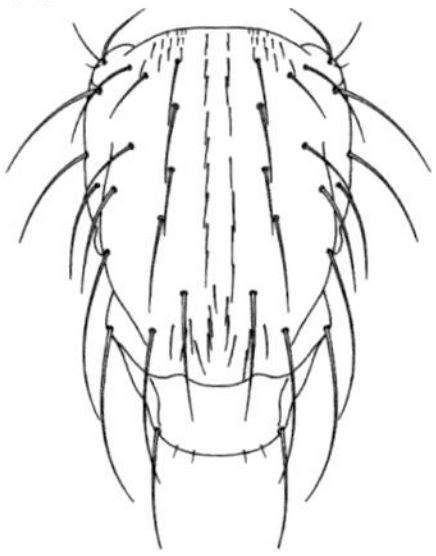

(b)

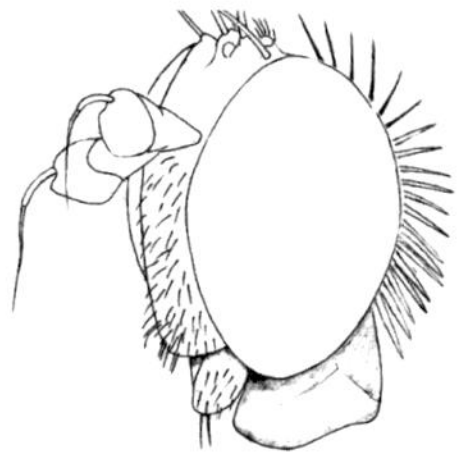

(c)

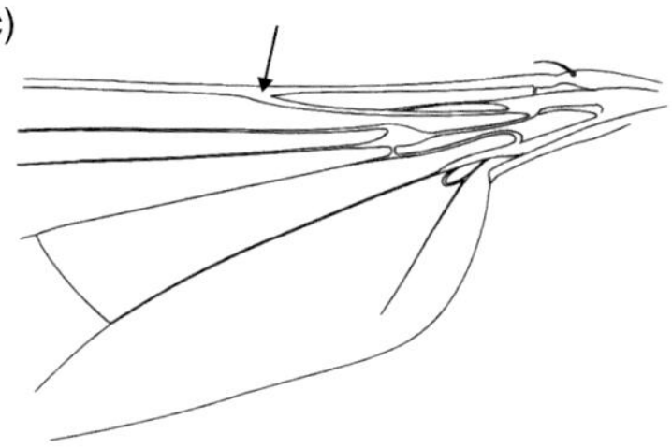

(cc)

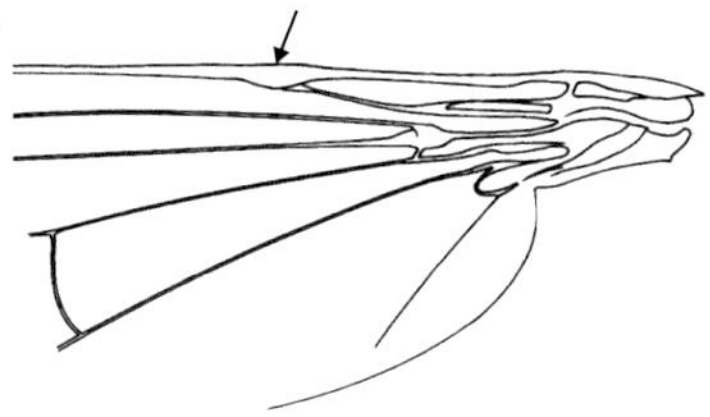

43(42) Calypter hairs white. Wing membrane brownish and darker in front of R_{4+5} than behind it. Apicoventral epandrial lobe large, pale and approximately oval, twice as high as wide, as large as cercus in lateral view (a).. ***nubilus***

– Calypter hairs black. Wing membrane almost uniformly pale grey. Apicoventral epandrial lobe smaller than cercus.. 44

(a)

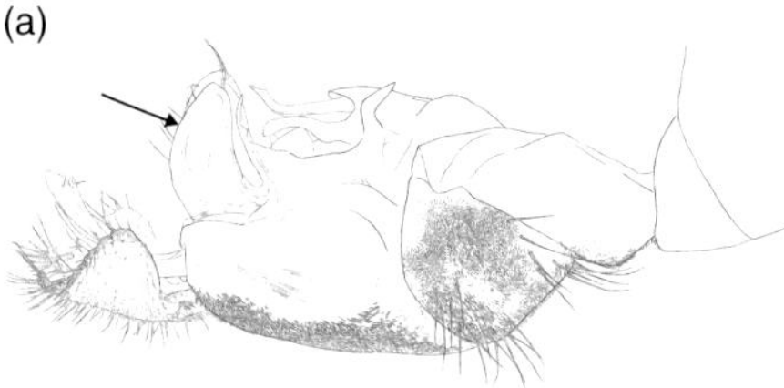

44(43) Front tibia with a long fine posteroventral seta at tip (a). Cercus large and approximately square with no extension on the inner corner (b); apicoventral epandrial lobe with a few conspicuous long hairs on margin (b). Front coxal hairs white on outer half, black on inner half (c)................ ***latilimbatus***

– Front tibia with short apical setae. Cercus triangular or semicircular with an extension on inner corner; apicoventral epandrial lobe with inconspicuous hairs. Front coxal hairs mainly black.. 45

(a)

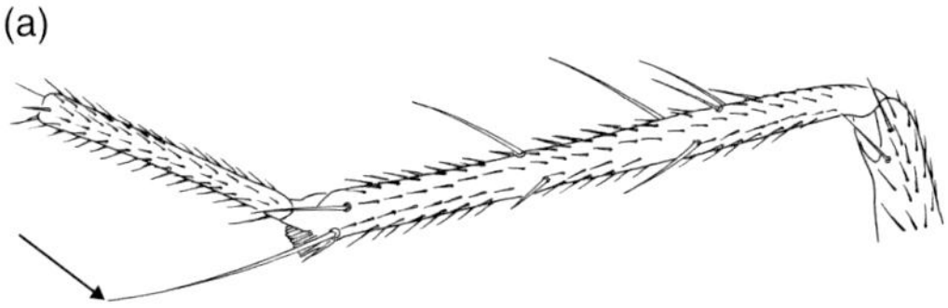

(c)

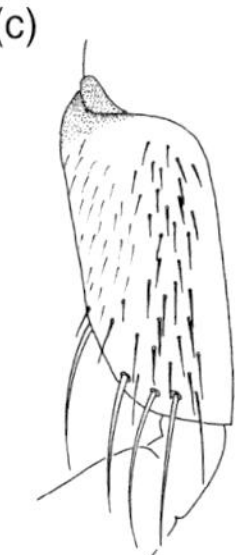

(b)

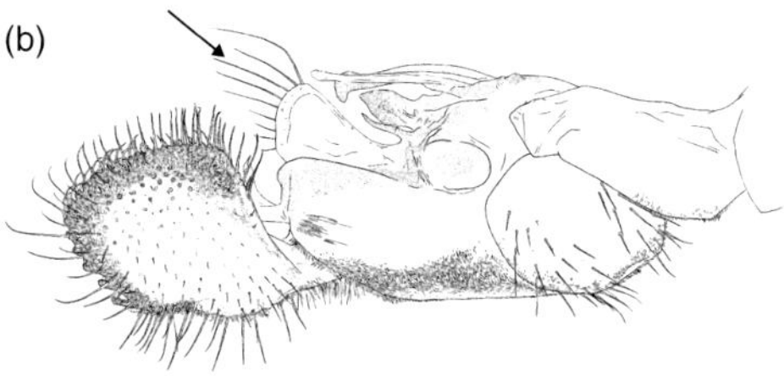

45(44) Front femur extensively dark dorsally and posteriorly. Hind femur with only a small or no dark mark at tip. Hind tibia yellow at tip. Hind basitarsus with more than 2 dorsal and more than 1 anterior setae. Wing margin smoothly rounded at anal corner. Cercus at inner corner with a tuft of short black hairs on a protrusion (b). Apicoventral epandrial lobe with a low triangular profile (c)...***andalusiacus***

– Front femur entirely yellow. Hind femur black from tip to preapical seta. Hind tibia black on apical sixth. Hind basitarsus with 2 dorsal and 1 anterior setae. Wing hind margin sinuate near base with a protruding anal lobe (aa). Cercus at inner corner with a ribbon-shaped projection (bb). Apicoventral epandrial lobe elongate-oval (cc). ....***excisus***

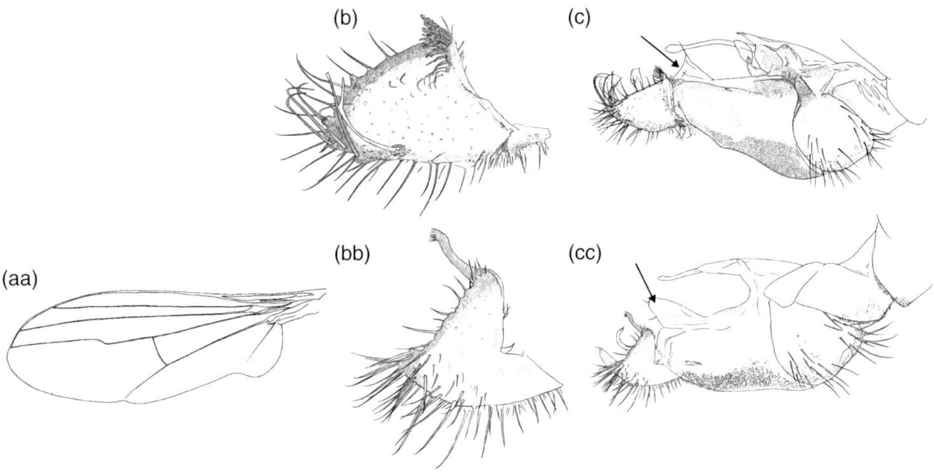

46(42) Face extending below eyes and widening on lower half (a, b). Stub-vein present on vein M_1 beyond dm-m (c). Front tibia with a very long posteroventral apical seta at apical three-quarters (d). Calypter hairs pale. Tergite marginal setae conspicuous, those on tergites 4 and 5 nearly as long as tergite. (See also couplet 15.)..***diadema***

– Face normal, tapered and not extending below eyes. No stub-vein on M_1. Front tibia with posteroventral setae at apical three-quarters either short or absent. Calypter hairs black. Marginal setae on tergites 4 and 5 about half length of tergites...47

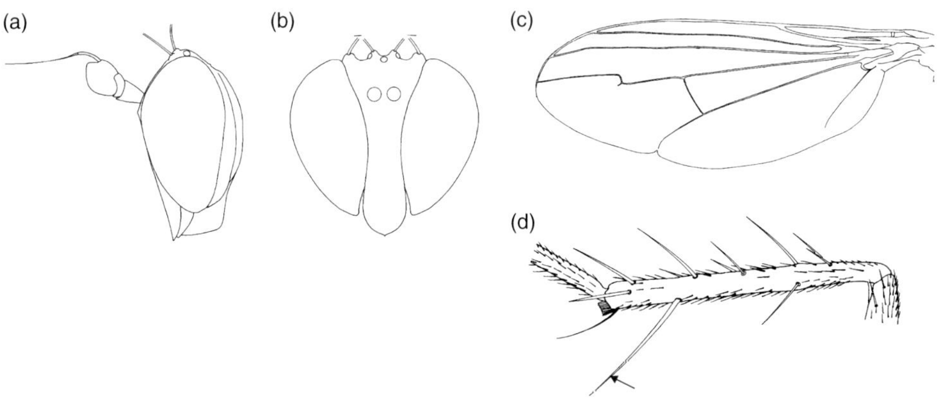

47(46) Front coxa entirely black. Lower postocular setae black. Cercus usually dirty white in centre and darkening gradually towards margin (a). Hind basitarsus with 3 dorsal setae. Femora often with dark areas on shaft. ..... ***rupestris***

– Front coxa partly or entirely yellow. Lower postocular setae pale. Cercus white with sharply demarcated black margin (aa). Hind basitarsus with 2 dorsal setae. Femora entirely yellow on shaft. 48

(a)

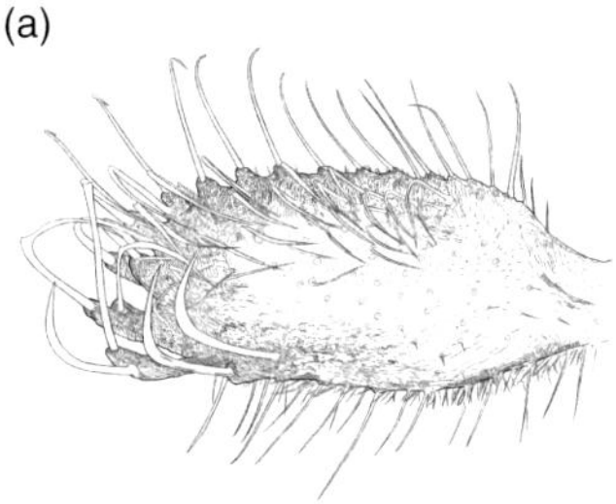

(aa)

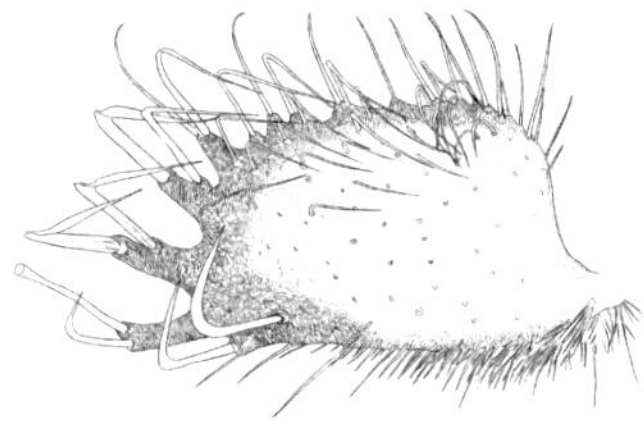

48(47) Hind tibia black on apical third (a). Face dusting yellow. Hind femur without a posteroventral fringe. Front tibia with a long fine anteroventral apical seta much thinner than other apicals (b). (See also couplet 58.) ..... ***notatus***

– Hind tibia black on apical sixth. Face dusting white. Hind femur with a posteroventral fringe of pale hairs about half depth of femur. Front tibia with all apical setae similar in structure. (See also couplet 56.) ..... ***caligatus***

(a)

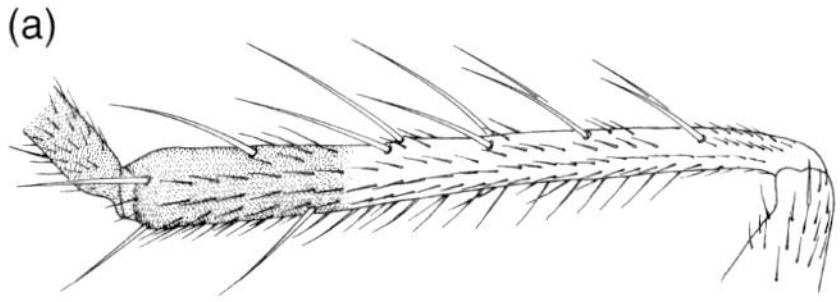

(b)

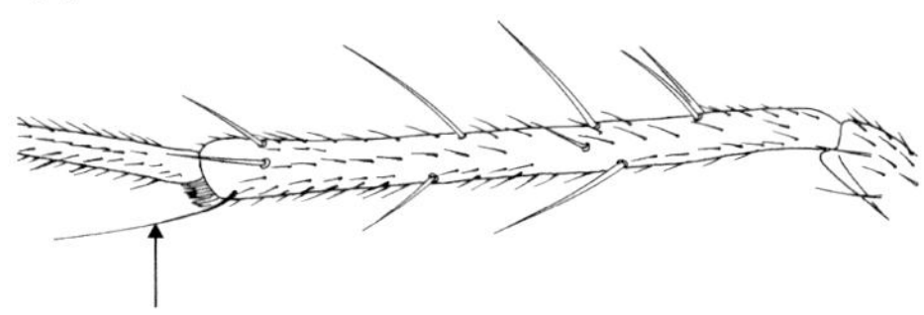

Yellow femora, unmodified tarsi, no fringe on hind femur, antenna partly yellow

49(40) Postpedicel elongate, at least 3 times as long as deep, tapering to a point. 50

– Postpedicel short, not more than twice as long as deep, at most slightly pointed but usually rounded or square at tip. 51

50(49) Pedicel and postpedicel partly yellow (a). Hind tibia entirely yellow, tip only slightly dusky. Front tibia with a long fine posteroventral apical seta (b). Hind tarsomere 2 clearly less stout than basitarsus. Apicoventral epandrial lobe low and square-ended (c). ..... ***longicornis***

– Pedicel and postpedicel entirely black (aa). Hind tibia black on apical sixth. Front tibia with all apical seta short. Hind tarsomere 2 as stout as and slightly longer than basitarsus (dd). Apicoventral epandrial lobe projecting, long and narrow, 4–5 times longer than wide (cc). ..... ***acuticornis***

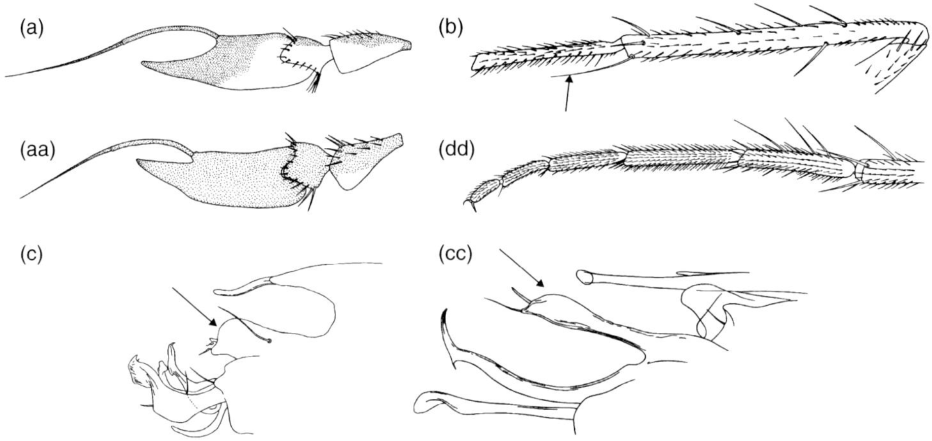

51(49) Front tibia with the 2–3 apical setae of similar structure (a)........52
– Front tibia with a finely tapered posteroventral apical seta, thinner and usually longer than adjacent apical setae (aa)........57

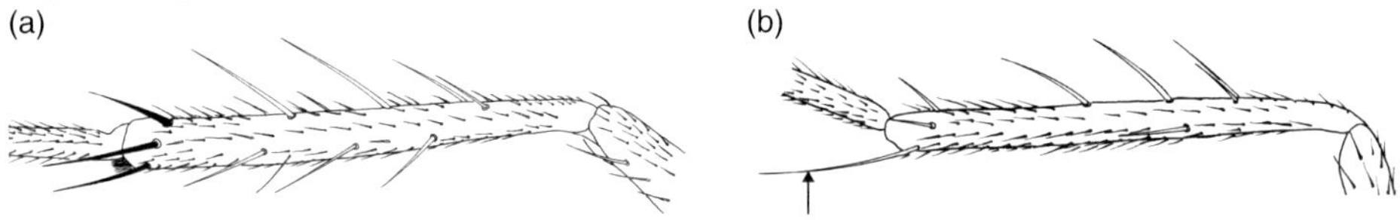

52(51) Postpedicel yellow at base ventrally, scarcely longer than deep. Calypter hairs pale. Cercus deeply split into a triangular lamella and a long narrow lobe (a). Hind femur with dark streaks along antero- and posteroventral surfaces (b)........***strigipes***
– Postpedicel entirely black, 1.5–2 times longer than deep. Calypter hairs black. Cercus oval, without a split (aa). Hind femur entirely yellow ventrally........53

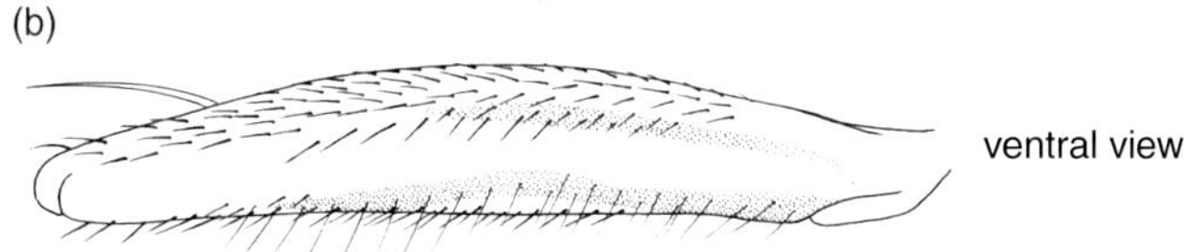

53(52) Costal stigma absent (a). Hind tibia slender, at its narrowest scarcely broader than tip of mid (not hind) tibia. Face pubescence pale yellow to whitish. 54

– Costal stigma present, short or long. Hind tibia stout, at its narrowest much broader than tip of mid tibia. Face pubescence either white or gold. 55

(a)

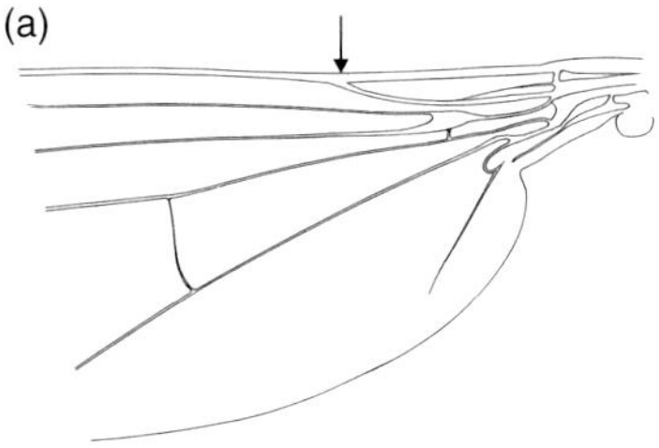

54(53) Scape entirely yellow. Postpedicel about 1.5 times longer than deep, arista inserted at about two-fifths from tip (a). Bend in vein M_1 moderate (b). Face dusting pale yellow. Cercus about 1.5 times longer than wide but sometimes shorter (fig. aa of couplet 52). .... ***simplex***

– Scape and pedicel broadly darkened dorsally. Postpedicel about twice as long as deep, arista inserted at about apical third (aa). Bend in vein M_1 slight. Face greyish white without yellow tinge. Cercus as long as wide (cc). .... ***mediicornis***

(a)

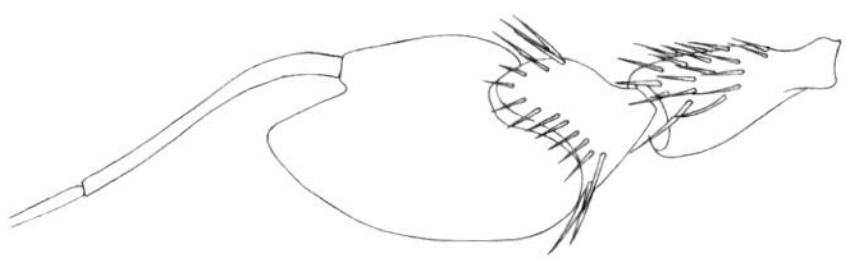

(b)

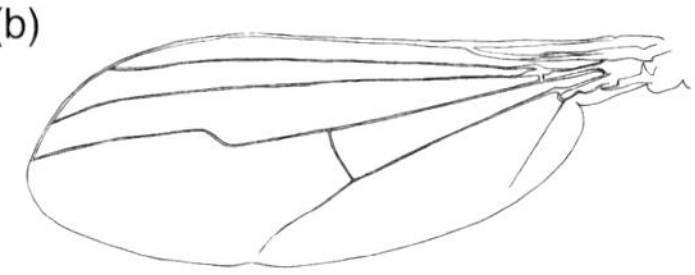

(aa)

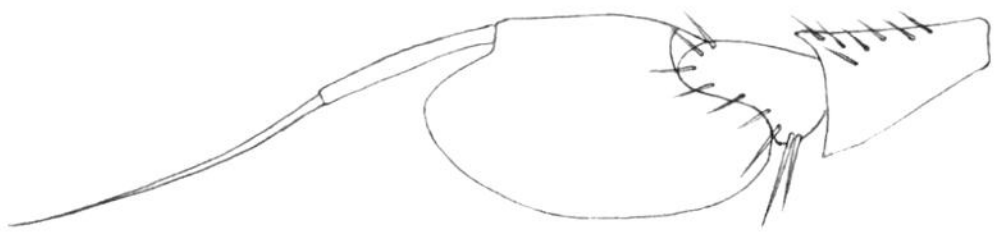

(cc)

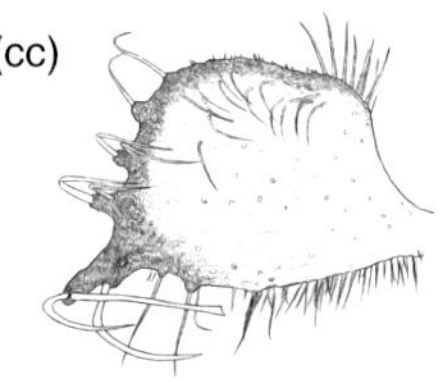

55(53) Hind basitarsus with 1 dorsal seta. Hind tibia black on apical third. Arista inserted at basal third of postpedicel (a). Hind tibia without a ciliolarium. Costal stigma long and tapered (b). .... ***agilis***

– Hind basitarsus with 2–3 dorsal setae. Hind tibia tip yellow or black on apical sixth. Arista inserted just before basal half of postpedicel (aa). Hind tibia with a ciliolarium of yellow pubescence on basal half. Costal stigma short (figs in couplet 56). 56

(a)

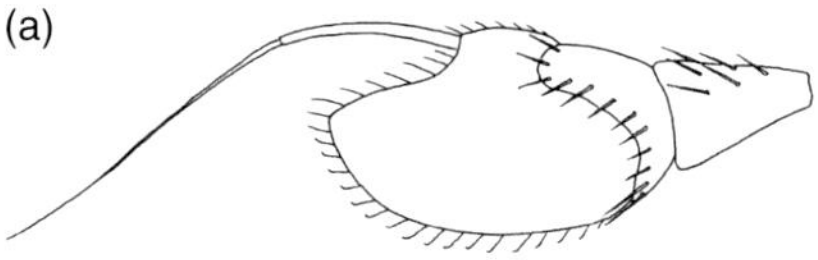

(b)

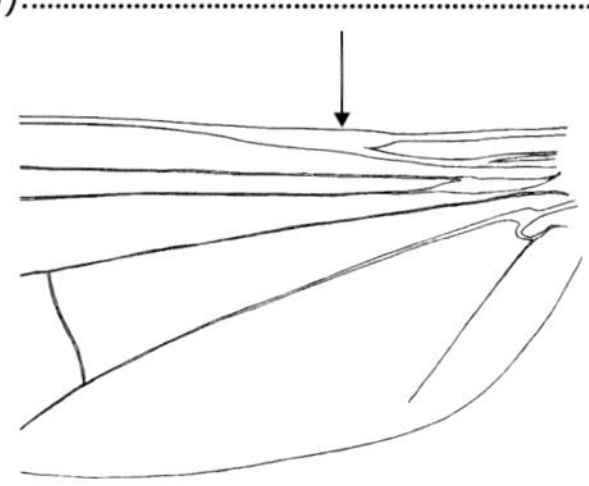

(aa)

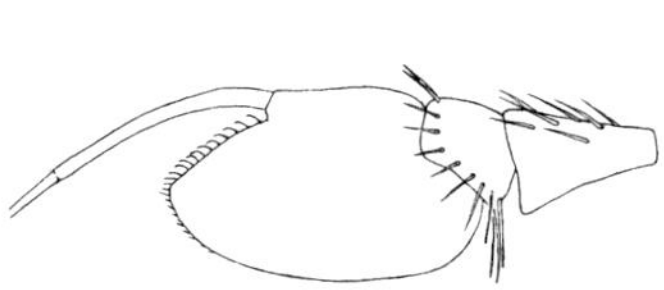

56(55) Face dusting white. Anal lobe of wing narrow, not projecting from smooth outline (a). Scape mainly or entirely black. Hind femur with a posteroventral fringe of pale hairs as long as black dorsal hairs on basal half (b). Ciliolarium not on a swollen area. (See also couplet 48.) .... ***caligatus***

– Face dusting golden. Anal lobe of wing broad and expanded (aa). Scape yellow with a narrow black streak dorsally. Hind femur without a posterodorsal fringe, only tiny pale hairs far shorter than dorsal hairs. Ciliolarium on a swollen area (cc)....................................***lineatocornis***

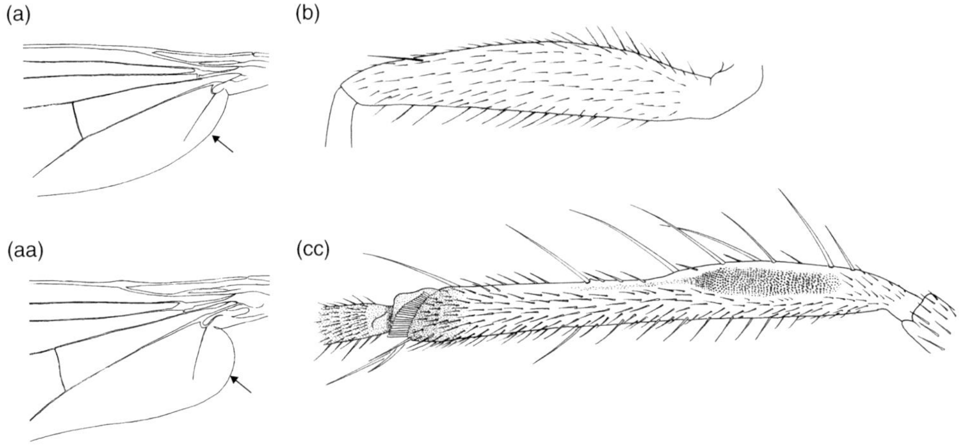

57(51) Hind tibia with its tip entirely encircled by black on apical quarter to third. Hind basitarsus black. Postpedicel entirely black.. 58

– Hind tibia either entirely yellow or dark at most on apical sixth but remaining paler or yellow on ventral or posterior face. Hind basitarsus yellow on basal third or more. Postpedicel usually yellow on at least basal quarter to third of outer face, the rest black..59

58(57) Hind coxae mainly black. Front coxa with hairs black from inner face to apical setae, pale on outer face. Scape black or at most paler on conus; pedicel black; postpedicel about twice as long as deep (a). Basiventral epandrial lobe short, its tip reaching less than halfway from its base to crest of apicoventral lobe (b). (See also couplet 48.)..***notatus***

– Hind coxae entirely yellow. Front coxa with hairs white, at most a few scattered black hairs. Scape yellow, at most slightly darker on dorsal surface; pedicel black dorsally and yellow ventrally; postpedicel less than 1.5 times as long as deep (aa). Basiventral epandrial lobe long, reaching half to two-thirds distance from its base to crest of apicoventral lobe and leaving a long narrow gap between them (bb)..***linearis***

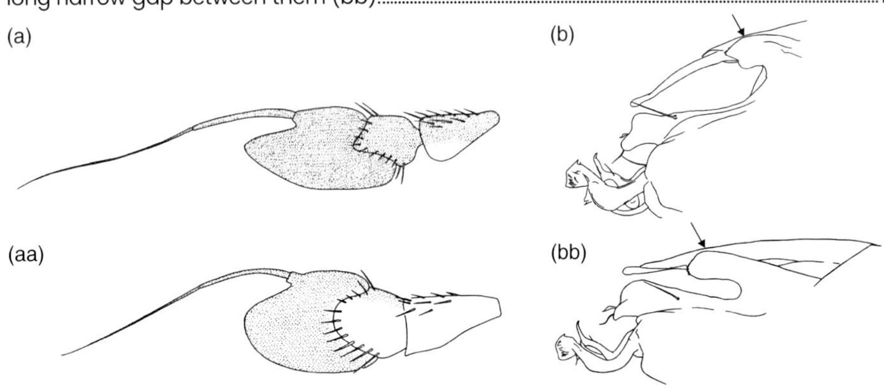

59(57) Wing not dark near tip. Vein M_1 with abrupt bends on apical section, often with a stub-vein (a). Cercus with exceptionally long apical hairs arising from short tubercles which are all equally small (b). Hind tibia without a ciliolarium. Larger species, wing length about 5–6 mm. .... ***nitidus***

– Wing darkened near tip (aa). Vein M_1 with smooth bends. Cercus with short apical hairs, none longer than width of cercus, some arising from long tubercles (bb). Hind tibia with a small ciliolarium. Smaller species, wing length about 4 mm. 60

(a)

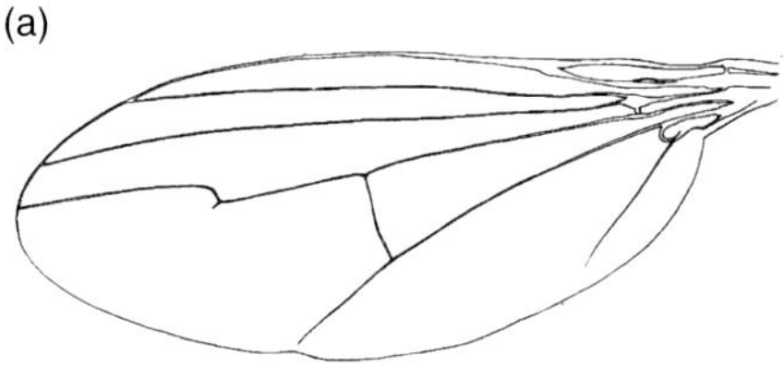

(b)

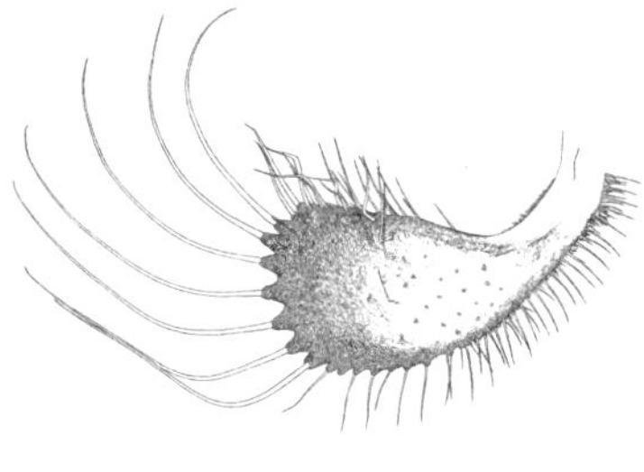

(aa)

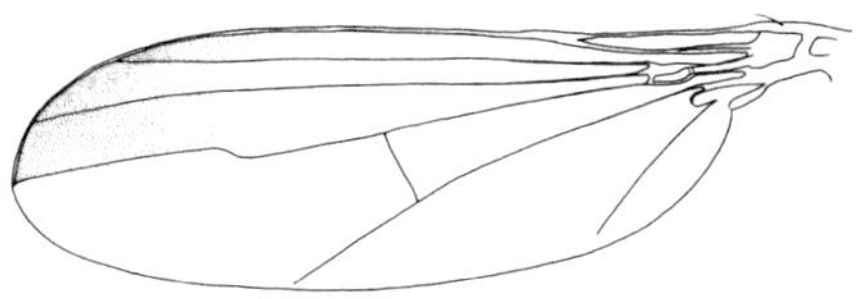

(bb)

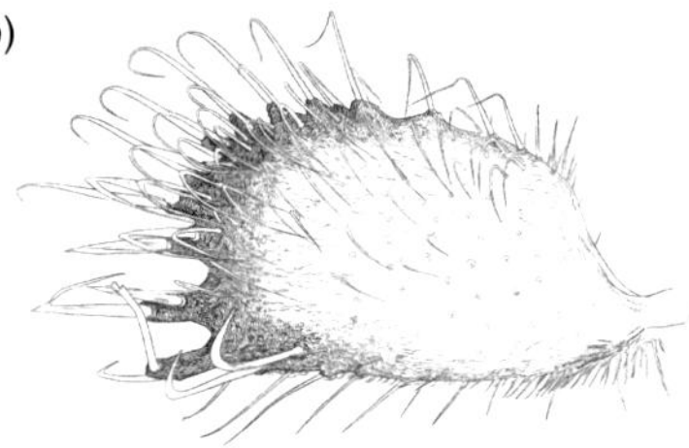

60(59) Face dusting white. Scape and pedicel black dorsally, yellow ventrally. Front coxa with a dark mark on outer side at base. Ciliolarium comprising a small oval patch of tiny dark points at basal sixth of tibia (a). .... ***sabinus***

– Face dusting pale yellow. Scape and pedicel entirely yellow. Front coxa entirely yellow at base. Ciliolarium comprising a narrow strip of dark points at basal one- to two-fifths of tibia (aa). .... ***calinotus***

(a)

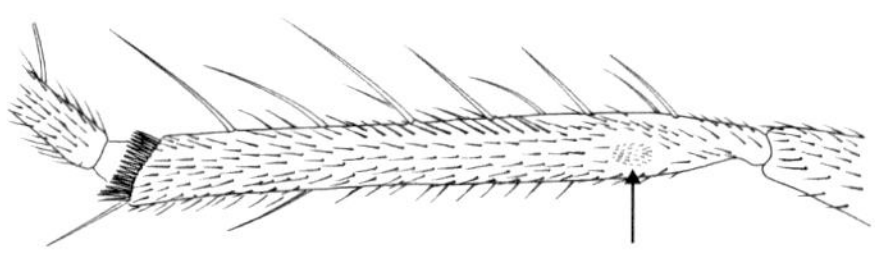

(aa)

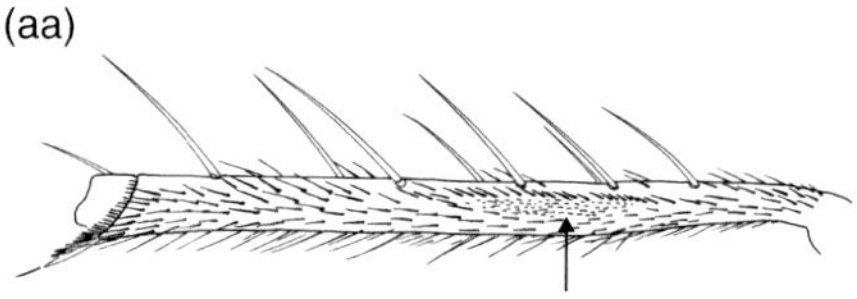

Key to species of *Dolichopus* (females)

1 All femora almost entirely black or dark brown, at least 1 pair completely encircled with black on at least part of their length, ignoring yellow tips and bases. 2

– Femora yellow, or if some are partly black then at least mid or hind femur mainly yellow... 14

2(1) All tibiae black or dark brown; in cases of doubt (*atripes*, *phaeopus*), arista arises further from tip of postpedicel, as in (a), not (b) 3

– Front and usually mid tibia entirely or mainly yellow. 9

(a)

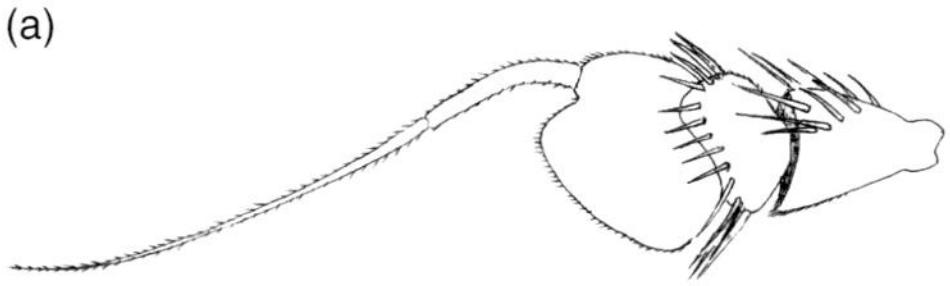

(b)

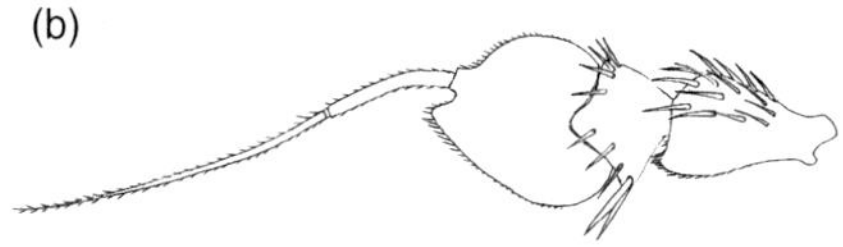

3(2) Lower postocular setae black.. 4

– Lower postocular setae white, lowest 1–3 may be black.. 8

4(3) Swollen base of costa black on leading edge (yellow ventrally), at low magnification (×10) appearing as dark as the tegula next to it (a). Mid basitarsus yellow on basal third beneath black vestiture. Upper proepisternal hairs black and white in about equal proportions. Anal lobe broad, as wide as maximum width of cell r_{4+5} with an acute anal angle (a).............. ***picipes***

– Base of costa yellow, clearly pale at low magnification compared to the black tegula. Mid basitarsus entirely black at base. Upper proepisternal hairs either all black or nearly all white. Anal lobe broad (a) or narrower, about two-thirds maximum width of cell r_{4+5} with an obtuse anal angle (aa).. 5

(a)

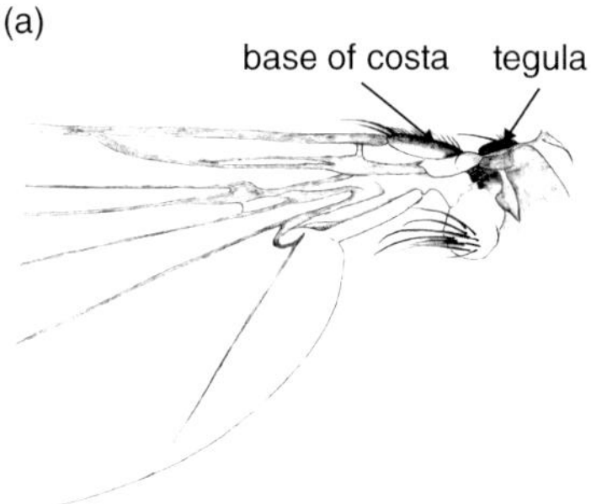

(aa)

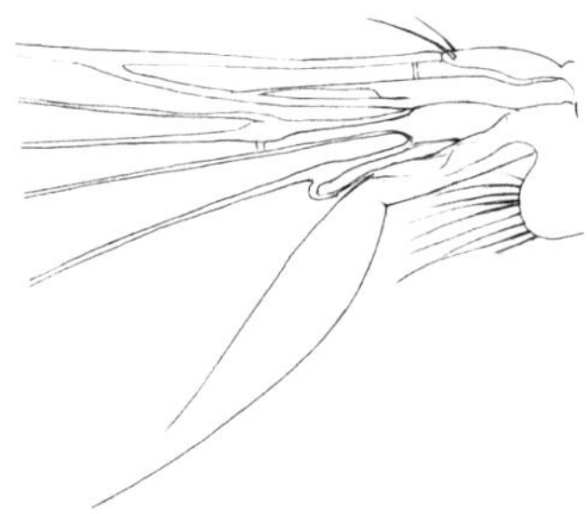

5(4) Cell c obviously pale compared to cell r_1, wing membrane usually dark on apical half compared to basal half, more obvious at low magnification (a). Front and mid knees black, rarely with a trace of yellow. Anal lobe narrow.. ***atratus***

– Cell c similar in shade to cell r_1, wing membrane not or scarcely darker on apical half but may be darker anteriorly or around crossvein dm-m (aa). Front and mid knees narrowly yellow.. 6

(a)

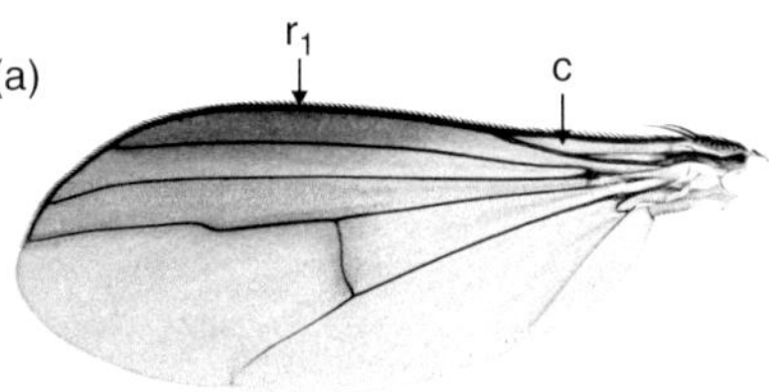

(aa)

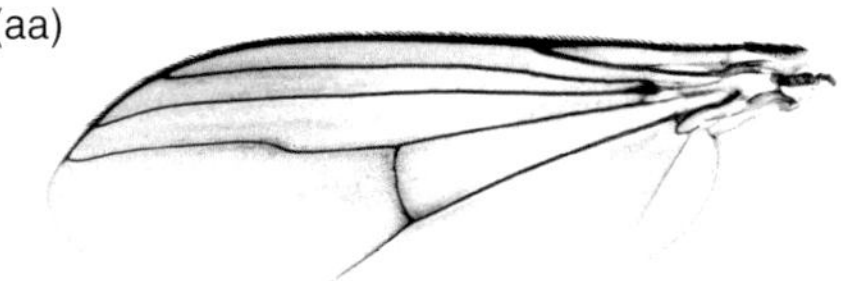

6(5) Mid and hind femora with 2–3 preapical setae (a). Mid tibia with 2–4 ventral setae (including pv and av). Wing membrane darker around crossvein dm-m, appearing as a spot at low magnification (fig. aa of couplet 5). Anal lobe broad. Rare montane species. .... ***maculipennis***

– Mid and hind femora with 1 preapical seta. Mid tibia with 1 ventral seta. Wing membrane not obviously darker around dm-m, not appearing as a spot at low magnification................. 7

(a)

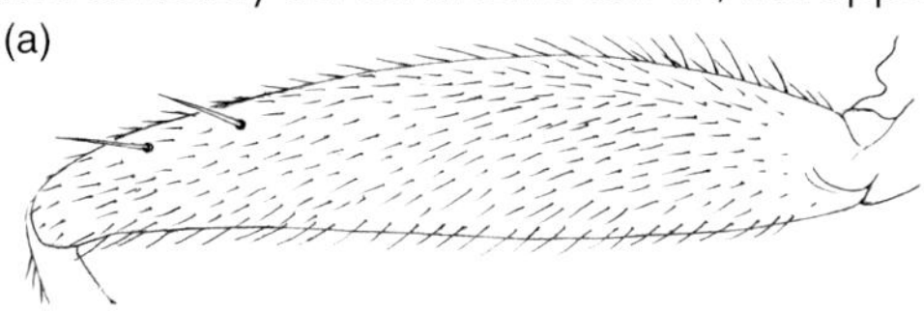

7(6) Mid and hind femora ventrally at base with fine pale hairs longer than half depth of femur (a). Front tarsomeres 1 and 5 black, 2–4 yellow. Front coxa yellow from tip to well above apical setae. Apical section of arista more than 3 times length of basal section. Clypeus without fine hairs. Frons violet or green. Rare fenland species. ***nigripes***

– Mid and hind femora ventrally at base with undistinguished black vestiture. Front tarsus entirely black. Front coxa with only extreme tip yellow. Apical section of arista twice length of basal section. Clypeus with short hairs at margin and at mouth edge. Frons green. Possibly extinct in Britain. ***melanopus***

(a)

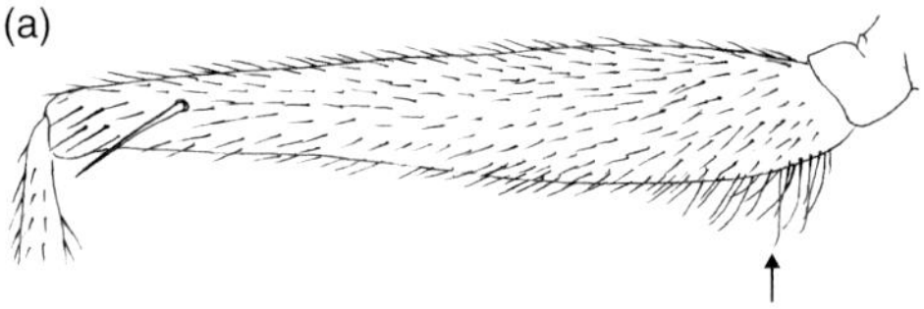

8(3) Swollen base of costa black or very dark on leading edge (yellow ventrally), at low magnification (×10) appearing as dark as the tegula next to it (fig. a of couplet 4). Front coxa with more numerous hairs (a, front view). Scutellum with fringe of moderately dense brown hairs on posterior margin (b). Face dusting brown-tinged. Vein M_1 with inflexion on apical section more marked (c); anal lobe slightly broader, anal angle about 90° (c). Larger species, wing length 4.1–4.6 mm. ***atripes***

– Base of costa yellow, clearly pale at low magnification compared to the black tegula. Front coxa with fewer hairs (aa). Scutellum with fringe of white hairs on posterior margin, sparser and often untidily aligned. Face dusting white. Vein M_1 with weaker inflexion (cc); anal lobe slightly narrower, anal angle more obtuse (cc). Smaller species, wing length 3.4–4.0 mm. ***phaeopus***

(a)

(aa)

(c)

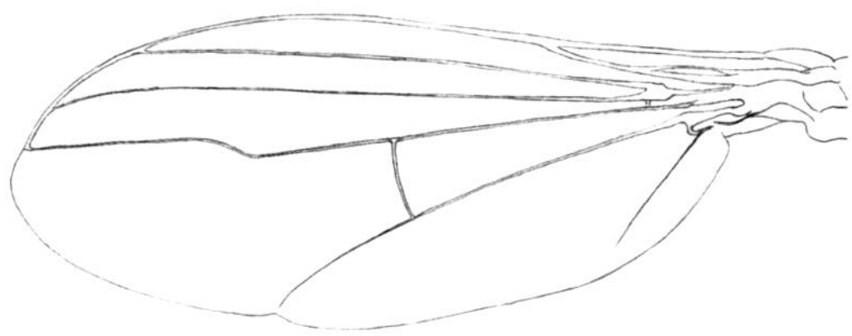

(b)

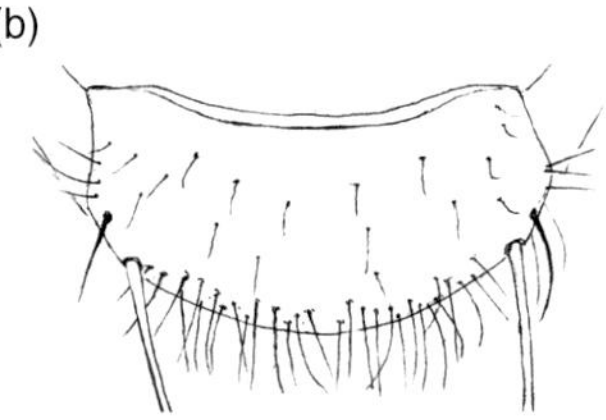

(cc)

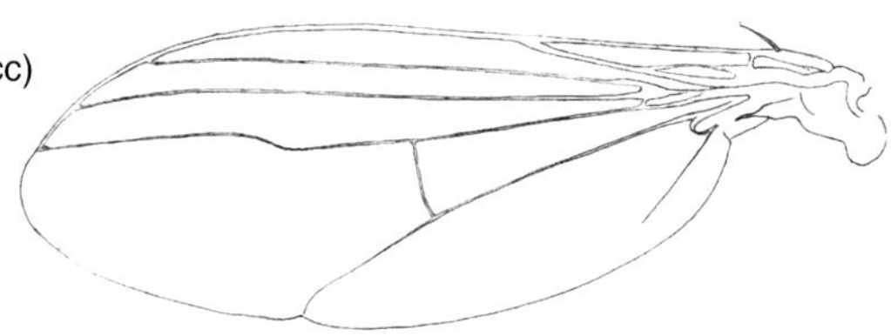

9(2) Lower postocular setae black. Large species, wing length at least 4.3 mm. 10

– Lower postocular setae white. Small species, wing length up to 4.3 mm. 13

10(9) Mid femur with 2 preapical setae (a). Anal lobe broad with angular outline, at least as wide as maximum width of cell r_{4+5} (b). Hind tibia yellow with a black tip. Face pale grey or yellow-tinged. .... ***campestris***

– Mid femur with only 1 preapical seta. Anal lobe margin smoothly rounded. Hind tibia either yellow with a black tip or mainly to entirely black. Face white.......... 11

(a)

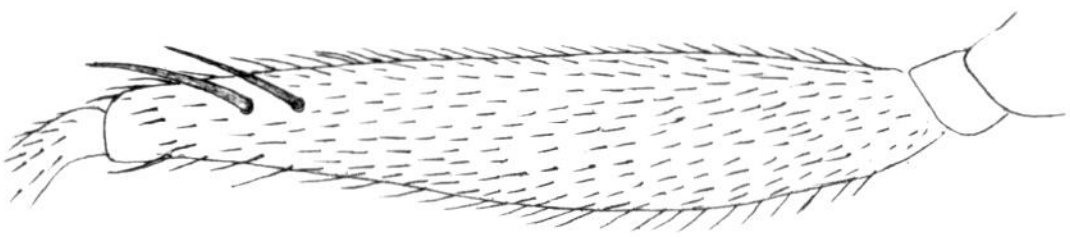

(b)

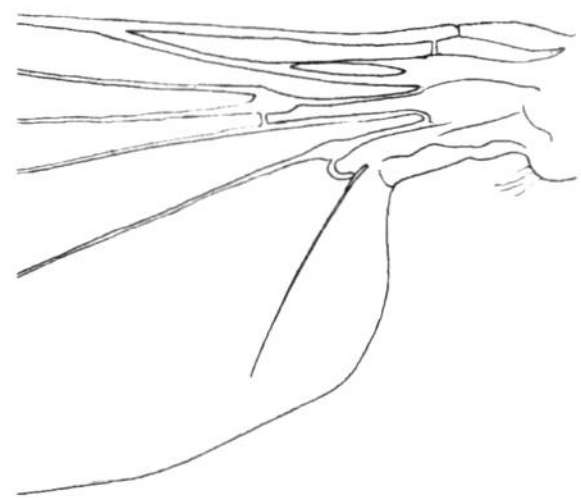

11(10) Mid tibia with at least 3 setae ventrally (av, v or pv) (a). Face narrower, its height at least twice its width (b). Postpedicel longer than deep (c); arista inserted close to tip of postpedicel, upper margin of postpedicel basal to arista several times its length beyond arista (c). Front tarsus yellow to tip of tarsomere 3, tarsomere 4 and 5 black. .... ***planitarsis***

– Mid tibia with only 1 ventral seta. Face broader, its height less than twice its width (bb). Postpedicel as long as deep, its upper margin basal to arista scarcely twice its length beyond arista. Front tarsus black from tip of basitarsus outwards. 12

(a)

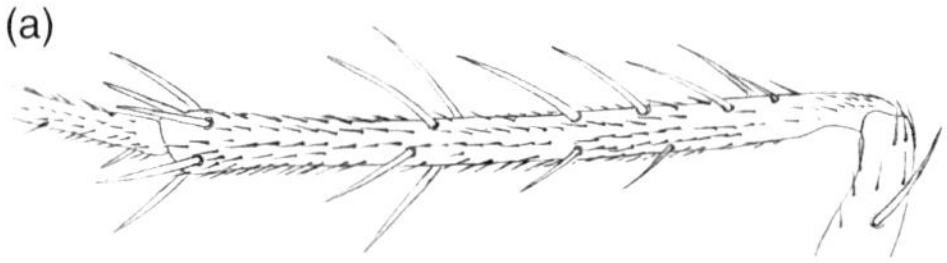

(c)

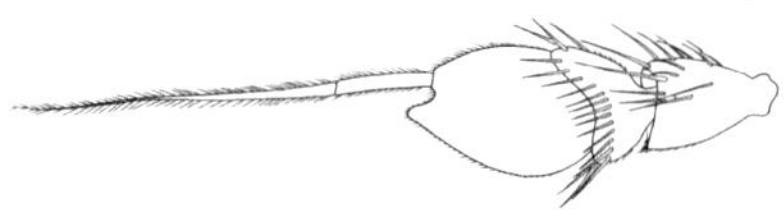

(b)

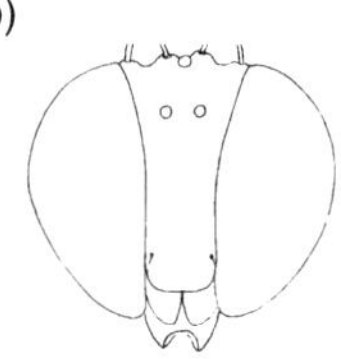

(bb)

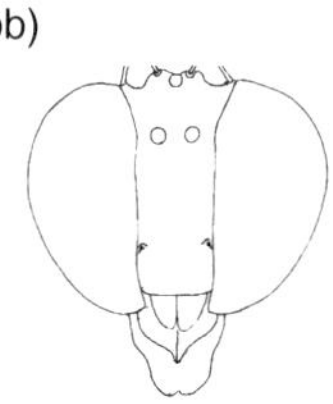

12(11) Scutellum with a marginal fringe of dense hairs regularly aligned (a). Mid tibia entirely clear yellow. Frons green or coppery. Mid coxa completely black or obscurely yellow at extreme tip.......... ***lepidus***

– Scutellum with only sparse hairs on apical margin. Mid tibia beyond yellow knee with slight to obvious darkening, often blotchy, on apical quarter or more. Frons blue or violet. Mid coxa with a clear yellow tip seen from front.......... ***laticola***

(a)

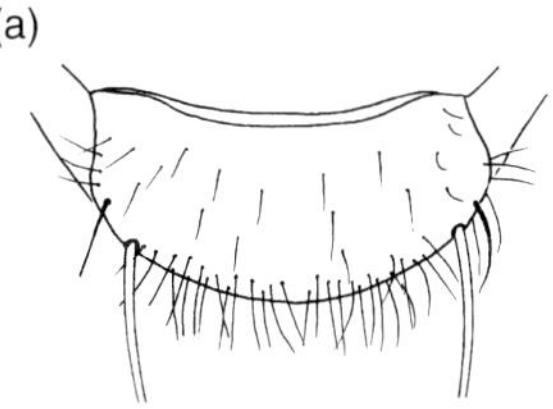

13(9) Hind basitarsus with 2 dorsal setae (rarely 1). Tips of coxae, trochanters and bases of femora black with only traces of yellow. Scape entirely black. Mid tibia with anteroventral seta placed beyond or opposite nearby antero- and posterodorsal setae. Hind basitarsus clearly shorter than tarsomere 2. [*atripes* and *phaeopus* with pale tibiae may run here, their arista arising about halfway along top of postpedicel, but closer to tip than base in *vitripennis* (a).] .... ***vitripennis***

– Hind basitarsus with 1 dorsal seta. Tips of coxae, trochanters and bases of femora usually yellow, often resulting in a conspicuous pale base to all legs. Scape usually yellow ventrally. Mid tibia with anteroventral seta placed proximal to nearby antero- and posterodorsal setae. Hind basitarsus approximately as long as or only slightly shorter than tarsomere 2. .... ***clavipes***

(a)

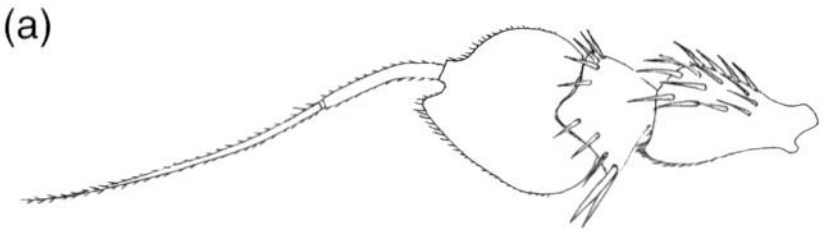

14(1) Hind femur with 2–4 anterior preapical setae (a) 15

– Hind femur with 1 anterior preapical seta. 19

(a)

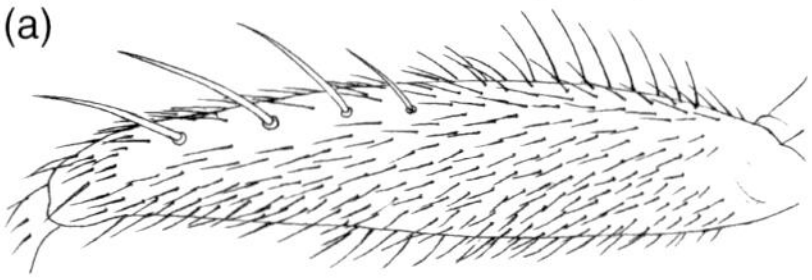

15(14) Mid femur with 2–4 anterior preapical setae (a). Mid tibia with several setae ventrally (antero- plus posteroventral) (b). 16

– Mid femur with 1 anterior preapical seta. Mid tibia with only 1 ventral seta. 17

(a)

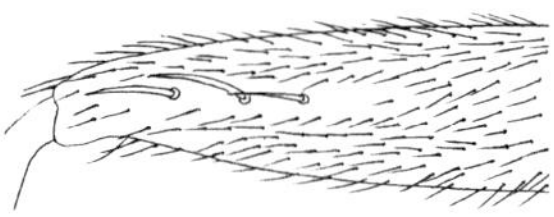

(b)

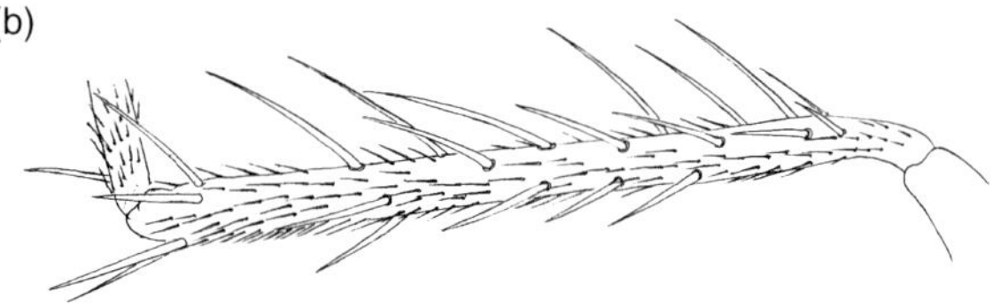

16(15) Front coxa black on about basal half, yellow apically. Mid tibia with 1 posteroventral seta, seen from behind (a, posterior view). Lower postocular setae entirely black. Palps black. Scape entirely black. Front tibia with many setae on posterior and posteroventral face (b, posterior view). .... ***ungulatus***

– Front coxa almost entirely yellow. Mid tibia with 2–3 posteroventral setae, seen from behind (aa). Lower postocular setae yellow or with only a few black setae at bottom of row. Palps yellow. Scape at least partly yellow ventrally. Front tibia with at most 3 setae on posterior and posteroventral face (bb) .... ***claviger***

(a)

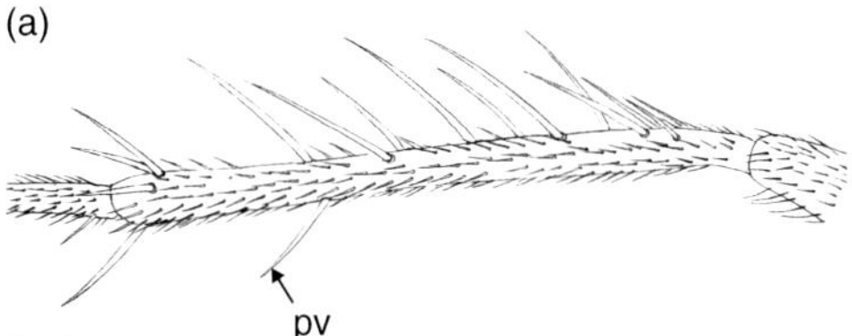

(b)

(aa)

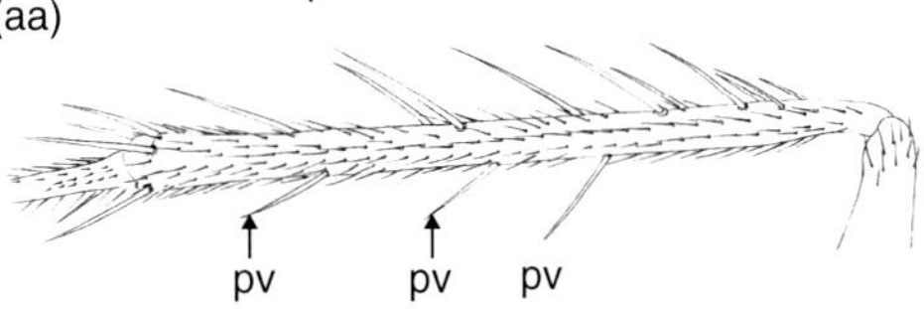

(bb)

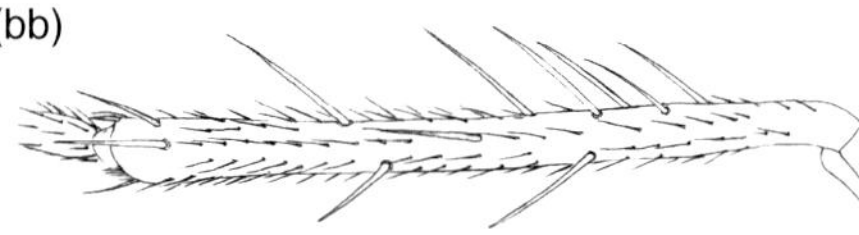

17(15) Hind tibia entirely or mainly black or dark brown, sometimes paler dorsally................***urbanus***
– Hind tibia entirely or mainly yellow. 18

18(17) Antenna mainly yellow with dorsal half of postpedicel black. Scape and pedicel shorter than postpedicel, which is slightly longer than deep. Frons green. Crossvein dm-m at midpoint between base and tip of wing. ....***popularis***
– Antenna entirely black. Scape and pedicel longer than postpedicel which is as long as deep (aa). Frons violet. Crossvein dm-m well beyond midpoint, almost at two-thirds from base. ....***latipennis***

(aa)

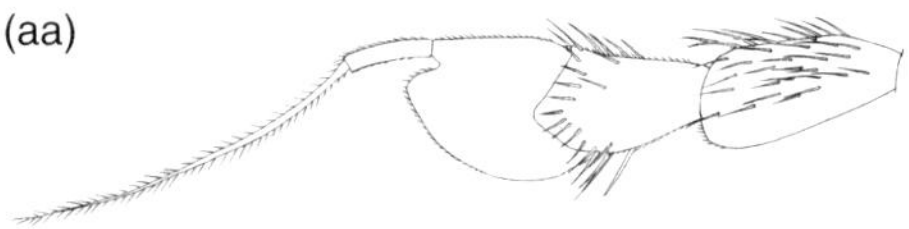

19(14) Mid tibia with 2–4 setae ventrally (a)..20
– Mid tibia with 1 seta ventrally (aa)..28

(a)

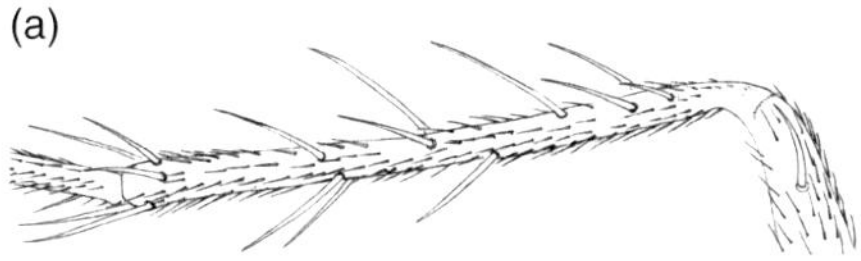

(aa)

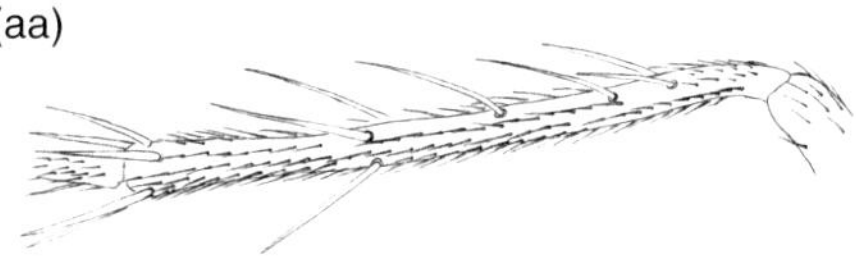

20(19) Postpedicel and pedicel entirely black, rarely with a tiny yellow dot at base of postpedicel ventrally. Mid tibia with at least 5 subequally strong anterodorsal setae, not counting a possible smaller seta near base (a). 21
– Postpedicel and pedicel partly yellow, sometimes only narrowly pale at base ventrally. Mid tibia with 4 equally strong anterodorsal setae, not counting a possible smaller seta near base (fig. a of couplet 19) or rarely a fifth strong seta near base forming a cramped group here..24

(a)

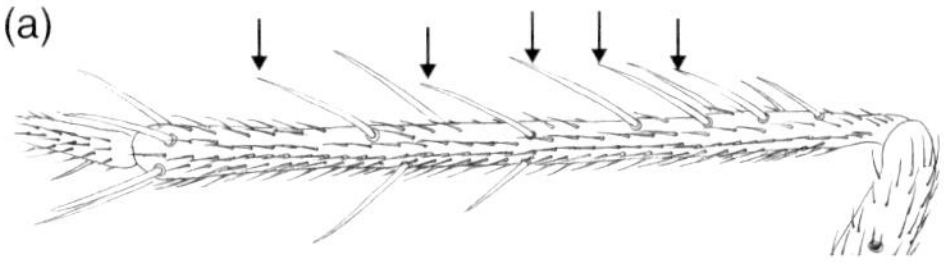

21(20) Front coxa black on outer face on at least basal quarter. Front coxa with hairs entirely black, apical setae black. Hind tibia with several strong ventral setae on apical half or more (fig. a of couplet 22). Scape usually entirely black..22
– Front coxa yellow on outer face with at most a dark spot at extreme base. Front coxa with hairs yellow on outer face on upper half, apical setae black (aa, front view). Hind tibia with numerous weak ventral setae no longer than width of shaft, apart from 1–2 long setae at three-quarters from base (bb). Scape yellow ventrally..23

(aa)

(bb)

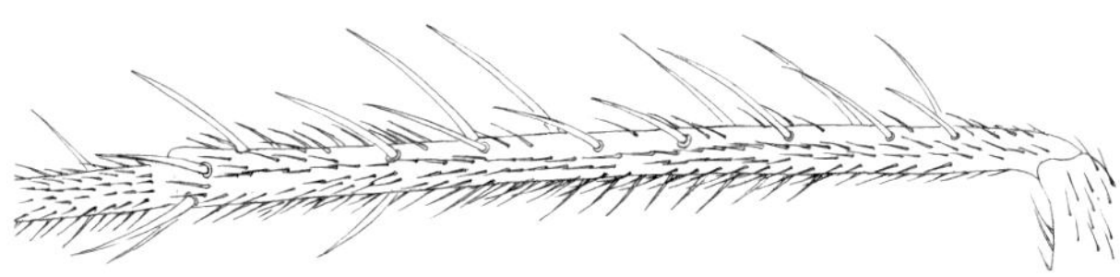

22(21) Lower postocular setae black. Hind tibia black at tip for more than 3 times width of shaft (a). Mid and hind coxae entirely black, colour continuous with dark base of trochanters. Front tibia with 2–3 strong posteroventral setae (b).. ***brevipennis***

– Lower postocular setae white. Hind tibia black at tip for about 1–1.5 times width of shaft (aa). Mid and hind coxae yellow at tip, continuous with yellow of trochanters. Front tibia with 1 strong posteroventral seta (bb)... ***migrans***

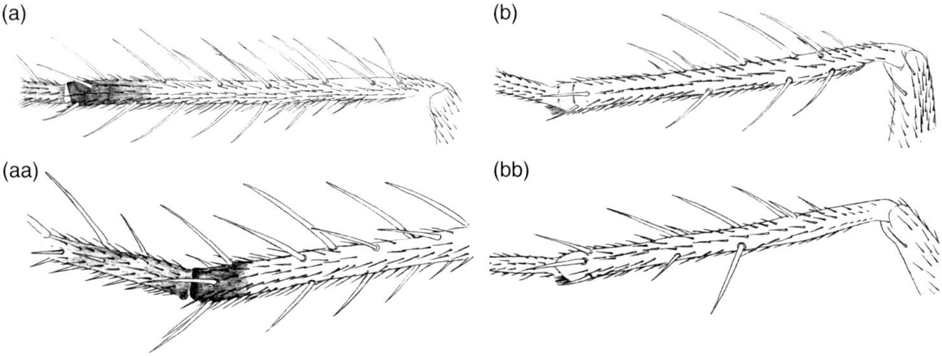

23(21) Vein M_1 with very weak inflexions on apical half, anal lobe broad, hind margin of wing with marked indentation at tip of M_4 (a, male shown). Mid basitarsus usually without a dorsal seta.. ***discifer***

– Vein M_1 with almost right-angled bends, anal lobe narrow, hind margin of wing scarcely indented at tip of M_4 (aa). Mid basitarsus with 1 strong dorsal seta................................ ***plumitarsis***

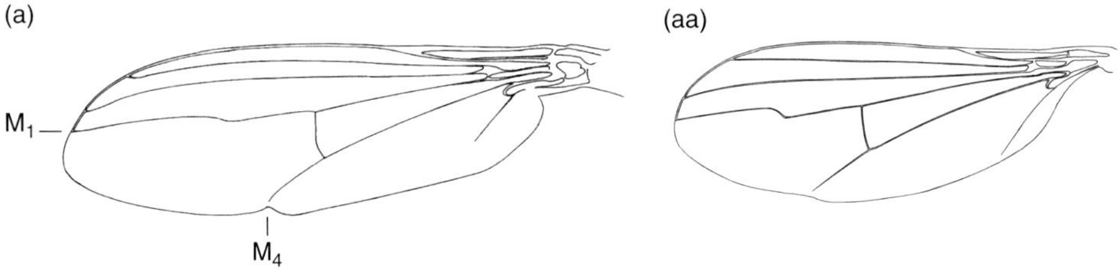

24(20) Hind coxa with a black wedge from base to the large lateral seta (a) or more extensively black. Mid coxa on outer face continuously black across most or whole of width from base to strongest outer seta (a). Front coxa with hairs usually mainly black, but variable. Face height twice its width.. 25

– Hind coxa mainly yellow, sometimes with a dark smudge near base rarely extending to lateral seta (aa). Mid coxa yellow on front margin above level of uppermost outer seta (aa). Front coxa with hairs mainly pale. Face height more than twice its width (about 2.3 times)........... 27

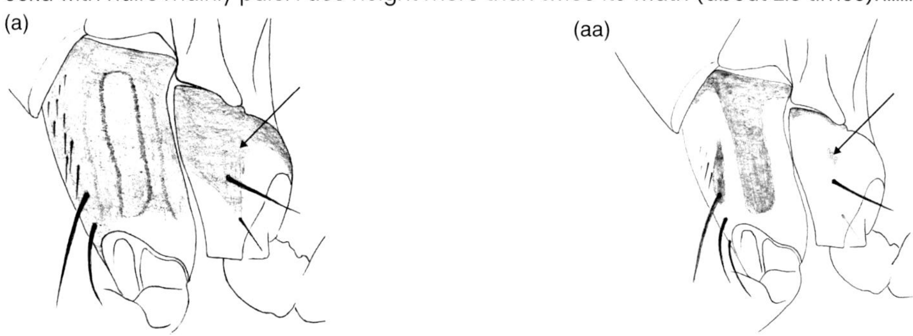

25(24) Antenna black along entire dorsal surface, obvious in dorsal view, postpedicel often mostly black on inner face. Front coxa usually with a dark smudge at outer corner at base. Smaller species, wing length 4.5–5.5 mm. .... ***trivialis***

– Antenna yellow, or at least not continuously black, on dorsal surface from scape to near insertion of arista. Front coxa usually entirely yellow at base. Larger species, wing length 5–6.5 mm. 26

26(25) Hind tibia dark at apex, especially on posterior face but sometimes rather indistinct, basitarsus entirely black at base (a). Basal section of arista less than one-third length of apical section (0.23–0.29 times) and slightly shorter than depth of postpedicel. .... ***festivus***

– Hind tibia entirely yellow, basitarsus very narrowly yellow at base. Basal section of arista more than one-third length of apical section (more than 0.35 times) and equal to or just longer than depth of postpedicel. .... ***cilifemoratus***

(a)

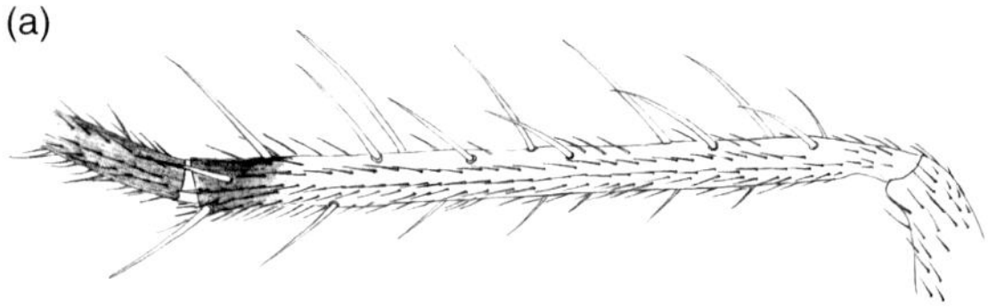

27(24) Hind basitarsus yellow with a narrow dark tip, leaving at least two-thirds as pale as tibia. Mid coxa yellow in front of ridge demarcating anterior margin of central depressed area, with a dark streak encompassing uppermost strong outer seta, only rarely more extensively dark (fig. aa of couplet 24) .... ***virgultorum***

– Hind basitarsus mainly black, even at the base much darker than tibia. Mid coxa black in front of ridge and scarcely separated from black area behind by a narrow yellow strip (bb). .... ***arbustorum***

(bb)

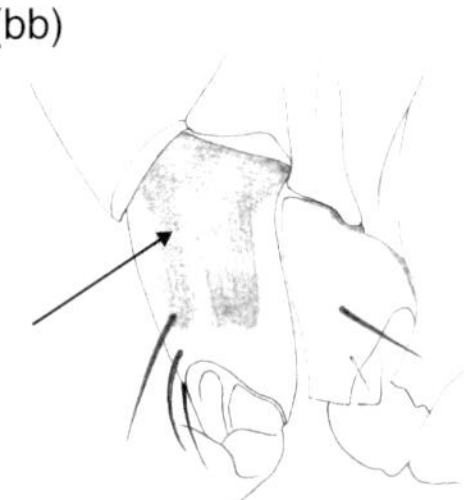

28(19) Rows of acrostichal setae splaying out into an unordered cluster in front of scutellum (a). Face covered with tiny white hairs (b). 29

– Rows of acrostichal setae strictly biserial posteriorly, not reaching beyond last dorsocentral seta. Face bare (except in *longitarsis*, with black lower postocular setae) 31

(a)

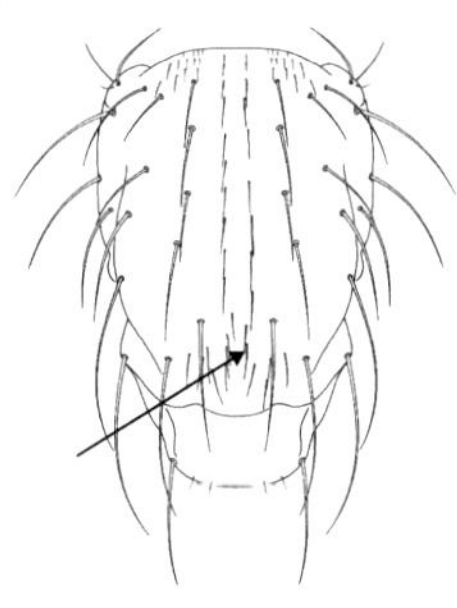

(b)

29(28) Hind tibia entirely yellow at tip. Scutum with small hairs in and beside dorsocentral row extending further forward than third dorsocentral seta (a). Front femur often slightly brownish dorsally, or ventrally near base, or more extensively (compare with clear yellow of hind femur).. ***andalusiacus***

– Hind tibia slightly to obviously darkened at tip, sometimes only a slightly darker shade of yellow. Scutum with additional small hairs not extending further forward than fourth dorsocentral seta (as fig. a of couplet 28). Front femur entirely yellow, same shade as hind femur.................................30

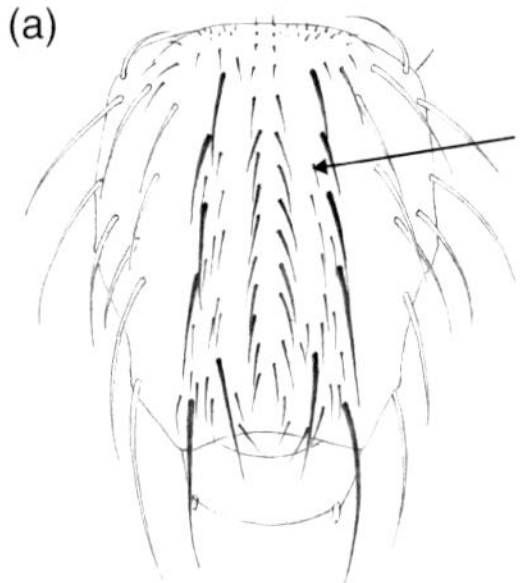

30(29) Front coxa black on outer face for more than half its length, from base to level with uppermost strong apical seta (a, lateral view). Hind tibia with tip black for at least twice width of shaft here, reaching to within a shaft's width of the long preapical ventral seta (b); this ventral seta almost as stout and long (about three-quarters) as adjacent antero- and posterodorsal setae (b). Margin of wing at anal lobe slightly bulging and sinuous (c)................................. ***excisus***

– Front coxa black only on upper quarter or less (aa). Hind tibia with tip yellow, or brownish to black for at most twice width of shaft here, rarely extending more than halfway to long preapical ventral seta (bb); this ventral seta clearly weaker and shorter (about half) than adjacent antero- and posterodorsal setae (bb). Margin of wing at anal lobe smoothly rounded.. ***latilimbatus***, ***nubilus*** (inseparable)

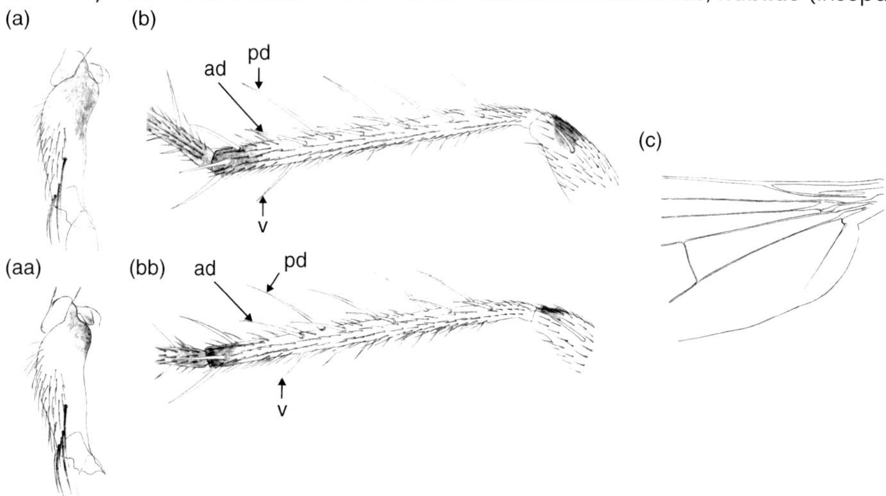

31(28) Wing with bends of M_1 almost square and usually with a stub-vein (a)...32

– Wing with bends of M_1 more gently curved and without a stub-vein. 34

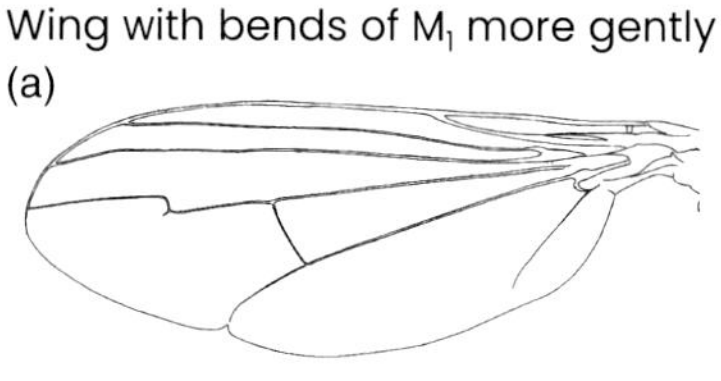

32(31) Face descending to bottom of eye or slightly beyond (a, b). Tergites with exceptionally long marginal setae, those on tergites 4 and 5 three-quarters to equal to length of tergite 5 (c). Front tibia with 2 strong posteroventral setae, the apical one often slightly longer (d). Scape entirely black. .... ***diadema***

– Face ending above lower margin of eyes. Tergites with short marginal setae, those on tergite 4 only rarely three-quarters length of tergite 5. Front tibia with 1 strong posteroventral seta on basal half, any seta on apical half much smaller. Scape yellow along whole ventral margin. 33

(a) (b) (c) (d)

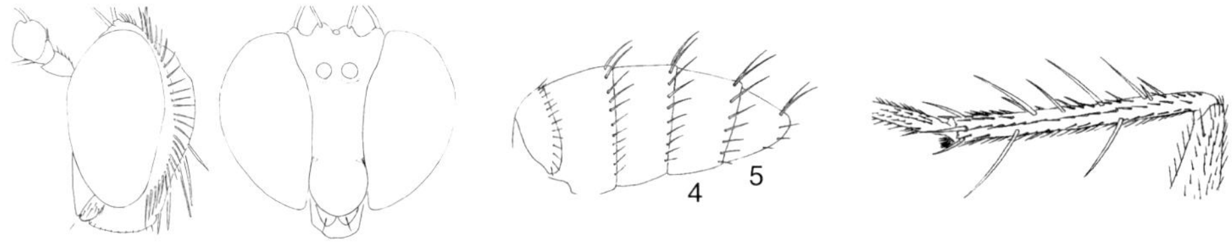

33(32) Hind basitarsus with 1 dorsal seta (a). Front coxa black on basal third on outer face (b, lateral view). Front coxa with hairs yellow on outer half, black on inner half. Antenna mainly black in ventral view, pedicel and postpedicel entirely black. Hind basitarsus black or dark yellow. .... ***griseipennis***

(Some *signifer* may fall here, with square bends on M_1 and 1 seta on hind basitarsus, but hind femur has a dark tip dorsally (not yellow) and hind tibia is true black on apical fifth almost to ventral preapical seta, not yellow or brownish.)

– Hind basitarsus with 2 dorsal setae. Front coxa entirely yellow, at most with a slight smudge at basal outer corner. Front coxa with hairs mainly black, yellow only on upper part of outer face. Antenna mainly yellow in ventral view, pedicel entirely yellow, postpedicel yellow on basal half. Hind basitarsus bright yellow with a black tip. .... ***nitidus***

(a) (b)

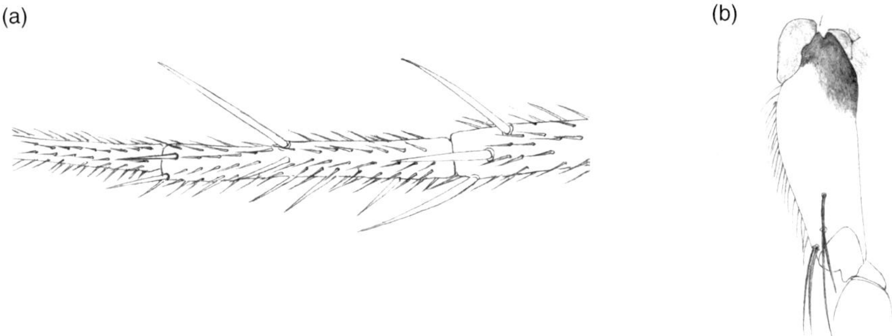

34(31) Front coxa dark on at least basal half. 35

– Front coxa entirely yellow or with at most a small dark spot or smudge on basal outer corner. 37

35(34) Hind basitarsus with 1 dorsal seta. Lower postocular setae white. Palps yellow. Front coxa with hairs white on outer half or more, black on inner half or less. Scape pale ventrally at tip. .... ***signifer***

– Hind basitarsus with 2–3 dorsal setae. All postocular setae black. Palps black below any pubescence with pale reflections. Front coxa with hairs mainly black, or pale only on upper half above uppermost apical seta. Scape entirely black. 36

36(35) All coxae with bright yellow of tips continuous with yellow of trochanters, leg bases yellow in ventral view. Scutellum posterior margin with a dense fringe of dark hairs (a). Mid tibia with 3 long anterodorsal setae in addition to 1 slightly shorter seta near base. Face with tiny white hairs (as fig. b of couplet 28)........ ***longitarsis***

– All coxae with black or dark yellow of tips continuous with partly dark colour of trochanters, leg bases dark in ventral view. Scutellum posterior margin with inconspicuous sparse short pale hairs. Mid tibia with 4 long anterodorsal setae in addition to 1 slightly shorter seta near base (as fig. a of couplet 19). Face bare........ ***rupestris***

(a)

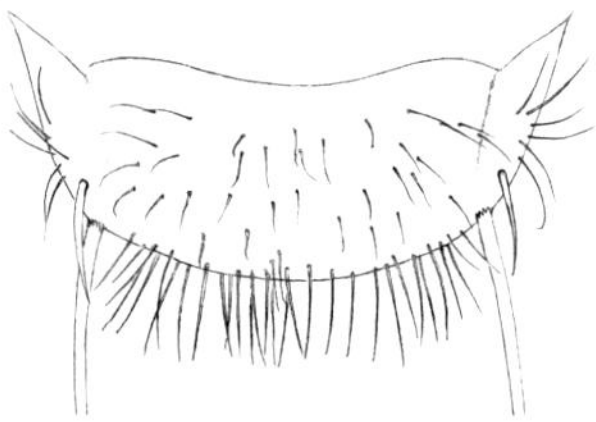

37(34) Hind tibia entirely or mainly black or dark brown, sometimes paler dorsally........ ***urbanus***

– Hind tibia mainly yellow........ 38

38(37) Postpedicel entirely black, only very rarely with a trace of yellow at base ventrally.39

– Postpedicel with at least some obvious yellow areas, if only at base ventrally. .... 51 or 57

39(38) Scape extensively black on dorsal surface, no hairs (or rarely 1) arising from the yellow area, most obvious on inside face when seen from above........ 40

– Scape entirely yellow or with only a narrow dorsal dark stripe under black hairs, some hairs arising from the yellow area........ 43

40(39) Postpedicel tip acutely pointed, lower margin slightly angled at apical third (a). Last (fifth) tergite with long marginal setae, almost equal to length of tergite. Hind coxa yellow with a dark wedge from base to lateral seta, or reduced to an isolated mark at seta (c)........ ***acuticornis***

– Postpedicel tip finely or bluntly rounded, lower margin smoothly curved (aa, bb). Last tergite without outstanding marginal setae. Hind coxa black on at least basal two-thirds (cc)........ 41

(a)

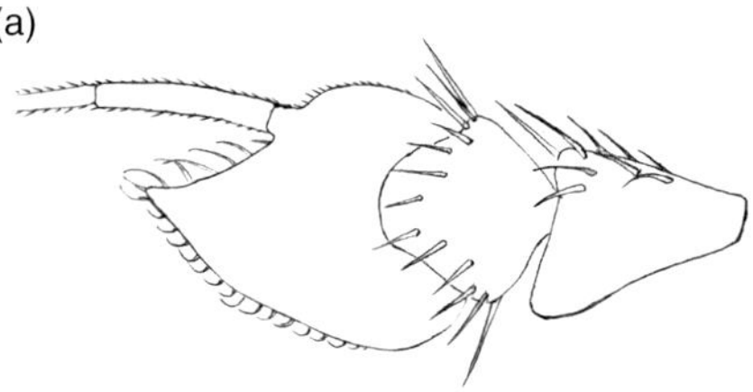

(c)

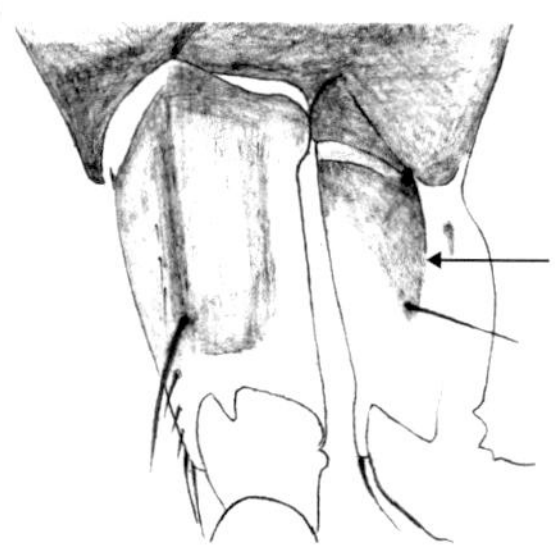

(aa)

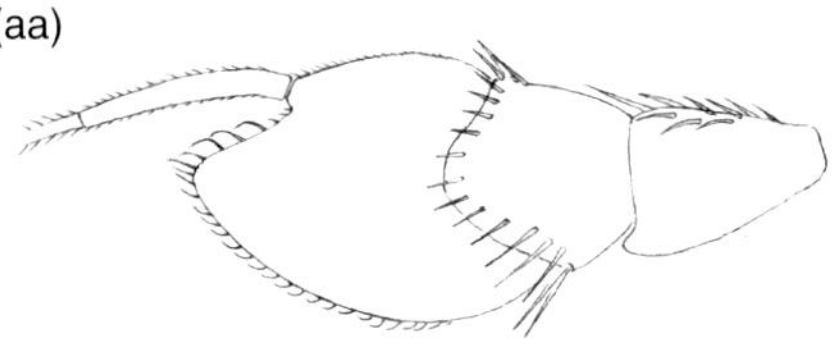

(cc)

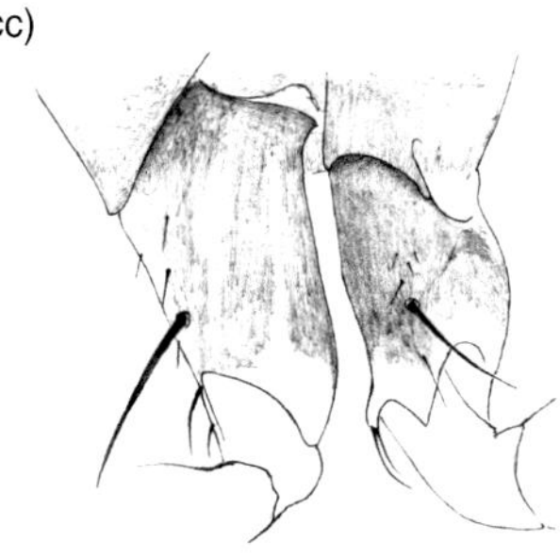

(bb)

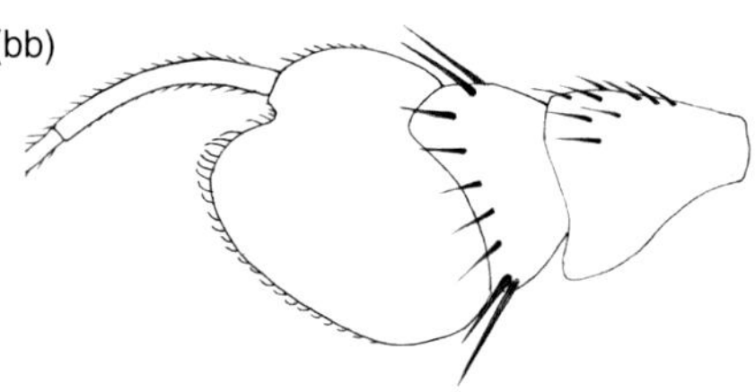

41(40) Hind basitarsus with basal dorsal seta in almost anterodorsal position, not aligned with distal dorsal seta, and only a tiny anterior seta close to distal dorsal seta (a, dorsal view). Hind tibia tip yellow or at most brown; hind basitarsus yellow on basal half. Pedicel partly yellow ventrally. (See also couplet 56.) (Rare specimens of *discifer* with only 1 ventral seta on mid tibia fall here – see couplet 23.) .... ***sabinus***

– Hind basitarsus with both dorsal setae in strict alignment, the distal setae accompanied by moderately long anterodorsal and anterior setae (aa, dorsal view). Hind tibia black on apical quarter, extending 3 times diameter of shaft to further than preapical dorsal seta; hind basitarsus entirely black. Pedicel entirely black. 42

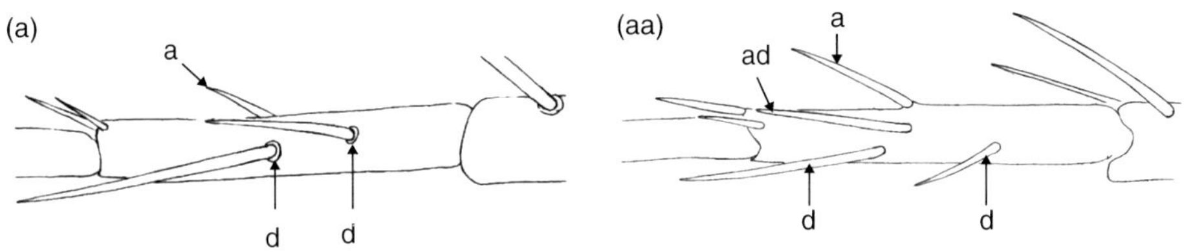

42(41) Face narrower, its height more than twice its width (2.2 times) (a). Front coxa entirely yellow at base. Postpedicel longer than deep (fig. aa of couplet 40). Mid tibia with 3 long anterodorsal setae (plus short basal one), ventral seta placed beyond the closest anterodorsal seta (b). Front basitarsus and often succeeding tarsomeres entirely yellow. .... ***notatus***

– Face broader, its height less than twice its width (1.8 times) (aa). Front coxa with a dark mark at base on outer side. Postpedicel rounded, as long as deep (fig. bb of couplet 40). Mid tibia with 3–4 long anterodorsal setae (plus short basal one), ventral seta about level with the closest anterodorsal seta (bb). Front basitarsus dark at tip, succeeding tarsomeres black. .... ***caligatus***

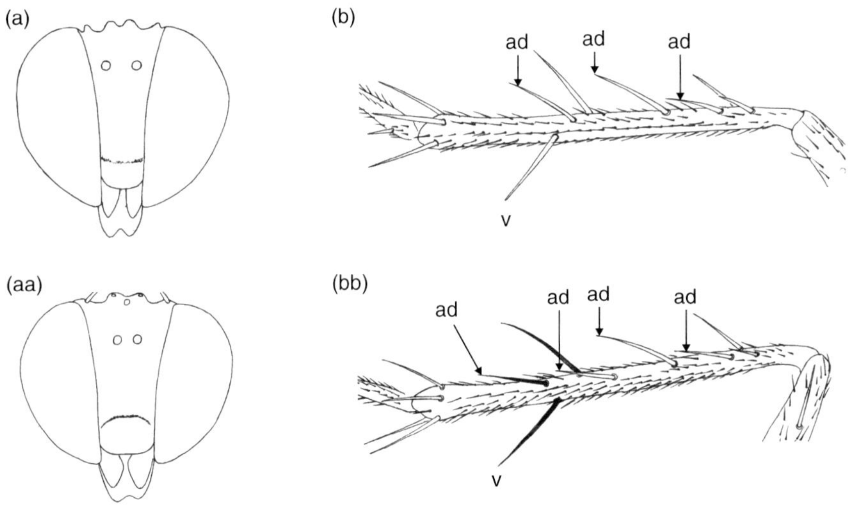

43(39) Hind basitarsus with 1 dorsal seta at midpoint. Arista inserted at about middle of postpedicel. .... ***agilis***

– Hind basitarsus with 2–3 dorsal setae. Arista inserted at middle or more apically. 44

44(43) Front coxa with black hairs extending at most to halfway between uppermost outer seta and coxa articulation with thorax, or with occasional black hairs above seta (a, front view). Face white or rarely slightly yellow-tinged below antennae. 45

– Front coxa with dark hairs reaching more than halfway between uppermost outer seta and coxa articulation (aa, bb). Face usually yellow-tinged. 48

(a) (aa) (bb)

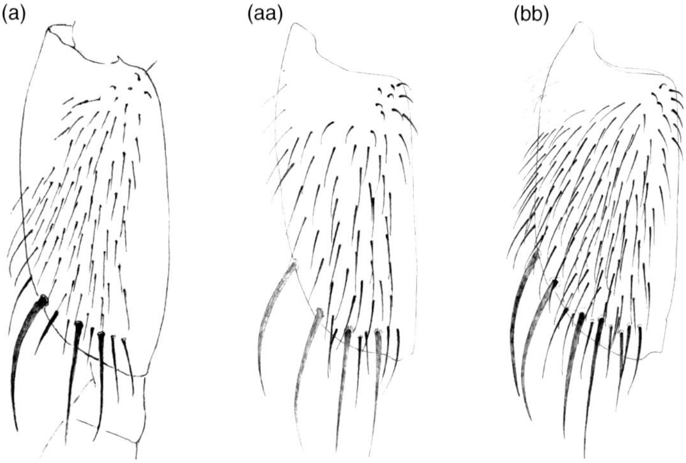

45(44) Hind coxa almost entirely yellow, mid coxa yellow on front and hind margins. Front coxa with mostly pale hairs. (See also couplet 54.) .... ***linearis***

– Hind and mid coxae mainly black. Front coxa with extensive black hairs. 46

46(45) Scape entirely yellow. Arista insertion intermediate between next two species (a). .... ***simplex***

– Scape narrowly black dorsally. 47

(a)

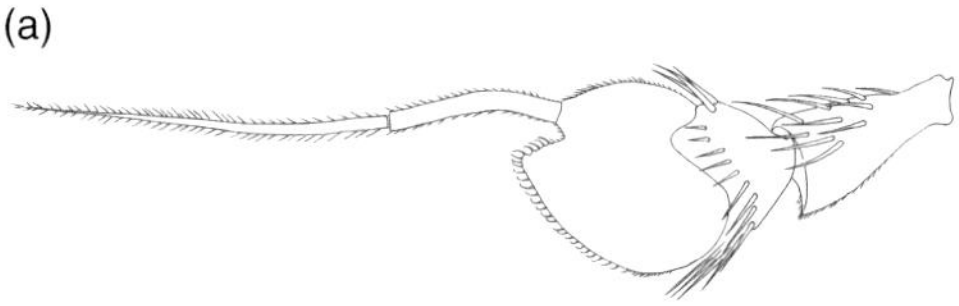

47(46) Arista inserted at about middle of postpedicel (a). Hind basitarsus with 3 dorsal setae, the basal one small (b). Mid basitarsus yellow with a black tip. .... ***lineatocornis***

– Arista inserted close to tip of postpedicel (aa). Hind basitarsus with 2 dorsal setae (bb). Mid basitarsus black. .... ***mediicornis***

(a)

(b)

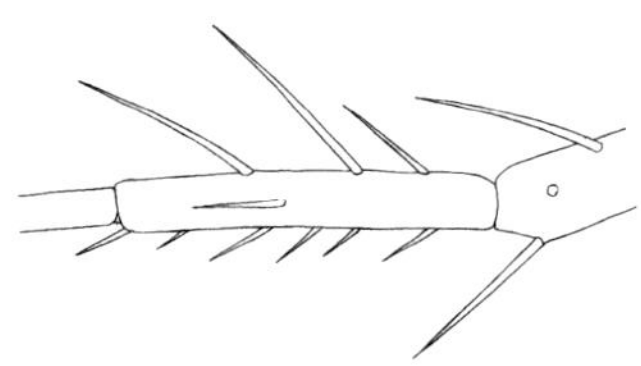

(aa)

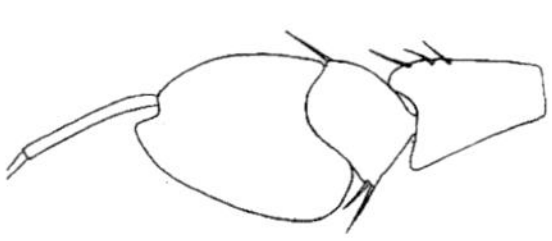

(bb)

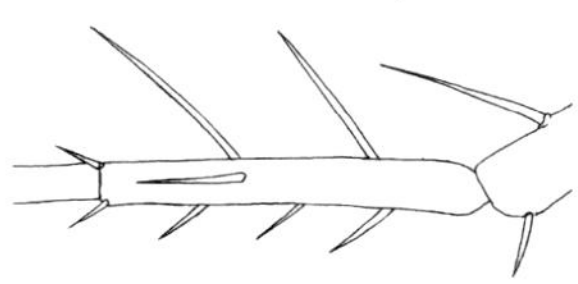

48(44) Front coxa with a black spot at base (a, lateral view). Hind tibia with black tip extending 2–3 times width of shaft (b). Mid tibia with 4 long + 1 short anterodorsal setae. Front coxa hairs sparse and longer (fig. aa of couplet 44). Mid tarsomeres 2 and 3 not or imperceptibly deeper than basitarsus (c). Proepisternal hairs white with usually 2–5 stouter black hairs on posterior upper corner 49

– Front coxa entirely yellow at base. Hind tibia tip yellow or slightly darkened on posterior side. Mid tibia usually with 5 long + 1 short anterodorsal setae. Front coxa hairs dense and shorter (fig. bb of couplet 44). Mid tarsomeres 2 and 3 slightly deeper than basitarsus, the effect accentuated by longer and denser ventral setulae and hairs (cc). Proepisternal hairs all white or with 1–3 weak black hairs in posterior upper corner 50

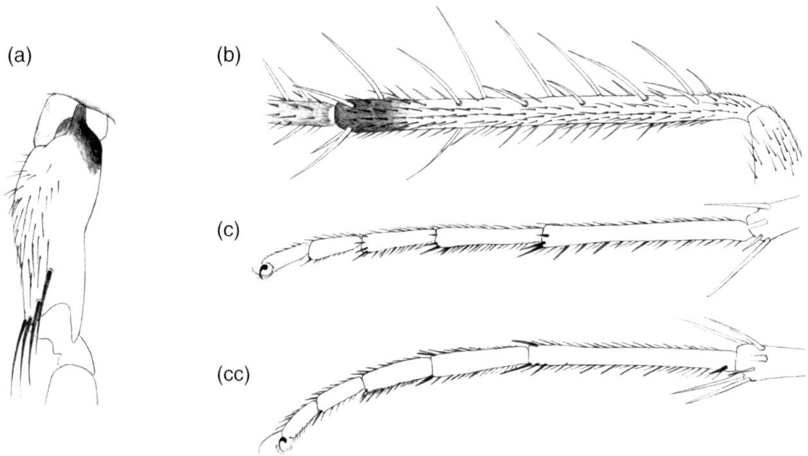

49(48) Arista inserted at about apical third of postpedicel (a). Hind basitarsus with 3–4 ventral setulae, the subapical one slightly thicker (b). Face greyish white when seen from above and illuminated from in front .... ***signatus***

– Arista arising near middle of postpedicel (aa). Hind basitarsus with subapical ventral setula not stouter than others. Face with pale golden yellow tinge, especially on upper part, when seen as above .... ***argyrotarsis***

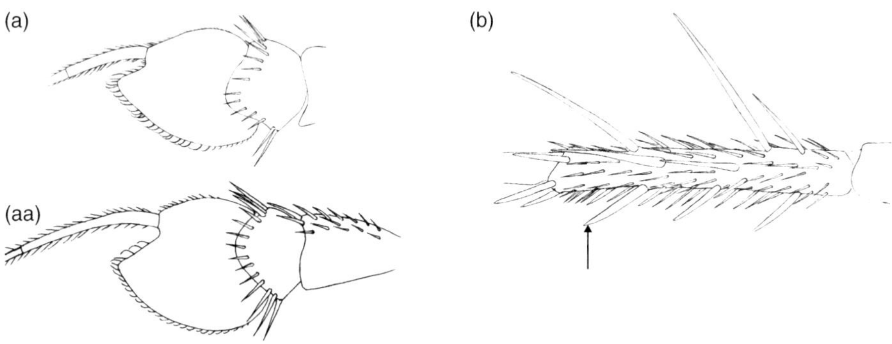

50(48) Scutellum with a posterior fringe of dense hairs in several rows, fringe almost half length of scutellum, top of scutellum with many almost equally long hairs.(a)................***pennatus***

– Scutellum without a dense fringe, only sparse hairs or a short feeble fringe whose length is no more than one-quarter length of scutellum...***subpennatus***

(a)

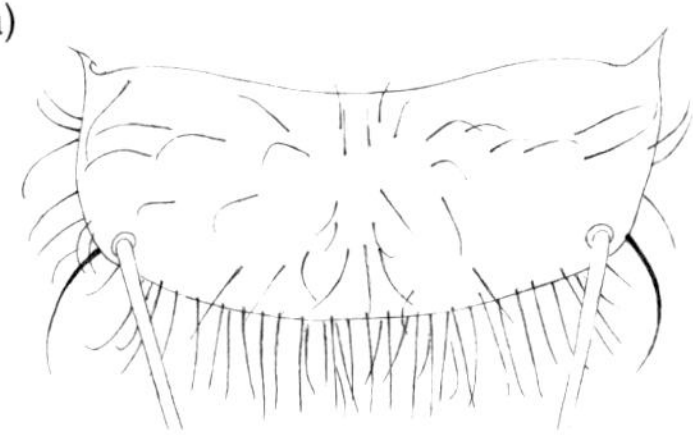

See alternative key at couplet 57 for these following seven species if couplet 51 proves difficult.

51(38) Hind basitarsus with distal dorsal seta equidistant from basal dorsal seta and tip of basitarsus (a, arrows), the 2 setae separated by at least 1.5 times width of basitarsus, usually clearly more; distal seta forming a close pair with a strong anterodorsal seta, these 2 setae separated at most by twice diameter of setae. 52

– Hind basitarsus with distal dorsal (sometimes posterodorsal) seta closer to basal dorsal seta than to tip of basitarsus, the 2 setae separated by less than twice width of basitarsus (aa, bb); anterodorsal seta absent or inconspicuous but if forming a pair with distal dorsal (or posterodorsal) seta then all 3 setae crowded together (cc)...55

(a)

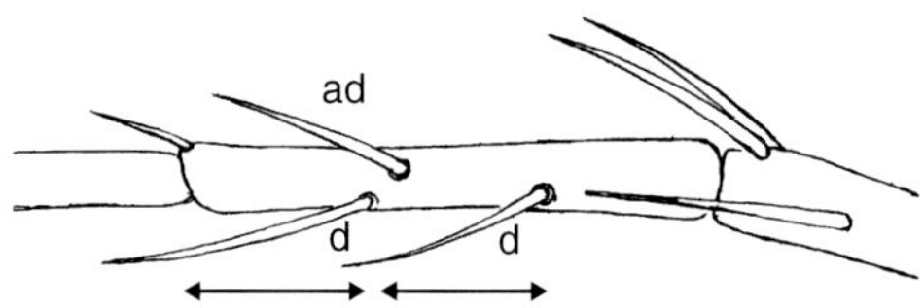

(aa)

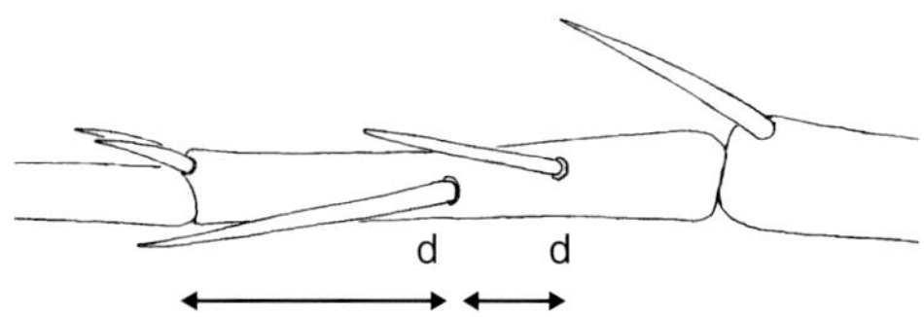

calinotus

(bb)

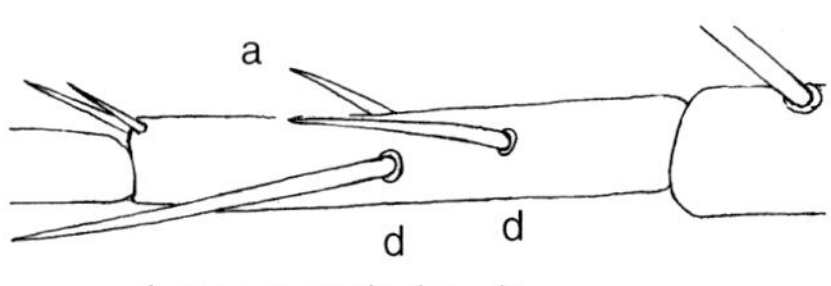

sabinus

(cc)

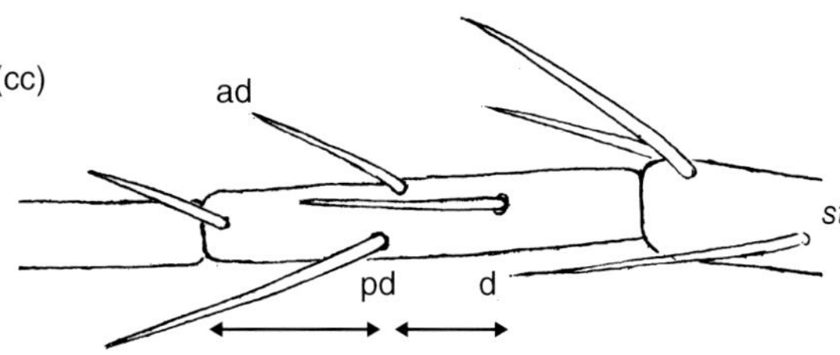

strigipes

52(51) Arista inserted at apical third of postpedicel, tip of postpedicel projecting only slightly beyond arista base (a). Front basitarsus yellow with a dark tip, tarsomere 2 black. Front coxa with black hairs usually reaching posterior (outer) margin above uppermost apical seta (b, front view). Clypeus bare.. ***plumipes***

– Arista inserted at middle of postpedicel, tip of postpedicel projecting well beyond arista base (aa). Front basitarsus entirely yellow, tarsomere 2 usually yellow. Front coxa with black hairs not or only feebly reaching posterior (outer) margin above uppermost apical seta (bb). Clypeus bare or with tiny hairs..53

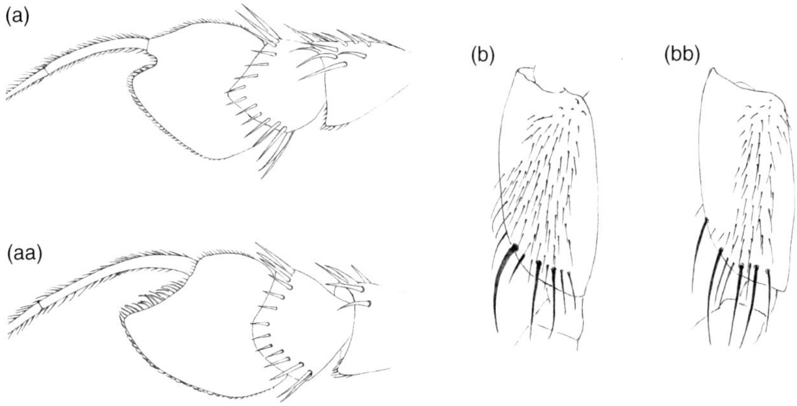

53(52) Hind coxa black with apical third yellow (a). Mid tibia with 4 long anterodorsal setae, ignoring any smaller basal seta. Scape and pedicel entirely yellow. Clypeus with tiny hairs. Anal lobe broad, as wide as cell r_{4+5} at dm-m (b).. ***wahlbergi***

– Hind coxa either entirely yellow or mainly yellow, including hind margin, with a black wedge descending from base to lateral seta (aa). Mid tibia with 3 long anterodorsal setae, ignoring any smaller basal seta. Scape dark dorsally. Clypeus bare. Anal lobe narrow, less than width of cell r_{4+5} at dm-m (bb).. 54

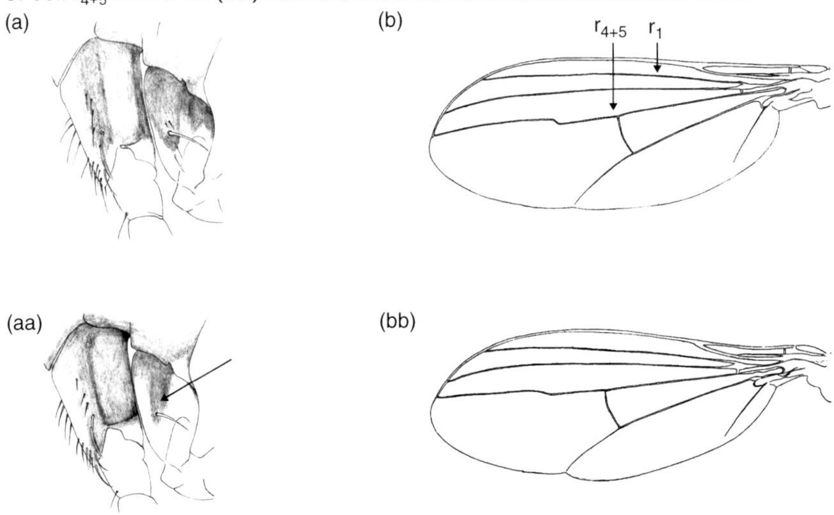

54(53) Postpedicel sharply pointed (a). Scape and pedicel black dorsally, yellow ventrally. Hind coxa with a black wedge, sometimes reduced to a streak at lateral seta; mid coxa mainly black. Anal lobe narrower than cell r_1 level with crossvein dm-m (fig. bb couplet 53). .... ***longicornis***

– Postpedicel more bluntly pointed (aa). Scape yellow dorsally. Hind coxa entirely yellow, mid coxa yellow on front and hind margins. Anal lobe slightly broader than cell r_1 level with crossvein dm-m (bb). (See also couplet 45.) .... ***linearis***

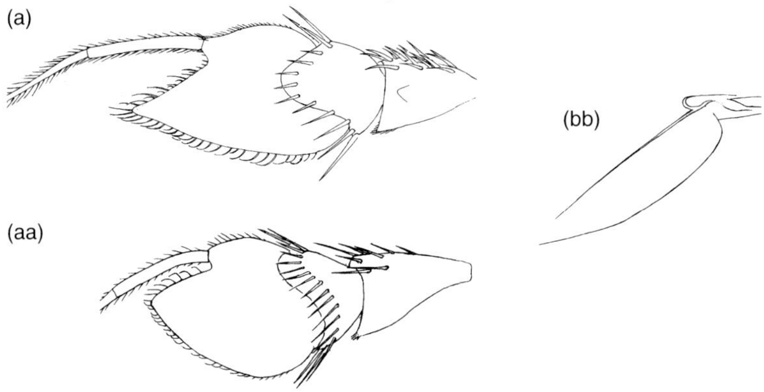

55(51) Tergite marginal setae conspicuously long, those on tergite 4 almost as long as tergite 5 (a). Frons violet or blue. Front tibia with 2–3 posteroventral setae (b). Hind femur with narrow dark streaks on antero- and posteroventral surfaces. Mid tibia with the long anterodorsal setae at two-fifths from base (c: i) obviously more dorsal in position than other anterodorsal setae (c: ii). Larger species on average, wing length 4.0–4.5 mm.......... ***strigipes***

– Tergite marginal setae short, those on tergite 4 less than half length of tergite 5. Frons usually green or dark brown with bronze or coppery tinge, rarely violet. Front tibia with 1 posteroventral seta, rarely with a second smaller seta. Hind femur entirely yellow ventrally. Mid tibia with the long seta at two-fifths (i) almost in line with other anterodorsal setae (ii), displaced by less than diameter of its socket (cc). Smaller species on average, wing length 3.5–4.2 mm.......... 56

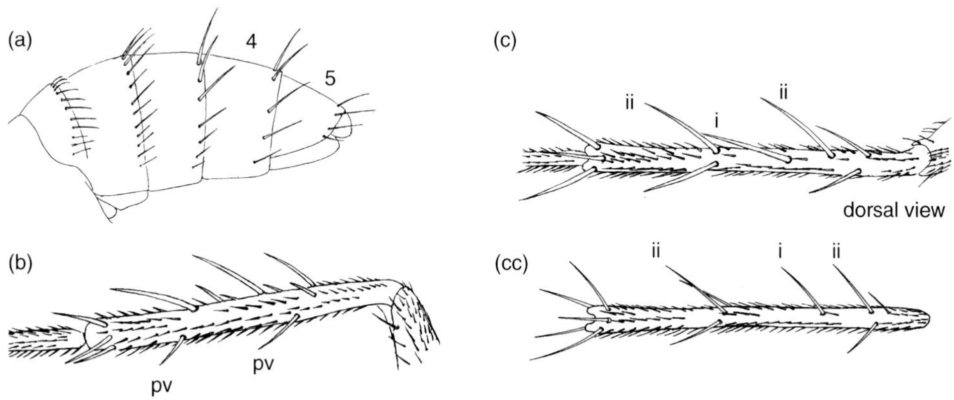

56(55) Scape and pedicel black dorsally, postpedicel in lateral view at most half yellow ventrally at base, whole antenna appearing more black than yellow. Front coxa with a small black spot or smudge at upper outer corner articulating with thorax. Hind basitarsus yellow basally and black on apical half or more, tarsomere 2 usually black. (See also couplet 41.).......... ***sabinus***

– Scape and pedicel entirely yellow, postpedicel in lateral view yellow for well over half its area, whole antenna appearing more yellow than black (aa). Front coxa entirely yellow at articulation with thorax. Hind basitarsus yellow with a black tip, tarsomere 2 yellow on basal half.......... ***calinotus***

(aa)

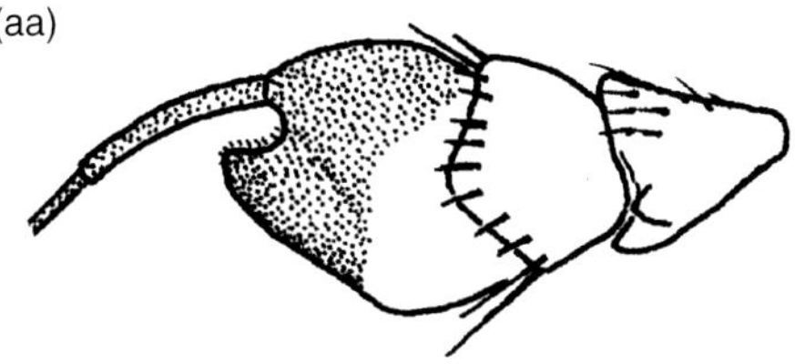

Alternative key to last seven species if couplet 51 proves difficult:

57(38) Arista inserted at about apical third of postpedicel. 58
– Arista inserted at about middle of postpedicel. .... return to 53

58(57) As couplet 55; first half .... ***strigipes***
– second half 59

59(58) Scape black dorsally .... ***sabinus***
– Scape entirely yellow 60

60(59) Mid tibia with 4 long anterodorsal setae in addition to a possible small basal seta. .... ***plumipes***
– Mid tibia with 3 long anterodorsal setae in addition to a possible small basal seta .... ***calinotus***

Dolichopus acuticornis Wiedemann, 1817

Males of this species and *longicornis* differ from all other British *Dolichopus* in having a long tapered postpedicel which is at least 3 times longer than deep. Other similarities are described under the account for *longicornis*. If the antennae are missing, *acuticornis* may be distinguished among yellow-legged *Dolichopus* with no obvious leg ornamentation in having a long fine posteroventral seta on the front tibia, a dark-tipped hind tibia and a pale hind coxa. The apicoventral epandrial lobe of the hypopygium is unique in being a long narrow projection rather than the usual low short lobe; the surstyli are also markedly long. Females are characterised by their acutely pointed and slightly upturned black postpedicel, largely pale hind coxa and long marginal setae on the fourth and fifth tergites but shorter on preceding ones. Wing length ♂: 3.4 (3.1–3.6) mm, ♀: 3.7 (3.3–4.0) mm.

This is an uncommon species with a mainly western distribution concentrated on the Welsh coast and in northwest England. In Scotland it is found mainly in the Western Isles with isolated records on the Sutherland coast, Strathspey and Firth of Forth. Coastal dunes are probably its principal habitat. Inland it is found on sandy or gravelly habitats such as shores of sandy rivers. It may not require wetland features within dunes, nor is it associated with woodlands as in northern Germany and Belgium (Emeis 1964, Maes and Pollet 1997). Flight: vi–viii.

Dolichopus agilis Meigen, 1824

Although relatively undistinguished, *agilis* is one of the few pale-legged *Dolichopus* with a slightly expanded hind tibia which, in combination with the lack of other modifications on the legs, requires comparison only with *longicornis* and *notatus* which differ in their far longer postpedicel, pale yellow face (not white) and possessing a long fine posteroventral seta on the front tibia (lacking in *agilis*). In both sexes, *agilis* is among the few species of *Dolichopus* with only one dorsal seta on the hind basitarsus, but care is needed with females not to confuse other species in which setae have been lost, leaving only one; in these cases the seta is not at the middle but closer to one end of the basitarsus. Characters that may avoid confusion with females of species with a similar antenna are the hind coxa being similar to *trivialis* (fig. (a) of couplet 24), the hind tibia having a narrowly black tip, the mid tibia with one short and four long anterodorsal setae, and the face being white. Wing length ♂: 3.6 (3.4–3.8) mm, ♀: 4.1 mm.

Its distribution is wide but sparse in England and Wales, reaching north to Yorkshire and with a concentration in East Anglia. It appears to be associated most often with dry grassland, dry heath and wetlands, including fens, bogs and grazing marsh, but also from dry woods and possibly coastal dunes. Emeis (1964) found it in saltmarsh in Germany. Flight: vi–viii.

Dolichopus andalusiacus **Strobl, 1899**

Dolichopus andalusiacus is one of the '*nubilus*' group discussed in the *nubilus* account. The unusual arrangement of acrostichal setae and presence of face hairs distinguishes *andalusiacus* from *rupestris, signifer* and *clavipes* which sometimes have partly dark femora like *andalusiacus*. Unusual features of the hypopygium are the triangular cercus with its basal projection surmounted by short black hairs, similar to that of the black-femured *vitripennis*, and the triangular basiventral apical epandrial lobe with a conspicuous seta on its crest. Females closely resemble those of *excisus* and the *nubilus*/*latilimbatus* pair but the hind tibia is bright yellow at the tip, which it rarely is in the other species, and the yellow often runs on to the base of the basitarsus. Specimens with entirely yellow tibiae but with only a few supernumerary setulae on the scutum (see below) are likely to be one of the *latilimbatus*/*nubilus* pair. Those with dark front femora are immediately recognised as *andalusiacus* but the colour may often scarcely differ from that of the hind femur. The acrostichal setae are numerous and more closely spaced than the distance between the rows (these distances are about equal in the other species) and the hairs in front of the scutellum extend forward along the dorsocentral row to nearly the second dorsocentral seta. Together, these hairs and the acrostichal setae make the dorsum particularly hairy. Wing length ♂: 4.3 (3.8–4.6) mm, ♀: 4.9 (4.1–5.4) mm.

The distribution is predominantly the south coast of England, particularly in Cornwall, and the south Welsh coast. Some inland records from southern England may be correct. Habitats are freshwater marshes and fens, including those on the coast, but *andalusiacus* does not appear to be associated with brackish conditions. Sites further inland include ponds within acid mire, gravel pits and washlands (not all confirmed). There may be a requirement for bare wet mud, sand or gravel, and a loose association with reed (Poulding 1992). Flight: v–x, peak in vi.

Dolichopus arbustorum **Stannius, 1831**

This species belongs to the group with a pale fringe on the hind femur and mainly pale hind coxa (the '*trivialis*' group) but is distinguished in lacking any modification in shape or hairing of the front tarsus. Other features are a mainly pale calypter fringe, very narrow anal lobe and a ciliolarium of yellow points on the basal half of the hind tibia which is not swollen. Females closely resemble those of *virgultorum* but are overall slightly darker, and may sometimes be indistinguishable. See the account for *virgultorum* for their separation. Wing length ♂: 4.8 (4.3–5.1) mm, ♀: 5.2 (4.8–5.7) mm.

Dolichopus arbustorum is an uncommon species distributed thinly in southern England and into south Wales, northwards to the southern fringes of Lancashire and Yorkshire. Its habitats include woodland, grassland on sand, pond margins and fen, which give little indication of exacting needs. An unconfirmed record has it reared from rotting wood. Flight: v–ix.

Dolichopus argyrotarsis **Wahlberg, 1850**

This is one of five species with silvery white tarsomeres on the mid leg; *argyrotarsis* is distinguished by having three (not just one or two) silver segments and with little difference in the thickness between all five tarsomeres. The ciliolarium is inconspicuous compared to others in this group, consisting of a narrow strip of yellow pubescence with no swelling of the shaft. Females are almost identical to those of *signatus*, differing as in the key and also by having a marginally narrower face although overlapping with the range in *signatus*. The extent of darkening at the tip of the hind tibia and the shape of the anal lobe of the wing overlap with the range found in *signatus* and not as described by d'Assis-Fonseca (1978). Wing length ♂: 4.9 (4.5–5.2) mm, ♀: 5.3 (4.6–5.7) mm.

This uncommon species is more frequent in Scotland than further south, with a concentration of records in the Highlands. In England and Wales it is found sparsely but widely away from the eastern lowlands. It is associated with exposed sediments of sandy and stony rivers and occasionally smaller streams. This requirement defines its range and explains apparently outlying occurrences in Hampshire and Sussex on shingly Palaeogene geology. Flight: v–viii, x, peak in vii.

Dolichopus acuticornis

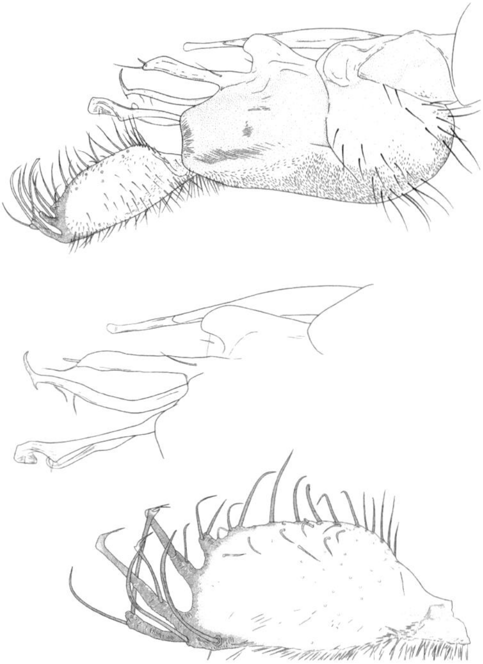

Dolichopus agilis

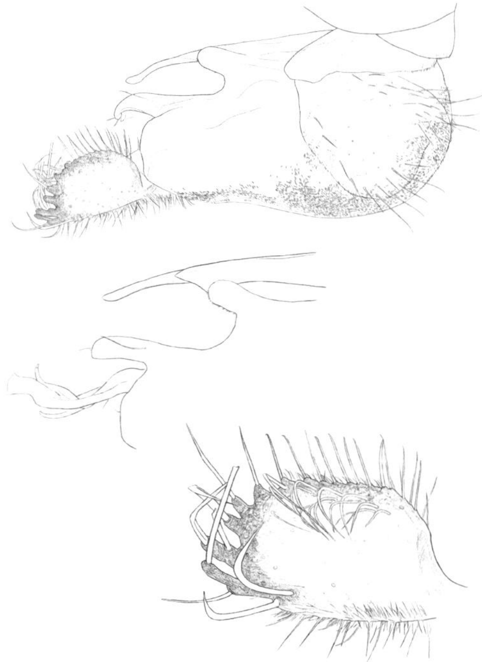

Dolichopus andalusiacus

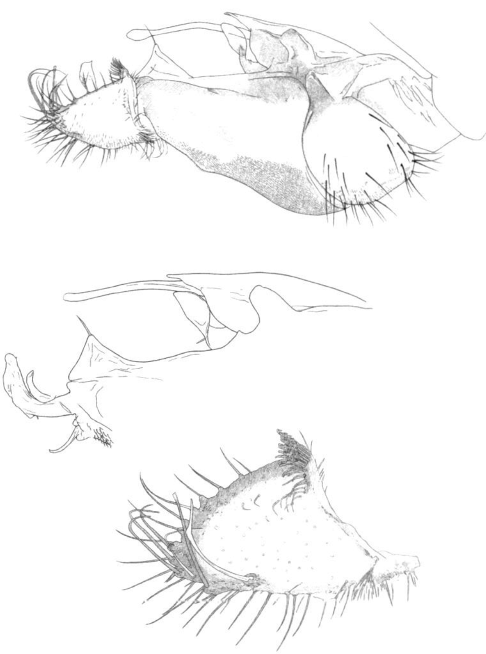

Dolichopus arbustorum

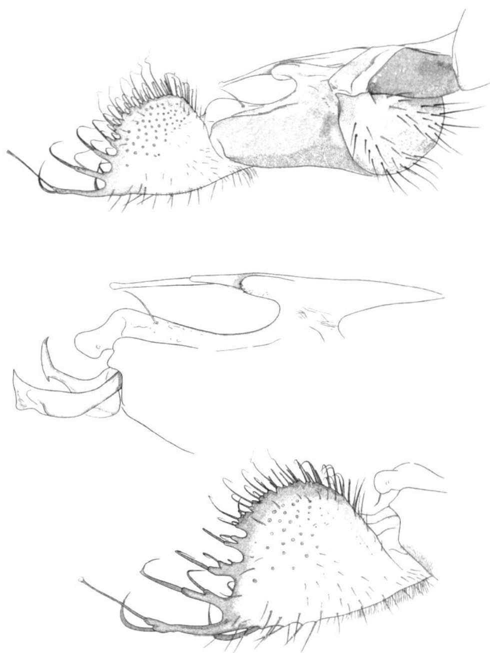

Fig. 16. *Dolichopus* hypopygia.

Dolichopus argyrotarsis

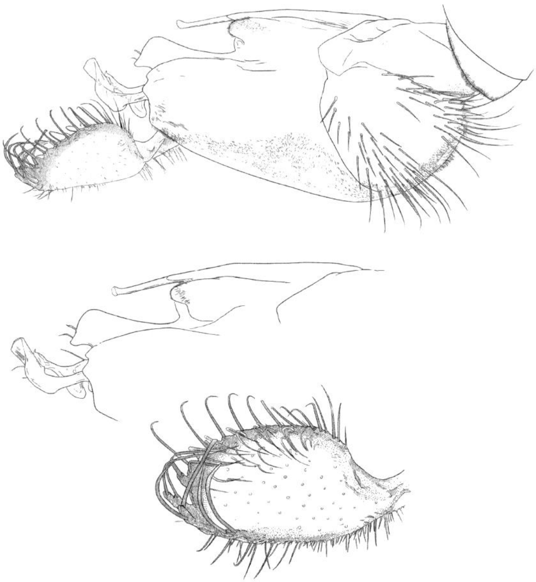

Dolichopus atratus

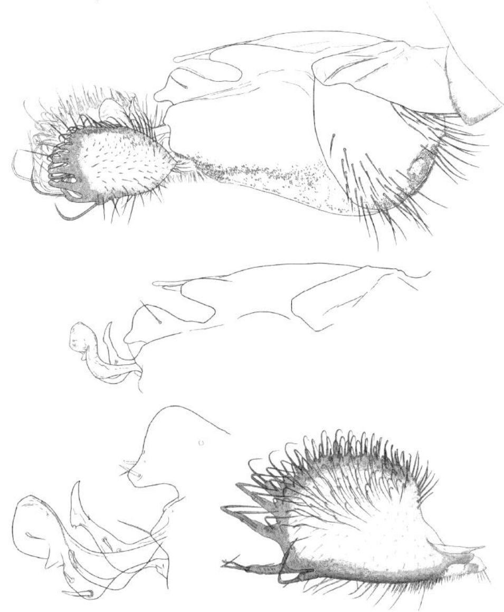

Dolichopus atripes

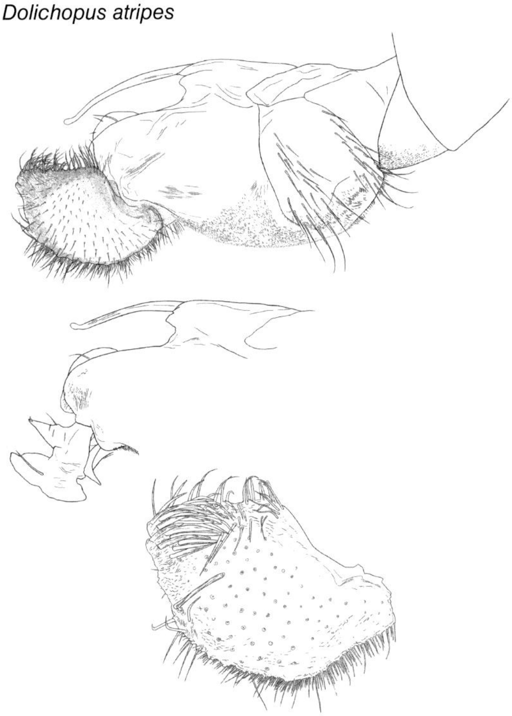

Dolichopus brevipennis

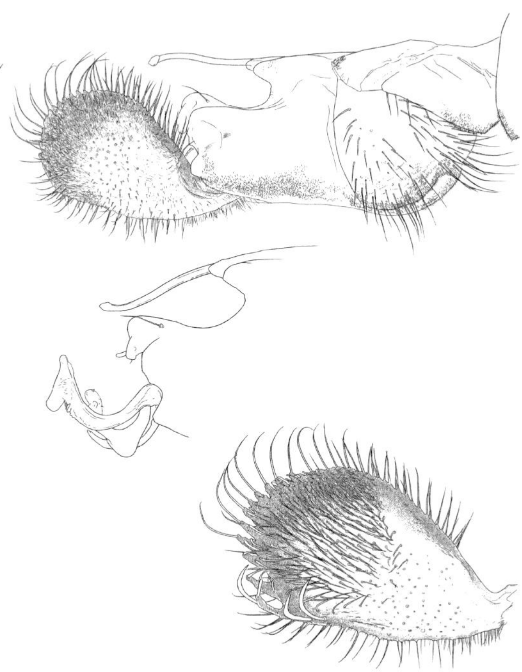

Fig. 16. (*Continued*)

Dolichopus caligatus

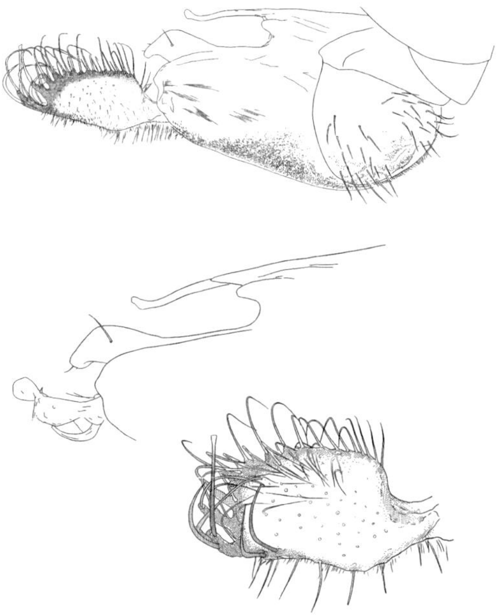

Dolichopus calinotus

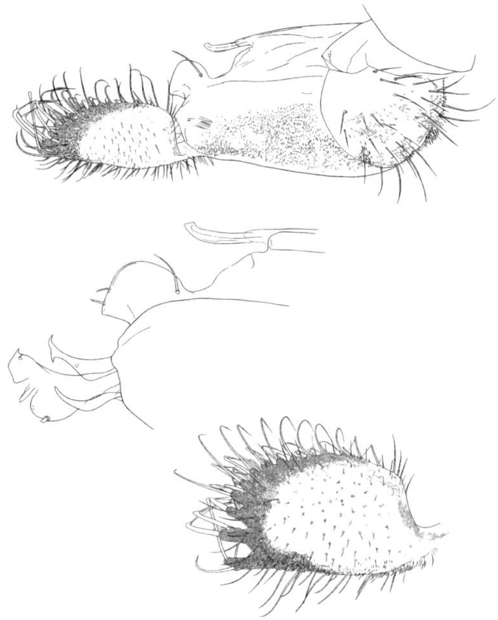

Dolichopus campestris

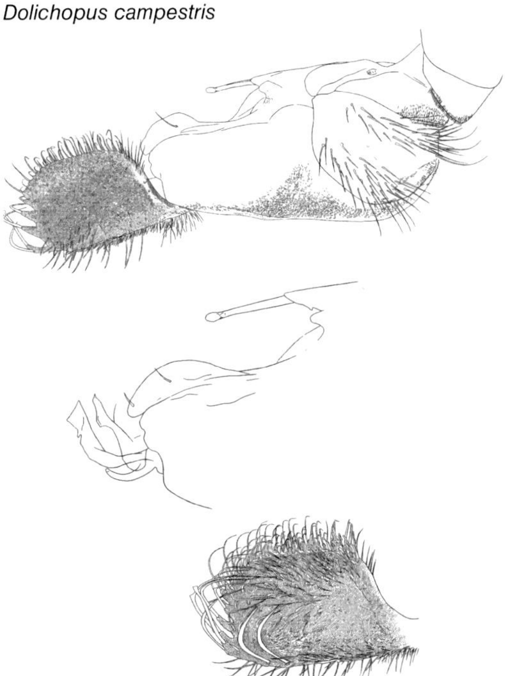

Dolichopus cilifemoratus

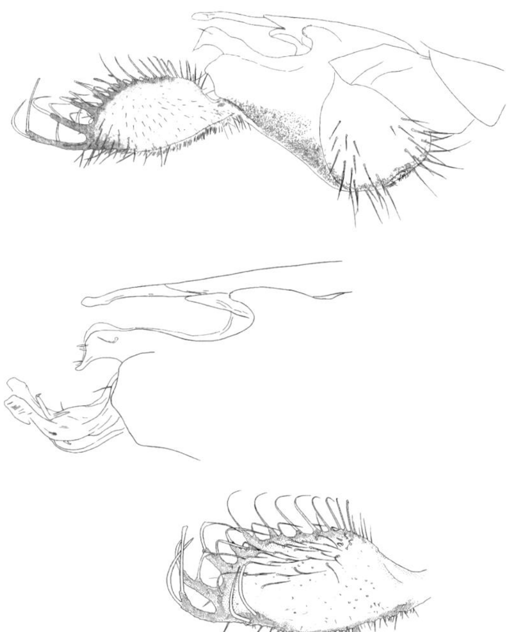

Fig. 16. (*Continued*)

Dolichopus claviger

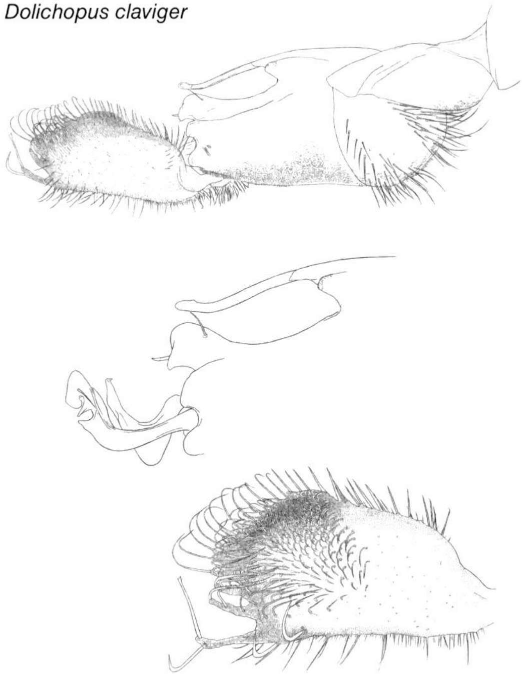

Dolichopus clavipes

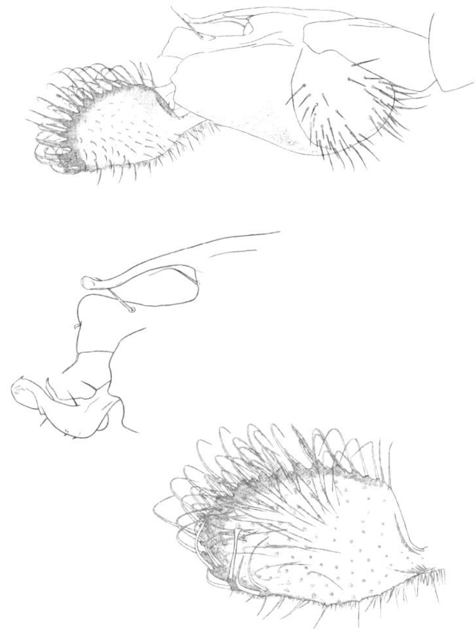

Dolichopus diadema

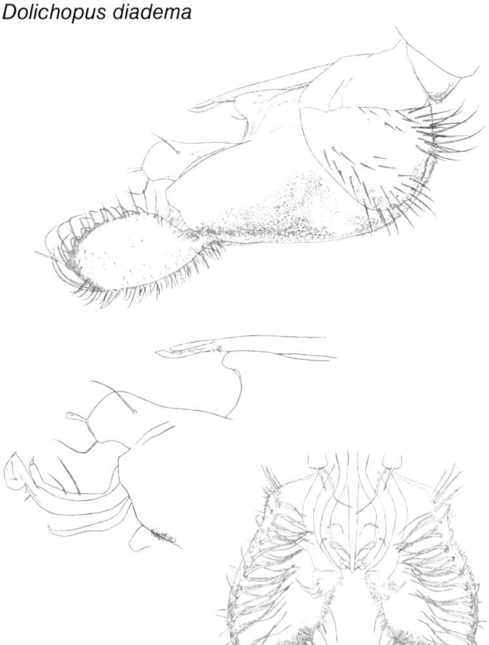

Dolichopus discifer

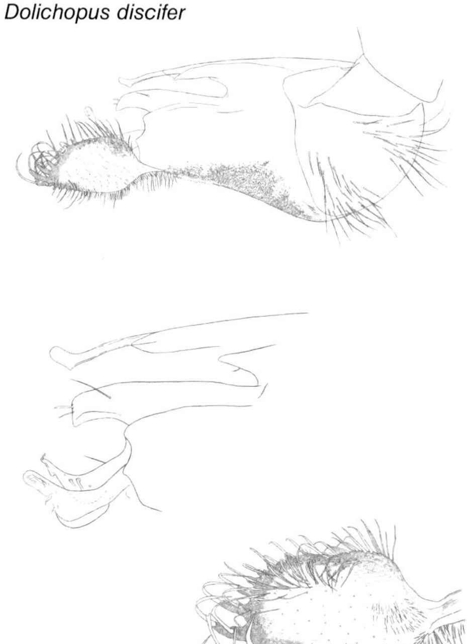

Fig. 16. (*Continued*)

Dolichopus excisus

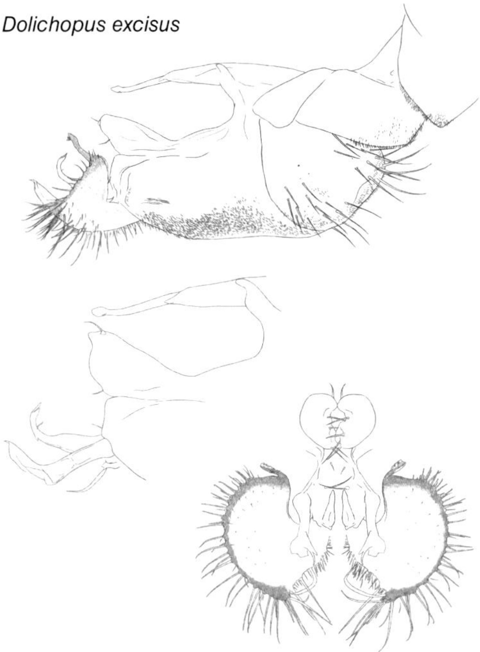

Dolichopus festivus

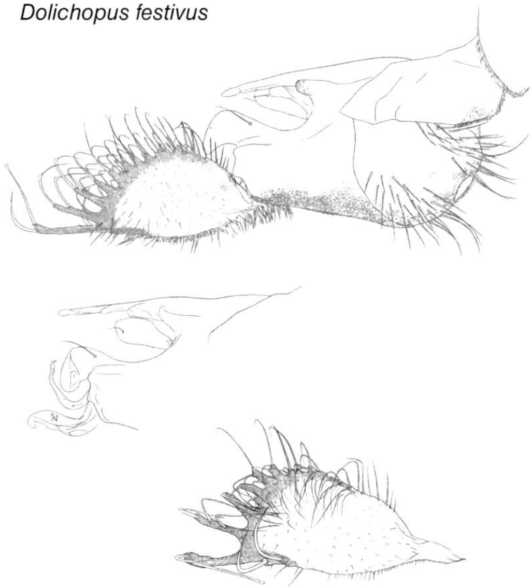

Dolichopus griseipennis

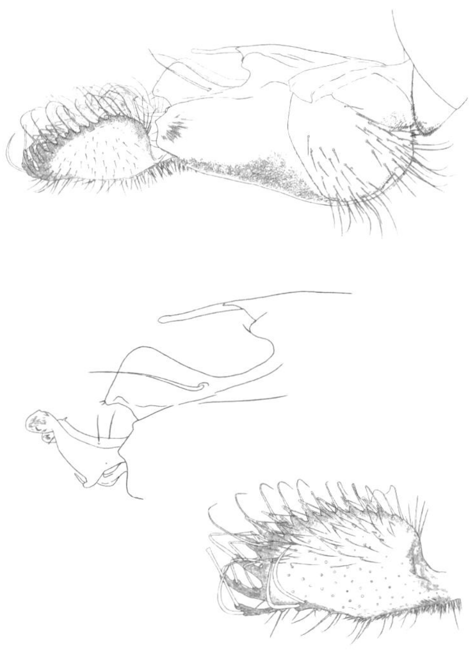

Dolichopus laticola

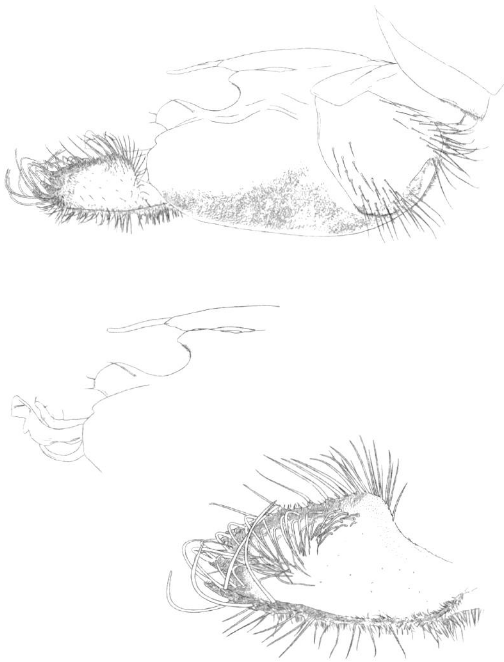

Fig. 16. (*Continued*)

Dolichopus latilimbatus

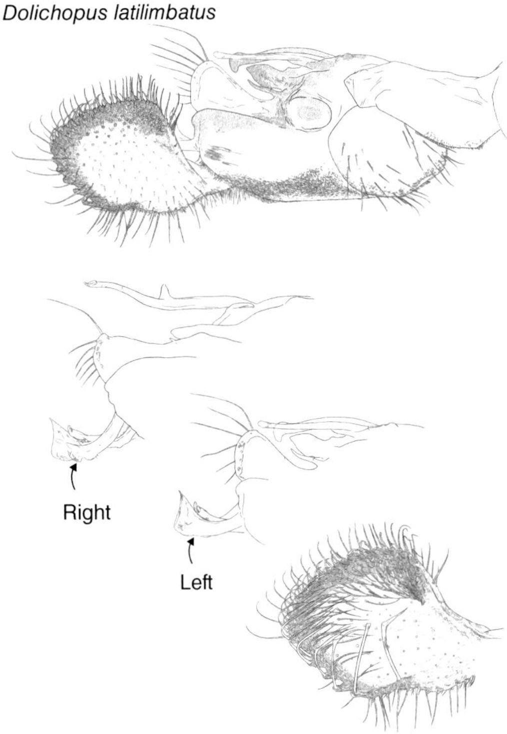

Dolichopus latipennis

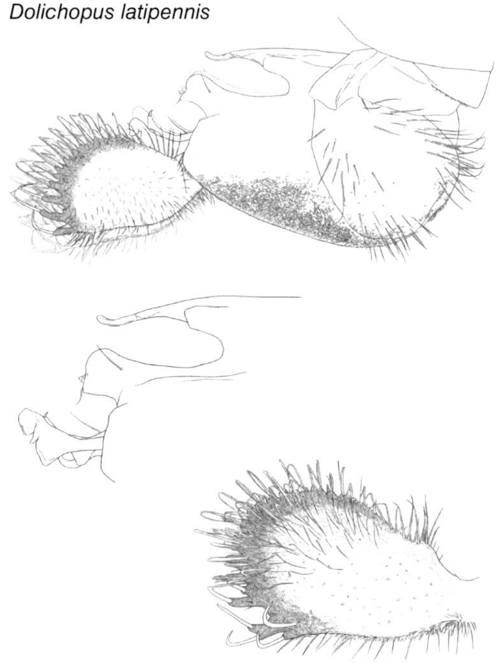

Dolichopus lepidus

Dolichopus linearis

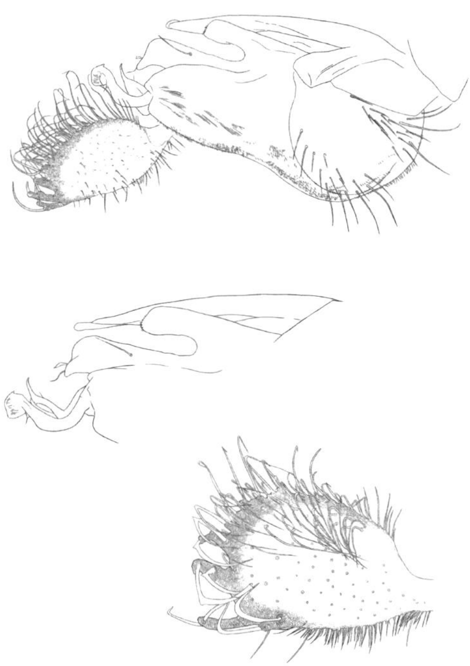

Fig. 16. (*Continued*)

Dolichopus lineatocornis

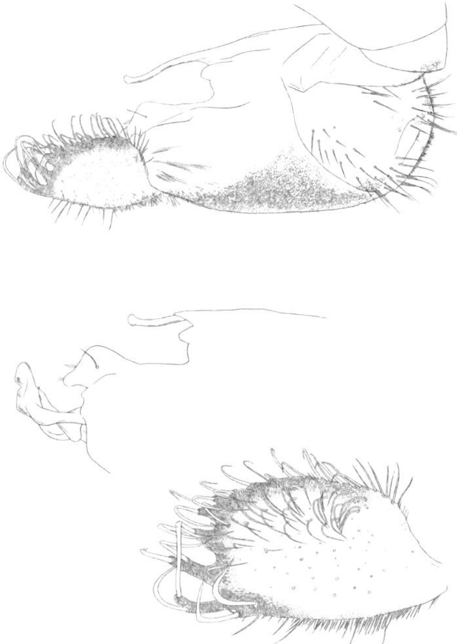

Dolichopus longicornis

Dolichopus longitarsis

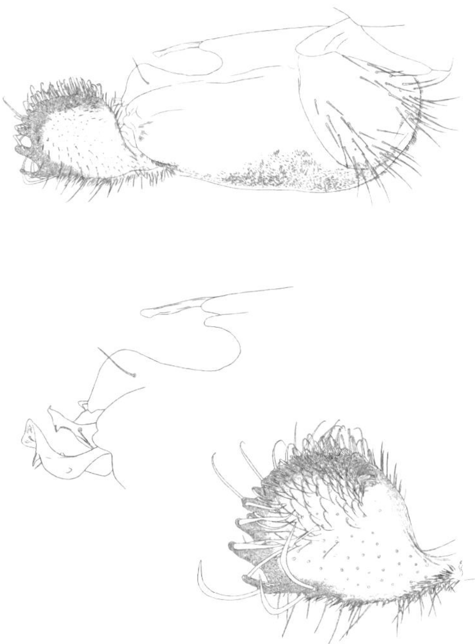

Dolichopus maculipennis

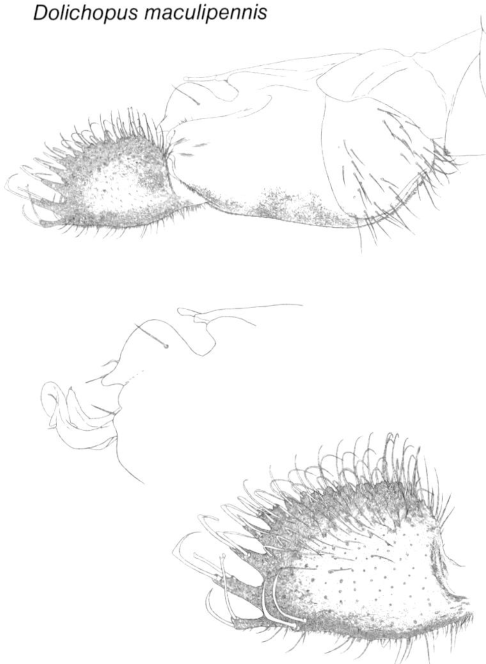

Fig. 16. (*Continued*)

Dolichopus mediicornis

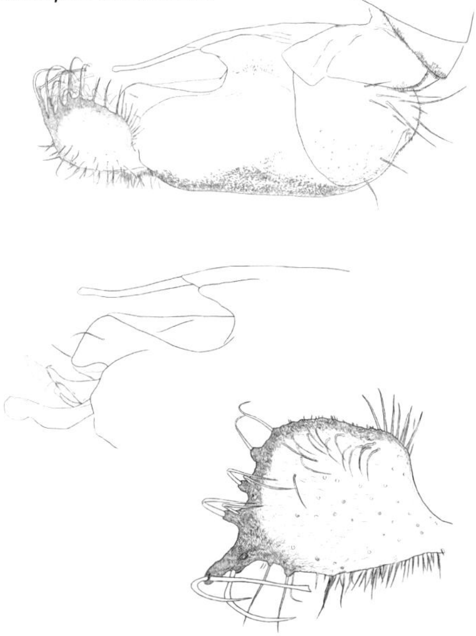

Dolichopus melanopus

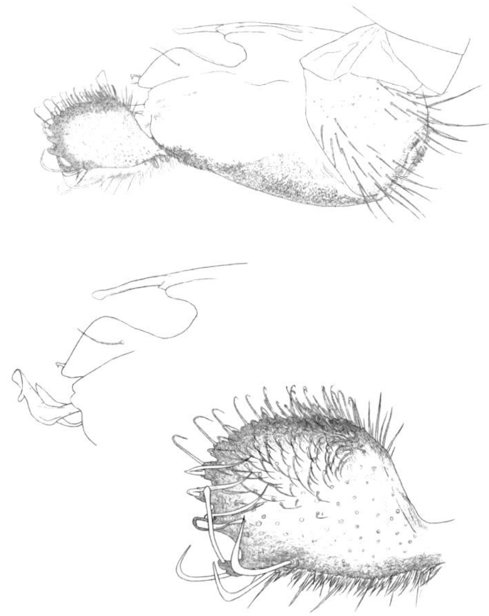

Dolichopus migrans

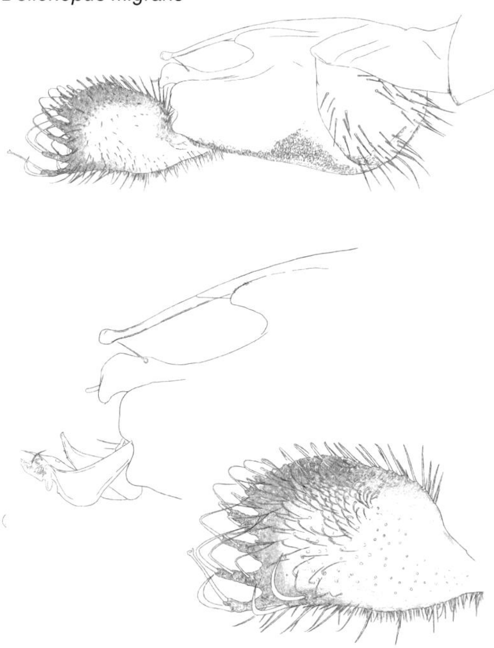

Dolichopus nigripes

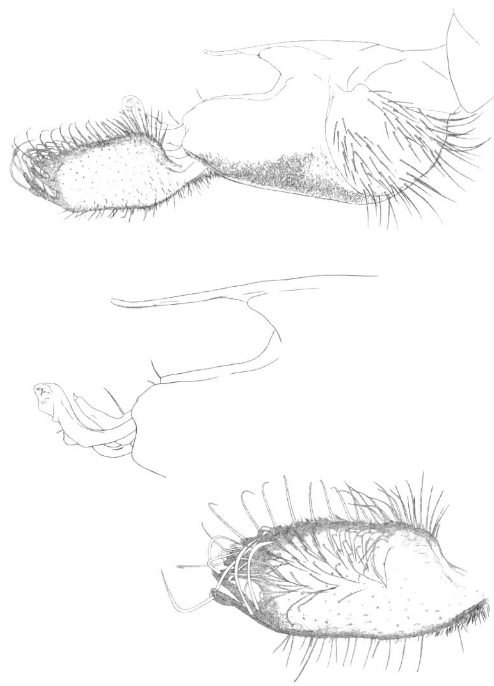

Fig. 16. (*Continued*)

Dolichopus nitidus

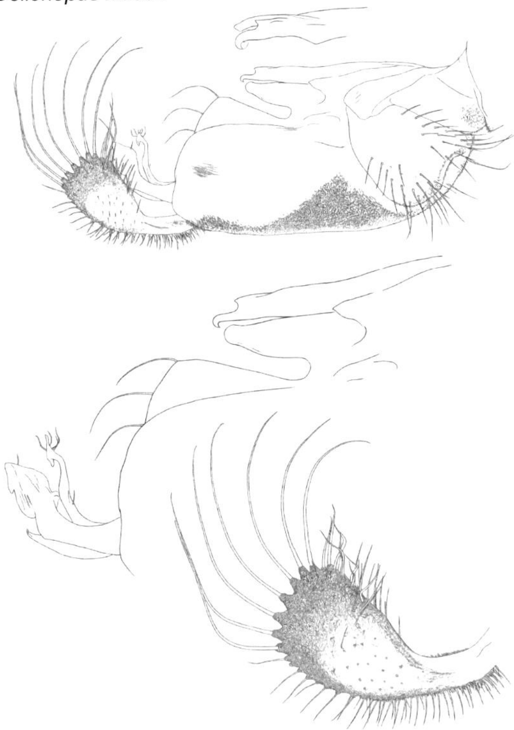

Dolichopus notatus

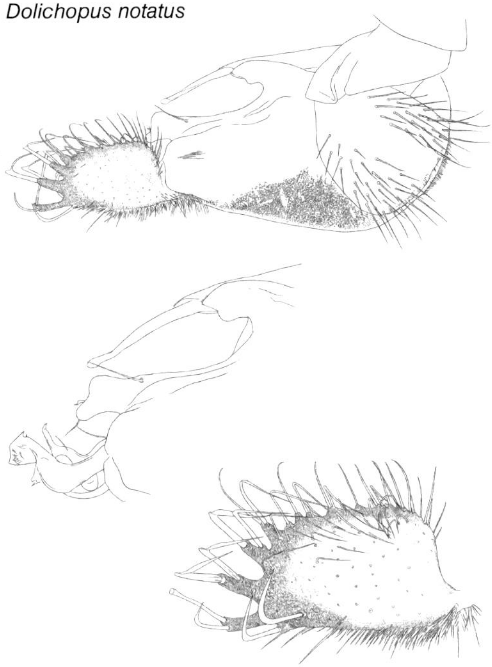

Dolichopus nubilus

Dolichopus pennatus

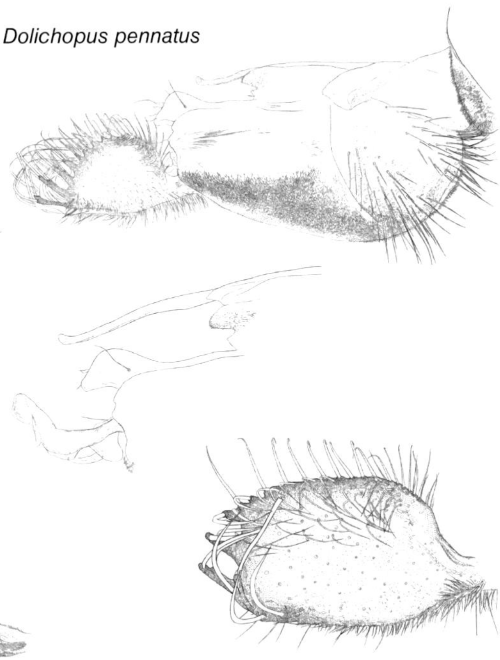

Fig. 16. (*Continued*)

Dolichopus phaeopus

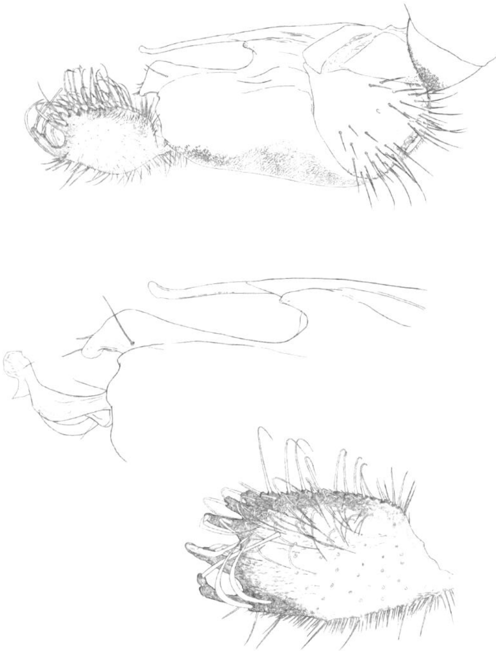

Dolichopus picipes

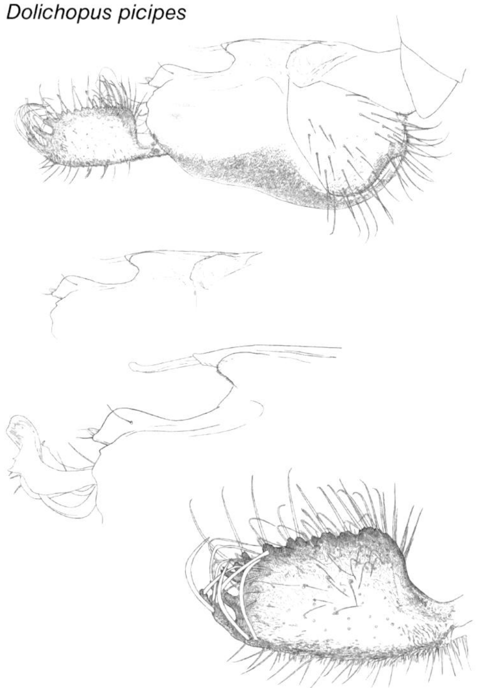

Dolichopus planitarsis

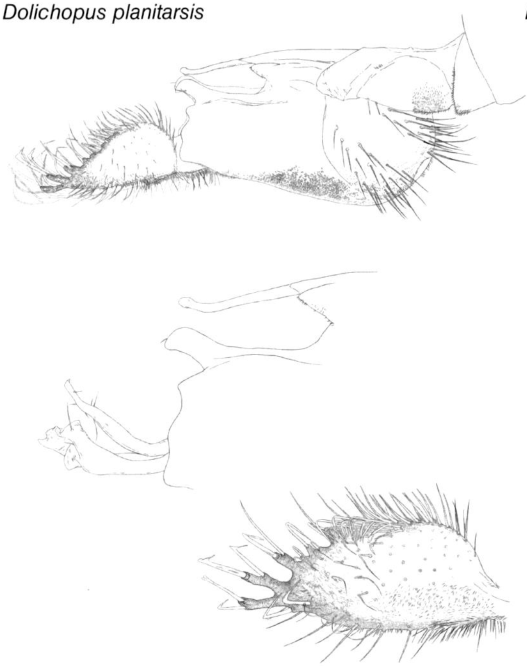

Dolichopus plumipes

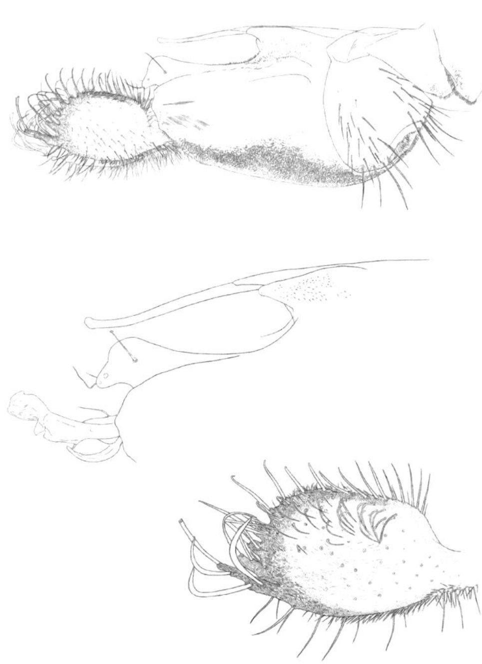

Fig. 16. (*Continued*)

Dolichopus plumitarsis

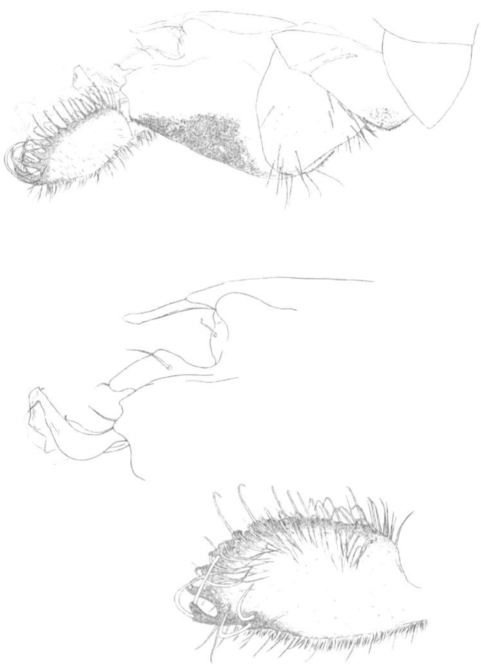

Dolichopus popularis

Dolichopus rupestris

Dolichopus sabinus

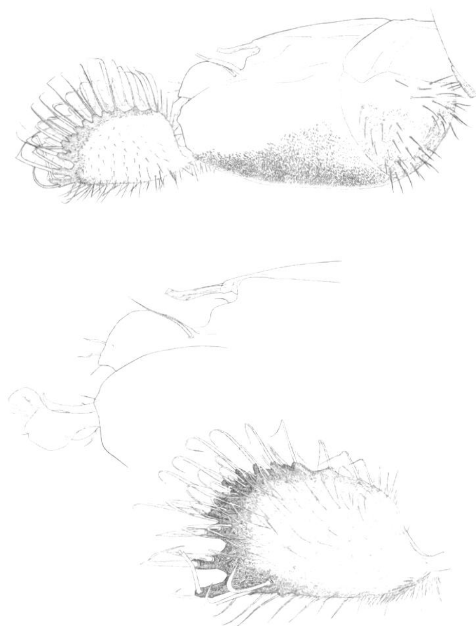

Fig. 16. (*Continued*)

Dolichopus signatus

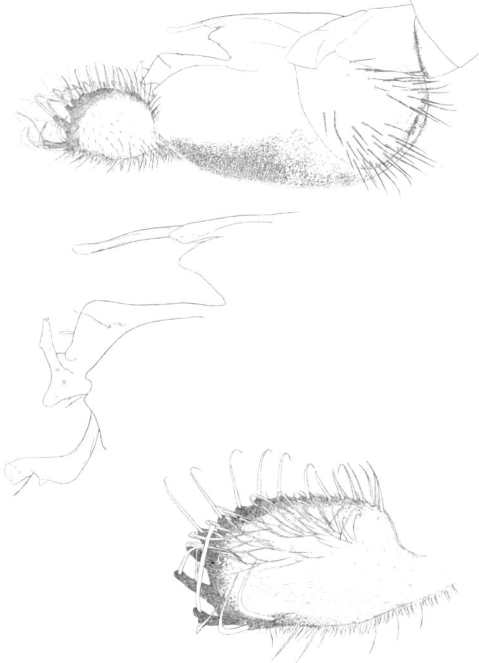

Dolichopus signifer

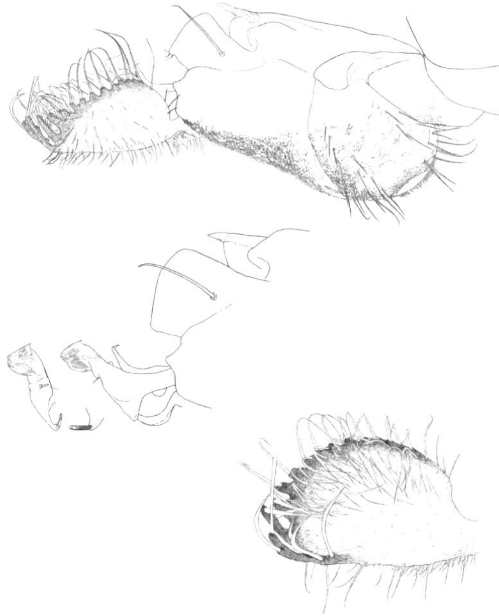

Dolichopus simplex

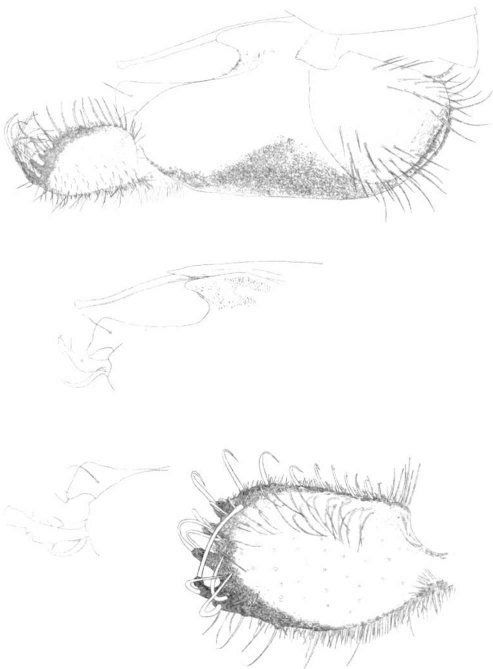

Dolichopus strigipes

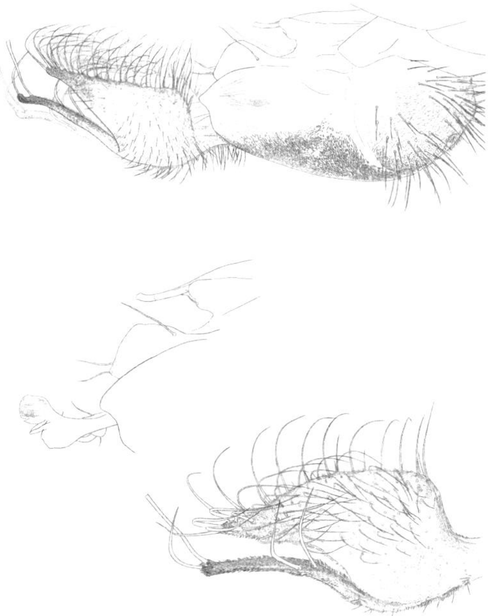

Fig. 16. (*Continued*)

Dolichopus subpennatus

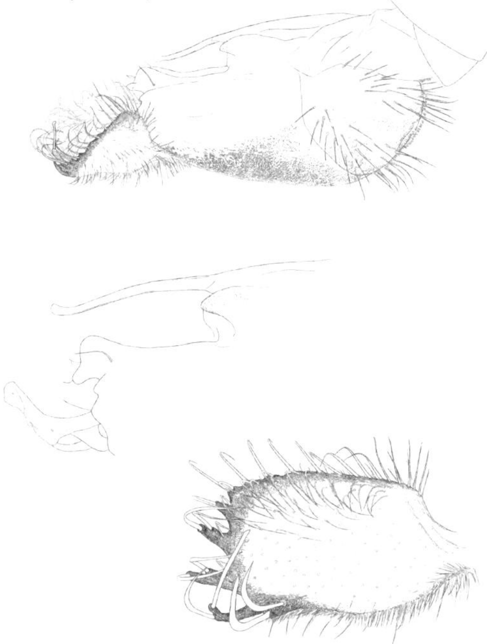

Dolichopus trivialis

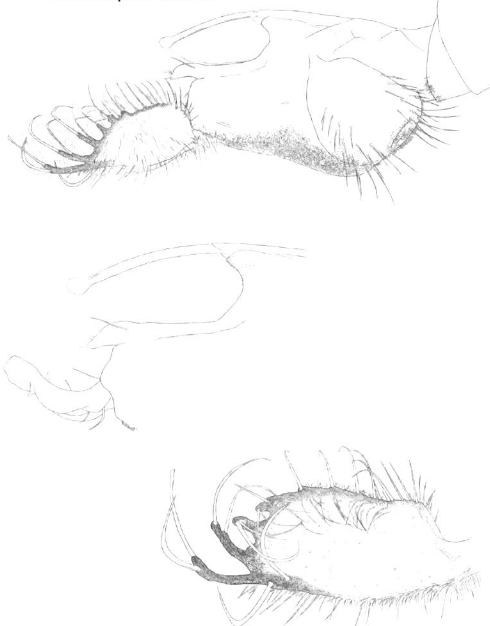

Dolichopus ungulatus

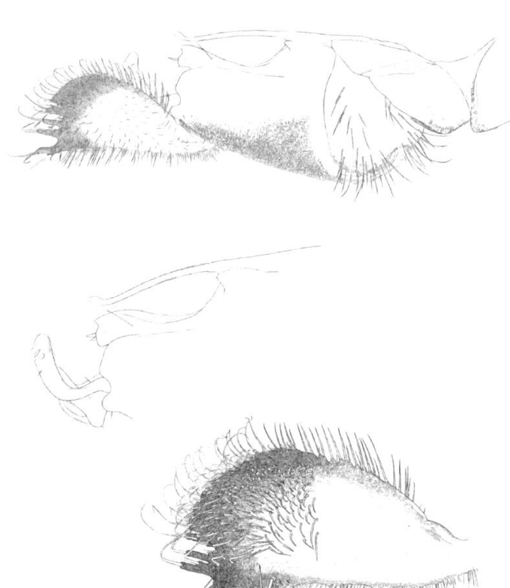

Dolichopus urbanus

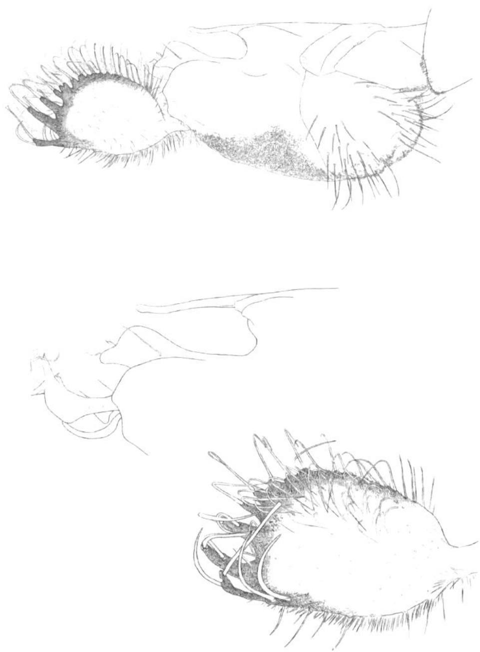

Fig. 16. (*Continued*)

Dolichopus virgultorum

Dolichopus vitripennis

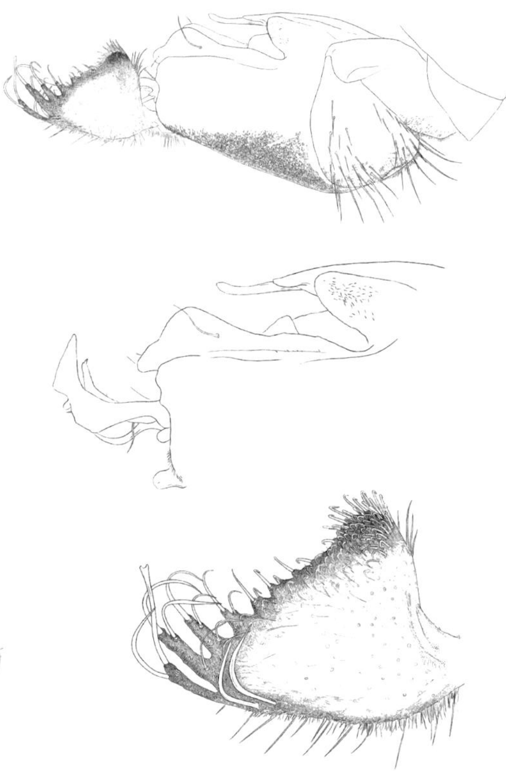

Dolichopus wahlbergi

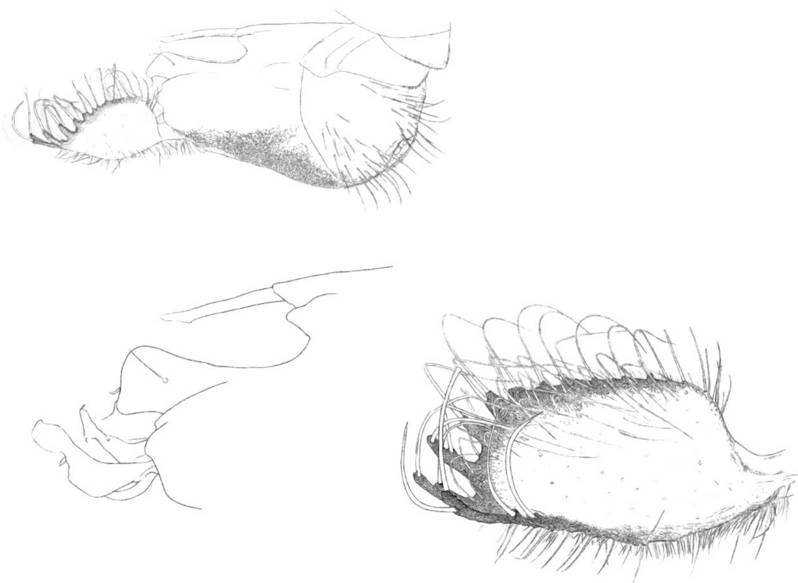

Fig. 16. (*Continued*)

Dolichopus atratus Meigen, 1824 (Plate 3)

Among *Dolichopus* with completely black legs, *atratus* is immediately recognised by the numerous setae on the hind basitarsus. Males may be recognised in the field by the dark outer half of the wings, a character shared only with *maculipennis*. In both sexes the legs are completely black with at most slight indications of yellow at the knees and 'ankles'. The basiventral epandrial lobe extends backwards leaving a long narrow gap before the apicoventral lobe. Females also have a darker apical half to the wing but less obviously than in males; when this distinction is slight, cell c in front of vein sc is almost colourless compared to the darker leading edge on the apical half of the wing. Wing length ♂: 5.3 (4.9–5.6) mm, ♀: 4.8 (4.7–5.0) mm.

This species is moderately common in western and northern Britain, and on acid soils of lowland England from Dorset to Norfolk, the Weald of Kent and East Sussex, but is absent from the limestone and chalk landscapes of lowland England. In upland areas it occurs where few other *Dolichopus* are found. It has a close association with acid mire and less so with wet heath. In the Norfolk fens it probably occurs only on small remnants of acid peat. Flight: iv–ix, xi, peak in vi.

Dolichopus atripes Meigen, 1824

This is one of the more robust black-legged *Dolichopus*. Males have a particularly short postpedicel and no ciliolarium. Of the entirely black-legged *Dolichopus*, only this species and *phaeopus* have pale lower postocular setae but *phaeopus* has a weak ventral fringe on the hind femur, lacking in *atripes*; unlike all but *melanopus*, it has a yellowish brown face. The almost semicircular cercus is characteristic and immediately distinguishes the species. Females may have their front and mid tibiae pale dorsally, although not bright yellow, and may run to *vitripennis* in the key; the arista of *atripes* arises about the middle of the postpedicel (much closer to the tip in *vitripennis*). *D. picipes* is the only other black-legged *Dolichopus* with a black leading edge to the base of the costa. In addition to the characters in the key, *atripes* females differ from *phaeopus* in having about ten black proepisternal hairs (4–7 in *phaeopus*) and the stem of the halter is completely yellow rather than partly darkened between a yellow base and knob. Wing length ♂: 4.4 (3.9–4.9) mm, ♀: 4.4 (4.1–4.6) mm.

Dolichopus atripes is a moderately common species on acid soils in northern and western Britain and in counties south of the Thames estuary, with a scatter of sites in the mires of Norfolk and Suffolk. It avoids the limestone and chalk of lowland England. It is found on acid mires and heaths, mainly at unshaded seepages and the margins of streams and pools. It is found less often in wet oak, alder or conifer woodlands. East Anglian sites appear to be neutral fens. Flight: iv–ix, peak in vii.

Dolichopus brevipennis Meigen, 1824 (Plate 3)

Among the yellow-legged species with a conspicuously large fifth tarsomere on the front leg, *brevipennis* is distinguished by its gold face, black lower postocular setae and ventral fringe on the hind femur. Its overall colour is a rather dark bronzy hue of green. Its ovate cercus with a wide dark apical margin, rather short marginal hair and dense black hair on the inner face is unlike other species. Females are characterised by having black postocular setae, only one preapical seta on the mid and hind femora and four ventral setae on the mid tibia. All its tibiae are well armed, the hind ones appearing particularly bristly as the ventral setae are as stout and numerous as the dorsals. The front feet look big as the fifth tarsomere is deeper than the preceding segments in an echo of the condition of the males. Wing length ♂: 5.4 (4.9–5.7) mm, ♀: 5.7 (5.2–6.1) mm.

This is a moderately common species in most of England and lower-lying areas of Wales but more sparsely distributed in Scotland where it is more often found near the coast, although not confined to it. It may be genuinely scarce in the extreme southeast of England. It is most often associated with unshaded margins of still water such as ponds, ditches and slow lowland rivers, but also by streams, stony rivers and in a variety of other wet places such as fens, seepages and brackish marsh. It is uncommon in woodlands. Unshaded water margins are probably the main requirement. Flight: iii, v–viii, peak in vi.

Dolichopus caligatus Wahlberg, 1850

Along with *strigipes*, *caligatus* has a weak posteroventral fringe on the hind femur, not long enough to be obvious at first glance but clearly differentiated from the vestiture. It differs from *strigipes* in having an entirely black postpedicel and lacking the deeply cleft cercus of that species. There is a short pronounced costal stigma. The hind tibial ciliolarium consists of tiny yellow points in a strip between the first dorsal seta to just beyond halfway. Females have almost entirely black antennae with only a narrow ventral pale strip on the scape, very short and rounded postpedicel, black-tipped hind tibia and a pronounced black mark at the base of the front coxa. Wing length ♂: 4.3 (4.0–4.7) mm, ♀: 4.4 (4.1–4.7) mm.

This is an uncommon species with a wide distribution from Devon to Sutherland although with more sites in Wales and Scotland than in England. There is little information on its habitat requirements; MacGowan (1987) regards *caligatus* as typically a moorland species. Sites also include lochs and streams in Scottish peat bogs, and fen and grazing marsh in Norfolk Broadland. Specimens from the isolated population in atypical habitat in East Anglia do not appear to differ from Scottish material so are unlikely to represent a different species. Flight: vi–viii.

Dolichopus calinotus Loew, 1871

This small species superficially resembles *sabinus* and *signifer* in having a dark mark at the wing tip but it is one of the few species among those with no leg ornamentation to have a golden face and mainly bright yellow antennae. Its inconspicuous ciliolarium consists of a narrow row of black points at a quarter to half the length of the hind tibia. The hypopygium has an unusually prominent and strongly rounded apicoventral epandrial lobe arising close to the basal lobe, as in *sabinus*. Females resemble a small *plumipes* with bright largely yellow antennae but have only three long anterodorsal setae on the mid tibia, a different arrangement of setae on the hind basitarsus and a narrow anal lobe on the wing. Wing length ♂: 3.9 (3.8–4.0) mm, ♀: 3.8 (3.8–3.9) mm.

This species was first found in Britain in 2016 on the east coast of Kent at upper saltmarsh (Drake and Pollet 2016). It appears to prefer saline habitat in northern Europe. Flight: vii.

Dolichopus campestris Meigen, 1824

This is the only *Dolichopus* with black femora and two or more anterior preapical setae on the mid femur. It shares almost entirely yellow tibiae with only the much smaller *vitripennis* and rare specimens of *clavipes*. The cerci are unusual in sometimes being dark brown, making them unique among *Dolichopus*, although more often they are white with a black margin. Females also have a ventral fringe on the hind femur whose hairs are as long as the dorsal hairs, whereas other black-legged *Dolichopus* have far shorter ventral hairs. The scutellum has a dense marginal fringe of white hairs, unlike the dark hairs found in the dense fringe on *lepidus* and most specimens of *atripes*. Wing length ♂: 4.8 (4.5–5.1) mm, ♀: 4.7 (4.5–5.1) mm.

In England *campestris* is moderately frequent but scarce in Wales and western English counties, and apparently absent from Scotland despite being found in Northumberland and Cumberland. It is found in a wide variety of wetland habitats including swamps, reedbeds, fen, margins of ponds and ditches in grazing marshes and washlands, and river shores. More sites are associated with still than flowing water and very few are wooded. Flight: v–ix, peak in vii.

Dolichopus cilifemoratus Macquart, 1827

This large species resembles *festivus* but differs in the conspicuously curved front tarsal segments when viewed from above or in front; like *festivus*, these tarsomeres have shiny black front faces and a fringe of fine short vertical hairs. Females also closely resemble those of *festivus*; the hind tibia is usually almost entirely yellow at the tip, as is the base of the basitarsus, but occasional *festivus* look similar. The ratio of the two sections of the arista can be difficult to use as the junction between them is often indistinct so small errors in measurement can lead to a big difference in the ratio. Wing length ♂: 5.3 (4.9–5.6) mm, ♀: 5.5 (4.3–6.3) mm.

Dolichopus cilifemoratus has a wide and scattered distribution in England, south Wales and Scotland but reliably identified records are probably confined to the eastern and southern parts of England, notably river catchments of the Yorkshire Ouse, the East Midlands Ouse and Nene, the Severn in Gloucestershire, and smaller floodplains in Sussex, Kent and Essex. It is most characteristically found in seasonally flooded wet grasslands next to ditches, drains and canals (Crossley 2014). Flight: v–viii.

Dolichopus claviger **Stannius, 1831**

Superficially this large long-legged species resembles *ungulatus*. Both have a similar hypopygium and cercus, two or more preapical setae on the mid and hind femora, ventral fringes on the front and mid femora and long crinkly hairs on the calypter. If the cercus and femoral setae and hairs are used as a first guess for the very common *ungulatus* then *claviger* could be overlooked, although there are many obvious differences. Females are distinguished by the combination of additional preapical setae on the mid and hind femora, and yellow scape, palps and front coxae which together readily separate *claviger* from *ungulatus*. Wing length ♂: 6.0 (5.2–6.7) mm, ♀: 6.4 (5.7–6.7) mm.

Dolichopus claviger has a wide distribution in Britain reaching northern Scotland but is more frequent on the east of the country, becoming scarce in western counties and in Wales. Its habitats are varied but more often deciduous woodland or scrub, or taller vegetation associated with a variety of wetlands and wet pasture. It does not appear to be specifically associated with wetlands. Flight: v–ix.

Dolichopus clavipes **Haliday, 1832**

This small species has variably coloured femora although in most specimens they are black. It is recognised by the combination of the long pale fringe on the hind femur and a single dorsal seta on the hind basitarsus. The hind tibia is unusual in being strongly thickened at the middle of the shaft and markedly bicoloured with the basal two-fifths yellow and the apical part black but sometimes entirely black in northern populations. The ciliolarium on the hind tibia is a particularly long field of yellow pubescence or short yellow points. Females are readily distinguished from other black-legged *Dolichopus* by the bright yellow bases to the legs, conspicuous in ventral view, and the pale palp and ventral side of the scape. The hind tibia is usually yellow with a black apex but can be entirely black, and the front and mid tibiae may sometimes be rather dark but not obviously entirely black. Wing length ♂: 3.8 (3.6–4.1) mm, ♀: 4.1 (3.7–4.4) mm.

Saltmarsh is the principal habitat of *clavipes*. It is distributed around much of the British coast in sheltered areas but is usually absent on shingly or rocky coasts. Flight: v–viii.

Dolichopus diadema **Haliday, 1832 (Plate 3)**

The combination of a long face that descends below the bottom of the eyes, a right-angled bend with a stub-vein on vein M_1 and a very long ventral seta at the apical third of the front tibia is unique in British *Dolichopus*; the last character is found in no other species. The rounded protruding apicoventral epandrial lobe of the hypopygium is unusual. Females share the face, wing and tibia characters although the ventral seta on the front tibia is not outstanding but is longer than the more basal seta. Long marginal setae on the tergites are found in only a few other *Dolichopus*, usually in other saltmarsh species. Unusually variable leg colour is the main reason for the convoluted wording of the first couplet that splits black-legged from yellow-legged species. Most individuals have obviously yellow femora but those from northern populations may have mainly black femora with narrow yellow streaks that interrupt the black on the anterior and posterior faces. Wing length ♂: 4.9 (4.7–5.2) mm, ♀: 5.2 (4.6–5.6) mm.

This saltmarsh species is found commonly around the English and Welsh coasts. It is infrequent on southern Scottish saltmarshes and has only scattered populations in the northern half of the country, including on the tiny rocky islands of Sule Skerry and Fair Isle. Although most numerous on upper saltmarsh, it can be found by brackish ditches and ponds. An inland site in Yorkshire has brackish influence. Flight: v–ix, peak in late vi–vii.

Dolichopus discifer **Stannius, 1831 (Plate 3)**

The particularly long slender front tarsus terminating in an expanded tarsomere make this species unique among British *Dolichopus*. Its hind tibia is also slender and no thicker than the mid tibia, unlike most species in this group except *migrans*. Entirely pale front coxal hairs are shared by *claviger* in this group but that species is amply distinct, e.g. by having ventral fringes on the front and mid femora. Both epandrial lobes are long, narrow and pointed, unlike other *Dolichopus*. Females are unique in having two ventral setae on the mid tibiae but no dorsal seta on the mid basitarsus. Rare specimens have a dorsal seta on the mid basitarsus but it is placed close to the base rather than on the apical half as in species of the '*trivialis*' group. The wings in both sexes are narrow and parallel-sided with venation resembling *Hercostomus* or *Poecilobothrus* with little inflexion on the apical section of vein M_1, and R_{4+5} and M_1 clearly converging towards the wing margin. Wing length ♂: 6.0 (5.7–6.3) mm, ♀: 5.8 (5.2–6.2) mm.

The name *nigricornis* Meigen, 1824 is used by some authors for this species.

Dolichopus discifer is a common and widespread species throughout Britain, including Shetland, although possibly more frequent in the north and west. It is a woodland species associated with wetland features such as streams, seepages or ponds. Outside woods, it may be found where trees, scrub or tall herbaceous vegetation provide shade. Flight: v–ix, xi, peak in iv.

Dolichopus excisus Loew, 1859

Dolichopus excisus is part of the '*nubilus*' group characterised by a finely hairy face and acrostichal setae that become numerous in front of the scutellum. It most resembles *andalusiacus* but is easily identified by the long ribbon extending from the cercus and the triangular apical epandrial lobe. Females closely resemble those of *andalusiacus* and the *nubilus*/*latilimbatus* pair but have more extensively dark tips on the hind tibia. The broadly dark front coxa resembles those found in some specimens of *andalusiacus* which is distinguished by its entirely yellow hind tibia and more extensive small hairs along the dorsocentral row of setae. The very slightly sinuous outline of the basal lobe of the wing is often difficult to see as the wing usually folds here. Wing length ♂: 4.3 (3.9–4.7) mm, ♀: 4.6 (4.3–4.8) mm.

Since its discovery in Britain in 2005 (Gibbs 2006) *excisus* has been found at near-coastal locations from Dorset to Norfolk, and at one inland location in Wiltshire. The habitats include freshwater ditches on coastal grazing marsh, reedbed, fen and heathland with pools. Flight: v–viii.

Dolichopus festivus Haliday, 1832

Among those with a pale fringe on the hind femora, this large species is distinguished by its mainly yellow antennae and dumpy postpedicel which is barely longer than deep and by the anterior apical setae on the second and third tarsomeres but not on the fourth on the front leg. These tarsomeres have shining black front faces as in *cilifemoratus*. On the hind tibia, the ciliolarium consists of tiny yellow points on the basal half on a slightly swollen section. Females closely resemble those of *cilifemoratus*; individuals whose hind tibia has an obviously dark tip are *festivus* but those with pale legs become difficult to distinguish. The arista character can be difficult to use as the junction between the sections is often indistinct, so small errors in this measurement can lead to a big difference in the ratio. The shape of the bends in M_1 are variable and overlap with the condition in *cilifemoratus*, and not as described by d'Assis-Fonseca (1978). Wing length ♂: 5.2 (4.7–5.5) mm, ♀: 5.7 (5.1–6.4) mm.

In lowland and southwest England, this is a common species which becomes scarce towards the northwest and in Wales. In Scotland it is frequent only in the low-lying catchments of the Forth and Clyde but does occur on the Outer Hebrides and Shetland. It is found mainly in drier habitat such as pasture and hedgerows in farmed countryside, where it appears to prefer ranker or tussocky vegetation. Porous soils such as gravels, sands and chalk probably favour it, and it is frequent in well-drained wetlands such as riverine gravels and disused quarries although it can be found in wetter places such as acid grasslands and grazing marsh with ditches. Most sites are unshaded. Flight: v–ix, peak in vii.

Dolichopus griseipennis Stannius, 1831

Both sexes are recognised by the unique combination of a tiny stub-vein at the inflexion of vein M_1 and only one seta on the hind basitarsus, both being relatively infrequent conditions. Males have a long fine posteroventral seta on the front tibia. Rare female specimens of *signifer* have a tiny stub-vein and one seta on the hind tibia and would run to *griseipennis* in the key but they have dark front coxae. *D. griseipennis* appears to have a larger size range than found in most *Dolichopus*. Wing length ♂: 5.3 (4.7–5.7) mm, ♀: 5.1 (3.9–5.9) mm.

The distribution is predominantly southern, and *griseipennis* is common in much of England north to Yorkshire, although not in East Anglia. While it is infrequent in inland Wales, northern England and Scotland, it is notably found on Scottish islands including Shetland, the Outer and Inner Hebrides and the tiny Ailsa Craig, North Rona and Sule Skerry. Presumably winter minimum temperature controls this distribution. It occurs with no apparent preference in a wide range of habitats, shaded to open, wet to moderately dry. Flight: iv–xi, peak in vii to early viii but with a long season into ix.

Dolichopus laticola Verrall, 1904

The scutum of this large species is uniformly black or almost black with scarcely a hint of green. The colour of the mid and hind tibiae is variable but the mid tibia usually has a slight dark ring just beyond the base, even if the rest of the shaft is blotchily yellow and black. The hind tibia may be entirely black or with a yellow dorsal face. The frons may be blue or green. The cercus unusually has long black hairs at the base on the ventral edge, where in most species these become short and pale. Females may be recognised by the violet disc of the frons and the unusual colouring of the mid tibia, as in the male; the inflexions in vein M_1 are very slight. Wing length ♂: 5.1 (4.9–5.3) mm, ♀: 5.6 (5.3–5.8) mm.

The distribution is confined to the fens of Norfolk where it is moderately frequent, at one pingo site in this county and at a few fens in Suffolk. Rich fens with high botanical diversity characterise the sites (Drake 2013, 2018c). Flight: v–viii.

Dolichopus latilimbatus Macquart, 1827

This is one of the group four species in the '*nubilus*' group. Males are immediately recognised by the large square cercus and large pale apicoventral epandrial lobe with several long hairs on its outer face. This species and *nubilus* have particularly complex hypandrium and basiventral epandrial lobes. There is no ciliolarium on the hind tibia. Females are indistinguishable from those of *nubilus*, as described under the account for that species. Wing length ♂: 4.3 (4.0–4.5) mm.

In lowland England *latilimbatus* is widespread and increasingly frequent towards the southeast, and moderately frequent in the southern half of Wales. It becomes scarce in northern England and rare in Scotland. Confusion with *nubilus* whose females are inseparable may account for inaccuracies in distribution, particularly around the coast. It is associated with water margins, especially those of ponds in a range of habitats, e.g. wet dune slacks, those on wide river shores and in grasslands. Other habitats include reedbeds, seepages running onto saltmarshes and, less often, stream margins. It shows a possible preference for areas with bare mud, sand and peat, e.g. in drying ponds. It is found far less often in shaded places. Flight: iv–xi, peak in late v to vi.

Dolichopus latipennis Fallén, 1823

Being large, having exceptionally long antennal scape and pedicel and a rounded wing that is only twice as long as wide, *latipennis* is unmistakable. The antenna appears particularly thick owing to long dense hairs on the first two segments and the costal stigma is unusually wide. Females also have an exceptionally long scape and pedicel and, although the wings are not particularly short, the position of dm-m well beyond the midpoint gives the impression of short wings. ♂: 5.1 (5.0–5.2) mm, ♀: 5.4 (5.3–5.5) mm.

Dolichopus latipennis has been found in only two areas, the northern Solway coast and marshes on or near the coasts of Norfolk and Suffolk, but has not been recorded from these latter sites for a considerable period. It may be associated with brackish marshes and reedbeds in Britain; Emeis (1964) classifies it as saline-associated. Flight: v–vii.

Dolichopus lepidus Staeger, 1842

Dolichopus lepidus is the only one among the large black-legged species with a long black fringe on the hind femur, black lower postocular setae and a rich golden face. The acrostichal setae are longer than normal, about two-thirds the length of adjacent dorsocentrals, and the scutellar hairs are more dense and conspicuous than in other species of the group. Females are distinguished by the dark fringe on the hind margin of the scutellum and entirely black proepisternal hairs. The hind tibia varies in colour but is nearly always yellow dorsally and black on the apical quarter. Wing length ♂: 4.7 (4.5–5.1) mm, ♀: 5.0 (4.7–5.3) mm.

This is a predominantly northern species common in Scotland and moderately frequent in northern England but scarcer in Wales and in the southern and eastern counties of England. Its distribution away from upland Britain is dictated by the occurrence of acid soils. It is closely associated with acid peat so occurs on mires, bogs, wet heaths and some fens where it can be found at pools and seepages. Some sites appear to be base-rich but these may be exceptions. The species is rarely found in shaded places. Flight: iv–viii, peak in vi.

***Dolichopus linearis* Meigen, 1824**

This is a moderately small species with particularly pale legs, the front and hind coxae being entirely pale and the mid coxa having only a dark streak. The hairs of the front coxa are almost entirely pale as found in only a few other species including *acuticornis* and *longicornis* which both have a very long postpedicel, and *arbustorum* and *virgultorum* which have a fringe of long hairs on the hind femur. The hind tibia is clearly black-tipped followed by a black tarsus contrasting with the almost entirely pale front tibia and tarsus. There is no ciliolarium on the hind tibia. The long narrow gap between the lobes of the epandrium is unusual. Females are distinguished by the combination of a single ventral seta on the mid tibia and entirely yellow scape and hind coxa; other species with mainly pale hind coxae are *arbustorum* and *virgultorum* which have more ventral setae on the mid tibia, and *acuticornis* and *longicornis* which both have a partly or entirely dark scape. Wing length ♂: 3.9 (3.6–4.2) mm, ♀: 4.2 (3.7–4.6) mm.

Dolichopus linearis is distributed widely but sparsely in England, in central to northern Scotland and south Wales. It is most often recorded by rivers and streams but also in wet grasslands, fen, wet woodland and carr and wet heath seepages. It may possibly be associated with more freely draining soils. Flight: vi–viii.

***Dolichopus lineatocornis* Zetterstedt, 1843**

The golden face and pronounced anal lobe separate *lineatocornis* from other species of *Dolichopus* with partly yellow antennae and no fringe on the hind femur. Its distinction is confirmed by the markedly swollen basal half of the hind tibia which is almost twice as thick as the mid tibia. The postpedicel is moderately long and pointed. Females may be difficult to identify as the key is based on very little material. Wing length ♂: 4.6 (4.3–5.0) mm.

Dolichopus lineatocornis is a rare species with its apparent centre of distribution in remnant wetlands of The Fens of Cambridgeshire and Huntingdonshire. There are also isolated sites in the southern counties of England and Yorkshire. It may be associated with seasonally flooded grasslands and mires. Flight: v–viii.

***Dolichopus longicornis* Stannius, 1831**

The exceptionally long and pointed postpedicel is found only in this species and *acuticornis*. These two species also have a long costal stigma which is about twice the length of dm-m, mainly pale hind coxa and yellow trochanters, which are all features found in few other *Dolichopus*. The antennae are also unusual in the group of *Dolichopus* with no fringe on the hind femora in being black above and yellow below on all segments, a combination found only in *sabinus* which has far shorter antennae. Females are easily distinguished by their long acutely pointed bicoloured postpedicel which is black above and yellow below for its entire length; the pale leg and coxa colour is as in the males. Wing length ♂: 3.7 (3.5–4) mm, ♀: 4.0 (3.6–4.5) mm.

Dolichopus longicornis is a moderately common species widely distributed in Britain. It is most strongly associated with river shores, particularly sandy ones, but is also found at the margins of streams, ponds and lakes, within carr woodland, seepages and some fens. Sites may be open or shaded. Flight: v–viii, peak in vii.

***Dolichopus longitarsis* Stannius, 1831**

Dolichopus longitarsis is included in the key under species either with or without modified tarsi as the slight expansion of fifth tarsomere may be overlooked. It is recognised by the combination of black lower postocular setae and a long black fringe on the hind femur. In addition, the scutellum has a dense fringe of fine hairs on its hind margin, the antennae are entirely black and the hind tibia is particularly stout, being twice as wide as the mid tibia and black on the apical two-fifths with an obvious ridge at its tip dorsally. Females are characterised by having black postocular setae, a long dark fringe on the hind margin of the scutellum and 2–3 setae on the hind basitarsus. They have the tiny facial hairs found in the '*nubilus*' group. Wing length ♂: 4.1 (3.9–4.5) mm, ♀: 4.1 (3.8–4.3) mm.

Dolichopus longitarsis is widespread although patchily distributed in Britain, reaching the north of Scotland. It is associated with particularly wet places such as springs, seepages, wet grassland with ditches, swamps, bogs and fens, but reaches its highest numbers in fens. There is no obvious preference for acidic or base-rich conditions. It is scarce in shaded sites, contrary to the observations of Emeis (1964) and Maes and Pollet (1997) who regard it as preferring shade. Flight: v–viii, peak in vi.

Dolichopus maculipennis Zetterstedt, 1843

This large species superficially resembles *picipes* and has similar conspicuous yellow knees. Among entirely black-legged *Dolichopus*, it is the only species having at least two anterior preapical setae on the mid and hind femora, and is the only one other than *atratus* having a strongly marked wing. Like *melanopus*, it has a fringe of long black hairs on the hind femur and a long ciliolarium of tiny black points on the slightly swollen basal half of the hind tibia. The cercus has a moderately dark lamella and wide black rim, and the epandrial lobes lie close to one another. Females are distinguished by the additional preapical setae on the mid and hind femora, several ventral setae on the mid tibia and a bluish frons. Confusion is most likely with *atratus* as both species have partly dark wings and occupy the same habitat but, apart from the characters in the key, *maculipennis* has black proepisternal hairs (not white) and a brown-tinged broad face whose depth is up to 1.5 times its width, compared to *atratus* with a white narrower face whose depth is at least 1.7 times its width. The scutum is shining black with no hint of green, unlike the obviously green scutum of *picipes* and *atratus*. *D. campestris* is similar in having two setae on the mid femur but only one on the hind femur; *planitarsis* also has several ventral setae on the mid tibia but only one femoral seta. Wing length ♂: 5.2 (4.9–5.4) mm, ♀: 5.2 (4.7–5.7) mm.

Dolichopus maculipennis is a rare submontane Scottish species occurring at altitudes of approximately 600–800 m. Originally found in the Breadalbane area of Mid Perthshire, subsequent records are from the Beinn Dearg area of Ross-shire and more recently from the southeast Cairngorms. MacGowan (1987) noted that it was associated with areas of base-rich soils derived from calcareous schists which are uncommon in highland Scotland. Habitats include peat pools, wet grassland and seepages. Flight: vi–vii.

Dolichopus mediicornis Verrall, 1875

Both sexes of this species are very similar to *simplex*, and females also to *lineatocornis*. The only characters that appear to differ are given in the keys but these may be unreliable as so few specimens of *mediicornis* and *lineatocornis* could be examined. ♂: 4.0 mm, ♀: 3.8 mm.

The only confirmed sites are in the New Forest, Hampshire, and adjacent heathlands of Dorset in England, and Dyfed in Wales. Its habitat is unknown. Flight: vi–viii.

Dolichopus melanopus Meigen, 1824

Only two *Dolichopus* with entirely black legs have modified front tarsi, so *melanopus* needs comparing only with *nigripes*. *D. melanopus* is smaller and has relatively shorter and much less bristly legs than *nigripes*, e.g. the hind tibia has only about seven short dorsal setae compared to about 14 long setae, each more than twice the shaft's width, in *nigripes*. The thick hind tibia has a ciliolarium consisting of minute points on a dull bare area in the basal half of the dorsal surface, contrasting with the shining apical half with normal black vestiture; at the tip on the dorsal side is an oblique furrow containing short flat white hairs. Females are unusual among the black-legged

Dolichopus in having tiny hairs on the clypeus and epistoma; they are otherwise rather undistinguished. Wing length ♂: 4.1 (3.9–4.3) mm, ♀: 4.3 (4.0–4.7) mm (non-British specimens).

Dolichopus melanopus is probably extinct in Britain. Two males were recorded from the New Forest, Hampshire, in 1872 (Verrall 1875). Flight: vi.

Dolichopus migrans Zetterstedt, 1843

This large slender-legged species resembles *discifer*. Among species with a modified front tarsus, it is the only one with black antennae with the scape yellow below and lacks a ciliolarium found in the others (indistinct in *plumitarsis*). Its scutellar fringe is dense, as in the much smaller *longitarsis*. The tibiae have more setae than usual; no other species in this group has four or more long ventral setae on the hind tibia. Females have a noticeably shiny green frons, compared to slightly dulled in most *Dolichopus*, and the dorsum is also unusually shiny; otherwise, there are no outstanding characters. Wing length ♂: 5.6 (5.4–6.0) mm, ♀: 4.9 (4.8–5.0) mm.

The species is reliably recorded only from the Brecklands of Norfolk and Suffolk and similar heaths in Lincolnshire and Yorkshire where it lives in dry grassland on sandy soil (e.g. Crossley 2014). In mainland Europe it is associated with sandy coastal habitats, including dunes and particularly where there is some scrub (Emeis 1964, Pollet and Grootaert 1996). Flight: v–viii.

Dolichopus nigripes Fallén, 1823

Only two *Dolichopus* with entirely black legs have modified front tarsi and a comparison with *melanopus* is given under that species. No other species has an apparently absent basiventral epandrial lobe and few have a relatively long rectangular cercus. Females resemble *picipes* in being similarly large, having conspicuous yellow knees and 'ankles' and wings having darker anterior cells. They are readily distinguished from other black-legged *Dolichopus* by the long fine pale hairs at the base of the mid and hind femora, and the unusual colour pattern of the front tarsus described in the key. The scutum is black with scarcely any hint of green, unlike the obviously green scutum of *picipes* and *atratus*. Vein M_1 has particularly weak inflexions. Wing length ♂: 4.9 (4.7–5.0) mm, ♀: 5.3 (5.0–5.4) mm.

The first record in Britain was from Dorset in 1839 but since then *nigripes* has not been found in that county and is now confined to a small area of the Norfolk fens. The habitat is fen with a more grassy than reedy component (Drake 2018c). Flight: vi–viii.

Dolichopus nitidus Fallén, 1823

This is a moderately large *Dolichopus* with largely pale legs which have entirely yellow tibiae and mainly yellow basitarsi. It can be recognised immediately by the fan of very long apical hairs on the cercus which are as long as the cercus itself; the epandrium is particularly stout and almost cylindrical. The right-angled bend of vein M_1, often with a tiny stub-vein, is otherwise found only in *diadema* and *griseipennis* which differ in having a ciliolarium and mainly black antennae compared to those of *nitidus* which have an entirely yellow scape and partly yellow pedicel and postpedicel. Females have the same distinguishing leg, wing and antennal features as the males; often the whole of each basitarsus and most of the second tarsomere are yellow so that the legs appear particularly bright. Wing length ♂: 5.1 (4.7–5.4) mm, ♀: 5.4 (5.0–5.6) mm.

This is an uncommon species with a wide but scattered distribution predominantly on or near coasts of England and Wales, and inland in northern England and Scotland. The habitats where it has been found suggest a wetland species of fens, bogs, seepages and marshes with no consistent factor emerging. Flight: vi–x.

Dolichopus notatus Staeger, 1842

This species is recognised by the combination of a long fine apicoventral seta on the front tibia, a long postpedicel that is about twice as long as deep and the hind tibia being strikingly black on the apical quarter to third. The antenna may be entirely black or the scape slightly orange at the tip below but never clear yellow, so *notatus* falls out twice in the key. Additionally, the hind tibia is markedly thickened and lacks a ciliolarium, and the wing is broadly rounded at the tip, unlike most

Dolichopus. Females have a slightly elongate postpedicel, the antenna is black except for the scape below, and the hind tibia is markedly black on the apical quarter as found in few other species that are amply distinct in other ways. Wing length ♂: 4.2 (4.0–4.5) mm, ♀: 4.2 (3.9–4.4) mm.

Dolichopus notatus is sparsely distributed on the English and Welsh coast from Lancashire to the Humber estuary. Records inland further north, including those in Scotland, may be unreliable. The main habitat is coastal wet dune slacks and saltmarsh transition to drier sand but it occurs inland at fens in East Anglia. Flight: v–viii, peak in vi.

***Dolichopus nubilus* Meigen, 1824**

Dolichopus nubilus is one of the small group of *Dolichopus* with tiny white facial hairs and acrostichal setae that extend beyond the last dorsocentral seta in a small cluster in front of the scutellum, a prominent apicoventral epandrial lobe and particularly complex basiventral lobes in some members. These features are sufficiently distinct that here this species together with *andalusiacus*, *excisus* and *latilimbatus* are referred to as the '*nubilus*' group. *Dolichopus nubilus* is a moderately small fly whose wings are usually brownish, contrasting with the pale grey or hyaline wings of other species with yellow femora and entirely black antennae. The exceptionally large apicoventral epandrial lobe is sufficient in itself to identify *nubilus*. Females are inseparable from those of *latilimbatus*; all characters form continua independently of one another so that, although some specimens agree with previously used combinations of states, there is no way of being certain that these combinations are unique to one or other species. Occasional individuals with entirely yellow tibiae and scarcely darkened hind femur resemble *andalusiacus* but have fewer supernumerary hairs along the row of dorsocentral setae. Wing length ♂: 4.2 (4.0–4.4) mm, ♀: 4.5 (3.8–4.9) mm (with male *nubilus*).

Dolichopus nubilus is a moderately common species found predominantly on British coasts. It is widespread inland in England but less so in Scotland and Wales. It occurs north to Orkney and on small remote Scottish islands. At the coast it is found in many types of wetland but principally saltmarsh where it can be abundant; it is also frequent at estuarine shores, brackish marshes, fens and reedbeds. Inland freshwater sites include reedbeds, fens, washlands and swamps, and there may be a slight preference for sites with patches of bare wet mud. Flight: iv–x, peak in vi.

***Dolichopus pennatus* Meigen, 1824**

This is a large robust species. The clearly thickened second and third tarsomeres of the mid leg are found only in this species and *subpennatus*, making them obvious although separating this pair of species requires some care. Among species with modified mid tarsi, these two species also have the most robust hind tibiae which are twice as thick as the mid tibiae. Selivanova *et al.* (2019) found that difference in the shape of the postpedicel is unreliable but no misfits have been found in British material. They suggest additional characters but those described by Pârvu (1997) appear to refer to *signatus*. Females resemble those of *argyrotarsis*, *signatus* and *subpennatus* but are immediately distinguished from all other large yellow-legged *Dolichopus* by the dense fringe on the hind margin of the scutellum. The slightly expanded second and third tarsomeres of the mid leg is a weak version of that found in the males and is also seen in females of *argyrotarsis*, *signatus* and *subpennatus*. Wing length ♂: 5.1 (4.7–5.6) mm, ♀: 5.4 (5.0–6.0) mm.

This species is common and widespread throughout Britain, reaching Orkney and the Outer Hebrides. It uses a wide variety of wetland habitats with no obvious preference for running or still water so it may be found at seepages, the margins of ponds, ditches and rivers and in fens, but avoids acid sites so is not often found on bogs or moorland. It will also stray from wet sites and may be found in hedgerows. Sites are usually unshaded although it can be found in woodland. Flight: iv–viii, peak in vi.

***Dolichopus phaeopus* Haliday in Walker, 1851**

This small black-femured *Dolichopus* may have its front tibia yellow and the mid and hind ones black, which is a unique combination, or all the tibiae may be black, and thus resembling the larger *atripes* and *picipes*. The lower postoculars are pale but some specimens have a few black setae below these. Females resemble those of *vitripennis* in their small size, and occasional specimens have pale front and mid tibia almost like *vitripennis* but they differ in the basitarsi of these legs being almost entirely black (yellow in *vitripennis*) and the arista arising at about the middle of the postpedicel (near the tip in *vitripennis*). Separation from *atripes* is given under that account. Wing length ♂: 3.7 (3.3–4.0) mm, ♀: 3.7 (3.4–4.0) mm.

Dolichopus tanythrix is one of the common species in nearby European countries and may yet be found in Britain. In the key it will run to *phaeopus* with black tibiae but differs in having a posterodorsal fringe of long hairs on the hind tibia, a long ventral fringe on the hind femur as in *lepidus* and white postocular setae.

The distribution is western and northern with a possibly higher density of records in Scotland. There are outliers in the Norfolk fens. It is found in unshaded wetlands and is most likely associated mainly with peat as its sites includes acid mire and seepages on uplands and patches of acid mire within fens in lowlands, and ditches in coastal grazing marsh close to acid heathlands. Flight: v–ix, peak in vii.

Dolichopus picipes Meigen, 1824 (Plate 3)

This is one of the largest of the entirely black-legged *Dolichopus* and is distinguished from those with no outstanding features on the wings or tarsi by the combination of narrowly pale knees and 'ankles', black postocular setae and an unremarkable rectangular slightly brown cercus. Females also have conspicuously yellow knees and 'ankles'. They differ from all but *atripes* among the black-legged species in having the leading edge of the costa near its base conspicuously dark, rather than entirely bright yellow like other veins at the wing base. The hind basitarsus usually has three dorsal setae rather than two as in most other species of this group. Wing length ♂: 5.2 (4.2–5.5) mm, ♀: 5.7 (5.3–6.1) mm.

Dolichopus picipes is a moderately common and widespread species found throughout Britain although yet to be recorded in the extreme north of Scotland. It is found in a wide variety of mainly unshaded wetlands, particularly in seepages in mire and valley fens, and beside small streams, but also in wetlands without water movement such as fens, swamps and lushly vegetated margins of ponds and ditches. Relatively few sites are wooded but it can be found in wet woodlands with seepages and streams. There appears to be no preference for acid or base-rich soils. Flight: v–ix, peak in vi.

Dolichopus planitarsis Fallén, 1823

This is the only black-legged *Dolichopus* with a modified mid tarsus. Its antennae are longer than in others of this group and the arista has conspicuous pubescence, unlike nearly all other *Dolichopus* in which it is almost bare. Several other features differ from most black-legged *Dolichopus*: the ground colour of the face is yellow rather than black below the white dusting, the front and mid coxal hairs are mainly white, the front tarsus is yellow to the third tarsomere, and the hind femur has a posteroventral fringe of short pale hairs that are about half the depth of the femur. The apicoventral epandrial lobe projects strongly and is about twice as long as wide. Females have a comparatively long postpedicel and rather short thick arista with obvious pubescence not found in other black-legged *Dolichopus*. Among this group, only *maculipennis* also has the additional ventral setae on the mid tibia. Wing length ♂: 4.9 (4.3–5.3) mm, ♀: 5.1 (4.7–5.5) mm.

Its distribution is wide, from Devon to Sutherland, but fairly sparse with a suggestion of greater frequency in northern England and Scotland than further south. However, its occurrence in central-southern England and East Anglia precludes it being characterised as 'northern and western'. Habitats include deciduous woodlands, wet grasslands including herb-rich grazing marsh, fen and reedbed. One recurring factor is underlying peat soil but this may not be essential. Populations appear to be small but stable over time. It flies earlier than most *Dolichopus*, peaking in mid-May to early June, so may be under-recorded. Flight: iv–viii.

Dolichopus plumipes (Scopoli, 1763)

The mid leg of *plumipes* is unusual in having both the tibia and basitarsus clearly modified. While the tarsal flag resembles that of *wahlbergi*, that species does not have the exceptionally flat mid tibia with a shiny dark dorsal stripe and white tip. The ciliolarium is an inconspicuous narrow line of tiny yellow points on the basal quarter. Deformities caused by parasitic nematodes give rise to the form *pectinitarsis* Stenhammar which is now a junior synonym of *plumipes* (Kahanpää 2008, Drake 2015). It has the fringed basitarsus of normal *plumipes* but the tibia shaft and setae resemble that of most other *Dolichopus*. Female *Dolichopus* with mainly bright yellow and rather short-tipped antennae and only one ventral seta on the mid tibia are usually *plumipes* although in coastal locations *calinotus*, *sabinus* and *strigipes* will need to be distinguished. Inland, *wahlbergi* is easily confused but has tiny hairs on the clypeus and its arista is placed obviously further from the tip of the postpedicel. Specimens of *plumipes* that have lost their antennae are indistinguishable from *simplex*. Wing length ♂: 4.2 (3.7–4.6) mm, ♀: 4.5 (3.8–4.8) mm.

This is the commonest British *Dolichopus*, occurring on several tiny offshore Scottish islands as well as Orkney. Apart from being a wetland species, it shows little habitat preference and may be found from saltmarsh to altitudes of over 1000 m in the Highlands, but is infrequent within deeply shaded sites. Flight: iv–x, peak in vi–vii.

Dolichopus plumitarsis Fallén, 1823

Among the yellow-legged *Dolichopus* with a modified front tarsus, *plumitarsis* differs in the tarsomeres being relatively slender compared to the conspicuous dorsal fringe; related species have relatively deeper segments and shorter fringes. The wings are noticeably broad and rounded with strongly angled bends on M_1. The hind tibia has no clear ciliolarium but a posterodorsal strip of fine pubescence narrows to a very thin line reaching almost to its tip. Females need comparing only with *brevipennis*, *migrans* and *discifer*. Apart from the characters used in the key, *plumitarsis* (and *discifer*) differ from the first two of these species in their hind tibia having a less obviously darkened tip, and *plumitarsis* (and *migrans*) differ from the first and last in the face being white rather than pale grey. Wing length ♂: 5.5 (5.3–5.8) mm, ♀: 5.6 (5.1–6.1) mm.

This species has been found only in a small area of Suffolk in species-rich fen meadow with areas of damp calcareous and neutral grassland where it may be associated with soligenous fen or with the margins of small streams. Flight: vi–vii.

Dolichopus popularis Wiedemann, 1817

This large species has a conspicuous mid-tarsal 'flag' unlike any other *Dolichopus* and which is clearly visible with the naked eye. Only a few species have more than one anterior preapical seta on the hind femur but among these only *popularis* has a modified mid leg. The hind tibia lacks a ciliolarium. Both sexes have the unique combination of additional preapical setae on the hind femur and largely yellow antennae. Wing length ♂: 5.1 (4.9–5.5) mm, ♀: 5.4 (4.5–5.8) mm.

Dolichopus popularis is a common species found throughout Britain including Orkney. It is found in a very wide range of wetland habitats and shows little preference. It is more frequently found in shaded habitat than open sites and may prefer flowing water, such as streams and seepages, to still water. Flight: iv–ix, peak in iv.

Dolichopus rupestris Haliday, 1833

Both sexes have a sombre dark brown scutum with scarcely a hint of green and are recognised by the combination of entirely black postocular setae and entirely black trochanters and coxae. This is one of the few species of *Dolichopus* whose femora are variably coloured from completely dark to partly yellow. The males of no other *Dolichopus* with partly dark front femora has a gold face, dense black hairs on the front coxa or a strong costal stigma. Its ciliolarium on the hind tibia is a narrow pubescent strip at about the basal quarter. The dirty white colour of the cercus and indistinctly demarcated black margin are unusual. Wing length ♂: 4.2 (3.7–4.5) mm, ♀: 4.3 (4.1–4.5) mm.

Dolichopus rupestris is a predominantly northern species. It is common from the Humber estuary to Scotland, including Shetland and St Kilda, frequent in Wales but scarce in southwest England. It is found most often on moorlands and mountains up to at least 950 m but can occur close to sea level on bogs and other acid sites. The habitat is usually acid mire, seepages, stream sides and wet grassland or montane heath, nearly always unshaded. Flight: vi–x, peak in late summer.

***Dolichopus sabinus* Haliday, 1838**

This small species resembles *calinotus* and *signifer* in having a dark-tipped wing. The antennae have the unusual pattern of all segments being yellow below and black above (as in *longicornis*) and an unusual arrangement of the epandrial lobes that touch at their bases (as in *calinotus*). The ciliolarium is a small oval tubercle with dark yellow pubescence but without a strip running down the dorsal edge. Females have a short and rather sparse but tidy fringe of hairs on the rear margin of the scutellum, not as long and dense as that of *longitarsis* or *pennatus* but more regular than in most *Dolichopus*. In most specimens the antenna is obviously yellow on the ventral half of all segments but the key accommodates those with rather dark antennae. Wing length ♂: 3.8 (3.6–4.0) mm, ♀: 3.9 (3.6–4.3) mm.

Dolichopus sabinus is a coastal species found moderately frequently on the English and Welsh coasts from the Humber to Cumbria but far more sparsely in Scotland on both the east and west coasts. It is found primarily on saltmarsh but also on brackish marshes, freshwater seepages running onto saltmarsh, wet dune slacks and lagoons. It may have a requirement for bare wet mud and water margins. Flight: vi–viii, peak in vii.

***Dolichopus signatus* Meigen, 1824**

Among males in the group with modified mid tarsi, this species is most likely to cause problems if the silver pubescence is overlooked since there is no marked difference in the thickness in the five tarsomeres. Conspicuous features are its markedly long and pointed postpedicel which is about 2.5 times longer than deep, the small oval ciliolarium on a tubercle which can be seen from behind without the need for a clear view of the posterior face, and the wing's simple hind margin at the anal lobe which in the other species with a modified mid tarsus is expanded so that the hind margin is conspicuously sinuate. The hind tibia is usually obviously black on the apical sixth, not just black-tipped. The epandrium is particularly large and its lobes form a sharp V-shape between them, unlike the usual rounded gap. Females closely resemble those of *argyrotarsis*, *pennatus* and *subpennatus* but are distinguished from all three by the arista arising closer to the tip of the postpedicel; *signatus* otherwise appears almost identical to *argyrotarsis* and differences are included in the account for that species. Wing length ♂: 4.2 (4.1–4.4) mm, ♀: 4.6 (4.1–5.0) mm.

The distribution suggests a common species, particularly in southern England, parts of northern England, Wales and the northern half of Scotland but there are large areas with no records in central England and from the northeast of England to southern Scotland. It is found in acid mires, poor fen and wet acid grasslands, usually in seepages or pool margins in unshaded areas but sometimes within woodlands. Flight: v–ix, peak in vi to mid vii.

***Dolichopus signifer* Haliday, 1838**

Dolichopus signifer is a small species whose scutum is dark brown to almost black with scarcely a hint of green. Both sexes are uniquely characterised by having dark front coxae and only one dorsal seta on the hind basitarsus. Both sexes also have variably coloured femora, the front one ranging from entirely yellow to entirely black, and the mid and hind femora from yellow to black with yellow streaks extending nearly their whole length. The tibiae may also be rather variable but the hind tibia is strikingly black on its apical fifth to quarter. Males are additionally distinct in having a dark wing tip which is usually visible to the naked eye, as in *calinotus* and *sabinus* which are also coastal species, a long fine apicoventral seta on the front tibia and a fringe of long black hairs on the hind femur. The ciliolarium consists of a long strip of dense yellow points starting on the posterior face near the base and twisting onto the dorsal face by the apical third. The epandrial lobes are unusual in being crowded one on the other. Wing length ♂: 4.5 (4.0–4.8) mm, ♀: 4.6 (4.1–5.0) mm.

Dolichopus signifer is an uncommon coastal species found in the southern half of Britain from Yorkshire around to Anglesey. It occurs on hard-rock cliffs at seepages and streams, pools in dunes and coastal gravels, and on upper saltmarsh although it does not appear to be a saltmarsh specialist. The requirement for saline influence is confirmed by its presence inland at a mildly brackish clay pit in Cambridgeshire (Drake 2002) and at saline colliery waste in Yorkshire. Flight: iv–x; Pollet *et al.* (2005) suggest two generations in Belgium.

Dolichopus simplex **Meigen, 1824**

Among species with no fringe on the hind femora and partly yellow antennae, *simplex* may be separated, together with *mediicornis*, by the complete absence of a costal stigma. Only a combination of character states then distinguishes it from the others species in the group: pale yellow face (never snow-white), entirely yellow scape, partly yellow pedicel and entirely black postpedicel, middle basitarsus completely black or dark brown (sometimes paler apically) and no ciliolarium on the hind tibia. Females closely resemble *plumipes* with the exception of their darker antennae but specimens with a tiny yellow spot on the ventral surface of the postpedicel are inseparable. Both these species usually have a dark mid basitarsus but in some specimens of *simplex* this basitarsus can be quite pale on at least the basal half, and the hind tibia of *simplex* is usually less strongly darkened at the tip. Wing length ♂: 4.5 (4.3–4.8) mm, ♀: 4.5 (4.0–4.8) mm.

Dolichopus simplex is a common species in the west and north of Britain, including Wales, much of Scotland north to Orkney and in some parts of eastern England. It is scarce in lowland England southeast of a line from Dorset to Lincolnshire with the exception of fens in East Anglia. Except in the east where it appears to be confined to fens, it occurs in no specific habitat so long as it is damp to wet; there is no preference for open or shaded sites, acid or base-rich, still or flowing water. Its distribution is probably determined by humidity. Flight: iv–x, peak in mid vi to early vii.

Dolichopus strigipes **Verrall, 1875**

Dolichopus strigipes is amply distinguished by several characters, the deeply split cercus being the most obvious. The narrow lobe of the cercus is, in nearly all other *Dolichopus*, the apical dorsal lobe that carries one curved and one straight seta at its tip and two curved setae usually near its base; in *strigipes* this lobe has become almost basal on the triangular lamella. The hind femur has a posteroventral fringe of short pale hairs, found otherwise only in *caligatus* among pale-legged *Dolichopus*; it is not immediately obvious as the hairs point somewhat backwards so are not always visible without turning the specimen. Other unusual characters found in both sexes are the violet frons, particularly strong marginal setae on the tergites and dark streaks on the antero- and posteroventral surfaces of the hind femur. Although these streaks are unique to *strigipes*, their position makes them difficult to see and they are sometimes faint. The apical section of vein M_1 appears to have a small dark spot owing to the microtrichia pointing into the wing boss and the halter has a small dark spot at its tip. Wing length ♂: 4.3 (4.1–4.6) mm, ♀: 4.4 (4.1–4.6) mm.

Dolichopus strigipes is a coastal species found from Anglesey to the Humber estuary with its greatest frequency in the East Anglia, and a confirmed record from Lothian. It is confined to saltmarsh but may stray into wet dune slacks, shingle lagoons and brackish grazing marsh. Flight: vi–viii, peak in vii.

Dolichopus subpennatus **d'Assis-Fonseca, 1976**

A comparison with the similar *pennatus* is made in the account for that species. Wing length ♂: 5.2 (4.7–5.5) mm, ♀: 5.5 (5.2–5.7) mm.

Dolichopus subpennatus is moderately common and widely distributed in Britain north to the Outer Hebrides. It appears to be strongly associated with rivers and larger streams with exposed sediments but is also found by drains and ditches on washlands and sometimes by ponds, including wet dune slacks which may have hydrological features in common with sandy and stony river shores. Most sites are not within woodland although they may be shaded by trees. Flight: v–viii.

Dolichopus trivialis **Haliday, 1832**

Among *Dolichopus* with a pale fringe on the hind femur, *trivialis* is most obviously distinguished by its moderately long postpedicel and largely black dorsal surface of the whole antennae, and more conclusively by only the front basitarsus having an apical anterior setae. The narrow, straight and pointed anteroventral epandrial lobe is easily seen and distinguishes *trivialis* from all other *Dolichopus*. Females are distinct in the '*trivialis*' group in having largely black antennae and a long postpedicel with its tip varying from pointed to rounded. Wing length ♂: 4.4 (4.1–4.6) mm, ♀: 4.8 (4.5–5.3) mm.

Dolichopus trivialis is a very common species throughout Britain, north to Sutherland and the Outer Hebrides. It occurs in a wide range of habitats, usually wetlands but by no means confined to these, and it may be found along hedgerows and in pastures in moderately dry landscapes as well as in true wetlands such as seepages, stream and pond margins, fens, bogs and wet woodland. It shows no preference for open or shaded sites. Flight: iii, v–x, peak in late vi–vii.

Dolichopus ungulatus **(Linnaeus, 1758)**

This was first described species of *Dolichopus* and the type species of the genus but is atypical among the British *Dolichopus* and consequently is readily identified. It is a large species with dark ventral fringes on all femora, two to five preapical setae on the mid and hind femora, black lower postocular setae and notably numerous setae on all tibiae. It superficially resembles *claviger* in the shape of the cercus and having several setae on the femora so *claviger* could be overlooked when these characters are used uncritically to identify *ungulatus*. Females, like the males, are unique in having additional femoral setae and black postocular setae, as well as bristly tibiae and black palps which are unusual among yellow-legged *Dolichopus*. In both sexes the rather sombre yellow of the legs is a useful indicator for this species. Wing length ♂: 5.5 (5.0–5.9) mm, ♀: 5.4 (5.0–5.7) mm.

Dolichopus ungulatus vies with *plumipes* as the commonest British *Dolichopus*. It is found throughout Britain but becomes infrequent in northern Scotland although it occurs on Orkney and the Outer Hebrides. It is found at a very wide range of wetland habitats including upper saltmarsh, and at wet features within dry landscapes. These are usually in open situations but also in woodlands. It is also frequent in hedgerows and scrub in moderately dry landscapes, sitting on sunlit foliage. Only extensive bogs and the centres of old fens appear to be avoided. Flight: v–ix, peak in vi–vii.

Dolichopus urbanus **Meigen, 1824 (Plate 3)**

The outstanding characters that distinguish this species from other *Dolichopus* are found in both sexes. They have an almost entirely dark hind tibia while the other tibiae and all femora are yellow, although some specimens of *diadema* may be similar. They have dark wings which are noticeably long relative to the abdomen and in both these respects *urbanus* resembles *Ethiromyia*. Among species in the yellow-legged group with modified mid tarsi, the face of both sexes is entirely white rather than having at least a hint of yellow and, in the male, the front coxal hairs are also mainly white, as found only in *wahlbergi*. The hind femur of both sexes may sometimes have two anterior preapical setae, as in *popularis*. The male has no ciliolarium on the hind tibia. Wing length ♂: 5.0 (4.4–5.5) mm, ♀: 5.3 (4.7–5.8) mm.

Dolichopus urbanus is a moderately common species distributed throughout Britain north to the Outer Hebrides. It is frequent in the north and west of Britain and in East Anglia but markedly scarce between these areas. It is a species of shaded wetland habitats such as seepages and streams within woodlands and scrubby area. It may be more closely associated with moving water than still-water features. Flight: v–viii, peak in vi although a 'spring' species.

Dolichopus virgultorum **Haliday in Walker, 1851**

This is a small and particularly pale species whose legs are almost white. Apart from part of the mid coxa, all legs are pale from the coxae to the basitarsi, including more tarsomeres of the front leg, and the front and mid coxal hairs are mainly white. The antennae are also rather pale with the postpedicel being brown rather than black above. This degree of pale colouration is unique among British *Dolichopus*. The fringe of tiny upright hairs on the front tarsal segments is inconspicuous compared to that of *trivialis*, *festivus* and *cilifemoratus*. The ciliolarium has dark yellow points on the basal third. The epandrium is narrowly pear-shaped with a concave dorsal surface, as in *arbustorum*. Females are not quite as pale as males but may still be picked out by their pale yellow legs and largely pale antennae. They resemble *arbustorum* in the pale yellow extending from the hind coxae onto the lower margin of the metepimeron. Darker examples may be distinguished by having a strongly defined black tip on the hind basitarsus and only a few or no black hairs on the front coxa, rather than small clusters at the inner upper corner and around the apical setae in *arbustorum*. Wing length ♂: 4.5 (3.7–4.9) mm, ♀: 4.7 (4.2–5.2) mm.

This was once an uncommon species restricted to southern England and Wales but after about 2015 it became markedly more frequent with a small northward expansion of its range. It remains one of the most strongly 'edge-of-range' British *Dolichopus*. It is found in lightly shaded sites, such as hedgerows, wood margins and scrubby grassland, and shows no particular association with wet habitats. Flight: v–x, peak in vii.

Dolichopus vitripennis Meigen, 1824

This is a moderately small species among those with black femora. All its tibiae are almost entirely yellow and it requires comparison with only the larger *campestris* which has additional setae on the mid femur, and dark specimens of *clavipes* and *signifer* which have a long ventral fringe on the hind femur. The white triangular cercus with its contrasting black basal corner and tuft of short hairs is sufficient to recognise the species; *andalusiacus* has a similar cercus but is a pale-legged species. Females are recognised by the combination of bright yellow tibiae, white postocular setae and two dorsal setae on the hind basitarsus. Rare specimens have postocular setae with dark reflections which can cause confusion with *lepidus* in which these are black, but *lepidus* is much larger, has three dorsal setae on the hind basitarsus and entirely black lower proepisternal hairs (not mainly white and inconspicuous). Wing length ♂: 3.7 (3.4–4.1) mm, ♀: 3.8 (3.4–4.2) mm.

Dolichopus vitripennis is a common species in the north and west of Britain and on acid soils in lowland England. Its habitat is wet and damp heath, moorland and the more mineral-based areas of acid or neutral mires and poor fen. Well-drained but constantly wet acid soil is probably the key requirement. It avoids shaded areas. Flight: v–ix, peak in vii.

Dolichopus wahlbergi Zetterstedt, 1843

Together with the common *plumipes*, this species has a distinctively fringed mid tarsus but unlike *plumipes* its tibia is cylindrical and entirely yellow. Like *plumipes*, the hind basitarsus has two dorsal setae; other species with a modified mid leg have at least three. The ciliolarium is an inconspicuous line of yellow points on the basal quarter. Females are unique in having tiny pale hairs on the clypeus and mainly pale antennae; the '*nubilus*' group and *longitarsis* also have facial hairs but on both the epistoma and the clypeus, and have entirely black antennae. The postpedicel is distinct from similar species as its apical part protrudes a long way in front of the arista insertion and is slightly concave dorsally, which gives it a characteristic upturned appearance. *Dolichopus polleti* Meuffels and Grootaert, 1989, recorded from Belgium, is similar but both sexes may be distinguished by the postpedicel being brownish yellow with a paler base, rather than dark brown to black on at least the apical third, the male mid tibia being hardly 3 times as long as the basitarsus and lacking a dorsal brown stripe (which distinguishes *plumipes*), and the female mid tibia having only three long anterodorsal setae. See the account for *plumipes* for the demasculinised form *pectinitarsis* that may run to *wahlbergi* in the current key to males. Wing length ♂: 4.6 (4.2–5.0) mm, ♀: 5.0 (4.3–5.4) mm.

Dolichopus wahlbergi is a common species in England, particularly in the south, and frequent in Wales but becomes scarcer northwards and is infrequent although widespread in Scotland. It is not a wetland species, although does occur in wet places, but lives mainly in moderately dry woods and often in hedges and scrub. Flight: iii, v–x, peak in late vi–vii with a moderately long autumn season.

ETHIROMYIA Brooks 2005 (Dolichopodinae)

Ethiromyia is a member of the Dolichopodinae whose characterisation is given under *Dolichopus*. It is distinguished from other genera in this subfamily by the combination of having wing veins M_1 and R_{4+5} straight and nearly parallel to one another, the scutum with violet reflections from a bluish black ground colour, a long apicoventral seta on the front tibia in the male, and a rounded cercus fringed with long fine marginal setae (Plate 4). The genus was erected for one Palaearctic species that had been included in *Hercostomus* and two Nearctic species previously in *Hercostomus* then *Gymnopternus* (Brooks and Wheeler 2005). It is phylogenetically related to *Dolichopus* and *Gymnopternus* (Brooks 2005, Bernasconi *et al.* 2007a) or to *Poecilobothrus* and some *Hercostomus* species (Pollet *et al.* 2010).

Ethiromyia chalybea (Wiedemann, 1817) (Plate 4)
Among larger British dolichopodids, the deep violet scutum is unique and, in combination with conspicuously long dark wings, pale legs and the long-haired cercus on a slender epandrium (Fig. 17), makes this species unmistakable. Wing length ♂: 5.1 (4.9–5.5) mm, ♀: 5.1 (4.6–5.3) mm.

Fig. 17. *Ethiromyia chalybea* hypopygium.

Ethiromyia chalybea is found in England and Wales north to Lancashire. The highest density of records is found in the southeast English lowlands. It lives in fens, reedbeds, marshes, wet grasslands and well-vegetated pond margins, although fen appears to be the habitat in which it reaches its greatest abundance. It appears to have become more frequent and is perhaps expanding its range since about 2020. Flight: vi–ix, peak in vi.

GYMNOPTERNUS Loew, 1857 (Dolichopodinae)

Species of *Gymnopternus* are medium-sized metallic dark green to bronze flies (Plate 4). They are members of the subfamily Dolichopodinae (see *Dolichopus* for its characterisation) and are distinguished from related British genera by the combination of having veins R_{4+5} and M_1 almost straight, parallel to one another beyond crossvein dm-m and barely converging, an anterodorsal comb of stout short closely spaced setae on the front tibia, and a cluster of tiny hairs on the pleuron in front of the hind spiracle. This last character is also found in *Dolichopus* and *Ethiromyia*. The arrangement of the acrostichal setae, which spread out into a cluster beyond the last dorsocentral seta, is found only in a few *Dolichopus* among the Dolichopodinae. The disc of the scutellum is also covered in tiny hairs. The male epandrium is typically large and approximately oval, bearing relatively small cerci fringed with short hairs, although similar in some species of *Hercostomus*. Females are often distinguished by their strongly protruding clypeus.

The following keys are based partly on Pollet (1990). Males have no secondary sexual characters with the exceptions of *cupreus* with a modified mid tibia and *aerosus* with a black face; the genitalia are relatively uniform but diagnostic for the limited British fauna. Females show even smaller structural differences. For instance, tiny differences in the shape of the postpedicel and pubescence of the arista are real but often difficult to assess. In both sexes it is necessary to rely strongly on small differences in leg colour which fortunately shows little variation within a species even when it is represented by only slight and rather nebulous duskiness. One constant morphological character is the tiny scattered hairs (pubescence of Pollet 1990) on the face, which are always present on the epistoma even if only represented by a single hair but on the clypeus are either present (*aerosus, angustifrons, brevicornis, celer, metallicus*) or absent (*assimilis, blankaartensis, cupreus, silvestris*). Some males lack these hairs when they are present in the female. The black dusting on the face of *aerosus* is diagnostic but this character cannot be relied upon for specimens in fluid or which are greasy as this causes the white faces of all other species to appear black.

Nine of the ten species of *Gymnopternus* known from Europe are recorded in Britain. The genus was treated in the Palaearctic as a subgenus of *Hercostomus* until Pollet (2003 (2004)) restored it to full generic rank. *Gymnopternus* is phylogenetically more closely related to *Dolichopus* and *Ethiromyia* than to most *Hercostomus* (Bernasconi *et al.* 2007a, Brooks 2005) although later analysis using molecular data suggests a less tidy relationship in which *Gymnopternus* becomes the sister group of the *H. nigrilamellatus* species-group (Pollet *et al.* 2010).

Gymnopternus are freshwater wetland flies. Many tolerate shaded sites such as woodland streams and seepages, and two are specialists of fenlands.

Key to species of *Gymnopternus* (males)

1 Femora mainly dark, at most yellow at extreme tip 2
– Femora mainly yellow, sometimes darkened on dorsal surface, mid femur always clearly yellow 3

2(1) Mid tibia swollen, slightly curved and armed ventrally with 3–4 short setae arising from small tubercles, and lacking ventral vestiture (a). Costa of wing conspicuously thickened beyond end of R_1. Hind tibia completely yellow. Larger species, wing length more than 3.5 mm .... ***cupreus***
– Mid tibia simple, bearing 1 anteroventral seta and normal vestiture. Costa not thickened as above. Hind tibia often darkened at tip, at least posteriorly. Smaller species, wing length less than 3.5 mm .... ***angustifrons***

(a)

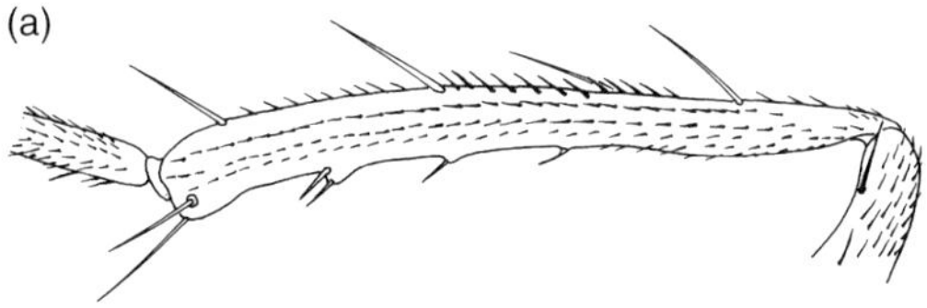

3(1) Cercus entirely or mainly dark 4
– Cercus entirely yellow 8

4(3) Wing with costa strongly thickened between humeral crossvein and R_1 (a) (in *silvestris* this section is sometimes slightly thickened) .... ***brevicornis***
– Above section of costa not thickened, distinctly thinner than costa beyond R_1 5

(a)

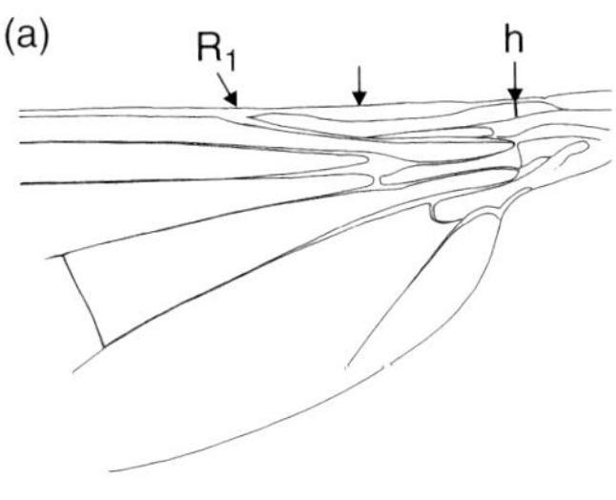

5(4) Larger species, wing length 4–5 mm. Face narrow and almost parallel-sided on its lower half, at its narrowest point not wider than separation of ocellar setae on their inner sides (a). Crossvein dm-m placed near middle of wing, with basal section of M_4 more than twice as long as apical section (2.5 times) measured from the tiny looped CuA near its base, and basal section of M_1 slightly longer than apical section (1.1 times or more) measured from its fork with M_4 (b). Cercus dark with pale base, all marginal setae similarly slender but differ in length........................ ***metallicus***

– Smaller species, wing length usually 2.5–4.0 mm. Face gradually tapered and broader at its narrowest point than distance between ocellar setae (aa). Crossvein dm-m slightly closer to base of wing, with basal section of M_4 not more than twice as long as apical section (1.6–2.0 times), and basal section of M_1 slightly shorter than apical section (0.85–0.90 times), both measured as above (bb). Cercus with some marginal setae thicker than others...................... 6

(a)

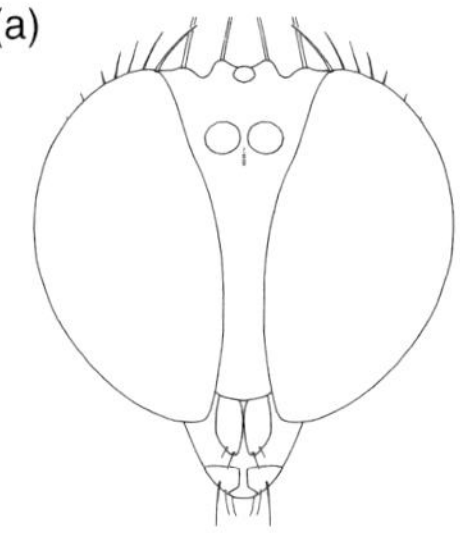

(b)

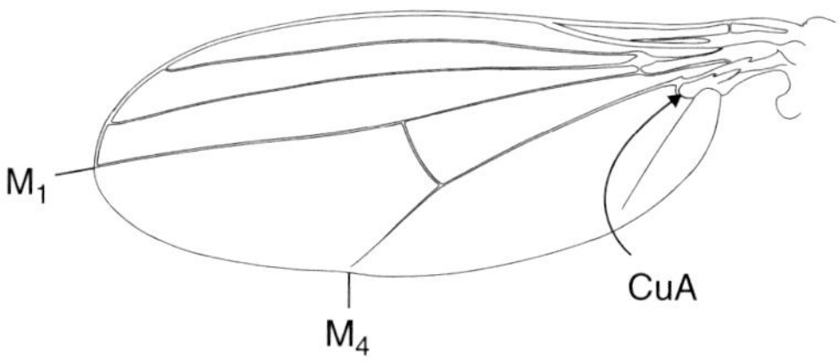

(aa)

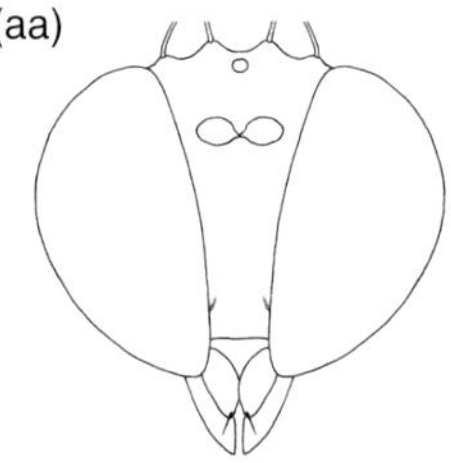

(bb)

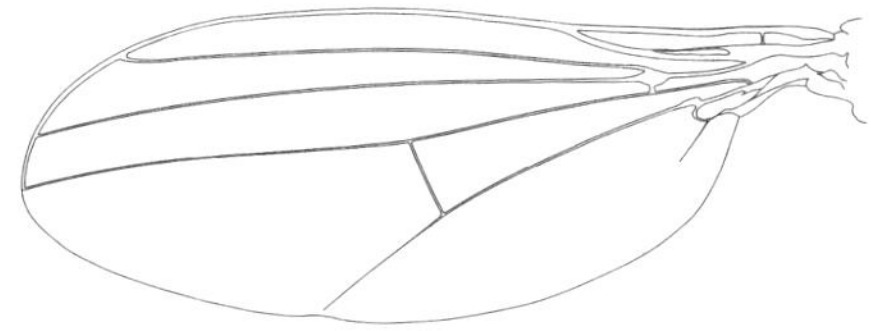

6(5) Face matt black viewed from all angles (when dry). Clypeus with short black pubescence (a). Wing clear with grey tinge. Front and hind femora usually clearly darkened along dorsal surface. Cercus relatively large (b). Basiventral epandrial lobe (bvepl.) slender, narrower than hypandrium (hyp.) (c).. ***aerosus***

– Face silvery white viewed from in front and above (but black when wet). Clypeus entirely bare although epistoma may have tiny hairs. Wing distinctly brownish. Front and hind femora either darkened dorsally or entirely yellow. Cercus relatively small (figs in next couplet). Basiventral epandrial lobe (bvepl.) at least as broad as hypandrium (hyp.) (cc).................. 7

(a)

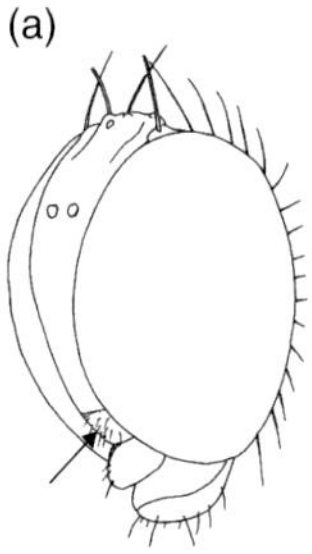

(b)

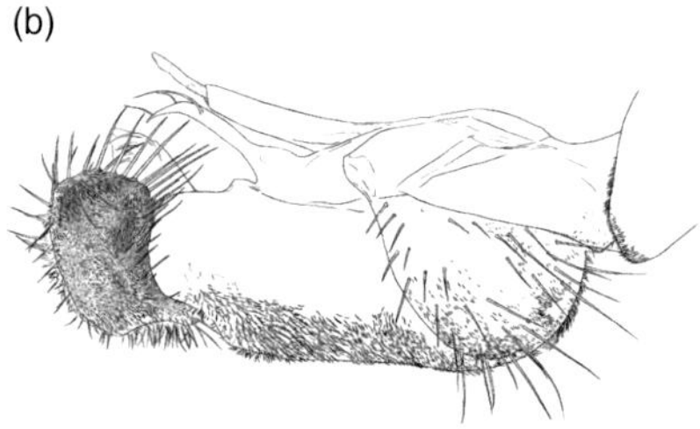

(c)

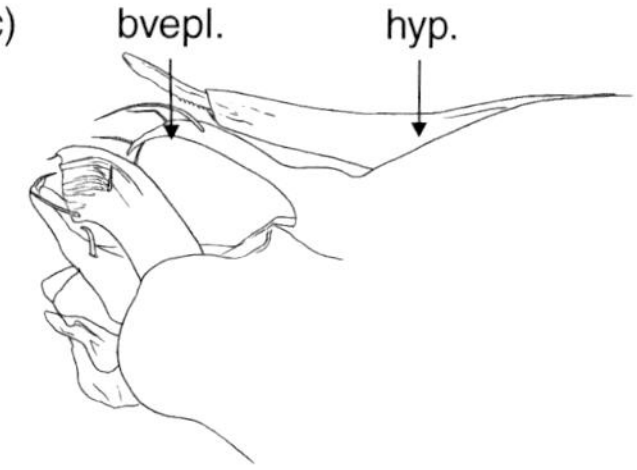

(cc)

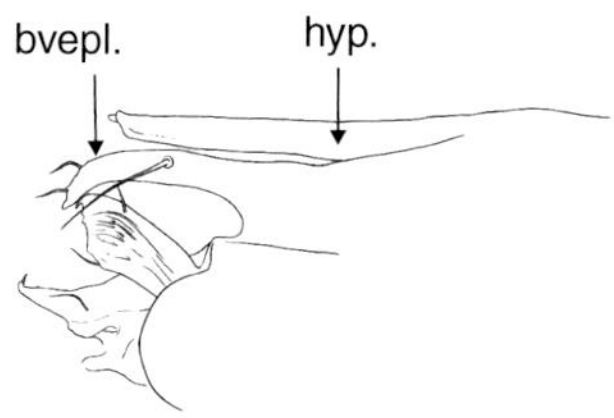

7(6) Hind tibia entirely yellow. Front and hind femora usually distinctly darkened at least posterodorsally. Epandrium stout and egg-shaped (a). Cercus dark with pale base, small and ellipsoidal, its depth just over half depth of black part of epandrium, inserted above dorsal (lower in figure) margin of epandrium so dorsal edges of cercus and epandrium not aligned, displaced by a small step (a)........ ***silvestris***

– Hind tibia slightly to distinctly darkened on apical quarter to third. All femora pale yellow. Epandrium moderately slender, its dorsal and ventral surfaces more parallel to each other in profile (aa). Cercus almost entirely dark, large and quadrate, its depth well over three-quarters depth of epandrium below the shining brown ridge, inserted almost at dorsal margin of epandrium so dorsal edges of cercus and epandrium are continuous without a step (aa)........ ***assimilis***

(a)

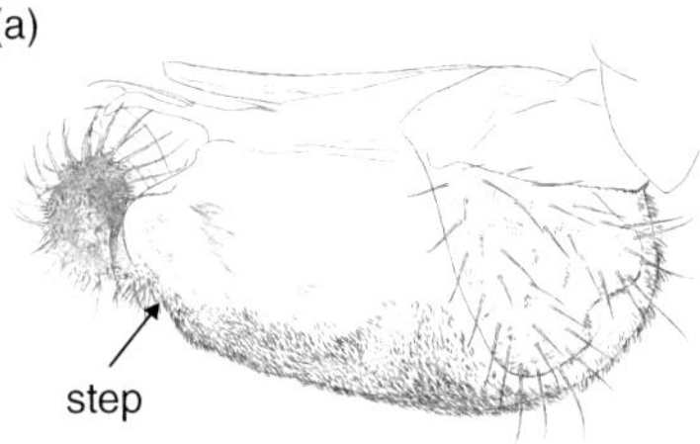

(aa)

8(3) Wing costa strongly thickened between humeral crossvein and R_1 (fig. a of couplet 4). All basitarsi yellow with tip dark, tarsomeres 2–5 dark........ ***celer***

– Above section of costa not thickened, distinctly thinner than costa beyond R_1. All basitarsi and usually all tarsomeres 2 entirely yellow........ ***blankaartensis***

Key to species of *Gymnopternus* (females)

1 Femora mainly dark on all faces, at most yellow at extreme tip........ 2

– Femora mainly yellow, sometimes darkened dorsally or rarely extensively, mid femur always almost entirely yellow........ 3

2(1) Mid tibia with 2 or more ventral setae, all short and two-thirds or less length of other tibial setae (a). Clypeus bare. Hind tibia usually entirely yellow or slightly dusky on apical fifth to quarter. Hind basitarsus mainly yellow. Larger species, wing length usually more than 3.5 mm........ ***cupreus***

– Mid tibia with only 1 anteroventral seta, as long as other tibial setae (aa). Clypeus with scattered tiny hairs (bb). Hind tibia usually dark to black on apical fifth to quarter. Hind basitarsus usually mainly black. Smaller species, wing length less than 3.5 mm. .... ***angustifrons***

(a)

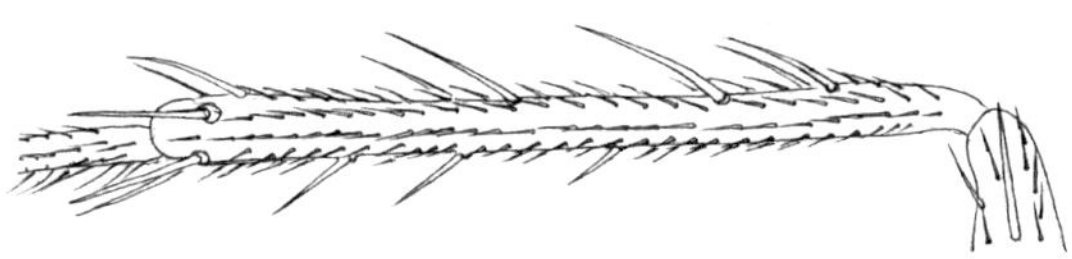

(aa)

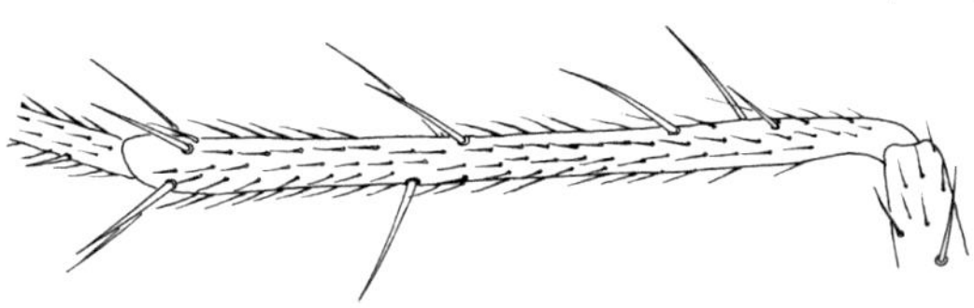

(bb)

3(1) Costa of wing just before vein R_1 markedly thinner than at middle of section from crossvein h to R_1, or than costa beyond R_1 (a). Clypeus with tiny hairs (as fig. bb of couplet 2).................. 4

– Costa between h and R_1 almost uniformly thick and similar to costa beyond R_1 (aa); in cases of doubt (*silvestris*), clypeus without tiny hairs.. 5

(a) (aa)

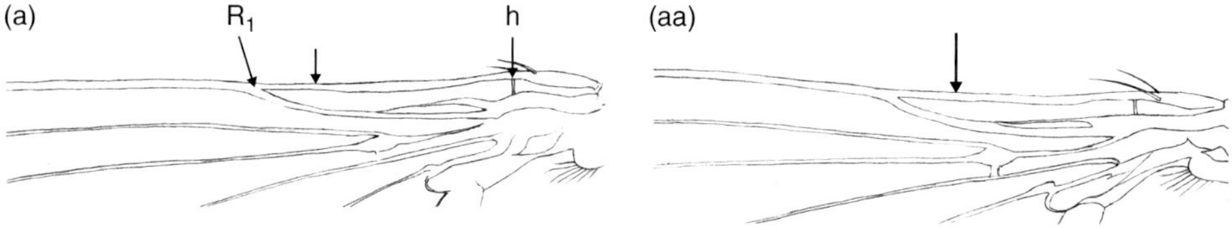

Note: The next couplet is unsatisfactory as many specimens have a mix of characters; those that do not fit well should be left undetermined.

4(3) Epistoma with a few short pale brown to white hairs. Clypeus with a few tiny pale to white hairs. Postpedicel slightly acute to square-tipped (a). Arista pubescence longer and sparser, longest hairs longer than basal thickness of arista. Front coxa dark basally and yellow usually on apical half to quarter. Front femur entirely yellow, rarely slightly dusky dorsally. Hind basitarsus yellow at least on basal half and usually as pale as middle of tibia shaft. Hind tibia at most slightly dusky on apical fifth.. ***celer***

– Epistoma with a few short black to brown hairs. Clypeus with a few hairs that are often dark and about half length of those on epistoma. Postpedicel usually rounded (aa). Arista pubescence shorter and denser, longest hairs hardly longer than basal thickness of arista. Front coxa mainly dark, yellow only on extreme apex (sixth or less). Front femur yellow, distinctly darkened posterodorsally along whole length, more obvious in lateral view. Hind basitarsus usually completely dark, more marked on posterior face, contrasting strongly with yellow at middle of tibia shaft. Hind tibia usually slightly to distinctly darkened on apical fifth.. ***brevicornis***

(a) (aa)

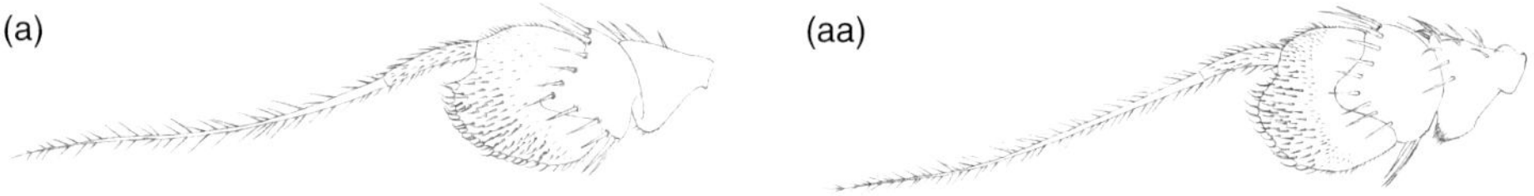

5(3) Clypeus with tiny hairs (as fig. bb of couplet 2; epistoma may also have hairs)................................... 6

– Clypeus entirely bare (epistoma may have hairs; ignore dense white dusting.)...................................... 7

6(5) Front and hind femora usually distinctly darkened at least dorsally, more obvious in lateral view. Wing clear, grey-tinged. Crossvein dm-m slightly closer to base of wing, with basal section of vein M_4 less than twice as long as apical section (1.5–1.9 times) measured from the tiny looped CuA near its base, and basal section of vein M_1 slightly shorter than apical section (0.8–1.0 times) measured from fork with M_4 (a). Hind tarsus dark from apical third of basitarsus to tarsomere 5. Clypeus hairs dark or pale. Smaller species, wing length 3.2–3.9 mm................ ***aerosus***

– All femora entirely pale yellow. Wing slightly but obviously brownish. Crossvein dm-m placed near middle of wing, with basal section of M_4 more than twice as long as apical section (2.1–2.5 times), and basal section of M_1 slightly longer than apical section (0.9–1.2 times) (aa). Hind tarsus dark from tarsomere 2 to tarsomere 5. Clypeus hairs pale. Larger species, wing length 4.1–4.8 mm.. ***metallicus***

(a) (aa)

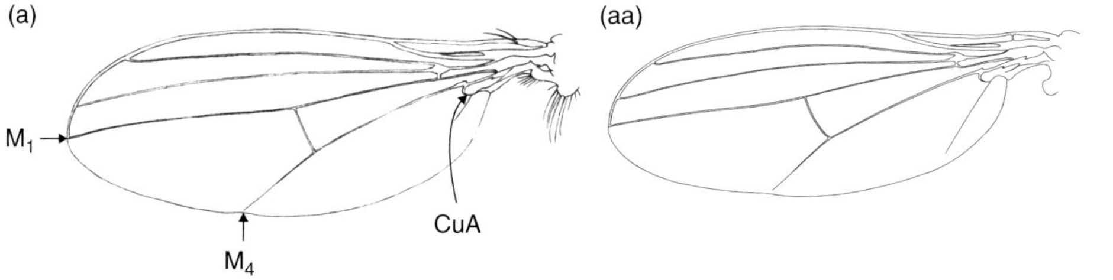

7(5) Front and often hind femora distinctly darkened, at least posterodorsally. Wing distinctly darkened .... ***silvestris***

– All femora entirely pale yellow. Wing slightly paler 8

8(7) Front coxa mainly dark, at least on outer face posterior to where vestiture begins. Hind tibia usually slightly dusky on apical quarter to third, basitarsus partly darker than basal half of tibia. Front and mid tarsi usually distinctly darkened from tip of basitarsus to tarsomere 5. Dust on frons pale grey, similarly coloured but thinner than dust on face. Arista pubescence fine and inconspicuous, individual hairs shorter than basal diameter of arista (a). Smaller species, wing length 3.1–3.6 mm .... ***assimilis***

– Front coxa usually mainly yellow, particularly on inner side underlying vestiture, but can be mainly dark. Hind tibia and basitarsus entirely yellow. Front and mid tarsi only feebly darkened. Dust on frons yellowish brown in contrast to white dust on face. Arista pubescence conspicuous, hairs longer than basal diameter of arista (aa). Larger species, wing length 3.5–4.3 mm .... ***blankaartensis***

(a)

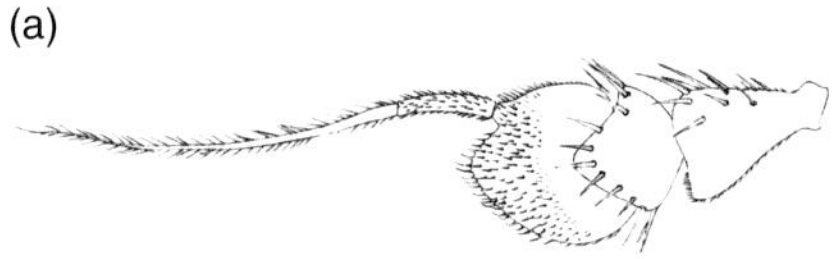

(b)

Gymnopternus aerosus **(Fallén, 1823)**

This is the only *Gymnopternus* with a matt black face. It shares with *brevicornis* and *silvestris* the dark shade along the dorsal surface of the femora. This shade may appear too diffuse to be a constant character but it is rarely absent and shows up in wet-preserved or greasy specimens when the face colour of all *Gymnopternus* becomes black like *aerosus*. Relative to the epandrium, the cercus is the largest of all *Gymnopternus* and the apical epandrial lobe one of the thinnest, and these are the easier characters to distinguish wet-preserved *aerosus*. Females are most easily identified by the dark streak on the front and hind femora in combination with the tiny hairs on the clypeus; *silvestris* has similar leg colour but differs in having slightly darkened wings, often a rather diffuse change from yellow to dark on the hind basitarsus, the tip of the postpedicel more obtuse, and slightly longer pubescence on the arista that more resembles that of *celer* than *brevicornis* (compare figures in couplet 4). Northern females may have almost completely black front femora and sometimes also the hind femora, causing confusion with *cupreus* and *angustifrons*, but in *aerosus* the mid femur remains mainly yellow. Wing length ♂: 3.6 (3.3–4.0) mm, ♀: 3.7 (3.2–3.9) mm.

Gymnopternus aerosus is a common and widespread species found equally frequently in Scotland, Wales and England but perhaps scarce on mesotrophic soils of the English Midlands. It may be found in most non-brackish wetland habitats where it shows a slight preference for more acidic sites but does occur in base-rich wetlands. Both shaded and completely open wetlands are used. Flight: iv–x, peak in vi.

Gymnopternus angustifrons **(Staeger, 1842)**

This is the only *Gymnopternus* with black femora and a simple mid tibia. Both sexes resemble *Hercostomus parvilamellatus* in their leg colour and size but are distinguished by the generic characters of veins R_{4+5} and M_1 being almost parallel to one another but converge slightly in *Hercostomus* and by the cluster of acrostichal setae in front of the scutellum. Wing length ♂: 3.2 (2.8–3.4) mm, ♀: 3.1 (2.9–3.4) mm.

The species is sparsely distributed in England and Wales northward to near the Scottish border. Records are mainly from north of a line from Lincolnshire to Dorset apart from a cluster in the Thames basin. Most records are from acid mire (Crossley 2014) but those from Devon are from wet woodlands. Flight: v–vii, ix, peak in vi.

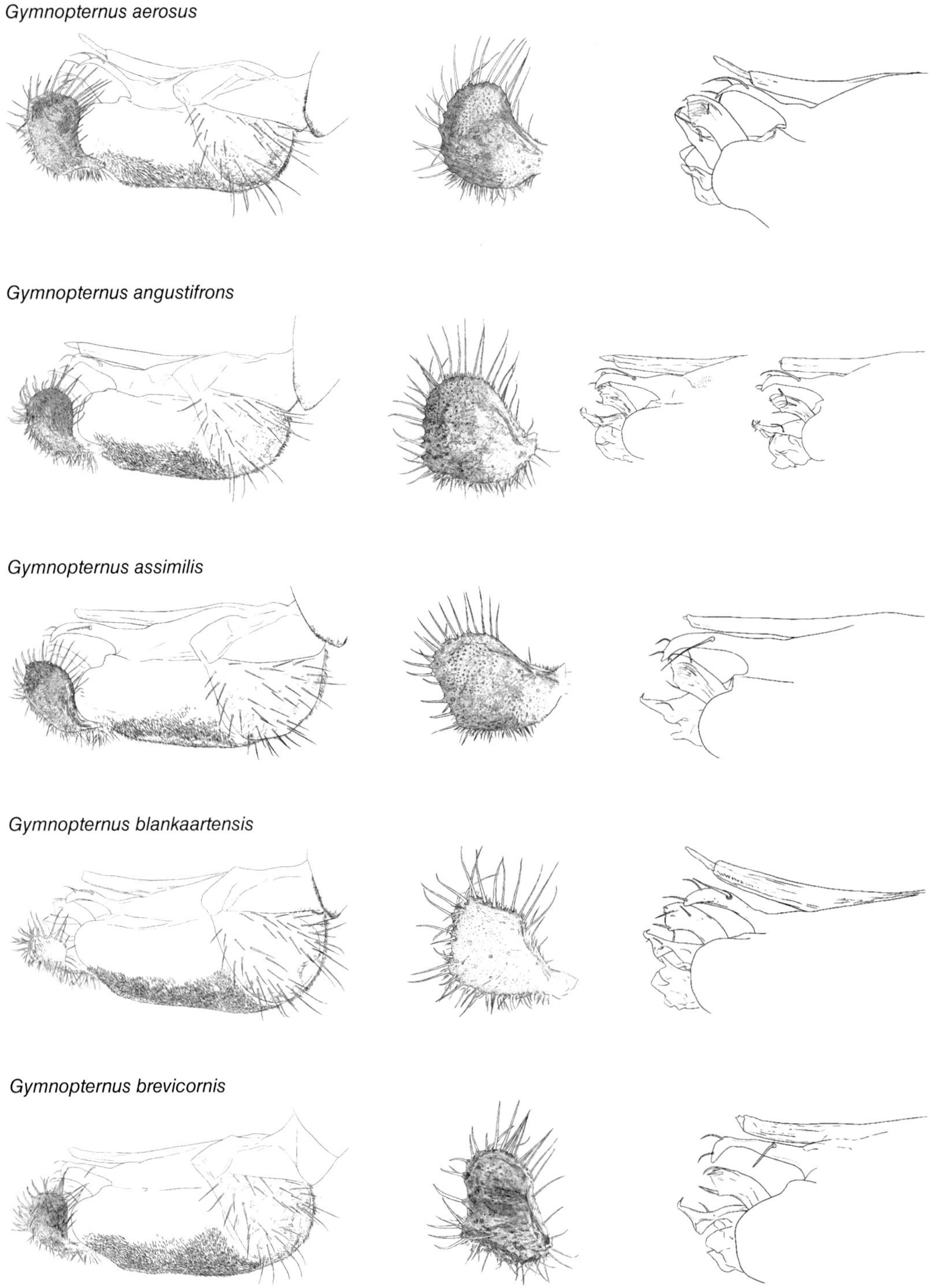

Fig. 18. *Gymnopternus* hypopygia.

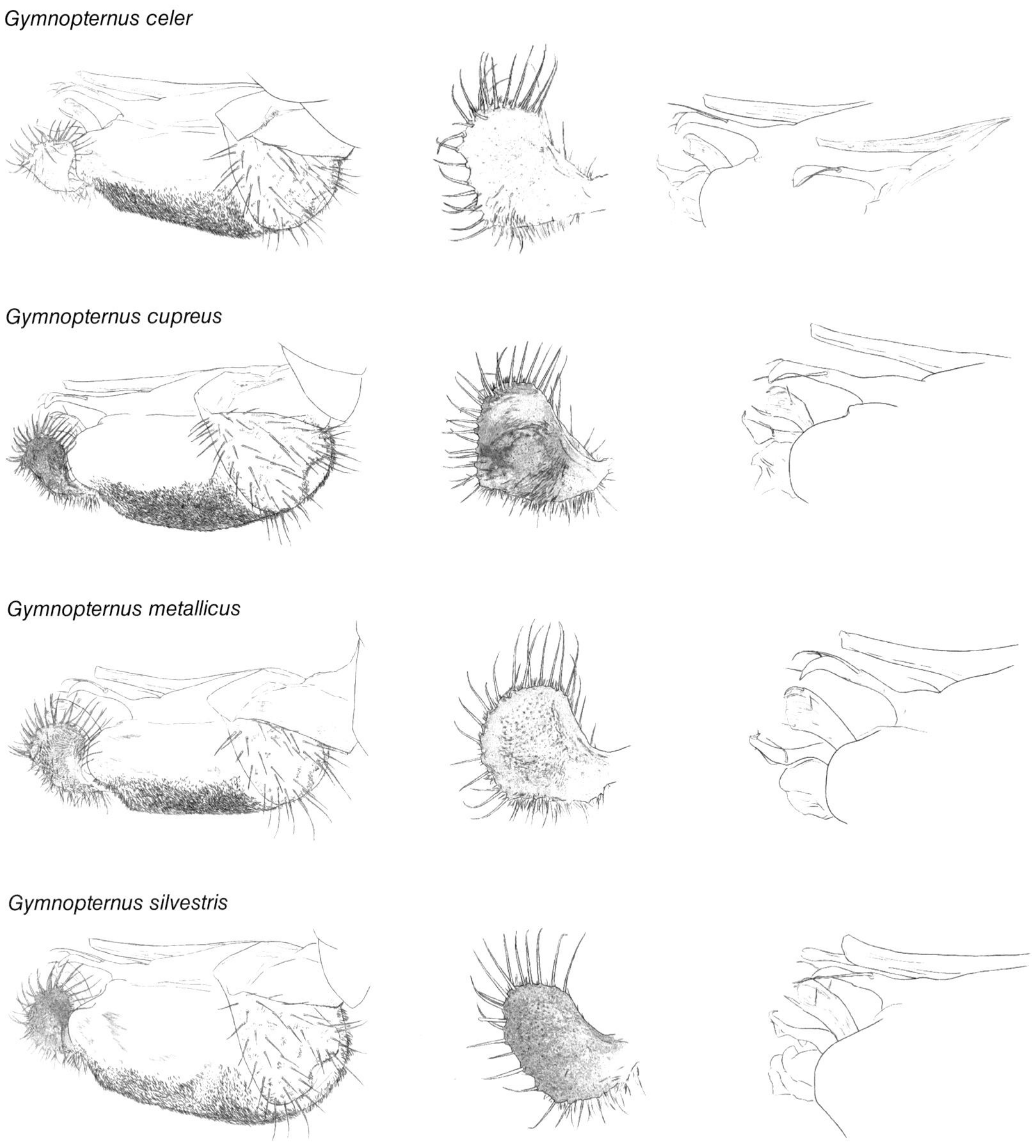

Fig. 18. Continued

Gymnopternus assimilis (Staeger, 1842) (Plate 4)

Gymnopternus assimilis, *silvestris* and *aerosus* are similar small species whose slight differences in leg colour do not always lead to a convincing identification but the relative sizes of the cercus and epandrium provide clear differences. In *assimilis* the cercus is large and nearly as broad as the depth of the epandrium at their articulation, similar to but slightly smaller than in *aerosus*, whereas in *silvestris* the cercus is clearly much smaller. The basiventral epandrial lobe is broader than in *aerosus* and *silvestris*. Females have rather variably coloured tibiae so may be confused with those of *blankaartensis* but their coxae and tarsi are nearly always darker than in *blankaartensis*. The arista has such short pubescence that it appears almost bare at a magnification of ×20 at which the pubescence is visible in other species. Even with several characters to compare, there can still be some doubt about the identification. Wing length ♂: 3.3 (2.8–3.9) mm, ♀: 3.3 (3.1–3.6) mm.

Gymnopternus assimilis has a wide distribution in Britain north to the Moray Firth but it is more frequent on the east side of England. Habitats are predominantly fens and reedbeds, but also acid mire, swamp, freshwater and occasionally brackish marsh, and ditch systems in wet grassland. Flight: v–viii, peak in vi.

Gymnopternus blankaartensis Pollet, 1990

This species and *celer* are the only *Gymnopternus* with yellow cerci. They differ in that *blankaartensis* lacks the swollen basal section of the wing costa found in *celer* and its mid basitarsus is entirely yellow rather than black-tipped. Females have particularly pale tarsi so resemble those of *metallicus* but specimens of *metallicus* in which the tiny hairs on the clypeus are rubbed off differ in being slightly larger, the apical section of M_4 being less than 1.7 times longer than the crossvein dm-m compared to about twice (1.7–2.4 times) in *blankaartensis*, and the arista pubescence being less conspicuous. Differences from *assimilis* are discussed under that species. The colour of the front coxa is not always constant; whereas specimens in which it is mainly yellow are *blankaartensis*, not *metallicus* or *assimilis*, some specimens have an extensively dark front coxa but all the other characters fit *blankaartensis*. Wing length ♂: 4.0 (3.8–4.2) mm, ♀: 3.9 (3.6–4.3) mm.

Since its recognition in Britain, *blankaartensis* has been recorded in southern lowland England and south Wales, with strong populations in East Anglian fens but relatively sparse elsewhere. It is almost restricted to fens or wetlands with fen-like characteristics including a withy bed (willow plantation) and emergent vegetation beside a canal. In the Norfolk fens, there appears to be a preference for tall or dense reed, sometimes with sallow scrub. Pollet *et al.* (1992) state that willows or poplars within fenland are essential for its occurrence. Flight: v–vii, peak in vi.

Gymnopternus brevicornis (Staeger, 1842)

Of the *Gymnopternus* with yellow femora, only this species and *celer* have an obviously swollen wing costa near its base and are distinguished by the colour of their cerci. Females closely resemble those of *celer* and may often be inseparable. All the key characters need to be assessed as they usually fail to agree entirely. In *brevicornis* the small hairs on the face are usually distinctly dark brown or black compared to pale or white in *celer*, the antennae are more rounded but some specimens may have a slightly square tip resembling the more obtuse end of the spectrum found in *celer*, and the leg features in the key are distinctly darker than in *celer* but rather variable although the dark hind basitarsus appears to be the most constant. Typical dark specimens may be confused with *aerosus* if the narrow section of the costa before vein R_1 is overlooked but *aerosus* usually has an entirely yellow hind tibia with the colour continuous with the yellow base of the basitarsus, rather than the tibia having a dusky tip running onto the dark basitarsus in *brevicornis*. Wing length ♂: 3.4 (3.1–3.8) mm, ♀: 3.5 (3.1–3.8) mm.

Gymnopternus brevicornis is widely distributed in Britain and is more frequent in upland areas or those with many streams, including southwest England, Yorkshire and the Scottish Highlands.

Streams and rivers in shade are its most frequent habitat but it also occurs occasionally at woodland seepages, shaded mire and lake margins, but only rarely in open wetlands. Flight: v–ix, peak in vii.

Gymnopternus celer **(Meigen, 1824)**

Only *celer* and *blankaartensis* have entirely yellow cerci; their separation is described under that species. Females closely resemble those of *brevicornis* and may sometimes be inseparable; see comments under that species. Wing length ♂: 3.8 (3.5–4.1) mm, ♀: 3.6 (3.4–3.9) mm.

Gymnopternus celer is widespread in Britain north to the Moray Firth, being more frequent in upland areas but scarce in central-southern England. It is closely associated with large streams and rivers, nearly always in shade or in woodland, those in the west and north of Britain often having gravelly shores. It is occasionally found in other wetlands such as fens and the ditches in wet grasslands but these habitats are rarely occupied in most of its range. Flight: v–ix, peak in vi–vii.

Gymnopternus cupreus **(Fallén, 1823) (Plate 4)**

This is the largest *Gymnopternus* and is the only Dolichopodinae with a thick curved mid tibia. Females are immediately distinguished by being the far larger of the only two *Gymnopternus* with black femora although rare individuals at the extremes in the size ranges overlap. It is the only *Gymnopternus* whose females have several small ventral setae on the mid tibia; all others have one long seta. Wing length ♂: 4.7 (4.3–5.2) mm, ♀: 4.0 (3.4–4.6) mm.

This is a common and widespread species found throughout Britain although apparently absent from the northern Scottish islands. It occurs in many wetlands but is particularly associated with the margins of running water, including seepages, and usually in woodland or under other shade-bearing vegetation. Flight: iii–ix, peak in late v–vi.

Gymnopternus metallicus **(Stannius, 1831)**

This is one of the largest species in the genus, differing from the others in its particularly narrow face which is narrowest at its lower third rather than at the mouth edge. The position of crossvein dm-m differs from other species although this is not easy to judge. It is placed further from the base and slightly nearer the middle of the wing so that the basal and apical sections of M_1 are about equally long (less than equal in other species) and, more obviously, the apical section of M_4 is clearly not even half the length of the basal section (at least half and often longer in other species). The paler base of the cercus and its fine marginal hairs distinguish males well although specimens with an unusually pale cercus may be mistaken for *blankaartensis* whose arista has obvious long pubescence rather than nearly bare in *metallicus*. Females are recognised by their large size, almost entirely yellow legs and fine hairs on the clypeus. The leg colour resembles that of *blankaartensis* which lacks the hairs on the clypeus. The venation is identical to that of males. Wing length ♂: 4.8 (4.3–5.1) mm, ♀: 4.4 (4.1–4.8) mm.

Gymnopternus metallicus is widespread in Britain, being common in much of England but becoming increasingly scarce going northwards, particularly on the west side of Scotland. Many wetlands support it, including wet woodlands, swamps, acid mires, fens, seepages, streams and pond margins. Sites are usually at least partly shaded, even if only by scrub or tall herbage. Flight: iv–ix, peak in vi.

Gymnopternus silvestris **Pollet, 1990**

Separation from the similar *assimilis* is discussed under that species. Females closely resemble *aerosus* in having a darkened dorsal face to the front femur; see the *aerosus* account for differences. The costa is sometimes narrowed just before vein R_1 so could be confused with *brevicornis* and *celer* but these species have tiny hairs on the clypeus. Wing length ♂: 3.4 (3.2–3.7) mm, ♀: 3.6 (3.3–3.9) mm.

This species is widespread across the southern half of England and Wales, north to South Yorkshire, with an isolated record from the Isle of Skye. It occurs in a range of wetlands that may be shaded or in the open, and which include pond and lake shores, acid mire and reedbed. Alder carr is better represented among its sites than for most dolichopodids. Flight: v–vii.

HERCOSTOMUS Loew, 1857 (Dolichopodinae)

Hercostomus has long been recognised as polyphyletic, having been treated as a waste-basket for members of the subfamily Dolichopodinae that do not fit the generic concept of *Dolichopus* (Bernasconi *et al.* 2007b). Consequently, even in the small British fauna there is considerable variation that makes it impossible to characterise the genus succinctly (Plate 4). Several taxa treated by d'Assis-Fonseca (1978) within *Hercostomus* have been removed from this genus: *Ethiromyia*, *Gymnopternus*, *Muscidideicus* and *Ortochile*. Attempts to find natural species-groups within the remaining *Hercostomus* have led to conflicting results. Brooks (2005) recognised three clades based on morphological characters, whereas Pollet *et al.* (2010) found seven clades based on molecular data, of which only some coincided with those of Brooks. These smaller clades do not help place the British fauna in useful groupings as most of the British species fall within six of the seven clades of Pollet *et al.* (four British species were not included in their analysis). These authors suggested that there was insufficient evidence to move species to other or newly proposed genera, so for convenience the British species are treated in the acknowledgedly polyphyletic genus *Hercostomus*. The genus *Sybistroma* is closely related to some '*Hercostomus*' so the characters used in the generic key to females have no phylogenetic basis but just happen to work for the small British fauna.

At least 66 species of *Hercostomus* are known in the Western Palaearctic of which 15 have been found in Britain. Several are widespread in Europe and may possibly occur in Britain. However, most British species also occur in northern countries including those in Scandinavia and bordering the Baltic Sea, whereas other widespread species are absent from these northern countries. On this criterion, only *Hercostomus vivax* (Loew, 1857) appears likely to occur in Britain. It resembles *argentifrons* in having entirely black legs, black postocular setae, a large cercus with a fringe of long hairs resembling that of *rusticus*, lacking the long apical seta on the front tarsus found in *nigrilamellatus* but has a shining frons and black knees (white-dusted and yellow, respectively, in *argentifrons*).

Habitat associations include a wide range from streams in woodlands (*chetifer*, *nanus*, *parvilamellatus*), drier woodlands (*argentifrons*, *fulvicaudis*, *nigrilamellatus*) and dry open sites (*gracilis*, *rusticus*), and unshaded wetlands (*nigripennis*, *nigriplantis*, *plagiatus*, *verbekei*). The sole upland representative is *sahlbergi*.

Key to species of *Hercostomus* (males)

1 Femora black 2
– Femora yellow 8

2(1) Hind tibia black 3
– Hind tibia partly or entirely yellow 6

3(2) Front and mid tibiae yellow. Wing membrane almost uniformly dark brown all over. Veins R_{4+5} and M_1 almost straight and parallel to each other, their widest separation near dm-m at most 1.5 times their separation at wing tip (a). Proboscis and palp long, pointing downwards (proboscis sometimes retracted but palp usually visible) (b) .... ***nigripennis***
– Front and mid tibiae black. Wing membrane browner in front, pale behind. Veins R_{4+5} and M_1 gently sinuous, converging towards wing tip, their widest separation near dm-m at least

twice their separation at wing tip (aa). Proboscis small and inconspicuous with its long axis horizontal; palp oval and inconspicuous. 4

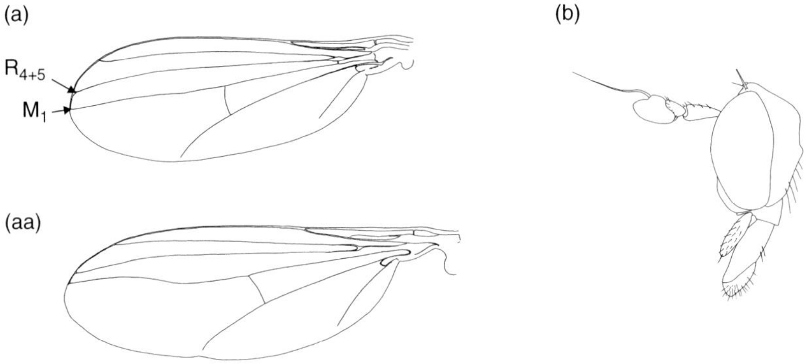

4(3) Small species, wing shorter than 3.5 mm. Cercus oval with its long axis in line with epandrium (a). Postpedicel elongate, about 2.5 times longer than deep (b). Lower postocular setae white. .... ***rusticus***

– Large species, wing at least 5.0 mm long. Cercus large, about twice as long as wide with its long axis at right angles to epandrium axis (aa). Postpedicel 1–1.5 times longer than deep (bb). Lower postocular setae black 5

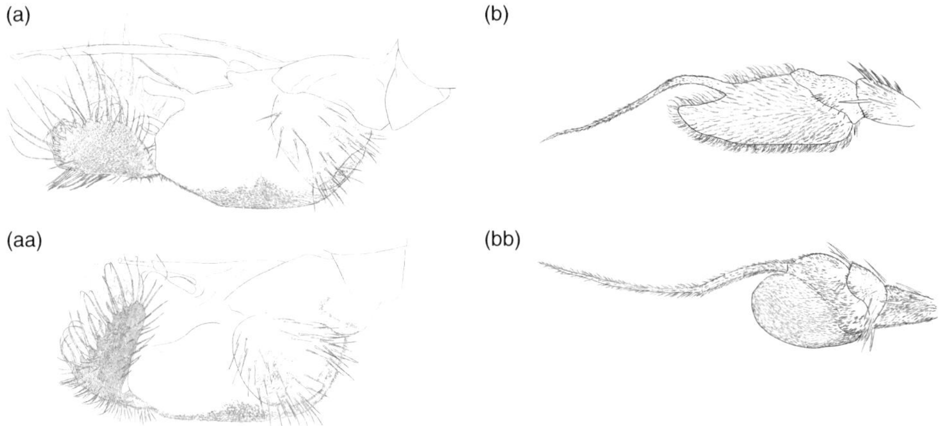

5(4) Front tibia with a long fine posteroventral apical seta about half length of basitarsus, and a second slightly shorter one (a). Mid tibia with 6 anterodorsal setae of varying lengths. Front and mid trochanters dark. Hind femur with ventral vestiture of fairly dense short pale upright hairs. Cercus with marginal hairs on long tubercles about 4–5 times longer than wide (b), base of cercus about one-third length of epandrium. .... ***nigrilamellatus***

– Front tibia with only short apical setae about 1.5 times width of tibial shaft (aa). Mid tibia with 3–5 anterodorsal setae. Front and mid trochanters and tips of coxae yellow. Hind femur with inconspicuous darker ventral vestiture. Cercus with marginal hairs on short tubercles about twice as long as wide (bb), base of cercus about one-quarter length of epandrium. .... ***argentifrons***

(a)

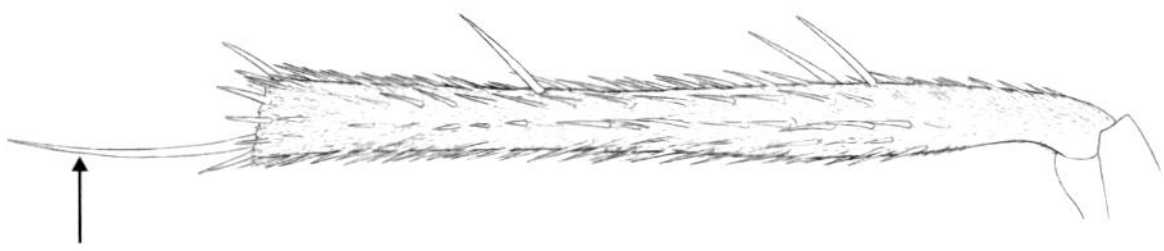

(b)

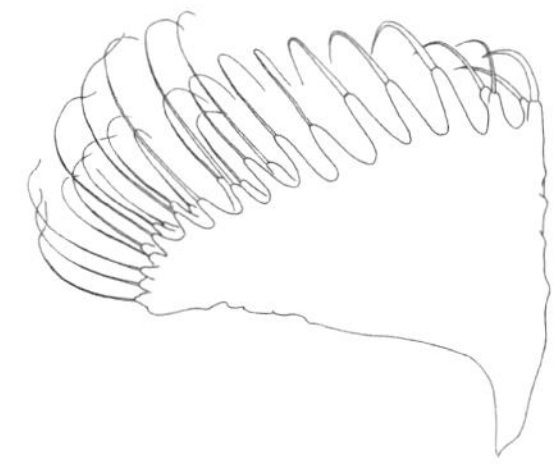

(aa)

(bb)

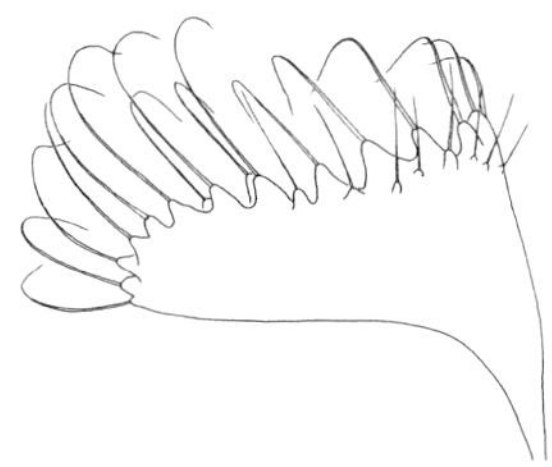

6(2) Small species, wing length 3–3.5 mm. Front tibia with a pair of tiny antero- and posterodorsal setae at basal quarter and no posteroventral apical seta (a). Front coxa with black vestiture. Cercus consisting of a very small pale ribbon scarcely extending beyond epandrium (b). .... ***parvilamellatus***

– Large species, wing length 5–6.5 mm. Front tibia with more than 2 setae on shaft, at least one clearly longer than width of shaft, and a long posteroventral apical seta at least one-quarter length of tibia. Front coxa with white vestiture. Cercus large and oval to rectangular with apical hairs almost as long as cercus. 7

(a)

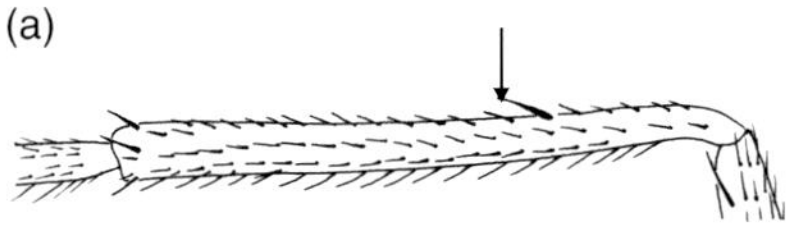

(b)

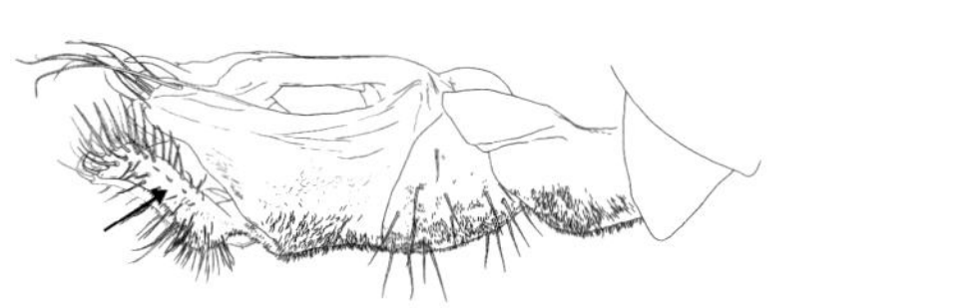

7(6) Front tibia with 3–6 setae on dorsal (antero- and posterodorsal) face and a very long ventral apical seta (a). Mid tarsomeres 2–4 slender and long, tarsomere 5 often flattened and wide, all setae tiny, basitarsus without ventral setae (b). Cercus with apical hairs as long as cercus length (c). .... ***nigriplantis***

– Front tibia with about 7–8 long setae on dorsal (antero- and posterodorsal) face, and ventral apical seta scarcely longer than dorsal setae on shaft (aa). Mid tarsus with shorter wider tarsomeres with distinct apical setae; basitarsus with several long ventral setae and 1 very long posteroventral seta as long as setae on tibia shaft (bb). Cercus with apical hairs about half length of cercus (cc). .... ***sahlbergi***

(a)

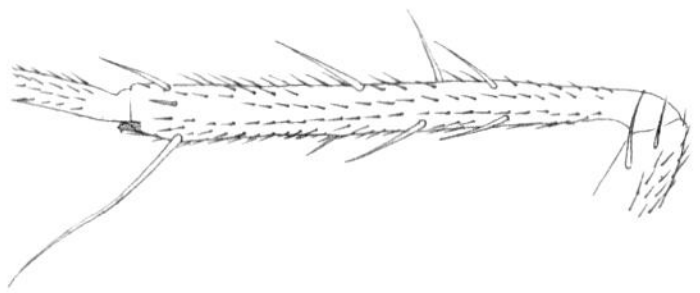

(aa)

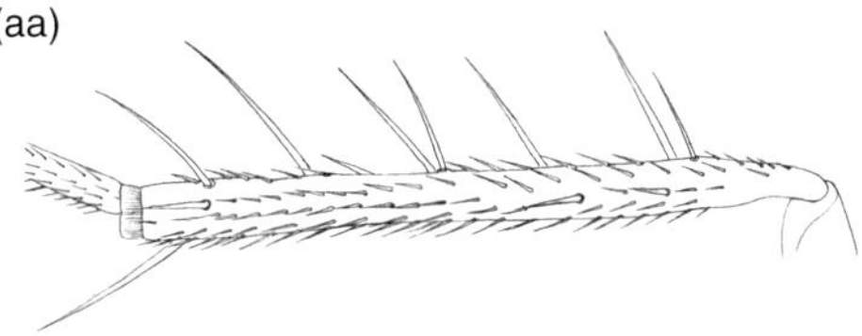

(b)

(bb)

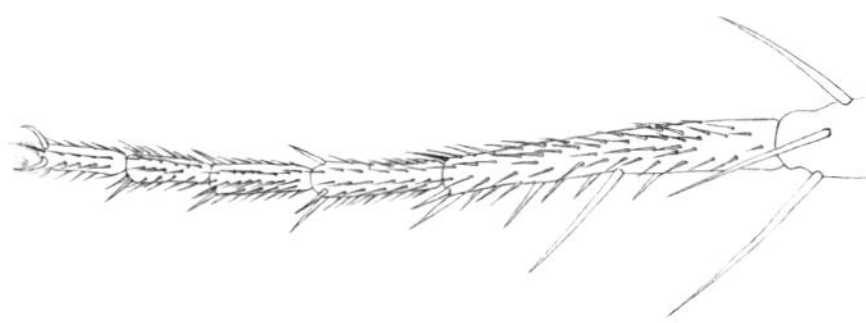

(c)

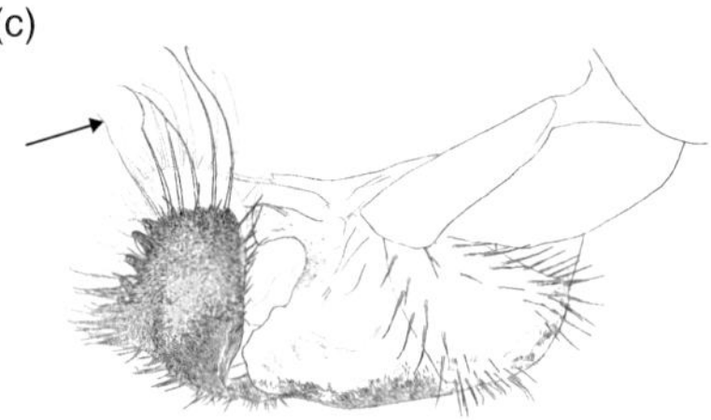

(cc)

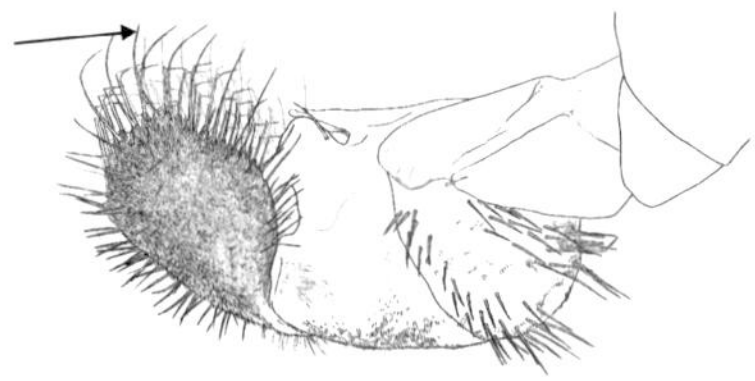

8(1) Metepimeron and all tergites dark. Mid coxa mainly dark, at least on outer face (compare with front coxa). Small to large flies, wing length 2.8–5.0 mm. 9

– Metepimeron and parts of tergites 1 and 2 yellow. Mid coxa yellow. Small flies, wing length 2.6–3.2 mm. 14

9(8) Hind coxa mainly dark. 10

– Hind coxa entirely or mainly yellow. 12

10(9) Large species, wing longer than 4 mm. Leading edge of wing conspicuously thickened and black (a). Mid tibia thickened and clearly wider than front tibia. Vestiture of front coxa white except for some stout short black hairs at base on inner side (apical setae black) .... ***gracilis***

– Smaller species, wing about 3–4 mm long. Leading edge of wing not thickened. Mid tibia similar in thickness to front tibia. Vestiture of front coxa mainly black. 11

(a)

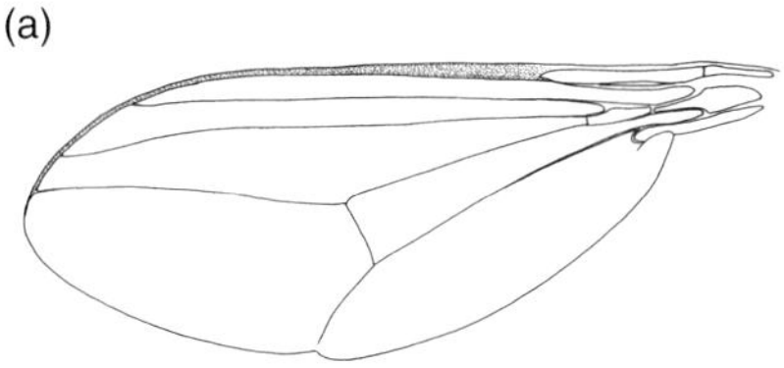

11(10) Face narrow, lower half clearly narrower than depth of postpedicel (a). Front tibia with a small pair of setae (antero- and posterodorsal) at basal third, 0 or 1 at apical third, and minute apical seta (b); front and mid tibiae without ventral setae. Lower postocular setae black. Epandrium smaller than head; cercus yellow and ribbon-shaped (c)........................***nanus***

– Face broad, at its narrowest near mouth as wide as length of postpedicel (aa). Front tibia with several stout dorsal setae (antero- and posterodorsal) and long posterior apical (bb), mid and hind tibiae with obvious ventral setae (av, pv). Lower postocular setae pale. Epandrium larger than head; cercus black and triangular (cc)..***germanus***

(a)

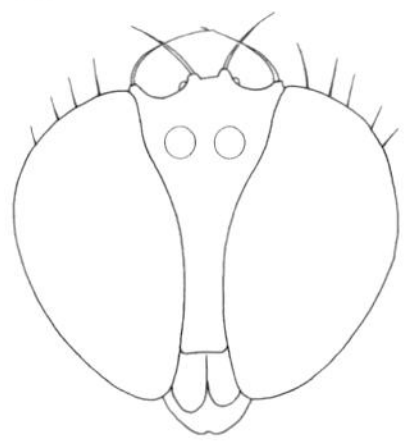

(b)

(c)

(aa)

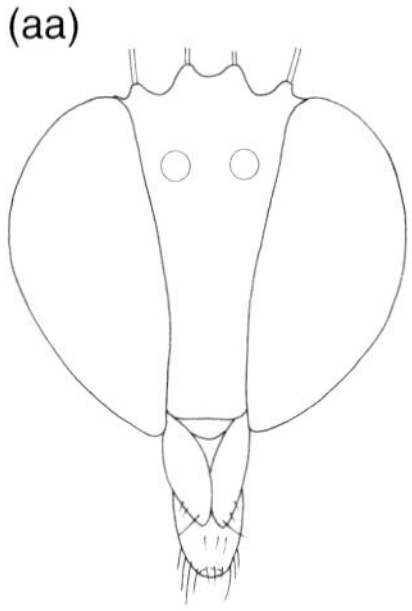

(bb)

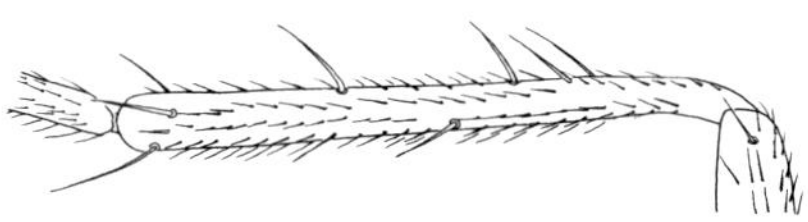

(cc)

12(9) Front tarsus conspicuously modified: tarsomere 5 flattened and white, tarsomeres 4 and 3 flattened, black with fringe of wide hairs, tarsomere 2 and basitarsus long and narrow (a). Antenna entirely black. Marginal setae of tergites obviously shorter than half length of segment. Front tibia with only 3 setulae, all shorter than width of shaft.................................***chetifer***

– Front tarsus not modified. Scape, pedicel and base of postpedicel yellow. Marginal setae on tergites 3 and 4 much longer than half length of segment. Front tibia with at least 1 seta twice as long as width of shaft...13

(a)

13(12) Postpedicel with an acute tip, about 1.5 times as long as deep (a) and longer than width of face. R_{4+5} and M_1 parallel for most of their lengths, converging near wing tip, ratio of sections of costa either side of R_{4+5} 2:1 (b). Cercus and basiventral epandrial lobe (bvepl.) both club-shaped (c). Hypandrium (hyp.) pale and almost straight (d)..***plagiatus***

– Postpedicel triangular, about as deep as long (aa) and shorter than width of face. R_{4+5} and M_1 starting to converge about halfway between dm-m and wing tip, ratio of sections of costa either side of R_{4+5} 3:1 (bb). Cercus small and ovoid (cc), basiventral epandrial lobe (bvepl.) only slightly swollen near tip (dd). Hypandrium (hyp.) dark and curved (dd)................................***verbekei***

(a)

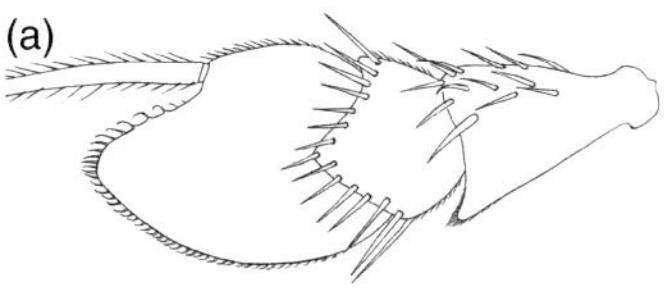

(b)

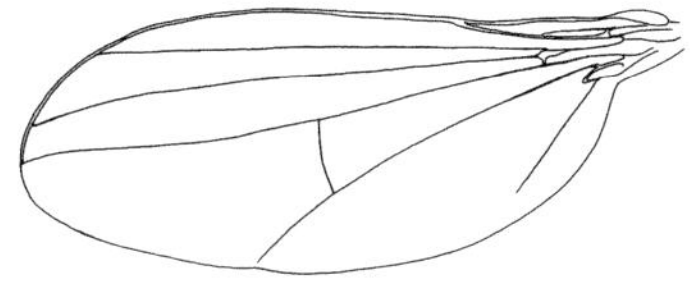

(aa)

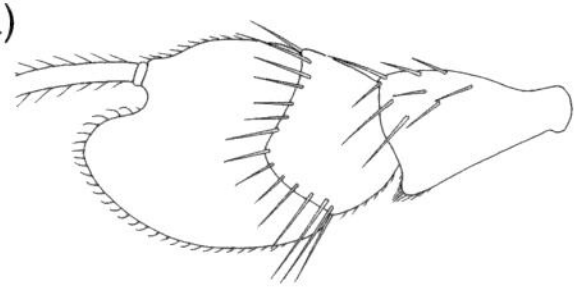

(bb)

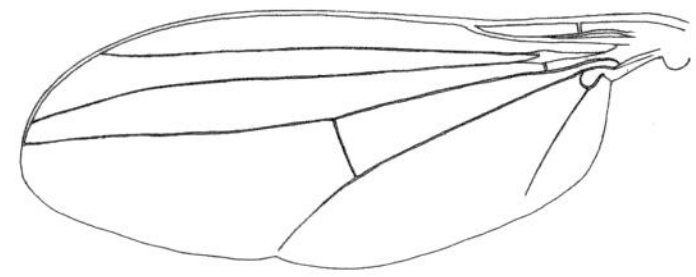

(c)

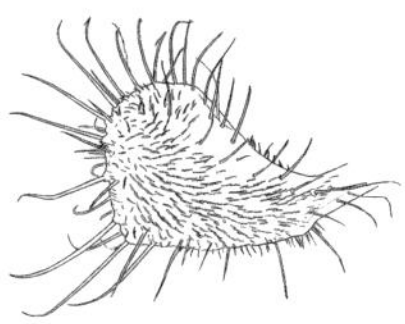

(d)

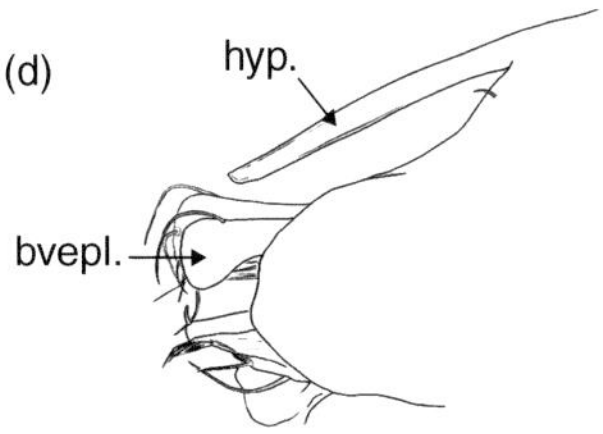

(cc)

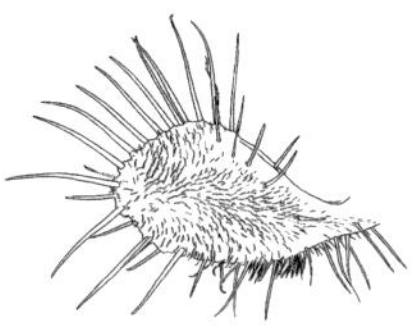

(dd)

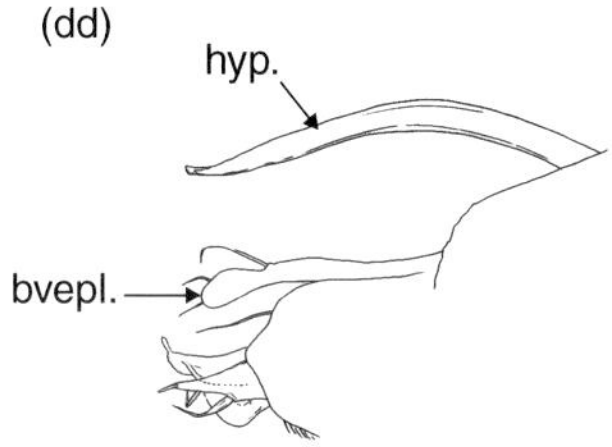

14(8) Front of thorax above coxa yellow from proepisternum to postpronotal lobe. Tergite 2 usually entirely yellow on anterior two-thirds. Postpedicel with tip right-angled to slightly rounded (a). Cercus very small, about one-quarter length of epandrium, widening gradually towards darkened tip into a long triangle with inconspicuous apical hairs (b). .... ***rothi***

– Front of thorax above coxa dark from proepisternum to postpronotal lobe, lower margin of proepisternum sometimes pale. Tergite 2 broadly dark on disc, usually to front margin, yellow laterally. Postpedicel with tip acute (aa). Cercus larger, almost half length of epandrium, strongly expanding into a broadly rounded tip with a conspicuous fringe of long hairs, brown or black with a yellow base (bb). .... ***fulvicaudis***

(a)

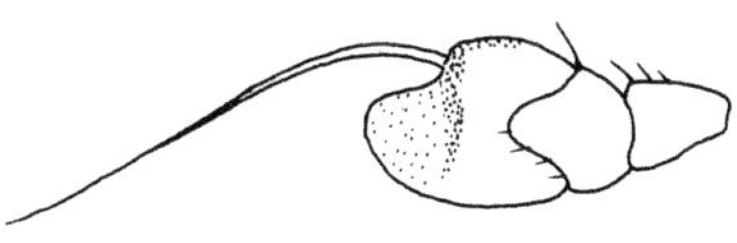

(b)

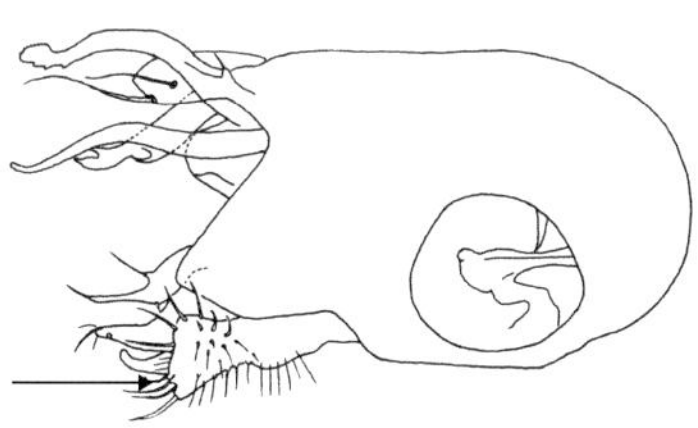

(aa)

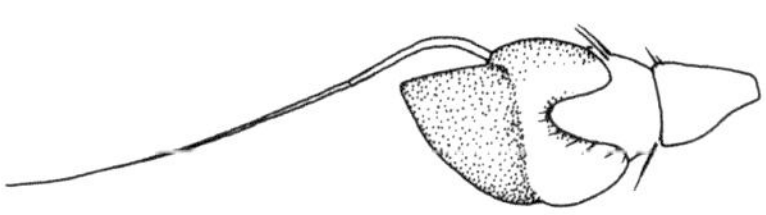

(bb)

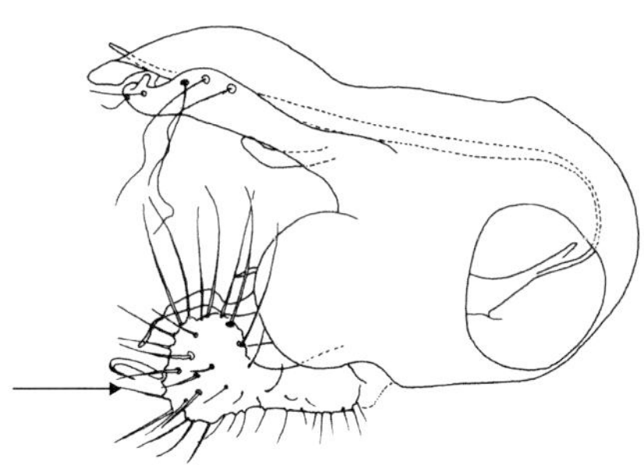

Key to species of *Hercostomus* (females)

1 Femora black. 2
– Femora yellow. 8

2(1) Hind tibia black. 3
– Hind tibiae entirely or mainly yellow. 6

3(2) Front and mid tibiae yellow. Wing membrane almost uniformly dark brown all over. Veins R_{4+5} and M_1 almost straight and parallel, their widest separation near dm-m at most 1.5 times their separation at wing tip (fig. a of male couplet 3). Proboscis and palp long, pointing downwards (proboscis sometimes retracted but palp usually visible) (a). .... ***nigripennis***
– Front and mid tibiae black. Wing membrane browner in front, pale behind. Veins R_{4+5} and M_1 gently sinuous, converging towards wing tip, their widest separation near dm-m at least twice their separation at wing tip (fig. aa of male couplet 3). Proboscis small and inconspicuous with its long axis horizontal; palp oval and inconspicuous (aa). 4

(a) (aa)

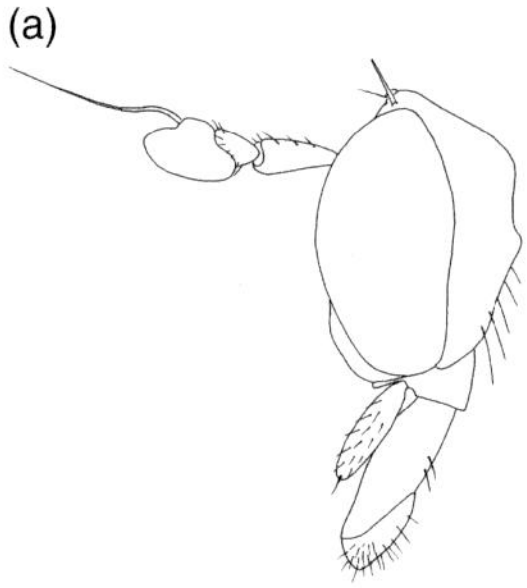

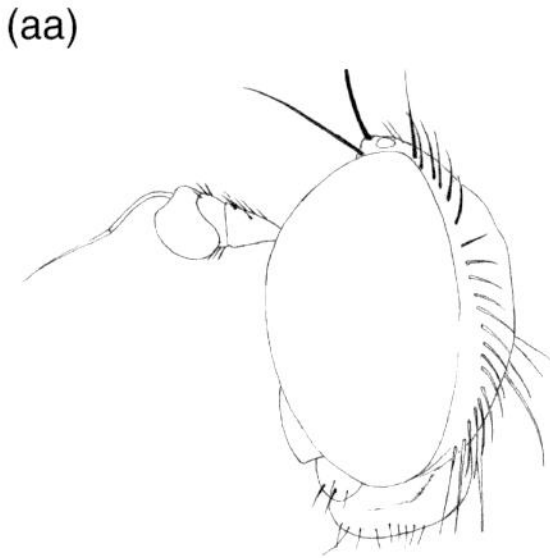

4(2) Small species, wing shorter than 3.5 mm. Lower postocular setae white. Postpedicel with tip acute. .... ***rusticus***
– Large species, wing longer than 5.0 mm. Lower postocular setae black. Postpedicel with point of tip square or rounded. 5

5(4) Front coxa with a narrow brown rim under apical setae, scarcely extending to first row of vestiture hairs. Mid coxa narrowly brown only under apical setae, not reaching vestiture hairs. Front trochanter brown or yellow. Front and mid knees black with yellow rim at tip, if present, not or scarcely reaching to first row of vestiture on anterior face. Postpedicel very slightly longer than deep, tip slightly angular, arista arising at apical third (a). Front and mid tibiae with 1 small + 4 long anterodorsal setae. .... ***nigrilamellatus***
– Front coxa with a yellow rim extending to first row of vestiture hairs which arise from the yellow area. Mid coxa tip yellow from tip to first or second row of vestiture hairs. Front trochanter yellow. Front and mid knees on anterior face of femur with yellow tip extending to third row of vestiture hairs on front femur and narrowly and obscurely to first row on mid femur. Postpedicel very slightly deeper than long, tip rounded, arista arising midway along dorsal margin (aa). Front tibia with 1 small + 3 long anterodorsal setae, mid tibiae with 1 small + 3 or 4 long anterodorsal setae. .... ***argentifrons***

(a) (aa)

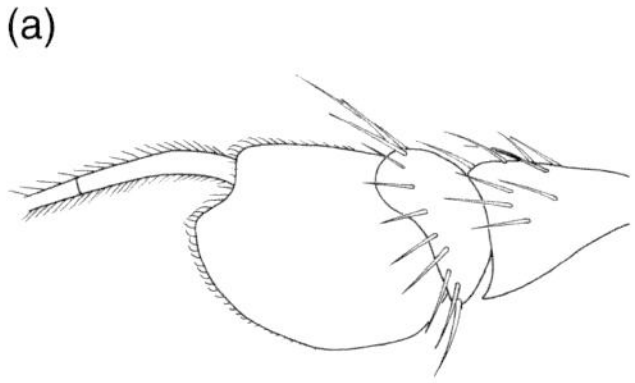

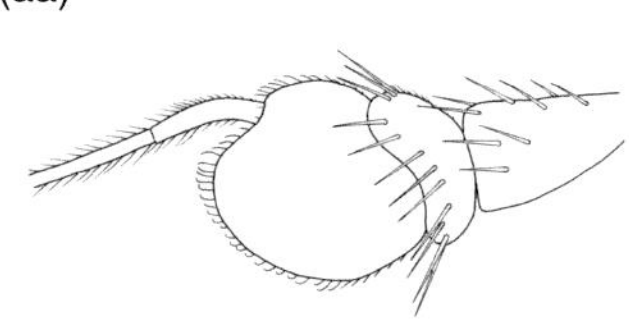

6(2) Hind tibia entirely yellow, colour continuing onto base of basitarsus. Front tibia with weak setae, 1 antero- and 1 posterodorsal at basal third, 1 posterodorsal at apical third (a). Smaller species, wing length 3.5–4 mm. Lower postocular setae black. .... ***parvilamellatus***

– Hind tibia black at tip, continuous with black base of basitarsus. Front tibia with at least 4 stout setae along shaft (aa). Larger species, wing length 4.5–6 mm. Lower postocular setae white. 7

(a)

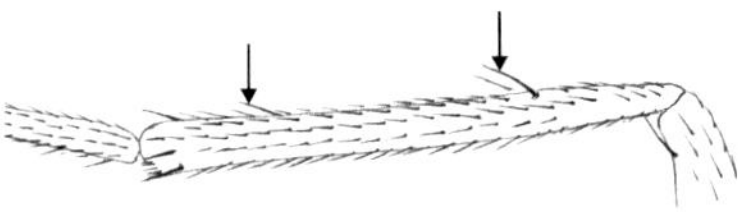

(aa)

7(6) Hind tibia yellow with dark tip. Front, mid and usually hind femora yellow at tip. Front tibia with 4–6 setae on shaft (as fig. aa of couplet 6). ..... ***nigriplantis***

– Hind tibia black at tip and base, yellow in between, sometimes only on dorsal face. All femora entirely black. Front tibia with 7–8 setae on shaft (aa). ..... ***sahlbergi***

(aa)

8(1) Metepimeron and base of abdomen dark. Mid coxa mainly dark, at least on outer face (compare with yellow front coxa). 9

– Metepimeron and parts of tergites 1 and 2 yellow. Mid coxa yellow. 14

9(8) Front tibia with a pair of tiny setae at basal third (antero- and posterodorsal) and none or 1 at two-thirds along shaft, apical setae absent or 1 tiny posterodorsal (a). Length of setae on hind tibia about twice width of shaft (b). Lower postocular setae black but sometimes pale in *nanus*. 10

– Front tibia with at least 1 seta twice width of shaft, and usually more than 3 on shaft and 3 obvious apical setae even if short (aa). Length of setae on hind tibia about 3 times width of shaft (bb). Lower postocular setae pale. 11

(a)

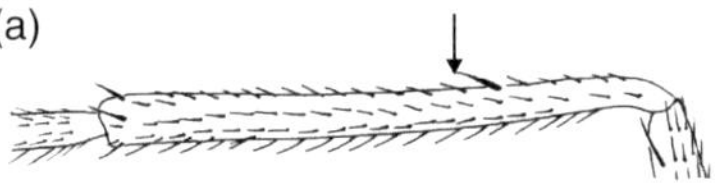

(b)

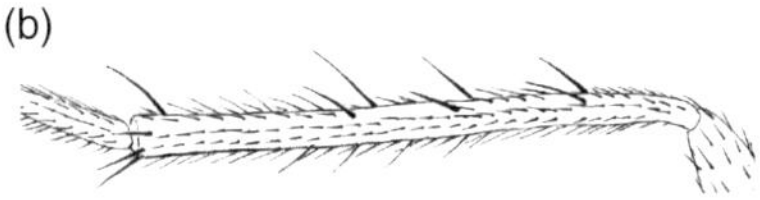

(aa)

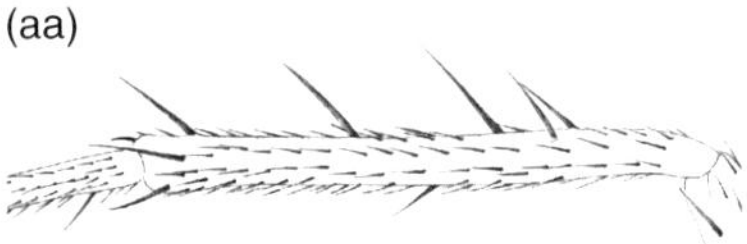

(bb)

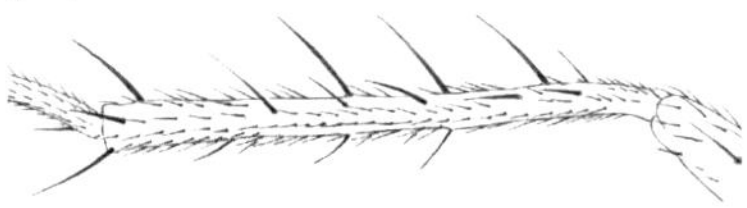

10(9) Scutum undusted shining blackish green. Proboscis and palp dark. Dusting of face white, dusting of frons pale grey and not continuing to shining margins. Hind coxa mainly dark. Scutellum with tiny hairs on disc. ..... ***nanus***

– Scutum thinly dusted on green ground colour. Proboscis and palp clear yellow. Dusting of face and frons silver over entire width. Hind coxa yellow. Scutellum bare on disc. ..... ***chetifer***

11(9) Large species, wing length at least 4.5 mm. Scutum vivid green. Vestiture of front coxa white except for some stout short black hairs at base on inner side. Calypter hairs mainly yellow. Mid tibia with only 1 seta ventrally (anteroventral). ..... ***gracilis***

– Smaller species, wing length up to 3.5 mm. Scutum dark green. Vestiture of front coxa mainly black. Calypter hairs dark. Mid tibia with 1–3 anteroventral and 1 posteroventral setae. 12

12(11) Antenna entirely black. Hind tibia black on apical quarter, basitarsus black. Hind and mid coxae mainly dark, yellow only at tips. Proboscis wedge-shaped and pointing slightly diagonally downwards, palp twice as long as wide (a). Front coxa without black hairs on upper outer corner above uppermost outer seta. .... ***germanus***

– Antenna mainly yellow, postpedicel black on apical half. Hind tibia yellow at apex, basitarsus yellow. Hind coxa mainly yellow, mid coxa yellow on entire anterior face. Proboscis short and horizontal, palp short (as fig. aa of couplet 3). Front coxa with many short black hairs on upper outer corner above uppermost outer seta. 13

(a)

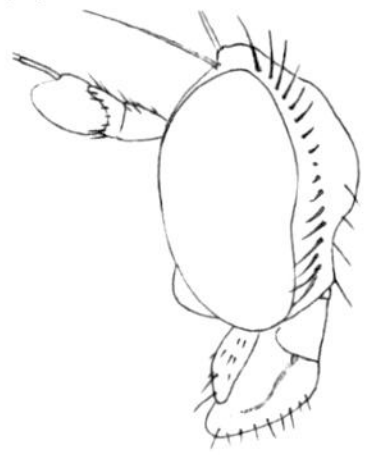

13(12) R_{4+5} and M_1 parallel for most of their lengths, converging near wing tip, ratio of sections of costa either side of R_{4+5} 2:1 (see fig. b of male couplet 13). .... ***plagiatus***

– R_{4+5} and M_1 starting to converge about halfway between crossvein and wing tip, ratio of sections of costa either side of R_{4+5} 3:1 (see fig. bb of male couplet 13). .... ***verbekei***

14(8) Front of thorax above coxa yellow from proepisternum to postpronotal lobe. Tergite 2 usually entirely yellow on anterior two-thirds, at most with a diffuse narrow central darkened zone on anterior third. Postpedicel with tip right-angled to slightly rounded (a). Smaller species, wing length 2.6–2.9 mm. .... ***rothi***

– Front of thorax above coxa dark from proepisternum to postpronotal lobe, lower margin of proepisternum sometimes pale. Tergite 2 broadly dark on disc, usually to front margin, yellow laterally. Postpedicel with tip acute (aa). Larger species, wing length 3.0–3.4 mm. .... ***fulvicaudis***

(a)

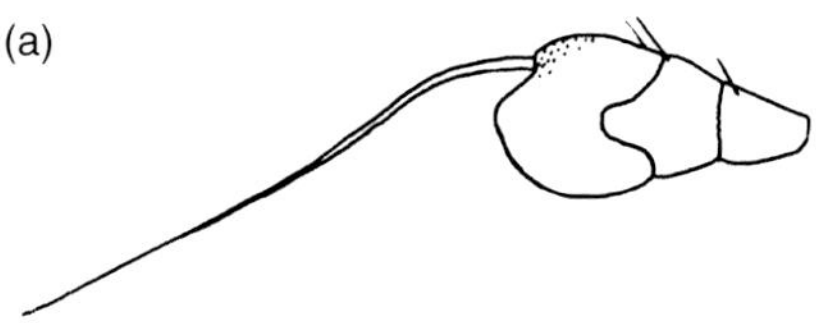

(b)

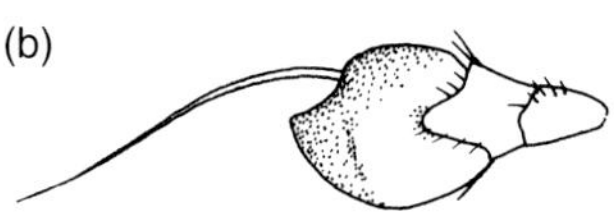

***Hercostomus argentifrons* Oldenberg, 1916**

Confusion with *nigrilamellatus* in British collections has led to this species being overlooked. As well as the characters given in the key, in comparison with *nigrilamellatus* both sexes of *argentifrons* are slightly smaller and have shorter and less dense vestiture on the coxae and legs. Wing length ♂: 4.7 (4.3–4.9) mm, ♀: 5.5 (5.3–5.7) mm.

Hercostomus argentifrons was recognised in Britain in 2022 (Godfrey and Drake 2024) and is known from southern counties of England but is likely to be more widespread. Its habitats include dry deciduous woodlands with streams or standing water. The larvae almost certainly develop in decaying wood of deciduous trees (Pollet and Hulcr 2007). In Britain it has been trapped over decaying alder and aspen. Flight: vi–viii.

***Hercostomus chetifer* (Walker, 1849) (Plate 4)**

The slender front tarsus with a black and white apical 'flag' immediately distinguishes *chetifer* from other British dolichopodids. The legs are generally rather thin, giving the fly a delicate appearance. The postpedicel is far shorter than in other *Hercostomus*, not as long as deep. Females are distinguished by their almost entirely yellow legs with only the mid coxae partly dark, and the bright yellow and rather large proboscis. Its entirely silver face and frons is also found in *fulvicaudis* and *rothi* which have partly yellow antennae (all black in *chetifer*) and a pronounced comb on the front tibia (absent in *chetifer*). Wing length ♂: 3.4 (3.0–3.7) mm, ♀: 4.1 (3.6–4.5) mm.

Hercostomus argentifrons

Hercostomus chetifer

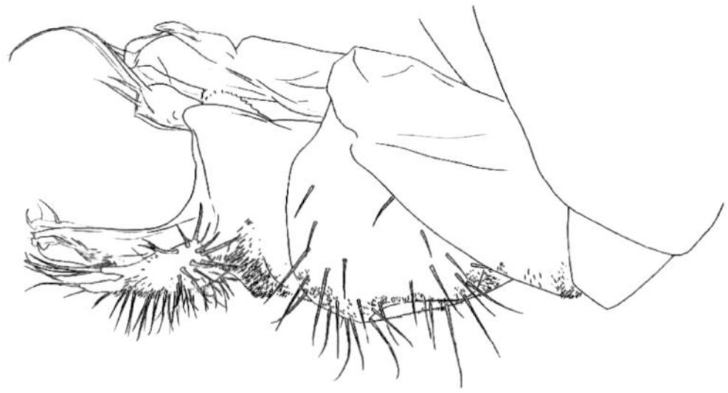

Hercostomus fulvicaudis

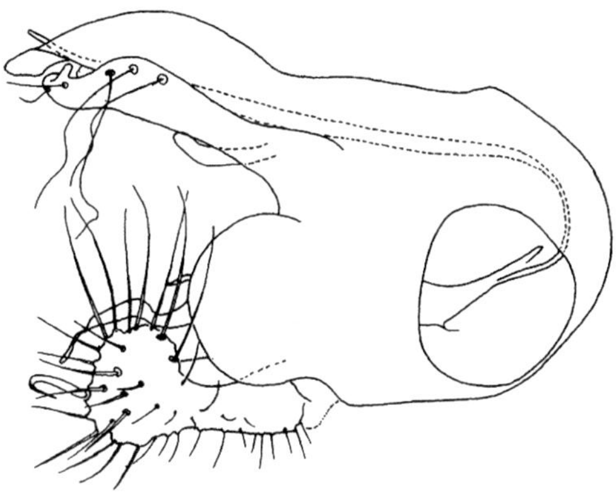

Hercostomus germanus

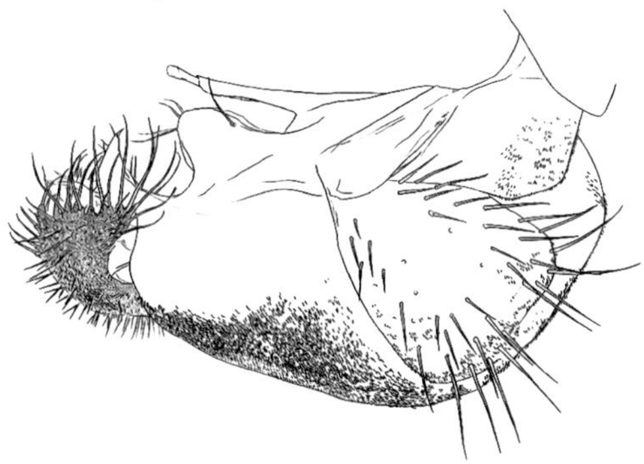

Hercostomus gracilis

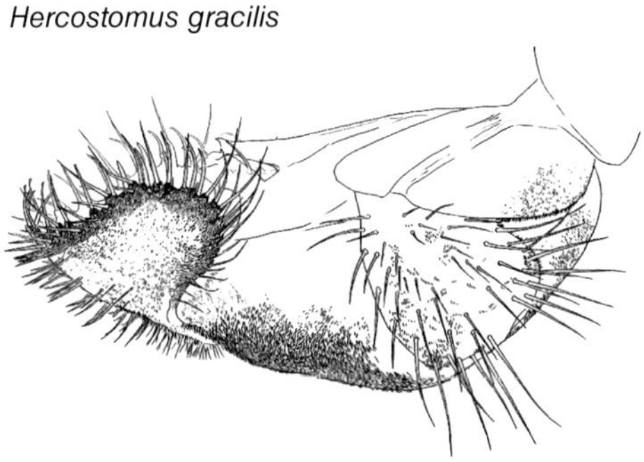

Fig. 19. *Hercostomus* hypopygia.

Hercostomus nanus

Hercostomus nigrilamellatus

Hercostomus nigripennis

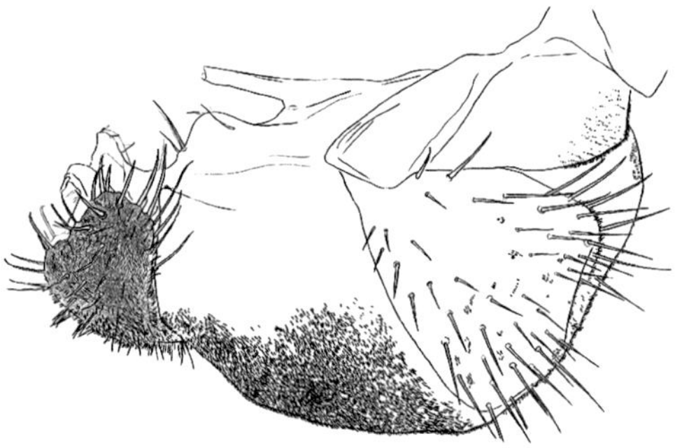

Hercostomus nigriplantis

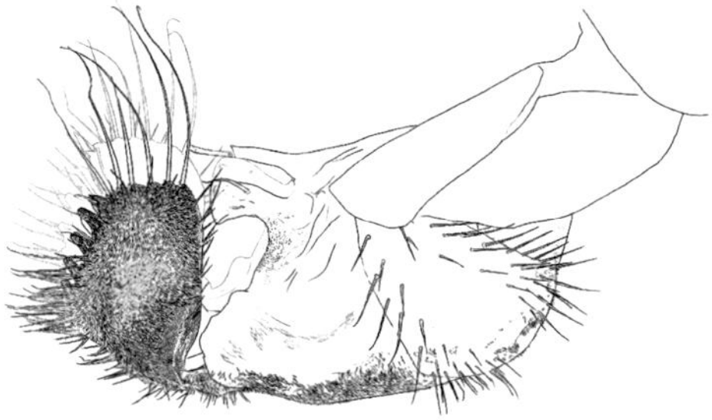

Hercostomus parvilamellatus

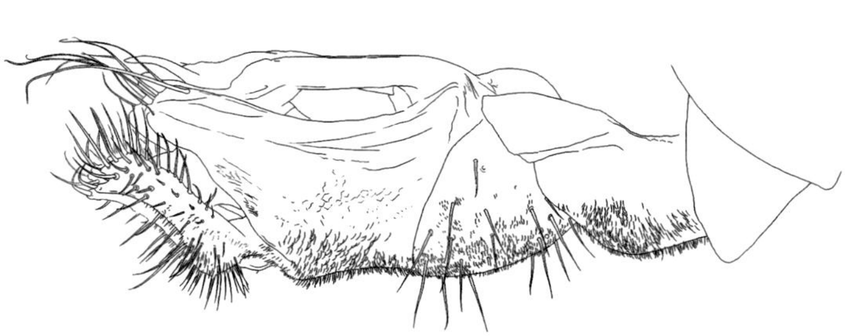

Fig. 19. Continued.

Hercostomus plagiatus

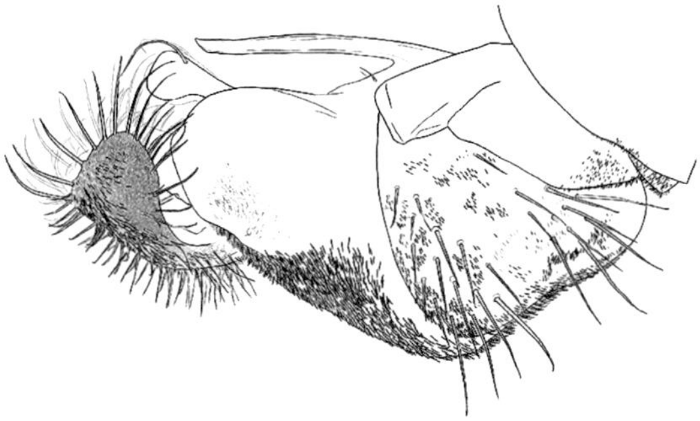

Hercostomus rothi

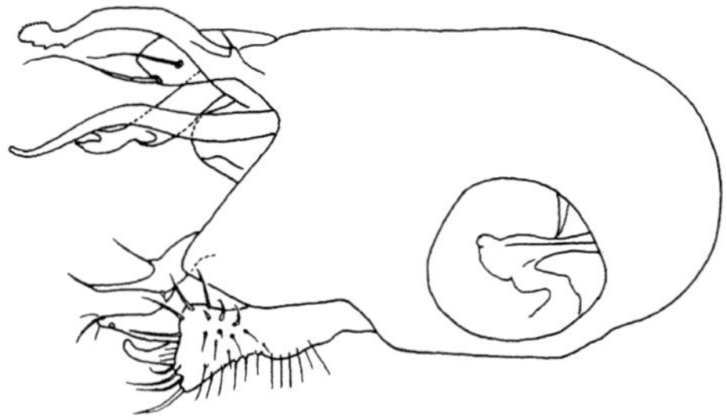

Hercostomus rusticus

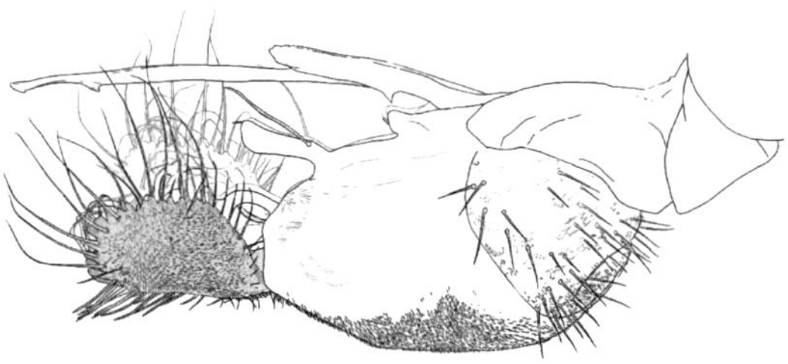

Hercostomus sahlbergi

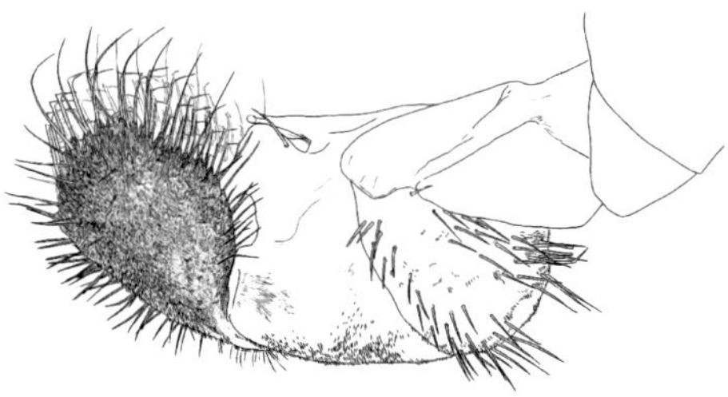

Hercostomus verbekei

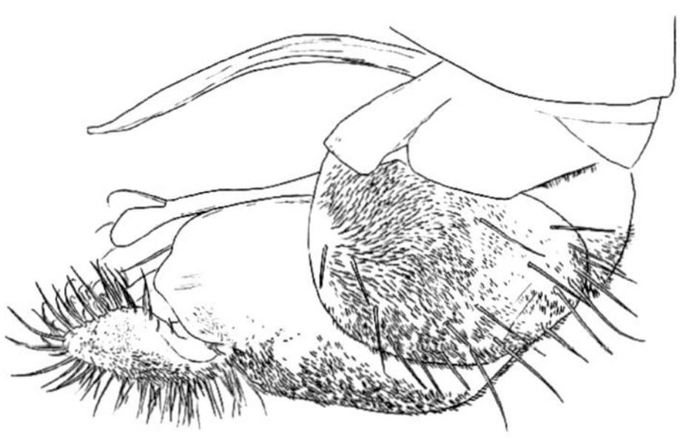

Fig. 19. Continued.

Hercostomus chetifer is found mainly in western and upland England and Wales with sparse records in southeast England and Scotland. It is closely associated with woodland streams, more rarely with unshaded streams, and occasionally with coastal seepages. Flight: v–ix, peak in vii.

Hercostomus fulvicaudis (Haliday in Walker, 1851)

Together with the very similar *rothi*, these are small rather pale flies, although *fulvicaudis* is less extensively yellow on the tergites and front of the thorax. They are the only *Hercostomus* showing a preponderance of yellow on the basal tergites, metepimeron, mid coxae and most tarsal segments and hypopygium, all contrasting with a blackish but densely dusted scutum. Also found only in this pair are white coxal apical setae on the front leg and entirely white postocular setae. The only other species of *Hercostomus* with partly yellow antennae are *plagiatus* and *verbeckei*. Females have the same unique colouring as males. Both *fulvicaudis* and *rothi* look similar to *Syntormon bicolorellum* and may be misidentified as *Syntormon* as the pedicel has the strong thumb-like extension into the postpedicel; *fulvicaudis* and *rothi* have biserial acrostichal setae and converging veins R_{4+5} and M_1 typical of *Hercostomus* (uniserial and parallel, respectively, in *S. bicolorellum*). Wing length ♂: 3.1 (3.0–3.4) mm, ♀: 3.5 (3.3–3.7) mm.

Records of *fulvicaudis* have been confused with those of *rothi* but as far as can be ascertained *fulvicaudis* may be either mainly western or perhaps near-coastal. It has been found in a variety of habitats including fens and woods, including moderately dry woods on limestone. Flight: vi–vii.

Hercostomus germanus (Wiedemann, 1817)

Among yellow-legged *Hercostomus*, this is the only species with its hind tibia being strongly black in the apical third; in *nanus* this tibia become gradually dark apically but not intensely black. The leg bristling is strong. *H. plagiatus* is superficially similar but has an entirely yellow hind tibia and partly yellow antennae. Females share the same distinguishing characters as the males and are additionally unique among British *Hercostomus* in having a slightly elongate proboscis that points diagonally forward and downward and comparatively long and narrow palps, although less conspicuously developed than in *nigripennis*. Wing length ♂: 3.6 (3.3–3.9) mm, ♀: 3.5 (3.3–3.8) mm.

Hercostomus germanus is widespread in Britain and shows no obvious trends in frequency. It occupies a very wide range of habitats including wet places such as acid moorland, valley mires and saltmarshes, and dry sites on chalk, limestone and dune grasslands. Nearly all sites are unshaded. It is one of the few dolichopodids frequently seen feeding at flowers. Flight: v–ix, peak in vii.

Hercostomus gracilis (Stannius, 1831) (Plate 4)

This beautiful large species is distinct from other British *Hercostomus* in numerous ways. Among the yellow-legged species it is by far the largest and has an unusual pea-green to sky-blue scutum and usually a gold face. The thick wing costa is unusual among British dolichopodids, and the numerous leg setae are matched among *Hercostomus* only by the dark-legged *argentifrons* and *nigrilamellatus*. Other unusual features include uncommonly stout mid tibia and tarsus and markedly long and dense proepisternal hairs. Females are similarly distinguished except for having a white face and the costa and mid tibia not being so markedly thick. Wing length ♂: 4.8 (4.3–4.9) mm, ♀: 5.2 (4.3–5.5) mm.

The distribution is mainly coastal areas of England and Wales where is it is found in dry places, mainly dunes and less often on cliffs, grasslands and coastal gravel. A large discrete inland population exits in Suffolk and Cambridgeshire on dry heathlands and in fens. Flight: vi–ix, peak in vii.

Hercostomus nanus (Macquart, 1827)

This is a small species whose small genitalia with short ribbon-shaped cerci quickly distinguishes it; *parvilamellatus* also has ribbon-shaped cerci but black femora. Its weak leg setae and almost black scutum also mark it out among the yellow-legged *Hercostomus* with dark mid and hind coxae. Females can be distinguished by the combination of very weak setae on the front tibia, the absence in most specimens of setae on the ventral face of the mid tibia, the gradual darkening of the hind tibia that runs into the black of the basitarsus, and a slight dark smudge at the tip of the hind femur. The lower postocular setae are black in males but either black or white in females. Wing length ♂: 2.9 (2.8–3.0) mm, ♀: 3.3 (2.8–3.6) mm.

In most of England this is a widespread species but in Wales it is infrequent and almost restricted to the coast, and in Scotland is very rare. It occurs most frequently at woodland streams and sometimes at springs, seepages and small rivers, but less often by standing water of ponds, ditches and pits. Flight: v–viii, peak in vi to early vii.

***Hercostomus nigrilamellatus* (Macquart, 1827)**

This species and *argentifrons* are the only large robust *Hercostomus* with entirely black legs, superficially resembling black-legged *Dolichopus*. The two species were confused until *argentifrons* was recognised (Godfrey and Drake 2024); as well as the characters in the key, *nigrilamellatus* differs in having longer and denser vestiture on the coxae and legs in both sexes. Wing length ♂: 5.2 (4.9–5.6) mm, ♀: 6.1 (5.7–6.5) mm.

Records are confused with those for *argentifrons*, and these combined records are mostly for England south of The Wash with outliers in Yorkshire and sparse records for Scotland. The habitats listed are mostly deciduous woodlands, usually wet or damp and occasionally dry, including those on chalk, and rarely wet grasslands and mires. Confirmed *nigrilamellatus* has been reared from decaying ash in England; Pollet and Hulcr (2007) additionally list elm, poplar and willow. Flight: v–viii.

***Hercostomus nigripennis* (Fallén, 1823) (Plate 4)**

This is the only small *Hercostomus* with black femora and hind tibia but with yellow front and mid tibiae, and can be recognised by its dark wings and long proboscis and palp. The leg colour and long proboscis bear a superficial resemblance to those of *Ortochile nigrocoerulea* but that species has a very long palp and more extensive yellow on the apical third of the mid femur and basal third of the hind tibia. *H. nigripennis* also bears a resemblance to *Gymnopternus* in the shape of the hypopygium and venation where veins R_{4+5} and M_1 are more parallel than in most *Hercostomus*, but no *Gymnopternus* has pale tibiae. Females have the distinguishing mouthpart, leg and wing features of males, and superficially resemble *Dolichopus vitripennis* in leg pattern and size but lack setae on the hind basitarsus. Wing length ♂: 3.8 (3.4–4.0) mm, ♀: 3.5 (2.9–3.9) mm.

This is a common species recorded widely in Britain but apparently absent in the outer Scottish islands. Its habitat is predominantly acidic examples of heath, mire, moorland and grassland. It can occur in moderately dry unshaded places, including dry limestone and chalk grasslands, perhaps associated here with de-calcified soils, and occasionally in woodlands. It is one of the few dolichopodids seen feeding on flowers. Flight: v–ix, peak in vii.

***Hercostomus nigriplantis* (Stannius, 1831) (Plate 4)**

Of the two large species of *Hercostomus* with black femora and yellow tibiae, this one is more likely to be found in Britain as the similar *sahlbergi* has been recorded only once. The mid tarsus is modified with an expanded fifth tarsomere, although the degree of expansion varies between populations from obvious to hardly noticeable. The form with a slender tarsomere was described by Verrall (1912) as *subsimplicipes* which Collin (1940) synonymised with *nigriplantis*. The thick costa approaches the more pronounced form seen in *gracilis*. Females are similarly distinguished by their large size and yellow knees on all legs (black in *sahlbergi*) and superficially resemble large black-legged *Dolichopus*. Wing length ♂: 5.8 (4.9–6.3) mm, ♀: 5.3 (4.5–5.7) mm.

Hercostomus nigriplantis has a mainly western distribution. It is locally frequent within a few kilometres of the coast from the Isle of Wight to Anglesey and more sparsely to Ayrshire but is scarce in eastern Britain. While it is most often found at the coast, it is not an obligate 'coastal' species, contrary to Pollet's early (1991) assertion, revised in Pollet and Grootaert (1996), as there are verified records well inland in Britain. It is found mainly by seepages, streams and pools on dunes, cliffs and in coastal woodlands, but more precise habitat requirements are unclear. Inland sites include bogs, riverbanks and fen but these appear to be unusual habitats for the species. It has been reared on several occasions from decaying wood or debris in tree forks (e.g. Dyte 1959). Flight: v–ix.

***Hercostomus parvilamellatus* (Macquart, 1827)**

Structurally this small species resembles *nanus* in weak sparse leg setae and a small ribbon-shaped cercus, but is the only small species with black femora and yellow tibiae, as in the much larger *nigriplantis* and *sahlbergi*. All *Dolichopus* species with similar leg colouration and dark lower postocular setae are large species. Wing length ♂: 3.2 (3.0–3.5) mm, ♀: 3.7 (3.4–3.9) mm.

This is an almost entirely English species found northwards to North Yorkshire but scarcely penetrating the Pennines, Wales or Devon in the west; within its range it is frequent. It is found by springs, seepages and headwater streams, often with mud or silt rather than fast stony ones, within woodland. Rare records are from unshaded ditches and ponds. It may be associated with calcareous sites (Crossley 2018). Flight: iii–ix, peak in vi.

***Hercostomus plagiatus* (Loew, 1857)**

Hercostomus plagiatus is a particularly bristly species, not just on its tibiae but its tergite setae are longer than half the segments' lengths. *H. germanus* is superficially similar in shape, being moderately strongly bristled and with a broad face but differs in having black antennae and a black apical third on the hind tibia, rather than entirely yellow as in *plagiatus*. In both sexes the partly yellow antennae are also unusual among *Hercostomus*, found only in *verbekei* and the small pale species *fulvicaudis* and *rothi*. The similar *verbekei* is reliably separated using the hypopygial characters, of which at least the cercus is easily visible, and in both sexes by the more strongly converging R_{4+5} and M_1. Wing length ♂: 3.6 (2.9–4.0) mm, ♀: 3.8 (3.7–4.0) mm.

This appears to be a lowland species with a distribution restricted to England south of the Humber and Wales. It is more frequent near the coast where it is found mainly at seepages on soft-rock cliffs and within reeds on upper saltmarsh. Inland it occurs in fens and reedbeds within which it is nearly found always close to water margins. The associations with reed and base-rich sites are strong. Sites are unshaded and any tree cover is incidental, such as invading willow scrub on fen, in contrast to its frequent occurrence in wooded sites in Belgium (Pollet 1993). Flight: v–ix, peak in late vi to early vii.

***Hercostomus rothi* (Zetterstedt, 1859)**

A comparison with *H. fulvicaudis* which it closely resembles is given in the account for that species. Wing length ♂: 2.7 (2.6–3.0) mm, ♀: 2.9 (2.7–2.9) mm.

Hercostomus rothi was first found in Britain in 2005 (Drake *et al.* 2013). The few subsequent records suggest that it occurs in the southern half of England with a concentration in fenland south of The Wash, and outliers in Somerset. Several records are from fens within an arable setting but its precise habitat requirements are uncertain. Flight: vi–viii.

***Hercostomus rusticus* (Meigen, 1824)**

This species resembles the larger *nigrilamellatus* in being completely black and the front tibia bearing a fairly long fine apicoventral seta about twice as long as the shaft width. No other British *Hercostomus* has such a long postpedicel with the arista placed at the apical third. Females are also completely dark and their postpedicel is pointed although not as long as that of the males, and can be separated from *nigrilamellatus* and *argentifrons* by their white postocular setae. Wing length ♂: 3.1 (2.8–3.2) mm, ♀: 3.2 (3.1–3.4) mm.

A strong population, thought to be a recent introduction, was found in London in 2017–2020 where it was living in a garden allotment (Drake and Painter 2023). It is a species of rather dry habitats. Flight: vi–ix, peak in late vi to mid vii.

***Hercostomus sahlbergi* (Zetterstedt, 1838)**

This is a large *Hercostomus* with black femora and all tibiae yellow, resembling only *nigriplantis* from which it may be distinguished by its more bristly tibiae and tarsi and relatively short hairs on the apical margin of the cercus. The front cells of the wing and cercus are conspicuously brown. Females are distinguished among *Hercostomus* with yellow tibia in having dark knees on all legs; its front tibiae have noticeably numerous setae. ♂: 5.7 (5.3–6.9) mm, ♀: 5.5 (5.3–5.8) mm (non-British specimens).

Only two specimens of *Hercostomus sahlbergi* are known from Britain, both collected on possibly the same occasion in Strathspey in 1938. Although this area has been well worked for Diptera, no further specimens have been found. It is a montane species in its European range. Ringdahl (1928) states that it occurs on gravel at the sides of streams and on nearby vegetation. Flight: vii.

Hercostomus verbekei **Pollet, 1993**

This species closely resembles *plagiatus* but is readily separated using the characters in the key. Wing length ♂: 4.5 (4.1–4.8) mm, ♀: 4.2 mm.

There is one record for Britain from Suffolk in 1987 (Hodge 1993a, Pollet 1993). The habitat was a sandy shore of a lowland river. In Belgium, Pollet (1993) describes its habitat as sedge and reed by ponds on clay soils, and similar to those of *plagiatus*. Flight: vii.

HYDROPHORUS Fallén, 1823 (Hydrophorinae)

Hydrophorus includes medium-sized to large slender species with a markedly robust, short and often armed front femur that contrasts with the long slender and weakly bristled mid and hind femora (Plate 4). Their wings are considerably longer than the body and this appearance is exaggerated by the very short abdomen. In both sexes, the genus may be recognised by the small pimple separated by a notch on the front edge of the postpedicel (similar in *Scellus*), having no seta on the hind coxa as in a few *Rhaphium* which have an apical rather than dorsal arista, and the arista terminating in fine hairs although these are difficult to see. *Hydrophorus* is unusual among dolichopodids in having a narrow or broad gena.

The two sexes of *Hydrophorus* appear very similar as the hypopygium is almost entirely encapsulated but the end of the abdomen does provide features for determining the sex. Males have a narrow sixth tergite whereas only five are visible in females, the male fifth tergite has abdominal plaques (oval black spots) near the lateral margin, which are absent from the fifth tergite in females, and the fourth sternite of the male often protrudes in profile as its hind margin is folded downwards to accommodate the hypopygium. Females are on average larger than males, as in most dolichopodids, but the difference is marked in *litoreus*, *praecox* and especially *oceanus*. While many show sexual dimorphism in the armature of the front legs, this varies inconsistently between species so cannot be used to determine the sex.

Leg armament consists of stronger setae and spines but whether these are setae or spines is not clear without careful examination (setae having a socket, spines are extensions of the cuticle) so, for simplicity, they are both called spines in most cases.

The genus has a predominantly boreal distribution with many European species occurring in Scandinavia and further east. Most of the nine species occurring in Britain are among the few that are widespread in Europe, with the exception of *pilipes*. Among species that may still be found in Britain, *borealis* Loew, 1857 and *rogenhoferi* Mik, 1874 are possible and their separation from similar species is discussed in those accounts.

All British *Hydrophorus* spend much of their adult life on water surfaces of small pools or still water by seepage, streams, rivers or gullies of intertidal mud. This habit makes them susceptible to water-trapping but often difficult to capture using a sweep-net as they fly very close to the water surface. Most species have a long flight period, some starting as early as February or March and extending to October or even November, and usually with no clear indication of a peak in activity; this suggests either continuous breeding, long overlapping generations or the ability to overwinter as adults. Dyte (1988a) described mate-guarding in *Hydrophorus oceanus* in which the male remains clinging to the female for a considerable time after copulation. *H. praecox*, whose males are also rather smaller than the females, almost certainly has the same behaviour as two accounts describe pairs remaining together for a long time (Emeis 1964, Pârvu 1984).

Key to species of *Hydrophorus* (males and females)

1 Wing unmarked 2
– Wing with marks, either mottled or with single spots on dm-m and apical section of M_1 6

2(1) Gena below eye very narrow, one-third or less depth of postpedicel, or not visible (a). Halter knob usually dark (variable in *oceanus*)....3

– Gena depth equal to or greater than postpedicel, and rarely narrower than palp (aa). Halter knob pale....4

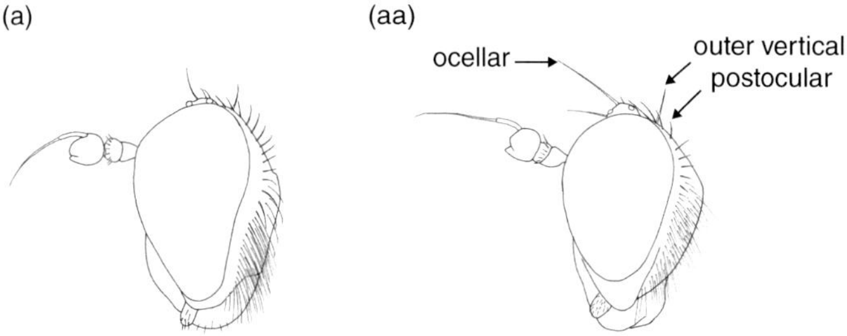

3(2) Scutum densely brown-dusted with a conspicuous pattern that shifts with varying angle of view. Scutellum with 1 pair of setae. *Male*: Front femur near tip beneath with only sparse short black setae (a). *Both sexes*: Front femur with antero- and posteroventral rows of setae separated. Front tibia with ventral spines on basal half not markedly weaker, although shorter, than on apical half (a)....***oceanus***

– Scutum disc almost uniformly shining blackish with a faint green or blue hue through very thin dust. Scutellum with 2 pairs of setae. *Male*: Front femur near tip beneath with 3 yellow flat club-shaped setae (aa). *Both sexes*: Front femur beneath with a single irregular row of setae varying in size. Front tibia with very short ventral spines, those on basal half reduced to tiny stubs (male) (aa) or about one-fifth of tibia width (female)....***litoreus***

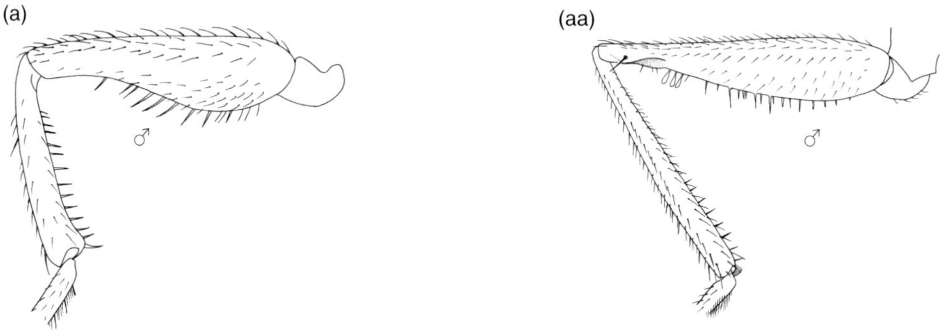

4(2) Epistoma glistening green or bronze, clypeus white-dusted (male) or grey-dusted with green ground colour showing (female). Front femur with a posteroventral row of long spines, in front of these a band of numerous small spines extending almost to tip, much closer together than their length (a). Outer vertical seta twice as long as postocular setae (as fig. aa of couplet 2). Scutum subshining dark or greenish without alternating coloured stripes and with only poorly defined dust stripes along rows of setae....***balticus***

– Epistoma and clypeus similarly white or yellowish brown (but if rubbed then metallic green). Front femur with 1 or 2 ventral rows of spines but without a cluster of numerous smaller spines, the sparse anteroventral spines on apical half separated by at least their length. Outer vertical seta not longer than postocular setae. Scutum with longitudinal stripes of different colours or dusting....5

(a)

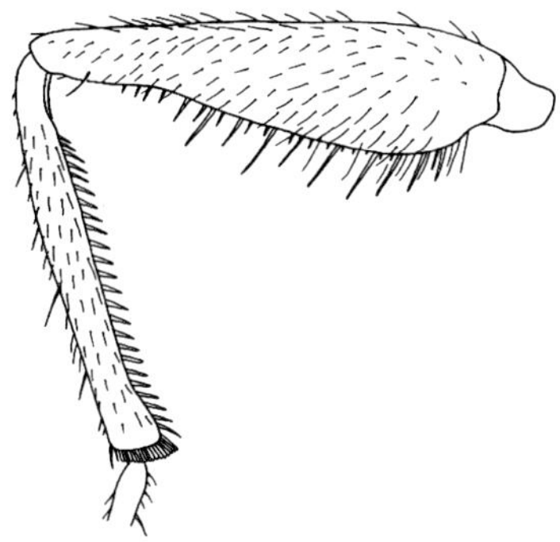

5(4) Front femur with spines near base shorter than width of tibia, continuing in 1 or 2 rows almost to tip of femur (a). Front tibia with distinct spines on apical half, about half width of tibia, the apical spine at tip longer, stouter and projecting beyond preceding small spines (a) although weaker in female. Ocellar setae not or scarcely longer than length of postpedicel. Dorsocentral setae increasing in length gradually, difference in size between last 2 setae not marked.. ***praecox***

– Front femur with 3–5 long splayed posteroventral setae near base, contrasting with anteroventral row of sparse short spines that become tiny points near tip (aa). Front tibia with tiny spines on apical half, the apical spine scarcely larger than preceding ones (aa). Ocellar setae 1.5 times length of postpedicel. Dorsocentral setae small and fine but last seta contrastingly long and strong, similar to scutellar setae.. ***viridis***

(a)

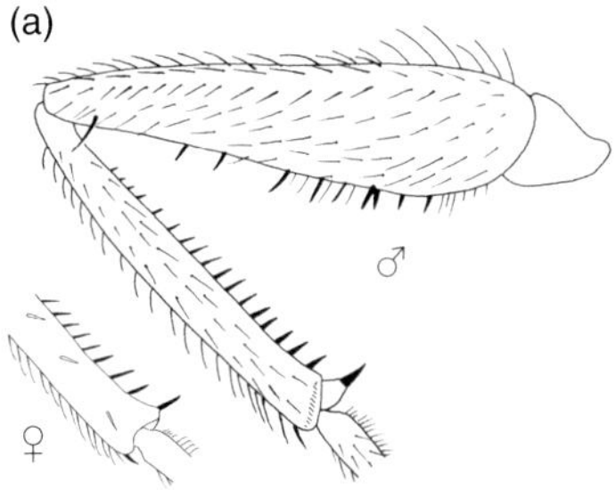

(aa)

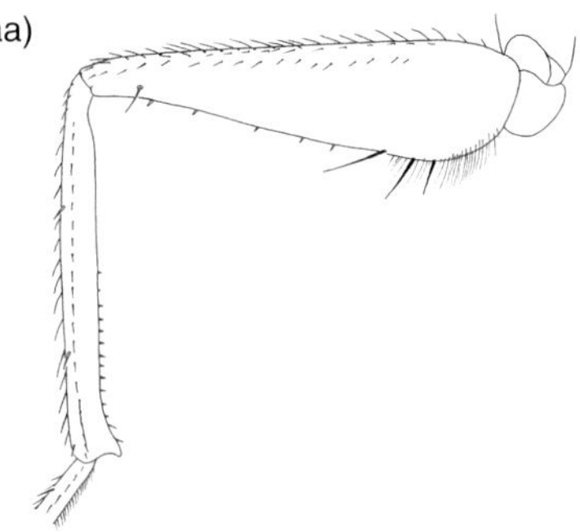

6(1) Wing with numerous dark spots and streaks (a). Small species, wing length 2.5–3.5 mm. Front coxa on outer edge with a row of 5–10 short thick black setae, sometimes interrupted in middle. Front tibia with an apicoventral curved spine as long as width of shaft (b).. ***nebulosus***

– Wing with 2 spots, on crossvein dm-m and on apical section of M_1 (aa). Large species, wing length 4.7–6.2 mm. Front coxa without short setae on outer edge, only black or pale hairs, but small setae present at tip. Front tibia without an apical spine. 7

(a)

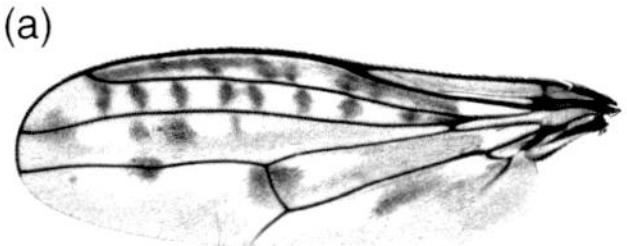

(aa)

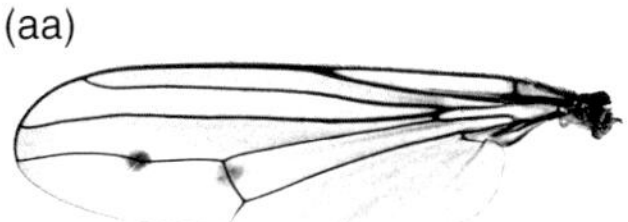

(b)

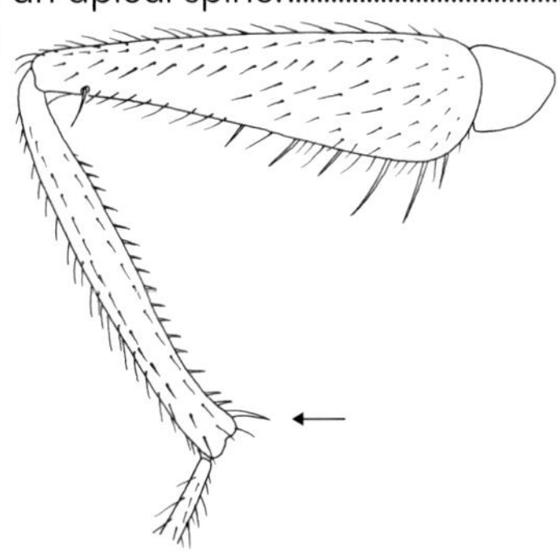

7(6) Epistoma shining green, undusted, contrasting with brown-dusted clypeus. Dorsocentral setae in middle of row finer and not longer than postocular setae near top of row (below topmost postoculars). Scutum when viewed from above and slightly anteriorly with distinct pale grey dust spots or short streaks along dorsocentral row more prominent than any dusted line along acrostichal row. Male front femur on apical half on posteroventral face with irregular rows of small spines that become tiny points apically (a)................................ ***bipunctatus***

– Epistoma and clypeus dusted white, gold or brownish yellow (but rubbed specimens may show green ground colour). Dorsocentral setae in middle of row at least as long and nearly as thick as postocular setae. Scutum when viewed as above with narrow pale-dusted line along acrostichal row more prominent than dusted line or spots on dorsocentral row. Male front femur on apical half with armament absent or more complex.. 8

(a)

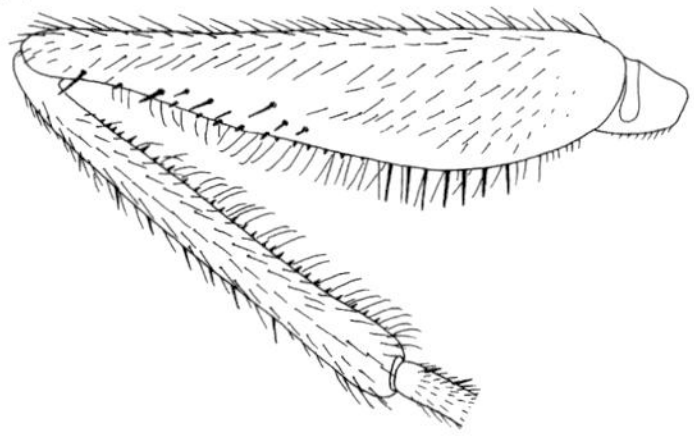

8(7) Males.. 9

– Females...10

9(8) Face entirely white. Front femur with armament along its whole length, on apical third with a cluster of 4–6 thicker irregularly curved spines and a low flange on anterior edge near tip, on basal half with short spines mixed with equally long dark hairs (a). Hind femur with ventral vestiture very short, less than one-quarter width of shaft, with several ventral setae (av, pv) 2–3 times longer than vestiture hairs.. ***albiceps*** ♂

– Face dusted brownish or gold. Front femur with about 7–8 short spines on basal third, no spines or points on apical half (aa). Hind femur with ventral vestiture long, about half width of shaft, with 1–2 ventral setae only about 1.5 times longer than vestiture hairs........... ***pilipes*** ♂

(a)

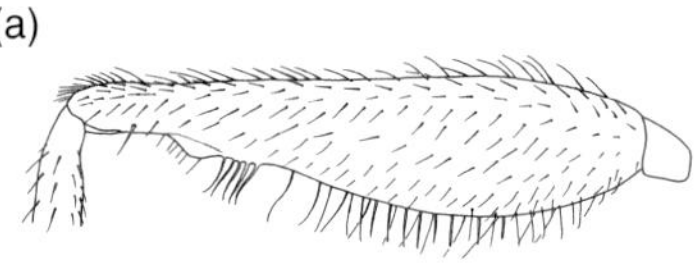

(aa)

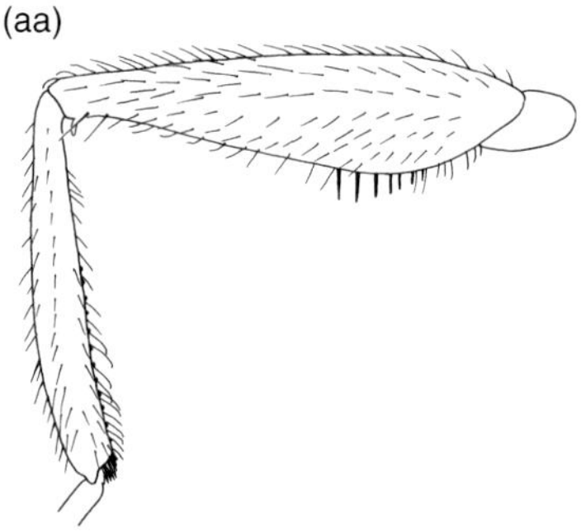

10(8) Front coxa with hairs mostly black, at least the longer ones on outer face. Dorsocentral and acrostichal setae short and fine, dorsocentrals (except last) clearly shorter than longest head setae. Hind femur with ventral vestiture very short, less than one-quarter width of shaft, with several ventral setae (av, pv) 2–3 times longer than vestiture hairs..................... ***albiceps*** ♀

– Front coxa with hairs pale with occasional black ones. Dorsocentral and acrostichal setae long and conspicuous, dorsocentrals about as long as longest head setae (although finer). Hind femur with ventral vestiture long, about half width of shaft, and with 1–2 ventral setae only about 1.5 times longer than vestiture hairs.. ***pilipes*** ♀

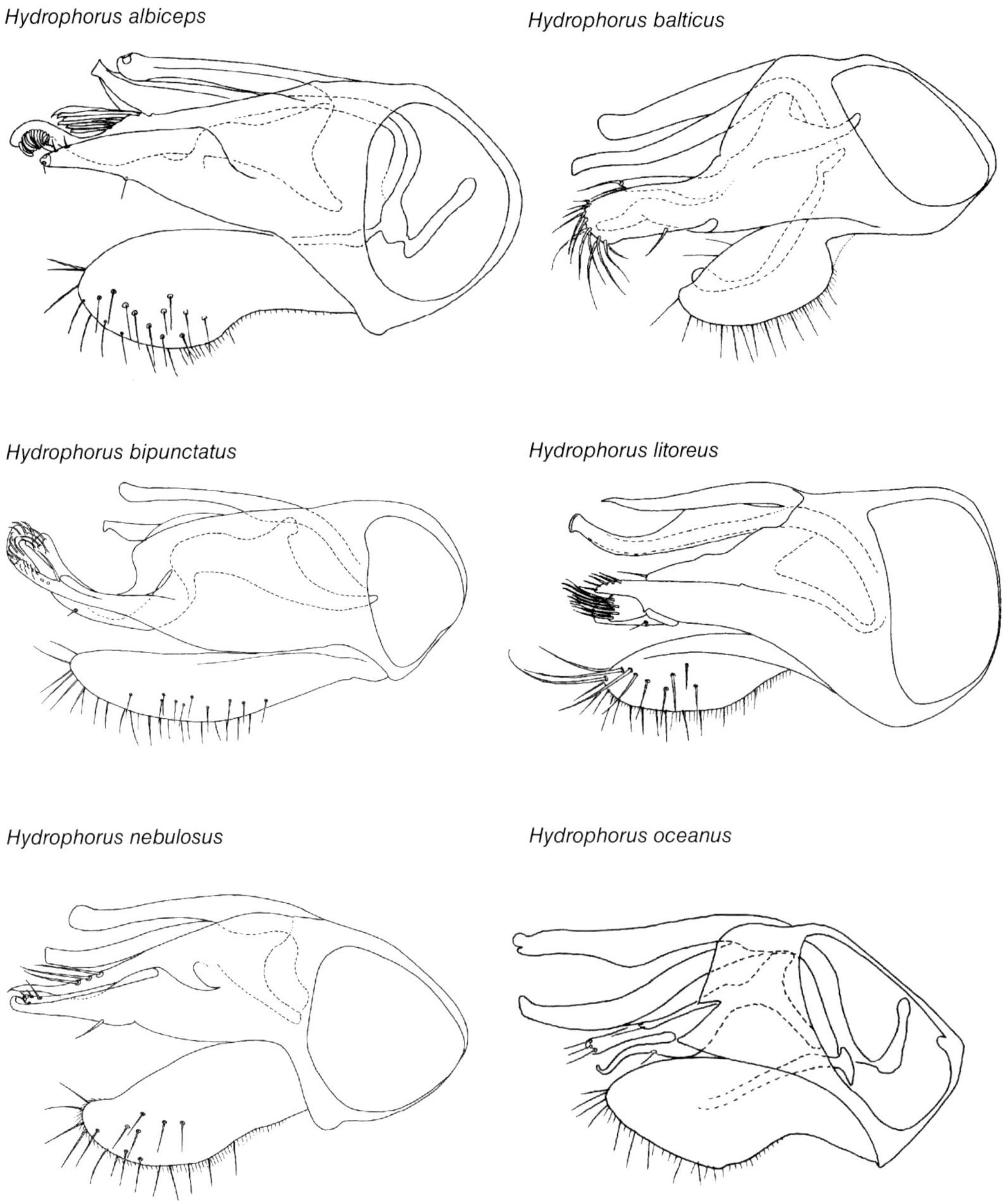

Fig. 20. *Hydrophorus* hypopygia.

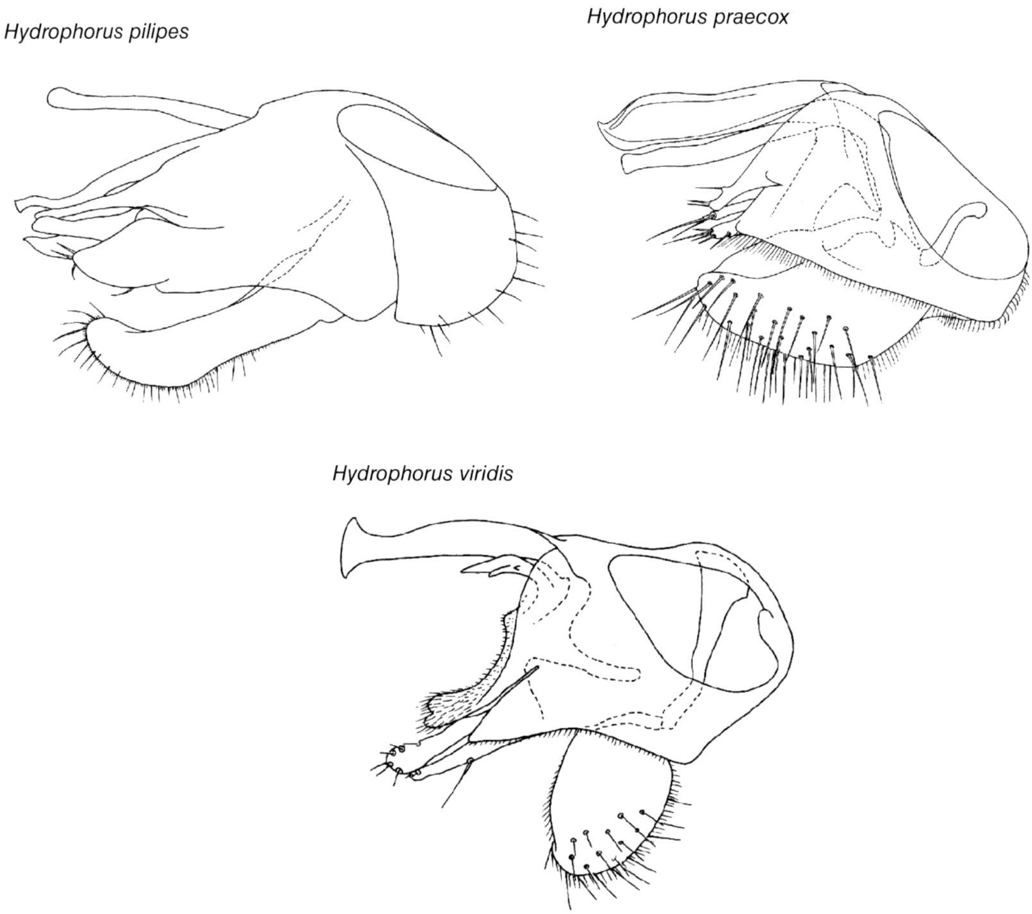

Fig. 20. Continued.

***Hydrophorus albiceps* Frey, 1915**

This is a large robust species with two wing spots and dark-shaded membrane on the front edge in females. The body is mainly dark green with an almost black shining scutum with very narrow continuous lines of dusting along the rows of acrostichal and dorsocentral setae, resembling *pilipes* in both sexes. Males are distinguished by the cluster of slightly wavy setae and an anteroventral flange near the tip of the front femur. Females differ from other British species apart from *bipunctatus* and *pilipes* in lacking femoral armature. Further features separating female *pilipes* are given in that account. *Hydrophorus borealis* is a possible species in Britain; it resembles *albiceps* in overall appearance and males have the same cluster of setae at the apical quarter of the front femur but lack the anteroventral flange. Wing length ♂: 4.9 (4.7–5.0) mm, ♀: 5.6 (5.0–5.9) mm.

Hydrophorus albiceps is a predominantly upland species found frequently in the Scottish Highlands, northern England and Wales but is scarce in southwest England. It is found mainly in bogs, acid heath and moorland where it lives on and next to pools and flushes, to an altitude of 950 m, but can occur in lakeside swamps and in bogs at low altitude. Flight: iii–x.

***Hydrophorus balticus* (Meigen, 1824) (Plate 4)**

Hydrophorus balticus is a dark brown or bronzy green species with a shining scutum. Both sexes in front view appear multi-coloured with a dark brown frons, shining green or sometimes bronzy green epistoma, white (in males) or greyish (females) clypeus and whitish grey coxae with the underlying green ground colour showing through. *H. bipunctatus* also has a shining green epistoma but its clypeus is slightly deeper than wide and has short grooves only at its upper margin, unlike *balticus* where the clypeus is only just as wide as deep and has grooves down its whole length. Thoracic and head setae are unusually long for the genus, e.g. the ocellars and longest dorsocentrals are no shorter than the greatest width of the front femur; in other *Hydrophorus* these setae are shorter than the femur's width. Both sexes have identical armature on the front femur and tibia. Wing length ♂: 4.5 (4.2–4.8) mm, ♀: 4.6 (4.2–4.9) mm.

This species is widespread in Britain and more frequent in eastern England. It occurs next to many types of unshaded water body including rivers, ponds, ditches and sometimes seepages either at coastal or inland locations, and saltmarsh. It is the characteristic *Hydrophorus* of pools on stony river shores. Flight: vi–x, peak in vii.

Hydrophorus bipunctatus **(Lehmann, 1822)**

Hydrophorus bipunctatus is a large robust dark brownish green species with a shining black scutum with dots, rather than lines, of paler dusting along the dorsocentral rows. It is easily distinguished by the entirely shining green or bluish epistoma, similar to that of *balticus* but clearly deeper than wide; in male *albiceps* the face is shining green only narrowly below the antennae. Females resemble *albiceps* and *pilipes* in overall appearance and in lacking armament on the front leg. The shining green face cannot be relied upon alone as rubbed specimens of these other two species may appear the same. *Hydrophorus rogenhoferi* Mik may occur in montane Britain; males resemble *bipunctatus* but have an entirely tawny face and very long ventral hairs on the front femur. Wing length ♂: 5.4 (5.1–5.7) mm, ♀: 5.9 (5.4–6.2) mm.

The species is widespread but not particularly frequent in Britain. It lives at the margins of flowing and still water where there is short dense vegetation, e.g. cattle-trampled grassland or swamp, or sparse fen, but is scarce at bare water margins. Flight: ii–xi, peak in late vi–vii.

Hydrophorus litoreus **Fallén, 1823**

This is the only clear-winged *Hydrophorus* with a narrow gena and a shining scutum almost lacking a dust pattern. In rear view the whole body is shining green to greenish black but in side view the grey dusting lightens the appearance to a rather pale green, resembling *praecox* and *viridis*. In front view the brown frons contrasts with the greyish white of the face and coxae, as in *praecox*. Male are distinguished by the three apical yellow spatulate setae on the front femur and the strongly thickened basal half of vein R_{4+5}. In females the feeble uniserial row of femoral spines distinguishes *litoreus* from other clear-winged *Hydrophorus*. Wing length ♂: 4.2 (3.8–4.6) mm, ♀: 4.9 (4.3–5.3) mm.

Hydrophorus litoreus is widely distributed in England but scarce in Wales and rare in Scotland. It lives at unshaded water margins, more often by still water of ditches and ponds than at rivers, and more often at muddy than stony places although it shows no clear preference. Flight: iii–xi.

Hydrophorus nebulosus **Fallén, 1823**

No other British dolichopodid has such strongly patterned wings. It is a dark bronzy green fly with contrasting pale grey pleura and is the smallest British *Hydrophorus*. The femoral armature is similar in both sexes, females having slightly longer spines. Wing length ♂: 3.0 (2.5–3.2) mm, ♀: 3.1 (2.9–3.5) mm.

In northern England, Wales and Scotland this is a frequent species, occurring at over 1200 m in the Highlands, but in southern England it is confined to a few sites. It is found only at acid habitats, mainly bogs but also wet heath, where it usually lives on small pools or sometimes at seepages. Flight: ii–xi.

Hydrophorus oceanus **(Macquart, 1838) (Plate 4)**

These are small greenish to brownish flies with strong dusting forming a pattern that shifts with the angle of view. No other *Hydrophorus* has such a densely dusted and patterned scutum, almost devoid of shine. In front view, the dusting of the frons, face and coxae is uniformly greenish grey in females and more shining white on the face and coxae in males. It is unique among British species in having only one pair of scutellar setae and two or three proepisternals. Females are markedly larger than males, with wings about 40% longer on average, and have stronger armature on the front legs and longer more numerous setae on other legs. Wing length ♂: 3.4 (3.0–3.8) mm, ♀: 4.9 (4.2–5.2) mm.

Hydrophorus oceanus is common on the coast of England and Wales, becoming scarce on the eastern coast and rarer in Scotland. It is confined to pools and creeks of saltmarshes and intertidal mudflats and on brackish pools on coastal dunes and grasslands. Flight: iv–x, peak in vii.

Hydrophorus pilipes Frey, 1915

This is a large robust species resembling *bipunctatus* and *albiceps*, with a dark brownish black scutum and abdomen which often shows little hint of green. The scutum has a very narrow line of paler dusting along the acrostichal row and dorsocentral setae arising from spots of dusting in the middle of the row. In front view, the male has a dark brown frons with yellow dust, a yellow to golden face and greenish grey coxae. The whole fly, particularly in the males, is noticeably hairier and has longer setae than any other *Hydrophorus*, although *balticus* has similarly long dorsocentral setae. Comparison is most needed with *albiceps* as they occur in similar habitats. Wing length ♂: 4.5 (4.2–5.1) mm, ♀: 4.6 (4.4–4.7) mm.

Hydrophorus pilipes is the only truly montane species found in Britain, occurring moderately widely in the Scottish Highlands. In the past it was found at lower altitudes but is now apparently restricted to above 900 m in the Cairngorms, although on mountains further west it will occur at lower elevations but rarely below 650 m. It lives at grassy flushes and on small peat pools (MacGowan 1987) often overlapping with *H. albiceps* in the lower part of its range. This species was known in Britain as *rufibarbis* Gerstäcker until MacGowan and Drake (2023) corrected the name. Flight: v–ix.

Hydrophorus praecox (Lehmann, 1822)

This is a small to medium-sized greyish green species with extensive whitish grey dusting masking a shining green ground, although metallic green and purple stripes can be seen along the dorsocentral rows. In dorsolateral view the scutum shows a clear dust pattern, in particular with streaks of white dust running up the thoracic suture, on the postpronotum and on the area immediately behind the field of setulae on the front of the scutum. The metepimeron does not completely cover the base of the first tergite, leaving at least one abdominal plaque visible (dark oval spots at lower margin of tergites); in most other *Hydrophorus* the abdomen is more cylindrical and the metepimeron envelops the base of the first tergite. Comparison is most needed with *viridis* which is the only other species with metallic green and purple stripes on the scutum but *praecox* usually lacks the narrow green stripe along the acrostichal row. The white hairs of the tergites are always obvious, more so than in *viridis*. In front view the frons, face and coxae are uniformly greyish white rather than yellowish as in *viridis*. The arrangement of spines on the front femur is rather variable. Wing length ♂: 4.3 (3.9–4.6) mm, ♀: 5.0 (4.3–5.5) mm.

Hydrophorus praecox is widely distributed around the English and Welsh coasts and inland mainly in England. It is rare in Scotland but recorded in Orkney and Fair Isle. It is found on the coast at saltmarsh, lagoons, dunes, soft-rock cliffs, and inland at puddles, ponds and seepages and river shingles with pools. Most inland locations are relatively bare with little vegetation and some have vestigial saline influence. It is almost always recorded skating on water. Flight: iii–x.

Hydrophorus viridis (Meigen, 1824)

This small dusted dark green fly resembles *praecox* with which it is compared under that species. The principal features are yellowish or pale brown dusting that is never glistening white, the frons, face and coxae in front view are uniformly pale yellowish white to pale golden, and the metallic green lines are developed equally along both acrostichal and dorsocentral rows. The femoral armature is the same in both sexes. No other British *Hydrophorus* has the cluster of slightly splayed long basal setae. Some characters used by d'Assis-Fonseca (1978) to distinguish females from those of *praecox* are misleading (both have yellowish veins at the wing base and small spines on the inner and outer surface at the base of the front coxa). Wing length ♂: 4.4 (4.4–4.5) mm, ♀: 4.7 (4.1–5.1) mm.

Hydrophorus viridis is rare, being found in a few places from Norfolk, along the south English coast to south Wales. It is mainly coastal, occurring on pools within seepages on soft-rock cliffs and at shallow brackish pools; one inland record is from a sandy Welsh river. It occupies a wider range of habitats in mainland Europe, e.g. Pârvu (2000) mentions small streams, swamps and marsh in upland Romania. Flight: vi–viii, xi.

LAMPROCHROMUS Mik, 1878 (Rhaphiinae)

Lamprochromus are tiny species characterised by a large matt black or dark brown, copper-edged spot on the side of the scutum and a smaller more shining purple one above the postpronotal lobe (Plate 5). In British species, all males and the female of one species have yellow second and third tergites; coupled with the unusual glossy violet frons, they thus resemble some small species of *Rhaphium* and *Syntormon*. The arista arises close to the base of the postpedicel in males. In addition, the acrostichal setae are biserial, there are three or four (rarely five) pairs of dorsocentral setae, the mid and hind femora have an anterior preapical seta, and the hind basitarsus is much shorter than the second tarsomere.

Germann *et al.* (2011a) found that *Lamprochromus* was more closely related to *Rhaphium* than to genera in the Sympycninae where it was previously included. Their conclusion is followed here although Grichanov and Ahmadi (2017) disagreed. There are nine species in the Western Palaearctic (Grichanov and Ahmadi 2017) of which three occur in Britain. Although unlikely, *Lamprochromus speciosus* (Loew, 1871) could occur here; males differ from the three British species in having a thickened dark hind tibia and pale scape and pedicel.

They are found in either fresh or brackish wetlands.

Key to species of *Lamprochromus* (males)

1 Lower half of pleura (katepisternum, meron) mainly yellow. Tergites 2 and 3 with yellow bands broadly interrupted by black on midline. Hind tibia often dark yellow, sometimes becoming black at tip and on basitarsus.................. ***kowarzi***

– Lower half of pleura entirely green with pale grey dusting. Tergites 2 and 3 entirely yellow on midline. Hind tibia and basitarsus yellow.................. 2

2 Postpedicel elongate-triangular with an acute tip, its pubescence sparse and straggly, longer than basal segment of arista (a). Scutum in dorsal view with a distinct more shining purple mid-stripe; no clearly distinct dark narrow stripe along row of anterior acrostichal setae, any darker patch here merging with anterolateral dark patches.................. ***bifasciatus***

– Postpedicel not markedly triangular, always broadly rounded at tip, its pubescence dense, even and shorter than basal segment of arista (aa). Scutum in dorsal view almost uniformly subshining grey-green, no differentiated mid stripe filling the space between dorsocentral rows; a distinct narrow dark mid-stripe along acrostichal rows from front of scutum to about second dorsocentral seta.................. ***semiflavus***

(a)

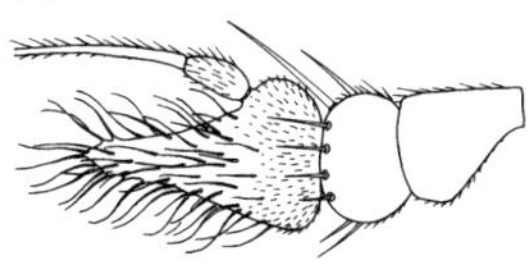

(aa)

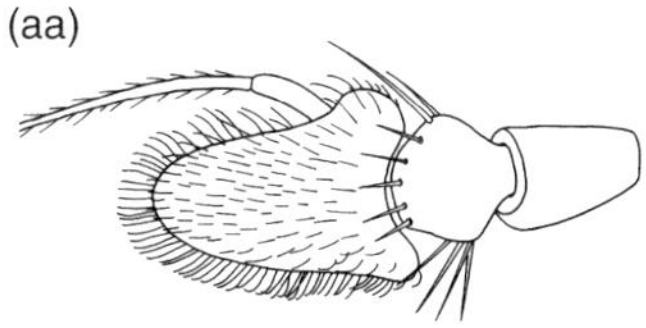

Key to species of *Lamprochromus* (females)

1 Tergites 2 and 3 and lower half of pleura (katepisternum, meron) mainly yellow.......... ***kowarzi***

– Tergites and pleura entirely green.................. 2

2 Occiput in posterodorsal view dark and more-or-less uniformly thinly dusted. Pubescence of postpedicel as long as basal segment of arista, longest hairs almost twice as long as width of this segment (a). Meron usually with a narrow triangular yellow streak running along anterior suture above base of mid coxa. Scutum as male. Front tarsus almost entirely yellow, tarsomere 5 sometimes darker.. ***bifasciatus***

– Occiput in posterodorsal view with 2 oval pale grey dust spots forming a 'V' above neck. Pubescence of postpedicel shorter than basal segment of arista, about equal to its width (aa). Meron rarely with a yellow anterior margin running along suture above base of mid coxa. Scutum as male. Front tarsus often with tarsomeres 3–5 dark but may be all pale. .... ***semiflavus***

(a)

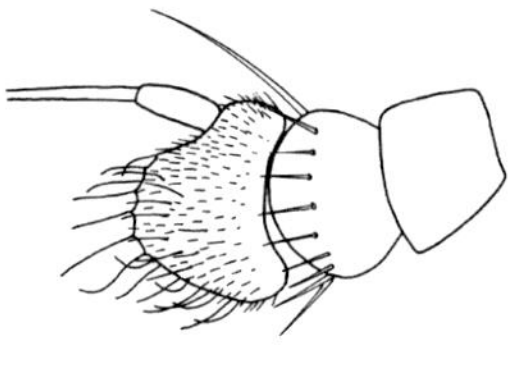

(aa)

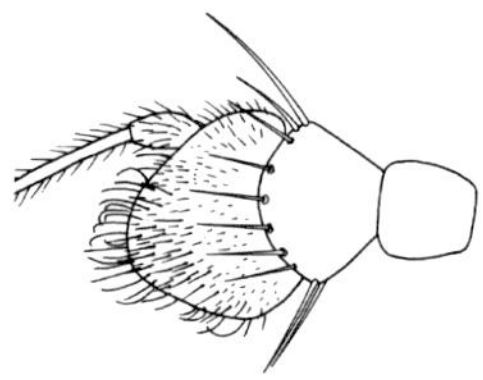

Lamprochromus bifasciatus **(Macquart, 1827)**

Males are readily identified by their long narrowly triangular and very hairy postpedicel. Females are more difficult to separate from those of *semiflavus* but the almost uniformly dusted occiput and broad glossy purple midline of the scutum are nearly always reliable features in clean dry specimens. Those in alcohol may be difficult to separate unless temporarily dried out. The apical setae of the front coxa are usually pale but may sometimes include black ones. Wing length ♂: 2.1 (2.0–2.3) mm, ♀: 2.4 (2.0–2.5) mm.

Lamprochromus bifasciatus is widespread in England north to North Yorkshire and Lancashire, with sparse Welsh records. An isolated record from north Scotland needs confirming. It is associated mainly with seepages, swampy water margins, tall reedbeds and fen scrub or carr. Many sites are shaded. Flight: vi–ix, peak in vii.

Lamprochromus kowarzi **Negrobov and Chalaja, 1988**

Among its Palaearctic congeners, this species is conspicuous in having the lower half of its thorax yellow, large yellow patches on the second and third tergites in the female as well as the male, and

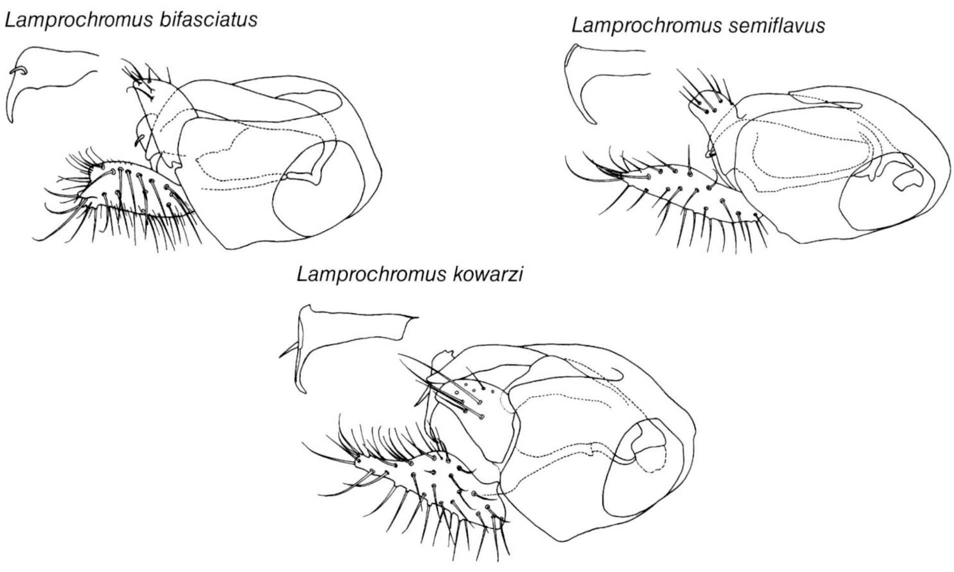

Fig. 21. *Lamprochromus* hypopygia and tip of phallus in lateral view.

the female having a partly or entirely pale antennal scape. The scutum has a vivid pattern of brilliant purple on the midline between the dorsocentral setae flanked by a subshining metallic green and purple-tinged line. The occiput has a pair of conspicuous white dust spots which in the female are more obvious than those of *semiflavus*. The strong apical setae on the front coxae are black, as in occasional *bifasciatus*. Wing length ♂: 2.2 (2.0–2.3) mm, ♀: 2.4 (2.2–2.6) mm.

Lamprochromus kowarzi is a rare species known from coastal sites in West Glamorgan, Devon to Hampshire and Suffolk to Norfolk. The habitat is the transition from upper saltmarsh to non-saline pasture where *kowarzi* is found at water margins of pools, ditches or remnant saltmarsh channels (Drake 2017). Most sites are next to estuarine saltmarsh but rarely on true upper saltmarsh. Flight: vi–viii, peak in vii.

***Lamprochromus semiflavus* (Strobl, 1880) (Plate 5)**

Males are readily identified by their large broadly rounded and evenly haired postpedicel. Females are very similar to those of *bifasciatus* but have two clear pale occipital dust spots, and usually the scutum has a less obvious shining purple midline stripe between the dorsocentral setae but instead has a narrow dark line along the acrostichal row in front of the suture. Wing length ♂: 2.3 (2.1–2.4) mm, ♀: 2.5 (2.3–2.7) mm.

This is a rare species of freshwater wetlands, recorded from Devon, Oxfordshire, Cambridgeshire and Sussex. It may be associated with freshwater reedbeds or lush water-side vegetation, although some sites are coastal. It was previously recognised in Britain under the name *strobli* Parent, 1925 (Hodge 1992). Flight: vi–vii, peak in vi.

LIANCALUS Loew, 1857 (Hydrophilinae)

Liancalus virens is the only species of the genus found in Europe. It is a large green slender-legged species with all femora equally narrow, and with weak short setae on the legs (Plate 5). The wings are narrow with distinctive markings. Crossvein dm-m is strongly oblique and M_4 very short. The thorax has pronounced purple or coppery stripes and each tergite is conspicuously banded, heavily dusted green in front and shining purple-coppery behind. It is distinct from other British dolichopodids in having three pairs of equally strong scutellar setae and a finger-shaped projection on the anepisternum behind the front coxa.

***Liancalus virens* (Scopoli, 1763) (Plate 5)**

This is one of the largest British dolichopodids. Wing length ♂: 6.8 (6.5–7.0) mm, ♀: 7.1 (6.7–7.4) mm.

It is widespread in Britain north to Orkney and moderately frequent in the west and north where rock exposures are more prevalent. It is strongly associated with trickles on wet rocks and waterfalls, and is often found at vertical rock-faces on the coast and at stony rivers and streams. On the soft geological formations of southeast England, it has been recorded at permanent trickles on urban brick-work (Hanson 1986, Sivell 2016). Its mating behaviour was described by Crossley (1988) who included two previous short accounts. Flight: i–xii, all year but more frequent in midsummer.

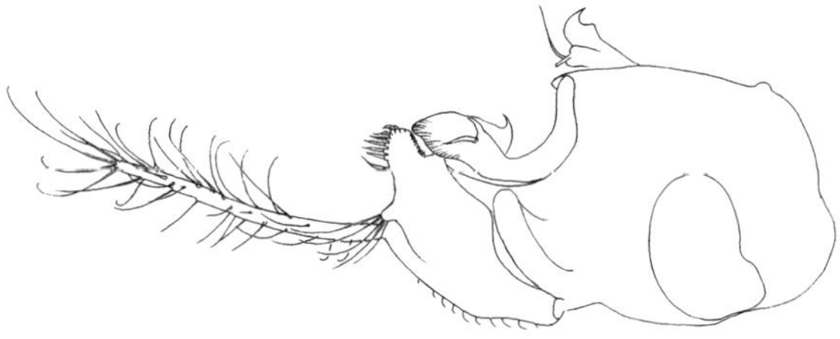

Fig. 22. *Liancalus virens* hypopygium.

MACHAERIUM Haliday, 1832 (Hydrophilinae)

The one British species of this genus is immediately recognised by its postpedicel having a bulbous base and exceptionally long tapered apical section that runs almost imperceptibly into the fine apical arista. Should the antennae be missing, *Machaerium* will be recognised by the combination of its large palps, robust femora with the hind one having two to three preapical setae and unusually long vestiture, and some ventral setae on the front tibia (Plate 5). The sexes are very similar externally as the hypopygium is almost entirely hidden; males have long pulvilli, those on the front leg being as long as the fifth tarsomere (much shorter in females), and the face is clearly narrow below the antennae and about 1.5 times wider at the mouth, compared to more parallel-sided in females and barely 1.25 times wider at the mouth than below the antennae.

Of the three Western Palaearctic species, *maritimae* is the only one found in the northern half of Europe.

Machaerium maritimae Haliday, 1832 (Plate 5)

This is a medium-sized shining green species with a purple or coppery hue, a conspicuous shifting dust pattern on the thorax, paired dust spots on each tergite, and almost entirely yellow legs. Wing length ♂: 4.1 (3.6–4.5) mm, ♀: 4.4 (3.5–5.0) mm.

It is common on the coasts of England and Wales but infrequent in the northern half of Scotland. It lives on upper saltmarsh and the margins of mudflats and estuarine creeks, and less often on other brackish sites such as lagoons and coastal grazing marsh. Flight: v–x, peak in vii.

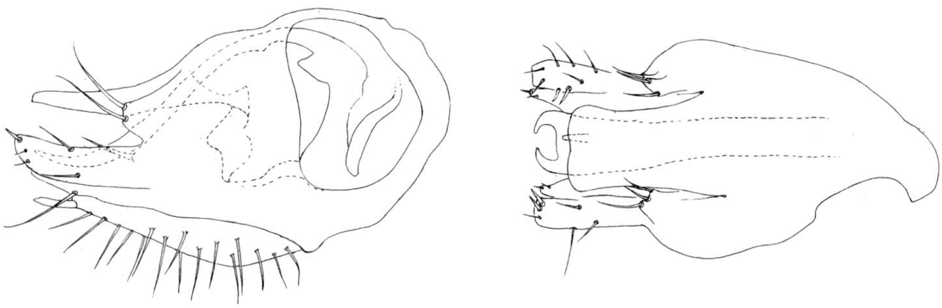

Fig. 23. *Machaerium maritimae* hypopygium, lateral and ventral views.

MEDETERA Fischer von Waldheim, 1819 (Medeterinae)

Medetera are tiny to medium-sized flies with black to green bodies, usually dulled by brownish dusting but sometimes shining (Plate 5). As in other Medeterinae, the prescutellar depression is large and well-marked, the occiput strongly concave so that the head fits closely to the thorax, and tergite 7 of males forms a peduncle carrying the large ellipsoidal genital capsule with elongate surstyli. *Medetera* differs from other Medeterinae in having black setae together with a distinctive venation where R_{2+3} and M_1 beyond dm-m are straight or gently bowed and run almost parallel to one another, with the intervening R_{4+5} running diagonally between them. The thorax is more rectangular than triangular in profile with the mid and hind coxae placed far back and arising under or behind the scutellum; this shape is probably related to the flies' posture. There are no preapical setae on the mid and hind femora but some species have a row of anterior setae on the hind femur which may be misconstrued as these. The tibiae have very few setae, apart from a pair (which may be absent) at the basal quarter of the mid tibia, some small dorsal setae on the hind tibia and small to tiny apical setae on all tibiae.

In the field the genus is easy to recognise owing to the flies' characteristic stance but identification to species is often difficult. The world fauna is at least 300 species of which about half are Palaearctic and many are still being described from Europe. The British fauna is poorly known, as shown by the rise in the number of species from 29 included by d'Assis-Fonseca (1978) to 43 species in the present work, although there is some doubt about at least two of these (*nitida*, *pseudoapicalis*). As many other species are probably awaiting discovery in Britain, the key is an interim attempt until a more detailed treatment of the Palaearctic fauna is published. Negrobov and Naglis (2016) provide the most recent key to the Palaearctic fauna based on the original detailed work of Negrobov and Stackelberg (1972, 1974a, b) and Negrobov (1977) which remains an essential companion to the updated keys as it includes more figures. Most male *Medetera* have no or only obscure secondary sexual characters and moderately variable colour but do have conspicuous genitalia whose examination often becomes essential for identification. The genitalia usually require mounting and viewing at high magnification. Sometimes both ventral and lateral views are needed so these should be examined before mounting the genitalia in a permanent medium. Males but not females are keyed here owing to the reliance on the male genitalia for identification and the poor knowledge of the females of many species, to which no adequate key has been written.

Adult *Medetera* of many species sit on vertical surfaces of tree trunks, rocks and walls where they are conspicuous by their outward-leaning posture, leading to a suggested common name of woodpecker flies (Bickel 1985). They are capable of running in any direction while still facing upwards. The adults prey on small arthropods whereas the larvae of many are known to be predators of the immature stages and emerging adults of scolytid bark beetles (Coleoptera: Curculionidae, Scolytinae) and have consequently been well studied as potential biological control agents (Beaver 1966, Grootaert *et al.* 1987, Hulcr *et al.* 2005). However, not all species of *Medetera* may feed on wood beetles as some, such as *belgica*, *micacea* and *petrophiloides*, are confined to open dry tree-less habitats such as coastal dunes and dry heaths. A wider range of beetle families and Nematocera is listed by Bickel (1985), while Pollet *et al.* (2011) suggest that the scolytid-feeding lifestyle may be restricted to the *signaticornis*–*pinicola* species-group of Bickel (1985) that is mainly associated with conifers, and that this may not be representative of many other *Medetera* species.

Key to species of *Medetera* (males)

1 Mid tibia without setae dorsally near base. Dorsal surstylus bent at almost a right angle (a), cercus with 2–3 apical projections (a). Hypandrium with paired outwardly pointing hooks (b). (Anterior dorsocentral setae short; halter with a dark knob; all postocular setae black.) 2

– Mid tibia with at least a posterodorsal seta near base. Dorsal surstylus not so strongly bent. Cercus usually with only 1 or no obvious apical projection. Hypandrium without paired hooks. 4

(a)

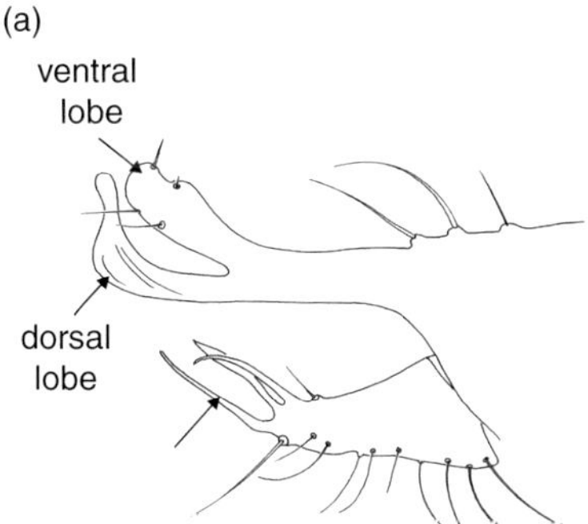

(b)

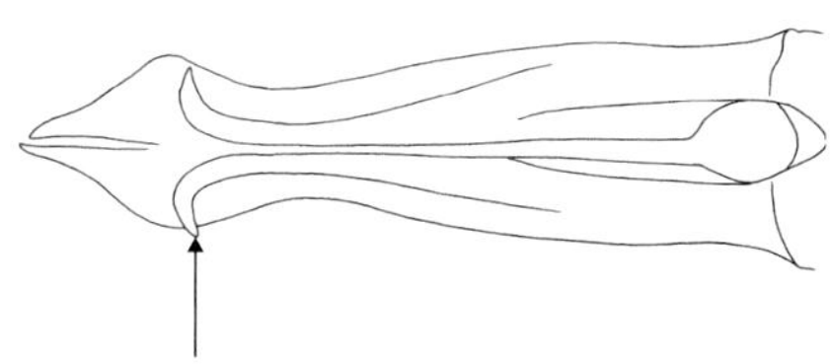

2(1) Phallus tip beyond sheath short, in lateral view kinked with a tiny tooth on dorsal edge (a). Cercus with 3 apical projections (b). Tibiae and basitarsi usually completely dark, but both can be pale, knees not or scarcely differentiated. .... ***belgica***

– Phallus tip long and slender. Cercus with 3 or 2 apical projections. Tibiae and basitarsi pale yellow, succeeding tarsomere becoming progressively darker, knees narrowly but distinctly yellow on femora. 3

(a)

(b)

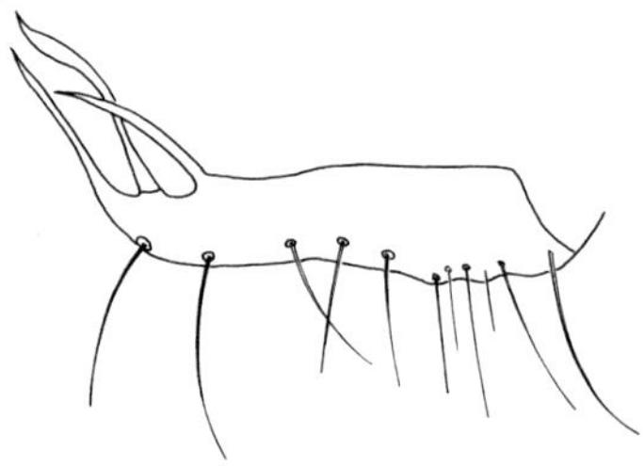

3(2) Sheath of hypandrium ending in a finger-shaped extension, phallus slender and unadorned where it emerges from sheath (a). Cercus with 3 apical projections (b). .... ***peloria***

– Sheath without an extension at tip, phallus with a tiny tooth where it emerges from sheath (aa). Cercus with 2 apical projections (bb). .... ***muralis***

(a)

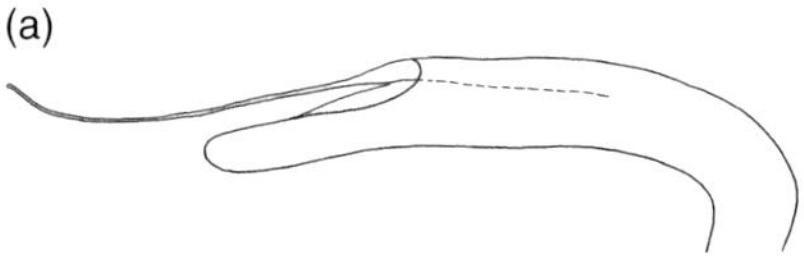

(aa)

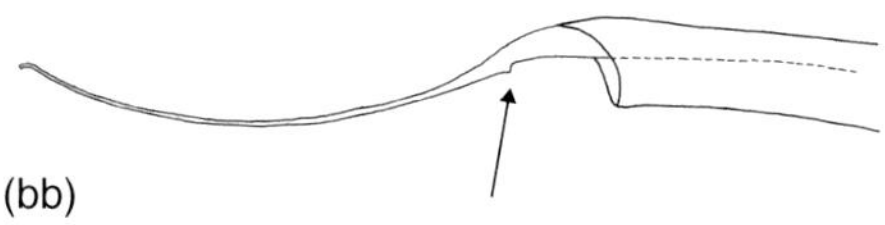

(b)

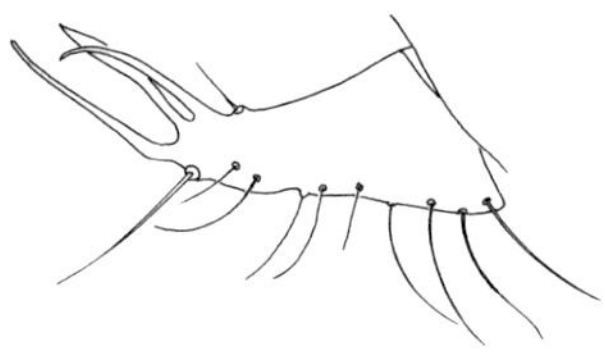

(bb)

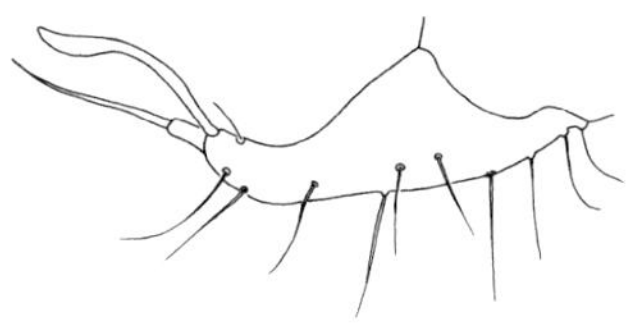

4(1) Scutellum with no small lateral seta, only 1 pair of strong apical setae. 3 dorsocentral setae. Epandrial lobe with 2 lanceolate setae, visible in dry material (figs in next couplet) 5

– Scutellum with 2 pairs of setae, the lateral ones may be very small. 3 to many dorsocentral setae. Epandrial lobe usually with simple setae, only rarely slightly broadened 6

5(4) Surstylus in ventral view on inner margin at its midpoint with an angular to pointed outline (a). .... ***micacea***

– Surstylus inner margin gently rounded (aa). .... ***mixta***

(a)

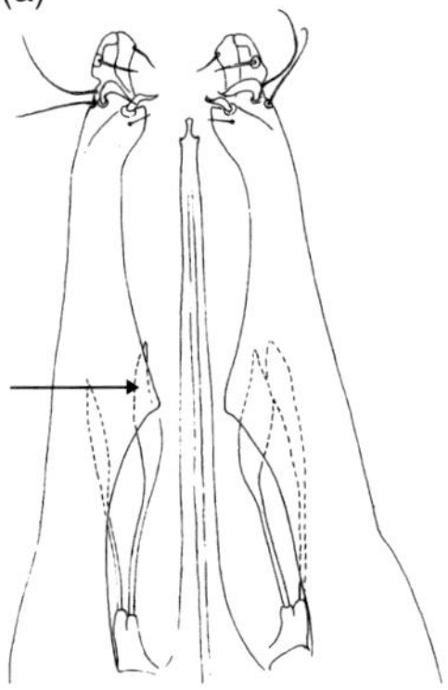

(aa)

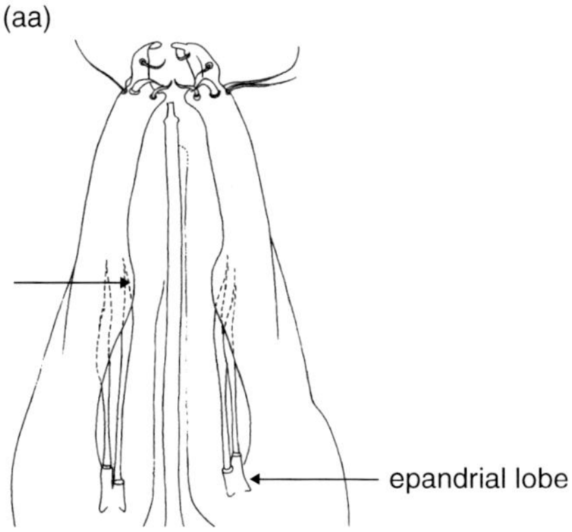

Note: These appear to be variation within a single species but are kept separate here in case further characters confirm their distinction.

6(4) Dorsocentral setae not decreasing in length towards front, the presutural seta strong (a) and rarely (*jacula*, *petrophila*) with a shorter strong seta in front of it (b); otherwise with marked contrast between the presutural dorsocentral seta and any tiny hairs in front of it. Acrostichal setae pale or at least not black, small and more-or-less equal in length................7

– Dorsocentral setae decreasing in length towards front, usually with no strong presutural dorsocentral (aa); if there is a small presutural seta it is close to suture and there are other still smaller setae in front of it. Acrostichal setae usually dark, usually increasing in size posteriorly. 13

(a)

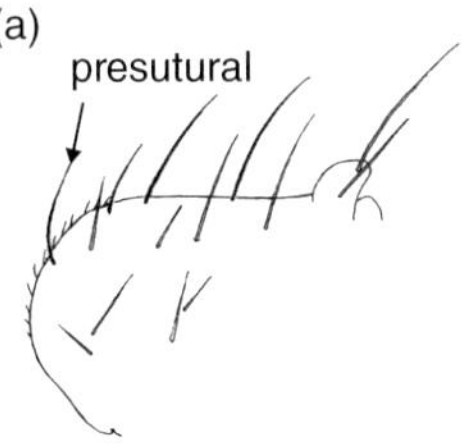

(b)

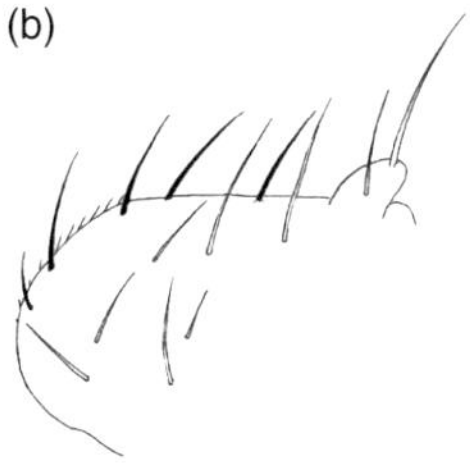

(aa)

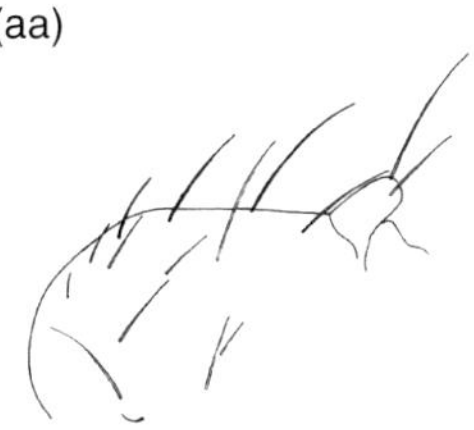

7(6) Legs almost entirely yellow, front coxa pale brown. Surstylus (a). If apical section of M_4 more than twice as long as dm-m, see *insignis*, couplet 17................ ***flavipes***

– Legs dark, at most knees yellowish................ 8

(a)

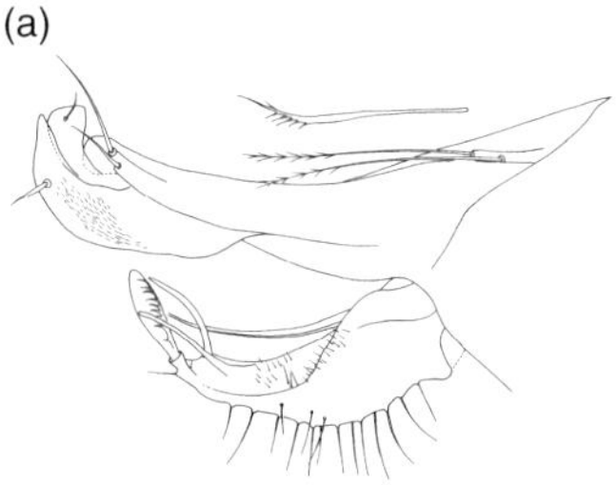

8(7) Surstylus elongate, two-thirds or more length of epandrium (a, b). Epandrium slender, its depth two-thirds or less length of surstylus (a, b)................ 9

– Surstylus short, less than half length of epandrium (aa, bb). Epandrium stout, deeper than length of surstylus (aa, bb)................10

(a, *truncorum*)

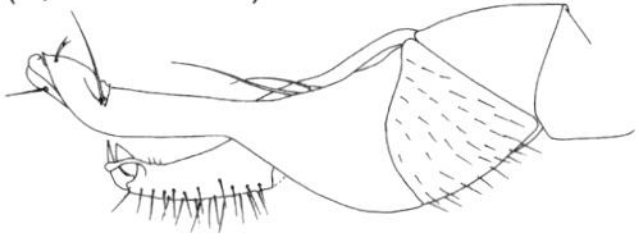

(b, *petrophiloides*)

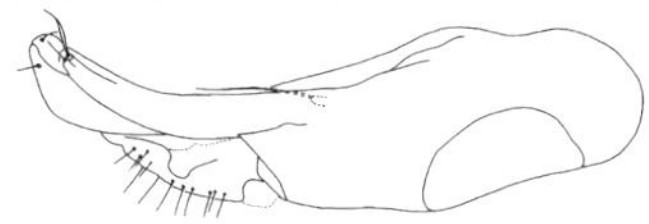

(aa, *jacula*)

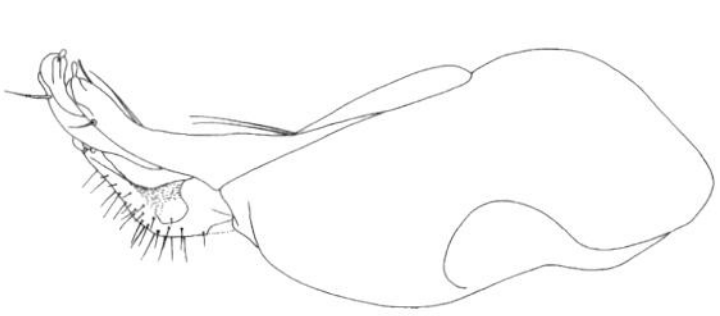

(bb, *saxatilis*)

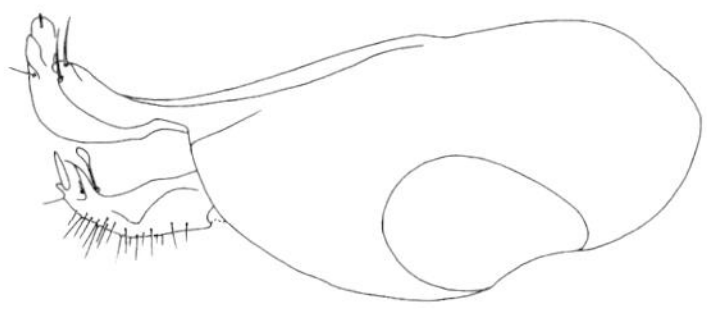

9(8) 3 long dorsocentral setae, with or without a clearly shorter second seta (a). Apical section of M_4 usually at least as long as dm-m (b). Clypeus green, shining on mid third, dusted on each lateral third. .... ***truncorum***

– 4 almost equally long dorsocentral setae (aa). Apical section of M_4 shorter than dm-m (bb). Clypeus shining or glossy violet or bluish green, dusted only at extreme sides. .... ***petrophiloides***

(a)

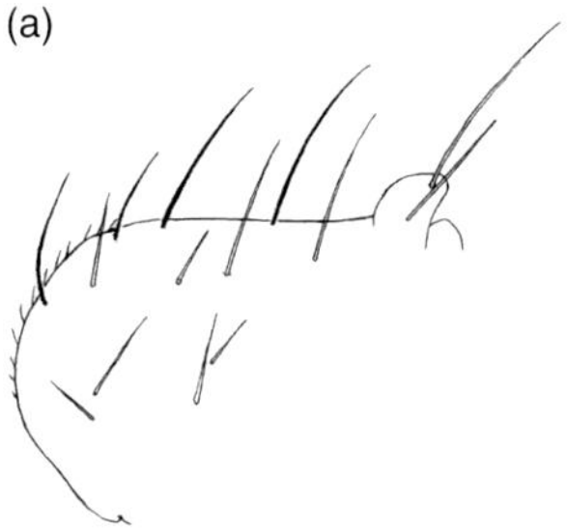

(b)

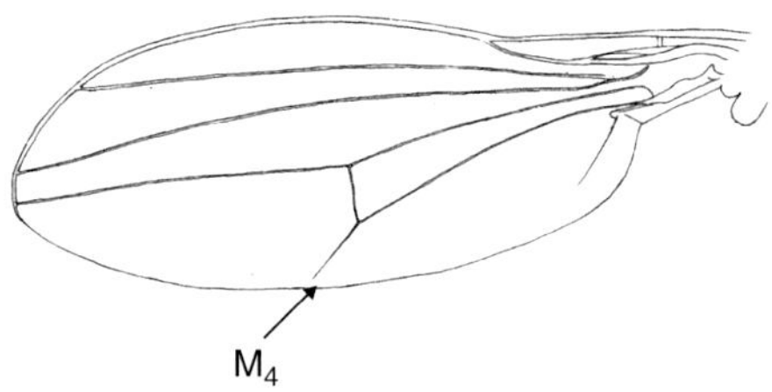

(aa)

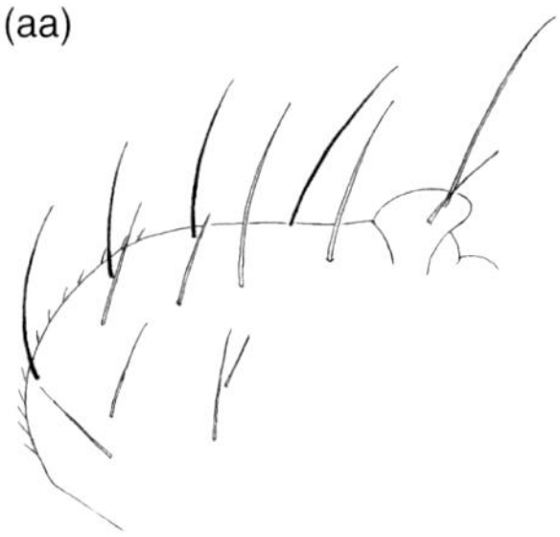

(bb)

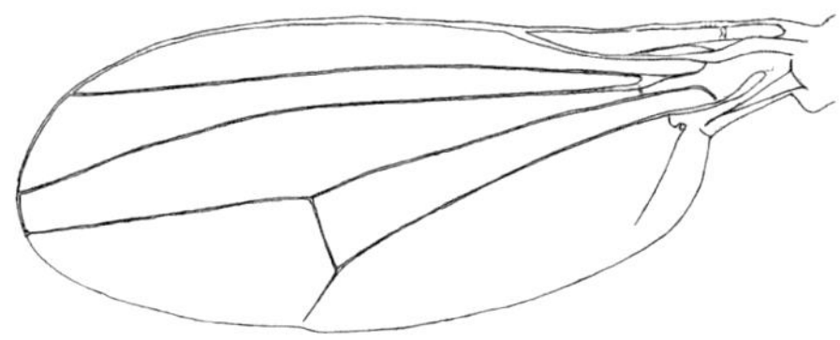

10(8) 1 presutural dorsocentral seta, with or without a small postsutural seta (a). 11

– 2 presutural dorsocentral setae, of which the front one may be small; all postsutural setae long (aa). 12

(a)

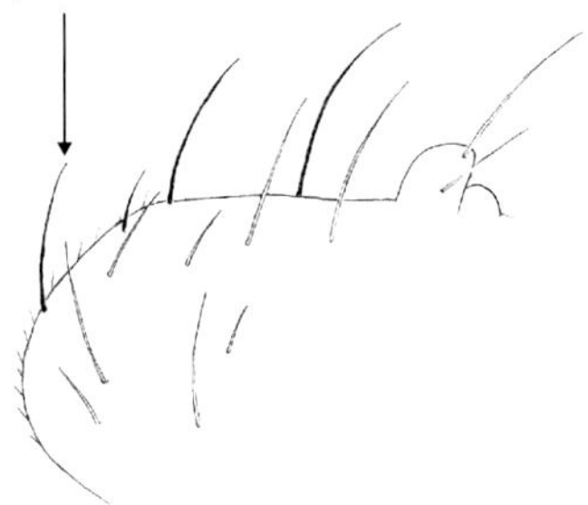

(aa)

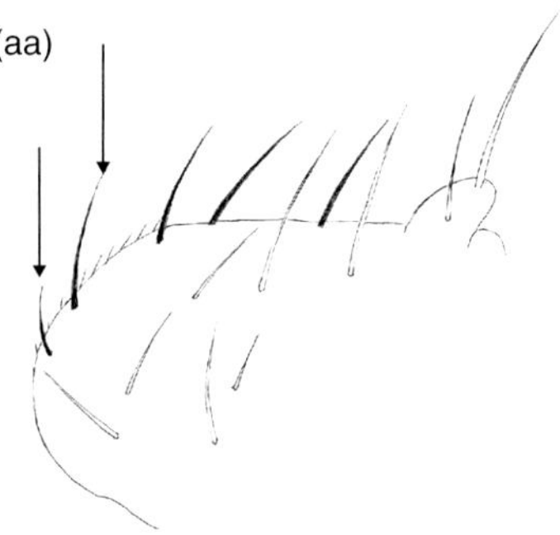

11(10) Cell r_{4+5} at crossvein dm-m about twice as wide as at its tip (a). Frons and upper part of epistoma dusted brownish. Hind femur with conspicuous white anteroventral setae near tip, as long as width of femur at its tip. Clypeus dusted to a third of its width each side. Postpedicel as long as deep. Hind basitarsus on posterior side at base with a brown flange rounded on apical margin (b). Surstylus with 2 long ventral setae on ventral lobe near tip (c). .... ***saxatilis***

– Cell r_{4+5} at crossvein dm-m about 3 times as wide as at its tip (aa). Frons and upper part of epistoma dusted pale grey. Hind femur with short anteroventral setae near tip. Clypeus only narrowly dusted at sides. Postpedicel 1.25–1.5 times longer than deep. Hind basitarsus on posterior side at base with a black flange evenly humped in profile (bb). Surstylus with 1 spatulate seta on ventral lobe near tip (cc[3]). .... ***dendrobaena***

---

3 Negrobov and Stackelberg 1974a. *Die Fliegen der palaearktischen Region* **4** (5), issue 302, Fig. 498.

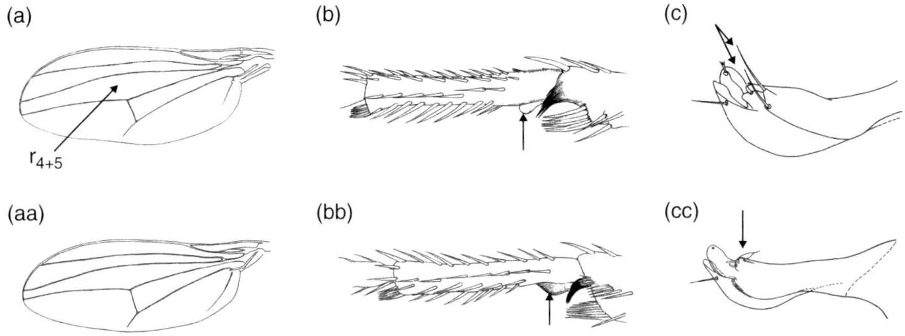

12(10) Hind femur with long pale anteroventral setae near tip (a). Arista subapical (b). Apical section of M_4 very short, angle between dm-m and basal section of M_4 more acute, R_{4+5} and M_1 more strongly converging at wing tip (c). Dorsal lobe of surstylus with a long apical seta, ventral lobe with only a short spatulate ventral seta (d)........ ***jacula***

– Hind femur with shorter anteroventral setae near tip, less than half depth of femur at their insertion and about as long as vestiture on anterior face, anteroventral hairs basal to these shorter than vestiture on anterior face (aa). Arista apical (bb). Apical section of M_4 relatively longer, angle between dm-m and basal section of M_4 approximately square, R_{4+5} and M_1 less strongly converging at wing tip (cc). Dorsal lobe of surstylus without a long apical seta, ventral lobe with a long ventral seta (dd[4])........ ***petrophila***

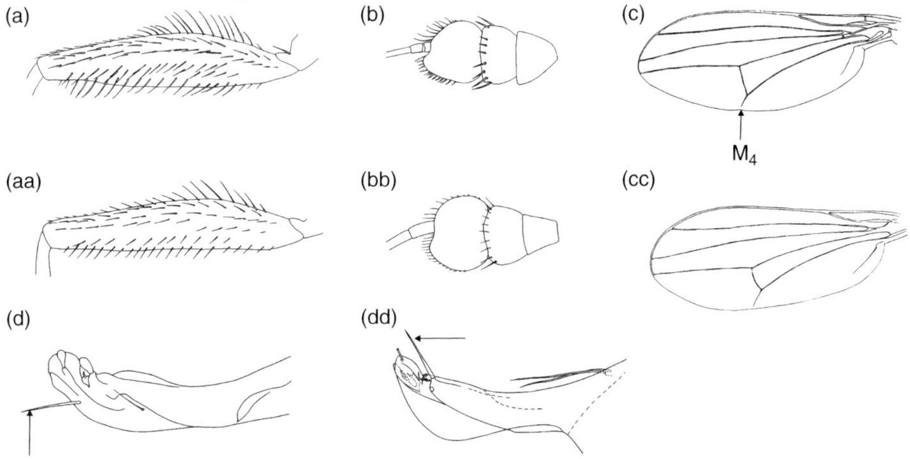

13(6) M_4 with a strong thickening at base (figs in next two couplets)........ 14

– M_4 not or only slightly thickened at base........ 16

[4] Negrobov and Stackelberg 1974b. *Die Fliegen der palaearktischen Region* **4** (5), issue 303, Fig. 749.

14(13) Thickening of M_4 with a blunter tip and about one-third length of basal section of this vein (a[5]). R_{4+5} evenly curved posteriorly near tip (a). Hypandrium narrow throughout and without spinules (b[6]). .... ***freyi***

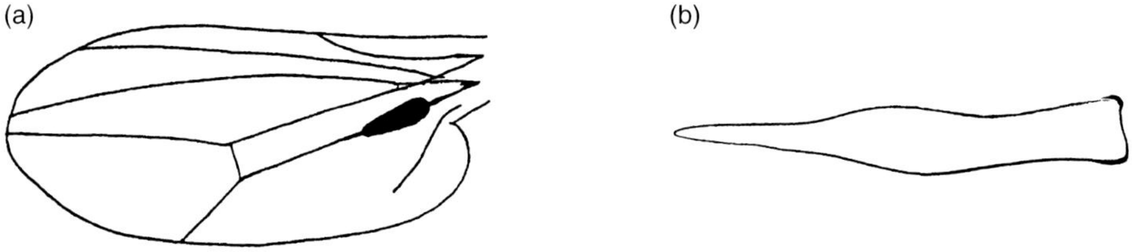

– Thickening of M_4 tapered and at least half length of basal section of this vein (figs in couplet 15). R_{4+5} straight or slightly curved anteriorly near tip. Hypandrium clearly broad at tip or middle (figs in next couplet). 15

15(14) Dorsal lobe of surstylus strongly angled (a[7]). Hypandrium with spatulate tip (b). Phallus slender, without spinules but with lateral spikes (c). R_{4+5} evenly curved to tip (d). Greyer species with pale tibiae, at least mid tibia yellow dorsally. Calypter margin yellow .... ***excellens***

– Dorsal lobe of surstylus almost straight and aligned with main axis of surstylus (aa[8]). Hypandrium greatly broadened on basal half, contracting towards tip, its sides covered in minute spinules (bb). Phallus with minute spinules (cc). R_{4+5} slightly recurved at tip (dd). Darker species, usually with dark tibiae. Calypter margin dark .... ***inspissata***

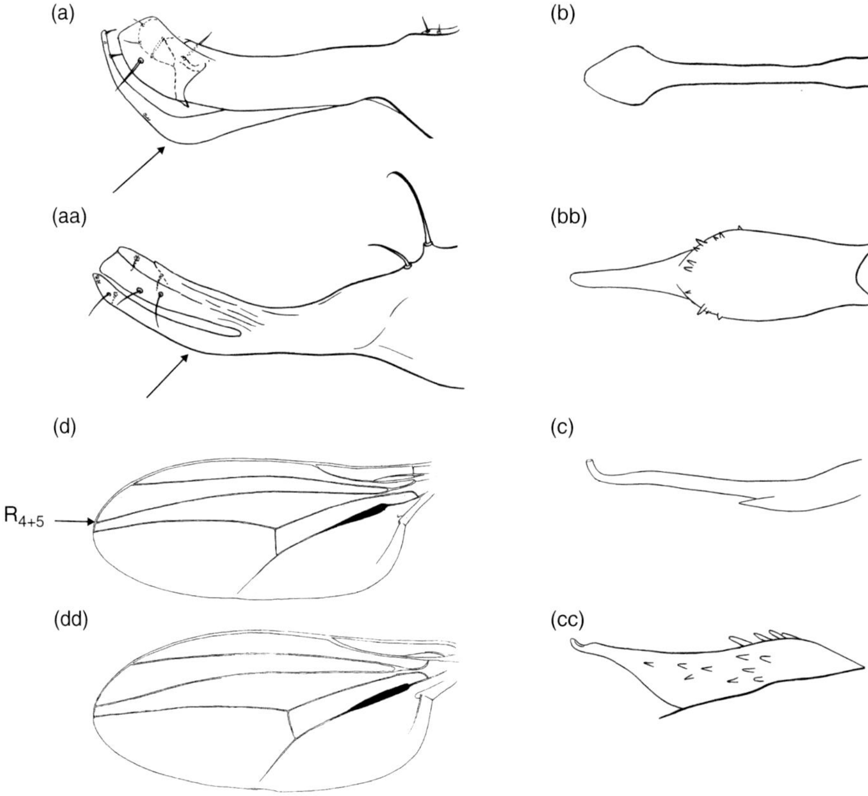

[5] MacGowan 2001. *Dipterists Digest (Second Series)* **8**, 85–90.
[6] Negrobov and Stackelberg 1974a. *Die Fliegen der palaearktischen Region* **4** (5), issue 302, Fig. 537 (part redrawn).
[7] Negrobov and Stackelberg 1974a. *Die Fliegen der palaearktischen Region* **4** (5), issue 302, Figs 511–513.
[8] Negrobov and Stackelberg 1974a. *Die Fliegen der palaearktischen Region* **4** (5), issue 302, Figs 588, 590, 592.

16(13) Scape and often pedicel yellow or reddish. 17
– Antenna entirely black. 23

17(16) Legs almost entirely yellow. Apical section of M_4 at least twice as long as dm-m. 2 pale proepisternal setae. [Surstylus broad with thick apical lobes (a[9]). Hypandrium gradually widening for half its length (b). Phallus gently curved at tip (c). Cercus with apical seta greatly expanded at tip (a).] .... ***insignis***
– Legs mainly black. Apical section of M_4 at most 1.5 times as long as dm-m. At least 4 pale or dark proepisternal setae. 18

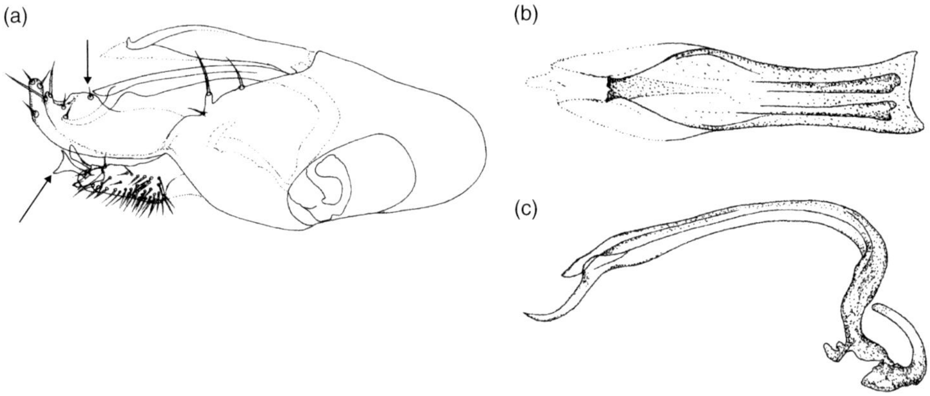

18(17) Halter dark, at least distinctly brownish. Apical section of M_1 almost straight. Hypandrium tip hammer-shaped (a, b[10]). Hind tarsomere 2 about 1.25–1.5 times as long as basitarsus. 19
– Halter yellow. Apical section of M_1 distinctly curved (cc). Hypandrium tip slightly swollen or parallel-sided but not strongly transverse. Hind tarsomere 2 about 1.75–2.0 times as long as basitarsus. 20

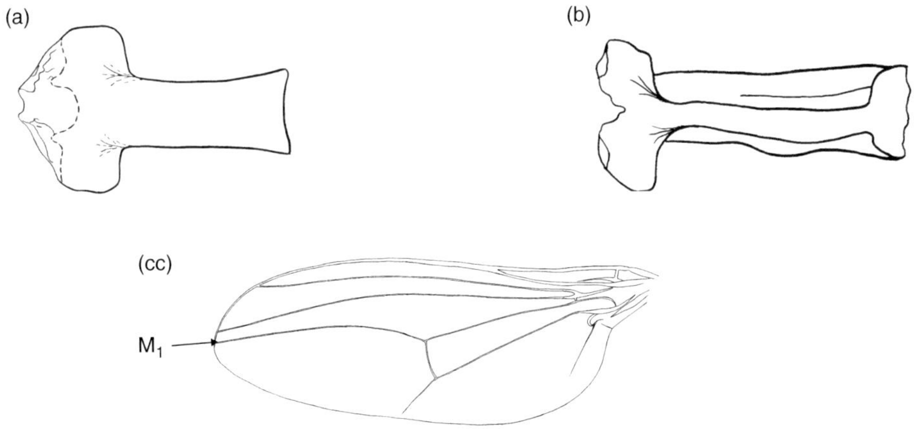

[9] Gibbs 2007. *Dipterists Digest (Second Series)* **14**, 95–97.
[10] Negrobov and Stackelberg 1974a, b. *Die Fliegen der palaearktischen Region* **4** (5), issue 302, Fig. 518; issue 303, Fig. 816.

19(18) Dorsal lobe of surstylus curved and longer than ventral lobe, visible without preparation (a[11]). Phallus with very shallow apical incision (b[12])..***fasciata***

– Dorsal lobe of surstylus straight and shorter than ventral lobe (aa). Phallus with deep incision at tip (bb)...***setiventris***

(a)

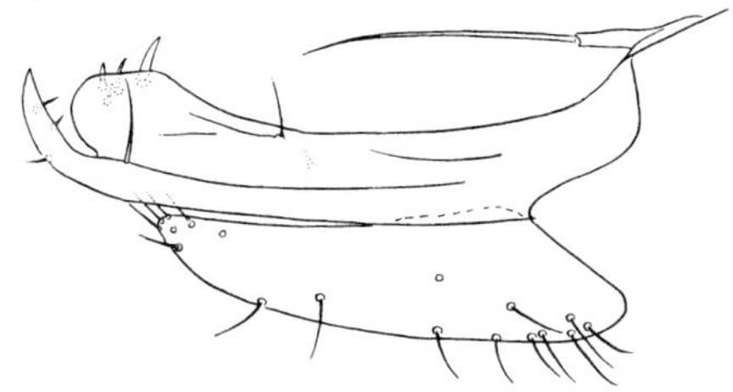

(b)

(aa)

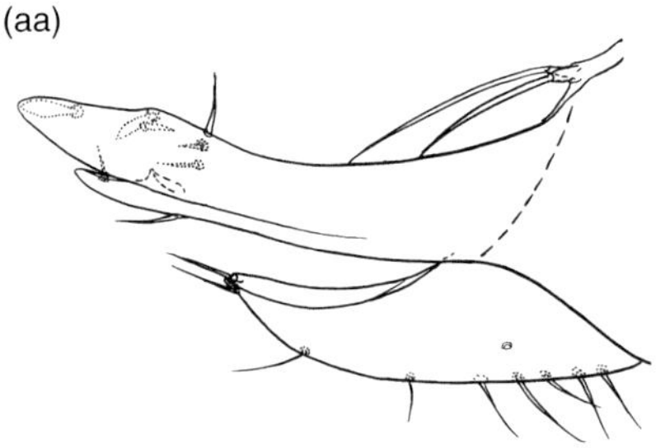

(bb)

20(18) Hind femur with anterior setae scarcely distinguished from adjacent vestiture. Front and mid tibiae with white vestiture, most leg setae pale. Hypopygium small, its maximum depth less than length of tergites 5 or 6 (a). Hypandrium broadly spatulate (b)...............................***pinicola***

– Hind femur with a row of long anterior setae at middle of anterior face, more easily seen in dorsal view (similar to fig. a of couplet 24). Front and mid tibiae with vestiture black and most leg setae black. Hypopygium larger. Hypandrium narrow or with only tip expanded............... 21

(a)

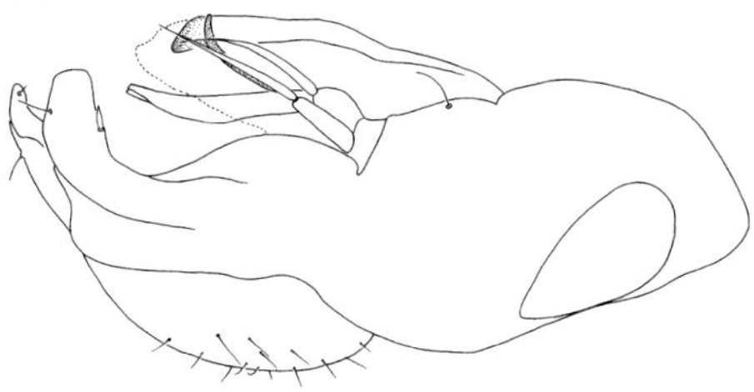

(b)

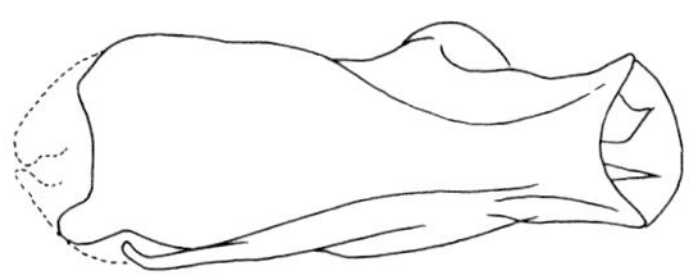

21(20) Hind tibia near tip with 2–3 long brown or pale dorsal setae, longer than width of tibia (a). Hind femur with no obvious anteroventral setae near tip, only short fine hairs. Front femur without posteroventral setae. Hind tarsomere 2 about twice as long as basitarsus. Arista length about two-thirds height of an eye. [Hypopygium very large, its maximum depth at least 1.5 times length of tergites 5 or 6. Surstylus base protruding ventrally as rounded hump (b). Hypandrium narrow and deep with a pair of asymmetrical horns at sides (c, d).]......................................***obscura***

[11] Negrobov and Stackelberg 1974a, b. *Die Fliegen der palaearktischen Region* **4** (5), issue 302, Fig. 514; issue 303, Figs 815, 817.

[12] MacGowan 2001. *Dipterists Digest (Second Series)* **8**, 85–90.

– Hind tibia near tip with only short black setae dorsally, as long as width of tibia (aa). Hind femur with long anteroventral setae near tip, almost equal to depth of femur near its tip. Front femur with long posteroventral setae, longer than width of tibia. Hind tarsomere 2 about 1.5 times as long as basitarsus. Arista almost as long as height of an eye. 22

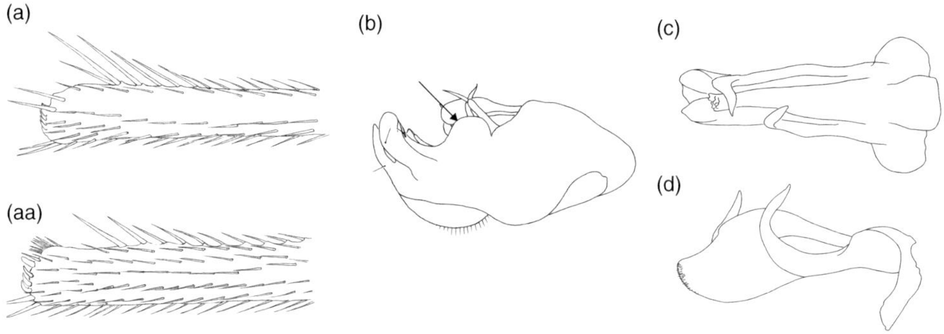

22(21) Hind tibia at tip anterodorsally with 2 short curved spine-like setae, close together (a). Face mainly shining with only narrow dusting at sides, at suture and below antennae. Surstylus with a strongly protruding dorsal lobe (b). Hypandrium tip rounded diamond-shaped (c). Phallus tip spear-shaped (d). .... ***bispinosa***

– Hind tibia at tip anterodorsally with 1 short strong curved spine. Face with conspicuous dusting at sides, suture and below antennae. Surstylus with apparently only 1 simple lobe (bb[13]). Hypandrium tip broad and rounded (cc). Phallus tip diamond-shaped (dd). .... ***nitida***

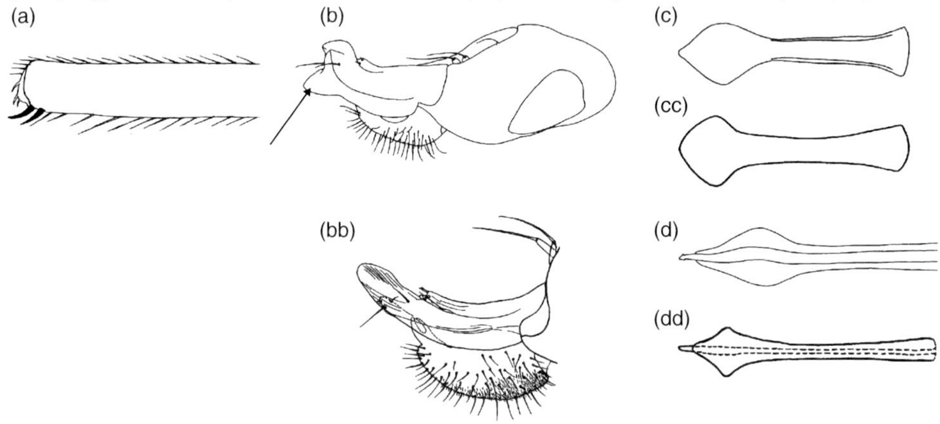

[13] Negrobov and Stackelberg 1974b. *Die Fliegen der palaearktischen Region* **4** (5), issue 303, Figs 691, 692, 699.

23(16) 1 notopleural seta. Mid tibia with 1 posterodorsal seta on basal quarter, no anterodorsal seta opposite it (very rarely present) (a). Tibiae yellow. Few acrostichal setae (4 or 5 pairs), almost no hairs on front of scutum, a few extra short hairs in front of anterior intra-alar seta. [Cercus with long apical seta-like extension (b). Hypandrium diamond-shaped at apex (c). Phallus tip with dorsal indentation (d).] .... ***unisetosa***

– 2 notopleural setae, the upper one often small. Middle tibia with both antero- and posterodorsal setae near base. Tibiae usually dark but sometimes yellow. Acrostichal setae more numerous, with many small hairs on front of scutum.......... 24

(a)

(c)

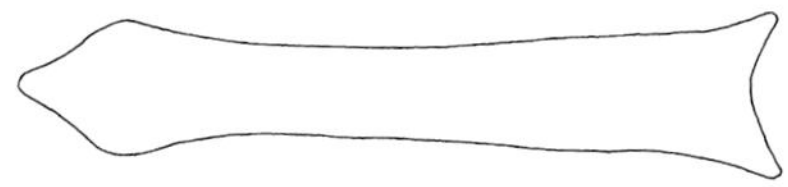

(b)

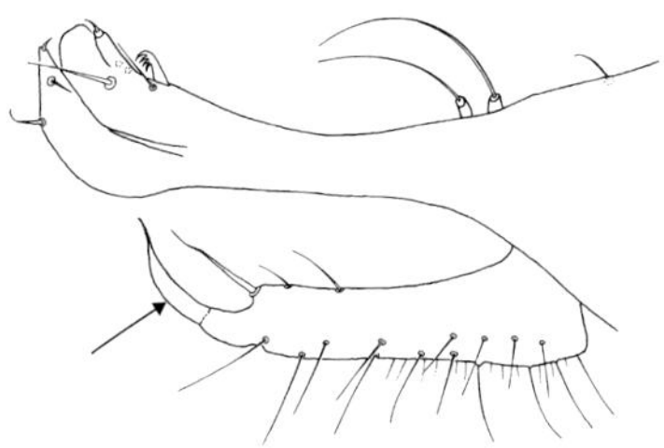

(d)

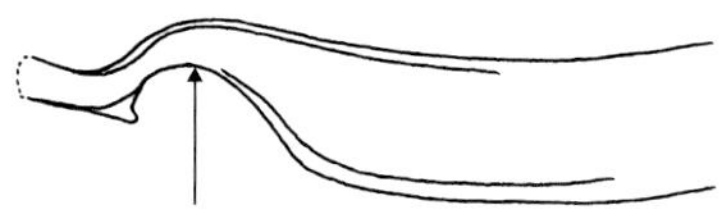

24(23) Hind femur with a row of about 6 long anterior setae on apical third, clearly distinct from surrounding vestiture, stouter and at least as long as anteroventral hairs near tip of femur (a), more easily seen in dorsal view. Apical section of M_4 usually more than 1.7 times as long as dm-m (b), but less in *feminina* (fig. a of couplet 25).......... 25

– Hind femur without differentiated anterior setae, all vestiture on anterior face similar but anteroventral hairs may be slightly longer. Apical section of M_4 not or scarcely more than 1.5 times as long as dm-m (fig. a of couplet 25).......... 34

(a)

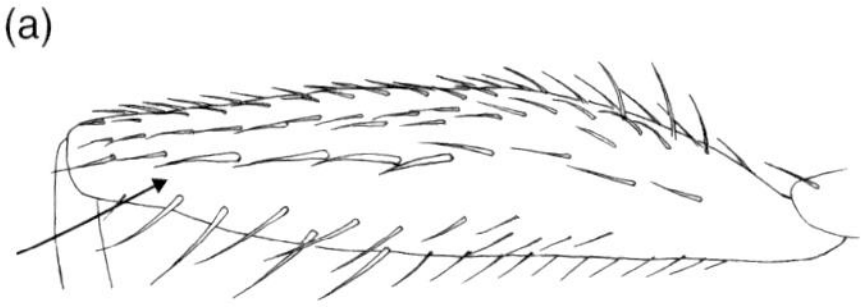

(b)

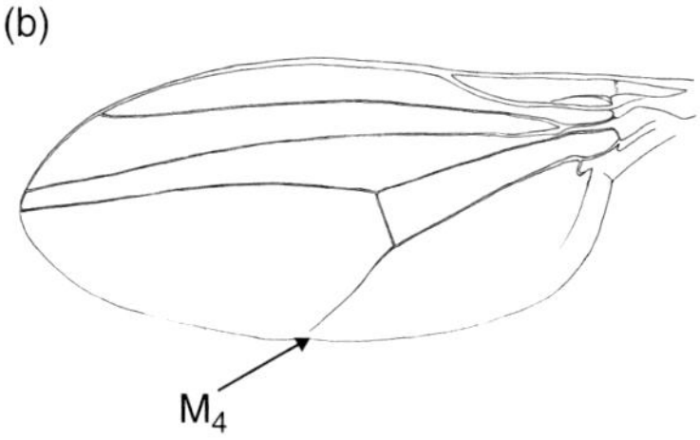

Note: Specimens of *insignis* with scape only obscurely pale fall between these options; go to couplet 17.

25(24) Apical section of M_4 short, about 1.0–1.5 times length of dm-m (a). [Setae of front coxa white; proepisternal seta white and conspicuous; tergite hairs on disc white; hypandrium in ventral view widening before tip and square-ended (b), surstylus and cercus (c, d).] .... ***feminina***

– Apical section of M_4 long, at least 1.75 times length of dm-m. Other characters variable.....26

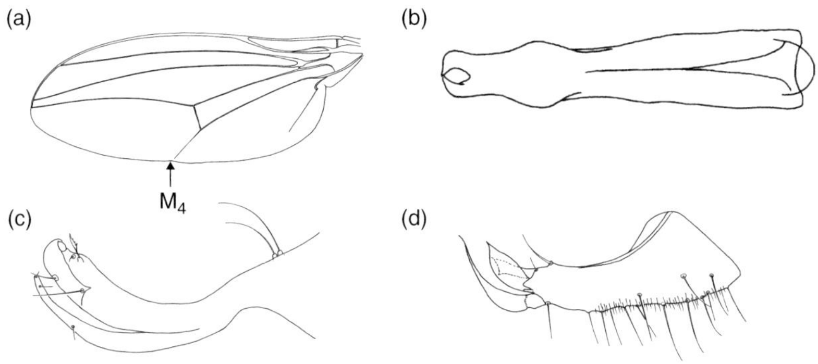

26(25) Setae and hairs on all coxae white. Tergite hairs white. At least mid and hind tibiae yellow or pale brownish dorsally. [Surstylus slender with fine apical lobes (a, c). Hypandrium diamond-shaped at apical quarter (b). Phallus abruptly curled at tip (c). Cercus apical seta long and finely pointed (d).] .... ***pallipes***

– Setae and hairs of coxae black. Tergite hairs black or dark with yellow reflections. Tibiae usually blackish or dark brown but sometimes as above 27

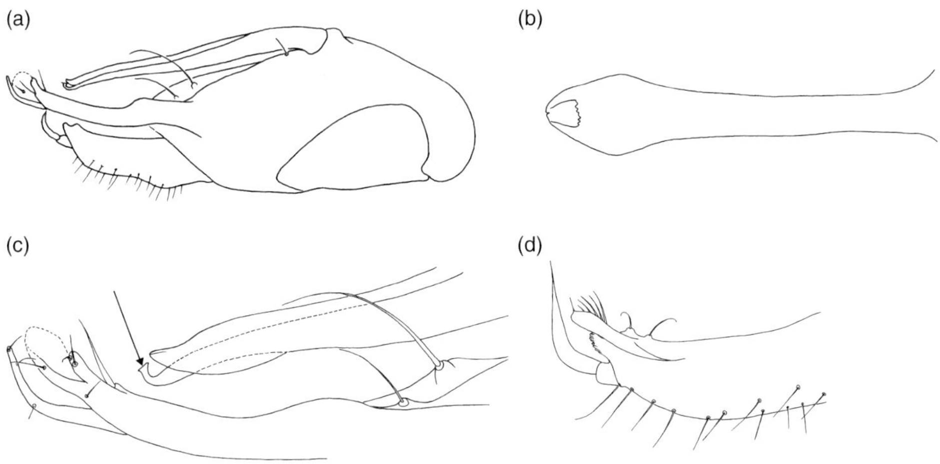

27(26) Halter knob dark, at least on outer face. Lower postocular setae dark. Hind femur with black ventral vestiture. Epistoma black. Calypter hairs dark 28

– Halter yellow. Lower postocular setae white or dark with yellow reflections. Hind femur with white ventral vestiture. Epistoma greenish black. Calypter hairs pale brown to white 29

28(27) Hind femur with both anteroventral setae near tip and anterior setae longer than width of tibia at basal quarter. Phallus forked (a). Hypandrium on apical half gradually widening before a short pointed tip (b), tapered in lateral view. Surstylus not markedly slender, ventral lobe broad and slightly clubbed (d).. ***tristis***

– Hind femur with both the anteroventral near tip and anterior setae shorter than width of tibia at basal quarter. Phallus simple, narrowing to fine curved tip in lateral view (aa). Hypandrium more-or-less parallel-sided in ventral view, tip truncate with a small dark toothed lip set back from apex (bb), visible in lateral view at low magnification (cc). Surstylus with dorsal lobe slender, ventral lobe moderately narrow, parallel-sided and not clubbed towards tip (dd).. ***nigrohalteralis***

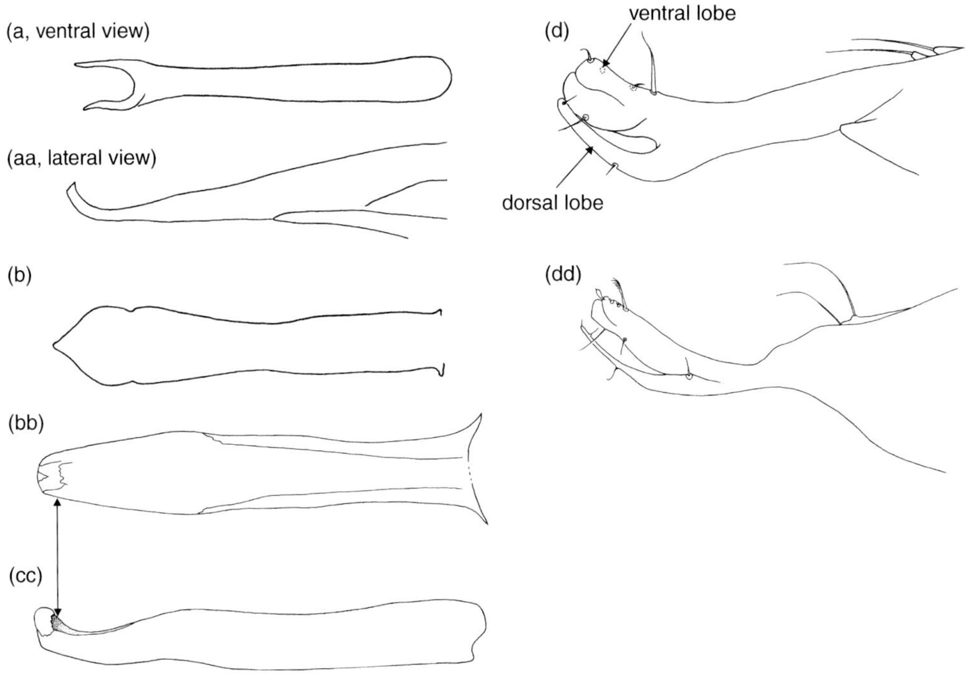

29(27) Hind femur on anteroventral apical quarter or less with short and inconspicuous setulae, vestiture continuing basally as tiny fine white hairs (a, *impigra*; b, *acanthura*)........... 30

– Hind femur on anteroventral apical quarter or more with long anteroventral setae (aa)...... 31

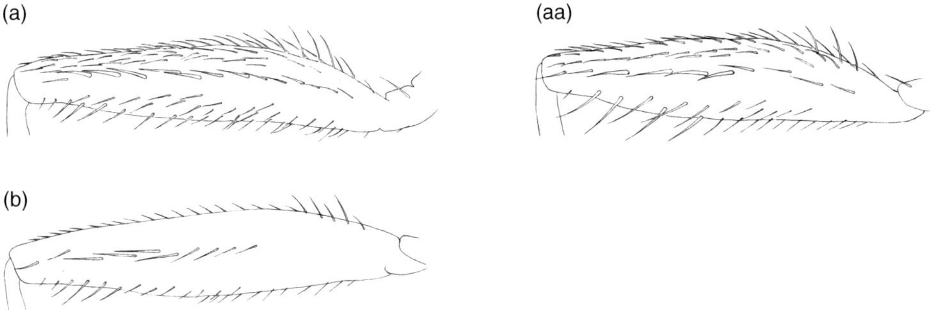

30(29) Mid femur with an untidy row of anteroventral setae, longest near middle of femur (a); if hairs short, possibly *apicalis* (Zetterstedt). Surstylus strongly bowed and with a distinct shoulder where joining epandrium (b). Epandrial lobes both consist of small tubercles scarcely higher than wide, inconspicuous in lateral view (b). [Hypandrium in lateral view with a hump two-thirds from base (c), in ventral view with apical third slightly broader (d).] .... ***impigra***

– Mid femur with anteroventral setae more even in size and spacing (aa). Surstylus gently curved with no shoulder at junction with epandrium (bb). Inner epandrial lobe long and conical, appearing almost as a spine at low magnification (bb). [Hypandrium (dd) as in *impigra*.] .... ***acanthura***

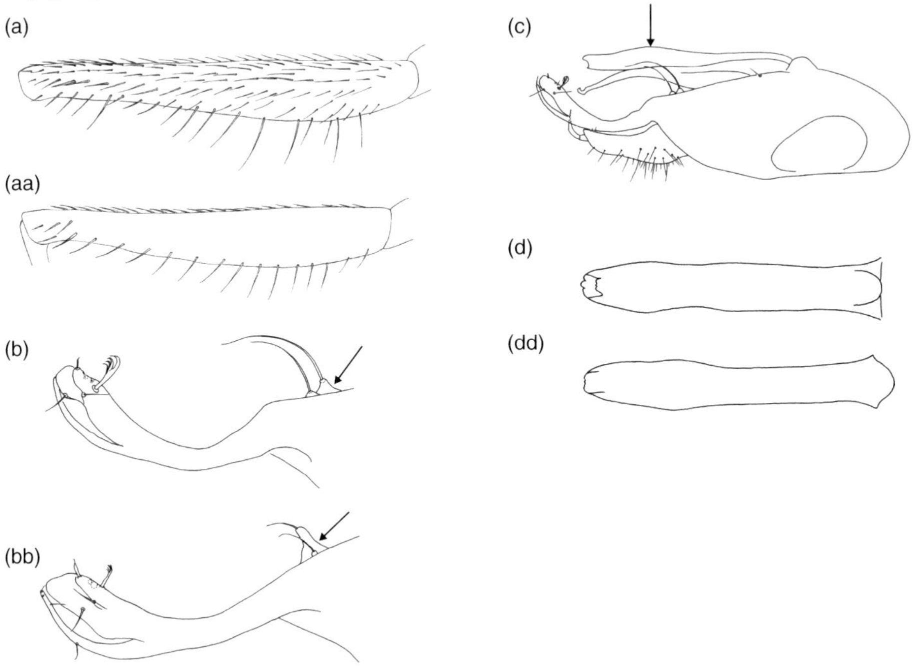

31(29) Lower postocular setae dark but often with yellow reflections........ 32

– Lower postocular setae pale, not just yellow reflection on black ground colour........ 33

32(31) Hypandrium tip spatulate with dark arc near tip (a, b), also visible laterally (c). [Surstylus (d).] .... ***abstrusa***

– Hypandrium parallel-side with an almost square tip without a subapical dark arc, as in *impigra* (fig. d, couplet 30)........ ***pseudoapicalis***

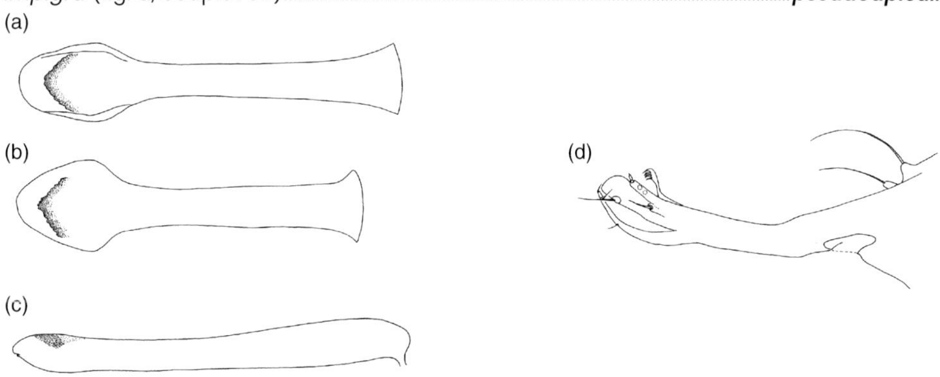

33(31) Tergites dark-haired. Hypandrium with broadly ovate tip (a), with upturned lateral margins appearing in lateral view as a hump at apical quarter (c). Surstylus gently bowed, dorsal lobe not so narrow (c). .... ***borealis***

– Tergites pale-haired. Hypandrium with ventral thorn (aa), seen as a dark spot in ventral view (bb). Surstylus strongly bowed, dorsal lobe slender (cc). .... ***jugalis***

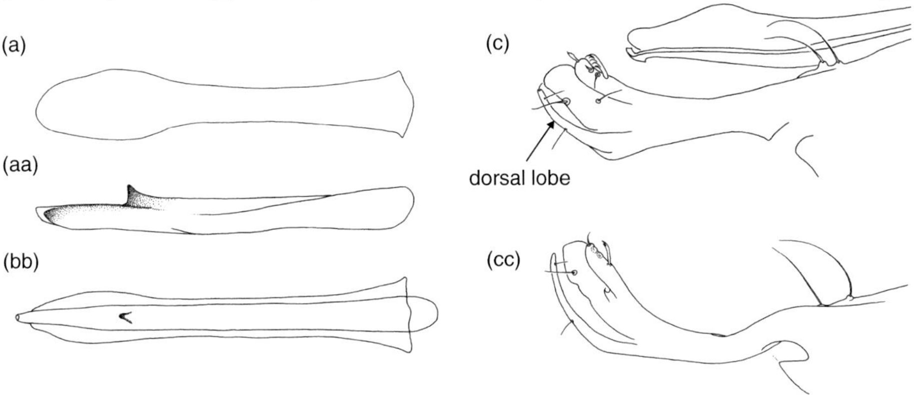

34(24) Clypeus glossy or strongly shining green, blue or violet, only narrowly dusted or dulled at sides. 35

– Clypeus black or dark green, dulled by dust or microsculpture but may be partly shining through microsculpture. 37

35(34) Apical section of M_4 very short, half length of dm-m. Separation of R_{2+3} and R_{4+5} about twice length of next costal section to M_1. Lower postocular setae conspicuous, white, long and dense, as long as antenna, their separation at insertions scarcely more than their width. Acrostichal setae white, small and numerous. Thorax conspicuously and broadly striped. Larger species, 4.2–4.5 mm. [Cercus with diamond-shaped apical seta (a). Phallus and hypandrium long and slender (b).] .... ***diadema***

– Apical section of M_4 at least as long as dm-m. Separation of R_{2+3} and R_{4+5} at least 2.5 times length of next costal section to M_1. Lower postocular setae unremarkable, shorter than antenna, pale to dark, much thinner than their separation at their insertions. Acrostichal setae pale to dark, not particularly small or numerous. Thorax shiny and unstriped. Smaller species, 3.0–3.5 mm. 36

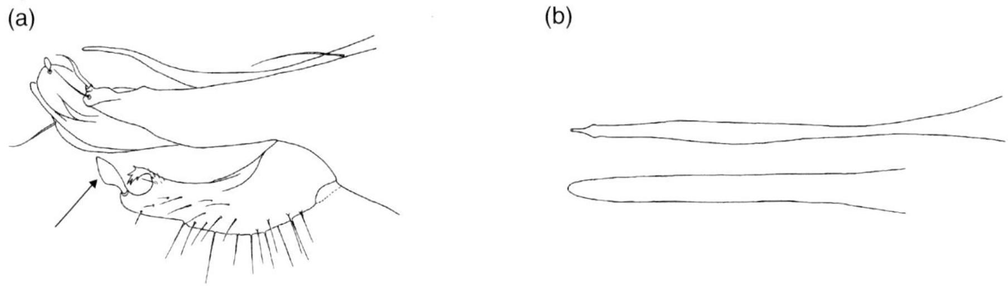

36(35) Apical section of M_4 as long as dm-m (as *infumata*, couplet 39). Separation of R_{2+3} and R_{4+5} about 2.5 times length of next costal section to M_1. Halteres black. Hind tarsomere 2 about 2.5 times length of basitarsus. Clypeus glossy violet. Postpedicel ovate, about 1.5 times longer than deep (b). Mid and hind femora almost devoid of hairs on ventral face.

Epandrium massive, about 3 times longer than surstylus (c). Phallus and hypandrium narrow and finely tapered (d, e), hypandrium tip slightly expanded at high magnification (e). Surstylus (f).. ***ambigua***

– Apical section of M_4 about 1.5 times length of dm-m (aa). Separation of R_{2+3} and R_{4+5} 5–6 times length of next costal section to M_1 (aa). Halteres pale. Hind tarsomere 2 about 1.75 times length of basitarsus. Clypeus green or coppery. Postpedicel shorter than deep (bb). Mid and hind femora with long white anteroventral hairs about as long as width of tibia. Epandrium about twice length of surstylus (cc). Phallus with dorsal forked flange (dd). Hypandrium broad, tapering to tip (ee).. ***parenti***

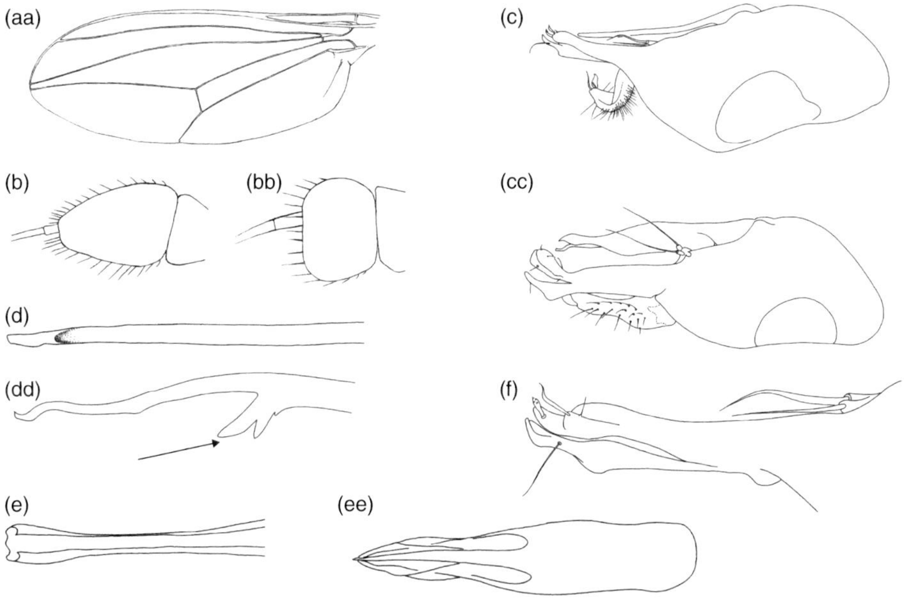

37(34) Legs mainly pale. Coxae and legs with pale setae. Proepisternal setae pale. [Hypopygium (a), tip of phallus (b), hypandrium ventral view (c).] .... ***grisescens***

– Legs black. Coxae and legs with dark setae. Proepisternal setae black...38

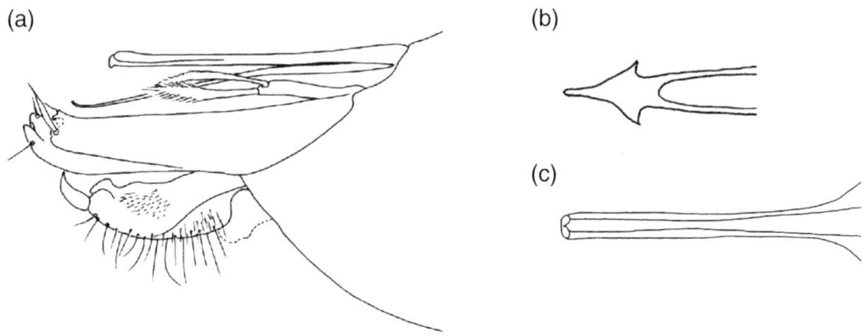

38(37) Halter knob dark or black. Apical section of M_4 slightly longer than dm-m (about 1.3 times). Knees entirely dark. 39

– Halter knob yellow. Apical section of M_4 about equal to dm-m. Knees yellow or reddish but may be narrow and indistinct. 40

39(38) Femora almost bare ventrally, hind femur without anteroventral setae, front femur without posteroventral setae. Arista longer than face, about equal to eye height. Proepisternum with 2–3 subequal setae. Cercus gently curved (a[14]). Hind basitarsus with a projecting transverse flange at base posteriorly beyond crown of black spines (b). Hypandrium narrow and parallel-sided for most of its length. [Apical section of M_1 practically straight (c).] .... ***infumata***

– Hind femur with distinct anteroventral setae at least as long as one-third depth of femur, front femur with distinct posteroventral setae. Arista scarcely longer than face, about 1.1 times. Proepisternum with 1 long seta and fine short hairs. Cercus strongly curved, cylindrical, terminating in a pale point (aa). Hind basitarsus with a brown triangular flange at base, no basal crown of black spines (bb). Hypandrium elongate bottle-shaped (dd[15]). [Phallus with paired lateral projections (ee).] .... ***cuspidata***

(a)

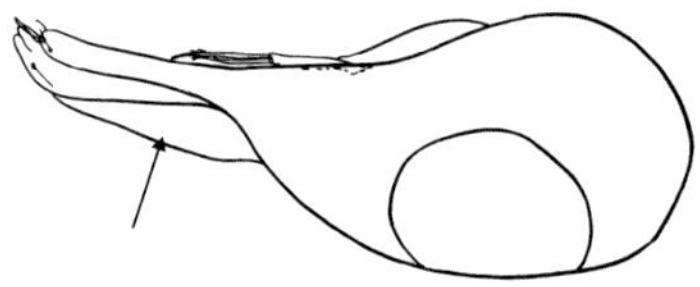

(c)

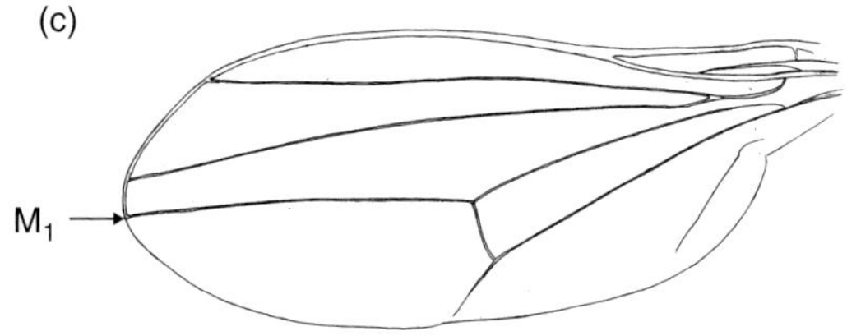

(aa)

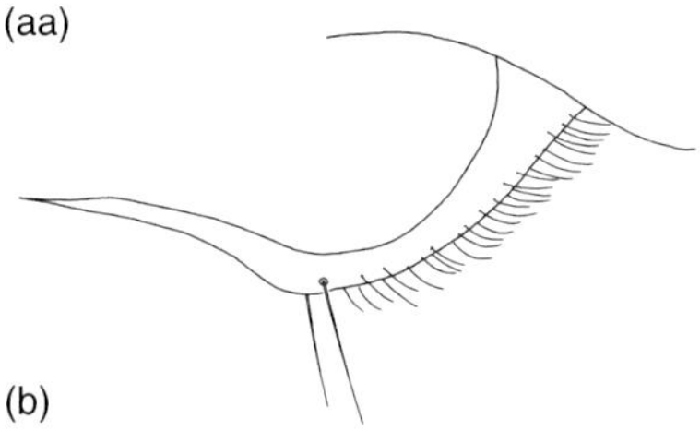

(dd)

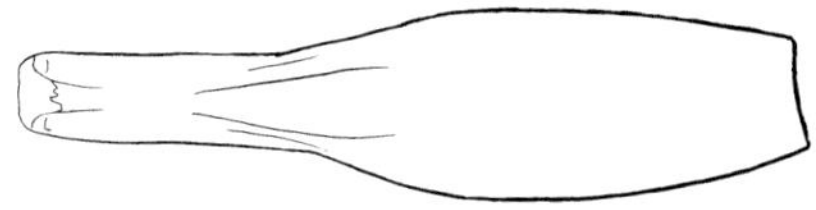

(b)

(ee)

(bb)

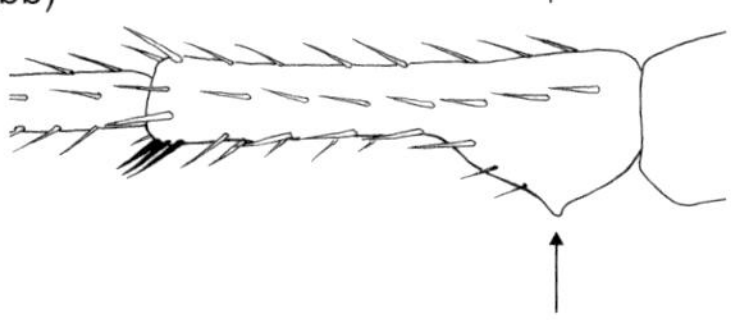

40(38) Apical section of M_1 strongly curved (a). Proepisternum above front coxa with 4 setae. Arista not more than 1.5 times length of antenna. Mid and hind femora with short anteroventral setae on basal half. Surstylus broad and as long as epandrium (b). Hind basitarsus without a flange on posterior side at base. Larger species, wing length about 3.0–3.5 mm. .... ***melancholica***

– Apical section of M_1 almost straight (as fig. c of couplet 39). Proepisternum with 2 setae. Arista more than 4 times length of antenna. Mid and hind femora almost bare ventrally. Surstylus narrow, about half length of epandrium (bb), its dorsal lobe thinner and longer than ventral lobe (cc). Hind basitarsus with a translucent flange on posterior side at base beyond a crown of black spines (dd). Smaller species, wing shorter than 3.0 mm. 41

[14] Negrobov and Stackelberg 1974a. *Die Fliegen der palaearktischen Region* **4** (5), issue 302, Fig. 579.

[15] Negrobov and Stackelberg 1974a. *Die Fliegen der palaearktischen Region* **4** (5), issue 302, Figs 486, 487.

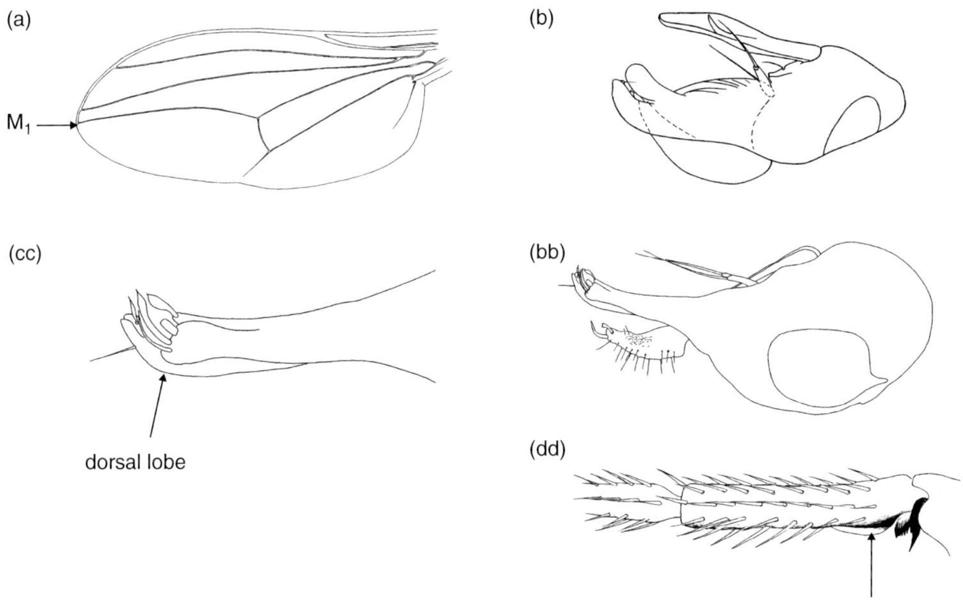

41(40) Setae of epandrial lobes bare, any minute spinules visible only at high magnification (a). .... ***bilineata***

– Setae of epandrial lobes with tiny perpendicular spinules (a[16]). .... ***veles***

Medetera abstrusa Thuneberg 1955

This is a moderately small shiny dark green species with a thinly dusted scutum and black legs with distinct yellow knees, resembling *impigra*. It is one of the few British *Medetera* with dark lower postocular setae but yellow reflections can mask the colour if not viewed from different angles. These setae are also dark in the *muralis* group, *cuspidata*, *nigrohalteralis* and *tristis*, which all have dark halteres rather than pale as in *abstrusa*. The hypandrium has a conspicuous crescent-shaped dark mark near its tip which can be seen at low magnification when the hypopygium is drawn away from the abdomen, providing unambiguous confirmation of the identification. Wing length ♂: 2.6 (2.2–2.8) mm.

Medetera abstrusa is widespread in Britain north to the Highlands, being moderately frequent in southern England but scarce in Scotland. Deciduous woodlands are its main habitat, including wood-pasture, carr woodland and scrubby habitats. It has been collected from the trunks of ash, birch, beech, oak and decaying sycamore. Flight: iv–viii, mainly vi–vii.

Medetera acanthura Negrobov and Thuneberg, 1970

Externally this species is similar to *impigra* but differs in the gently curved surstylus (more arched in *impigra*) and in having a long inner epandrial lobe that appears as a spine at low magnification. The setae of the two epandrial lobes lie at almost the same level along the surstylus; in *impigra*, the inner

[16] Cole 1989. *British Journal of Entomology and Natural History* **2**, 115–118.

seta base is a short cone, not much larger than the base of the nearby outer seta. Specimens with their hypopygium agreeing with that of *acanthura* have either obviously long or inconspicuous short anteroventral setulae on the hind femur, and may represent different species. Wing length ♂: 3.0 (2.8–3.2) mm.

Medetera acanthura was first found in Britain in Perthshire and Moray (Drake 2024). The sites were a wet alder wood and dry birch wood where a male was taken from a birch trunk. Pollet *et al.* (2009) suggest that it prefers deciduous trees. Flight: v–vi.

Medetera ambigua Zetterstedt, 1843

Among British *Medetera* with a glossy clypeus, *ambigua* is distinct in being entirely black, including the halter although this may be partly pale on the stem and sides of the knob, and the knees which are at most indistinctly brown. It has a shiny black scutum with a dusted prescutellar depression, shiny black tergites with a conspicuous anterolateral wedge or band of pale dust, and an elongate-oval postpedicel about 1.5 times longer than deep. The femora appear at first devoid of hairs as the vestiture is minute and pale while the only setae are dorsals at the base of the hind femur. The hypandrium and phallus taper to very fine tips. Wing length ♂: 3.3 (3.1–3.5) mm.

Medetera ambigua is widely distributed in Britain north to the Moray Firth but more frequent in eastern England. Many sites are heathlands, either wet or dry, and several records refer to captures on Scots Pine or unspecified conifers, e.g. Collin (1941); Kenis *et al.* (2004) add pines and spruce. Flight: v–viii.

Medetera belgica Parent, 1936

This is a tiny dark species with shining bluish to violet pleura and notopleuron, largely dark legs, a dark halter knob, black postocular setae, and dorsocentral setae increasing in size from front to back with several tiny ones in front of the suture. The epistoma is dull black, not shining as stated by Negrobov and Stackelberg (1972) and Negrobov and Naglis (2016). Examination of the phallus is essential but it is particularly difficult to see without damaging the hypandrium which must either be removed completely or pushed to one side in wet material. Only a very short section of the phallus projects beyond the sheath of the hypandrium. This species was confused in Britain with *muralis* and *peloria* (Drake 2024). Compared with these species, *belgica* is darker-legged, although the basitarsus can be pale, and the wing costa is marginally darker and thicker. Slightly paler-legged specimens of *belgica* may appear to be *muralis* or *peloria*. This concept of *belgica* follows Negrobov and Stackelberg (1972) but, as Parent described *belgica* from a female, it is possible that the dark-legged form is the true *muralis*, as suggested by Jonassen *et al.* (2013) although this is unlikely as Meigen (1824) clearly described *muralis* as a paler-legged species. Wing length ♂: 2.3 (2.1–2.4) mm.

Many records of *muralis* refer to *belgica* which is a common and widespread species, becoming less frequent in Scotland where it may be more confined to coastal areas. It is found mainly on isolated trees in open habitats or in hedgerows and at woodland edge, and in both wet and dry deciduous woodlands. Some records are for boulders and vertical surfaces other than tree trunks. Flight: vi–vii.

Medetera bilineata Frey, 1915

Medetera bilineata is a small mainly black species with brown dusting forming stripes on the scutum. Although specimens agreeing with *bilineata*, rather than the almost identical *veles*, have been found in Britain, *bilineata* may be a junior synonym of *veles*, as proposed by Bickel (1985). The separation is based on the absence of tiny upright hairs on the setae of the epandrial lobe, which are present in *veles*. Bickel's illustration of the whole hypopygium differs from the single specimen attributed to *bilineata*, having a strongly angled junction of the surstylus with the epandrium rather than smoothly joined (fig. bb of couplet 40). Wing length ♂: 2.6 mm.

A specimen of *bilineata* with almost bare epandrial setae was found in Devon from dry moorland grassland and gorse although deciduous trees were present within a few hundred metres. An apparently identical specimen was found in Argyllshire, Scotland, but was included in *veles* var. *scotica* by Cole (1989). Flight: vi.

***Medetera bispinosa* Negrobov, 1967**

This large almost black species is comparable to *obscura* and *pinicola*, with a blue- or green-tinged scutum and green-tinged shining abdomen, but differs in having the first two tergites thinly dusted in contrast to the shining following tergites. The black proepisternal setae form a dense narrow cluster with the setae in 2–3 rows, differing from a single row in *obscura* and *pinicola*. The dorsal lobe of the surstylus is thin and leaf-like and can be seen easily if the hypopygium is drawn away from the abdomen. Wing length ♂: 4.9 (4.8–5.0) mm.

Confusion with *nitida* was recognised by Dyte (1996) and it is possible that all British records of *nitida* refer to *bispinosa*. Records of confirmed *bispinosa* are thinly spread in lowland England with an isolated record in Yorkshire, while records attributed to *nitida* show a similar but slightly wider distribution that includes Wales. Information on the larvae is summarised by Dyte (1996); they are associated mainly with elm but also oak (Lloyd 1944, Beaver 1966, as *nitida*; Beaver gives a detailed account of larval and adult biology). Flight: v–vii.

***Medetera borealis* Thuneberg 1955**

This member of the *apicalis* group most resembles *jugalis* in having pale postocular setae, as usual in most species of *Medetera*, but the obviously black tergite hairs of *borealis* separate these two species. In profile, the hump near the tip of the hypandrium is discernible when the hypopygium is extended, differing from the straight profile of *abstrusa* and *jugalis* (which has a tooth on this surface). Wing length ♂: 2.5 (2.4–2.8) mm.

It is likely that many records will be incorrect. The few verified males are known from Devon to Yorkshire so it is presumably widespread but probably uncommon. Two records are from acid mire with birch and another from fen with carr. Flight: v–vii.

***Medetera cuspidata* Collin 1941**

This is one of the few *Medetera* with dark halteres and closely resembles *infumata*. The curved rod-like cercus is not found elsewhere in the genus and has an unusual arrangement of hairs on the dorsal (lower) side, with a row of short hairs from the base terminating in a pair of long stronger hairs before the bare pale pointed tip. Wing length ♂: 2.5 mm.

Apart from the type specimens from the Clyde area (1906), *cuspidata* is known from a single record in the Highlands. Flight: vi.

***Medetera dendrobaena* Kowarz, 1877**

This species is recognised by the combination of strongly converging veins R_{4+5} and M_1 and one long anterior presutural dorsocentral seta. A small second dorsocentral seta may be present, as in *saxatilis*. Wing length ♂: 3.0 (2.8–3.2) mm.

Although widely distributed in England where there is a greater concentration of records in the south, in Wales it is found only in the south and it is apparently absent from Scotland. It is found on tree trunks but with no well-defined habitat associations, being found within woodlands and on isolated trees in grasslands and heathlands. The trees frequently include oak, beech, birch and Scots Pine. Flight: vi–x, peak in vii.

***Medetera diadema* (Linnaeus, 1767)**

This is one of the largest British *Medetera* and clearly distinguished by its striped scutum, dark femora that grade to yellow or pale brown about halfway with the rest of the legs yellow or sometimes entirely yellow (entirely dark in females), a glossy green clypeus, an exceptionally short apical section of M_4 less than half the length of dm-m, and all leg setae short except the dorsals on the basal half of the hind femur. Wing length ♂: 4.5 (4.2–4.9) mm.

Its distribution is mainly in the southern and eastern lowland quarter of England north to the Humber, with scarce records in the extreme south of Scotland. For a large and easily identified species, there are few recent records which may indicate a decline in status. Habitats are mainly lowland and coastal floodplains although *diadema* has been recorded at drier sites, including sparse deciduous woodland. It has been found on or close to willow, poplar and pine. Flight: vi–viii.

***Medetera excellens* Frey, 1909**

This is a small pale grey-dusted species that resembles *inspissata* in the shape of its tapered swelling on vein M_4 but is separable under low magnification when the hypopygium is drawn away from the abdomen as the tiny points on the hypandrium are visible. The colour of the tibiae is variable but the mid tibia, at least, is yellow dorsally (dark in *inspissata*). The colour of the abdominal hairs, as used by d'Assis-Fonseca (1978), is unreliable. Wing length ♂: 2.5–2.6 mm.

The Scottish Highlands is the only region where *excellens* has been reliably recorded. It has been reared from Scots Pine, spruce and larch (MacGowan 2001). Flight: iii, vi–ix.

***Medetera fasciata* Frey, 1915**

Only the bright orange scape and pedicel and the more reddish brown tips to the surstyli differ in colour from the entirely shining black body, legs and most of the genitalia, where even the cerci are black; only the scutum is brown-dusted with or without a distinct pattern. The mid femur has a long anteroventral fringe along almost its whole length, with the longest hairs about two-fifths to two-thirds the depth of the femur at their insertion, whereas the hind femur has only short fine ventral vestiture and no longer anteroventral setae. As in *setiventris*, the pedicel is very short, as is the arista which is not more than 1.5 times as long as whole antenna. The curved tip of the longer ventral lobe of the surstylus may be curled strongly around the tip of the shorter dorsal lobe and can be seen without a genitalia preparation. This is the *striata* of earlier British authors, corrected by MacGowan (2001). Wing length ♂: about 3.0–3.5 mm.

This species is known only from the Scottish Highlands (MacGowan 2001); a female under the name *striata* from Hampshire has not been checked. *Medetera fasciata* has been reared from dead Scots Pine. Flight: v–vii.

***Medetera feminina* Negrobov, 1967**

Medetera feminina is larger than most related species in the *apicalis* group, has pale lower postocular setae, pale setae and hairs on the front coxa, a long white proepisternal seta and 2–3 hairs, pale hairs on the tergite disc but longer-than-average black marginal seta on the sides, and only 2–3 anteroventral setae at the tip of the hind femur that are scarcely half the femur's depth (more similar to *impigra* than others in the *apicalis* group). The hypandrium is slightly widened before contracting again at the tip but its ventral surface is uninterrupted by undulations as seen in several others in the *apicalis* group. Wing length ♂: 3.2 (3.1–3.4) mm. This is the senior synonym of *oscillans* Allen, 1976 (Drake *et al.* 2024).

Records of reliably identified specimens fall within a small area from Kent to Oxfordshire and Cambridgeshire and an outlying record from Cheshire. Allen (1976) collected adults from Lombardy poplar, on dead and decaying cherry and English elm. It has been reared from the decaying subcortical layer of a fallen grey poplar. Flight: vi–vii.

***Medetera flavipes* Meigen, 1824**

The single long presutural dorsocentral and almost entirely yellow legs characterise this large species; other pale-legged species such as *diadema* and *grisescens* have dorsocentrals increasing in size towards the rear, although in *insignis* the presutural seta is long enough to cause confusion.

The face is entirely grey-dusted with a green or violet ground colour showing through on the clypeus. The ventral surstylus has two apical setae, one smaller than the other, and the dorsal surstylus has a small subapical seta (fig. a of couplet 7). Wing length ♂: 3.6 (3.5–3.7) mm.

Medetera flavipes is widely distributed in the southern half of England north to Cumbria but with most records from the southeast lowlands; it is scarce in Wales. There are relatively few recent records and this probably indicates a decline as the species is large and unmistakable. It appears to be associated mainly with dry grasslands and scrubby places. It has been reared from material collected from a rot-hole. Flight: vi–ix.

Medetera freyi Thuneberg, 1955

The rather short blunt-ended thickening of M_4 is diagnostic of *freyi* which is a tiny dark species with a subshining scutum and black legs with yellow knees and basitarsi. Wing length ♂: 1.8 mm.

This species is known only from Inverness-shire where it was reared from material collected from under the bark of a fallen and decaying aspen (MacGowan 2001). Flight: vi.

Medetera grisescens de Meijere, 1916

It is possible that the species recorded in Britain as *grisescens* is not this species as the British specimens do not agree with the description in Bickel (1987); in particular, the scutum and face of British specimens are black with dense brown and grey dusting with no hint of metallic green, and the cercus and surstylus more resemble the figure by Bickel (1987) for the much smaller *opaca* de Meijere, 1916. British specimens are medium-sized flies with slender yellow legs that are darker in basal half of the femora and towards the tip of the tarsi. Until the identity is confirmed, British material is best recorded as *grisescens* to prevent confusion. Wing length ♂: 3.3 (2.9–3.7) mm.

Medetera grisescens is an introduced species recorded in winter from heated greenhouses in Surrey (Halstead 2003) and Berkshire. It was described from Java and has a wide distribution from the Indian subcontinent, through southeast Asia to Australia and Pacific islands, spreading as a tramp species (Bickel 1987).

Medetera impigra Collin, 1941 (Plate 5)

This is a dark species with a blue-black tinge to the pleura and dark legs with narrowly yellow knees. The rather untidy row of long anteroventral setae on the mid femur is unusual in this species-group. The lower postocular setae appear to be yellowish, with the lowest one or two long setae more obviously dark, but the colour is difficult to judge owing to pale reflections obscuring the dark ground colour. The long low hump on the hypandrium seen laterally is a good indicator for this species. *M. impigra* can be confused with the very similar *acanthura* as both differ from others in the *apicalis* group in having inconspicuous anteroventral setae at the tip of the hind femur; the difference in their epandrial lobes can be seen at low magnification. Very similar species are *pseudoapicalis* – see the account for that species – and *M. curvipyga* Naglis and Negrobov, 2014 which differs in the hypopygium being slightly more tapered at the end and the seta on the ventral lobe of the surstylus being feathered, but *impigra* has a feathered seta too, not simple as inaccurately illustrated by Negrobov and Stackelberg (1974a). Wing length ♂: 3.1 (2.7–3.3) mm.

Medetera impigra is widespread and moderately frequent in England, and scarce in Wales and Scotland but found on the Sutherland coast. It is found mainly in deciduous woodland and copses, more often in damper sites but also in dry ones, and occasionally on isolated trees in open habitats. It has been reared from the decaying wood of aspen, beech, birch, horse chestnut, larch, poplar, sycamore and oak. Beaver (1966) gives a detailed account of the larval and adult biology. Flight: iv–ix, peak in late vi–vii.

Medetera infumata Loew, 1857

This is a small almost black species, including the halteres, with a thin fairly even coat of brown dust on the scutum. The pale knees are scarcely differentiated. Black halteres are found in few other species but only *cuspidata* has similar venation. Apart from the long dorsal hairs on the basal half of the hind femur, the femoral hairs are otherwise short and almost absent ventrally. Fine microsculpture dulls the black clypeus which is dusted only at the margins; it is slightly shining but not glossy and contrasts with the duller and blacker epistoma. Wing length ♂: 2.5 (2.3–2.6) mm.

The species is known only from Strathspey apart from an old unverified record from the Clyde area. Adults are found most commonly in association with Scots Pine (MacGowan 1987) and have been found on a decaying Scots Pine log (Cole 1989); Kenis *et al.* (2004) also list spruce. Flight: v–vii.

Medetera insignis Girschner, 1888

This species is distinct from other British *Medetera* in having clear pale yellow femora and tibiae, a reddish scape and the apical section of M_1 being about twice the length of dm-m. Problems may arise as the presutural dorsocentral seta is moderately long so the setae in the row do not increase in size gradually, leading to *flavipes* in the key, and the pale colour of the scape may be obscure and overlooked so that the specimen fails to fit anywhere. Other species with a reddish scape have black legs. Pale *pallipes* may be confused but its veins R_{4+5} and M_1 end close together, rather than further apart as in *insignis*, and the has slightly darkened femora. Wing length ♂: 2.4 mm.

A male was found in Somerset in a cutting through limestone shaded by deciduous trees (Gibbs 2007). Flight: vii.

Medetera inspissata Collin, 1952

This species resembles *excellens* and they are the only known British species with a tapered thickening on M_4. They are both small and dark with pale dusting, although *inspissata* appears darker overall. The tiny spines on the hypandrium are visible at a magnification of ×40 if the hypopygium is drawn away from the abdomen, without the need for a preparation. Wing length ♂: 2.5 (2.3–2.6) mm.

Medetera inspissata is widely but sparsely distributed in England and Scotland with an apparent concentration of records in the aspen woodlands of Speyside. It is associated with aspen and other poplars and has been reared from material collected from under the bark of these trees (MacGowan 2001). Flight: v–vi. viii–ix.

Medetera jacula (Fallén, 1823) (Plate 5)

As one of the larger species in the group with a large presutural seta, *jacula* should be clearly separable from related species using the characters in the key. The scutal stripes are well developed, with pale grey-dusted stripes between the acrostichal and dorsocentral rows contrasting with dull coppery brown of the rest of the scutum. As in *petrophiloides*, a few upper postocular setae and the seta on the mid and hind coxae are sometimes dark. *Medetera meridionalis* Negrobov, 1967 is very similar to *jacula* but lacks the smaller second anterior dorsocentral seta and has a straighter surstylus; Kejval and Pollet (2023) discuss other small differences in the hypopygium and face. Wing length ♂: 2.9 (2.7–3.1) mm.

Medetera jacula is widespread in England, more sparsely distributed in Scotland and apparently confined to coastal areas in most of Wales. The sites are usually dry, including grasslands, heathland and dunes, but occasionally wetter, and it is usually found on trunks of both deciduous and coniferous trees. Flight: v–xi, mainly late iv-mid vii.

Medetera jugalis Collin, 1941

This is a fairly dark species with only the knees narrowly yellow or brown. The long anteroventral setae of the mid femur are almost regularly spaced and even in size. Obvious white setae and hairs include the lower postoculars and postgenals, the tibial vestiture, those on the front and mid coxae, and the tergite discs and their margins. The spine on the ventral edge of the hypandrium, not far from its tip, can be seen at low magnification when the hypopygium is drawn away from the abdomen. Wing length ♂: 2.8 (2.5–3.0) mm.

The distribution is mainly in the southern half of England, with a few isolated Scottish records in the Highlands and Skye. There is a recurring association with wood-pasture. Some records were made from poplar. Flight: v–viii, x.

Medetera melancholica **Lundbeck, 1912**

Medetera melancholica is a large black species with a grey-dusted scutum flanked by browner dusting outside of the dorsocentral setae, and with numerous although not particularly long acrostichal setae. There are 3–4 long black proepisternal setae. The arista is markedly short, about equal to the height of the face. Femoral hairs are very short and almost absent ventrally, leaving long hairs only dorsally at the base of the hind femur. Unlike many species, the hind basitarsus has no small tooth or flange on its posterior side at the base. Wing length ♂: about 3.0–3.5 mm.

This species is sparsely distributed in the Scottish Highlands with isolated records that need confirming from Yorkshire, Derbyshire and Surrey. Kenis *et al.* (2004) associate it with alder and pine, and it has been reared from ash (MacGowan 1987). Flight: vi–viii.

Medetera micacea **Loew, 1857**

This tiny species has a greyish striped scutum and only three equally large dorsocentral setae. It thus resembles occasional specimens in the *truncorum* group. Many other Palaearctic species have only one pair of scutellar setae, like *micacea* and *mixta*, so it is possible that others may be overlooked using the key here. *Medetera micacea* differs from these others (except *mixta*) by the combination of the clypeus being shiny centrally and the epistoma entirely dull, femora dark, proepisternal setae white, and both setae of the epandrial lobes lanceolate, these setae being distinct in dry specimens. However, *mixta* is identical externally and in having these setae, differing only in having a smoothly undulating profile to the inner margin of the surstylus, which in *micacea* is more angular or almost pointed. Wing length ♂: 1.9 (1.8–2.0) mm.

Medetera micacea is widespread in England and Wales, with isolated Scottish records. Habitats are invariably dry, unshaded and with short herbs, and include coastal, sandy and chalk grasslands, coastal dunes and shingles, and heathlands. There is no association with trees. Flight: v–x, mainly late vi to mid vii.

Medetera mixta **Negrobov, 1967**

This species is almost identical to *micacea* – see the account for that species. Wing length ♂: 1.9 mm.

Specimens that fit the concept of *mixta* were found at two dunes on the Norfolk coast, in company with *micacea* (Drake 2024). Flight: vii.

Medetera muralis **Meigen, 1824**

This small species closely resembles *belgica* and *peloria*; it cannot be separated from *peloria* externally. Most specimens of *belgica* have darker tibiae and tarsi although pale-legged *belgica* are inseparable without examining the hypopygium. Wing length ♂: 2.1 (2.0–2.2) mm.

Medetera muralis appears to be less frequent than either *belgica* or *peloria*. It is likely to be widespread but the only confirmed records are from Gwynedd in mixed conifer and deciduous woodland at an altitude of about 300 m and Inverness-shire from deciduous woodland at 400 m. Flight: vi–vii.

Medetera nigrohalteralis **Pollet and Drake, 2025**

This species is entirely dark with only the knees slightly paler than the rest of the fly. All setae, calypter hairs and both the stem and knob of the halter are dark to black. The postpedicel is slightly triangular with a pointed tip below the arista. It had been confused in Britain with *tristis* which, in comparison, has more obviously yellow knees, paler calypter hairs and a more rounded postpedicel although these differences cannot be relied upon alone. The most definitive differences are found in the hypandrium and phallus tip. The hypopygium resembles that of *impigra* but the tip of the phallus lacks the tiny dorsal flange of that species. Wing length ♂: 2.5 (2.2–2.8) mm.

Confirmed records of *nigrohalteralis* and *tristis* show the former to be more frequent so it is likely that many British records of *tristis* refer to *nigrohalteralis*. Its distribution is therefore probably wide in Britain, reaching Skye and the northern Highlands, and shows no obvious gradient in frequency in the country. It is recorded mainly in dry and wet deciduous woodlands or in areas with trees in otherwise open habitat, with perhaps a greater frequency in more acid sites. Adults have been collected from ash, beech, birch and oak, and in birch-sallow carr woodland. Flight: v–viii.

Medetera nitida **(Macquart, 1834)**

This species may not occur in Britain as all specimens examined are *bispinosa* – see the account for that species for more information. Wing length ♂: 4.0 mm (French specimen).

Medetera obscura **(Zetterstedt, 1838)**

This large shining black fly resembles *bispinosa* and *pinicola* although the tergites are less shining than those of *bispinosa*. The clypeus is almost glossy, dulled only by minute punctuation. The hypopygium is exceptionally large, almost as long as the abdomen and much deeper than the lengths of the last few tergites. Unusually it has a black cercus, as in *pinicola*, with only sparse tiny hairs near its base. Wing length ♂: 3.9 (3.7–4.0) mm.

Medetera obscura appears to have a wide distribution in Britain from the southern coast to the Scottish Highlands and including Wales. It is associated with pines (Kenis *et al.* 2004, Negrobov and Stackelberg 1971); Collin (1941) reared it from material under pine bark. Flight: iv–viii.

Medetera pallipes **(Zetterstedt, 1843)**

This is a small species with largely yellow or pale brownish mid and hind tibiae that contrast strongly with the black femora; when these two tibiae are darker, their dorsal faces remain pale. The apical section of M_4 is at least twice the length of dm-m, and R_{4+5} and M_1 end very close together at the wing tip (costal ratio either side of R_{4+5} at least 4.0). The posterodorsal seta of the mid tibia is weak, about half the length of the anterodorsal seta. The anteroventral hairs towards the tip of the hind femur are moderately long, about half the femur's depth, and those of the row above these are much longer, the last three being at least as long as the width of the femur in *dorsal* view, and are usually dark or have dark reflections. *Medetera unisetosa* has similar leg colouration but a shorter apical section on M_4, only one notopleural seta and usually only a posterodorsal seta near the base of the tibia. *Medetera lorea* Negrobov, 1967, which could occur in Britain, is also similar to *pallipes* and is differentiated by the dorsal lobe of its surstylus being longer and extending beyond the ventral lobe, although in *pallipes* the oval apical seta on the ventral lobe becomes transparent when cleared in a mount so it appears that the dorsal lobe extends well beyond the darker tip of the ventral lobe. Wing length ♂: 2.3 (2.1–2.5) mm.

Medetera pallipes is widespread and frequent in southern England but scarcer towards the north and west, with few Welsh and a single old Scottish record. It is associated with trunks of deciduous trees, mainly isolated or sparse trees in grasslands, and more rarely in closed woodland. Flight: v–viii, peak in vi.

Medetera parenti **Stackelberg, 1925**

This is a shining dark greenish black fly with black legs and yellow knees. The shining coppery- or blue-green face is unusually narrow, about 4 times higher than its minimum width, and the tiny palps are the same glossy green colour as the clypeus. Both the mid and hind femora have long white anteroventral hairs that are unusual in being longer on the basal half or mid third than near the tip. The vestiture of the hind tibia is about twice as long ventrally as dorsally. The hypandrium has a forked thorn on its ventral (the more hidden) surface which can be seen if the hypandrium is lifted away from the epandrium. As Cole (1989) showed, the number and relative sizes of the proepisternal setae and smaller hairs vary and make unreliable key characters, as used by Negrobov and Naglis (2016). Wing length ♂: 3.0 mm.

The few records are from central-southern England and include a preponderance from managed landscapes, including parks and a cemetery. Records include rearing from material collected under poplar bark, including grey poplar (Cole 1989); Negrobov and Stackelberg (1971) give both aspen and oak. Flight: vi–vii.

Medetera peloria Negrobov, 1967

This tiny dark species closely resembles *belgica* and *muralis*, with bluish to violet pleura and notopleuron, a dark knob to the halter and dorsocentral setae increasing in size from front to back. The epistoma is dull black, not shining as given by Negrobov and Stackelberg (1974b) and Negrobov and Naglis (2016). It differs from *belgica* in having obviously yellow knees and paler tarsomeres that become gradually darker to the black fifth segment, but is inseparable from *muralis* externally. See the account for *belgica* for more details. The cercus has three equally large apical projections, not two as given by Negrobov in 1967 and by Negrobov and Naglis (2016). Wing length ♂: 2.2 (2.0–2.3) mm.

Medetera peloria has been recorded in southern England from Devon to Cambridgeshire and from south Wales, suggesting a possibly widespread species (Drake 2024). It has been found in deciduous woodland, wood-pasture and scrub, and taken from tree trunks and a limestone rock face. Flight: vi.

Medetera petrophila Kowarz, 1877

This species resembles the widespread *jacula* from which it is best separated by its shorter anteroventral setae on the hind femur; the anteroventral hairs before the last two to three longer ones are shorter than the vestiture on the front face, but about equal or longer in *jacula*. Both species have a similar postpedicel but in *jacula* the arista appears to be more dorsal as the tip of the postpedicel projects slightly, but less so in *petrophila*. The clypeus is subshining but not glossy on the central third and not markedly different from rubbed specimens of *jacula*. Wing length ♂: 3.1 (3.0–3.2) mm.

The distribution is uncertain as many records require confirming. d'Assis-Fonseca (1978) may be correct in stating that it is known only from Scotland where there are coastal records clustered on the Moray Firth and scattered records around the northern and western coasts. There are no reliable data on its habitat preference. Flight: vi–viii.

Medetera petrophiloides Parent, 1925

Confusion is most likely with *truncorum* which has a similar slender epandrium and long surstyli but the arrangement of the dorsocentral setae will separate most specimens (the second dorsocentral setae may be slightly shorter in some specimens of *petrophiloides*); the colourful scutal stripes are particularly obvious in *petrophiloides* and it usually has an almost glossy violet to turquoise clypeus rather than green with wide dusted margins as in *truncorum*. Features that will help separate *petrophiloides* from other *Medetera* with large presutural dorsocentrals are a few black setae among the lateral setae on the first tergite, dark setae on the mid and hind coxae, and usually one to a few black upper postocular setae; these setae are usually pale in similar species. The non-British *Medetera tenuicauda* Loew, 1857 has the same arrangement of dorsocentral setae and similar narrow hypopygium as *petrophiloides* but lacks scutal stripes. Wing length ♂: 2.9 (2.7–3.0) mm.

Medetera petrophiloides is found around most of the British coast although more sparsely in Scotland, and sparingly at inland sites in England. This is the commonest *Medetera* found on dry coastal dunes, shingles and sandy soft-rock cliffs. There are reliable inland records from heathlands, e.g. the Breckland of Norfolk and Suffolk, but many inland records not from heathlands are probably errors. It appears to have no association with trees in Britain. Flight: v–ix, peak in vi.

Medetera pinicola Kowarz, 1877

Among species with yellow basal antennal segments and pale halteres, this species resembles the similarly large *obscura* and *bispinosa* but has mainly pale setae and hairs on the coxae, legs, proepisternum and tergites. *Medetera adjaniae* Gosseries, 1989 is a similar species that could occur in Britain and is distinguished by its epandrium being compact and shorter than the surstylus. Wing length ♂: 3.4 (3.2–3.6) mm.

Medetera pinicola is widespread in the northern half of Scotland, with sparse scattered records in Wales and England to the south coast. It has been reared from Scots Pine (Rotheray and Shaw 1989) and spruce (Kenis *et al.* 2004). Flight: v–vii.

Medetera pseudoapicalis **Thuneberg, 1955**

As this book was being finalised, *pseudoapicalis* was recognised from DNA in several widely spaced samples in England and Scotland in the BOLD database (Ratnasingham and Hebert 2007). It has been incorporated into the key using information from Negrobov and Stackelberg (1974a) but the key has not been tested on specimens. Its hypandrium appears to be almost identical to that of *impigra*.

The records in BOLD are from Devon, Oxfordshire, Yorkshire and southwest Highlands.

Medetera saxatilis **Collin, 1941**

Except for the hypopygium, *saxatilis* closely resembles *truncorum*, as discussed under that species. The number of dorsocentral setae varies from three to four, sometimes in the same individual, and those with only three setae will run to *turkestanica* Stackelberg, 1926 in Negrobov and Stackelberg (1972) and Negrobov and Naglis (2016). Wing length ♂: 2.9 (2.7–3.0) mm.

Medetera saxatilis is common and widespread in England although becoming less frequent towards the north, apparently confined to the coast in most of Wales and very scarce in Scotland where records may be based on females which are usually indistinguishable from those of *truncorum*. It is found in open habitats that are usually dry, including grasslands, heathlands, dunes and river shingles, but sometimes in wetlands settings; it is very rarely found within woodlands. Although adults can be found on tree trunks, it is more often found on the ground. Flight: v–x, peak in vii.

Medetera setiventris **Thuneberg, 1955**

See the account for *fasciata* which resembles *setiventris*.

The species is known in Britain from a single male from Strathspey, Scotland (MacGowan 2001). It is associated with pines and spruce (Kenis *et al.* 2004). Flight: vii.

Medetera tristis **(Zetterstedt, 1838)**

As discussed under *nigrohalteralis*, these two almost entirely black species have been confused in Britain. The forked tip of the phallus is unique among British *Medetera*. Wing length ♂: 2.4 (2.4–2.5) mm.

Owing to the confusion with *nigrohalteralis*, the distribution is unclear but *tristis* appears to be the less frequent of the two. Confirmed records are from Hampshire, Wiltshire, Norfolk and Suffolk. One record was from carr woodland edge in fen. Flight: iv–vii.

Medetera truncorum **Meigen, 1824**

This species shares the same arrangement of dorsocentral setae found in many specimens of *saxatilis* and *dendrobaena*, in which the second (first postsutural) is smaller than those either side. Unlike these two species, this small second seta is always present in *truncorum* and is usually at least half the length of the third (usually clearly shorter than half in the other two species, sometimes absent). Vein R_{4+5} is slightly closer to M_1 than to R_{2+3} in *truncorum* compared to wider apart in *saxatilis*, and the apical section of M_4 is usually slightly longer than dm-m in *truncorum* but equal or shorter in *saxatilis*, as illustrated in the key couplets 9 and 11, although variation makes this last ratio unreliable as a key character to separate these two species. The narrow epandrium and long slender surstylus is also found in the non-British *M. murina* Becker, 1917 and *M. tenuicauda*, all three having four dorsocentral setae about the same length. Wing length ♂: 2.8 (2.5–3.0) mm.

Medetera truncorum is the commonest species in the genus. It is found throughout England but apparently confined to coastal areas in Wales, which is probably an artefact of low recording, and is relatively sparsely distributed in Scotland although reaching the Outer Hebrides and Fair Isle. It is found in a wide range of habitats wherever trees are present although it may be seen on walls, rocks and vertical surfaces other than tree trunks. Flight: iv–x, peak in vii.

Medetera unisetosa Collin, 1941

This is a fairly small species with yellow-tipped black femora and mainly pale tibiae, notably the mid tibia being entirely yellow whereas the front and hind ones are often slightly darkened. The single posterodorsal seta on the mid tibia and single notopleural seta are unusual in *Medetera*. However, a small proportion of specimens have a tiny to large anterodorsal seta on the mid tibia, making this character unreliable on its own. There are rather few tiny hairs on the front of the scutum which is often densely haired in other species. All femora have only inconspicuous short pale anteroventral hairs and the hind femur has no row of stouter anterior hairs. *M. unisetosa* bears a superficial resemblance to *pallipes*, as discussed under that species. Wing length ♂: 2.7 (2.5–2.8) mm.

It is likely that *unisetosa* has been overlooked or confused with another species as there are few records pre-dating 2000 although inspection of collections shows it to be more frequent. The apparent distribution is mainly southwest Britain from the Solent to north Wales, with a single Scottish record on the west Highland coast. It has been recorded from deciduous woodlands and from deciduous trees, including oak and ash, in pasture. Flight: v–viii.

Medetera veles var. _scotica_ Cole, 1989

It remains unclear whether this variety is distinct from *bilineata* or indeed from the nominal species, *veles* Loew, 1861, as mentioned in the account for *bilineata*. Cole's (1989) figure of the surstylus is very similar to that of a specimen from Devon attributed to *bilineata*, and unlike that of Bickel (1985), whereas the Cole's figure of the epandrial lobes show clear spinules on the setae which is characteristic only of *veles*.

Medetera veles var. *scotica* is known from three sites in Highland and one from north Wales whose identification has not been verified. Two Scottish sites were pine woods, one with calcareous flushes, and one specimen was collected from a large barkless pine log. The Welsh site was wet acid heath or bog. Flight: vi–vii.

MELANOSTOLUS Kowarz, 1884 (Diaphorinae)

These are tiny rather stocky species resembling *Chrysotus* and *Diaphorus nigricans*, with a dark subshining green scutum that is often slightly blue-tinged on the pleura and tergites, and entirely dark legs, setae, hairs and halteres (Plate 5). Males of the genus are distinguished, along with only *Asyndetus* and *Diaphorus*, in lacking claws and having large pulvilli, particularly on the front tarsus. *Asyndetus* is clearly distinguished by its wing venation and *Diaphorus* by its eyes touching above the antennae rather than being clearly separated by a wide frons in *Melanostolus*. Additionally, males differ from *D. nigricans* in the wing's hind margin being rounded and unremarkable compared to the pronounced square anal lobe of *nigricans* (Fig. 14). Males have two pairs of conspicuous setae on sternite 8 at the tip of the abdomen, as in several other genera of Diaphorinae. From the few other small and almost entirely black dolichopodids, the biserial acrostichal setae distinguish *Melanostolus* from *Acropsilus* (no setae) and *Sympycnus cirripes* (uniserial) and the completely black knees distinguish it from *Campsicnemus pusillus*, *Medetera nigrohalteralis* and *M. tristis* whose knees are narrowly pale.

In Britain only *melancholicus* is known. Of the remaining two European species, *nigricilius* (Loew, 1871) may occur in Britain although it is rare within its range. Males of *nigricilius* are likely to be mistaken for a species of *Chrysotus* with pale front and mid tibiae but differ in having conspicuous setae on sternite 8, a broad face and a slightly clubbed hind femur with a ventral fringe of very long hairs.

Melanostolus melancholicus (Loew, 1869) (Plate 5)

Differences between this species and *Diaphorus nigricans* are covered above. The genitalia of these two species are identical. Females are also very similar to those of *D. nigricans* and do not show the marked difference in wing shape seen in males but, apart from having black halteres (yellow in *nigricans*), *melancholicus* has a slightly green-tinged face and frons which are both similarly thinly dusted (entirely dark ground colour with darker dusting on the frons than on the face in *nigricans*), the clypeus and epistoma are about equal in height and the clypeus is about 1.5 times broader than high (clypeus clearly shorter than the epistoma and about twice as wide as high in *nigricans*) and the frons and eye margins converge slightly (almost parallel in *nigricans*). Wing length ♂: 2.3 (2.1–2.8) mm, ♀: 2.5 (2.1–2.9) mm.

Records are thinly scattered from Yorkshire southwards, including Wales. The two habitats used are sparsely vegetated seepages on soft-rock cliff and in quarries, and bare or thinly vegetated exposed riverine sediments. Feature in common are sparsely vegetated sandy or fine-grained substratum, and perhaps being weakly base-rich. There are rare records from other wetland habitats, including bogs, although these may include errors for other small dark diaphorines if based on females.

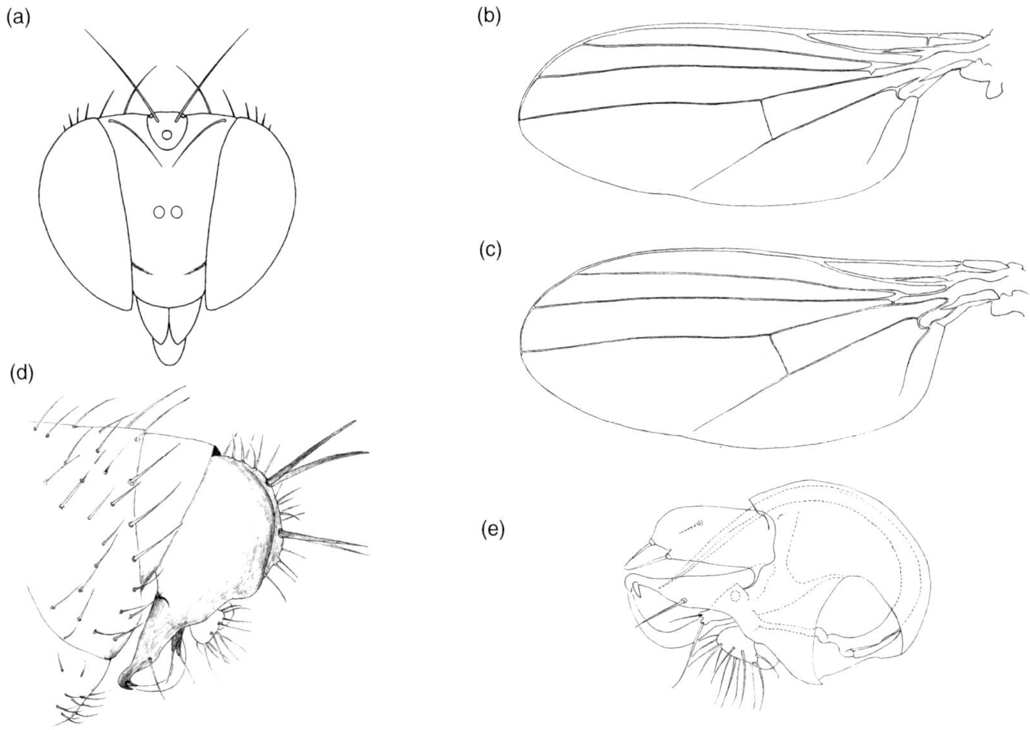

Fig. 24. *Melanostolus melancholicus* head of female (male is similar) (a), wing of male (b) and female (c), end of male abdomen (d) and hypopygium (e).

MICROMORPHUS Mik, 1878 (Peloropeodinae)

Micromorphus are tiny stocky flies, dull dark grey or brown with no hint of green, and with strong dusting (Plate 5). There are no acrostichal setae or lateral seta on the scutellum, the prescutellar depression is well-marked, the mid and hind femora each have an anterior preapical seta, veins R_{4+5} and M_1 are almost straight and parallel beyond dm-m, and dm-m is much shorter than the apical section of M_4. This combination of characters is found in *Chrysotimus*, also in the Peloropeodinae, but that species is bright green and lacks the matt black epistoma of *Micromorphus*. Additionally, in both sexes the ocellar triangle is exceptionally short, the front tibia has no setae and the mid tibia has only a strong anterodorsal and weaker posterodorsal at its basal third.

There are over 30 species of *Micromorphus* globally (Bickel 2022) and four in Europe (Pollet 2011). Three clearly distinct species are present in Britain but they cannot be matched unequivocally with described Palaearctic species or by using the key of Negrobov (2000). They are therefore referred to as species A, B and C, using letters designated by David Gibbs in an unpublished manuscript. Cole (1987) recognised two species and Gibbs (*in litt.*) a third. They have previously been recorded in Britain under the one name *albipes* Zetterstedt, 1843, making the distribution and habitat affinities of each uncertain. One species is readily identifiable in both sexes but the separation of the other two requires examination of the hypopygium at high magnification so no key to females is given.

Micromorphus species are found in unshaded wetlands, including saltmarsh and brackish marsh. Being tiny species, they are easily overlooked in sweep-net samples although their stance is typical of a dolichopodid. Flight times are given for confirmed identifications; a few records of '*albipes*' have been made in April and October outside the dates given for the three types.

Key to species of *Micromorphus* (males)

1 Mid and hind coxae dark, often all femora dark too. Phallus tip shaped like a crochet hook (a). Process arising near base of phallus with a terminal spherical cluster of dense black hairs, most easily seen in ventral view (b). Dorsal arm of surstylus straight with upturned claw (c) .... Species A

– All coxae and femora yellow. 2

(a)

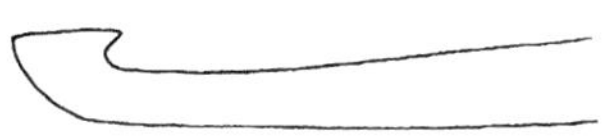

(b)

(c)

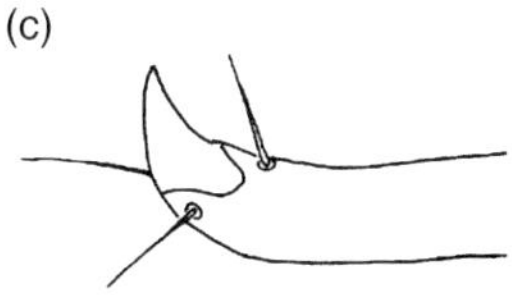

2 Hypopygium with the dorsal arm of surstylus broad, strongly shouldered and angled ventrally away from cercus (a). Tip of phallus rounded (b). Process arising near base of phallus with terminal brush of many rows of short black hairs when seen in ventral view (c). .... Species B

– Hypopygium with the dorsal arm of surstylus slender and straight, with a slightly upturned claw (aa). Tip of phallus shaped as a weak crochet hook (bb). Process of phallus without a terminal cluster of hairs. .... Species C

(a)

(b)

(c)

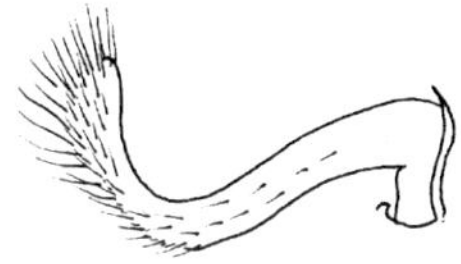

(aa)

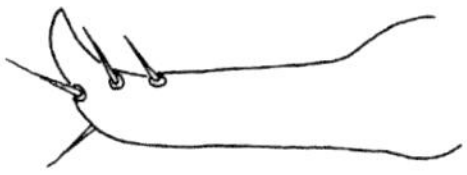

(bb)

Micromorphus Species A of Gibbs (Plate 5)

This species is usually entirely dark-legged, in particular the mid and hind coxae are nearly black even when the front coxae and all femora are pale brown. The lowermost parts of the pleura adjoining the coxae are equally dark. Sp A is therefore readily distinguished externally from the two paler species. The scutellum is completely dark without the yellow rim present in most specimens of Sp B and Sp C. Dark-legged females are presumably this species. Wing length ♂: 1.6 (1.5–1.7) mm.

Sp A agrees most closely with *claripennis* Strobl, 1899 and *paludicola* (Karl, 1921), particularly in having dark coxae and femora and also in other parts of their descriptions although differing in some minor characters. A figure of the hypopygium that agrees closely with that of Sp A is attributed to *albipes* (Grichanov *et al.* 2011a) but the true identity of *albipes* remains uncertain as the type specimen is a damaged female.

Confirmed records of Sp A occur widely and commonly in England north to Stirling in Scotland, and in Wales. It is found in freshwater marshes, water margins and fens in unshaded places, and occurs together with Sp B in brackish coastal sites. Flight: v–ix.

Micromorphus Species B of Gibbs

This species and Sp C of Gibbs are apparently indistinguishable externally. They are both yellow-legged, including all coxae, together with the lower parts of the pleura and metepimeron, and most specimens have a yellow rim on the scutellum. At low magnification (×40) neither arm of the surstylus of Sp B extends far beyond the epandrium which consequently looks approximately spherical, whereas in Sp C both arms and the epandrial lobe close to them clearly protrude as an apparently single rod. At high magnification (×100) the dorsal arm of the surstylus of Sp B is strongly hooked with an almost right-angled shoulder close to its tip, which is sometimes visible in profile at low magnification in dry specimens. Wing length ♂: 1.7 (1.4–2.0) mm.

Zetterstedt's (1843) description of *albipes* fits either Sp B or Sp C since these have pale tarsi along with the rest of the legs and some specimens have a pale rim on the scutellum.

This is an apparently obligate coastal species of saltmarsh or brackish marsh. Confirmed records indicate a frequent occurrence in southern England; the most northern is from Cumbria. It also occurs in Wales. Scottish coastal records of '*albipes*' may refer to this form, and if true then it occurs frequently on the west coast, Orkney and the outlying St Kilda. Flight: v–viii, mainly vi–vii.

Micromorphus Species C of Gibbs

See Sp B for differences between these two very similar species. Wing length ♂: 1.6 (1.4–1.7) mm.

Species C may be *mesasiaticus* Negrobov, 2000 described from Tajikistan, as there is good agreement with the description and figure, as well as that given by Grichanov and Vicklund (2007) for their interpretation of this species. The ratio of the lengths of the sections of costa either side of R_{4+5} disagrees with that for Sp C but it may have been inverted in the description. Grichanov and Vicklund (2007) suggest that *mesasiaticus* is one of the already-described species *albipes*, *claripennis* or *paludicola*.

The few records attributable to this species are from southern England northwards to Cambridgeshire, and south Wales. It is a freshwater species of unshaded water margins of ponds and marshes. The distinction between the habitats of Sp C and Sp A is unclear. Flight: v–ix.

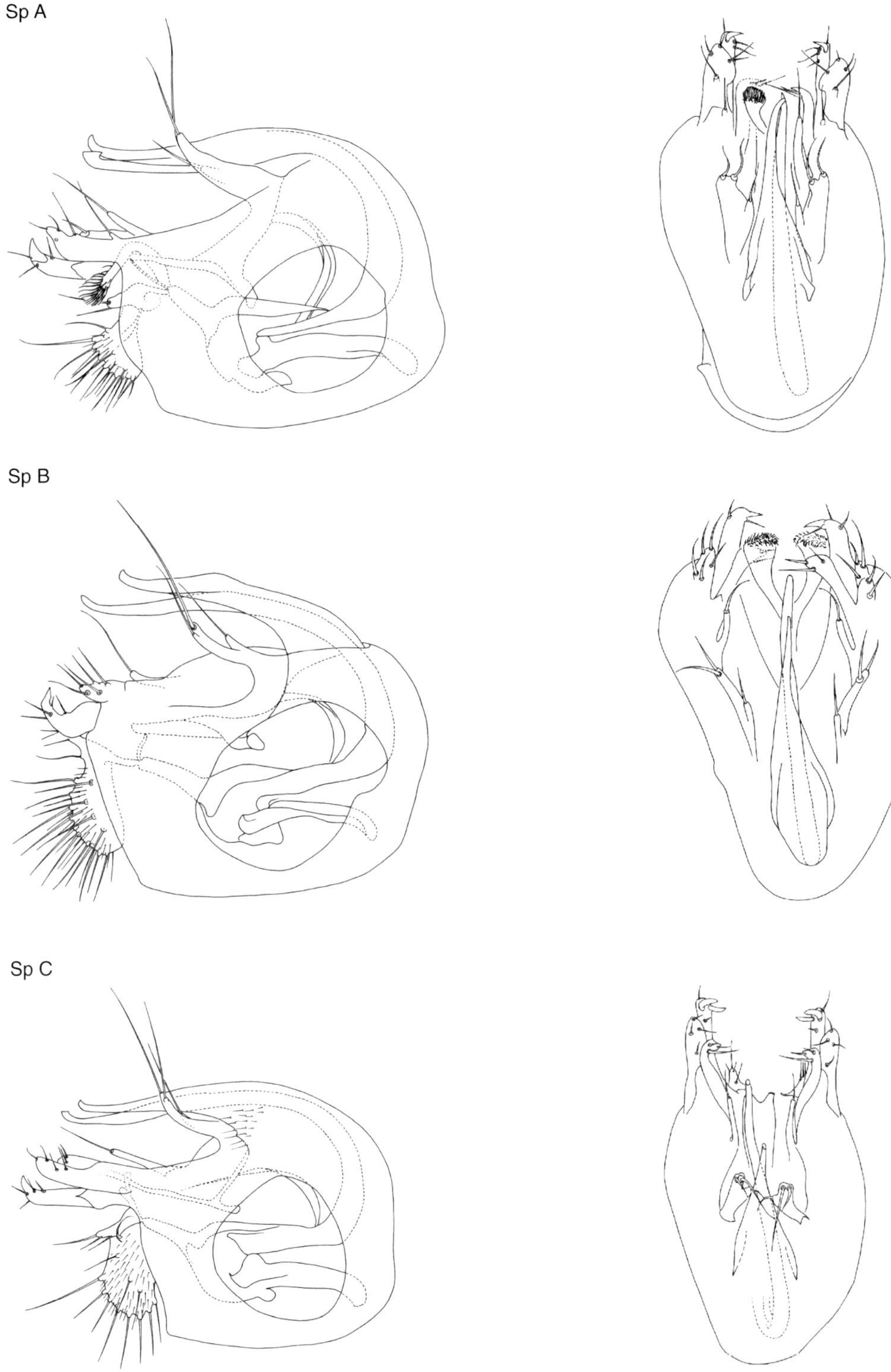

Fig. 25. *Micromorphus* hypopygia, lateral and ventral views.

MICROPHOR Macquart, 1827 (Microphorinae)

Species of *Microphor* are small to tiny black or dark grey flies with no hint of metallic green or blue, their legs entirely black or at most with brownish knees and tibiae (Plate 5). The wing has three veins stemming from the discal medial cell (dm), instead of the usual two veins in all other British dolichopodids, and has a dark pterostigma at the end of R_1 which is not found in any other genus. *Microphor* more resemble some small ocydromiinine Hybotidae and *Atelestus* (Atelestidae) than other British dolichopodids but the reflexed genitalia show a closer affinity to the dolichopodids than to those other groups. There are superficial resemblances with the all-black *Diaphorus nigricans* in the holoptic male eyes and wide base to the wedge-shaped wing but *Diaphorus* has the habitus of a typical dolichopodid.

In the Western Palaearctic 12 species are known of which five occur in Britain (Barták and Kubík 2022). These authors provide a key and detailed accounts for all these species, correcting some errors in Chvála's (1986) revision. As Barták and Kubík (2022) found wide variation in the numbers of acrostichal and dorsocentral setae of males, Chvála's (1986) key to females based on these setae is unreliable. The key provided here includes a larger number of characters than usual as some appear to work in combination but overlap when taken individually. The genitalia provide unambiguous characters and their examination is essential to reliably separate the three species of '*anomalus*' group.

Most records refer to moderately wet places including a range of wetlands although *holosericeus* is often found in drier habitat such as meadows, deciduous woodland and scrub. *Microphor anomalus* and *crassipes* have been recorded swarming (Collin 1961, Empidid and Dolichopodid Recording Scheme). A male of the *anomalus* group ('*pusillus*') was reared from soil under beech trees (Beling 1882). Female *crassipes* have been recorded feeding on insects caught in spiders' webs (Laurence 1948) and they are known to be both predaceous and flower-feeders (Brooks and Cumming 2022).

Key to species of *Microphor* (males)

1 Hind tibia and basitarsus slender (a). Dorsocentral setae uniserial, fewer and more conspicuous, those on the front half about half length of last 1–2 setae; together with acrostichals forming an untidy group with uneven lengths (b – lateral view, c – front view). 2

– Hind tibia and basitarsus strongly dilated (aa). Dorsocentral setae irregularly biserial, more numerous and smaller, those on front half one-quarter or less the length of last 1–2 setae; together with acrostichals, forming a neat group with equal lengths (bb, cc). 3

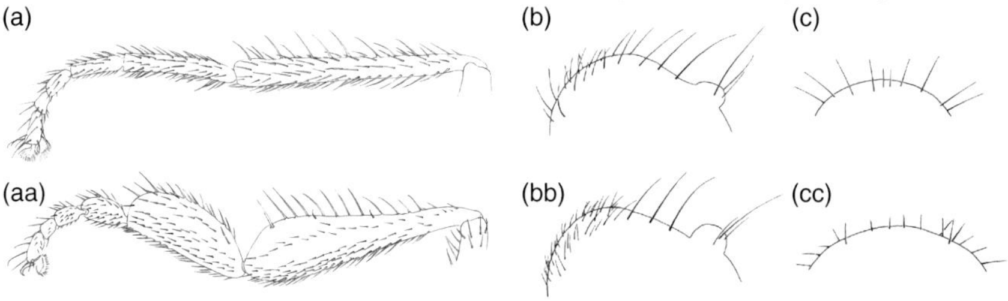

2(1) Larger species, wing longer than 2 mm. Arista shorter than postpedicel (a). Postpedicel less abruptly narrowed on dorsal margin just beyond its base (a). Scutum in front view with broad black stripes along rows of setae, separated by narrow pale grey stripes. Tergite discal hairs conspicuous, as long as tergite length, lateral hairs much longer than tergite. Mid femur with a posteroventral fringe of long hairs not clearly distinct from long hairs on posterior face. Palp with dense stout setae (b). Phallus tip broad, conical and only slightly curved (c). .... ***holosericeus***

– Small species, wing length about 1.7 mm. Arista as long as or longer than postpedicel (aa). Postpedicel abruptly narrowed on dorsal margin just beyond its base (aa). Scutum uniformly subshining black with scarcely discernible stripes. Tergite discal hairs less than two-thirds tergite length, lateral hairs about as long as tergite. Mid femur with posteroventral fringe of long hairs in a single row clearly distinct from short posterior vestiture. Palp with fewer shorter setae. Phallus tip slender and strongly curved (cc). .... ***strobli***

(a)

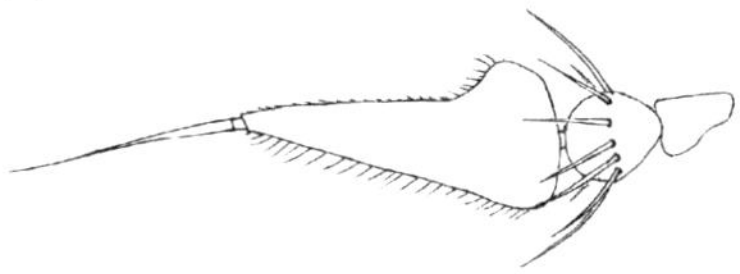

(b)

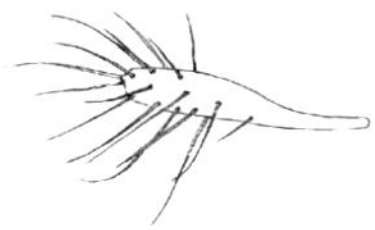

(c)

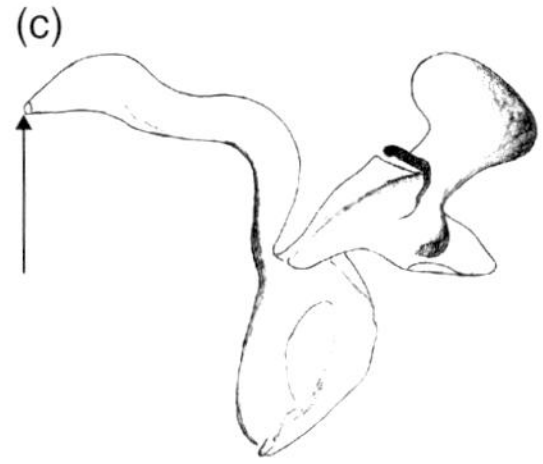

(aa)

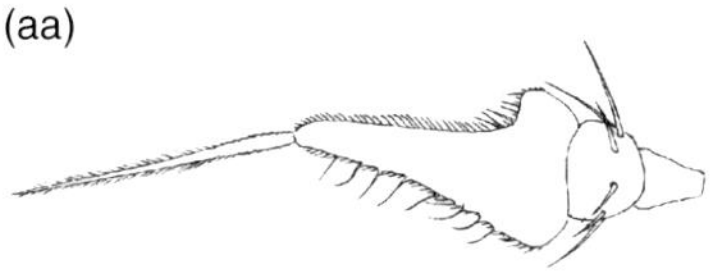

(cc)

3(1) Sternites 2–4 each with at least 3 pairs (rarely fewer) of approximately equally long straggling setae, sternite 1 with 1–3 pairs of setae, usually curved forward but sometimes apparently absent when tucked between hind coxae (a). Acrostichal setae short, on front half only slightly longer than distance between the rows of setae. Dorsal surstylus lobe narrow, slightly bowed and cylindrical with a few tiny hairs along shaft (b). Hind basitarsus as wide as broadest part of tibia (fig. aa of couplet 1). .... ***anomalus***

– Sternites 2–4 each with 1–2 pairs of long straggling setae and some shorter setae, sternite 1 without setae (aa). Acrostichal setae longer, on front half at least 1.5 times as long as distance between rows of setae. Dorsal surstylus lobe either broader with a small knob supporting a single hair or cylindrical with a cluster of terminal hairs but none on shaft (figs a and aa of couplet 4). Hind basitarsus as wide as or narrower than broadest part of tibia. 4

(a)

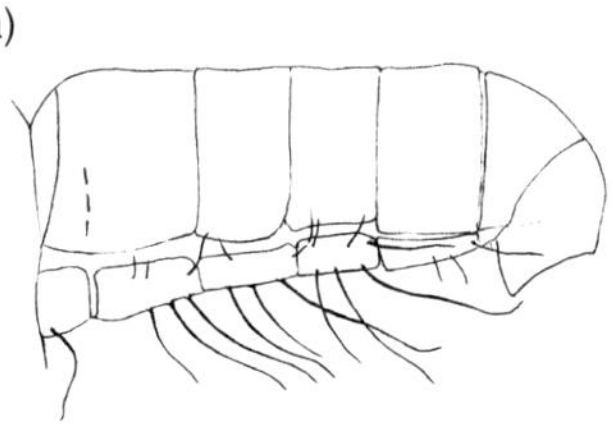

(aa)

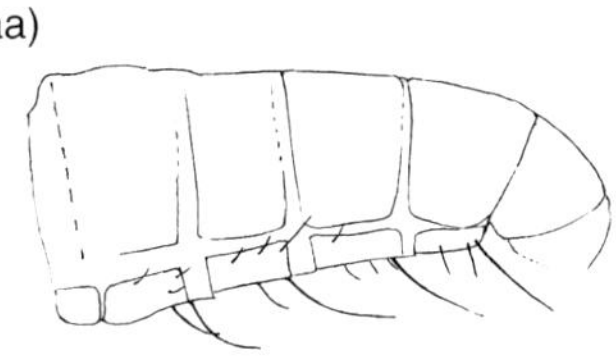

(b)

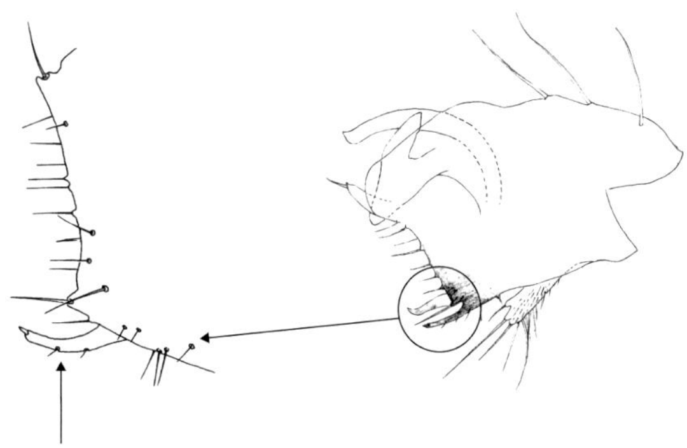

4(3) Dorsal surstylus lobe of epandrium curved with a tiny dorsal hair at about halfway and another near tip (a, i). Epandrium posterior margin without a differentiated tubercle just ventral to surstylus-bearing corner (a, ii). Phallus and phallic process gently curved, with rounded tips (b).
[Supplementary characters that may be variable: Mid femur with posteroventral hairs longer, those in middle about 1.6–2.2 femur depth (c). Mid tibia with 2–4 strong dorsal setae at least twice as long as shaft width, arising along whole length of shaft (d). Hind basitarsus narrower than broadest part of tibia (0.6–0.75 times) (e). Front tibia with a posterior ciliation of fine hairs, on basal half at least as long as width of shaft, usually obviously longer.] .... ***crassipes***

– Dorsal surstylus lobe straight with a cluster of tiny hairs at tip (aa, i). Epandrium posterior margin with a distinct tubercle longer than remaining hair-bearing knobs just ventral to surstylus-bearing corner (aa, ii). Phallus tip upturned, phallic process narrowly tapered to pointed tip (bb).
[Supplementary characters: Mid femur with posteroventral hairs shorter, those in middle about 1.3–2.0 times femur depth (cc). Mid tibia with 1–3 weak dorsal setae up to 1.5 times as long as shaft width, often on only basal and apical quarters (dd). Hind basitarsus as broad as broadest part of tibia (0.9–1.2 times) (ee). Front tibia with a posterior ciliation of fine hairs whose length on basal half is less than width of shaft.] .... ***intermedius***

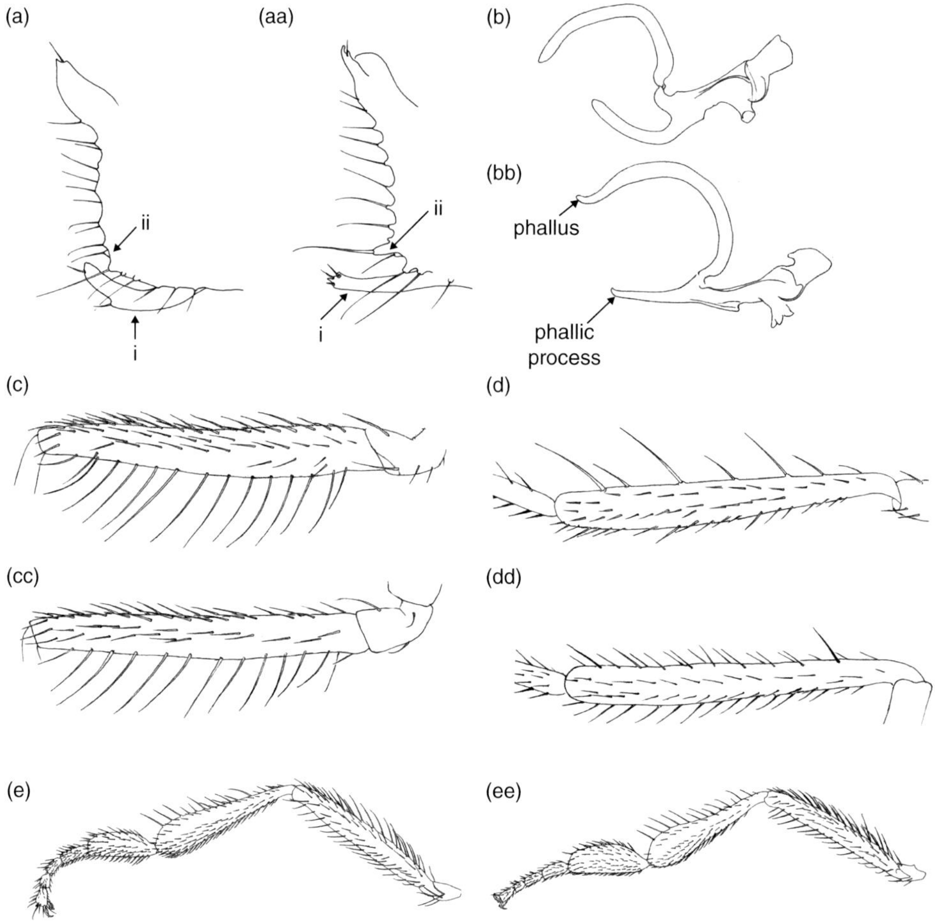

Key to species of *Microphor* (females)

1 Acrostichal and dorsocentral setae larger and less numerous, those on front half about half length of last 1–2 dorsocentrals, forming an untidy group with uneven lengths (figs b and c of male couplet 1). Dorsocentral setae strictly uniserial. 2

– Acrostichal and dorsocentral setae smaller and more numerous, those on front half about one-quarter or less length of last 1–2 dorsocentrals, forming a neat group with equal lengths (figs bb and cc of male couplet 1). Dorsocentral setae in 2–4 rows (rarely nearly uniserial anteriorly). .... ***anomalus*, *crassipes*, *intermedius***

2 Larger species, wing length at least 2 mm. Arista shorter than postpedicel (fig. a of male couplet 2). Scutum grey-dusted with distinct darker brown longitudinal stripes along lines of setae. Acrostichal setae quadriserial. Notopleuron with 3–5 hairs in addition to 2 setae. .... ***holosericeus***

– Small species, wing length about 1.7 mm. Arista longer than postpedicel (fig. aa of male couplet 2). Scutum uniformly light grey-dusted, unstriped. Acrostichals biserial. Notopleuron with 1–2 hairs in addition to 2 setae. .... ***strobli***

Microphor anomalus (Meigen, 1824) (Plate 5)

Microphor anomalus can be recognised by the combination of many long setae on the sternites, the hind basitarsus being as broad as the widest part of the hind tibia, and neat short acrostichal setae on the front half of the scutum. It resembles *intermedius* in the structure of the hind leg, relatively longer arista and weaker dorsal setae on the mid tibia but usually has fewer and shorter hairs in the posteroventral fringe of the mid femur where the maximum length of these hairs is usually clearly less than 1.5 times the depth of the femur. The scutum pattern does not give an unambiguous separation but, in anterior view, most specimens are pale grey with clear narrow dark lines along the acrostichal and sometimes also the dorsocentral rows; as the specimen is tilted backwards, the whole scutum becomes uniformly black. In dorsolateral view narrow grey lines can often be seen between the acrostichal and dorsocentral rows. In comparison, *crassipes* and *intermedius* are usually more evenly coloured in anterior view with less distinct or no dark lines below the rows of setae. The phallus is evenly curved and nearly parallel-sided to its tip, as in *crassipes*, and the phallic process is tapered to a fine upturned tip, similar to *intermedius* but with a broader base. Females of these three species appear to be inseparable although they are clearly distinct from the strongly striped *holosericeus*. Wing length ♂: 2.6 (2.5–2.8) mm, ♀: 2.3 (2.1–2.5) mm (females measured from specimens taken with males).

Microphor anomalus is widespread and moderately frequent in the southern half of England, becoming scarce in northern England. It has been recorded very rarely in southern Wales and central Scotland. Its habitat includes water margins, fens, meadows and deciduous woodlands usually with some wetland feature. Flight: v–viii.

Microphor crassipes Macquart, 1827

This species is recognised by the combination of sparse setae on the sternites, the hind basitarsus being relatively narrow compared to the widest part of the hind tibia, and strong long dorsal setae on the mid tibia. From *anomalus*, it is additionally distinct in having a relatively shorter arista which is about half or less the length of the postpedicel (two-thirds in *anomalus*), and a long dense posteroventral fringe on the mid femur with hairs averaging twice the femur's depth and the range of values scarcely overlapping with the sparse shorter hairs of *anomalus*. *M. crassipes* is distinguished from *intermedius* by many characters that overlap but are on average different, and for this reason many features are used in the key. In anterior view the scutum has darker stripes below the rows of acrostichal setae and sometimes below the dorsocentral row. Externally, the distinctive shape of the dorsal surstylus may sometimes be distinguished without the need for a genitalia preparation. The phallus and phallic process are about the same width with rounded tips and are evenly curved; in

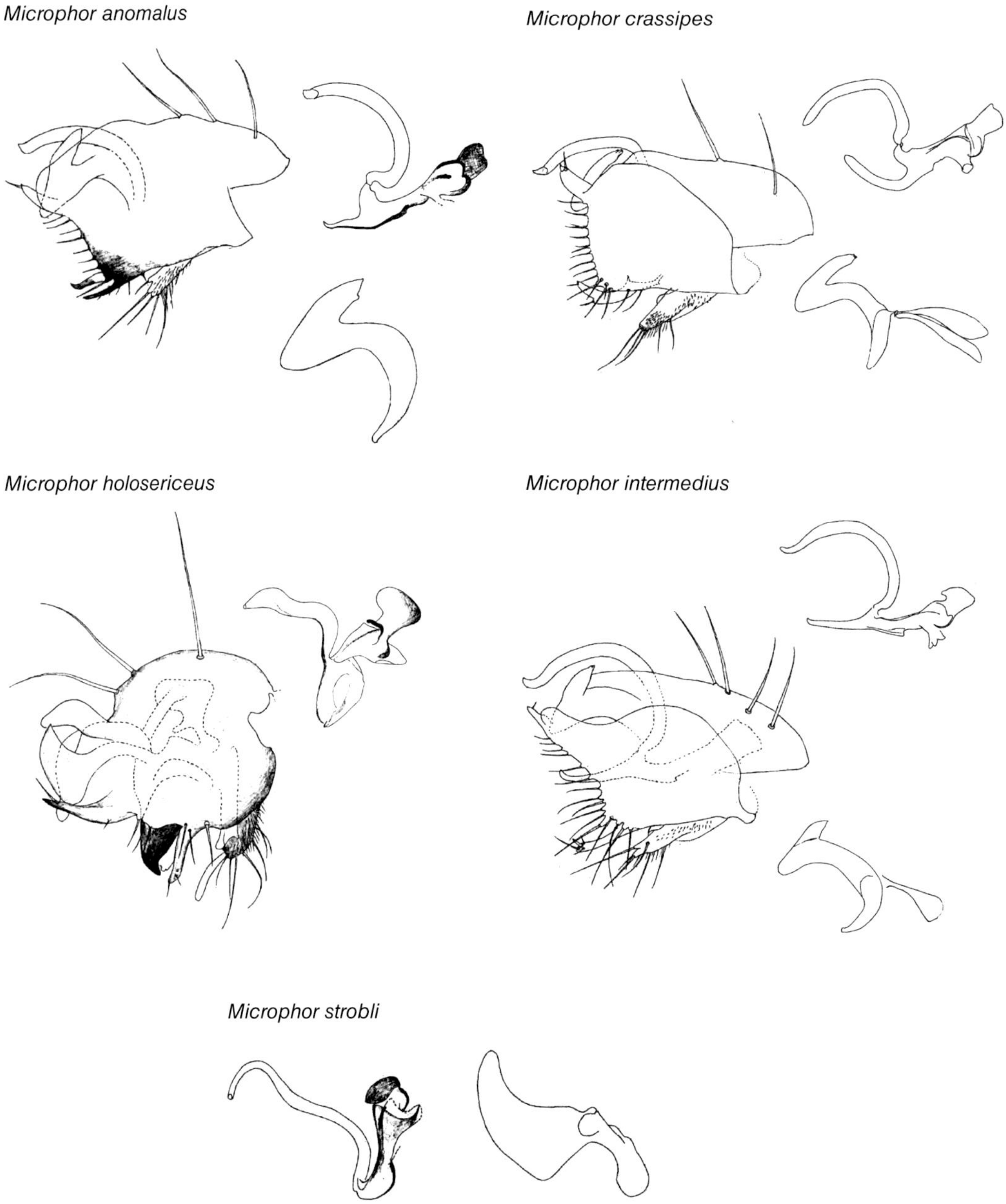

Fig. 26. *Microphor* hypopygia, phalli and postgonites.

comparison, the phallus of *intermedius* has a slightly upturned tip and the phallic process is slender, straight and tapered. Wing length ♂: 2.6 (2.5–2.8 mm).

As this species has been confused with *intermedius*, the records probably apply to both species as they appear to be equally frequent in southern England. The species pair is widespread and moderately frequent in England and south Wales but with only two Scottish record from Perthshire.

It has been recorded mainly from woodlands, scrub and less frequently meadows, and usually associated with streams, rivers and sometimes standing water. Flight: v–viii – combined *crassipes* and *intermedius*.

Microphor holosericeus **(Meigen, 1804)**

This species resembles *strobli* in having slender hind legs but is distinguished by its dull black scutum with paler narrow grey lines and being considerably more hairy. Females are clearly striped grey and brown and are more obviously patterned than in *anomalus*, *crassipes* and *intermedius*. Wing length ♂: 3.0 (2.5–3.4) mm, ♀: 3.0 (2.6–3.4) mm.

The species is widespread and moderately frequent in England and Wales but with only rare records in Scotland. Its wide range of habitats include deciduous woodland, wood-pasture, meadows and pasture, and various water-margin situations where the water or wetland may be incidental. Flight: iv–viii, probably peaking in May.

Microphor intermedius **(Collin, 1961)**

This species closely resembles *anomalus* and *crassipes* but may be recognised by the combination of sparse hairs on the sternites (similar to *crassipes*) and a broad hind basitarsus (similar to *anomalus*). The scutum is usually entirely grey in anterior view with no dark stripes below the acrostichal or dorsocentral rows although indistinct dark stripes are evident when the specimen is tilted back. When the hypopygium is pulled away from the abdomen, it is sometimes possible to discern the tiny tubercle with its hair on the posterior margin of the epandrium close to the dorsal surstylus; this tubercle is absent in *crassipes*. See the accounts for *crassipes* and *anomalus* for further distinctions. Wing length ♂: 2.4 (2.3–2.5) mm.

The few confirmed records of *intermedius* suggest that it is probably as frequent as *crassipes*, at least in the southern half of England, and it has been recorded northwards to Nottinghamshire. It has been found next to rivers and canals. Flight: v–viii – combined *crassipes* and *intermedius*; confirmed *intermedius* vi–vii.

Microphor strobli **Chvála, 1986**

The small size and dull black almost unstriped thorax make *strobli* particularly undistinguished. Females are uniformly pale grey with any darker brown lines being barely discernible. In both sexes the steep step on the postpedicel between the swollen base and tapered section differs from the more gently sloping profile of *holosericeus*. Wing length ♂: 2.0 (2.0–2.1) mm, ♀: 2.1 (2.0–2.2) mm.

Records are concentrated in northern England with one in the southwest but this reflects a low level of recording. The habitat where known includes calcareous fens, reedbeds, post-industrial flower-rich coastal grasslands, scrub and scrubby woodland, and riverside exposed cobbles and sand (Plant and Cole 2005). Flight: vi–vii.

MICROPYGUS Bickel and Dyte, 1989 (Sympycninae)

These are small brown flies resembling *Campsicnemus* in their partially constricted face and overall appearance with long wings and hind femora and relatively short abdomen (Plate 5). *Micropygus* differs from *Campsicnemus* in lacking the strong proepisternal seta, although Parent (1933) states that it is present in his generic description, and the hind basitarsus is noticeably thicker and slightly shorter than the second tarsomere compared to much shorter in *Campsicnemus*. Females have six obvious tergites compared to five in *Campsicnemus*. There are no acrostichal setae, five pairs of long dorsocentral setae, and an anterior preapical seta on the mid and hind femora. The end of the abdomen looks superficially similar in both sexes but males, while they have no secondary sexual characters, are recognised by their narrower face which is scarcely wider at the constriction than the front ocellus compared to obviously wider than the ocellus in females.

The genus *Micropygus* is endemic to New Zealand from where the single species recorded in the British Isles was presumably introduced.

Micropygus vagans Parent, 1933 (Sympycninae) (Plate 5)

Although rather drab and with mainly yellow simple legs, this species is readily distinguished by the small pale spot over crossvein dm-m which itself is almost transparent, with the adjacent membrane being clear and contrasting with the remaining uniformly pale brown wing. In both sexes, the occiput, scutum and postnotum have a marked shifting tessellated pattern of pale brown dust and almost black areas with no hint of green. Chandler (1999) provides a detailed description. Wing length ♂: 2.7 (2.6–3.0) mm, ♀: 2.9 (2.7–3.1) mm.

It was first found in Ireland in 1971 and since 1995 has been recorded widely in Scotland, particularly in the southern half, and bordering English counties (Chandler 1999, Chandler and Smith 2005). It is found in damp or dry woods or beside ponds and streams in shaded places, and rarely in open habitats. Flight: vi–x, peak in vii.

MUSCIDIDEICUS Becker, 1917 (Dolichopodinae)

There is only one species in this genus worldwide. It is distinguished by the combination of dorsal hairs on the scape and seven dorsocentral setae. Among Dolichopodinae it is distinct in have strong acrostichal setae that are each about two-thirds the length of the closest dorsocentral seta, long white hairs on front coxa, dense numerous white hairs on the proepisternum, R_{4+5} and M_1 almost parallel and both strongly bowed (less pronounced in the female), and a dorsoventrally flattened abdomen (Plate 6). It stands out from similar-sized dolichopodids because the head and front coxa appear under-sized and the front tarsus is noticeably shorter than the tibia. Genera in the subfamily Dolichopodinae usually have an obvious anterior preapical setae on the hind femur but in *Muscidideicus* this seta is weak and sometimes lost among quite long vestiture.

Muscidideicus praetextatus (Haliday, 1855) (Plate 6)

This is a medium-sized green species with yellow legs; males have brilliant green eyes in life. Wing length ♂: 4.6 (4.3–5.1) mm, ♀: 5.1 (4.8–5.6) mm.

It is a rare species found intermittently on the English coast in the vicinity of The Wash and along the south coast, in Wales from Carmarthenshire to Anglesey, and rarely in Scotland. It is confined to saline conditions and is found most often at upper saltmarsh but may occur on nearby dunes or estuarine mud. Flight: vi–viii.

Fig. 27. *Muscidideicus praetextatus* hypopygium.

NEMATOPROCTUS Loew, 1857 (Rhaphiinae)

Nematoproctus are medium-sized flies resembling *Rhaphium* in their overall slender shape and shining green colour (Plate 6). The genus is distinguished from all others by the combination of biserial acrostichal setae, six pairs of dorsocentral setae, no anterior preapical seta on the mid or hind tibiae, and a purple spot at the side of the scutum. Males additionally have a bare sixth tergite and dense postgenal whiskers, and females have robust deep mouthparts. With the exception of the absence of a preapical seta on the mid femur, these characters are the same for *Rhaphium*.

Two of the three Western Palaearctic species occur in Britain. *Nematoproctus longifilus* Loew, 1857 could occur here; it is recognised by its cercus being far longer than that of *distendens*, reaching forward to the third abdominal segment, and by its almost entirely yellow legs like those of *praesectus*. In females, both the antennal scape and palps are partly yellow; in neither of the other two species are both these parts partly yellow at the same time.

They are rare lowland wetland species found mainly by standing water or in swamp.

Key to species of *Nematoproctus* (males)

1 Hind leg mainly black, yellow only on basal half of femur. Cercus long ribbon-shaped, conspicuous, projecting far beyond surstylus (a). Mid femur with ventral hairs as long as depth of shaft (b). Front tarsus with a ventral fringe as wide as shaft only on basal two-thirds of basitarsus, becoming shorter apically (c)........ ***distendens***

– Hind leg mainly yellow, becoming black only at tip of tibia and on tarsi. Cercus triangular, inconspicuous, scarcely projecting beyond surstylus (aa). Mid femur with ventral hairs short, less than half depth of shaft (bb). Front tarsus with ventral fringe as deep as width of shaft on tarsomeres 1–3, becoming shorter apically (cc)........ ***praesectus***

(a)

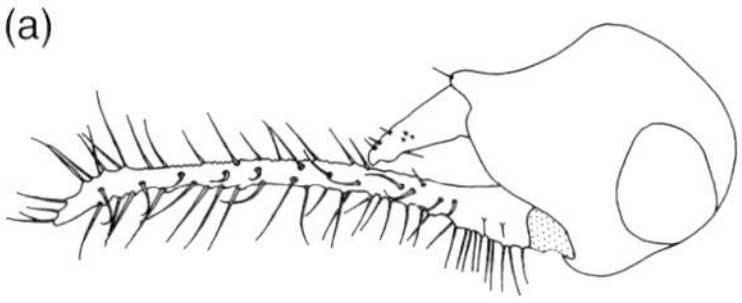

(aa)

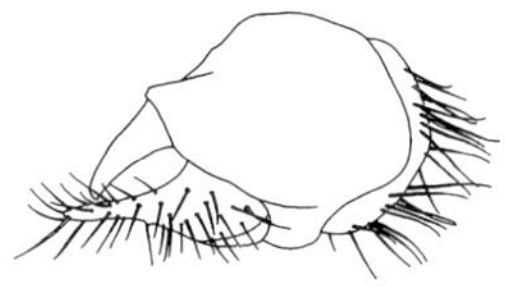

(b)

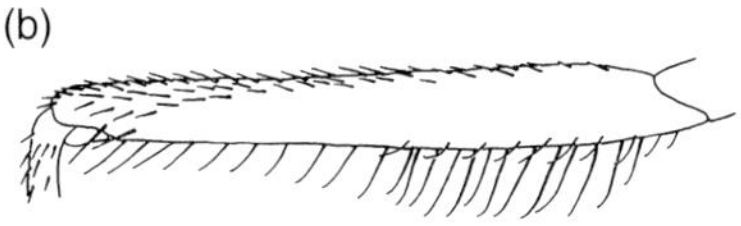

(bb)

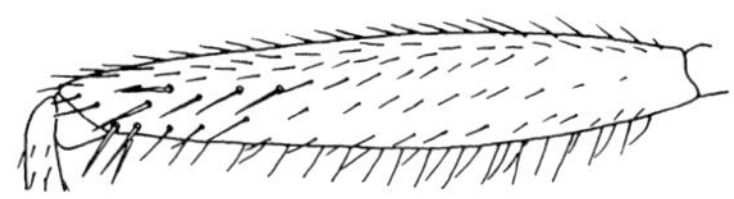

(c)

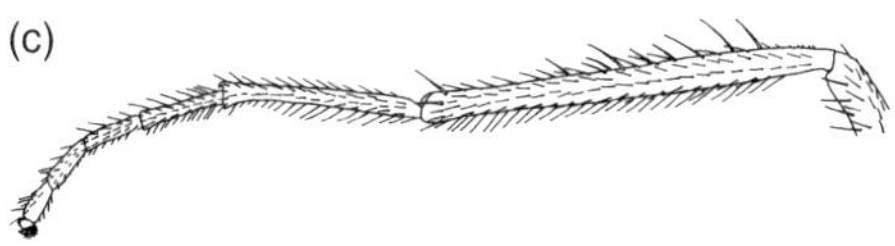

(cc)

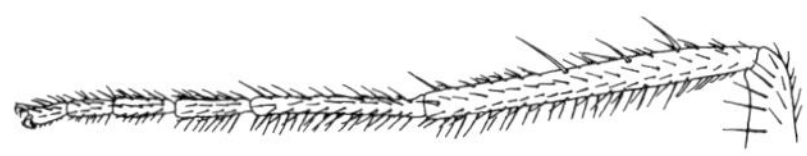

Key to species of *Nematoproctus* (females)

1 Front coxa black at base. Hind tibia dark to black on at least apical half. Wing darker anteriorly. Tergites 2 and 3 with long marginal setae, half length of tergites. .... ***distendens***

– Front coxa entirely yellow. Hind tibia dark on apical quarter. Wing uniformly grey. Tergites 2 and 3 with short inconspicuous marginal setae, one-third length of tergites. .... ***praesectus***

Nematoproctus distendens (Meigen, 1824) (Plate 6)

The long ribbon-shaped cercus and dense whiskers resemble those of several species of *Rhaphium*, but male *distendens* are distinguished by the short triangular postpedicel with a dorsal arista. The hind legs are deep black from the middle of the femora to the tips of the tarsi. The femora have long fine hairs, particularly the front one posteriorly and mid one ventrally. In both sexes the wings are clearly darker anteriorly. In females, the tergite hairs are long, particularly the marginals, forming a lateral 'fan' on the first tergite. Wing length ♂: 4.5 (4.4–4.6) mm, ♀: 4.6 (4.0–5.0) mm.

This is a rare species found sparsely in England from the south coast to Yorkshire. Its habitat appears to be lowland swamp, either in partial shade or unshaded, and perhaps with bare or sparsely vegetated substratum. Flight: v–vii.

Nematoproctus praesectus **Loew, 1857**

This almost entirely yellow-legged species has inconspicuous cerci in the male. The femora have conspicuous hairs only anteriorly on the front femur and only towards the tip on the mid femur. The ventral fringe of the front tibia is dense and even, continuing onto the tarsi and fading out on the fourth tarsomere. The female's tergite marginal setae are inconspicuous and there is no clear lateral 'fan' on the first segment. Wing length ♂: 3.5 mm, ♀: 4.4 (4.3–4.4) mm.

The species was recorded for the first time in Britain together with *distendens* in West Sussex (Drake 2019) at pond margins and swamp. Pollet *et al.* (1988) recorded it from river banks with sparse vegetation. Flight: v.

NEURIGONA Rondani, 1856 (Neurigoninae)

Species of *Neurigona* are moderately large and very long-legged flies, their front coxae over-reaching the tip of the katepisternum, often slender-bodied in the male in contrast to more compact females, and unusually with conspicuous yellow colouration, especially on the tergites (the thorax may be mostly dark grey) but with no trace of green (Plate 6). The genus is recognised by the combination of a dorsal to subapical arista, biserial acrostichal setae, a strongly marked prescutellar depression, no anterior preapical setae on the mid and hind femora, and dark setae and hairs on the front and hind coxae. *Anepsiomyia*, *Xanthochlorus* and some *Achalcus* are also partly yellow flies without any green marks but are smaller, and *Xanthochlorus* has pale seta on the coxae. Males often have large, somewhat lumpy male genitalia carried on a peduncle and with reduced cerci.

Neurigoninae is a small family of two Palaearctic genera of which only *Neurigona* occurs in Britain. There are 11 European species (Pollet 2011), six in Britain. It is considered unlikely that others will be found here.

The British species of *Neurigona* are somewhat more xerophilic than most dolichopodids and many have a strong association with tree trunks. Kowarz (1869) was probably the first author to describe the flight of males up and down tree trunks. d'Assis-Fonseca (1955) describes the behaviour of male *pallida* making zig-zag flights up smooth trunks of ash and sycamore with a narrow range of diameters of about 12–18 cm; Smith (1959) recorded a similar preference for trunks of small diameter for *quadrifasciata* whereas *suturalis* used much larger trunks. Smith (1959) made detailed observations of the courtship of *quadrifasciata*, and of particular interest is that male stands behind the female while reaching forward with his exceptionally long front legs, and waving his broadly feathered fore tarsi over the female's eyes. *Neurigona quadrifasciata* has been reared from soil in a beech wood (Beling 1882). Insufficient is known of their immature stages to be certain whether it is only the adult that make use of tree trunks or whether larvae develop in decaying wood as well as in soil.

Key to species of *Neurigona* (males)

1 Abdomen with wide dark bands 2
– Abdomen entirely yellow, without dark bands or with only very narrow weak bands 5

2(1) Scutum almost entirely yellow with a narrow dark mark in front of scutellum (a). Tergites scarcely longer than wide (a). Dark bands on tergites 2–4 at least half length of tergite on midline (a). Eyes touching on face. Front tarsomeres 2–5 very short, altogether only two-thirds length of basitarsus (b). .... ***erichsoni***
– Scutum mainly or entirely grey. Tergites at least 1.5 times longer than wide, with dark bands substantially less than half length of tergites on midline. Eyes separated by about width of front basitarsus, face parallel-sided and silver. Front tarsomeres 2–5 combined at least as long as basitarsus 3

(a)

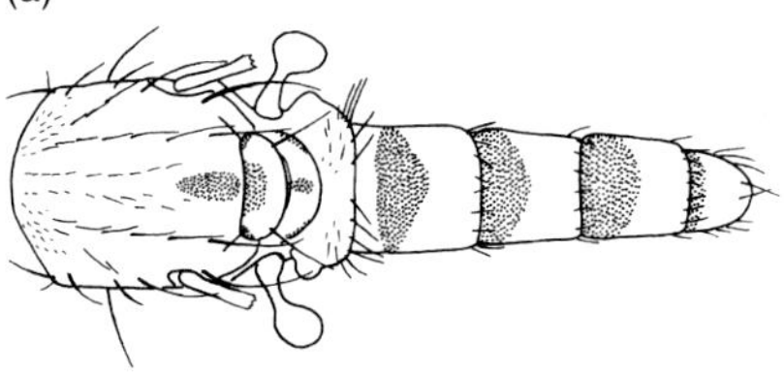

(b)

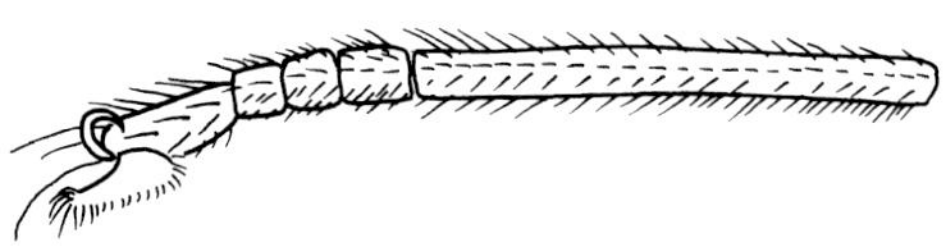

3(2) Front tarsus pale and slender with all hairs short and simple (a). Vein M_1 gently bowed, not sinuous (b). Anal lobe absent. Wing slightly darkened anteriorly. .... ***suturalis***

– Front tarsus black or bicoloured with modified shape, hairs or setae. Vein M_1 sinuous (bb). Anal lobe present. Wing membrane uniformly clear. 4

(a)

(b)

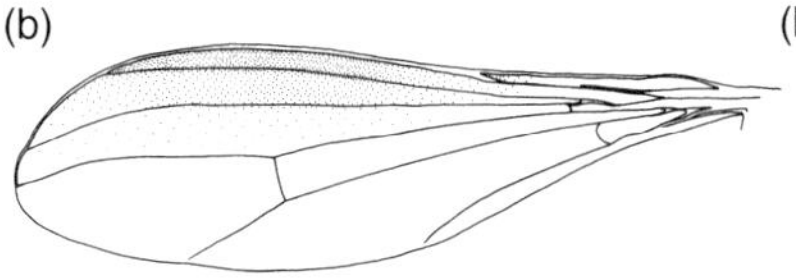

(bb)

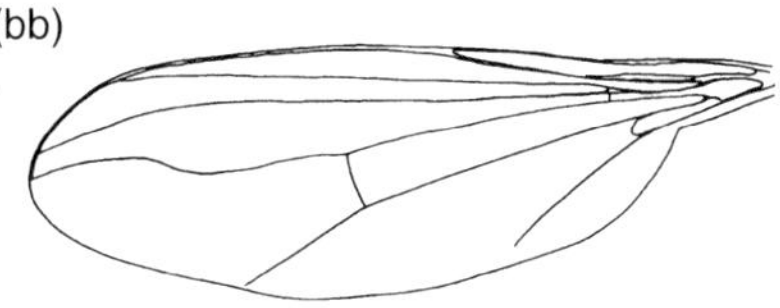

4(3) Front tarsomeres 3 and 4 with conspicuous lateral feathering of flat black setae contrasting with white tarsomeres 1, 2 and 5; tarsomeres 1 and 2 very long and slender (a). Last large (fifth) tergite pale, posterior margin bearing long curled pale setae. .... ***quadrifasciata***

– Front tarsus black and uniformly thick, almost as wide as front tibia and clearly thicker than mid tarsus; tarsomeres 3–5 with dense ventral hairing making segments appear stout (aa, lateral view). Last large (fourth) tergite unremarkable, bearing only short marginal setae. .... ***biflexa***

(a)

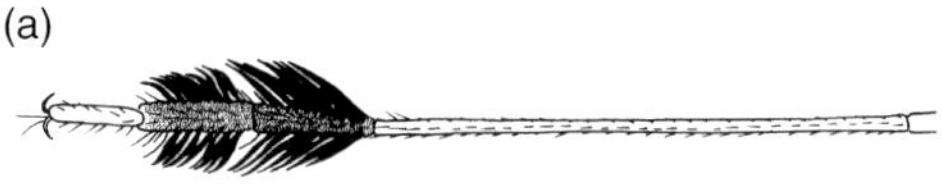

(aa)

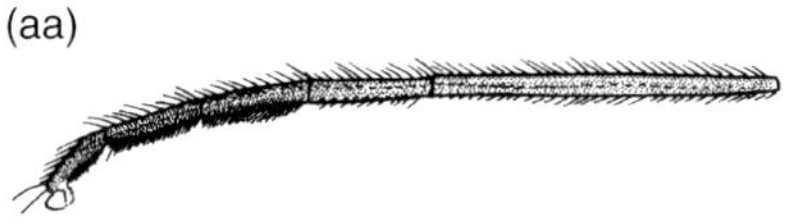

5(1) Scutum entirely yellow. Mid femur beneath at base with 3–4 short flat setae (a). Front tarsus almost unmodified, tarsomere 4 slightly clubbed (b). Large almost entirely yellow fly, wing longer than 5 mm. Tergites 4 and 5 and genitalia projecting ventrally. .... ***pallida***

– Scutum grey. Mid femur without flattened setae at base. Front tarsomere 5 with short coarse anterior spines and enlarged anterior sinuous claw (bb). Smaller fly, wing shorter than 4 mm. Tergites 4 and 5 and genitalia not deeper than tergite 2. .... ***abdominalis***

(a)

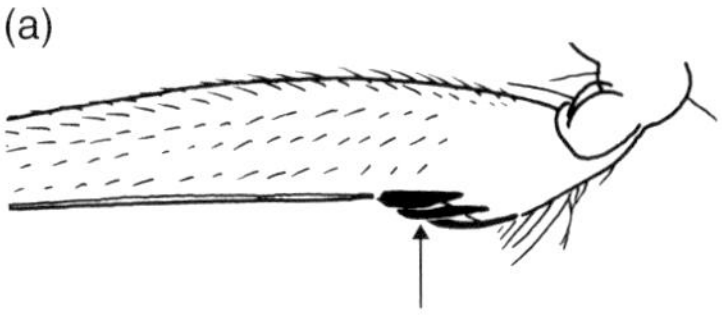

(b)

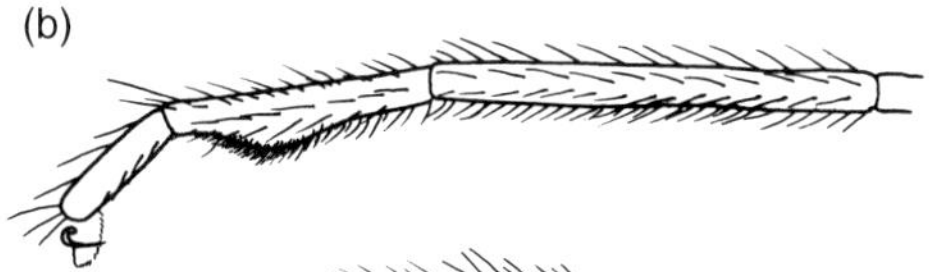

(bb)

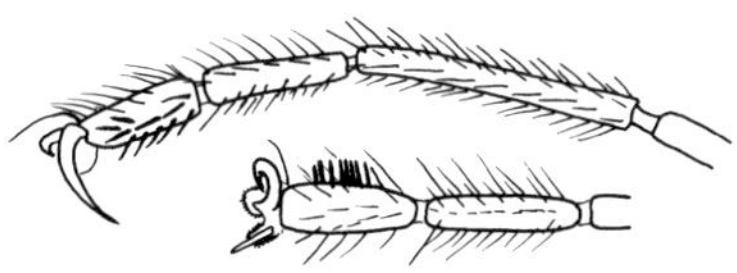

Key to species of *Neurigona* (females)

1 Abdomen with dark bands. 2

– Abdomen entirely yellow, without dark bands. 5

2(1) Scutum, at least in front half, predominantly yellow. 3

– Scutum entirely grey. 4

3(2) Scutum dark on hind third between dorsocentral setae and with 3 narrow dark lines extending forward along dorsocentral and acrostichal rows (a); postnotum almost entirely dark. Dark bands on tergites 3 and 4 strongly indented on midline (a)................ ***quadrifasciata***

– Scutum and postnotum almost entirely yellow (aa). Dark bands on tergites 3 and 4 lunulate (aa)............ ***erichsoni***

(a)

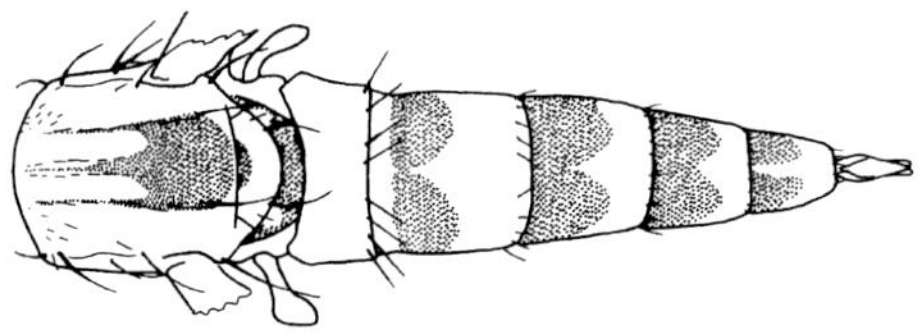

(aa)

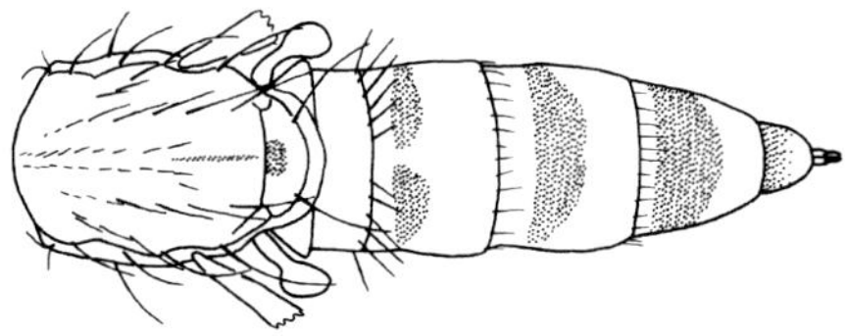

4(2) Vein M_1 very gently bowed, meeting costa almost at wing tip (a). Wing slightly darkened anteriorly. Hind tibia darker than yellow hind femur............ ***suturalis***

– Vein M_1 strongly sinuous, converging closely with R_{4+5}, meeting costa well before wing tip (aa). Wing entirely clear. Hind tibia and femur equally pale............ ***biflexa***

(a)

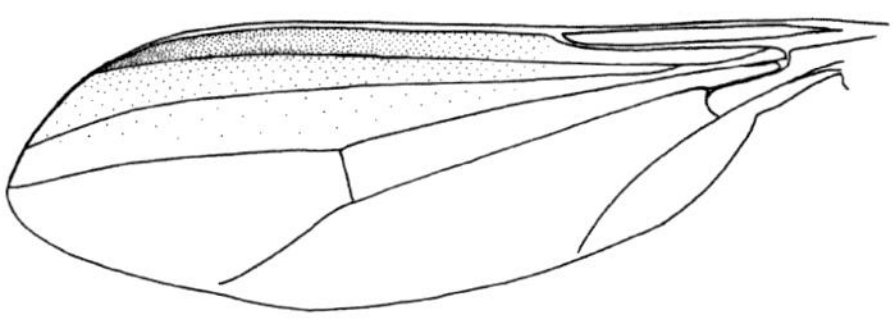

(aa)

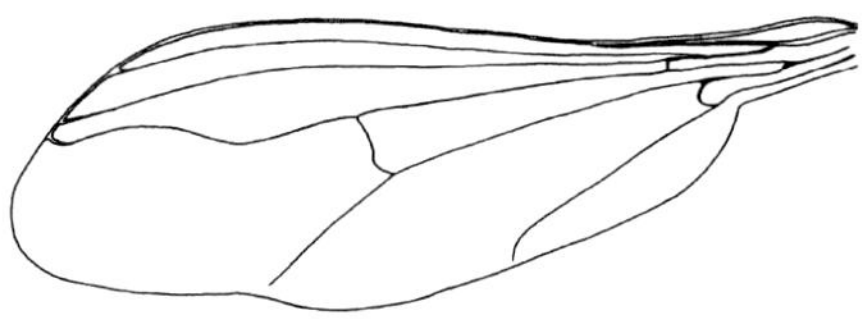

5(1) Scutum mainly yellow or pale brown; pleura yellow on posterior half. Large species, wing length 6–7 mm, rarely smaller............ ***pallida***

– Scutum and most of pleura dark grey. Smaller species, wing length 4–5 mm..... ***abdominalis***

Neurigona abdominalis (Fallén, 1823)

As the only British species with a grey scutum and no bands (or very narrow bands) on the abdomen, *abdominalis* is distinctive. Should the tergite bands be wide enough to cause confusion with *suturalis*, whose bands may be quite narrow, males can be separated by their different front tarsus and claw, and by the colour of the scutellum which is completely yellow in *abdominalis* but grey on the basal half in *suturalis*. Females of *abdominalis* are very similar to those of *biflexa*, except for the tergite bands, but may also be distinguished by the venation; in *abdominalis*, R_{4+5} and M_1 gradually converge at their tips whereas in *biflexa* they are parallel for the last 0.5 mm (about one-tenth of wing length) before the wing tip. Wing length ♂: 3.6 mm, ♀: 4.5–5.1 mm.

This is a rare species recorded at single sites in Kent, Essex, Hertfordshire, Cambridgeshire and Norfolk. It appears to be a species of woodland edge or light woodland, including suburban gardens (Bowden 1998, Laurence 1993). Flight: vi.

Neurigona biflexa Strobl, 1909

This species superficially resembles *quadrifasciata* but male *biflexa* are easily separated by their simpler front tarsi and absence of long marginal hairs on the last long tergite. Females resemble those of *abdominalis* – see the account for that species. Both sexes have a strongly sinuous vein R_{4+5} (male shown in female key but nearly identical). Wing length ♂: 3.9 (3.7–4.3) mm, ♀: 4.2 mm.

Neurigona biflexa has been recorded from only one site in Anglesey in 1987 where a female was found in scrub at the centre of a coastal dune system (Cole 1991). Flight: vii.

Neurigona erichsoni (Zetterstedt, 1843)

The almost entirely yellow scutum and pleura and lunulate bands on the tergites distinguish this particularly pale species. Males are also distinct in having a short abdomen compared to others in the genus. Wing length ♂: 3.8 (3.6–4.0) mm, ♀: 4.7 mm.

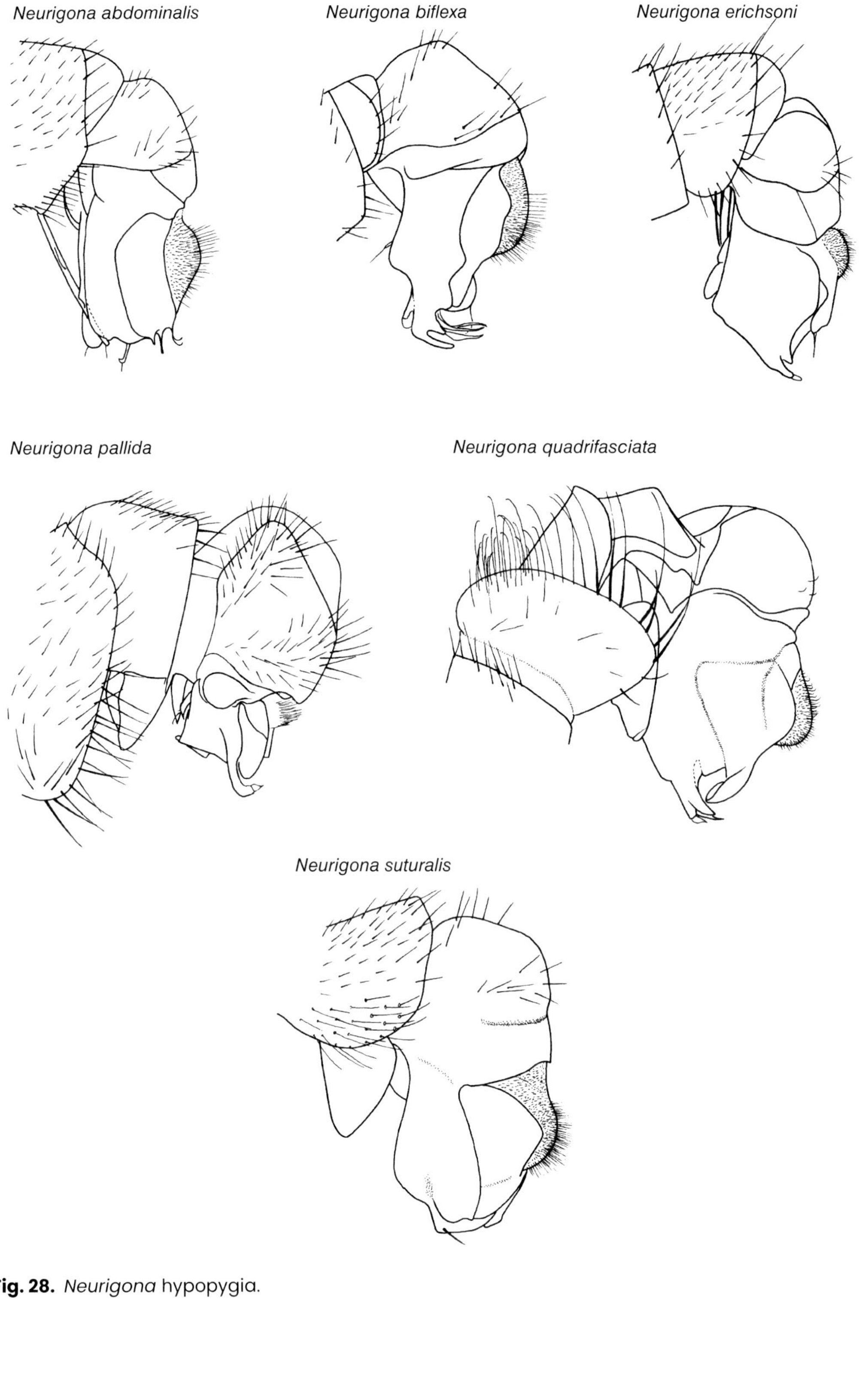

Fig. 28. *Neurigona* hypopygia.

Since its discovery in Britain in 2012, *erichsoni* has been recorded in Kent, Surrey, West Sussex and Dorset in deciduous woodlands (Drake and Hunnisett 2014). Flight: vii–viii.

Neurigona pallida (Fallén, 1823) (Plate 6)

This is a large robust and almost entirely yellow species, turning pale brown in dried specimens, with obvious dull black areas on some of the pleura. The male's leg ornamentation is restricted to the black flat setae at the base of the femur and an apical 'spur' of white hairs on the mid coxa. Wing length ♂: 5.6 (5.4–5.7) mm, ♀: 5.8 (4.7–6.4) mm (rare individuals can be smaller).

Neurigona pallida is widespread in England north to Cumbria and Yorkshire although apparently rare in the East Midlands and Wales. There are no Scottish records. It is usually found in damp and dry deciduous woodland, frequently oak woods but rarely in coniferous woodland, and very often close to streams. Flight: v–x, mainly vi–vii.

Neurigona quadrifasciata (Fabricius, 1781)

The conspicuous feathering of the male's front tarsi is unique to this species and is visible with the naked eye so the species can be identified in the field. If these legs are lost, *quadrifasciata* may be distinguished from the superficially similar *suturalis* by its inflected vein M_1 and the presence of long curled white hairs on the hind margins of the last large tergite. In females, the strongly indented tergite bands and tridentate scutum mark are unique among British species. Wing length ♂: 4.5 (4.2–4.9) mm, ♀: 4.6 (4.1–5.3) mm.

The species is frequent in southern and eastern England north to South Yorkshire, but almost absent in upland England north of the West Country. There are no Welsh records but an isolated population exists in Strathspey. It occurs most often in usually damp deciduous woodland, but also on isolated large hedgerow trees. Many records are by streams or other wetland features but this may be recording bias. Flight: iv–ix, mainly vi–vii.

Neurigona suturalis (Fallén, 1823) (Plate 6)

This slender species may be distinguished by the smoothly curved vein M_1, with no trace of a sudden inflexion, and the slightly darkened leading edge of the wing. Wing length ♂: 5.0 (4.6–5.2) mm, ♀: 4.6 (4.1–4.9) mm.

This is a lowland species found sparsely but widely in England, approximately south and east of a line from Dorset to South Yorkshire, rarely in Wales and with an old record from the Highlands in Scotland which needs confirming. It is found in dry woodland and scrub, and is perhaps associated with limestone and chalk. Flight: v–vii, peak in vi.

ORTHOCERATIUM Schrank, 1803 (Hydrophilinae)

Orthoceratium are moderately large, slender-legged species superficially resembling *Liancalus* (Plate 6). The single British species is unique in both sexes in having a long black seta at about the basal third of the front coxa, standing out against the white hairs. The combination of uniserial acrostichal setae, seven dorsocentral setae, a clearly demarcated prescutellar depression and strongly excavated vertex either side of the ocellar triangle is found in only a few other hydrophiline genera.

Palaearctic *Orthoceratium* includes two species of which *O. lacustre* (Scopoli, 1763) occurs in the Mediterranean basin and is unlikely to be found in Britain (Pollet and Stark 2018). The British species, *sabulosum*, had previously been misidentified as *lacustre*.

Orthoceratium sabulosum (Becker, 1907) (Plate 6)

Orthoceratium sabulosum is shining bright green all over, including the legs except for their narrow bright yellow knees; in other dark-legged dolichopodids the green on the legs is rather dark. Males may be immediately recognised from all other dolichopodids by the very pronounced bends on vein M_1, and their front femora having a narrow posteroventral depression near the base covered in dense pale pubescence. Wing length ♂: 5.6 (5.4–6.0) mm, ♀: 6.0 (5.5–6.3) mm.

Orthoceratium sabulosum is an uncommon coastal species recorded most frequently on the south and east English coasts and the south coast of Wales. Its distribution extends north to Cumbria on the west coast and to the Moray Firth on the east. Records are mainly from saltmarsh or coastal dune slacks and brackish marsh, with a few from coastal freshwater pools, gravel pits and small reed-fringed lakes (Pollet *et al.* 2017). Clemons (2003) lists drier sites. Flight: v–x.

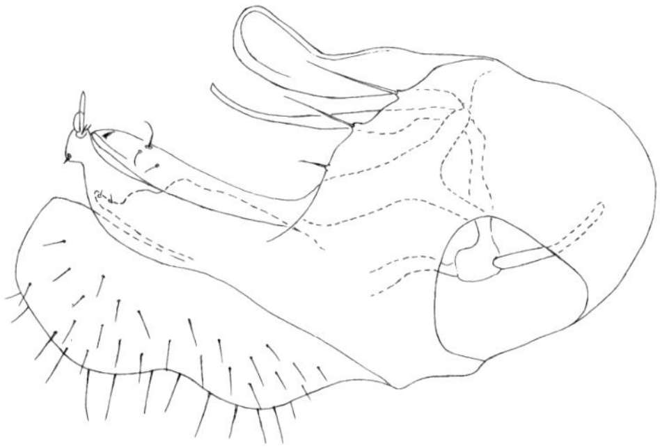

Fig. 29. *Orthoceratium sabulosum* hypopygium.

ORTOCHILE Latreille, 1809 (Dolichopodinae)

Ortochile has the features that characterise Dolichopodinae (scape hairy, dorsal arista, biserial acrostichal setae, and mid and hind femora each with an anterior preapical seta) and is distinguished from other genera in the subfamily by the very long mouthparts. There is one British species; two others occur in southern Europe.

Ortochile nigrocoerulea Latreille, 1809 (Plate 6)

This is a small fly whose long mouthparts and a sometimes strikingly bicoloured mid femur, with its apical half to quarter yellow, make it easily distinguished from other British species. *Hercostomus nigripennis* has moderately long mouthparts but entirely black femora. Wing length ♂: 2.9 (2.6–3.2) mm, ♀: 3.1 (2.9–3.3) mm.

This species was once distributed moderately widely in southeast England and with outlying populations in Dorset and Cornwall. There has been no reliable record since 1960. Dyte (1988b) suggests that it may have been a migrant that established temporary populations. Its habitat requirements are unknown. The long mouthparts are an adaption for feeding at flowers (Andrews 1939, Drake 1999). Flight: vi–viii.

Fig. 30. *Ortochile nigrocoerulea* hypopygium.

POECILOBOTHRUS Mik, 1878 (Dolichopodinae)

These are moderately large metallic green species (Plate 6). The genus is in the Dolichopodinae so possesses the characters of that subfamily (scape hairy, dorsal arista, biserial acrostichal setae, mid and hind femora each with an anterior preapical seta) and is recognised within the subfamily by the combination of the scutum having a dark purplish black spot above the notopleuron just behind the suture and usually a strong posterior preapical seta on the mid femur in addition to the anterior one (weaker in *chrysozygos*). A purple thoracic spot is also found in *Dolichopus ungulatus* and Rhaphiinae. In British species the tibiae have many strong setae, with at least nine on the mid leg and 12 on the hind leg, veins R_{4+5} and M_1 converge towards the wing margin where they reach the costa well in front of the wing tip and, except in *chrysozygos*, the arista is conspicuously pubescent with individual hairs being as long as the basal diameter of the arista. The cercus is characteristically long-triangular with long curved setae set on finger-shaped lobes along the whole of the outer edge.

There are at least nine species of *Poecilobothrus* in the Western Palaearctic of which five occur in Britain. *Poecilobothrus regalis* (Meigen, 1824) is widespread in central Europe and may occur in Britain; it is distinct in having a dorsal seta on the hind basitarsus, like *Dolichopus*, but has the venation typical of *Poecilobothrus* and would run to *majesticus* in this key. Small but consistent differences in the hypopygium, particularly in the easily seen basiventral epandrial lobe, may help identify species not yet known in Britain, along with *majesticus* whose genitalia have not been figured as the hypopygium of the type specimen is distorted.

Key to species of *Poecilobothrus* (males)

1 Wing strongly darkened on apical half with a white spot at tip (a) .... ***nobilitatus***
– Wing without strongly contrasting shades on apical half 2

(a)

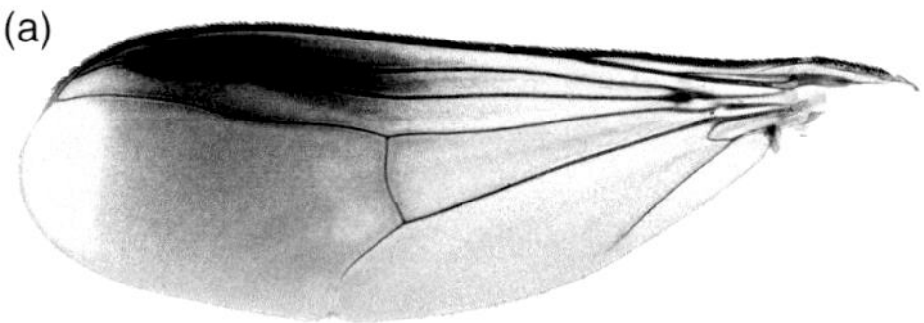

2(1) Face dusting golden or yellow. Hind tibia black on at least apical fifth 3
– Face dusting silver or white. Hind tibia entirely yellow or dark only at tip 4

3(2) Front tarsus black with a white band on basitarsus and tarsomere 2 (a). Hind tibia black on about apical half. Antenna mainly yellow, postpedicel black on apicodorsal half. Hind femur ventrally with unremarkable setulae .... ***chrysozygos***
– Front tarsus dark from tip of basitarsus outwards. Hind tibia black on apical quarter to fifth. Antenna entirely black. Hind femur posteroventrally at base with a few hairs that become bi- to triserial rows of closely spaced and very fine black setulae reaching tip of femur .... ***ducalis***

(a)

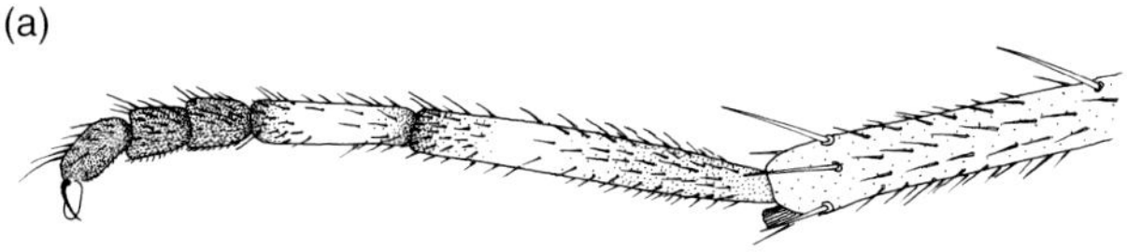

4(2) Hind coxa yellow. Antenna mainly yellow, postpedicel black in apical half. Hind femur at base with a single posteroventral row of short setae that become smaller towards tip, ending as tiny points............ ***principalis***

– Hind coxa black. Antenna black apart from underside of scape. Hind femur ventrally as *ducalis*............ ***majesticus***

Key to species of *Poecilobothrus* (females)

1 Postpedicel entirely black. Mid coxa black with a yellow tip. Palp black, dark brown or with only apical half yellow............ 2

– Postpedicel yellow in basal half. Mid coxa yellow on inner face or more extensively yellow, as well as at tip. Palp bright yellow............ 3

2(1) Front coxa with hairs on inner half black and short, on outer half pale and fine with a small proportion of black hairs mixed with them (a, front view). Front coxa entirely yellow except at extreme base. Scape entirely black. Mid tibia with 2 or more true ventral setae (b). Hind tibia black on apical fifth or less (c). .... ***nobilitatus***

– Front coxa with hairs dense and nearly all black (as fig. b of couplet 3). Front coxa with a dark mark at outer setae or more extensively dark, and on extreme inner (hairless) side. Scape black dorsally, yellow ventrally. Mid tibia with only 1 true ventral seta (bb). Hind tibia black on apical quarter to third (cc)............ ***ducalis***

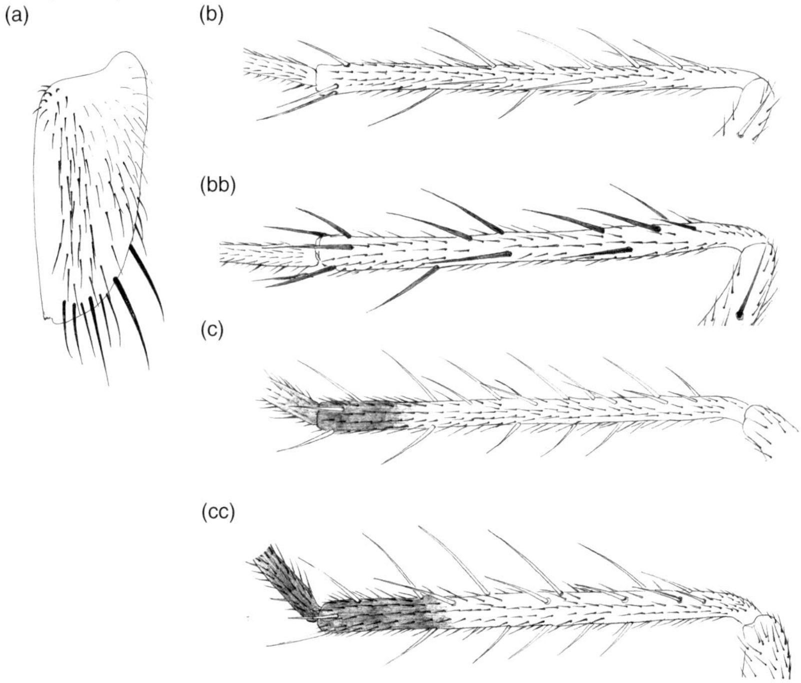

3(1) Hind tibia black on about apical half (a). Hind femur with a black dorsal spot at tip (a). Front coxa with vestiture dense and almost entirely black with only a few fine pale hairs on upper outer corner, black hairs above outer seta reaching most or all the way to articulation with thorax (b, front view). Crossvein dm-m placed at about middle of wing so apical section of M_4 is almost twice as long as dm-m (c) and apical section of M_1 is about equal to basal section measured from fork of M_1 and R_{4+5} (level with crossvein h). Vein M_1 with distinct double inflexion on apical section (c). .... ***chrysozygos***

– Hind tibia yellow with at most one-sixth black at tip on posterior side. Hind femur entirely yellow at tip. Front coxa with vestiture sparse and black on inner and lower halves, pale and fine on upper outer quarter, black hairs above outer seta reaching at most halfway to articulation with thorax (as fig. a of couplet 2). Crossvein dm-m placed beyond middle of wing so apical section of M_4 is about equal to dm-m (cc) and apical section of M_1 is noticeably shorter (85–90%) than basal section measured from fork of M_1 and R_{4+5}. Vein M_1 with inflexion scarcely marked (cc). .... ***principalis***

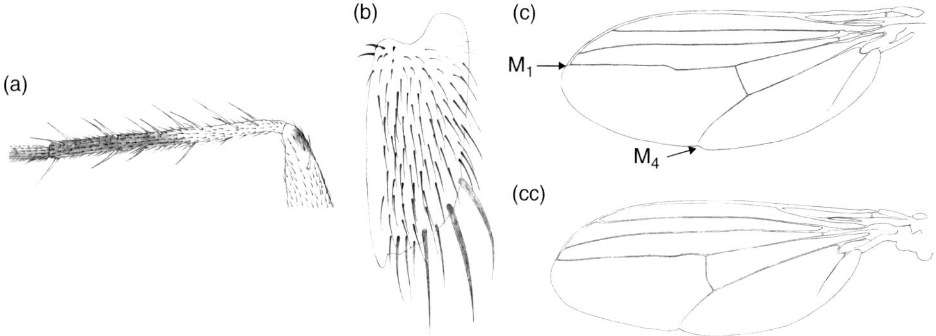

Poecilobothrus chrysozygos (Wiedemann, 1817) (Plate 6)

No other British dolichopodid has such conspicuously annulated front tarsi. Males are recognised by a yellow face and mainly yellow antennae. Females are recognised by their more extensively black hind tibia, the tiny black spot on the tip of the hind femur and having a distinct inflexion on the apical section of vein M_1, almost as in *Dolichopus*. Wing length ♂: 4.6 (4.2–4.9) mm, ♀: 4.8 (4.5–5.2) mm.

Poecilobothrus chrysozygos is found in lowland England northward to South Yorkshire and reaching south Wales but with its greatest frequency in southeast England and East Anglia. Its habitat is swamp, reedbed, fen and well-vegetated margins of ponds and ditches on wet grasslands, nearly always unshaded. There may be a preference for more nutrient-rich sites as it is scarce in apparently suitable habitat of Norfolk fens.

Poecilobothrus ducalis (Loew, 1857)

A yellow face and black antennae distinguish male *ducalis*. Females have partly black front coxae, even if represented by only a small spot by the outer seta and on the less easily seen inner surface, rather than being entirely yellow. The black marking at the tip of the hind tibia is usually wider than in *nobilitatus* but can be too similar in the two species to be reliable alone. The extensive black hairs on the front coxa are as in *chrysozygos* and are markedly different from the condition in *nobilitatus* and *principalis*. Wing length ♂: 5.7 (5.2–6.5) mm, ♀: 5.5 (4.7–6.0) mm.

This is a coastal species found on the English coast from Suffolk to Hampshire and in Somerset with most records from the Thames estuary in Essex and Kent. It lives on upper saltmarsh and brackish marsh. Flight: v–viii.

Poecilobothrus majesticus d'Assis-Fonseca, 1976

This species is known only from the male type specimen which is somewhat crumpled but is still in moderately good condition. It closely resembles *ducalis* and differs only in having a white face, entirely yellow hind tibia and yellow scape. Wing length ♂: about 4.5 mm.

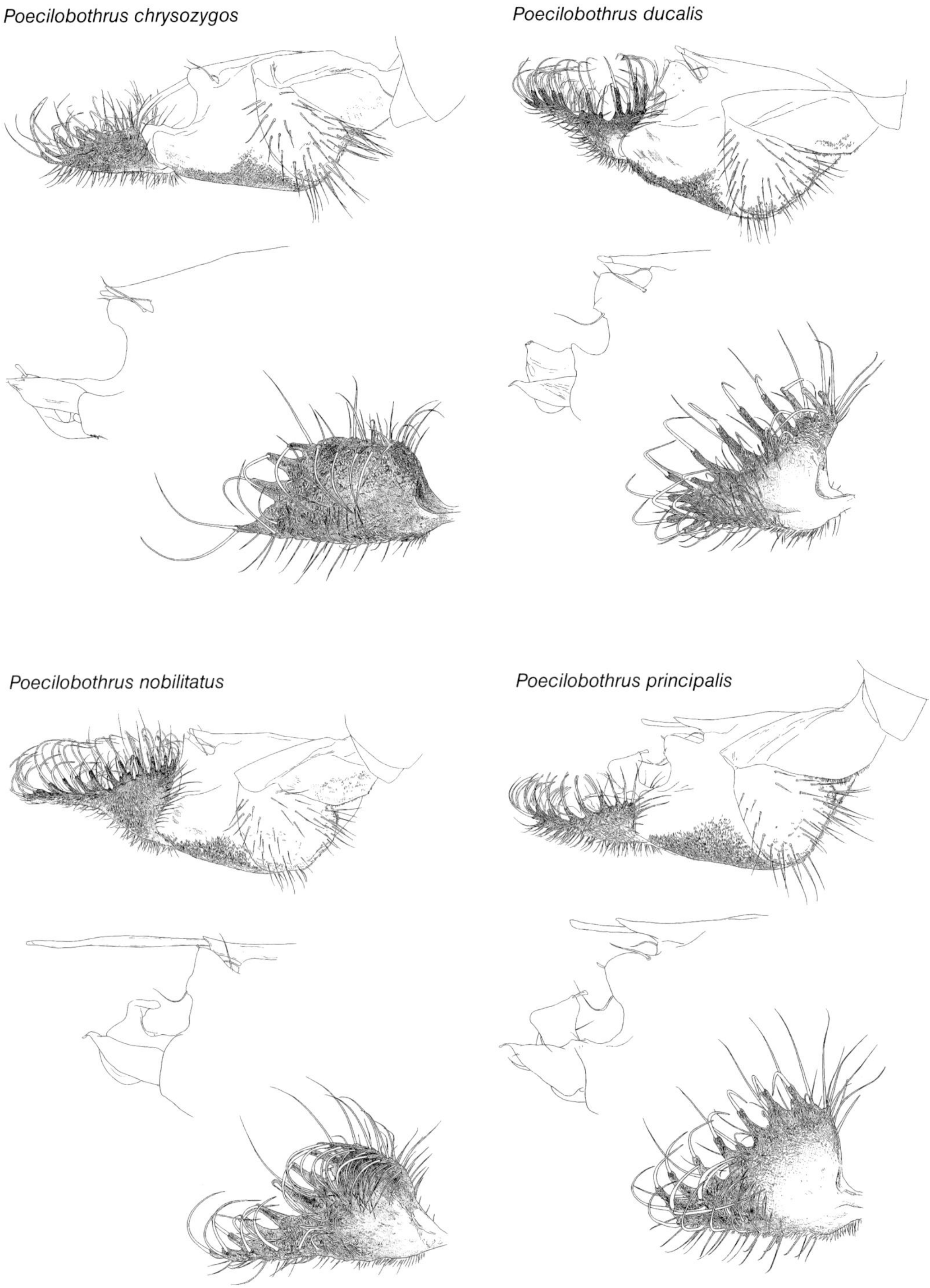

Fig. 31. *Poecilobothrus* hypopygia, excluding *majesticus*.

A single male collected from Walton-on-the-Naze, Essex in 1909, presumably from a coastal habitat, remains the only record of this species worldwide. Flight: viii.

Poecilobothrus nobilitatus **(Linnaeus, 1767) (Plate 6)**
The male's conspicuous wing pattern makes this species instantly recognisable in the field and cannot be confused with any other British dolichopodid. It has a white face and entirely black antennae. Females also have entirely black antennae, unlike other *Poecilobothrus*. In samples of mixed female *nobilitatus* and *Dolichopus* species, *nobilitatus* is quickly separated by veins R_{4+5} and M_1 converging closely towards the wing tip, compared to remaining nearly parallel to one another beyond the inflexion in *Dolichopus*. Wing length ♂: 6.1 (4.9–6.8) mm, ♀: 5.4 (4.1–6.1) mm.

Poecilobothrus nobilitatus is a common and widespread species in the southern half of Britain and frequent in lowland areas of Wales, becoming less frequent in the north of the country. In Scotland it was unknown until 2015, since when it has been found in the central lowlands and the Aberdeenshire coast. Almost any type of water body will support adults, usually in unshaded places. Although lekking adults are a common site at water margins, it is not clear where larvae develop. Courtship has been described in detail (Land 1993a, b, Lunau 1992, 1996, Lunau *et al.* 2006) although mating is rarely observed (Noble 2016). Flight: v–xi, peak in vii.

Poecilobothrus principalis **(Loew, 1861)**
A white face and mainly yellow antennae distinguish *principalis*. Females have almost entirely yellow hind tibiae that are only slightly dark at the tip, unlike the obviously black-tipped tibia of the other *Poecilobothrus*, and the hind coxa is largely yellow rather than black as in the other species. Wing length ♂: 5.5 (4.8–6.0) mm, ♀: 5.4 (4.8–5.8) mm.

It is distributed on the English and Welsh coasts south of a line from the Humber estuary to Pembrokeshire with the greatest concentration of records from the Thames estuary in Essex and Kent. It is found on saltmarsh and brackish marshes though may stray onto adjacent dunes. Adults form leks on pools, like those of *nobilitatus*. Flight: vi–ix, peak in vii.

RHAPHIUM Meigen, 1803 (Rhaphiinae)

Rhaphium are small to large metallic green to dark brown flies with a dark abdomen except in *fasciatum*. The combination of a purple spot at the side of the scutum, an apical or subapical arista on a short to long-triangular postpedicel, and the lack of a prescutellar depression will distinguish both sexes of *Rhaphium* from all other genera (Plates 6 and 7). The purple spot is also found in *Nematoproctus*, *Poecilobothrus* and *Dolichopus ungulatus*, and this area of the scutum is sometimes purplish, black or coppery is a few other genera. Males of several genera have antennae similar to those of *Rhaphium*, and of these *Achalcus*, *Argyra*, *Syntormon* and *Systenus* are most likely to be confused with only cursory inspection. One group of *Rhaphium* species has a distinctive strap-shaped postpedicel that is not found in other genera. The cercus and surstylus in many species are rarely simple and the cercus, at least, can usually be seen without having to extend the hypopygium. Females have robust deep mouthparts and usually conspicuous pale dust patches on the sides of the tergites, as in *Argyra*. In addition, both sexes have biserial acrostichal setae (rarely uniserial), the mid femur has an anterior preapical setae, and the vertex is strongly excavated either side of the ocellar triangle.

There are about 50 species in Europe (Pollet 2011) of which half (26 species) occur in Britain. Several species occur in nearby mainland European countries but perhaps only *discigerum* Stenhammar, 1850 is likely to be found in Britain; like *antennatum*, it has a swollen tip on the arista but has black rather than yellow femora.

Negrobov (1979b) and Negrobov *et al.* (2012) did not accept Collin's (1940) suggested use of the names of *riparium* for *praerosum* (Loew), *caliginosum* for *lanceolatum* and *appendiculatum* for *macrocerum* (Meigen). As a result, the species illustrated here for British specimens under the first name in each of these pairs are illustrated by Negrobov under the second name.

Rhaphium includes species that were originally placed in three genera, *Rhaphium*, *Porphyrops* and *Xiphandrium*. The synonymy of these has long been accepted (e.g. d'Assis-Fonseca 1978) but in the limited British fauna the names *Porphyrops* and *Xiphandrium* do conveniently identify two clearly distinct groups. In the species accounts, the more immediately recognisable of these is referred to as the '*appendiculatum*' group which includes species previously in *Xiphandrium* characterised by possessing a seta on the outer face of the hind coxa and the males usually having a long ribbon-shaped postpedicel and short arista. Here these are *appendiculatum*, *albomaculatum*, *auctum*, *brevicorne*, *caliginosum*, *lanceolatum* and *monotrichum*.

Within females, two pairs and one trio of species appear to be impossible to separate reliably: *appendiculatum* and *caliginosum*, *lanceolatum* and *auctum* + *monotrichum*, *riparium* and *penicillatum*. Females often predominate in catches of some of these species so a large proportion of records may be based on suspect identifications.

All species of *Rhaphium* are associated with wet habitats, with *appendiculatum* and *caliginosum* living at the drier end of the spectrum. Some show a strong affinity for particular habitats: acid mires and wet heaths support *fascipes*, *longicorne* and possibly *lanceolatum*, river shores support *fractum*, *gravipes*, *penicillatum*, *rivale* and *suave* as well as several other species that also live by smaller streams, while *consobrinum* is a saltmarsh species and *elegantulum* is mostly found at lake shores.

Key to species of *Rhaphium* (males)

1 Antenna very long and thin, combined with arista more than twice as long as front tibia (a). Scape stout and broader than widest part of postpedicel (a). Cercus clothed in dense short upright hairs (b). Hind coxa with 2 black lateral setae. .... ***longicorne***

– Antenna and arista combined much shorter than twice the length of front tibia. Scape conical and clearly narrower than widest part of postpedicel. Cercus clothed in hairs of uneven lengths, often straggly. Hind coxa with 0 or 1 lateral seta 2

(a)

(b)

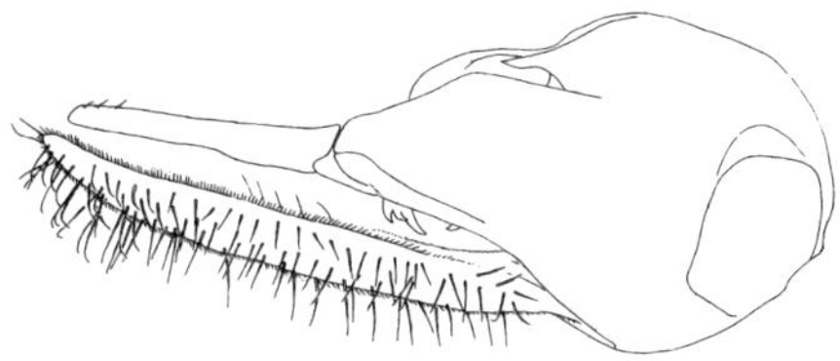

2(1) Tergites 2 and 3 mainly yellow. All legs yellow from coxa to basitarsus. [Front tibia with only 1 stout dorsal seta followed by a row of 5 or more setulae (a). Small delicate species.] .... ***fasciatum***

– All tergites dark. At least part of one coxa dark... 3

(a)

3(2) Mid coxa with a pale apical spur of coalesced or tightly clustered white hairs (a). Postpedicel in most species elongate, ribbon-shaped and 2 or more times longer than arista (b), less in *brevicorne* (c); *laticorne* with a black coxal spur is included in second half of couplet (cc)..4

– Mid coxa either with apical spur black or with all apical hairs distinctly separate and not forming a spur (aa). Postpedicel in most species long-oval and usually shorter than arista (bb) (not in *laticorne*, cc). 10

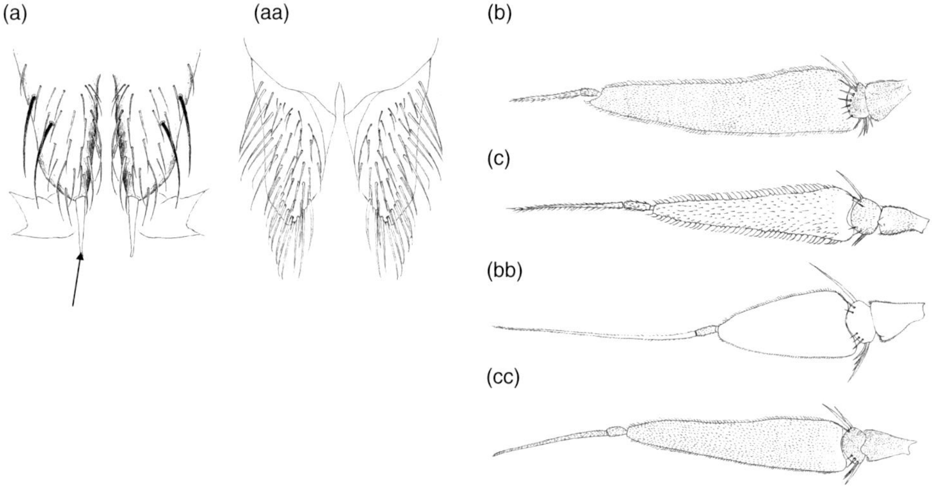

4(3) Frons shining violet or bluish black at its centre. Face narrow, at mouth equal to or less than depth of postpedicel (one- to two-thirds), width at mouth one-fifth or less its height (a). Mid and hind coxae with lateral seta black. Front tibia with ventral vestiture much shorter than dorsal vestiture (b). 4 subequal postsutural dorsocentral setae. Hind tibia usually black and contrasting strongly with yellow femur. 5

– Frons green with dense dust on its centre. Face wider, equal to depth of postpedicel and width at least one-third its height (aa, two examples). Mid and hind coxae with lateral seta pale, scarcely distinguishable from white hair. Front tibia with ventral row of upright hairs clearly longer than dorsal vestiture (bb, two examples). 3 strong postsutural dorsocentral setae, the second postsutural fine and short. Hind tibia and femur usually with no strong contrast in colour. 9

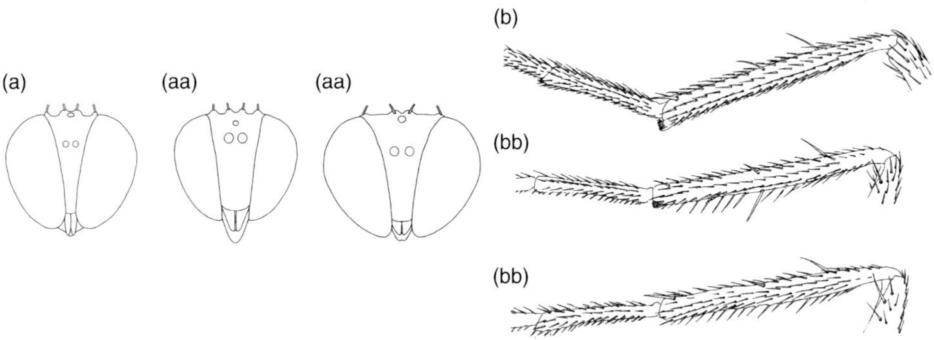

5(4) Cercus long ribbon-shaped or tapered triangular, at least as long as epandrium, extending beyond surstylus (ignoring hairs) (a, two examples). 6

– Cercus rounded or short triangular, not longer than epandrium, not extending beyond surstylus (ignoring hairs) (aa, two examples). 8

(a)

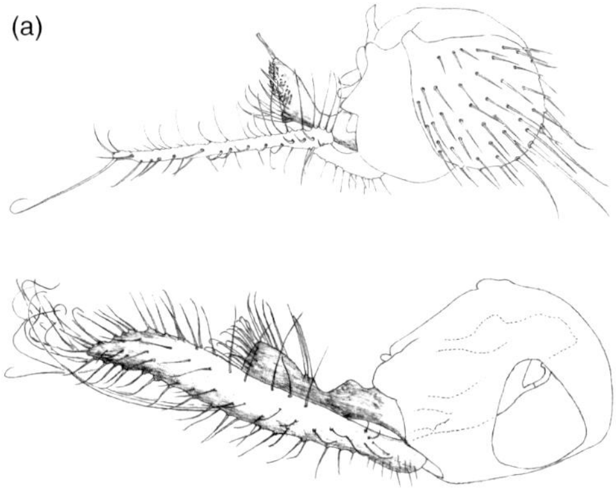

(aa)

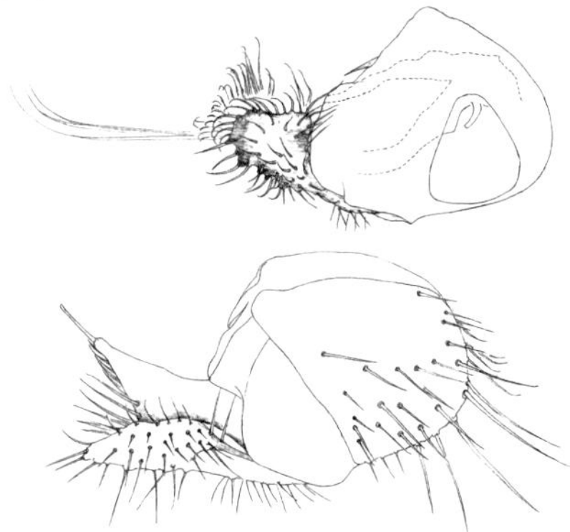

6(5) Cercus shaped as a parallel-sided ribbon (a). Front tarsus with vestiture clearly far shorter than width of shaft, at most slightly longer dorsally than on other faces (b). Hind femur without an anterior preapical seta. Sternite 8 with short hairs (c).......... ***appendiculatum***

– Cercus shaped as a long tapered triangle. Front tarsus with vestiture of dorsal hairs equal to or longer than width of shaft and at least twice as long as vestiture on other faces (figs b and bb of next couplet). Hind femur with an anterior preapical seta. Sternite 8 with long hairs (cc). 7

(a)

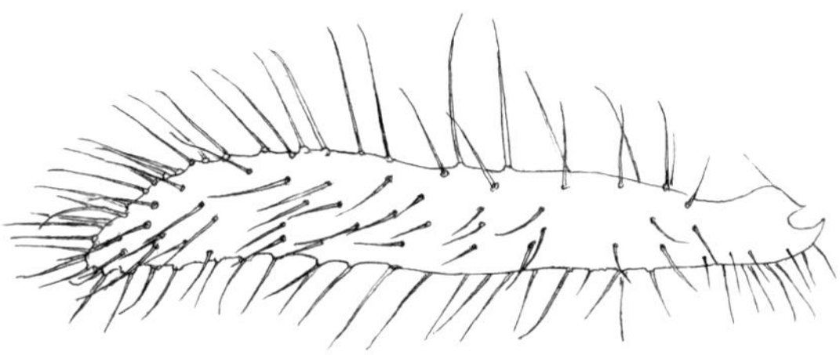

(c)

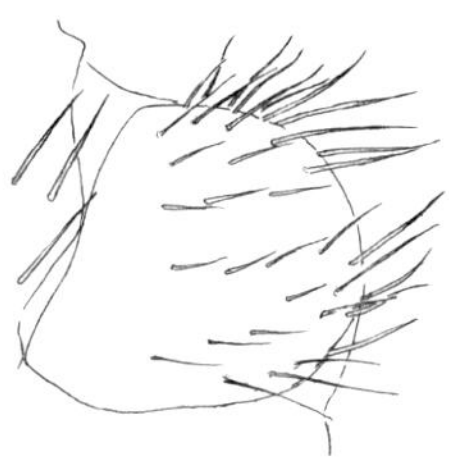

(b)

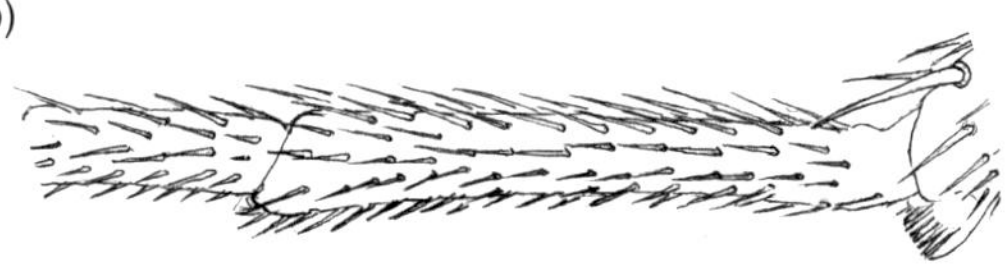

(cc)

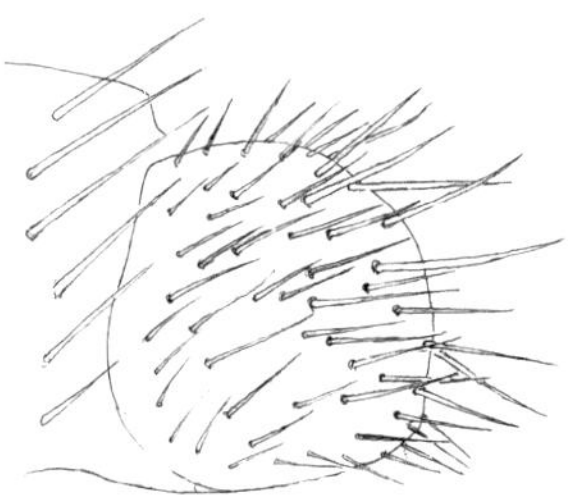

7(6) Cercus with a single outstanding apical hair, all other hairs weaker and similar to each other (a). Front coxa with apical setae not differentiated from dense white hairs. Front basitarsus with a conspicuous anterior to dorsal fringe, hairs 3 times length of fine short posterior and ventral hairs (b). Anal angle of wing very weak, anal lobe narrow, its width about one-third length of dm-m (c). .... ***monotrichum***

– Cercus with several outstanding black hairs on inner margin, thicker and longer than other hairs and similar to apical hair (aa). Front coxa with about 3 strong black apical setae. Front basitarsus with dense anterior and dorsal hairs about twice length of posterior and ventral hairs, and not conspicuously differentiated (bb). Anal angle of wing distinct, width of anal lobe half length of dm-m (cc). .... ***auctum***

(a)

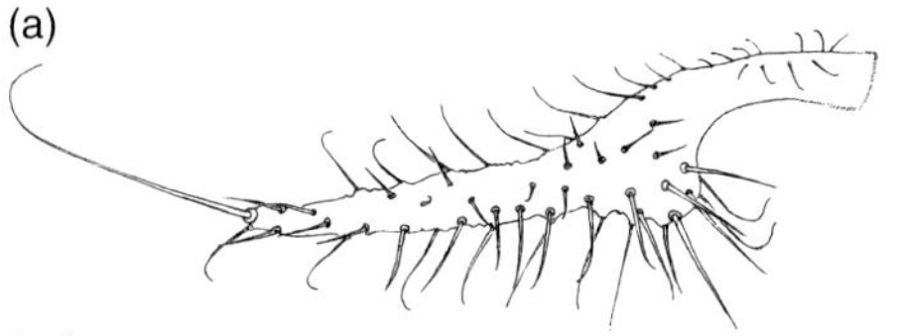

(aa)

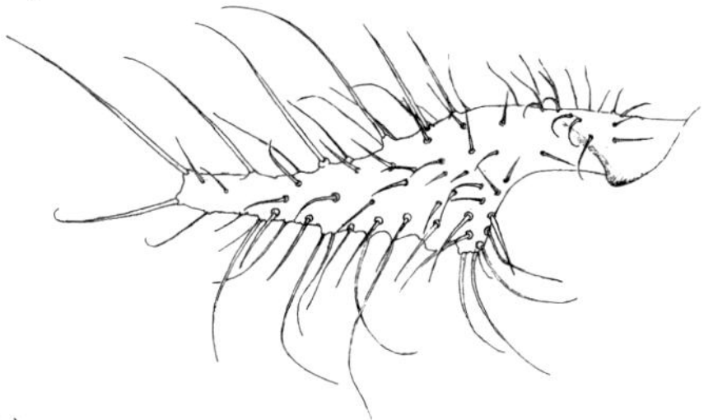

(b)

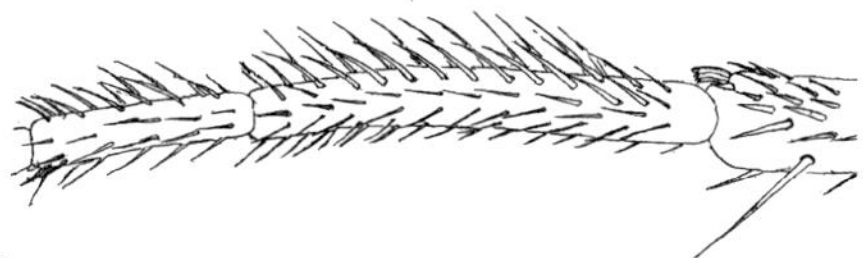

(bb)

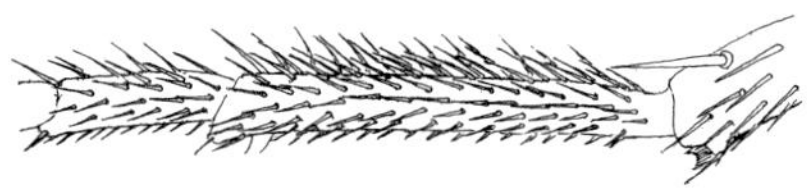

(c)

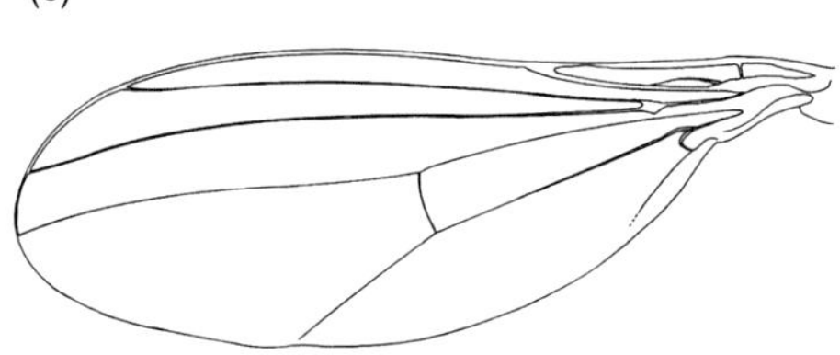

(cc)

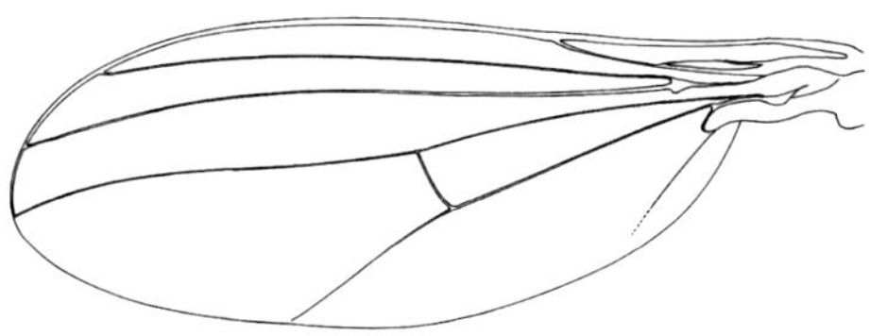

8(5) Arista length equal to postpedicel width and one-fifth its length (a). Surstylus with an apical tuft of long hairs extending well beyond square cercus (b). Hind femur without an anterior preapical seta. .... ***caliginosum***

– Arista length twice postpedicel width and half its length (aa). Surstylus with a fan of hairs that does not extend beyond tip of triangular cercus (bb). Hind femur with an anterior preapical seta. .... ***lanceolatum***

(a)

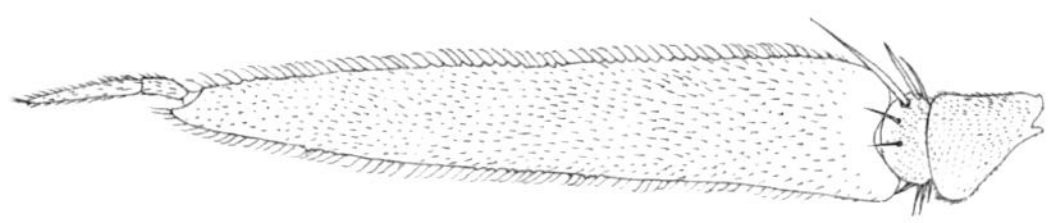

(b)

(aa)

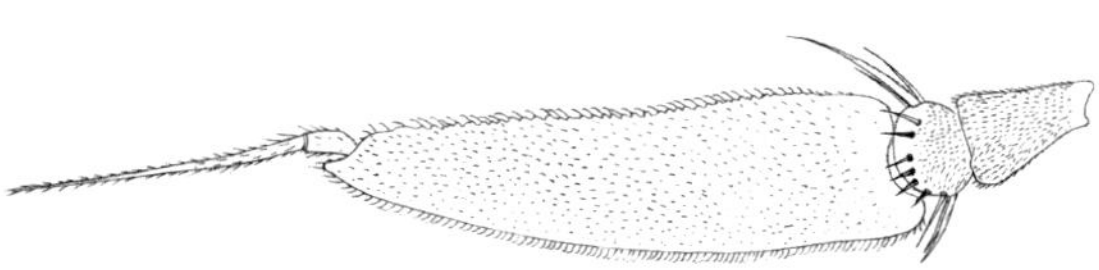

(bb)

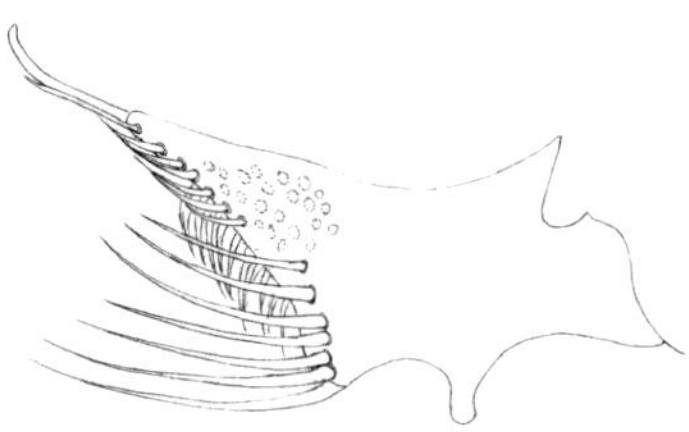

9(4) Cercus lobes shorter than epandrium, both lobes about equal in length (a). Postpedicel about 1.5–2 times length of arista (b). Hind basitarsus shorter than tarsomere 2.. ***brevicorne***

– Cercus inner lobe longer than epandrium, ribbon-shaped, much longer than inconspicuous outer lobe (aa). Postpedicel about 4 times length of arista (bb). Hind basitarsus as long as tarsomere 2. ***albomaculatum***

(a)

(b)

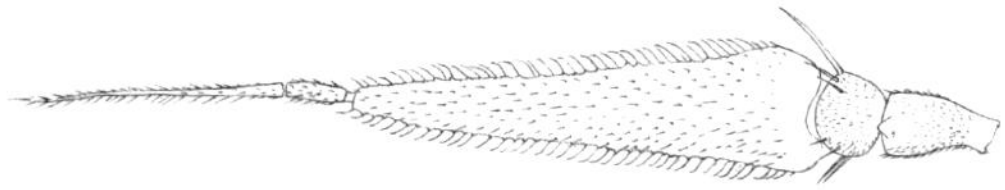

(aa)

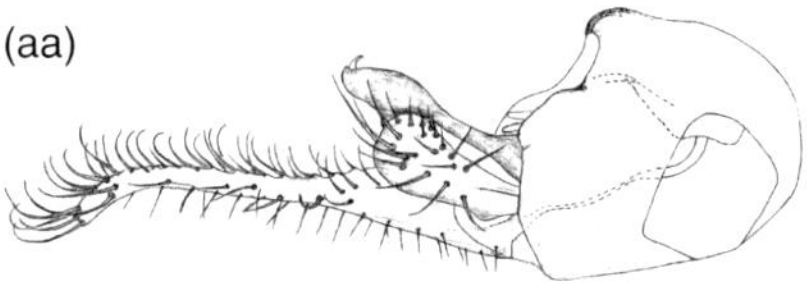

(bb)

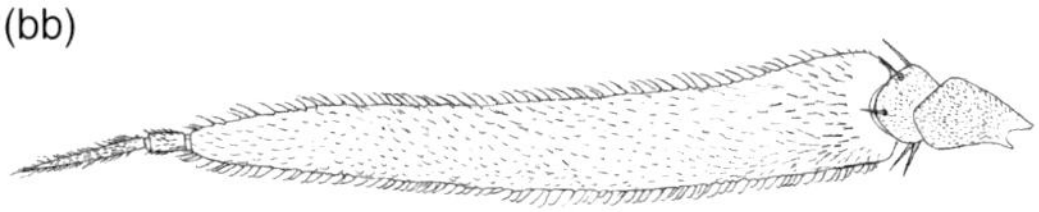

10(3) Front coxa with black hairs. Face black except in *micans* which is white when viewed obliquely from above. 11

– Front coxa with pale hairs. Face white. 16

11(10) Mid coxa without an apical spur. Front basitarsus shape modified, partly swollen, constricted or spindle-shaped, not simply cylindrical, clearly wider than second tarsomere. 12

– Mid coxa with an apical spur. Front basitarsus shape simply cylindrical, scarcely wider than second tarsomere, but may have an apical cluster of hairs. 15

12(11) Hind femur entirely black. Front basitarsus swollen ventrally at tip. 13

– Hind femur yellow on about basal half. Front basitarsus either spindle-shaped or constricted in middle. 14

13(12) Hind tibia gradually widening towards tip, clearly wider at three-quarters along shaft than at a quarter (a). Hind tarsomeres becoming progressively narrower from basitarsus to tarsomere 5 (a). Front tarsomere 2 slightly swollen apically (b). Arista about as long as postpedicel (c). Cercus simple, long and tapered (d). ***fractum***

– Hind tibia more-or-less cylindrical, almost equally wide at three-quarters as at a quarter along shaft (aa). Hind tarsus changing little in width between tarsomeres 2 and 3 (aa). Front tarsomere 2 with an anterior apical peg (bb). Arista about 1.5 times longer than postpedicel (cc). Cercus forked, with a short outer lobe (dd). ***nasutum***

(a)

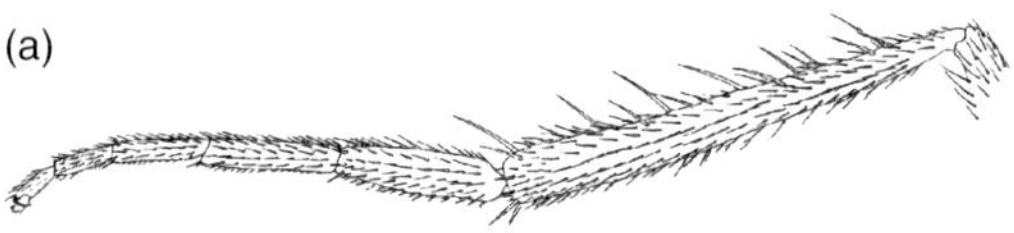

(c)

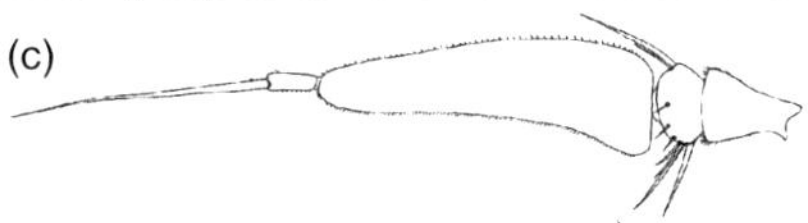

(aa)

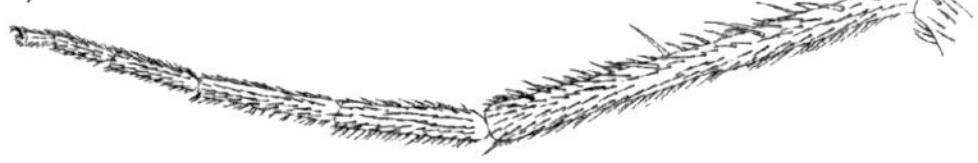

(cc)

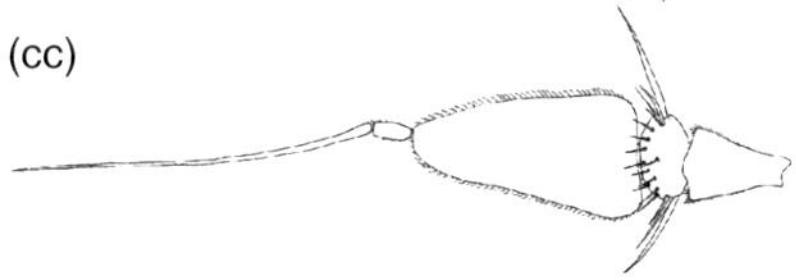

(b)

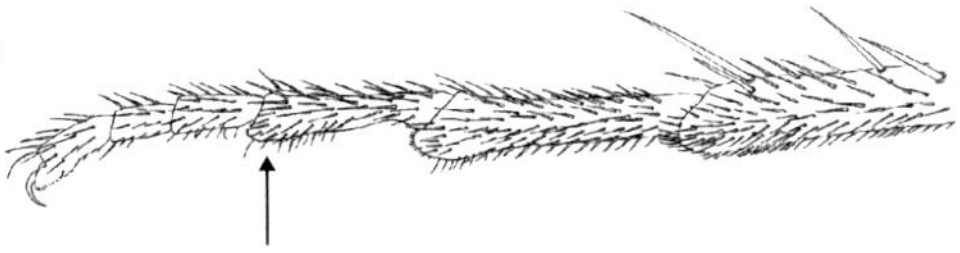

(d)

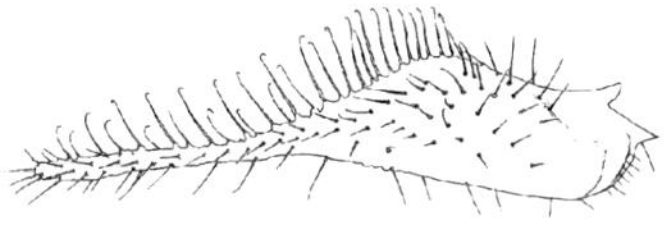

(bb)

(dd)

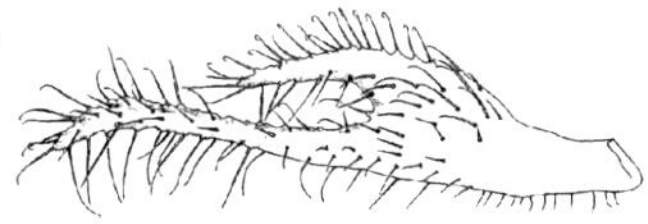

14(12) Front basitarsus with a row of short thick ventral hairs (a). Palp with 2 long blunt-tipped hairs (b). Front femur on basal third with posteroventral hairs but no setae. Mid basitarsus with 1 long ventral seta at base, as strong as largest tibial apical seta (d). Face silver when viewed from above. Whiskers white. Surstylus triangular, much shorter than cercus (e).. ***micans***

– Front basitarsus with dense short vestiture ventrally (aa[17]). Palp with normal fine hairs. Front femur on basal third with a posteroventral row of long thick setae (cc). Mid basitarsus with 3–4 long ventral setae on basal quarter (dd). Face black. Whiskers black. Surstylus rod-shaped, about as long as cercus (ee[18]).. ***pectinatum***

(a)

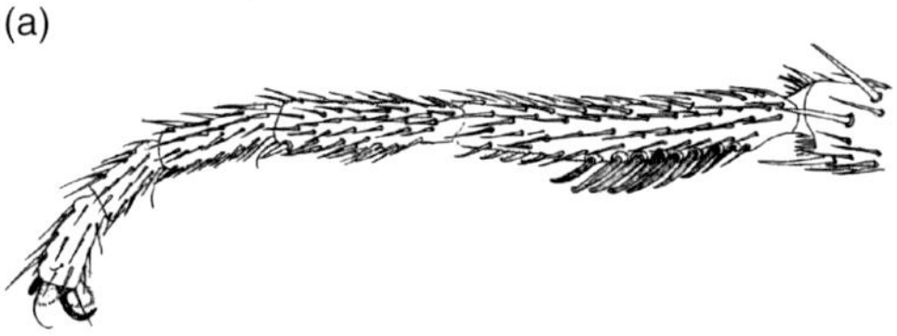

(b)

(cc)

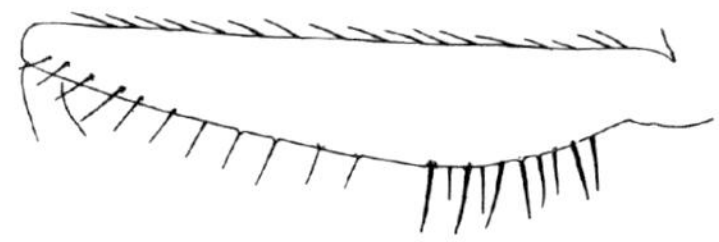

(aa)

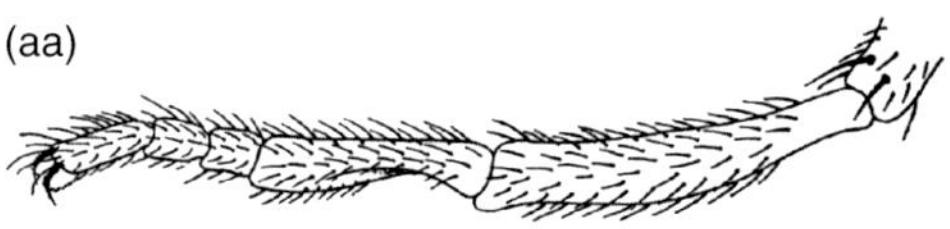

(d)

(e)

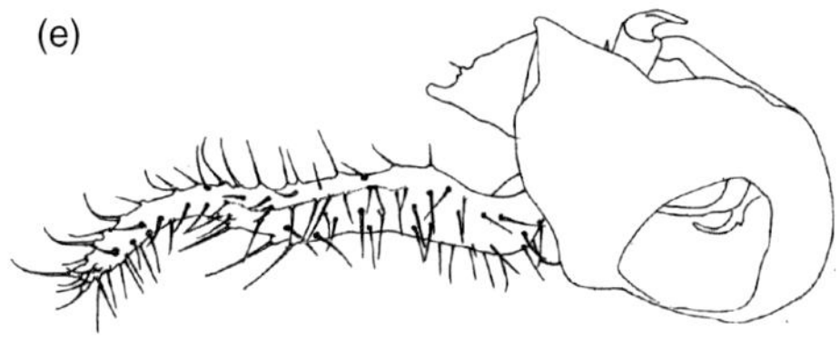

(dd)

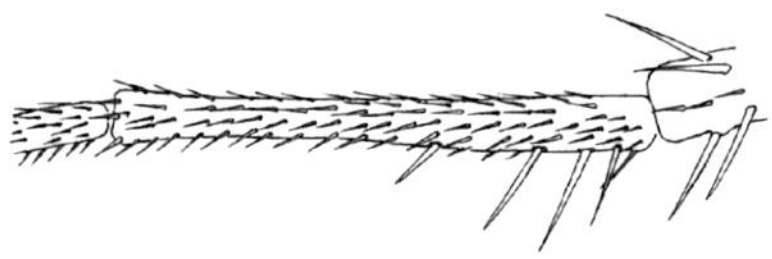

(ee)

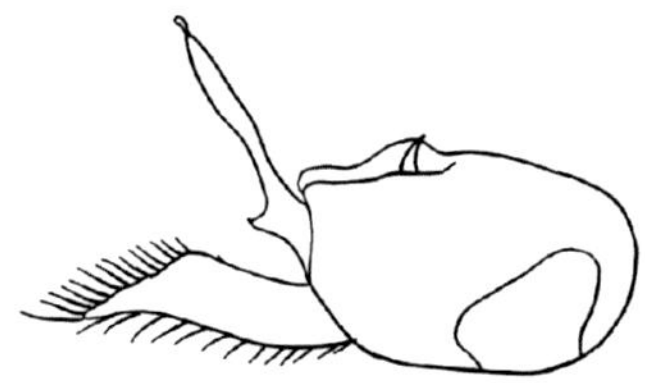

15(11) Hind tibia strongly thickened apically, clearly wider at three-quarters along shaft than at a quarter (a), yellow on basal half and abruptly black on swollen apical half. Front basitarsus about half length of tibia (b) and with no differentiated apical tuft of hair. Mid femur with ventral hairs simple and short, all hairs curved towards tip. Front femur with posterior fringe becoming gradually longer towards tip where hairs are about twice width of shaft. Cercus simple, tapered (e).. ***fascipes***

– Hind tibia cylindrical, not swollen, scarcely broader than mid tibia (aa), yellow grading to black at base and tip. Front basitarsus two-thirds length of tibia (bb) and with a posterior tuft of long hairs at tip (cc). Mid femur with a ventral fringe of black hairs, those on basal quarter curved towards base, the rest curved towards tip (dd). Front femur with posterior fringe abruptly much longer on apical third where hairs are 3 times width of shaft. Cercus forked (ee).. ***commune***

[17] d'Assis-Fonseca 1978. *Handbooks for the Identification of British Insects* **9**(5), Fig. 146.

[18] Negrobov 1979. *Die Fliegen der palaearktischen Region* **4**(5), issue 322, Fig. 1859.

(a)

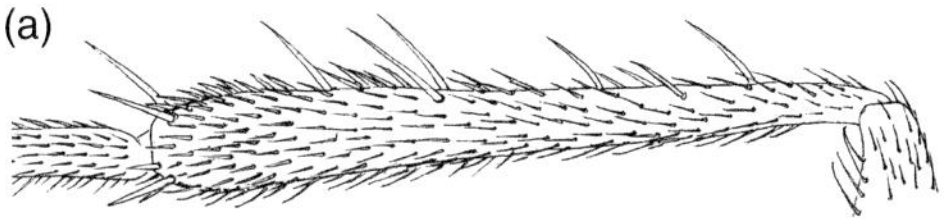

(aa)

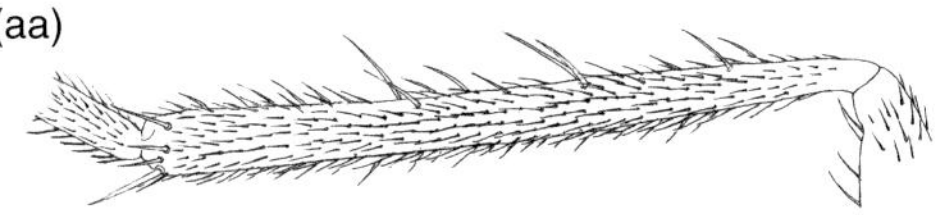

(b)

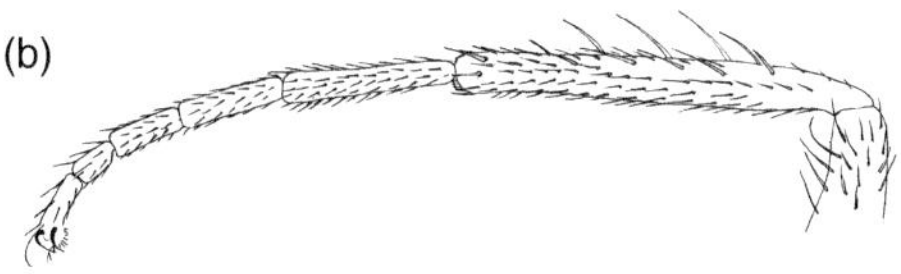

(bb)

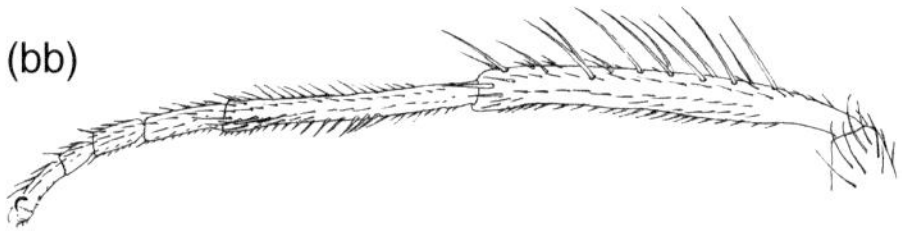

(cc) dorsal

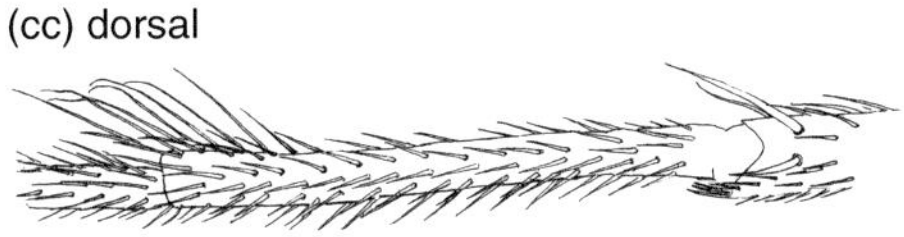

(e)

(ee)

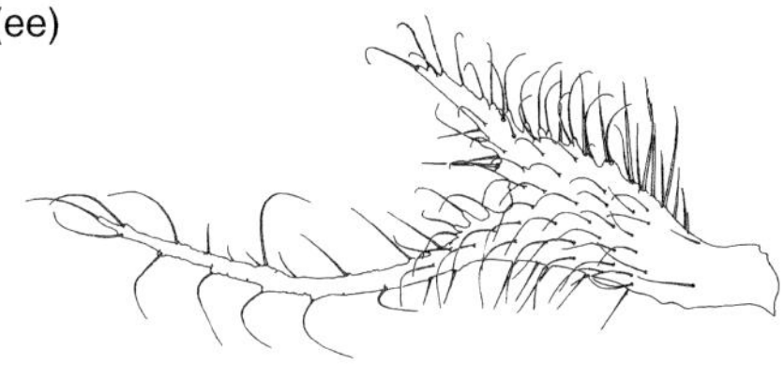

(dd)

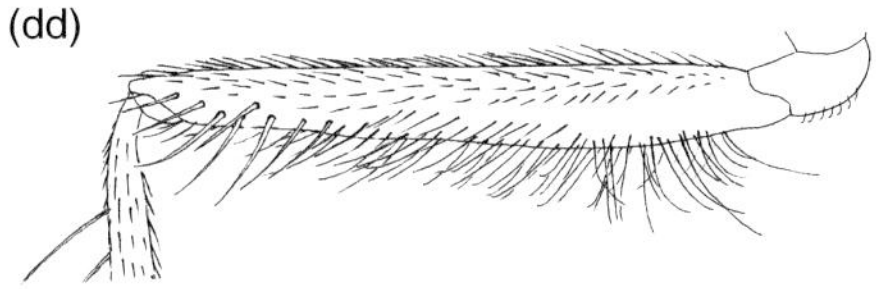

16(10) Mid tarsomeres 4 and 5 strongly expanded and black, tarsomeres 1–3 pale and slender (a). Front tarsomeres 2–5 combined clearly shorter than basitarsus. Hind coxa with an apical anterior fringe of long dense hairs extending to tip of trochanter. Cercus long with a lateral fork at halfway (b). Surstylus shaped as a narrow curved and tapered rod (c). Scutellum with several fine lateral hairs (d).. ***crassipes***

– Mid tarsus unmodified, segments cylindrical. Front tarsomeres 2–5 combined at least as long as basitarsus. Hind coxa with at most a few sparse short hairs. Cercus long or short, simple or forked near base. Surstylus shape various, not a finely tapered rod except when straight (*patulum*). Scutellum with 1–2 fine lateral hairs anterior to large apical seta, except in *patulum* which has many fine lateral and apical hairs.. 17

(a)

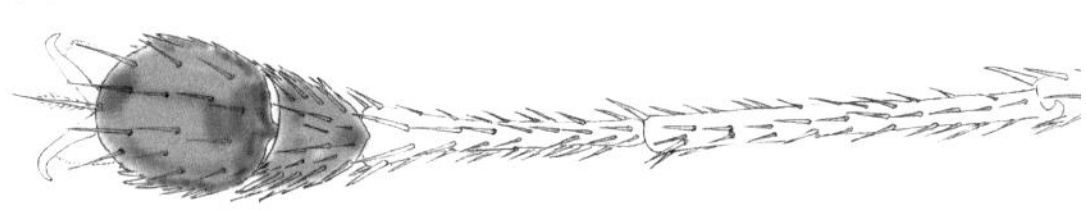

(c)

(b)

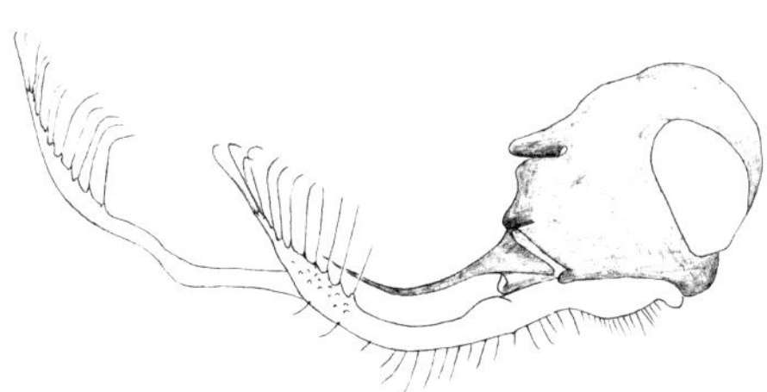

(d)

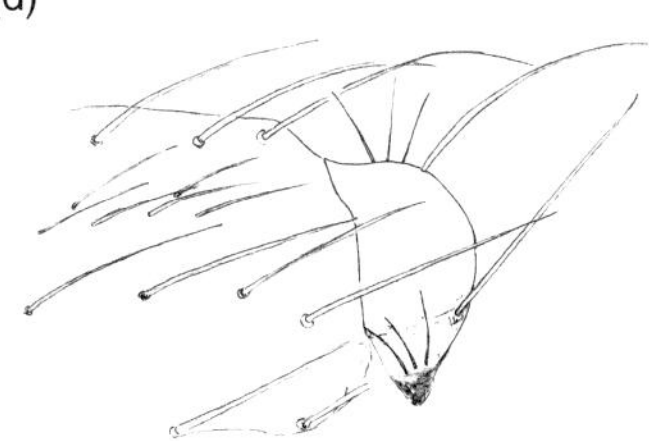

17(16) Mid coxa with an apical spur of coalesced or strongly converging hairs, spur 2–4 times thicker than individual adjacent hairs; spur may be split into two. Cercus shaped as a long tapered ribbon (a)......18
– Mid coxa without an apical spur, all apical hairs thin and clearly separated. Cercus long (as above) or short (aa, 2 examples)......22

(a)

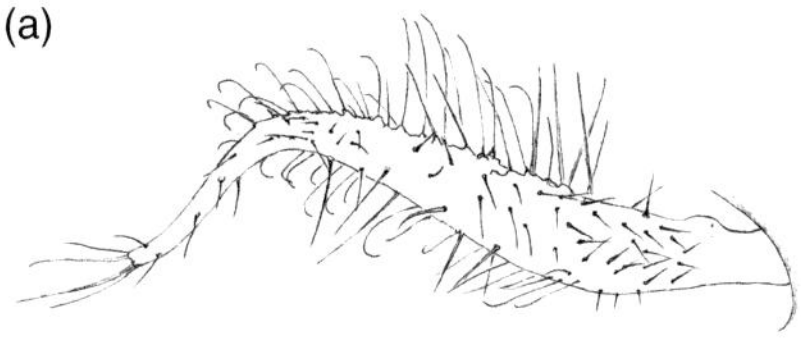

(aa)

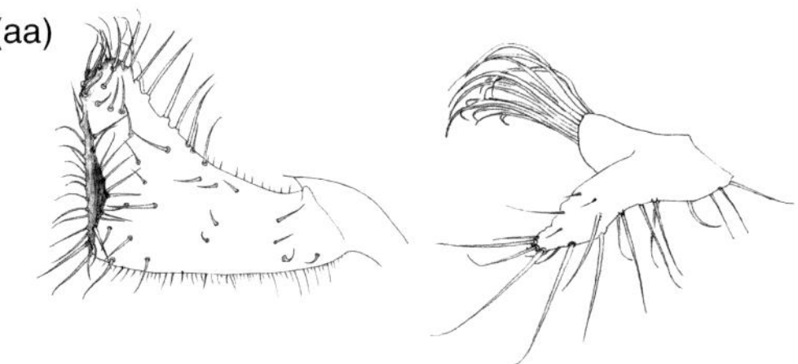

18(17) Front coxa mainly yellow, mid coxa yellow on apical quarter of anterior face. Arista with a swollen tip (a). Front tibia with 2–3 weak setae on dorsal surface (b). Palp yellow. Mid femur with only 1 preapical seta on each side. ......***antennatum***
– Front and mid coxae mainly black. Arista tip simple or at most slightly thickened by dense pubescence. Front tibia with at least 4 strong setae on dorsal side. Palp black. Mid femur usually with at least 2 preapical setae on posterior side and often 2 on anterior side......19

(a)

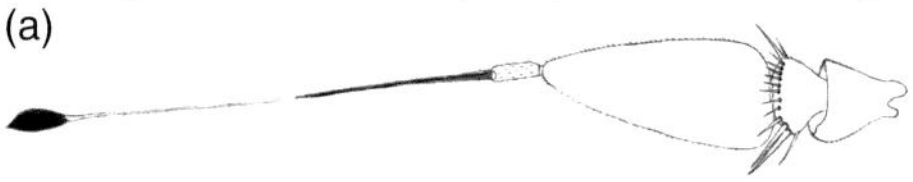

(b)

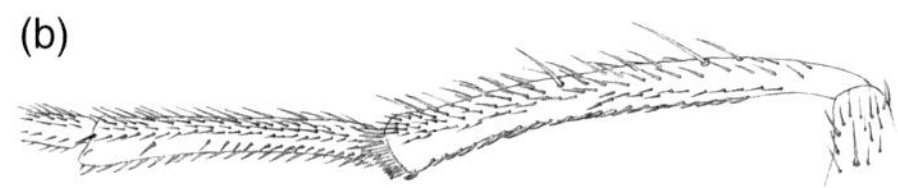

19(18) Front coxa at tip with black setae among white hairs. Mid tibia with antero- and posteroventral setae at two-thirds along shaft (a, ventral view). Hind femur yellow on basal half or more. Hind tibia partly or mainly yellow. Hind tibia with 4–5 anterodorsal setae. Large species, wing length at least 4.5 mm......20
– Front coxa at tip with only white hairs. Mid tibia with only 1 anteroventral seta at two-thirds along shaft, no posteroventral seta (aa). Hind femur mainly black. Hind tibia entirely or mainly black. Hind tibia with 2 anterodorsal setae. Small species, wing length less than 3.5 mm......21

(a)

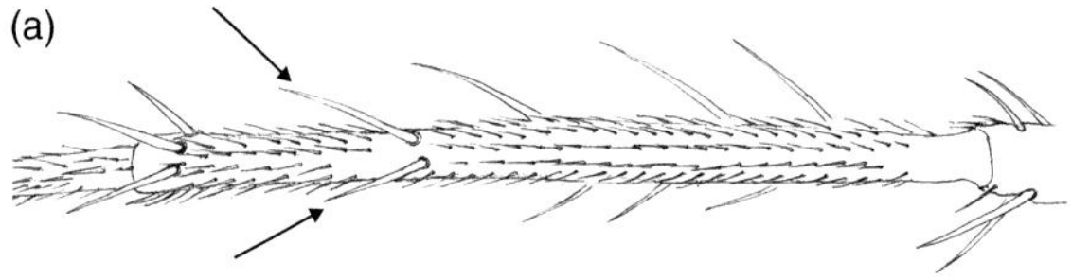

(aa)

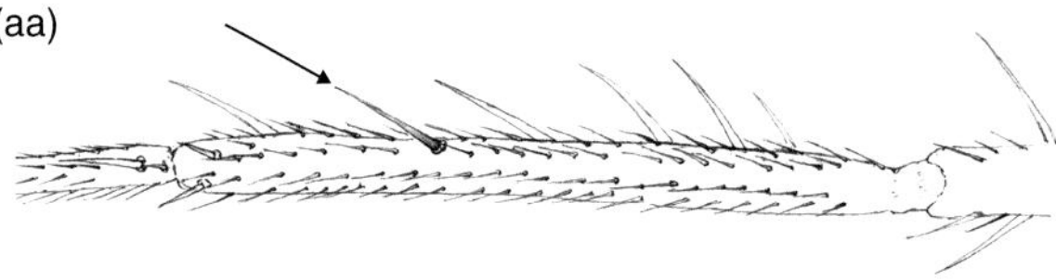

20(19) Tip of arista finely tapered (a). Tergite 6 bluish compared to coppery or greenish preceding tergites. Mid femur yellow. Front femur with posterior hairs inconspicuous, fine, short and pale. Hind femur at base with short black vestiture on all faces. Scutellum with 1–2 small lateral hairs anterior to large apical seta. Surstylus broad and elongate with a rounded tip (b)......***elegantulum***

– Tip of arista slightly swollen due to thicker pubescence, about 3 times wider than stem immediately before it (aa, enlarged tip below). Tergite 6 with colour same as preceding tergites. Mid femur mostly black. Front femur posterior fringe white and as long as depth of femur. Hind femur at base with long white posteroventral hairs, 2–3 times length of black vestiture. Scutellum with few to many small hairs anterior to large apical seta and along posterior margin. Surstylus a narrow pointed rod (bb). .... ***patulum***

(a)

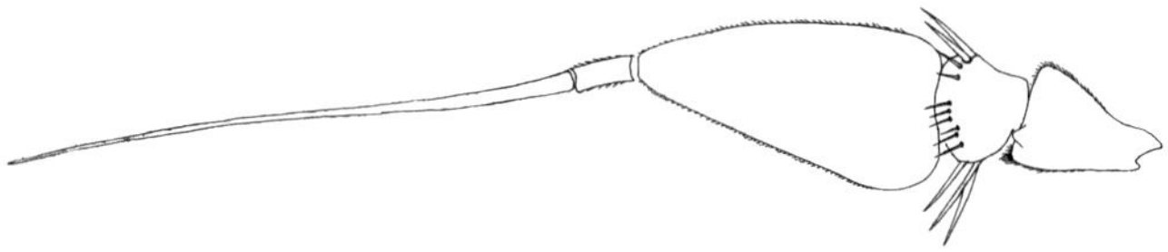

(b)

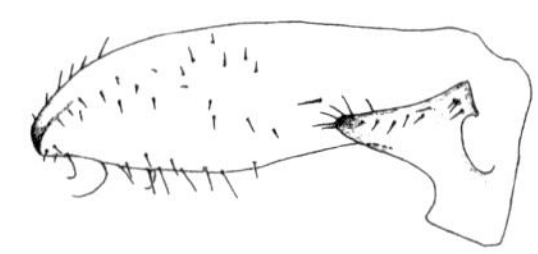

(aa)

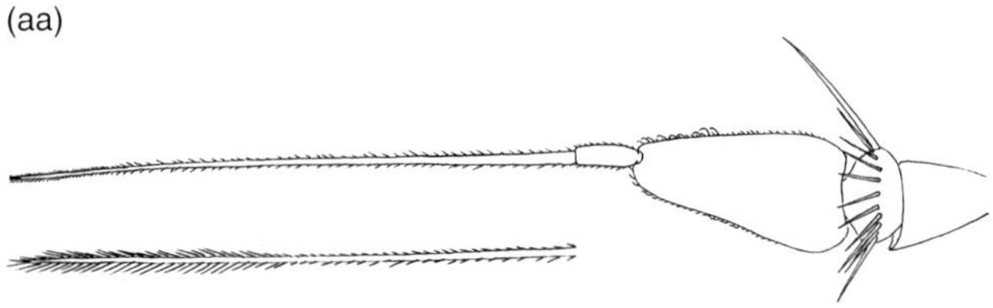

(bb)

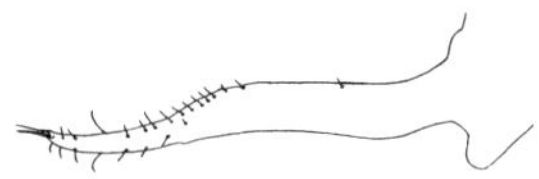

21(19) Postpedicel several times longer than deep, and 1.5 times longer than arista (a). Front tarsomere 5 with normal short dorsal hairs, basitarsus with long dorsal vestiture (c). Front tibia with many (7–9) anterodorsal setae (c). Cercus long and tapered, without a basal lobe (d). Surstylus broad (e). .... ***laticorne***

– Postpedicel scarcely twice as long as deep, and only half to two-thirds as long as arista (aa). Front tarsomere 5 with very long dorsal hairs (bb), basitarsus with dorsal vestiture scarcely longer than ventral hairs (cc). Front tibia with fewer (up to 5) long anterodorsal setae (cc). Cercus long and tapered with a small basal lateral lobe (dd). Surstylus narrow (ee)..... ***rivale***

(a)

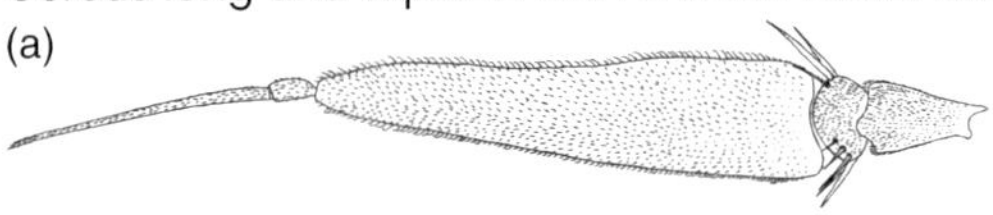

(aa)

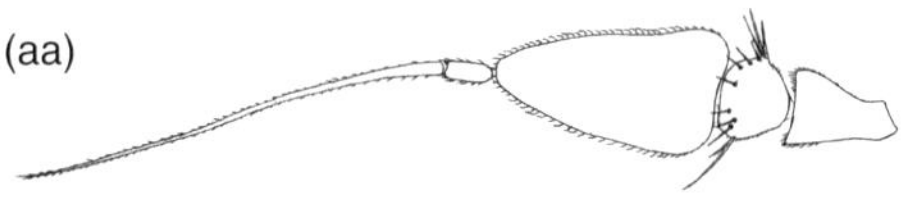

(bb)

(c)

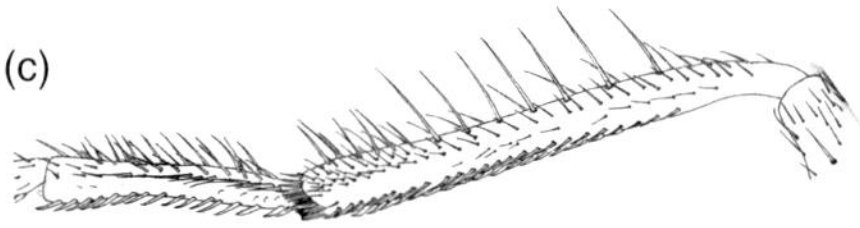

(cc)

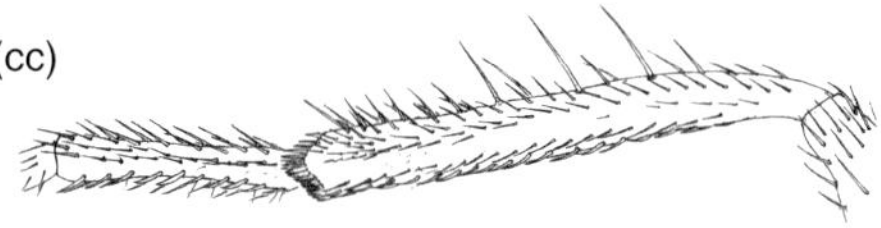

(d)

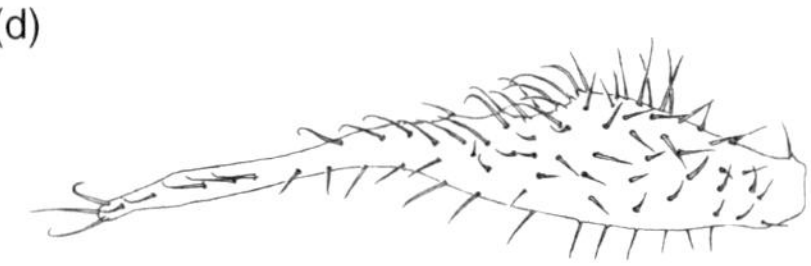

(dd)

(e)

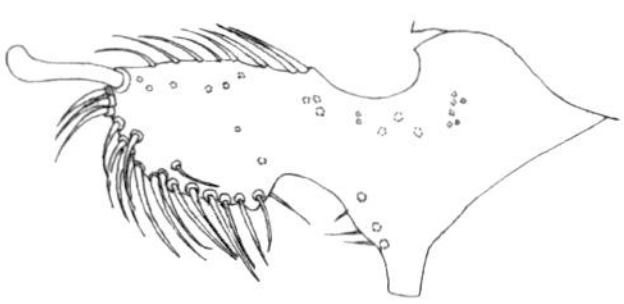

(ee)

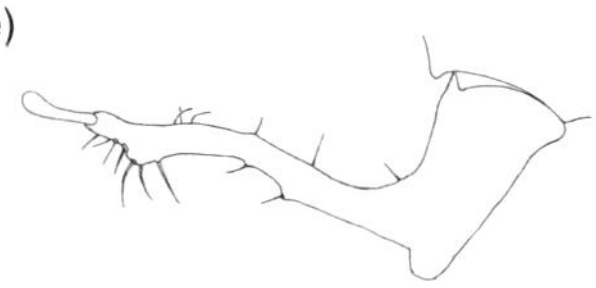

22(17) Front femur with a ventral fringe of black setae as long as depth of femur, posteriorly with long white hairs (a). Mid tibia with anteroventral setae at one-third and two-thirds along shaft, but no or a tiny posteroventral seta (b, ventral view). Front basitarsus cylindrical to tip (c). Smaller species, wing length 3.2–4.0 mm. ***consobrinum***

– Front femur without conspicuous ventral setae, with ventral to posterior long white and black hairs. Mid tibia with paired antero- and posteroventral setae at two-thirds along shaft (as fig. a of couplet 19). Front basitarsus swollen ventrally at tip (cc). Larger species, wing length 3.9–5.1 mm. 23

(a)

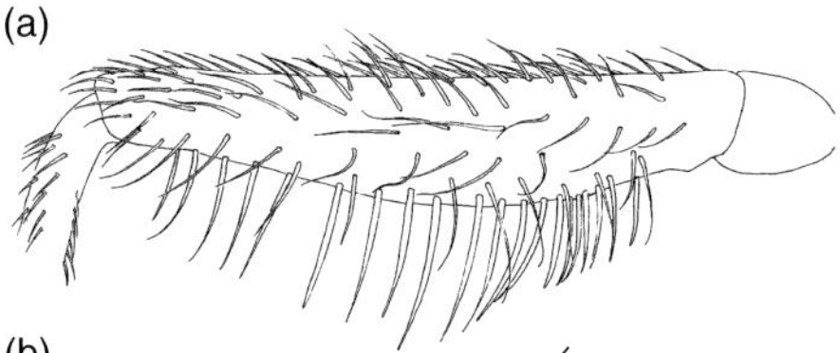

(c)

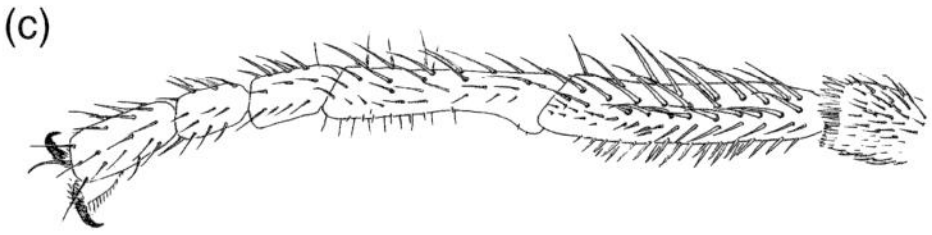

(b)

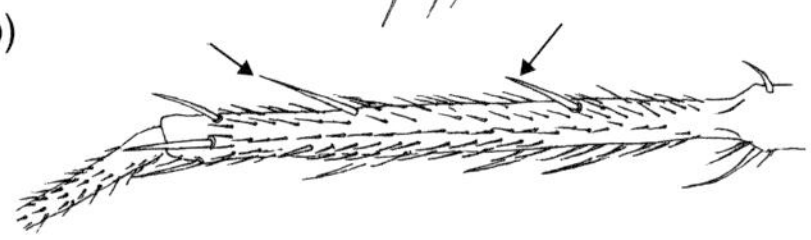

(cc)

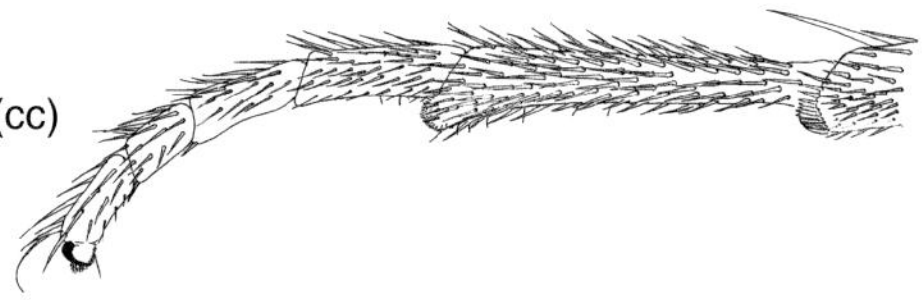

23(22) Cercus long, twice as long as epandrium and projecting far beyond surstylus (a). Front tibia with ventral vestiture of more upright hairs (60° to shaft), rising to three-quarters width of shaft and individual hairs almost as long as shaft width (b) (if no ventral fringe then try *patulum*). Front basitarsus relatively short, about three-quarters length of tarsomeres 2–5 combined (c) and 0.4 times length of tibia. Face wider, one-quarter to one-third its height and almost as wide as depth of postpedicel. ***gravipes***

– Cercus short, not longer than epandrium and not or scarcely projecting beyond surstylus. Front tibia with short ventral vestiture lying at 45° to shaft, rising to less than one-quarter width of shaft (bb). Front basitarsus about as long as tarsomeres 2–5 combined. Face narrow, width about an eighth its height in middle and about half depth of postpedicel. 24

(a)

(c)

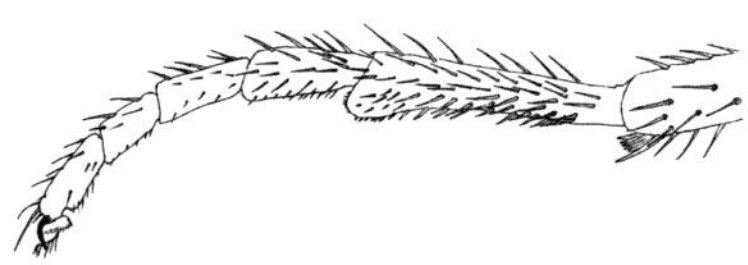

(b)

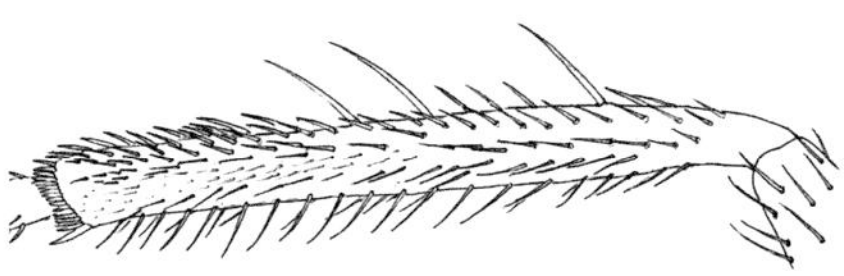

(bb)

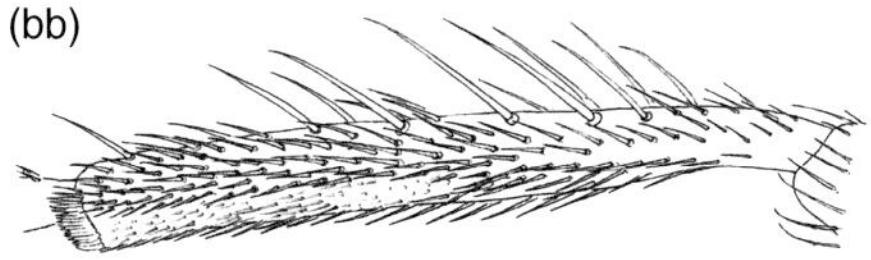

24(23) Hind femur almost entirely black, any yellow at base extending to only one-third along shaft. Front basitarsus with dorsal vestiture rising to depth of shaft, some hairs longer than shaft width, much longer than ventral vestiture (a). Coxae with hairs often yellowish. Front coxa usually with 1–3 black setae among hairs at tip. Cercus bilobed, the lobes projecting sideways and backwards (b). Surstylus swollen on apical section (c). ***penicillatum***

– Hind femur yellow with apical third to quarter dark on anterior face. Front basitarsus with dorsal vestiture only one-third width of shaft, individual hairs half width of shaft and not much longer than ventral vestiture (aa). Coxae with hairs usually pure white. Front coxa with no black setae at tip. Cercus either broadly triangular or very narrow. Surstylus narrow on its apical section (cc). 25

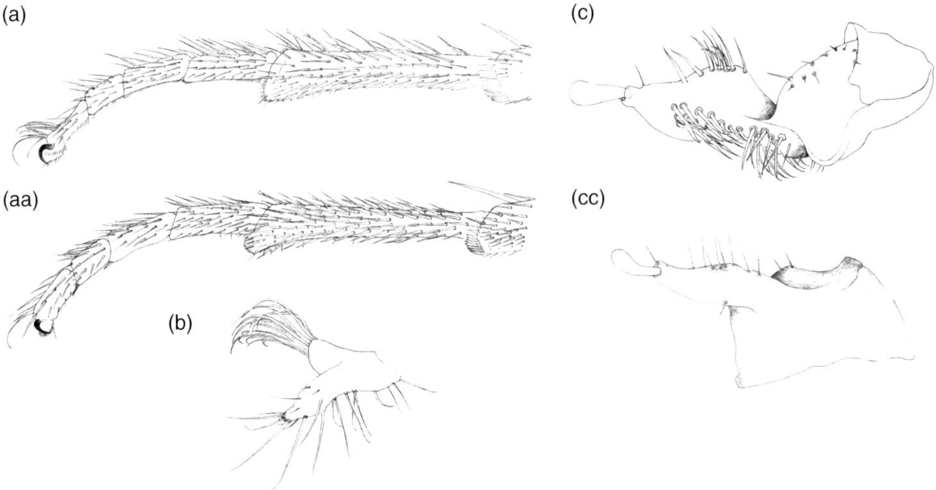

25(24) Cercus broad, flat and roughly triangular with hairs scattered over whole surface (a). Hind tibia often entirely black or sometimes entirely yellow but in all cases having the same shade along whole length except on dorsal surface which is often paler basally. Front femur black almost to tip where yellow extends to at most twice width of shaft here, usually just reaching first strong preapical seta (b, anterior face). Mid basitarsus with basal ventral seta usually not much different in size to more apical setae (c). Postpedicel tip usually slightly more broadly rounded (d).. ***riparium***

– Cercus apically reduced to a narrow ribbon bearing an apical tuft of strongly bent short hairs (aa). Hind tibia either entirely yellow or with a dark apical quarter. Front femur either black with yellow tip extending 3–4 times width of shaft here, usually proximal to strong preapical seta, or almost completely yellow. Mid basitarsus with basal ventral seta usually 1.5–2 times longer than any more apical setae (cc). Postpedicel tip usually slightly more pointed (dd).. ***suave***

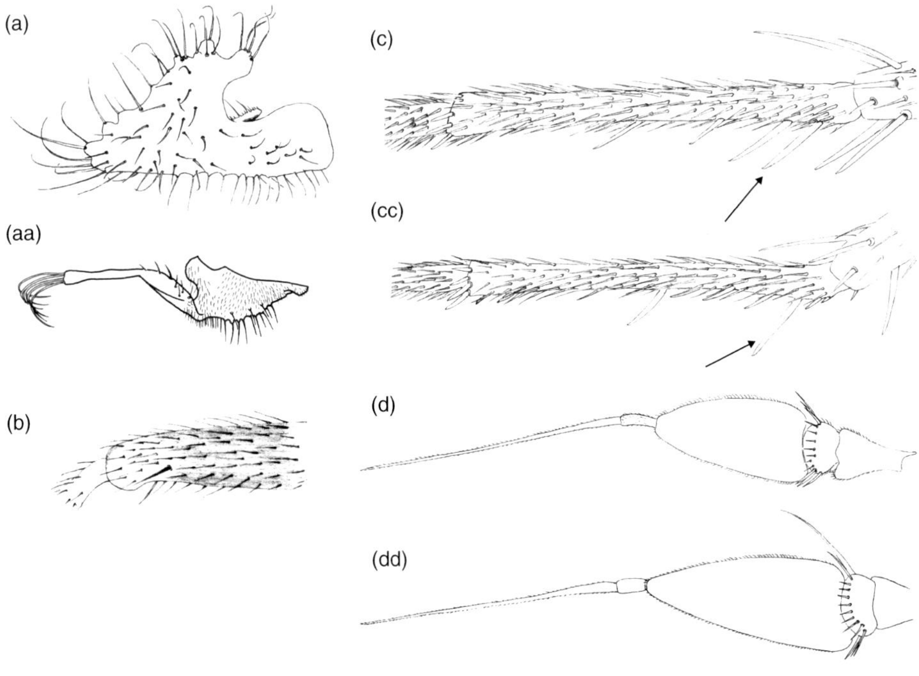

Key to species of *Rhaphium* (females)

Not included: *suave* and *pectinatum*.

1 Antenna, excluding arista, distinctly longer than length of head. Postpedicel tapering into thick arista with little discontinuity in their widths (a). Scape ovoid and broader than widest part of postpedicel. Proepisternum with numerous short black setae in front of spiracle. Hind coxa with 2 lateral setae (lower one weaker) AND frons glossy violet. .... ***longicorne***

– Antenna, excluding arista, far shorter than length of head. Arista seta-like and joining postpedicel with an abrupt change in width (aa). Scape conical and clearly narrower than widest part of postpedicel. Proepisternum with only pale hair. Hind coxa with 0 or 1 lateral seta (if more than 1 seta then frons dusted green).. 2

(a) (aa)

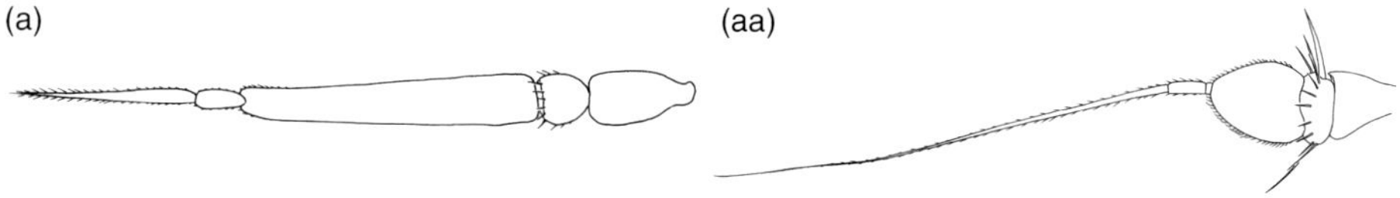

2(1) Hind coxa with a strong white or black lateral seta in addition to hairs.. 3

– Hind coxa without a lateral seta, only white hairs.. 11

3(2) Frons shining violet, rarely greenish but then shining.. 4

– Frons dusted on green or bronze ground. 7

4(3) All legs yellow from coxa to basitarsus. Small delicate species, wing length up to 2.8 mm. 4 dorsocentral setae..***fasciatum***

– At least part of one coxa dark. Larger, wing length at least 3.2 mm. 5–6 dorsocentral setae........ 5

5(4) Hind femur without a preapical seta (a)...................................... ***appendiculatum*** and ***caliginosum***

Note: These species cannot be reliably separated; large specimens with wing longer than 3.5 mm are likely to be *appendiculatum*, smaller specimens are likely to be *caliginosum* but this distinction is not reliable.

– Hind femur with a preapical seta (aa).. 6

(a) (aa)

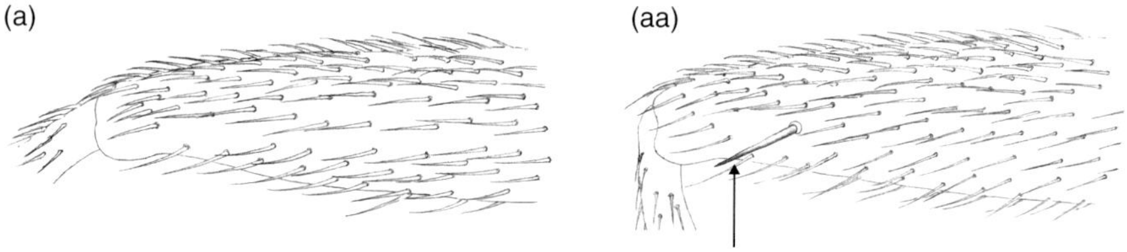

6(5) Hind tibia yellow with black apical quarter to third. Arista apical or almost so (a, two examples). Postpedicel with dorsal and ventral margins both more-or-less straight on apical half, forming a triangular shape (a). R_{4+5} more arched so that costa section R_{2+3} to R_{4+5} is at least twice section R_{4+5} to M_1 (b), and maximum distance between R_{4+5} and M_1 shortly beyond dm-m usually more than 1.3 times minimum distance near wing tip. Front coxa with black apical setae not much stouter or longer than apical white hairs. .... ***monotrichum***

– Hind tibia with the same darkish shade along entire length (possibly not applicable to *lanceolatum*). Arista subapical, dorsal to tip of postpedicel (aa, two examples). Postpedicel with dorsal margin more strongly curved than ventral margin on apical half, forming a more rounded shape (aa). R_{4+5} less arched so that costa section R_{2+3} to R_{4+5} up to twice section R_{4+5} to M_1 (bb), and maximum distance between R_{4+5} and M_1 usually less than 1.4 times minimum distance near wing tip. Front coxa with apical black setae stouter and longer than apical white hairs........................ .... ***auctum*** and ***lanceolatum***

These species appear to be inseparable; *auctum* is slightly larger (wing length 3.8–4.3 mm) than *lanceolatum* (wing length 3.4–3.9 mm), and in *lanceolatum* the maximum distance between R_{4+5} and M_1 usually less than 1.25 times minimum distance near wing tip, but widely variable in *auctum*.

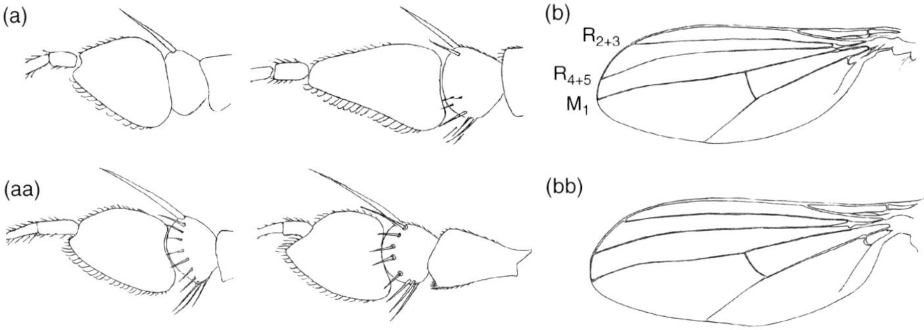

7(3) All coxae with pale setae (and white hairs). 4 strong dorsocentral setae. Small species, wing length 2.8–3.6 mm. 8

– All coxae with black setae (and white hairs). 5–6 strong dorsocentral setae. Large species, wing length 4.2–5.5 mm. 9

8(7) Front coxa yellow on at least apical third. Femora mainly or entirely yellow. Hind basitarsus slightly shorter than tarsomere 2 (a). Front tibia with a distinct ventral seta, usually as long as width of shaft and about twice length of ventral vestiture (but sometimes very small) (b). .... ***brevicorne***

– Front coxa usually mainly dark, yellow on apical quarter (but variable). Femora mainly dark, at least front and hind femora dark dorsally. Hind basitarsus and tarsomere 2 equal in length (aa). Front tibia with ventral seta usually indistinct, shorter than width of shaft and about 1.5 times length of ventral vestiture (but can be as long as in *brevicorne*) .... ***albomaculatum***

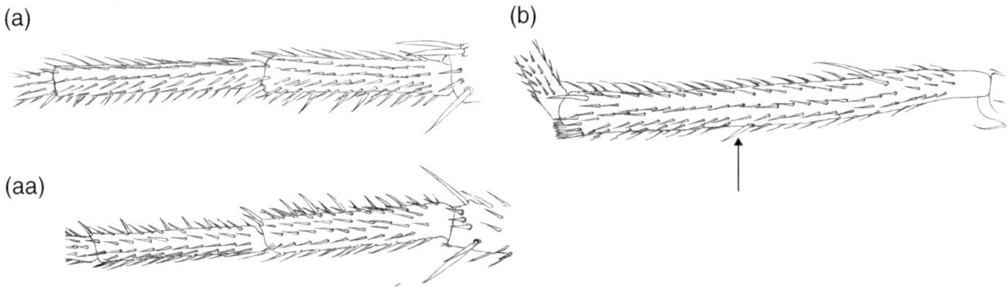

9(7) Front leg more-or-less continuously yellow from apical half of coxa plus trochanter to tip of femur. Mid femur yellow. Hind femur with a weak anterior preapical seta (a). (See more characters in couplet 20.) .... ***micans***

– Front leg black from coxa plus trochanter to near tip of femur. Mid femur mainly dark, not clear yellow. Hind femur with no or a weak anterior preapical seta. 10

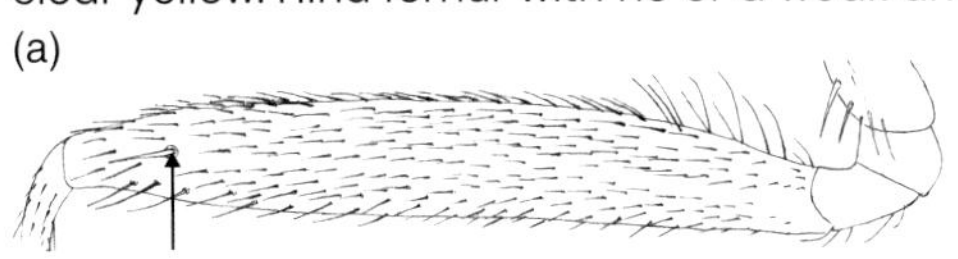

10(9, 21) Frons dusting pale grey on green ground. Mid tibia with 2 or more anteroventral setae (a). Front basitarsus nearly two-thirds length of tibia and as long as tarsomeres 2–5 combined (b). Vein M_1 with a marked inflexion (c). Front coxa with several black setae as long as adjacent white hairs.......... ***commune***

– Frons dusting brownish on brown ground. Mid tibia usually with 1 (sometimes 2) anteroventral seta. Front basitarsus half length of tibia and shorter than tarsomeres 2–5 combined (bb). Vein M_1 with a weak inflexion (cc). Front coxa with a few weak black setae.......... ***fascipes***

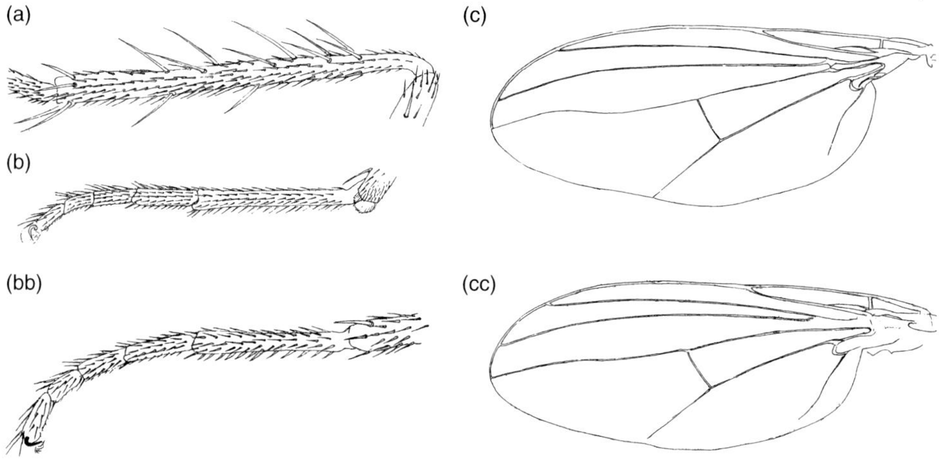

11(2) Mid femur yellow or dark but not intensely black on all faces, some pale streaks present.... 12

– Mid femur almost entirely intensely black on all faces, with a narrow yellow tip and sometimes partly yellow base.......... 21

12(11) Mid tibia with 1 strong posteroventral seta nearly as stout as and almost paired with anteroventral seta (a). Mid coxa with setae and hairs on apical rim mostly black. 2 presutural dorsocentral setae, front one may be shorter. No setulae between dorsocentral and intra-alar rows. Large species, wing length 5.0–6.5 mm.......... 13

– Mid tibia without a strong posteroventral seta, but 1–3 weak setae half to two-thirds length of anteroventral seta may be present, usually beyond anteroventral seta. Mid coxa with setae and hairs on apical rim usually mainly white with at most some scattered black hairs, but mainly black in *micans* and some *fractum*. Usually only 1 strong presutural dorsocentral seta. Setulae absent or present between dorsocentral and intra-alar rows (fig. aa of couplet 17). Smaller species, wing length 3.0–5.0 mm.......... 16

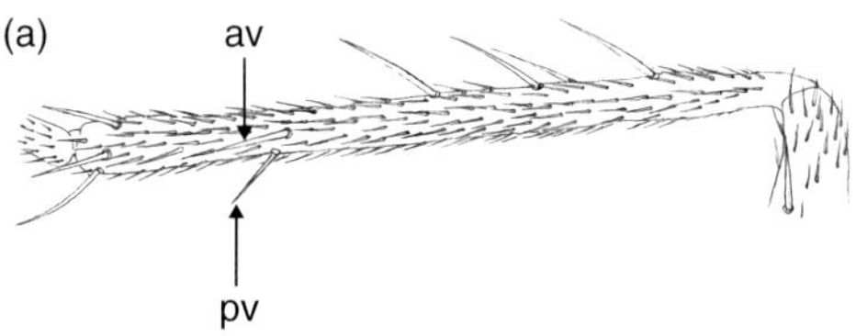

13(12) Tibial setae weak and sparse, only 2 anterodorsal setae (rarely 3–4) on hind tibia (a). Scutellum with 3–5 small diverging lateral setae anterior to long apical seta (b). Palp ground colour yellow at tip (view from below so not confused with pale marginal pubescence). Front femur without a posterior fringe of black and white hairs but only fine short white hair. Vein M_1 strongly sinuous, maximum distance between R_{4+5} and M_1 1.7–2.0 times shortest distance near wing tip (c). Mid femur usually with only 1 posterior preapical seta.......... ***crassipes***

– Tibial setae strong and more numerous, at least 3 anterodorsal setae on hind tibia (aa). Scutellum with 1 small lateral seta (sometimes 2) anterior to long apical seta (bb). Palp ground colour black at tip (viewed as above). Front femur with a posterior fringe of long black and white hairs, as wide as shaft of front tibia (not femur) (dd, dorsal view). Vein M_1 only gently sinuous, maximum distance between R_{4+5} and M_1 less than 1.7 times distance near wing tip (cc). Mid femur with 2 posterior preapical setae. 14

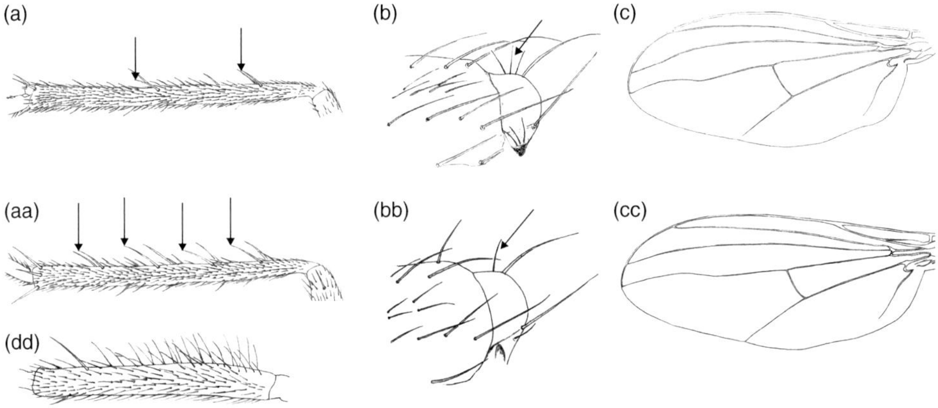

14(13) Scutum shiny with scarcely any pattern. Hind femur and tibia yellow, at least on anterior face. Front tibia with 4–5 posterodorsal setae (a, anterior view). Postpedicel short- to long-triangular (b). .... ***elegantulum***

– Scutum with clear dark lines between acrostichal and dorsocentral rows that fuse to form 1 dark line when viewed from behind. Hind femur yellow with tip black or most of shaft black, hind tibia usually yellow with black apical half or less (but may be entirely yellow). Front tibia with 2–3 posterodorsal setae. Postpedicel rounded to short triangular. 15

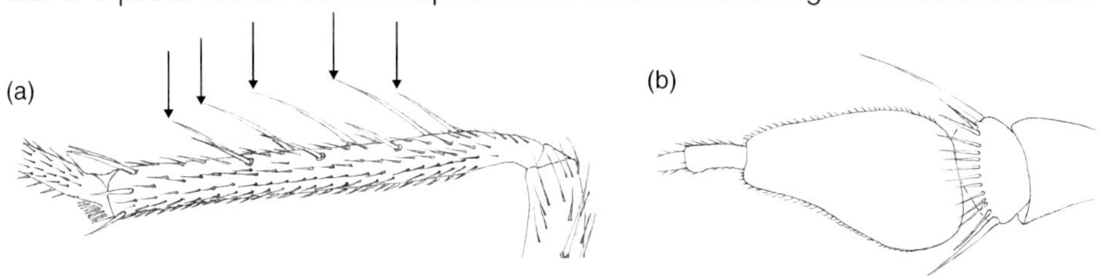

15(14) *Three options*:

– Hind femur yellow with black tip. Front femur usually yellow with dark dorsal smudge. Front tibia usually with 3 strong and 1 weak anterodorsal setae, the most distal seta usually before two-thirds distance along shaft (a, posterior). Arista strictly apical (b, c). ....typical ***riparium***

– Hind femur black with base also black or partly yellow with colour extending to basal third. Front femur usually mostly black. Front tibia with 4–6 strong anterodorsal setae, the most distal seta placed at about three-quarters distance along shaft (aa, posterior). Arista just subapical (bb, cc). .... ***penicillatum***

– Hind femur yellow on about basal half, black on apical half. Front femur mostly dark. Other characters variable. ....either ***riparium*** or ***penicillatum***

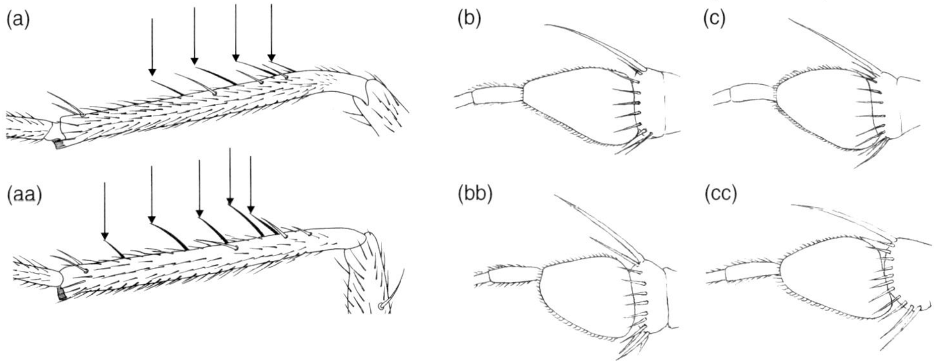

16(12) Front coxa usually mainly black. Face dusting grey or tinged with yellow or brown. Scutum seen from behind with a central dark stripe between dorsocentral rows flanked by strongly contrasting pale-dusted sides. Setulae absent or present between dorsocentral and intra-alar rows (fig. aa of couplet 17)............ 17

– Front coxa yellow on at least apical half. Face dusting silver-white. Scutum seen from all angles usually with scarcely differentiated mid and lateral areas, at least in front half, all shiny except on front margin. No setulae between dorsocentral and intra-alar rows............ 20

17(16) Scutum between dorsocentral and intra-alar rows in front of suture with no additional setulae. Hind tibia with 2 anterodorsal setae (b). Front tibia with setae in anterodorsal series more numerous and stronger than in posterodorsal series (c, posterior). (Mid femur usually with 1–2 posterior preapical seta. Mid tibia with 1 anteroventral seta.)............ ***laticorne***

– Scutum between dorsocentral and intra-alar rows in front of suture with 1–4 setulae as long as acrostichal setae (aa). Hind tibia with 3–5 anterodorsal setae. Front tibia with setae in anterodorsal and posterodorsal series not markedly differing in number and size (cc). (Mid femur usually with 2–3 posterior preapical setae. Mid tibia usually with 2 anteroventral setae.)............ 18

(b)

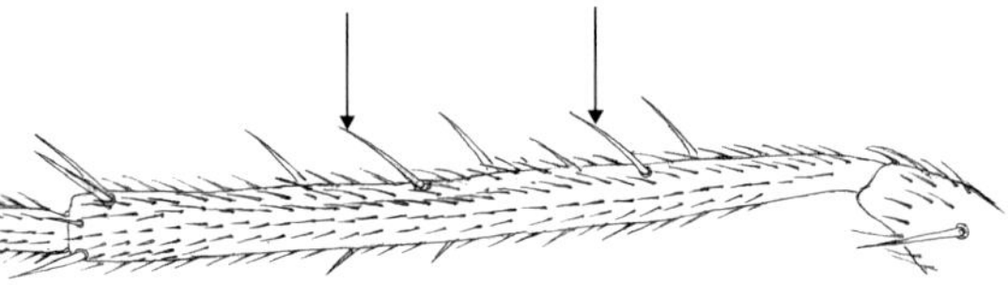

(aa)

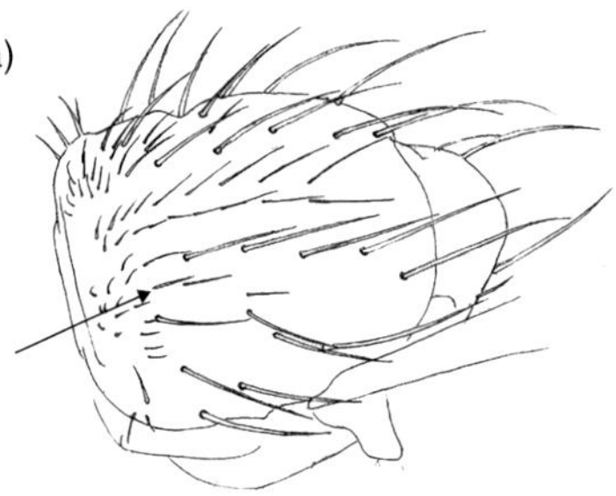

(c)

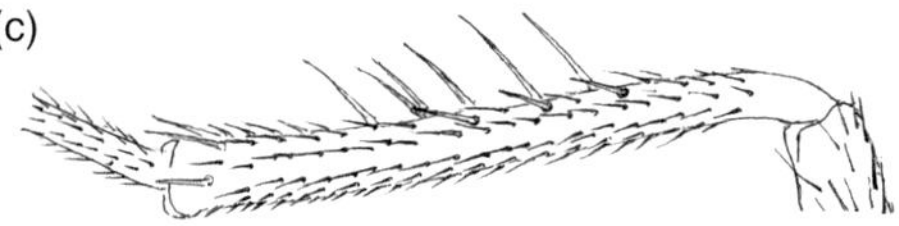

(cc)

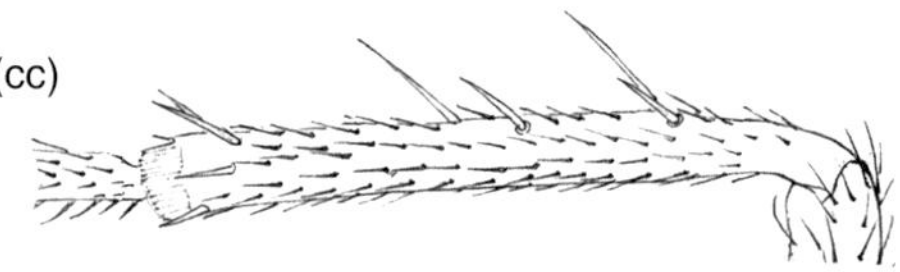

18(17) Hind femur without or with a very weak anterior preapical seta; dorsal hairs on basal quarter twice as long as more distal hairs and 3 times as long as vestiture on anterior face (a). Frons and face with same intensity of grey or brown dusting. Scutum, when viewed from in front, grey-brown with 2 black undusted lines, but when tilted back the paler mid line becomes divided by a third black line, or whole series of lines becomes confused. Front femur black on basal half or more............ ***fractum***

– Hind femur with an anterior preapical seta; dorsal hairs at base not markedly longer than other vestiture (aa). Frons subshining, less densely dusted than face which is grey or tinged with yellow. Scutum with central pale-dusted line along acrostichal row not divided by a narrow undusted dark line. Front femur entirely yellow, at least dorsally............ 19

(a)

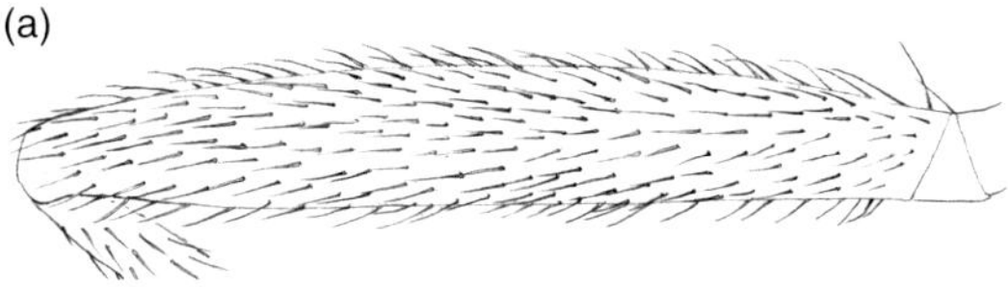

(aa)

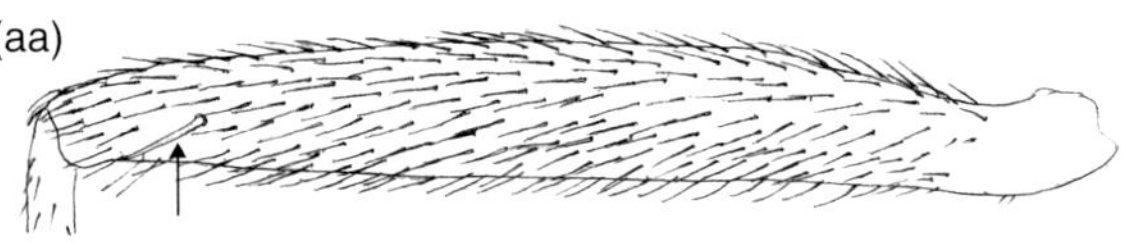

19(18) Epistoma as high as wide (a). Front tibia at least twice as long as basitarsus (2.1–2.3 times). .... ***consobrinum***

– Epistoma slightly higher than wide (aa). Front tibia less than twice as long as basitarsus (1.8–1.9 times).. ***nasutum***

(a) (aa)

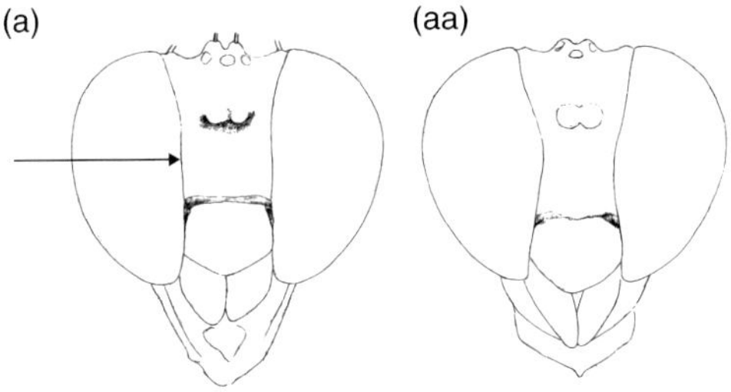

20(16) Front coxa with several strong black setae at apex. Mid coxa black, with several black setae and many black hairs. Hind femur with a dorsal fringe at base (a). Palp ground colour black at tip (view from below to avoid confusing with pale marginal pubescence). Front tibia with 3–5 anterodorsal setae (b). Face narrower, epistoma slightly narrower than high (c). Front basitarsus relatively shorter, about three-quarter length of tarsomeres 2–5 combined (0.7 times). Hind coxa usually with 1–2 strong black lateral setae. (See also couplet 9.)............ .... ***micans***

– Front coxa with entirely white apical hairs and setae. Mid coxa yellow at tip, with 1–2 black setae and all hairs white. Hind femur without a dorsal fringe at base. Palp ground colour yellow at extreme tip (viewed as above). Front tibia with only 2 anterodorsal setae. Face broader, epistoma slightly wider than high (cc). Front basitarsus relatively longer, nearly equal to tarsomeres 2–5 combined (0.9 times). Hind coxa without lateral seta........................... ***antennatum***

(a) (b) (c) (cc)

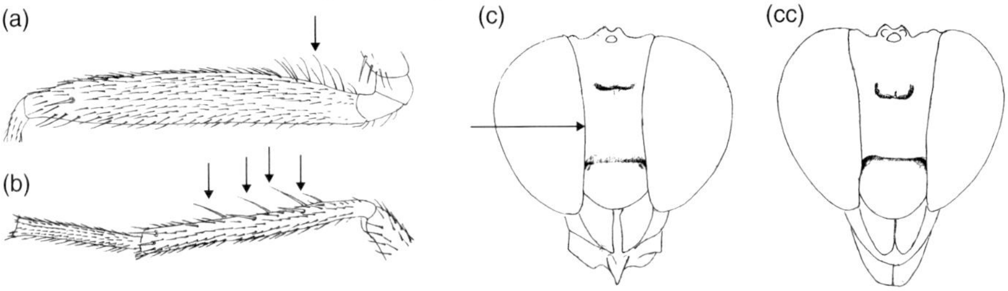

21(11) Scutum between dorsocentral and intra-alar rows in front of suture with 1–3 setulae as long as acrostichal setae (fig. aa of couplet 17). Hind femur with an anterodorsal fringe at base. Mid coxa with apical hairs mainly black........................... return to 10 (*commune, fascipes*)

– Scutum between dorsocentral and intra-alar rows without additional setae. Hind femur without an anterodorsal fringe at base. Mid coxa with apical hairs mainly white...................... 22

22(21) 5 dorsocentral setae, only 1 presutural or with small anterior seta half or less length of first strong seta. Hind tibia with 2–3 anterodorsal setae. Scutum with wide dark shiny stripe between dorsocentral rows flanked by paler brown either side, especially when viewed from behind. Frons bronze. Small species, wing length 3.1–3.4 mm....................................... ***rivale***

– 6 or more dorsocentral setae, of which 2–3 are presutural. Hind tibia with 4 anterodorsal setae. Scutum usually with 2 narrow dull black stripes between acrostichal and dorsocentral rows separated by a wider shiny pale stripe along acrostichal row when viewed at all angles, and usually not fusing to a single wide black stripe when viewed from behind. Frons black or dark bronze. Large species, wing length 4.5–5.6 mm.. 23

23(22) Hind femur almost entirely black, at most with a yellowish patch at base beneath. Face greyish white, often with faint yellowish reflections. 6 dorsocentral setae............. ***gravipes***

– Hind femur yellow on basal half or more. Face more brown-dusted with slight bronze reflections. Usually 7 or more dorsocentral setae.. ***patulum***

Rhaphium albomaculatum

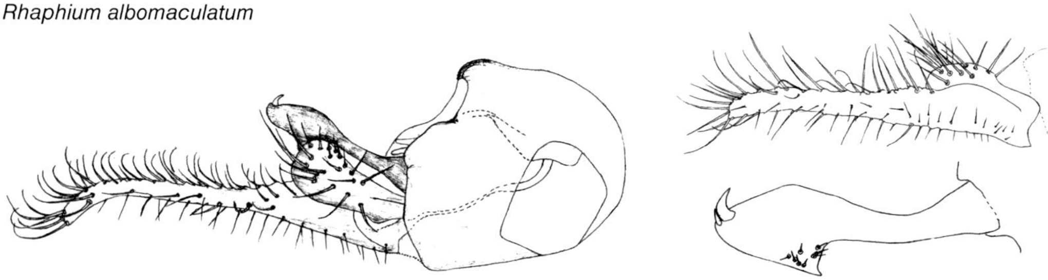

Rhaphium antennatum

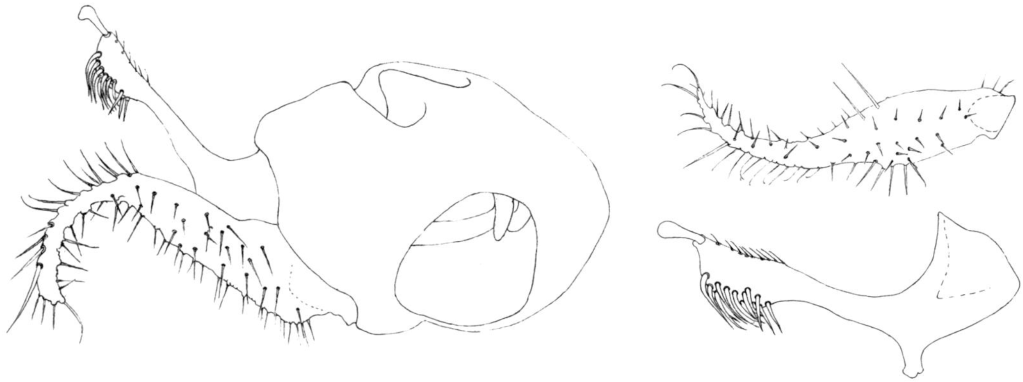

Rhaphium appendiculatum

Rhaphium auctum

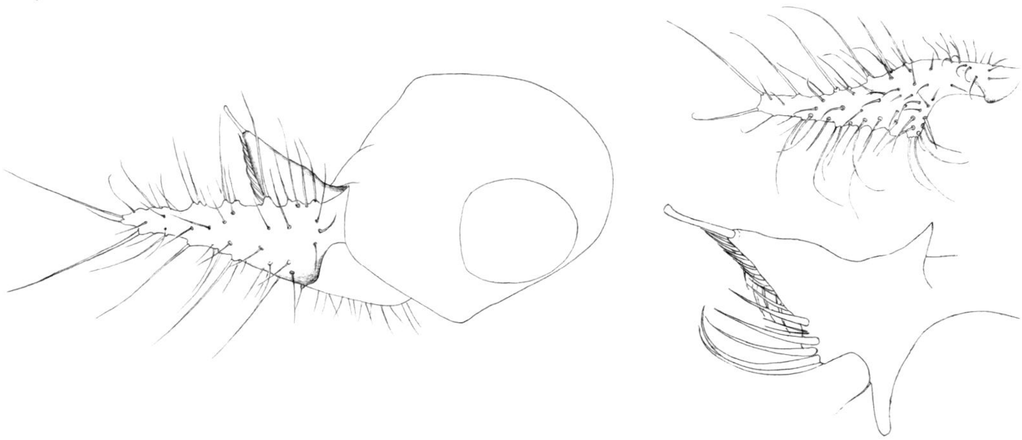

Fig. 32. *Rhaphium* hypopygia, and cercus and surstylus angled to show their maximum breadth.

Rhaphium brevicorne

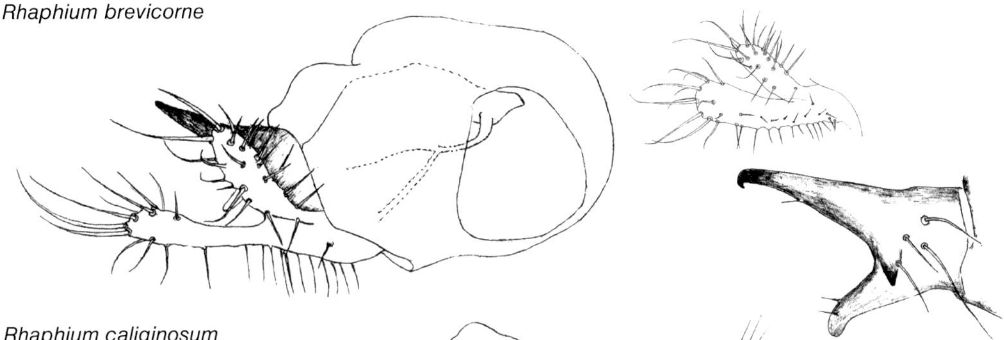

Rhaphium caliginosum

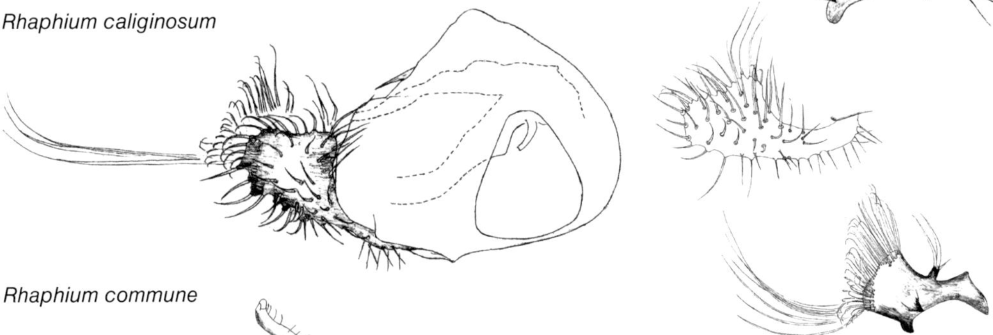

Rhaphium commune

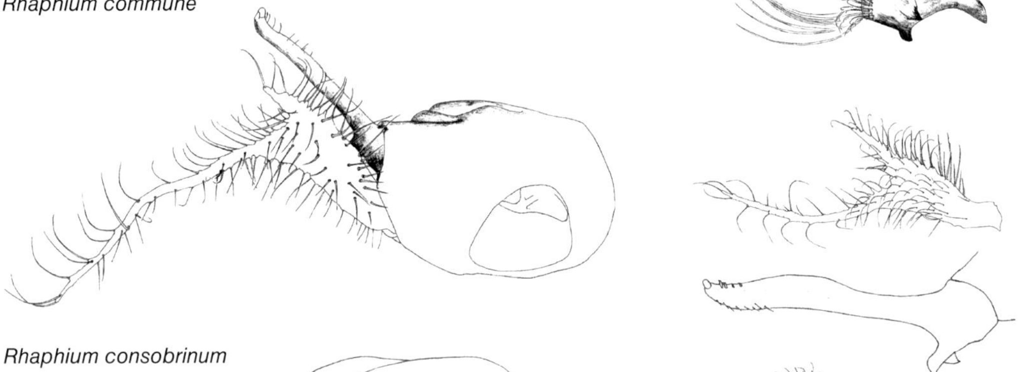

Rhaphium consobrinum

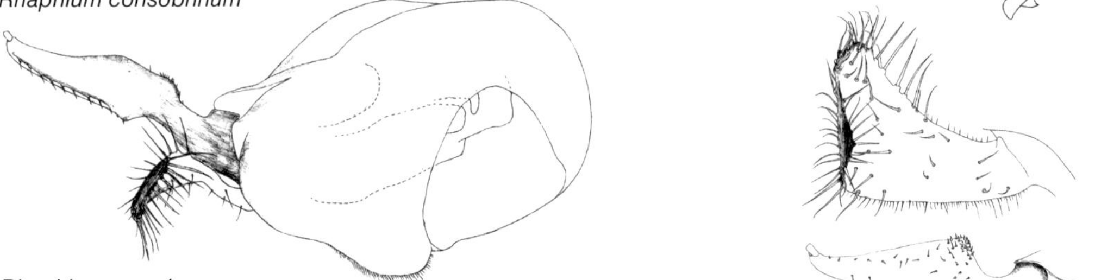

Rhaphium crassipes

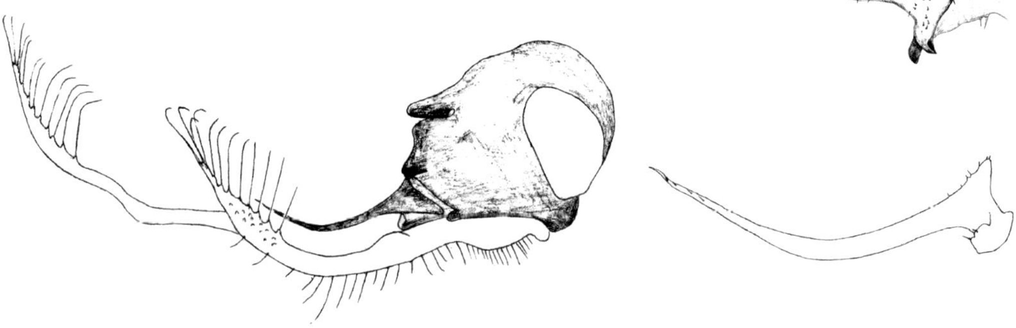

Fig. 32. (*Continued*)

Rhaphium elegantulum

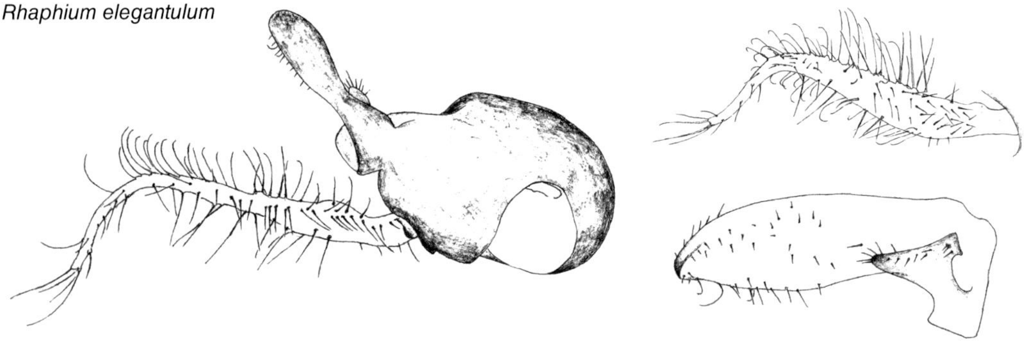

Rhaphium fasciatum

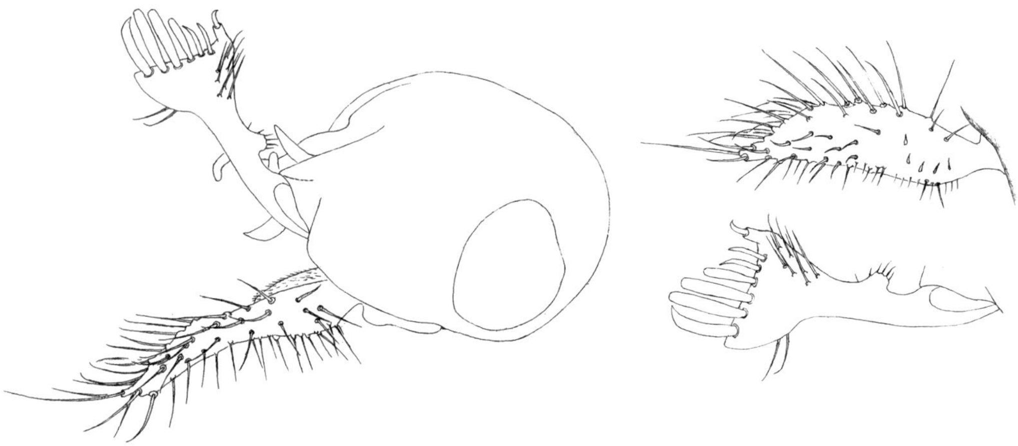

Rhaphium fascipes

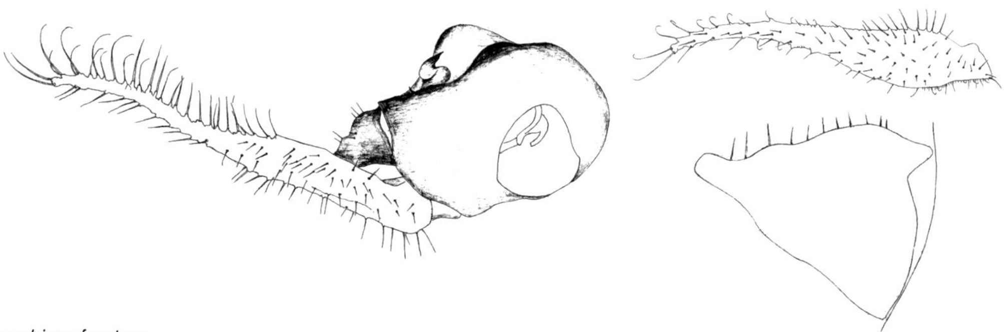

Rhaphium fractum

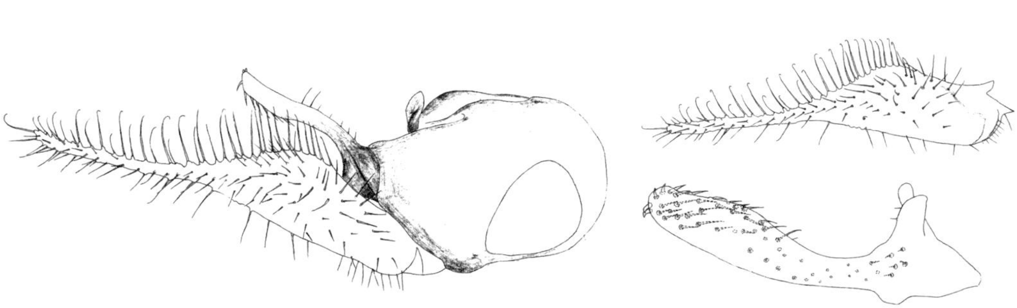

Fig. 32. (*Continued*)

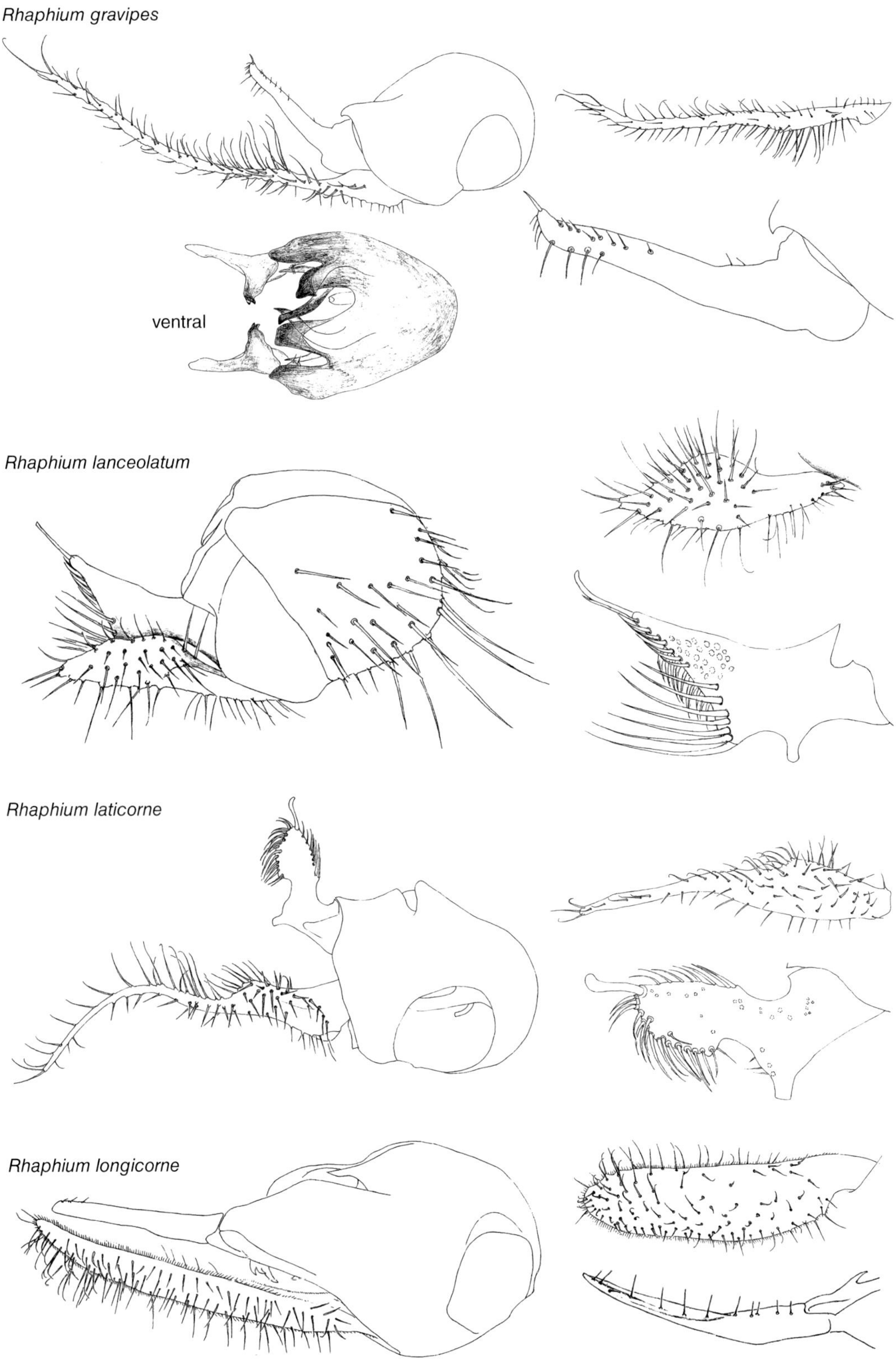

Fig. 32. (*Continued*)

Rhaphium micans

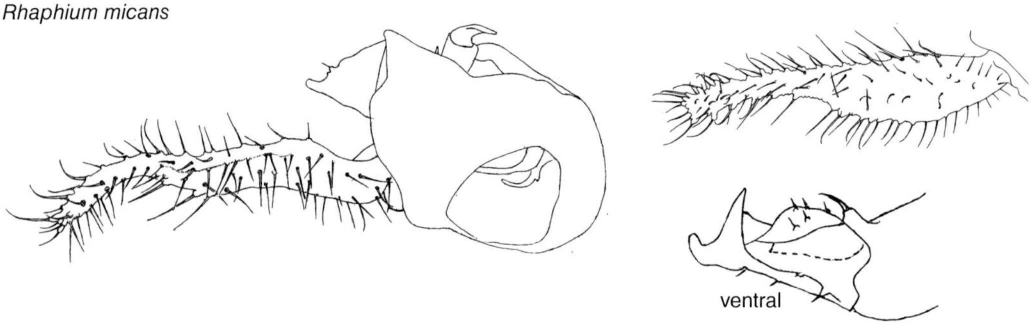

Rhaphium monotrichum

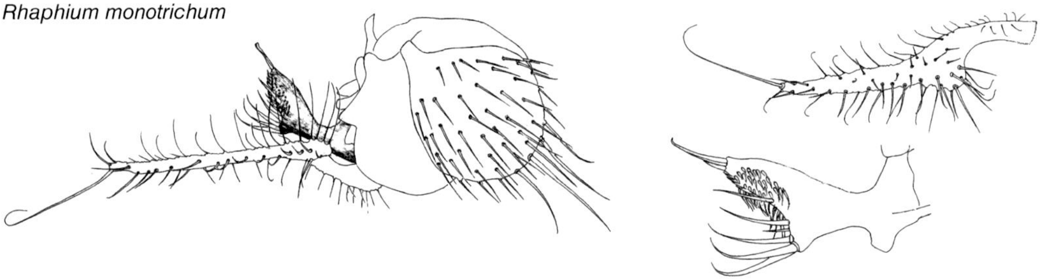

Rhaphium nasutum

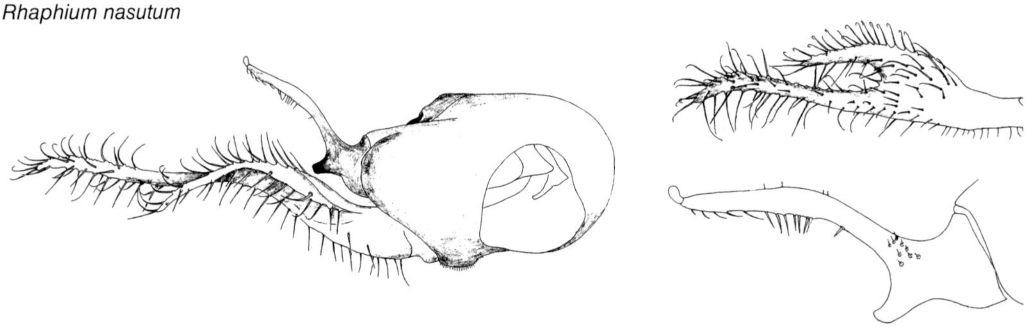

Rhaphium patulum

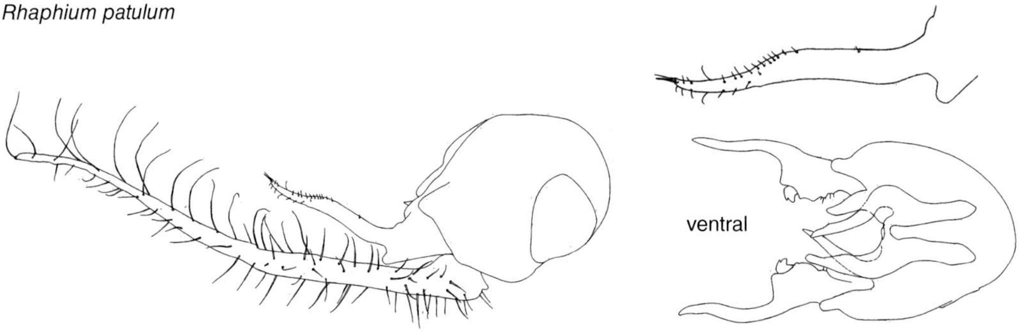

Fig. 32. (*Continued*)

Rhaphium penicillatum

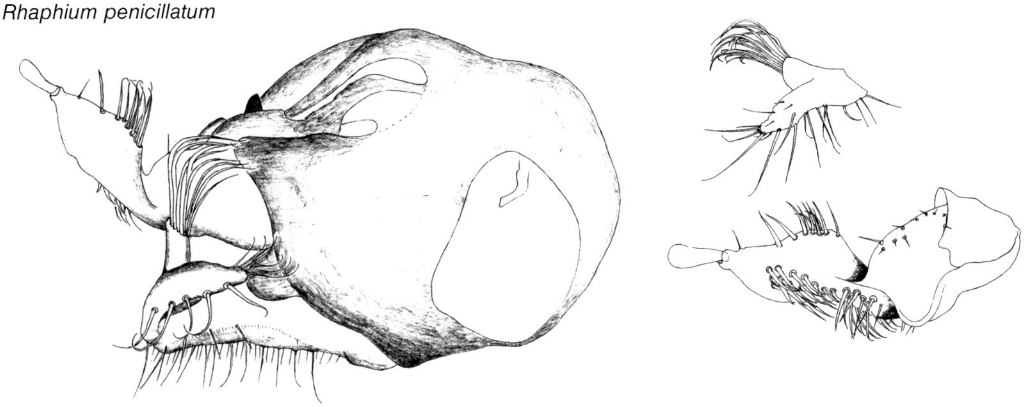

Rhaphium riparium

Rhaphium rivale

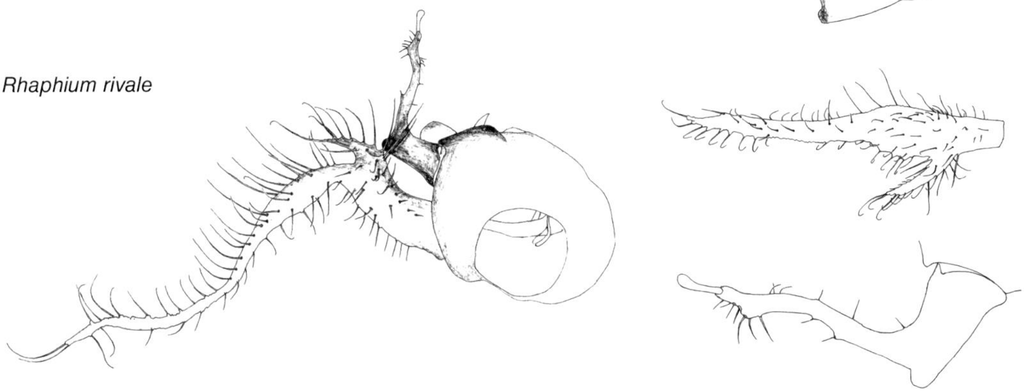

Rhaphium suave

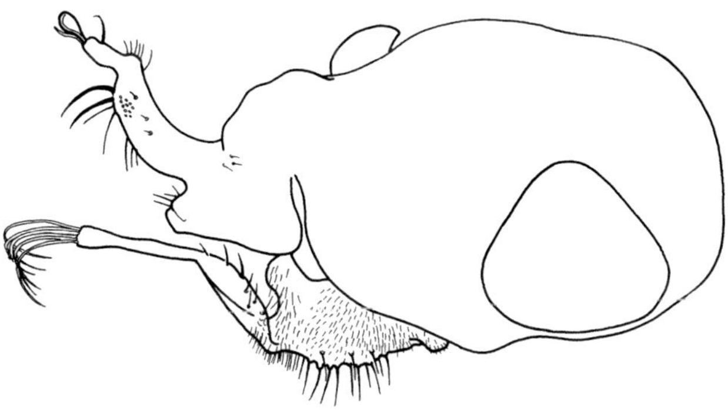

Fig. 32. (*Continued*)

Rhaphium albomaculatum (Becker, 1891)

Males resemble *appendiculatum* in having long ribbon-shaped cerci and a short arista but differ in many ways including a green dusted frons, no black setae on the front coxa and short stout ventral setae on the front tibia. Its leg colour is particularly variable, ranging from mainly yellow to nearly entirely black but always with the mid tibia and knees and the apical half of the front coxa obviously paler than the remaining parts of the legs. Females need comparing only with *brevicorne* as the only two species with pure white setae on the coxae. The coxae and femora are usually completely black though pale-legged specimens resemble *brevicorne* but in these specimens the yellow colour is somewhat darkened compared to the clear yellow of *brevicorne*. The ventral seta of the front tibia sometimes approaches the size in *brevicorne*. Wing length ♂: 2.7 (2.4–2.9) mm, ♀: 3.1 (3.0–3.4) mm.

This species has a clear northern and western distribution delimited by a line from the Severn estuary to the Humber where is may be frequent in more upland localities. It is usually found by streams and rivers, usually in woods but not confined to them, and less often by seepages and lakes. Some records are from sites approaching 1000 m altitude. Flight: v–vii.

Rhaphium antennatum (Carlier, 1835)

The black 'flag' at the end of the arista, beyond a white section shortly before the flag, makes this species unmistakable; by comparison, *patulum* has longer pubescence at the tip of its arista which makes it look swollen but the whole arista is black. Should the arista be missing, the yellow front coxae and palps are unique among the group of species with no lateral seta on the hind coxa, and its face is also wider than in most of this group, about as wide as the depth of the postpedicel, although these characters are common in the *appendiculatum* group. Females also have almost entirely yellow legs including all or most of the front coxa. The yellow at the tips of the mid and hind coxae runs continuously through the trochanters to the femora, making them far paler than most *Rhaphium*, but similar to *micans* which has several strong black setae at the tip of the front coxa. Wing length ♂: 3.2 (2.9–3.5) mm, ♀: 3.9 (3.6–4.3) mm.

The species is sparsely distributed mainly in lowland England with records north to Cumbria and into Wales but is more frequent in eastern England. Records are concentrated on coastal marshes and river floodplains in the east. Its habitat is mainly margins of still water bodies, particularly ditches and ponds, including slightly brackish water and sometimes reedbeds. Flight: v–ix, mainly vii–viii.

Rhaphium appendiculatum Zetterstedt, 1849 (Plate 6)

Males closely resemble *caliginosum*, both being the only species in the *appendiculatum* group to have a very short arista that is not or scarcely longer than the depth of the postpedicel, and lacking a preapical seta on the hind femur. The ribbon-shaped cercus resembles that of *albomaculatum* but lacks a small basal lobe; other differences are mentioned in the account for that species. Females with no anterior preapical seta on the hind femur and a shiny violet frons are immediately recognisable as either this species or *caliginosum* but the two cannot be separated. Large specimens are probably *appendiculatum* (since their males are larger than those of *caliginosum*) but this cannot be used as a reliable character. Wing length ♂: 3.4 (3.1–3.8) mm.

This is one of the commonest British *Rhaphium*, found in much of Britain. It occupies a wide range of wet habitats but is found most often by shaded seepages or stream margins. Flight: iv–x, with a long flight season.

Rhaphium auctum Loew, 1857

This is one of the largest species in the *appendiculatum* group. Its cercus resembles that of *monotrichum* but has several long hairs on the outer margin; their surstyli are very similar. Females are readily distinguished from the smaller *monotrichum* as the difference in the colour of the hind tibia appears to be constant, but probably not from *lanceolatum*. The position of arista and shape of the postpedicel vary but give a good initial guide; likewise, venation overlaps between *auctum* and *monotrichum*. More comparisons are given in the accounts for *monotrichum* and *lanceolatum*. Wing length ♂: 3.7 (3.6–3.8) mm, ♀: 4.1 (3.8–4.3) mm.

The distribution is wide in England but with a marked absence in eastern central counties, sparse in Wales, and in Scotland apparently confined to the central lowlands and southern

Highlands. It is found mainly in seepages and fens, by streams and in wet woodland. It is often but not exclusively associated with base-rich sites (Crossley 2020). Flight: v–ix, mainly vi–vii.

***Rhaphium brevicorne* Curtis, 1835**

Rhaphium brevicorne is a small dark brown or greenish black species. Its legs are mainly yellow with contrasting darker mid and hind tibiae but they can vary from entirely yellow to nearly entirely black apart from the knees and the apical half of the front coxa. The relative lengths of the arista and postpedicel vary from the arista being much shorter to slightly longer than the postpedicel. The illustration of the surstylus by Negrobov *et al.* (2013) differs slightly from that illustrated here for a specimen collected in southern England not far from Curtis's type locality. Pale-legged females are distinctive but northern darker-legged specimens resemble *albomaculatum*. Wing length ♂: 2.7 (2.5–3.0) mm, ♀: 3.3 (3.1–3.6) mm.

The species is widespread in Britain and although scarce in Scotland, where it is almost confined to the coast, it is recorded from Orkney and Shetland. It shows a high concentration of records in coastal locations but inland there may be a possible association with headwater streams. It is almost always found by running water, from tiny trickles to river margins, and usually in open or only lightly shaded places. It is particularly characteristic of seepages and tiny streams on coastal cliffs. Flight: iv–x, mainly vi–vii.

***Rhaphium caliginosum* (Zetterstedt, 1843)**

This species closely resembles *appendiculatum* but is easily separated using the genitalia and in having a dark shade along the dorsal side of the front femur which is always yellow in *appendiculatum*. It is almost identical to *lanceolatum* apart from the genitalia, but has a relatively shorter arista and lacks a preapical seta on the hind femur. Females with no anterior preapical seta on the hind femur and a shiny violet frons are usually this species or *appendiculatum* but the two cannot be separated. Small specimens are probably *caliginosum* (since their males are smaller than those of *appendiculatum*) but this cannot be used as a reliable character. The smaller *fasciatum* also has the same frons and hind femur characters but differs in having completely yellow coxae. Wing length ♂: 3.0 (2.9–3.2) mm.

Rhaphium caliginosum is a common and widespread species in England and Wales but is scarce in Scotland, reaching the Inner Hebrides. It is found in almost any wetland although usually by flowing water, from seepages to rivers, and less often by standing water bodies including slightly brackish places. Sites may be shaded or in the open. Flight: iv–x, long flight period.

***Rhaphium commune* (Meigen, 1824)**

The legs of this large species are usually black apart from the hind tibia which is yellow with a black base and tip, and with ill-defined boundaries where the colour changes; the front and mid tibiae and basitarsi may sometimes be paler. Its legs are particularly hairy and bristly, with at least ten setae on the dorsal surface of all tibiae. The diverging dense black ventral hairs on the mid femur are not found in other species, in which any ventral hair-fringe is pale and relatively sparse. The posterior hair tuft at the tip of the front basitarsus is unique. Females share with *fascipes* the leg pattern of black coxae, trochanters and femora with a strongly contrasting yellow basal half to the hind femur. The lateral seta on the hind coxa is usually weak and sometimes absent. Wing length ♂: 4.8 (4.2–5.4) mm, ♀: 5.2 (4.5–5.5) mm.

Rhaphium commune is found sparsely in the southern half of Britain and is very scarce in Wales and Scotland north to the Moray Firth. Its habitat appears to be seepages or margins of running water, most often in woodland, and there may be an association with base-rich sites. Occasional records are from ponds and in open habitats. Flight: v–viii, mainly v–vi.

***Rhaphium consobrinum* Zetterstedt, 1843**

Males are immediately recognised by the ventral fringe of long seta-like hairs on the front femur. The front and mid tibiae each have many more dorsal setae than in most species. Females are moderately large flies whose yellow legs are only rarely slightly browner in parts, and strongly resemble the slightly smaller *laticorne* which lacks setulae between the acrostichal and intra-alar rows, and *nasutum* which may at times be inseparable from it. Wing length ♂: 3.7 (3.2–4.0) mm, ♀: 4.3 (3.9–4.8) mm.

This species occurs on the coast around much of Britain but is scarcer in Scotland. It is a saltmarsh species but there are rare inland populations, particularly by large rivers in Yorkshire. With the exception of these inland sites, the habitat is upper saltmarsh and brackish grazing marsh, where it can be numerous. Flight: iv–ix, peak in vii.

Rhaphium discolor Zetterstedt, 1838 may be the senior synonym of *consobrinum*. The type specimen is a female whose identity needs to be confirmed by examination and bar-coding.

Rhaphium crassipes (Meigen, 1824)

Males are strikingly different from all other species in many ways. Apart from the characters used in the key, which are nearly all unique to *crassipes*, it has conspicuously long black hind legs that contrast with the mainly pale yellow front and mid legs. The leg colour varies slightly but the front femora are always black on the basal half. Females are in the group of large species with a strong posteroventral seta on the mid tibia. Its legs are nearly always mostly yellow and it is distinct from other British *Rhaphium* except *patulum* in having a cluster of lateral hairs on the scutellum and yellow-tipped palps as otherwise found only in *antennatum*. Vein M_1 is not always strongly sinuous and can resemble some *riparium*. Wing length ♂: 4.9 (4.7–5.1) mm, ♀: 5.6 (5.0–6.1) mm.

Rhaphium crassipes is widespread in Britain reaching the north of Scotland but scarce in Wales and East Anglia. Its habitat is usually seepages and margins of streams and rivers, often in woodland or partial shade, although also in more open places. It has been recorded at 450 m altitude. Flight: iv–vii (viii?), mainly v–vi.

Rhaphium elegantulum (Meigen, 1824)

This is the largest British *Rhaphium*, with a shining scutum that has scarcely any dust pattern and usually mainly yellow legs. The contrasting colour pattern of the tergites is found only in *longicorne* in which the last two, not just the last, large tergites differ in colour from those in front of them. The near absence of posterior hair fringes on the front femur is unique among the group of British species with no lateral seta on the hind coxa. Females are robust flies with mainly yellow legs and a shining scutum and tergites with shades of violet, blue and green but without the male's conspicuously differentiated colour of the last tergite. The wings are brownish with the anterior cells r_1 and r_{2+3} darker still. Wing length ♂: 5.1 (4.9–5.3) mm, ♀: 6.3 (6.1–6.5) mm.

The species is moderately widespread in England but scarce in Wales and in Scotland mainly from the central belt and Strathspey. It has a clear affinity with large ponds and lakes, and particularly artificial water bodies including reservoirs, pits and ornamental ponds which are over-represented in the records. The common feature may be bare shores with fine sediment exposed during summer draw-down, adjacent to emergent vegetation. Flight: v–ix, peak in vi.

Rhaphium fasciatum Meigen, 1824

This is the smallest British *Rhaphium*, a delicate pale species with an almost black shining scutum with no dust pattern, pale legs, coxae and metepimeron, only four dorsocentral setae, almost straight long veins and a barely detectable preapical seta on the hind femur. The conspicuously yellow-ringed tergites and entirely yellow coxae of the males are unique among British species. All these features make it superficially resemble *Lamprochromus* or sympycnines such as *Anepsiomyia* or *Syntormon* although in the males the antennae are typical of the *appendiculatum* group to which it is probably only distantly related. Wing length ♂: 2.6 (2.5–2.8) mm, ♀: 2.7 (2.6–2.8) mm.

Rhaphium fasciatum is widespread in Britain although slightly more frequent in the north and west. It is found mainly by flowing water, from trickles to river margins, by the well-vegetated margins of ponds and lakes, in fens and reedbeds with tall vegetation and carr woodland. There may be a minor requirement for light shade. Flight: iv–ix, mainly v–vi.

Rhaphium fascipes (Meigen, 1824)

The clubbed hind tibia which is black on the swollen half and abruptly yellow on the narrow basal half is unique in British *Rhaphium*; the effect is exaggerated by dense hairs on the swollen half.

Several other species have slightly swollen hind tibiae, particularly *fractum*, *laticorne* and *rivale* which all have a broad face and relatively short arista but lack the strongly contrasting tibial colour pattern. Females have dark coxae and femora with the basal part of the hind femur contrastingly yellow, as found in *commune*. As with that species, the hind coxa usually has a lateral seta but the key allows for occasional specimens in which it is absent. Wing length ♂: 3.7 (3.6–3.9) mm, ♀: 4.1 (3.8–4.3) mm.

This species appears to be confined to the south of England and Wales; it has not been found in Scotland since 1906. The sites are wet heathland and wet acid grassland, sometimes in sparse woodland. Flight: iv–ix.

***Rhaphium fractum* Loew, 1850**

Rhaphium fractum is a large dark subshining species with mainly black legs apart from parts of the tibiae and tarsi. The legs and scutum are particularly bristly, approaching the condition in *commune*, and the tibiae, although variable in colour, are usually dark. The short arista, scarcely as long as the postpedicel, is unusual among the group with no lateral seta on the hind coxa; *laticorne* has an even shorter arista but is a smaller shining green species. A character used in previous keys is the tiny spot near the inflexion of vein M_1 but this is not always clear and may be absent from one wing while present on the other, or entirely absent. Females have yellowish legs although the colour is slightly variable and may approach the condition in dark *riparium* from which they are distinguished by having setulae between the dorsocentral and intra-alar rows and numerous small acrostichal setae. The short row of dorsal hairs at the base of the hind femur is an uncommon feature. Wing length ♂: 3.9 (3.7–4.1) mm, ♀: 4.4 (4.1–4.7) mm.

The species is found in the west and north of Britain. Nearly all sites are large rivers, particularly those with exposed gravel or stony sediments on their shores. Flight: v–viii.

***Rhaphium gravipes* Haliday, 1851**

Among species with mainly dark femora and a swollen tip on the front basitarsus, *gravipes* may be distinguished by the rather short front basitarsus which is about three-quarters the length of the rest of the tarsus rather than about equal or longer in all other *Rhaphium*. The row of long ventral hairs on front tibia is unique to this species. The genitalia are very similar to those of *patulum* which differs in having a spur on the mid coxa. Females have almost entirely dark femora and their scutum in most specimens has an unusual pattern of two narrow dull dark stripes that remain separated by a broader pale stripe at all angles of view; in other species with a patterned scutum, the pair of lines fuse when viewed from behind, except in *patulum*. Specimens with only 6 dorsocentral setae, an entirely black hind femur, pale grey face and the mid coxa without obvious black apical setae are *gravipes*, but variation in some of these characters can approach the condition in *patulum* (see that account for these problematic specimens). Wing length ♂: 4.1 (3.8–4.4) mm, ♀: 5.1 (4.4–5.6) mm.

The distribution of *R. gravipes* is northern, with records mainly from the Scottish Highlands and the northwest Pennines in England; there is a single record from Gwent. The records appear to be entirely from large stony rivers. It has been reared from larvae found in wet sandy sediments on a river shore. Flight: iv–viii.

***Rhaphium lanceolatum* Loew, 1850**

This small species resembles *appendiculatum* and *caliginosum* but has a distinctly shorter postpedicel relative to the arista, the hind femur has a preapical seta and the surstylus resembles those of *monotrichum* and *auctum*. Females appear to be inseparable from those of *auctum* as there is considerable variation in *auctum* that overlaps with the condition in probable *lanceolatum* found together with males. Veins R_{4+5} and M_1 are marginally more parallel to each other near the tip but the condition in *auctum* overlaps with this. Size appears to be the only distinction, *lanceolatum* being the smaller species. Wing length ♂: 3.3 (3.1–3.5) mm, ♀: 3.8 (3.4–4.0) mm.

This species is principally recorded from the northern half of Scotland where it is widespread, but there are confirmed records from North Yorkshire and the New Forest, Hampshire. The habitat includes acid valley mire with seepages and possibly river and lake margins but information is scarce; MacGowan (1987) refers to several captures beside acid upland bogs, and Pollet *et al.* (1988) describe the habitat as water-meadows neighbouring deciduous woodland. Flight: v–viii.

Rhaphium laticorne **(Fallén, 1823) (Plate 7)**

This small species has a long postpedicel resembling that of the *appendiculatum* group but otherwise there is no similarity. The front tibia often has many almost evenly spaced anterodorsal setae which is unusual in the genus. A resemblance with *consobrinum* is found in the similar front basitarsus and second tarsomere which articulate with mismatched profiles and their basitarsi have long upright 'prickly' vestiture. Females have yellow legs that are only rarely dark brown in parts, so resemble *consobrinum* and *nasutum*; like *consobrinum*, the vestiture of the femora is particularly short and even on all faces but, except for the anterodorsal row on the front tibia, *laticorne* has fewer setae. Wing length ♂: 3.2 (3.0–3.4) mm, ♀: 3.6 (3.1–3.9) mm.

Rhaphium laticorne is found widely in England north to Yorkshire and Cumbria, south Wales and rarely on the east coast of Scotland. It is found at the margins of a wide range of flowing and still water bodies, usually those with sandy or stony shores, in unshaded or lightly shaded sites. Flight: iv–ix, mainly vi–vii.

Rhaphium longicorne **(Fallén, 1823) (Plate 7)**

This is a large robust dark metallic green to almost black species with outstandingly long antennae. Even if the antennae are missing, both sexes are distinguished from other British *Rhaphium* by the combination of their large size, black coxa and femora (except the extreme tips), glossy violet frons and two strong black lateral setae on the hind coxa. Wing length ♂: 4.4 (4.1–4.6) mm, ♀: 4.9 (4.7–5.1) mm.

This is a locally common species dependent on acid peatlands so that it is frequent in the north and west of Britain. There are isolated populations scattered across lowland southern England where this habitat occurs. It is found at seepages and pools on peat up to 930 m, often where purple moor-grass or *Sphagnum* are dominant, and only infrequently within woodland. Flight: iv–viii, peak in vi.

Rhaphium micans **(Meigen, 1824)**

Males are distinguished from other British *Rhaphium* by the dome-shaped ventral fringe of small setae on the front basitarsus, the long thick setae on the palps and the combination of black hairs on the front coxa and white hairs on the occiput. Females have mainly yellow legs including at least half of the front coxa, which is a combination found in *antennatum* and pale-legged *auctum* which both have strong black setae on the front coxa (none in *micans*). The scutum has a poorly developed pattern and is shiny from most angles of view, rather like most in the *appendiculatum* group. Wing length ♂: 3.8 (3.3–4.0) mm, ♀: 4.5 (4.1–4.9) mm.

The species is sparsely distributed in England and Wales, north to about the Humber. There are unconfirmed and unlikely records from the west coast of Scotland. It is found at the margins of large sandy and stony streams and rivers, and, especially in the east, ponds, lakes and flooded gravel pits. There appears to be an association with finer sandy sediments, often with emergent vegetation. Sites are usually unshaded. Flight: v–ix.

Rhaphium monotrichum **Loew, 1850**

Apart from the distinctive genitalia, males are easily separated from other species in the *appendiculatum* group by the long dorsal vestiture on the front basitarsus and very narrow anal lobe, which are unique to *monotrichum*. The surstylus illustrated by Negrobov (1979b) differs from British specimens. Typical female specimens are distinctive in having a moderately narrow triangular postpedicel, widely spaced veins R_{4+5} and M_1, and a markedly black apical quarter to third on the hind tibia, compared to being similar in shade along the entire length in *auctum*, although often paler ventrally than dorsally, and variable in possible *lanceolatum*. The first two characters can overlap with those of *auctum* and possible *lanceolatum* and the venation needs to be measured carefully. Other constant features are the rather weaker apical setae on the front coxa compared to quite stout setae in *auctum* (but intermediate in possible *lanceolatum*) and the extreme tip of hind femur is not darker behind than in front (darker behind in *auctum*, variable in possible *lanceolatum*). Wing length ♂: 3.2 (2.8–3.3) mm, ♀: 3.6 (3.3–4.0) mm.

This is one of the more widespread and common species of *Rhaphium* found north to the Outer Hebrides and north Scottish coast, and is apparently as frequent in Scotland as in most of England. It is found in a wide range of wetlands, including brackish sites, with no apparent specificity but, as with most *Rhaphium*, may require moving ground-water as it is rarely found in floodplain and coastal fens and marsh. In contrast, Pollet *et al.* (2009) state that it has an affinity for swamps, and reed marshes in particular. Flight: iv–x, peak in vi.

Rhaphium nasutum (Fallén, 1823)

All the tibiae are usually yellow in this fairly shiny dark green species. The ventral projection at the tip of the second tarsomere of the front leg is unique among male British *Rhaphium*. Females are yellow-legged but the colour is very variable, sometimes including a dark front femur and dark dorsal surface on the hind femur. Typical pale-legged specimens closely resemble *consobrinum* in having a shiny scutum lacking a distinct pattern and often violet on the rear half; distinguishing the two species requires careful measurement of the key characters. Vein dm-m is usually slightly less angled than in *consobrinum*. Wing length ♂: 3.9 (3.7–4.2) mm, ♀: 4.3 (4.0–4.5) mm.

Rhaphium nasutum is widely but sparsely distributed in Britain. Its habitat is unclear but it is most frequently found at the unshaded margins of rivers, but also at river washlands, lakes and pits. Pollet *et al.* (2009) give its habitat as well-developed riparian vegetation by lakes and large rivers. Flight: iv–ix.

Rhaphium patulum (Raddatz, 1873)

Two characters distinguish *patulum* from other British species: the very slightly swollen tip of the arista (not strongly swollen as in *antennatum*) and the numerous small hairs on the margin of the scutellum (not just in front of the large apical setae as in *crassipes*). It closely resembles *gravipes* which differs in having long ventral hairs on the front tibia, absent in *patulum*. These two species also have very similar hypopygia; in *patulum* the tiny hairs at the tip of the surstylus are strictly apical whereas in *gravipes* they are just subapical. In ventral view, all the lobes of the epandrium are similar in width but in *gravipes* some are far wider than others. Females closely resemble those of *gravipes* and the characters taken from d'Assis-Fonseca (1978) appear to be the only useful distinctions. Even these have intermediates, particularly the hind femur of *gravipes* may be up to a third yellow at the base on the ventral and posterior faces so approaches the condition in *patulum*. Differences in the tinge of the dust of the face and frons of these two species are often scarcely detectable and the colour of hairs and setae on the mid coxa are not always consistent: *patulum* has more black hairs and setae, and the hairs tend to be more yellowish, whereas *gravipes* has fewer or no black hairs and dark setae, but there is too much overlap for the distinction to be reliable with isolated individuals. Wing length ♂: 4.8 (4.3–5.6) mm, ♀: 6.1 (6.0–6.2) mm.

This species is found mainly but infrequently in Scotland with very few records in north England and Wales. Its sites include rivers with shores of mud, sand and stones, and riverside reedbed and fen. Flight: iv–viii.

Rhaphium pectinatum (Loew, 1859)

While resembling others in the small group of species with black hairs and setae on the back of the head and front coxa and lacking a spur on the mid coxa, this species is distinct in its ventral setae on the front femur, the cluster of long white hairs at the base of the mid femur, and its hypopygium. A female has not been seen so is not included in the key. d'Assis-Fonseca (1978) used the presence of extra setulae between the dorsocentral and intra-alar setae (as in *consobrinum*), a relatively elongate postpedicel, almost entirely yellow tarsi, and the crossvein dm-m being less than twice as long as the apical section of M_4. Wing length ♂: 4.2 mm.

The only records of males are one from the River Thames in London (1868) and another from River Exe in Devon (2015) (Wolton and Drake 2015). The Exe site was reedbed by the river's estuary. Flight: vi–vii.

***Rhaphium penicillatum* Loew, 1850**

This large species superficially resembles *riparium* but males usually have entirely black femora which are largely yellow in most specimens of *riparium*, the hairs of the front coxa and occiput have a yellow tinge, unlike the white of *riparium*, and the front fifth tarsomere has an apical cluster of several strong hairs, which are less distinct in *riparium*. Typical female specimens are dark-legged, particularly on the front face and apical half of the hind femur, and can be confidently named, but paler examples taken in the company of male *penicillatum* look identical to darker-legged *riparium* and there appears to be no way to distinguish them. Wing length ♂: 4.4 (4.1–4.9) mm, ♀: 5.5 (5.1–5.7) mm.

The species is sparsely distributed in the west of England, Northumberland and Wales, with one record near the Forth in Scotland. It is found on the shores and associated vegetation of sandy and stony rivers. Flight: iv–vii, peak in v.

***Rhaphium riparium* (Meigen, 1824)**

This large species is similar to *penicillatum* (see that account) and more particularly to *suave* which differs on small and perhaps unreliable external differences. All three of these species are best separated using their distinctive genitalia. Females are large and pale-legged with a distinctly dark tip on the hind femur, but sometimes the front femur and the apical half of the hind femur can be entirely dark, so resembling *penicillatum*. Veins R_{4+5} and M_1 can sometimes be as widely spaced as in *crassipes* so this character should not be relied upon to separate these two species. Wing length ♂: 4.6 (4.1–5.1) mm, ♀: 5.6 (5.0–5.9) mm.

Rhaphium riparium is widespread and common in the west and north of Britain but scarce in Wales. Apart from a population in Surrey–Sussex area, elsewhere in lowland England there are only sparse records. It is found mainly by running water in woodlands or under tree cover, but rarely by seepages. Flight: iv–ix, mainly v–vii.

***Rhaphium rivale* (Loew, 1869)**

This is a small species whose long hairs on the last tarsomere of the front leg are unique among male British *Rhaphium*; the other tarsomeres are cylindrical with unremarkable hairing. Even if the front tarsi are missing, then the only similarly undistinguished species with black femora and a wide face are *gravipes* and *patulum* which have a far longer undivided cercus and surstylus with only a tiny apical seta. Females have a dark greenish brown scutum and mainly black femora but the hind one may be slightly paler behind at the base, so approaching the condition in the rather larger *commune* and *fascipes*. Wing length ♂: 2.9 (2.4–3.1) mm, ♀: 3.5 (3.1–3.6) mm.

The distribution is western and northern, from south Wales northwards. It is found almost exclusively by rivers with stony or sandy shores. Flight: v–viii, peak in v.

***Rhaphium suave* (Loew, 1859)**

Males are externally almost identical to *riparium* and the small non-genitalia differences used in the key may be unreliable. In addition, the postpedicel is more obviously pointed at the tip compared to a more rounded tip in *riparium*, but with overlap between the two species. The narrow cercus with its apical 'claw' of hairs is unmistakable. Even though Loew (1859) described the female, neither Parent (1938) nor Negrobov (1979b) included it in their keys, and their example is followed here. Loew's description closely fits typical pale-legged *riparium* which may be similar in size and overall appearance, so is compared here with Loew's description of *suave*, with the condition in *riparium* in brackets: face dusting greyish yellow (white), arista 1.75 times the length of the antenna (at least twice in other larger species), mid coxa with all hairs white (black apical hairs), hind basitarsus paler at base (usually all black), front femur with fairly long but extremely delicate white hairs near the base on the ventral surface, longer than those on the mid femur (similarly short on both femora), front basitarsus shorter than tarsomeres 2–5 combined (slightly shorter). Wing length ♂: 4.1 (3.9–4.4) mm.

Males have been recorded from three rivers in Gwent, Cheshire and Northumberland (Drake 2007b). The habitat is sandy shores of rivers, sometimes with dense herbaceous vegetation or willow scrub. Flight: vii.

SCELLUS Loew, 1857 (Hydrophilinae)

Scellus are large dark species with stout front legs armed with setae, spines and projections, and long slender mid and hind legs with only a few weak setae (Plate 7). They have many of the characters found in many hydrophilines such as a face that widens towards the mouth, whiskers on the postgena, and weak marginal setae on the tergites, but are distinguished by the scape being clearly longer than the pedicel, the dorsocentral setae reduced to numerous tiny hairs, one notopleural seta, three stout proepisternal setae, and the subcosta ending free in the costal cell. Males are clearly distinguished by possessing a conspicuous bright yellow flag, the cingulum, arising between the fifth and sixth tergites. Doane (1907) describes the courtship of a North American species (*S. virago* Aldrich, 1907) in which females take the initiative; this agrees with the suggestion that the cingulum disperses pheromones rather than acts as a visual signal to females (Runyon 2008).

Five species occur in the Western Palaearctic with one in Britain. Although their occurrence in Britain is unlikely, the males of three species found in Scandinavian countries may be distinguished from *notatus* by lacking the apical extension of the front tibia (*dolichocerus* Gerstäcker, 1864), or having dense hairs on the ventral surface of the mid femur, which is bare in *notatus* (*gallicanus* Becker, 1909, *spinimanus* (Zetterstedt, 1843)) (Negrobov 1978).

Scellus notatus (Fabricius, 1781) (Plate 7)

Scellus notatus is a large coppery green and moderately shiny species with similarly coloured legs lacking pale areas. Both sexes have similar conspicuously marked wings (figure female genus key couplet 13). Males exhibit secondary sexual characters on the front tibia, notably a broad two-pronged ventral peg, which in females is replaced by a row of long setae (figures genus key couplet 4 for males, 13 for females) and on the mid tibia which has a cluster of tightly curled hairs at its tip. Wing length ♂: 5.9 (5.5–6.2) mm, ♀: 6.2 (4.6–6.9) mm.

It is frequent throughout most of England but scarce in the southwest, Wales and Scotland where it occurs north to the Moray Firth; the distribution may be limited by high rainfall in the west and north. Its habitat does not appear to be clearly defined. It is more frequently recorded in dry places such as heathland, chalk grassland and sites with bare soil such as quarries and brown-field sites, scrub and woodland, but also some wet habitats including reedbeds, wet grasslands and upper saltmarsh. Flight: v–x, with extended flight period showing little indication of a peak.

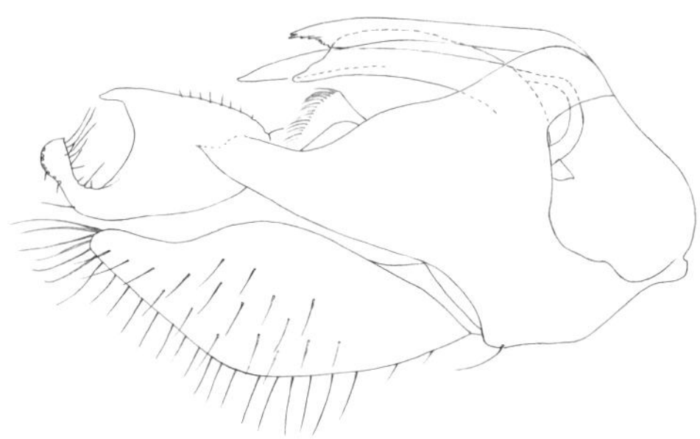

Fig. 33. *Scellus notatus* hypopygium.

SCHOENOPHILUS Mik, 1878 (Hydrophilinae)

There is only one species in the genus worldwide although some authors (e.g. Grichanov 2017) treat *Schoenophilus* as a subgenus of *Thinophilus*. It is somewhat nondescript but can be recognised by the combination of a single notopleural seta, large palps and the absence of postgenal whiskers, and in males by tergite 6 being not visible (Plate 7). The combination of only 4 dorsocentral setae and no acrostichal setae is found in only a few other genera. Confusion is most likely with sympycnines but these have a dorsal arista and almost all have two notopleural setae.

Schoenophilus versutus (Haliday in Walker, 1851) (Plate 7)

Schoenophilus versutus is a very small dull-dusted greyish brown fly with at most a hint of olive-green on the scutum. All the legs are moderately slender; the femora are dark and the tibiae dull yellow to brown with weak sparse setae only on the mid and hind tibiae and particularly sparse hairs on the front coxa (about ten). The wing has a weak small dark spot at the wing boss and on dm-m.

The almost round postpedicel is often dull yellow on the basal half, and the arista is placed just dorsal of the midpoint. The front eye facets are particularly large, about 1.5 times the diameter of those next to the frons. Most of the hypopygium is hidden but the long ribbon-shaped phallus is wrapped around the capsule externally, but shown extended in Fig. 34. Females are similar to the males apart from the genitalia. Wing length ♂: 2.0 (1.8–2.1) mm, ♀: 2.2 (2.1–2.4) mm.

Schoenophilus versutus is sparsely distributed in England and Wales, mostly around the coast but with some inland locations. There are no Scottish records. It is a wetland species of seepages on coastal cliffs, water margins of dune slacks, ponds and ditches, and water margins within both calcareous fens and acid mires. Flight: v–ix, mainly vi.

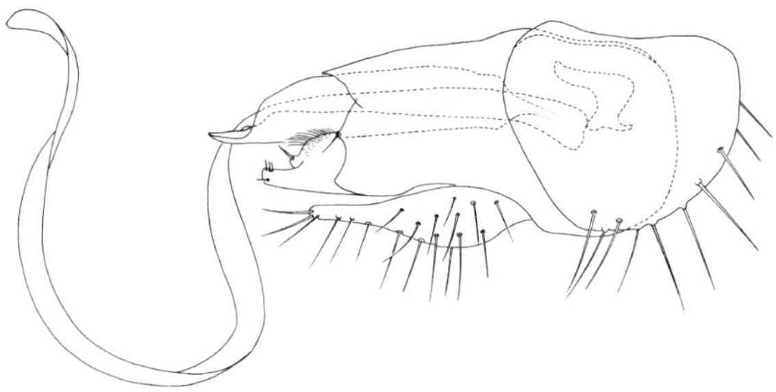

Fig. 34. *Schoenophilus versutus* hypopygium.

SCIAPUS Zeller, 1842 (Sciapodinae)

Sciapus are medium-sized to large green flies with long very slender pale legs and the males with a slender abdomen that is at least twice as long as the thorax (Plate 7). It is the only British genus in which vein M is prominently forked with the anterior branch M_1 at right angles to M_2 and converging strongly with R_{4+5} near the wing tip; M_2 reaches at least halfway to the wing margin. M_2 is sometimes reduced to a fold but is still visible when viewed at an angle to the plane of the wing. A few species of *Dolichopus* have a rudimentary fork in M. Females of all species and the males of a few have a 'cage' of strong spines on the front femur and coxa. In life, males hold their wings apart, delta-fashion, rather than folded over the abdomen as in most other subfamilies.

Ten species occur in Britain. Among the 40 European species, more may be found here. *Sciapus albifrons* (Meigen, 1830) resembles *heteropygus* in having largely yellow tergites and entirely yellow antennae, but differs in the hind basitarsus of the male being longer than the second tarsomere (similar in length in *heteropygus*); males of *lobipes* (Meigen, 1824) and *nervosus* (Lehmann, 1822) resemble those of *wiedemanni* in having a lobed front tarsus, but in *lobipes* the tarsus is entirely black rather than with yellow basal tarsomeres as in *wiedemanni*, and *nervosus* lacks acrostichal setae which are biserial in all British species; male *spiniger* (Zetterstedt, 1859) resembles *platypterus* in having a narrow face but differs in having long ventral spines on the front femur and no modification of the mid tarsus. Meuffels and Grootaert (1990) provide detailed figures of male and female genitalia and their work has been used in this key. Grichanov and Negrobov (2014) provide a key to all Palaearctic species. Bickel (1994) provides a wealth of information about the subfamily.

In contrast to many dolichopodids, most species of *Sciapus* are found in dry open habitats such as coastal dunes, heaths, soft-rock cliffs, cliff-top grasslands and other dry grassland types. The most widespread species, *platypterus*, is found in woodland and scrub habitats. Two species may require wet areas in otherwise dry habitats: *laetus* is found at wet slacks and ponds on dunes and other coastal sites, and *basilicus* has been recorded only at stony river shores (Crossley 2009).

Key to species of *Sciapus* (males)

1 Wing broadest well beyond middle, anal lobe undeveloped (a). Mid tarsomeres 3 and 4 slightly deeper than other tarsomeres, milky white except apical half of tarsomere 4 (b). Face narrow, its width at middle about one-eighth its height (c). Front femur with a dense anteroventral fringe of long pale hairs, apically up to twice as long as width of shaft (d). Hypopygium black with only cercus projecting, other appendages tiny.................... ***platypterus***

– Wing shape normal. Mid tarsus unmodified. Face broad, width at least half its height (cc). Front femur without a differentiated anteroventral fringe but may have ventral setae near base. Hypopygium with several long appendages visible, usually pale...2

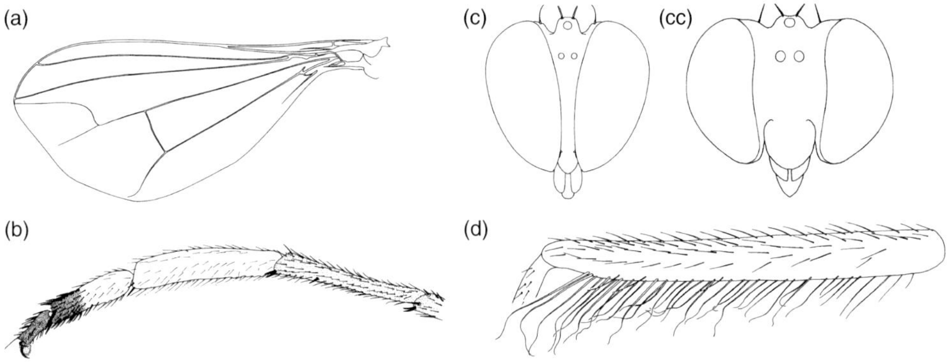

2(1) Scutum, tergites and frons brilliantly glossy green without dusting. Antenna entirely black. Mid tibia at base with 1–3 thin setae either side (a, dorsal view). Small species, wing length 2.4–3.0 mm. .... ***laetus***

– Scutum dull-dusted, at most slightly shining through dust. Antenna partly to entirely yellow. Mid tibia with only 1 anterodorsal seta at base, with or without others further along shaft. Small to large species, wing length at least 3.0 mm...3

(a)

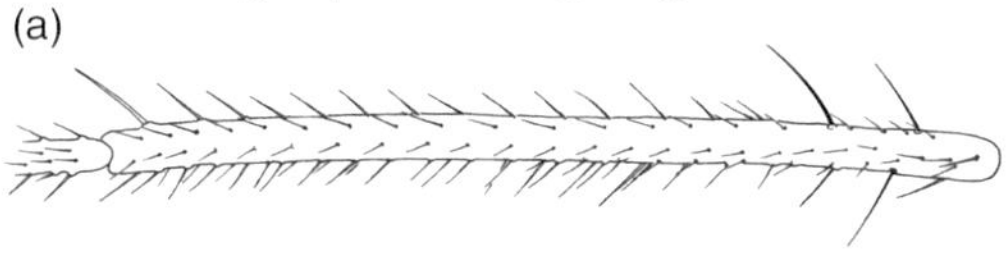

3(2) Tergites largely yellow with narrow dark bands on anterior and posterior margins. Postpedicel entirely yellow. Surstylus with long dorsal appendage, organ X strongly curved (a)..................... ....***heteropygus***

– Tergites entirely green or with lateral pale patches that do not meet dorsally on midline except on tergite 1. Postpedicel partly to entirely dark. Surstylus not forked, organ X square at base with finger-shaped appendage or fused with cercus.. 4

(a)

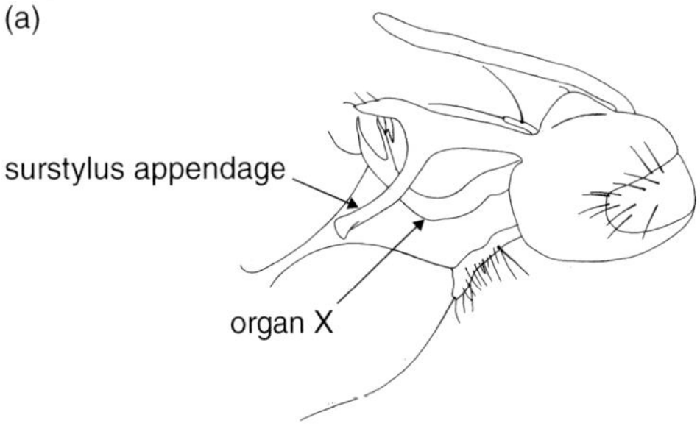

4(3) Front tarsomere 4 modified in shape and colour compared to preceding segments. Tergites 2 and 3 (and sometimes 1 and 4) with yellow lateral spots... 5

– Front tarsomeres not modified although tarsomere 5 may be slightly wider than others. Tergites with pale spots or entirely dark.. 6

5(4) Front tarsomere 4 with a black dorsal projection extending over tarsomere 5 (a). Front femur near base with a ventral row of fine setae longer than depth of femur. Front coxa with long apical setae about 1.5 times as long as width of coxa. Veins R_{4+5} and M_1 converge at a wide angle towards wing margin forming a triangular apex to cell r_{4+5} (b). Hind basitarsus longer than tarsomere 2 (about 1.3 times). .... ***wiedemanni***

– Front tarsomere 4 slightly deepened and white with mainly white pubescence, contrasting with black tarsomere 5 (aa). Front femur completely bare ventrally. Front coxa with dense apical hairs but no setae. Veins R_{4+5} and M_1 converge gradually, nearly parallel for several times their separation at wing tip (bb). Hind basitarsus as long as tarsomere 2. .... ***pallens***

(a)

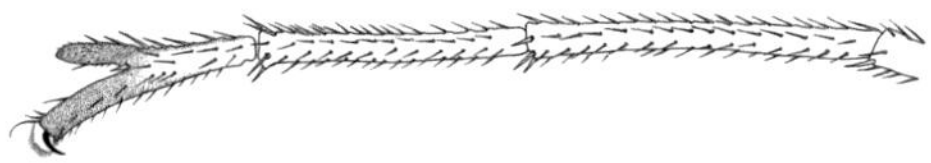

(b)

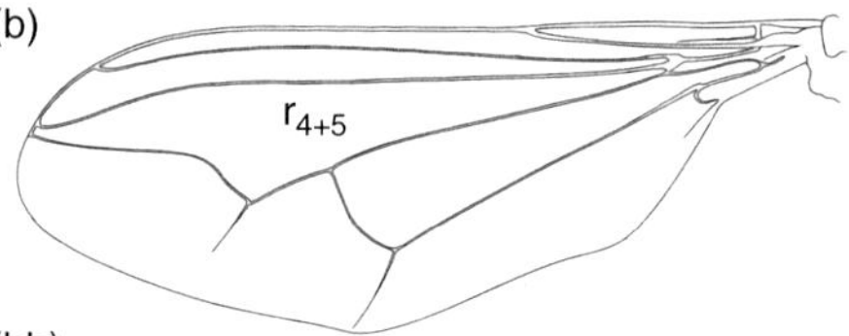

(aa)

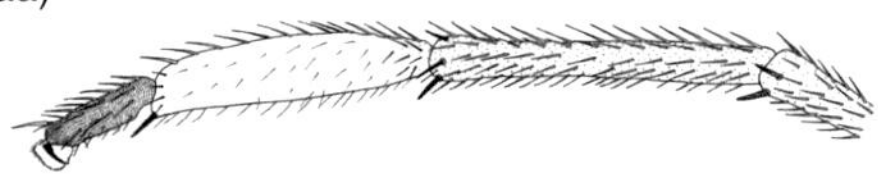

(bb)

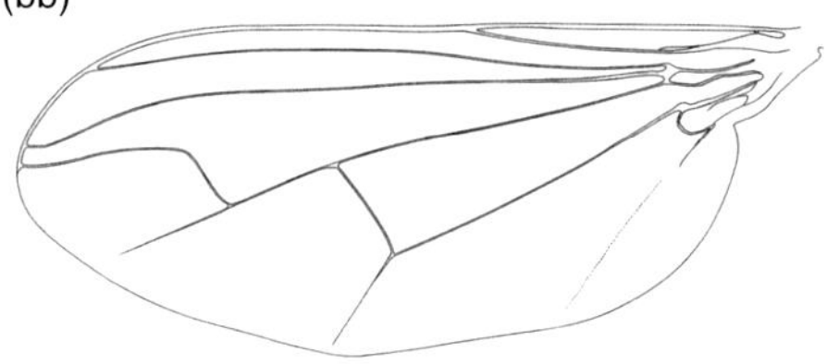

6(4) Front femur with a ventral row of setae on basal third to half, longest seta at least as long as width of shaft (a, b). 7

– Front femur without ventral setae but may have short hairs or an isolated short seta on basal third to half. 8

(a)

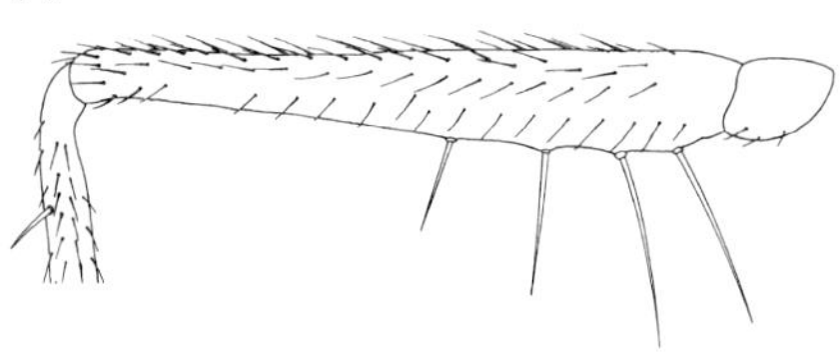

(b)

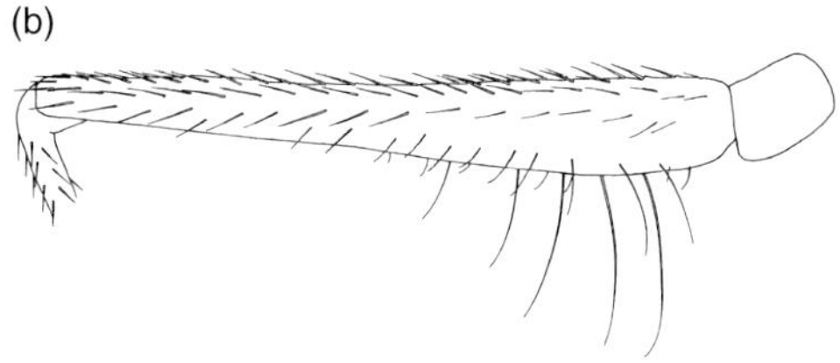

7(6) Scutum slightly shining when viewed from above. Mid and hind coxae and usually metepimeron mainly dark. All tergites dark. Cercus with long hairs extending well beyond its tip if imaginarily straightened (a). Mid tibia and basitarsus in dorsal view with a spiky appearance due to vestiture being longer and upstanding (60° to shaft), differing from vestiture on front and hind legs (b). Hind basitarsus and tarsomere 2 about equal in length, basitarsus at most 1.15 times longer. Face wider, width at middle 0.7 times its height (c). .... ***longulus***

– Scutum dull-dusted. Mid and hind coxae and usually metepimeron pale. Tergites 2 and 3 with yellow lateral marks. Cercus with shorter hairs barely surpassing tip of organ X (aa). Mid tibia and basitarsus in dorsal view with unremarkable hairs lying closer to shaft (30°) and the same on all legs (bb). Hind basitarsus longer than tarsomere 2, about 1.33 times. Face narrower, width at middle 0.6 times its height (cc). .... ***contristans***

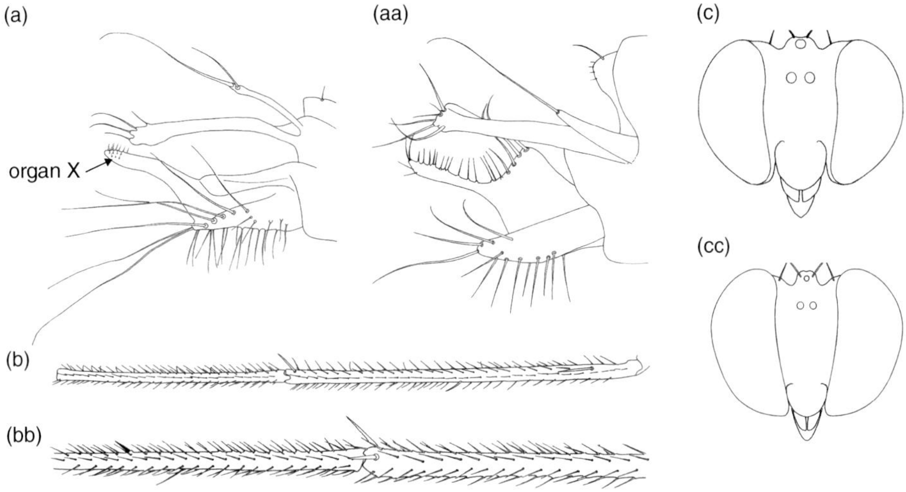

8(6) Mid tibia with 2–3 anterodorsal setae all about twice width of shaft (a). Hind tibia with 3 anterodorsal setae clearly distinct from hairs (b). Metepimeron mainly yellow. Mid coxa usually entirely yellow. Uppermost postocular setae usually all white.............................***maritimus***

– Mid tibia with only 1 long anterodorsal seta near base, seldom with more very small setae (aa). Hind tibia with setae scarcely differentiated from hairs or at most only 1 anterodorsal seta (bb). Metepimeron dark. Mid coxa dark at base. Uppermost 6–10 postocular setae dark. 9

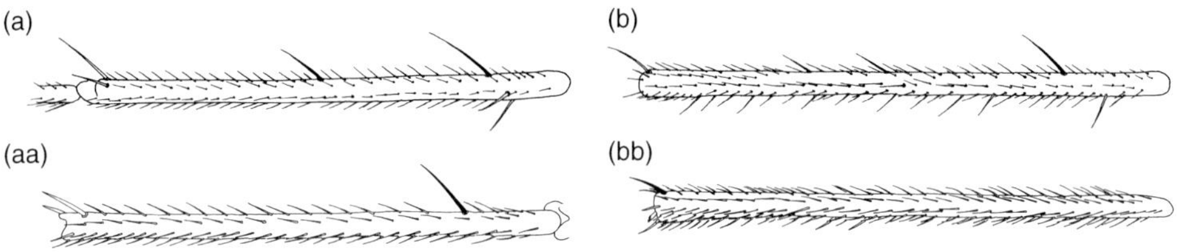

9(8) Front basitarsus with 1–3 small posteroventral setae as well as a small basal posteroventral (a). Hypopygium with tip of surstylus lacking a ventral flange in lateral view (b).....***zonatulus***

– Front basitarsus with only 1 small basal posteroventral seta. Hypopygium with tip of surstylus with a flange on ventral surface forming distinct bulge (bb). ....***basilicus***

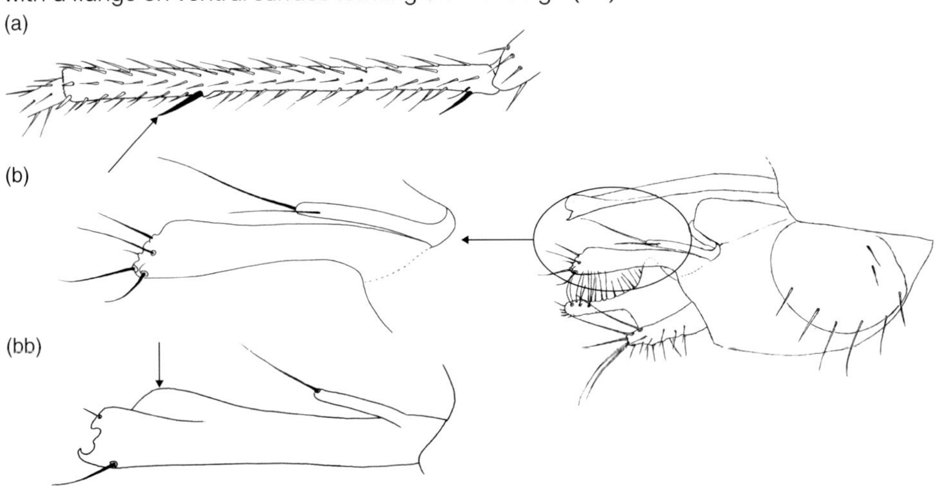

Key to species of *Sciapus* (females)

1 Face narrow, width one-third or less its height (about 0.29–0.35 times) (a). Hind basitarsus well over 1.5 times length of tarsomere 2 (about 1.8 times) (b).......... ***platypterus***

– Face broader, width two-thirds its height. Hind basitarsus and tarsomere 2 about same length. 2

(a)

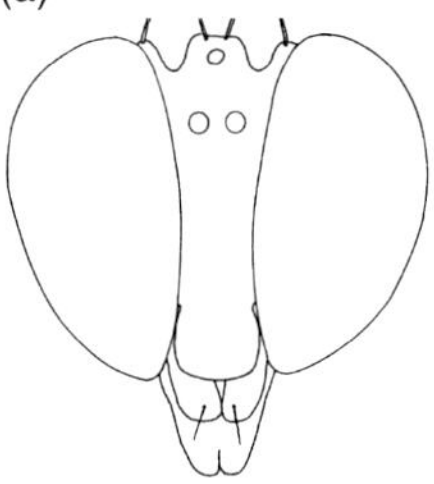

(b)

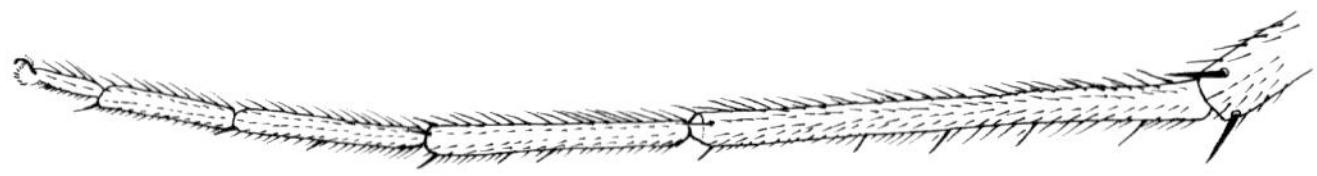

2(1) Scutum, tergites and frons brilliantly glossy green without dusting. Antenna entirely black. Small species, wing length 2.8–3.1 mm.. ***laetus***

– Scutum dull-dusted, at most slightly shining through dust. Antenna partly to entirely yellow. Small to large species, wing length 2.8–5.0 mm and more than 3.5 mm in most species.. 3

3(2) Metepimeron yellow, or rarely paler grey than pleura. 4

– Metepimeron dark.. 9

4(3) Tergites extensively yellow, tergite 1 entirely, tergites 2 and 3 broadly at sides (at least) and tergite 4 narrowly but entirely yellow on hind margin. Mid femur with a stout anterior preapical seta (a, dorsal view). Front basitarsus about twice as long as tarsomere 2 (b). Hind femur with ventral vestiture of fine yellow hairs on basal half. Large species, wing length at least 4.0 mm.. ***heteropygus***

– Tergites dark or with only small lateral pale marks. Mid femur without or with an anterior preapical setula. Front basitarsus at least 2.5 times as long as tarsomere 2. Hind femur with ventral vestiture entirely or mostly small black hairs. Small to large species.. 5

(a) dorsal view

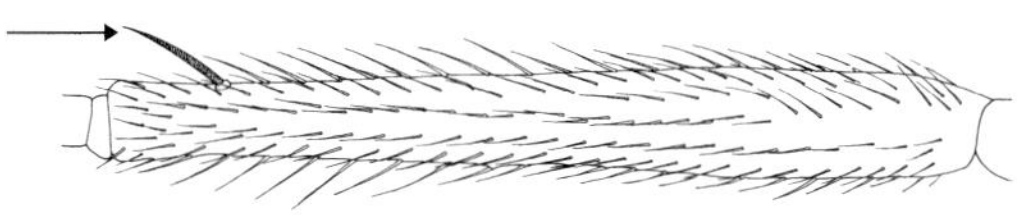

(b)

5(4) Small slightly shining species, frons notably shiny compared to dusted face, wing length 2.8–3.4 mm. Acrostichal setae sparse, rarely extending beyond fourth dorsocentral seta, small to tiny, slightly shorter than the small uppermost postocular setae (ignoring two long setae at top of row). Mid coxa usually dark. Acanthophorite almost flat in profile (a). [Acanthophorite spines slender, outer ones about equal to width of preapical seta on hind femur (b). Cercus with a very long pale hair at tip, similar long dark hairs on lobes of acanthophorite (b).].. ***longulus***

– Larger dull-dusted species, frons and face similarly densely dusted, wing length at least 3.9 mm and usually rather longer. Acrostichal setae well developed, extending to fifth (penultimate) dorsocentral seta, slightly or obviously longer than the small uppermost postocular setae. Mid coxa at least half yellow. Acanthophorite strongly arched in profile (figs in next 2 couplets)..6

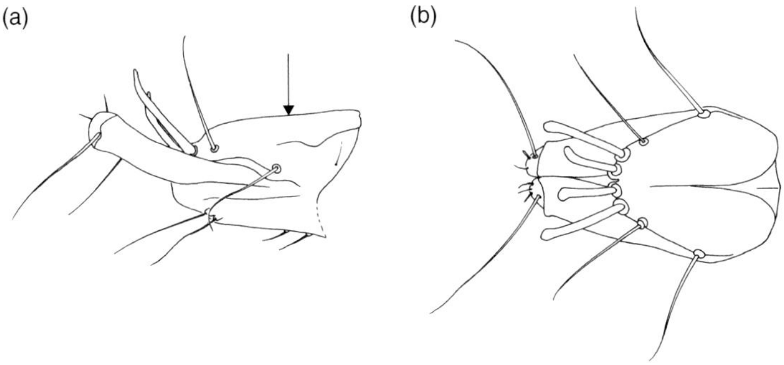

6(5) Postpedicel almost entirely yellow, brown only at tip and dorsally. Acanthophorite spines slender...***pallens***

– Postpedicel entirely brown. Acanthophorite spines slender or stout...7

7(6) Tibiae with more numerous and stronger setae, e.g. front tibia with 2–3 strong anterodorsals and both posterodorsals as long as width of shaft (a, dorsal view), hind tibia with 3 strong anterodorsals. Front basitarsus with a few ventral setulae differentiated from vestiture hairs. Front tibia 1.5 times longer than basitarsus. [Acanthophorite spines stout, not expanded at tip (b). Hairs at tip of cercus short (c).]...***maritimus***

– Tibiae with fewer and weaker setae, e.g. front tibia with only 1 strong anterodorsal and both posterodorsals weak and short, hind tibia with at most 2 strong anterodorsals. Front basitarsus with ventral setae absent or scarcely differentiated from vestiture hairs. Front tibia at most 1.25 times longer than basitarsus. 8

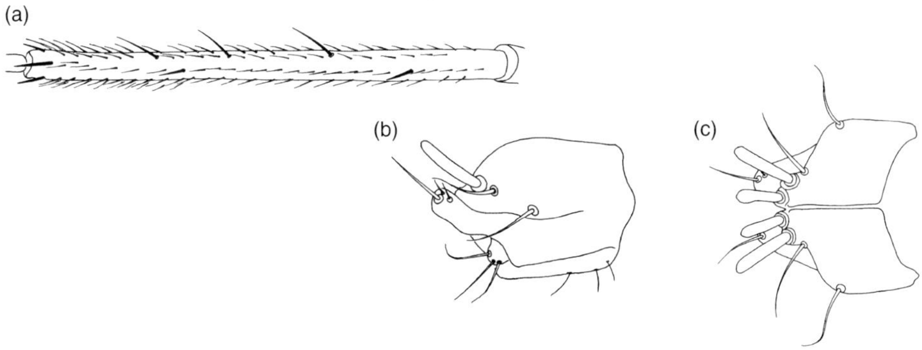

8(7) Acanthophorite spines stout, slightly clubbed towards tip (a, b). M_1 with 'upright' section from fork to sharp bend about equal to distance from sharp bend to R_{4+5}; these veins converging at a wide angle (c)........ ***wiedemanni***

– Acanthophorite spines long and thin, not wider at tip (aa, bb). M_1 with 'upright' section from fork to bend longer than distance from bend to R_{4+5}; these veins converging at a shallow angle so are more nearly parallel (cc)........ ***contristans***

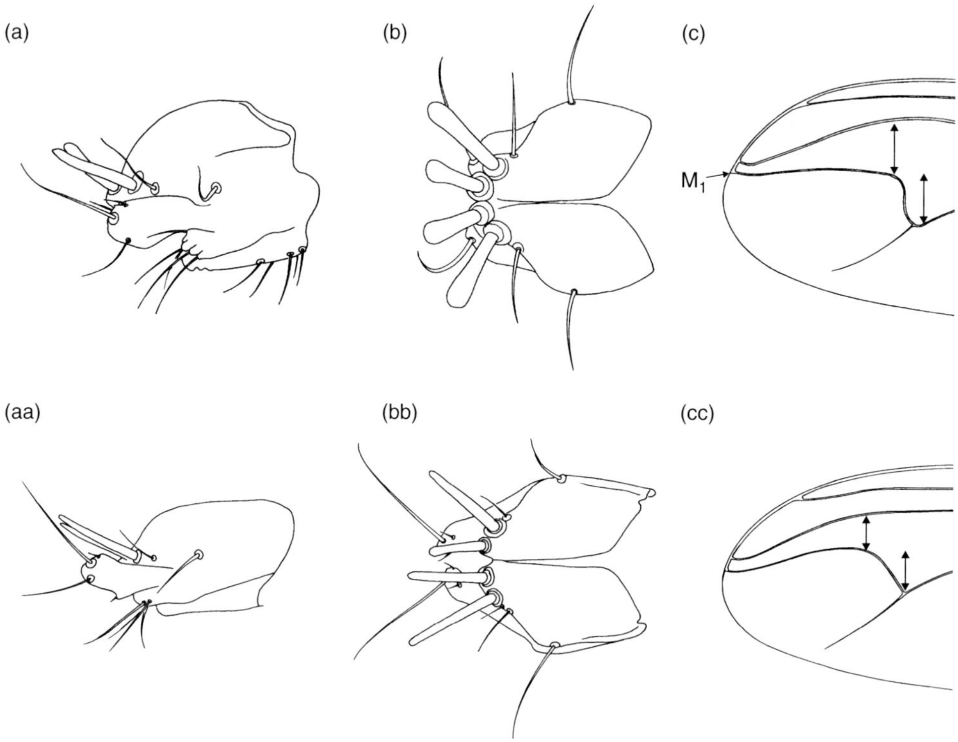

9(3) Front basitarsus ventrally with some short hairs slightly longer than others, giving an impression of minute spinules (a). Mid and hind tibiae bearing several short setae (b, dorsal view)........ ***zonatulus***

– Front basitarsus ventrally with all hairs equally short. Mid and hind tibiae very poorly bristled, usually with only 1 anterodorsal seta near base........ ***basilicus***

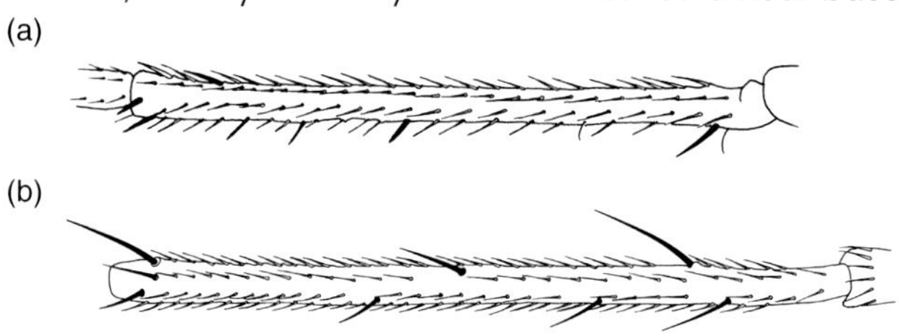

Sciapus basilicus

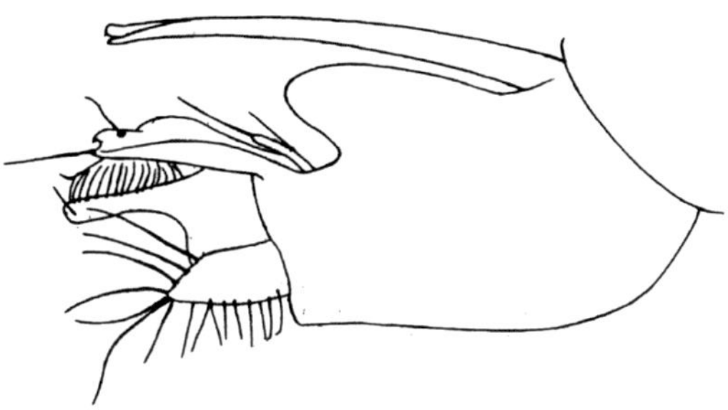

Sciapus contristans

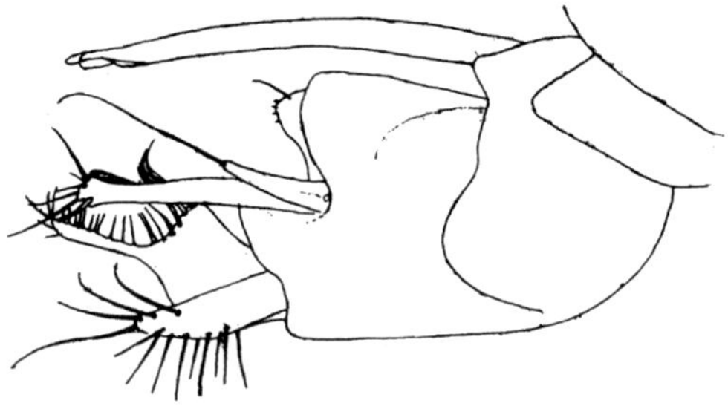

Sciapus heteropygus

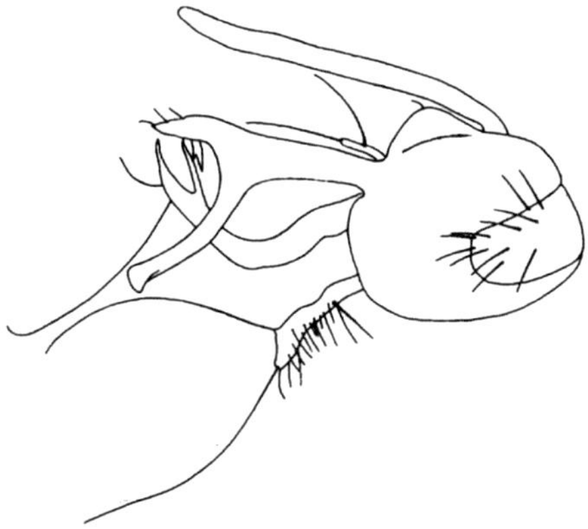

Sciapus laetus

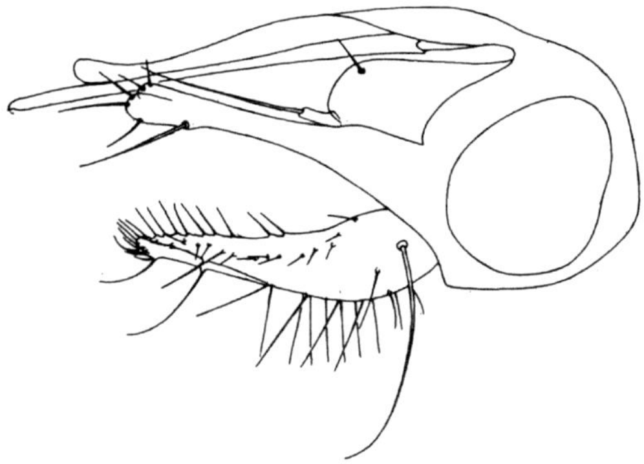

Sciapus longulus

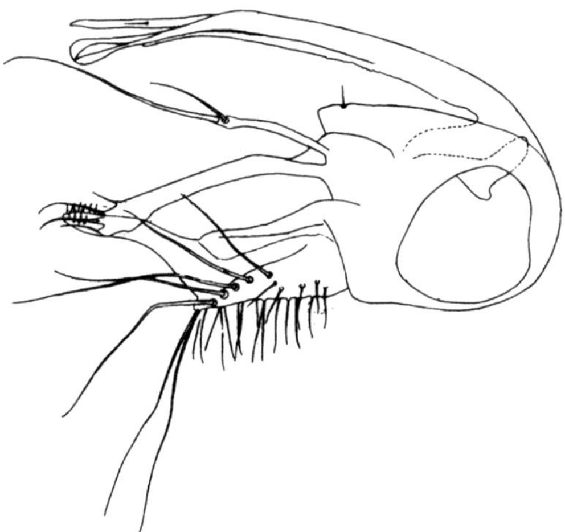

Sciapus maritimus

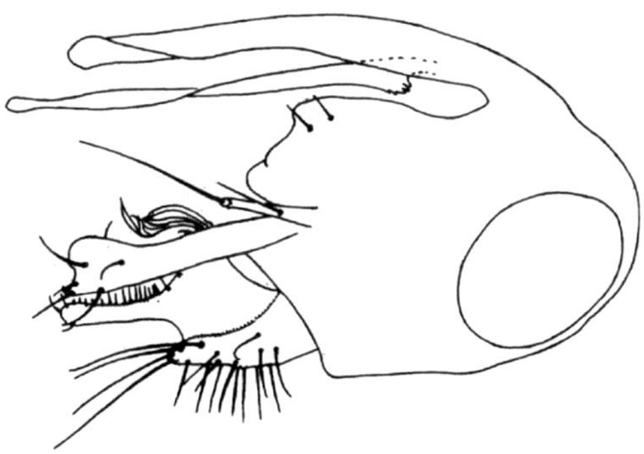

Fig. 35. *Sciapus* hypopygia.

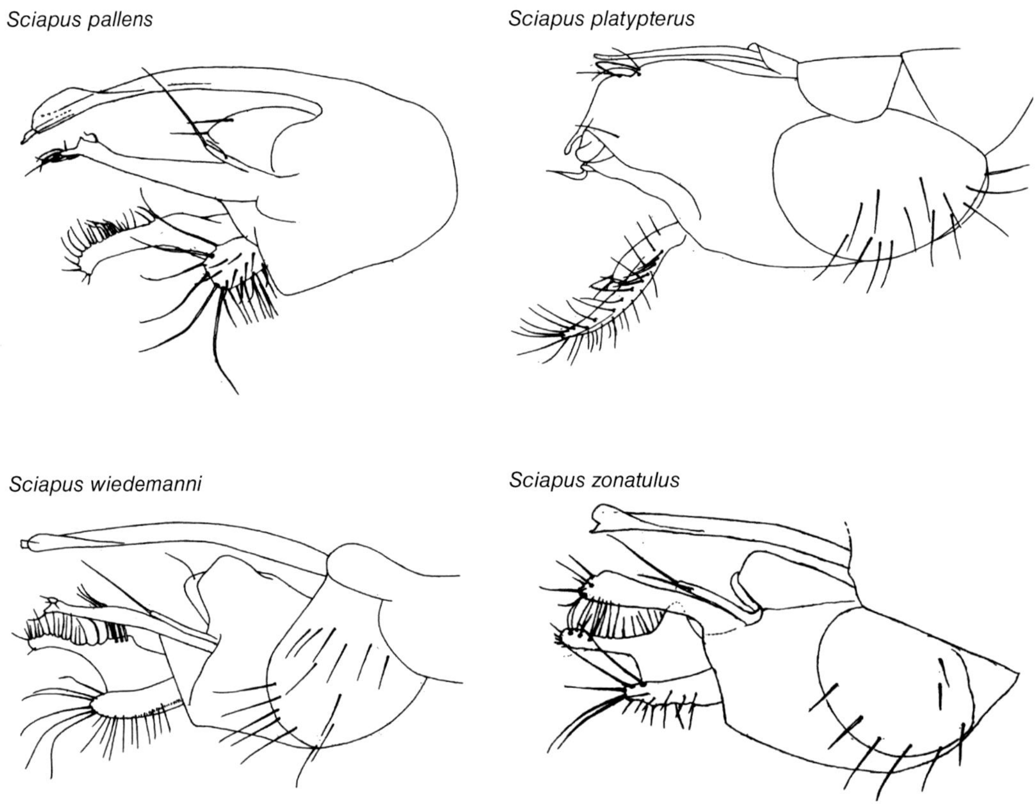

Fig. 35. Continued.

Sciapus basilicus Meuffels and Grootaert, 1990

This is one of two species in the '*contristans*' group (see account for that species) with a dark metepimeron, differing from *zonatulus* in minor details given in the key. The distinguishing feature of the hypopygium is the small flange on the inner side of the epandrial lobe which in side view appears as a ventral bulge (uppermost in the illustration) near the tip. Wing length ♂: 3.5–4.1 mm (Meuffels and Grootaert 1990), ♀: 3.8 (3.6–4.0) mm.

Records are from the River Usk in Gwent, River Swale in Yorkshire and the River Feshie in Strathspey. These sites are large stony rivers with exposed sediments where the flies have been found at partially vegetated shingle bars (Cole 1998, Plant 2006). The type series was collected from coastal dunes in the Netherlands (Meuffels and Grootaert 1990) and elsewhere in Europe it has been found in marshland and in sparsely vegetated or bare ground in a sand-pit (Roháček *et al.* 2021). Flight: vi–vii.

Sciapus contristans (Wiedemann, 1817)

This is probably the commonest of four closely related and often confused species that comprise the '*contristans*' group (along with *basilicus*, *maritimus* and *zonatulus*) which are moderately large dull-dusted species having similar hypopygia. *S. contristans* is the only one of the group whose male has ventral setae on the front femur, a character also found in *laetus*, *longulus* and *wiedemanni*, and its hind basitarsus obviously longer than the second tarsomere (equal in length in the other three species). Females resemble those of *wiedemanni* but are reliably separated by their thin parallel-sided acanthophorite spines. Wing venation varies and is only partially helpful. Wing length ♂: 4.3 (4.2–4.5) mm, ♀: 4.1 (4.0–4.4) mm.

The distribution is mainly southeast England, becoming sparser towards the north and west in Wales, with the northmost record from Yorkshire apart from a possible outlier in the Outer Hebrides. The habitat is probably mainly dry heaths, but also dunes, coastal soft-rock cliffs and dry grasslands. Flight: vi–viii.

***Sciapus heteropygus* Parent, 1926**

This is a large pale species, in the female with large yellow lateral marks on the second and third tergites occupying more than half their lengths in side view and reaching the midline on the rear margins, unlike the small pale lateral spots found in a few other species. The antenna of females is not always entirely bright yellow as previous keys have suggested. Wing length ♂: 4.5 mm, ♀: 4.2 (4.1–4.3) mm.

Confirmed records of this rare species are from Devon, Dorset and Gloucestershire, with an unverified record from Buckinghamshire. Sites for the three southwestern records are at or close to the coast, two from cliff grassland and one of these on limestone. One population was found in an urban garden (Woollatt 1972). Flight: vi–viii.

***Sciapus laetus* (Meigen, 1838) (Plate 7)**

This species is markedly distinct from other British *Sciapus* in both sexes being almost entirely glossy green with dense white pubescence restricted to the lower half of the face and middle of the occiput. It is the smallest in the genus. Wing length ♂: 2.7 (2.4–3.0) mm, ♀: 3.0 (2.8–3.1) mm.

Sciapus laetus is found only in southern England and south Wales where it is restricted to the coast. It is found on dunes and estuarine upper saltmarsh. There appears to be a requirement for brackish or fresh water in pools, ditches or seepages running over the shore. Flight: vi–ix, peak in vii.

***Sciapus longulus* (Fallén, 1823)**

This is a small species differing from all others in being only thinly dusted, although not glossy as in *laetus*, notably on the frons where the green ground colour shines through in contrast to the dusted face. Males have particularly dark bodies compared to most species except *laetus*, including a dark metepimeron and mid and hind coxae. The cercus and organ X appear to be fused, giving a characteristic shape like a hand with a pointing finger; the very long hairs on the cercus are unique although difficult to see. The front tibia has a small anterodorsal seta on the basal quarter, usually absent in *contristans*. In females the acrostichal setae are small to minute, or even apparently absent, and thus resembling the condition in the non-British *nervosus* which has pale setae on the pedicel (black in *longulus*) and its hind basitarsus is much longer than the second tarsomere (similar in length in *longulus*). The mid and hind tibiae of *longulus* have only one strong anterodorsal seta in the basal quarter and rarely with a short one midway, thus resembling *basilicus* but not *maritimus* or *zonatulus*. Wing length ♂: 3.3 (3.1–3.6) mm, ♀: 3.1 (2.8–3.4) mm.

Sciapus longulus is widespread in lowland England, reaching coastal south Wales, but scarcer towards the northwest; it is recorded just over the Scottish border in Berwickshire. It is found mainly in unshaded dry rank grasslands including chalk grassland, cliff-tops, dunes, coastal grazing marsh and tall-herb fens. Some sites are more obviously wetland in character. Flight: v–viii, peak in vii.

***Sciapus maritimus* Becker, 1918**

This is one of the '*contristans*' group; see the account for that species. In both sexes *maritimus* has markedly more spinose tibiae than other species in this group (or most other *Sciapus*) and its legs are short compared with most other *Sciapus*. In males, front tarsomeres 2–5 are finely spiky due to the vestiture standing more upright than usual and the basitarsus in both sexes has isolated posteroventral setulae which, in the males, are longer than the shaft's width, as in *zonatulus*. British specimens have some dark upper postoculars but Meuffels and Grootaert (1990) give them as all white in their diagnosis. Wing length ♂: 4.0 (3.9–4.3) mm, ♀: 4.2 (4.1–4.3) mm.

This is a rare species with confirmed records from one site in Ayrshire, although some records in Kent and Hampshire may be correct (Crossley 1998). It is associated with coastal dunes in Britain and in northern Europe (Meuffels and Grootaert 1990). Flight: vi–viii.

Sciapus pallens (Wiedemann, 1830)
This is a large dull-dusted species readily distinguished in males which have a conspicuously modified front tarsus and much paler antenna than in most British species. They also have particularly long strong thoracic setae, for instance the acrostichal row extends to the last dorsocentral seta where they are about as strong as the postpronotal seta. The male's cercus is particularly short relative to organ X with which it is fused. Female differs from species other than *heteropygus* in having a mainly yellow postpedicel, but otherwise it closely resembles *contristans* in having narrow acanthophorite spines and the front tibia, basitarsus and tarsomeres 2–5 being about the same length. Wing length ♂: 5.1 (4.7–5.8) mm, ♀: 4.7 mm.

The species was first recorded in Britain from a large population in the garden of the Natural History Museum, London (Sivell 2019) and has since been recorded in Bedfordshire. Flight: v–vii.

Sciapus platypterus (Fabricius, 1805) (Plate 7)
Both sexes of this large species have narrow faces unlike any other British species; males are easily distinguished by their broad wings, modified mid tarsus and ventral hair-fringe on the front femur. Females are unique in their hind basitarsus being much longer than the second tarsomere. Wing length ♂: 4.9 (4.0–5.2) mm, ♀: 4.6 (4.1–4.9) mm.

This is the commonest and most widespread species of *Sciapus* in Britain although not recorded from the Scottish outer islands. It is found mainly in deciduous woodland, both wet and dry, but also in the shade of hedges and isolated trees. Flight: v–x, mainly vi–vii.

Sciapus wiedemanni (Fallén, 1823)
Sciapus wiedemanni is a fairly large, weakly shining species whose males are easily distinguished by their modified front tarsus and both sexes by the more triangularly wedge-shaped tip of cell r_{4+5}; in other species, veins R_{4+5} and M_1 are parallel near their tips for a distance of several times their separation at the costa. It can be grouped with the 'long-legged' members of the genus (*longulus*, *contristans*), a feature that is obvious in comparison but not easy to quantify. The female's acanthophorite spines are dumpy with slightly expanded tips, and this character reliably distinguishes *wiedemanni* from *contristans* which has slender spines; wing venation is variable and not always reliable. Wing length ♂: 4.6 (4.3–4.9) mm, ♀: 4.5 (4.2–4.8) mm.

The species is widespread but not common in Britain, being more frequent in lowland England and scarce in Scotland. It appears to be confined to the coast in Wales and southwest England. Its habitat preferences are unclear; in the west of Britain it is found mainly on dunes and soft-rock cliffs, and inland and eastern sites include many heathlands and some dry grasslands. Flight: v–viii, x, peak in vii.

Sciapus zonatulus (Zetterstedt, 1843)
Sciapus zonatulus and *basilicus* are the only species with an entirely dark metepimeron in both sexes although in *longulus* males it is mainly grey. Males have spiky upstanding vestiture on the front tibia, as in *maritimus*, and both sexes have a few sparse longer setae below the front basitarsus, as in *maritimus*, whereas most other species have no differentiated spinules. Meuffels and Grootaert (1990) state that, in males, the hind margin of wing is irregularly curved with a bulge at the tip of M_4 but this is difficult to appreciate except when the wing is perfectly flat. Wing length ♂: 3.4 (3.1–3.8) mm, ♀: 4.0 (3.9–4.2) mm.

Confirmed records are patchily and sparsely distributed in England from Cornwall to Yorkshire, and possibly south Wales and Moray Firth in Scotland. Confusion with *maritimus* and *contristans* makes a definitive statement of distribution impossible. Confirmed records are from coastal soft-rock cliffs and dry heaths. It has been found in dry wooded habitats and riverside dune in Belgium (Pollet 2002), in dry habitats including gardens in the Netherlands (Meuffels and Grootaert 1990) and on coastal dunes in Malta (Ebejer 2021). Flight: v–ix.

SYBISTROMA Meigen, 1824 (Dolichopodinae)

These are slender, long-legged species possessing the usual characters of dolichopodines (hairy scape, dorsal arista, biserial acrostichal setae, mid and hind femora each with an anterior preapical seta). The genus is not clearly defined but males of the few British species can be recognised by tergites 6 and 7 being modified as a short stalk bearing the hypopygium, and the greatly elongated apicoventral epandrial lobe of the hypopygium bearing very long setae (Plate 7). These epandrial lobes may be mistaken for the cerci which are relatively inconspicuous in British *Sybistroma*. Veins R_{4+5} and M_1 converge towards the wing margin where they meet the costa at the wing tip. Setae (av, v, pv) are absent or tiny on the ventral face of the front and mid tibiae, and weak on the hind tibia. *Sybistroma* is closely related to some species of *Hercostomus*, and this probably accounts for the difficulty in finding characters that convincingly separate females of the two genera.

Three of the six species of Western Palaearctic *Sybistroma* occur in Britain. Two found in central Europe could yet be found in Britain. Males of *sphenopterum* (Loew, 1859) resemble *discipes* in having an elongate postpedicel but the last tarsomere of the front leg is only slightly broadened, about twice as long as deep, and the apicoventral lobe of the genitalia is strongly expanded. *S. sciophilum* (Loew, 1869) is probably less likely to occur in Britain; it closely resembles *obscurellum* but its hypopygium is far shorter, scarcely reaching the middle of the fourth tergite. Grichanov and Kazerani (2014) provide a key to males of an expanded concept of *Sybistroma* in the Western Palaearctic fauna.

Sybistroma are species of deciduous woodland, the adults possibly associated with the canopy.

Key to species of *Sybistroma* (males)

1 Front tarsus with no modifications (a). Separation of veins R_{4+5} and M_1 at wing tip about as wide as cell r_1 level with crossvein dm-m. Front tarsus less than 1.5 times length of tibia. .... ***obscurellum***

(a)

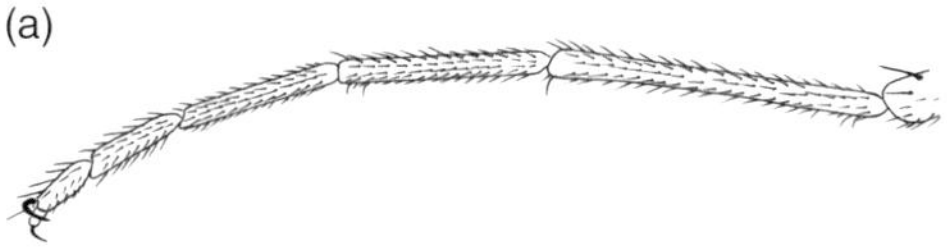

– Front tarsus with modified structure or chaetotaxy (figs in couplet 2). Separation of veins R_{4+5} and M_1 at wing tip less than width of cell r_1. Front tarsus nearly 2 or 3 times length of tibia. ...2

2 Tarsomere 5 of front leg black, greatly expanded and flattened; basitarsus without long hairs (a). Metepimeron, most of hind and mid coxae and epandrium yellow. Postpedicel long and pointed (b). .... ***discipes***

– Tarsomere 5 of front leg white on anterior face, slightly expanded, about 2.5 time deeper than tarsomere 4 (aa); basitarsus with an anterior row of about 8 long hairs decreasing in size apically (bb, dorsal view). Metepimeron, most of hind and mid coxae and epandrium dark. Postpedicel rounded, not longer than wide (bb).......... ***crinipes***

(a)

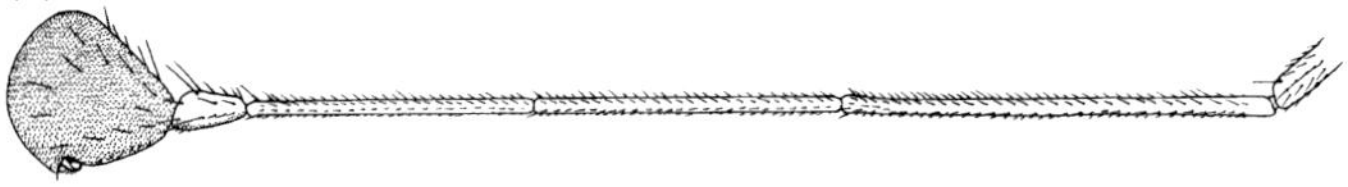

(b)

(aa) lateral and dorsal views

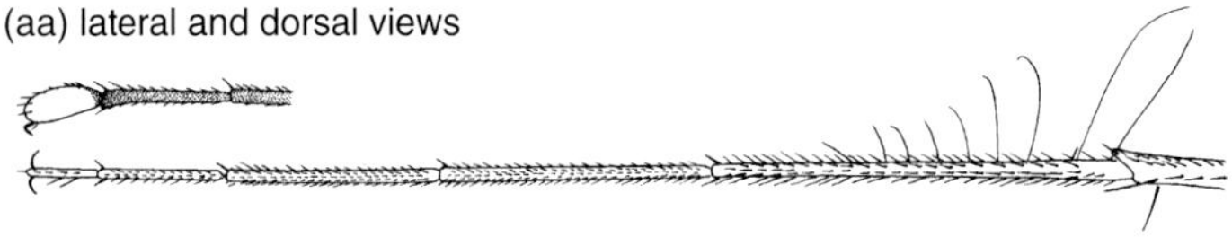

(bb)

Key to species of *Sybistroma* (females)

1 Mid and hind coxae and metepimeron mainly dark (a). Mid tibia with 2 strong anterodorsal setae, each loosely or closely paired with a posterodorsal seta; ignore any weak anterodorsal seta close to the basal strong seta (b, dorsal view). Front coxa with hairs on inner face mainly black on most of its length. Postpedicel rounded, not longer than deep, with scarcely differentiated tip (c).. ***crinipes***

– Mid and hind coxae and metepimeron mainly yellow (aa). Mid tibia with 3 strong anterodorsal setae, the distal 2 setae often equidistant from distal posterodorsal seta; ignore any weak anterodorsal seta close to basal strong seta (bb, dorsal view). Front coxa with hairs on inner face mainly or entirely pale. Postpedicel slightly longer than deep with bluntly pointed tip (cc).. 2

(a)

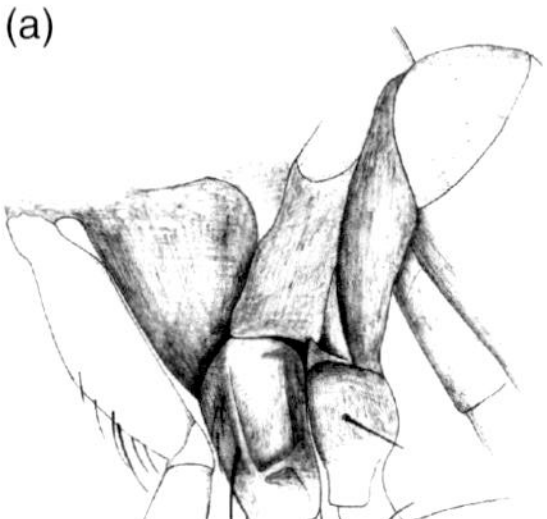

(b)

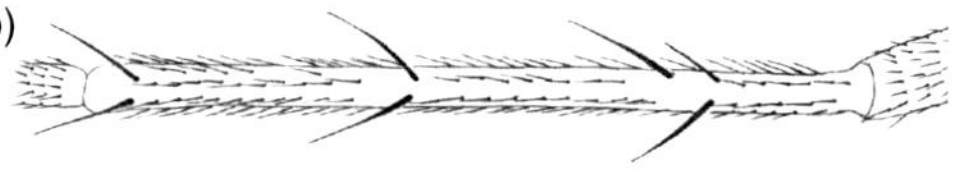

(c)

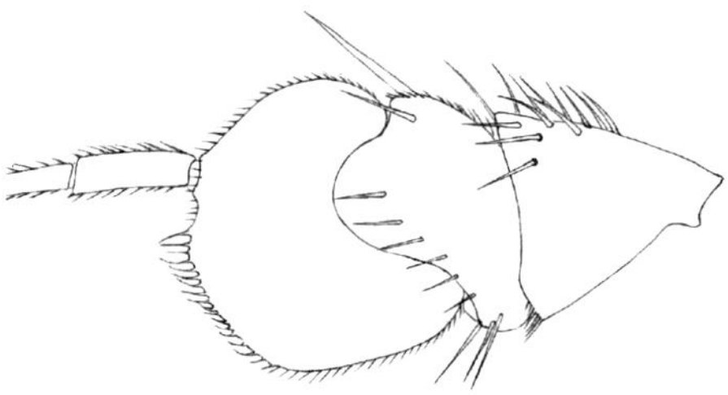

(aa)

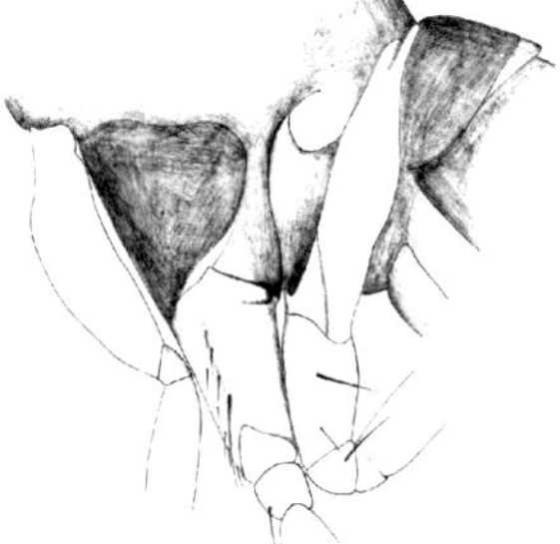

(bb)

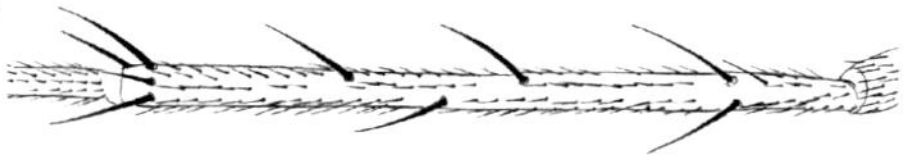

(cc)

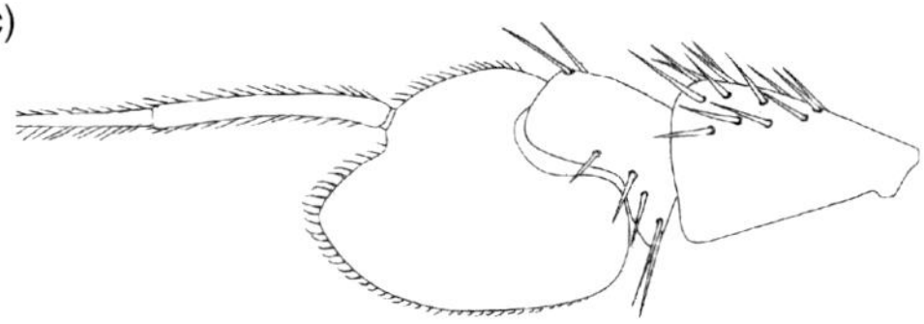

2 R_{4+5} and M_1 converging less strongly at their tips, ratio of costal sections either side of R_{4+5} about 3.0–4.5; greatest separation of R_{4+5} and M_1 near middle of wing about 2.0–2.5 times their separation at wing tip (a). Apical section of M_4 at least twice as long as dm-m (usually more than 2.3 times), measured as if extended to wing margin (a). Leg segments relatively shorter and stouter, e.g. hind basitarsus about 6 times longer than wide and just over half length of front coxa (b). Mid tarsus about 1.1–1.3 times length of mid tibia. Arista slightly longer than face. Smaller species, wing length 3.3–4.5 mm. .... ***obscurellum***

– R_{4+5} and M_1 converging more strongly, ratio of costal sections either side of R_{4+5} about 4–5; greatest separation of R_{4+5} and M_1 near middle of wing nearly 3 times their separation at wing tip (aa). Apical section of M_4 clearly less than twice as long as dm-m (about 1.8 times), measured as above (aa). Leg segments relatively longer and more slender, e.g. hind basitarsus about 10 times longer than wide and equal to length of front coxa (bb). Mid tarsus 1.3–1.4 times length of mid tibia. Arista clearly longer than face. Larger species, wing length 4.4–5.2 mm. .... ***discipes***

(a)

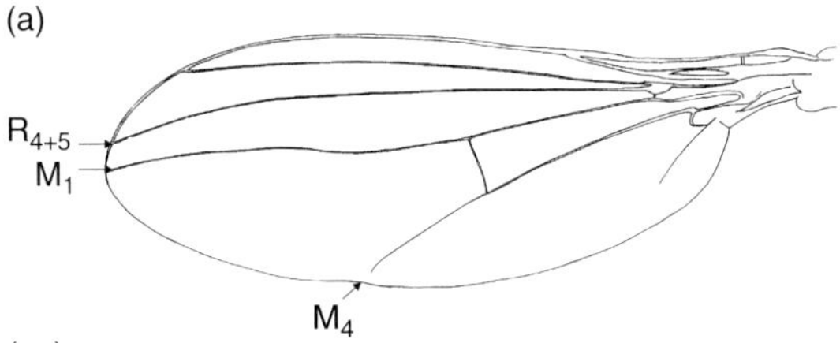

(b)

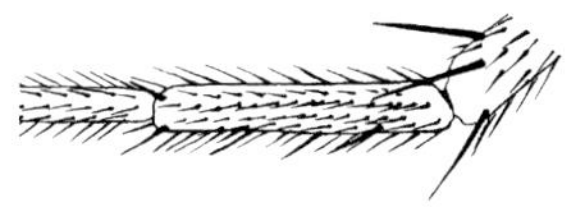

(aa)

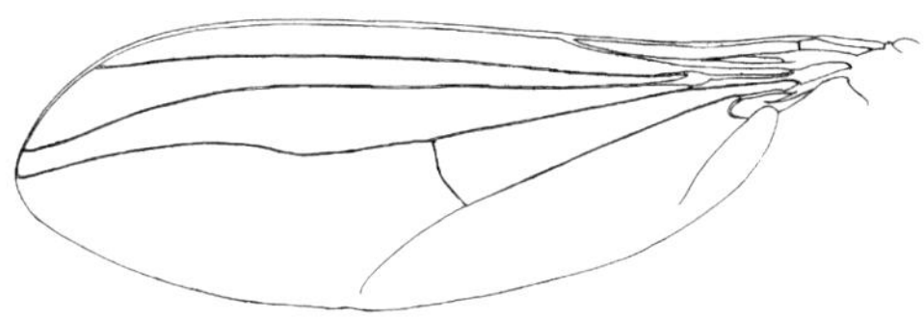

(bb)

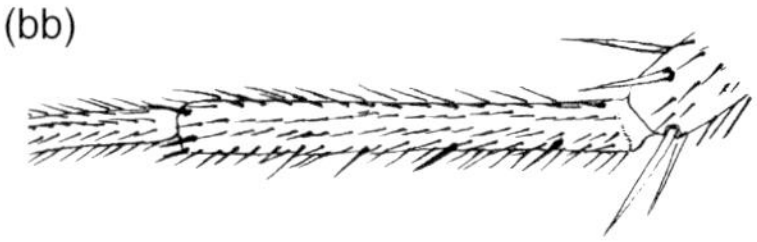

Sybistroma crinipes Staeger, 1842

The sparse fringe of long fine hairs on the front basitarsus coupled with a white front face of tarsomere 5 are unique in British dolichopodids. Both sexes are distinct from the other two species by their darker coxae and metepimeron, and very short rounded postpedicel. Compared with the other two species, the male's arista is nearly uniformly thick with a truncated blunt tip rather than slightly tapered, and the epandrium is entirely black rather than yellow on the apical half. Wing length ♂: 4.2 (4.1–4.5) mm, ♀: 4.4 (4.2–4.8) mm.

Sybistroma crinipes is widely distributed in Britain but moderately uncommon, being particularly sparse in lowland England and showing a possible increase in frequency in hilly areas in the north and west. It is found most often in wet deciduous woodland but sometimes in dry woods and occasionally in open wet pasture or moorland. Flight: iv–ix, peak in vi.

Sybistroma discipes (Germar, 1821)

The greatly enlarged fifth tarsomere on the front leg superficially resembles that of *Dolichopus brevipennis* and *D. migrans* which are larger species with the typical neat cercal setae of *Dolichopus* rather than the long somewhat straggly setae of the epandrial lobe of *discipes*. The antennae are moderately long and the first section of the arista is twice as long as the apical section. Females resemble those of *obscurellum* but have more sweepingly curved R_{4+5} and M_1 veins that converge closely at the wing tip, as in *crinipes*, and are more long- and slender-legged flies. Wing length ♂: 4.8 (4.4–5.3) mm, ♀: 4.9 (4.4–5.2) mm.

Sybistroma discipes is widely but sparsely distributed in England and Wales, but rare in Scotland. It is more frequent in the west and north of its range, being scarce in lowland England. Most records are from deciduous woodland but it may be associated with isolated trees rather than woodland itself as it is sometimes found in shaded places away from woods. Flight: vi–viii.

Sybistroma obscurellum (Fallén, 1823) (Plate 7)

Both males and females resemble *discipes* in the yellow colouring of the mid and hind coxae, metepimeron and epandrium, unlike the dark colour of these parts in *crinipes*. Females resemble those of *discipes* but are smaller with relatively shorter leg segments, and veins R_{4+5} and M_1 are

Sybistroma crinipes

Sybistroma discipes

Sybistroma obscurellum

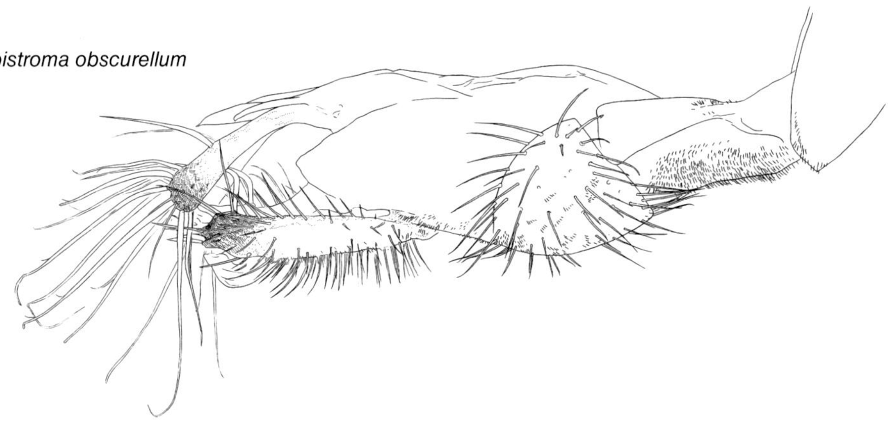

Fig. 36. *Sybistroma* hypopygia.

straighter and lie further apart towards the wing margin. Variation in shape of wing veins and relative lengths of leg segments make it difficult to pick on just one character than most reliably separates this species from *discipes*. Wing length ♂: 4.0 (3.8–4.2) mm, ♀: 4.0 (3.3–4.5) mm.

Sybistroma obscurellum is a common and widespread species in Britain with the most northern record from Orkney, although it appears to be relatively uncommon in East Anglia. It is found in mainly woodlands and shaded sites, showing no particular preference for wet or dry places, and can occur in many other situations. Flight: iv–x, peak in vii with long late season.

SYMPYCNUS Loew, 1857 (Sympycninae)

These are small dull dark green to dark brown flies with moderately slender legs and a long abdomen extending about as far as the tip of the hind femur (Plate 7). The genus is readily recognised in males as the hind tarsus has modified setae (Fig. 37), although this is less obvious in *aeneicoxa*. Apart from this character in males, neither sex can be recognised as belonging to this genus using just a few characters. The options are reduced to a few other sympycnine genera using the combination of being small species with a wing length less than 3.5 mm, the postpedicel being triangular with the arista inserted on the basal half or midway, the pedicel having a simple straight apical margin, the face tapering to the mouth edge, having six dorsocentral setae, no proepisternal seta, and the mid femur having an anterior preapical seta. *Teuchophorus* males have this combination but are readily distinguished by their thick black costa and ornamented hind tibia. *Lamprochromus* females may also have this combination of characters but are distinguished by their black marks on the side of the scutum.

Five of the 14 Western Palaearctic species occur in Britain. Among the remainder, *brevimanus* Loew, 1857 and *simplicipes* Becker, 1908 are sufficiently widespread to be possible future additions to the British fauna. They are included in accounts of *aeneicoxa* to which they would run in the key.

Key to species of *Sympycnus* (males)

1 Legs almost entirely black. Hind tarsomere 3 with a posterior coarse fringe of long stiff hairs (Fig. 37)........ ***cirripes***

– Legs mainly yellow, at least mid tibia yellow dorsally in dark specimens. 2

2(1) Front coxa and femur mainly black. Hind femur black on about apical two-fifths on all surfaces except ventrally. Hind tarsomeres 3 and 4 with a conspicuous posterior fringe........ 3

– Front coxa and femur mainly yellow. Hind femur black or dark only dorsally on apical half, or entirely yellow. Hind tarsomere 3 with an inconspicuous posterior fringe or an apical cluster of large setae. 4

3(2) Mid tibia with a short posteroventral seta at apical third (rarely absent). Hind tarsomere 3, when viewed from above, even in width with 2 long fine basal hairs shorter than length of segment (about 0.8 times) and not much longer than other hairs in posterior fringe (Fig. 37). Hind tarsomere 2 slightly longer than tarsomere 3. .... ***pulicarius***

– Mid tibia without a posteroventral seta. Hind tarsomere 3, when viewed from above, markedly concave on posterior margin, wide at base and narrowing on apical two-thirds, with 2 very long fine basal hairs 1.2 times length of segment and 2–3 times longer than other hairs in posterior fringe (Fig. 37). Hind tarsomeres 2 and 3 equal in length. .... ***septentrionalis***

4(2) Front tarsomeres 2–5 short and stout, clearly wider than basitarsus, with spiky hairs (a). Hind tarsomere 3 with an inconspicuous posterior fringe of short upright hairs not longer than width of shaft (Fig. 37). Eyes almost touching on lower third of face, separated at most by the width of an ocellus (b). Front tibia without a comb of small setae. .... ***aeneicoxa***

– Front tarsomeres 2–5 more slender, not wider than basitarsus (aa). Hind tarsomere 3 at its tip with a posterior cluster of 1 large flat seta and 2–3 smaller setae (Fig. 37). Eyes well-separated by twice the width of an ocellus (bb). Front tibia with an anterodorsal comb on apical half........ ***spiculatus***

(a)

(aa)

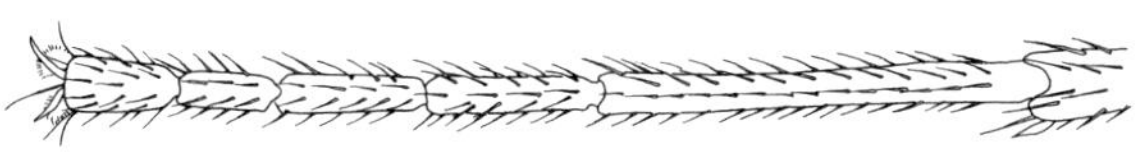

(b)

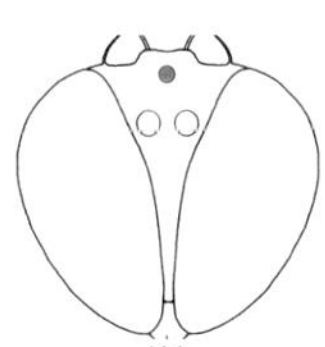

(bb)

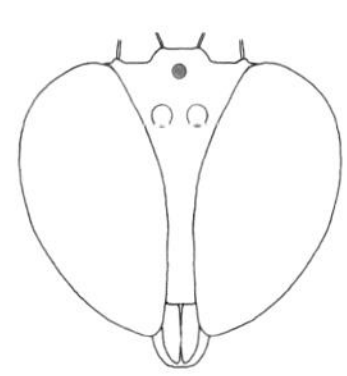

Key to species of *Sympycnus* (females)

1 Legs almost entirely black. Front tibia with anterodorsal row of close-set setae but no clearly defined comb of small setae (a). .... ***cirripes***
– Legs mainly yellow. Front tibia with comb of longer setae (aa). 2

(a) (aa)

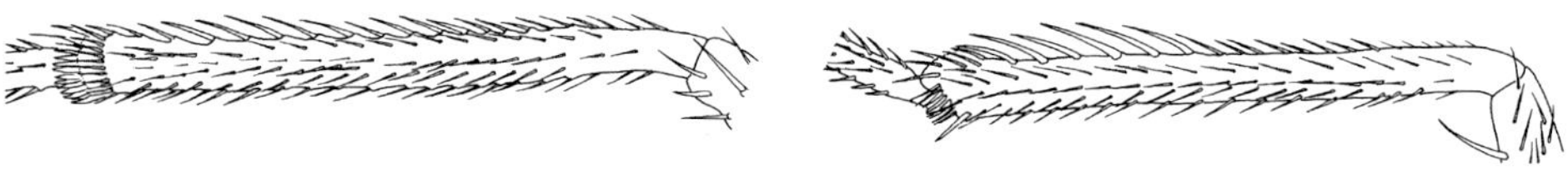

2(1) Front femur entirely encircled in black on at least basal third, usually on basal two-thirds. Front coxa largely black. Postpedicel short, not longer than deep, tip rounded or with right-angled point. .... ***pulicarius*** and ***septentrionalis***
– Front femur either entirely yellow or darkened dorsally on whole length and basally, not entirely encircled in black but with dull yellow encroaching on at least posterior face. Front coxa either largely dark or yellow. Postpedicel short or longer than deep. 3

3(2) Hind tibia usually conspicuously dark on apical half. Hind femur darkened as a ring or dorsal saddle near tip only, not along whole length. Hind basitarsus as long as tarsomere 2. Postpedicel small and rounded, not longer than wide. .... ***aeneicoxa***
– Hind tibia almost entirely yellow. Hind femur with dorsal surface dark along whole length, rarely less. Hind basitarsus slightly longer than tarsomere 2. Postpedicel as long as wide, more triangular and pointed at tip. .... ***spiculatus***

Sympycnus aeneicoxa (Meigen, 1824)

In both sexes, this species has particularly pale legs. The front coxa, all trochanters and the bases of all femora are clear yellow without the smudges or dark areas found in other yellow-legged species, and any darkening being restricted to the hind leg where the femur is darkened at the tip and most of the hind tibia is black, along with the tarsus. Males are distinguished from other British species by the eyes almost touching (well-separated in other species) and by the short stocky front tarsomeres 2–5 contrasting with the slender basitarsus, rather than simple as in other species. The posterior

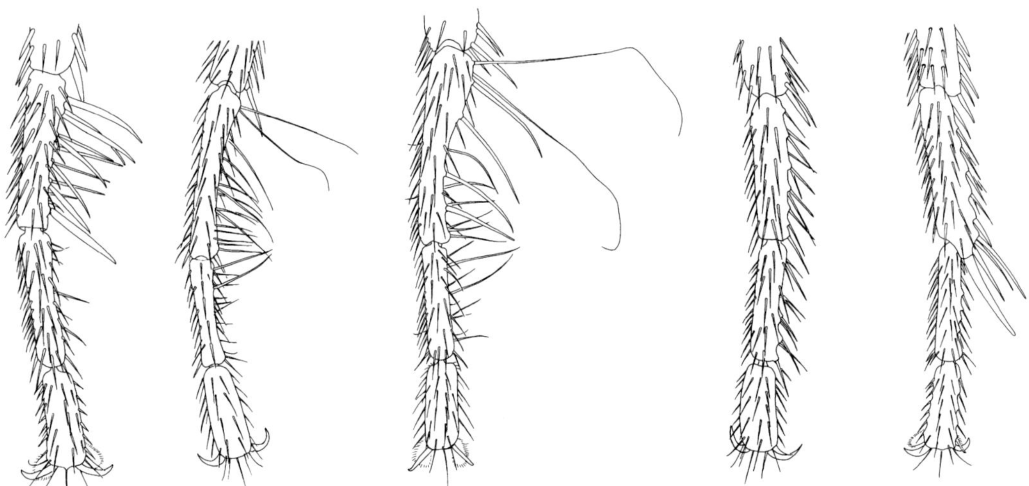

Sympycnus cirripes *Sympycnus pulicarius* *Sympycnus septentrionalis* *Sympycnus aeneicoxa* *Sympycnus spiculatus*

Fig. 37. *Sympycnus* male hind tarsomeres 3–5, ordered as in key.

fringe on the hind tarsus is inconspicuous, thus differing from the obvious ornamentation of other species. The European species *simplicipes* runs to *aeneicoxa* in the key but in both sexes has yellow patches on the abdomen, a yellow metepimeron and a brilliantly blue frons, which in British species are respectively dark green, black and dull-dusted. A widespread European species, *brevimanus*, has male front tarsi similar to those of *aeneicoxa* but black femora. Wing length ♂: 2.5 (2.4–2.7) mm, ♀: 2.7 (2.5–3.0) mm.

Sympycnus aeneicoxa is widespread in Britain but more frequent from northern England to Sutherland in Scotland, and in the dry lowland belt of England from Hampshire to East Anglia, but scarce in Wales and southwest England. It appears to be an unspecialised wetland species found in fens, bogs, wet grasslands, lushly vegetated water margins of rivers, pond and seepages, and usually in unshaded areas. Flight: v–ix, mainly late vi to early vii.

Sympycnus cirripes (Haliday in Walker, 1851)

As the only almost entirely dark *Sympycnus* known in Britain, this species is readily identified in both sexes. Other characters are unnecessary for identification but it is worth noting that both sexes have a particularly dark wing membrane and intensely black basicosta, and no clearly differentiated comb on the front tibia. Wing length ♂: 2.6 (2.5–2.8) mm, ♀: 3.0 (2.7–3.3) mm.

Sympycnus cirripes has a northern and western distribution in upland Britain, being absent from England southeast of a line from the Severn to the Humber estuaries. It is common in Scotland at altitudes of up to 1000 m, Wales and northern England. It occurs mainly in acid wetlands including moorlands, mires, streams, rivers and seepages, and occasionally in base-rich sites on limestone. Most sites are unshaded but some are within woodlands. Flight: vi–ix, peak in vii.

Sympycnus pulicarius (Fallén, 1823) (Plate 7)

The largely black front coxa and femora distinguish both sexes of this species and the very similar *septentrionalis* from other yellow-legged species. Separating the two similar species is straightforward in males but no characters work consistently for females, so consequently they are best treated as the very common *pulicarius* if occurring inland or, if on a saltmarsh, ignored since both species occur together in this habitat. Wing length ♂: 2.5 (2.4–2.7) mm, ♀: 2.9 (2.8–3.1) mm.

Sympycnus pulicarius is a very common and widespread species recorded from southern England to Shetland and small offshore islands in Scotland. It has no obvious habitat preference beyond wet or damp places. Flight: iii–x, long season, no distinct peak.

Sympycnus septentrionalis Pollet, 2015

Sympycnus septentrionalis closely resembles *pulicarius* but males are readily distinguished by the concave hind tarsomere 3 and much longer hairs here. Females are indistinguishable from those of *pulicarius*; the differences suggested by Pollet *et al.* (2015) do not work for British material. Wing length ♂: 2.5 (2.2–2.6) mm, ♀: 2.8 (2.7–3.1) mm.

Records of *septentrionalis* are scattered around the British coast with a slight preponderance in the west. It is a species of upper saltmarsh. Flight: vi–vii, mainly vi.

Sympycnus spiculatus Gerstäcker, 1864

This is the largest British species. Both sexes differ from other species in having an entirely yellow hind tibia (other species have at least some dark areas) and the front and hind femora usually being dark along the whole dorsal surface rather than as a partial or complete ring of dark colour, which is best appreciated when viewed from the side. Also in both sexes, the postpedicel is relatively longer and more pointed than in the other species. The male's hind tarsus has a conspicuous flattened seta at the tip of tarsomere 3, along with two to three smaller setae in a cluster. Wing length ♂: 2.9 (2.7–3.2) mm, ♀: 3.2 (2.9–3.5) mm.

Sympycnus spiculatus is widely distributed in Britain with more records in upland areas and few (or erroneously recorded) in lowland southeast England. It is strongly associated with limestone in England and Wales, or perhaps with other base-rich conditions in Scotland, and this may account for Crossley (2014) associating the species with dog's mercury which is often favoured by base-rich conditions. It is often recorded by streams, rivers and seepages but frequently in dry places without obvious wetland habitat; nearly all sites are woodland or shaded. Flight: v–ix, mainly vi–vii.

SYNTORMON Loew, 1857 (Sympycninae)

Syntormon are small to medium-sized metallic green flies, sometimes with yellow abdominal marking (Plates 7 and 8). The genus is characterised by the articulation of the antennal segments in which the pedicel extends as a peg into a cavity on the inner face of the postpedicel. This protrusion is less pronounced in *bicolorellum*, in which it resembles the condition in some *Gymnopternus*, *Hercostomus* and *Thinophilus*. In males, the genus can be distinguished from most others by the combination of an anterior preapical seta on the mid femur, the arista being apical or subapical, and the genitalia completely encapsulated so that no appendages are visible. They often have modifications of the legs and long tapered postpedicels with an apical or subapical arista that superficially resemble those of *Rhaphium* and *Systenus*. Recognising the genus in females relies strongly on the pedicel shape as many other characters vary widely between species. Those of all species except *bicolorellum* stand out by having a strongly protruding clypeus, as in *Gymnopternus* and *Telmaturgus*. Some species of *Syntormon* have dorsal hairs on the scape, which is unusual in Sympycninae (only *Anepsiomyia*) although is a characteristic of all Dolichopodinae and most *Argyra*. Wing veins R_{4+5} and M_1 are usually almost straight and nearly parallel to one another, or M_1 has only a slight inflexion; crossvein dm-m is about equal to or rarely shorter than the apical section of M_4.

There are 41 species of *Syntormon* in the Western Palaearctic (Grichanov 2013) of which 16 occur in Britain, plus another one if *setosum* is regarded as a valid species (Drake 2021a). One other species, *S. punctatum* (Zetterstedt, 1843), could be found in Britain; its male is distinguished by having two long setae near the base of the hind basitarsus, illustrated by Becker (1918b) under the name *bisetosum* Becker, while the female resembles *aulicum*. *Syntormon luteicorne* Parent, 1927 is another possible addition but it is known only from females and is discussed in the account for *bicolorellum* which it resembles in having yellow antennae. *Bathycranium*, which included *bicolorellum*, was synonymised with *Syntormon* by Speight *et al.* (1995). The aedeagus and lateral view of the hypopygium is illustrated for British species by Drake (2021a, 2023).

All British *Syntormon* are wetland species often associated with swamps, fens, seepages, wet woodlands and water margins. *S. bicolorellum* seems to be the least confined to such wet conditions. *S. filiger*, *mikii* and *pseudospicatum* are found on saltmarsh and brackish sites and only rarely inland, while *sulcipes*, *tarsatum* and *zelleri* are restricted to acid sites.

At the time of writing, there are conflicting opinions on whether the gender of the genus name is neuter or masculine. Neuter is used here, following Chandler (2024). An Opinion arguing that the gender should be neuter is awaiting a response from the ICZN (Drake and Welter-Schultes, 2025). Drake (2021a) explains why, accepting that neuter is the correct gender, why *submonilis* remains in the masculine while *monile* is neuter, inevitably leading to confusion.

Key to species of *Syntormon* (males)

1 Metepimeron (above hind coxa), tergite 1 and at least base of hind coxa dark. Tergites 2–4 in side view not continuously yellow along lower margins, except in *macula* with an indistinct spot on apical section of M_1. 2

– Metepimeron, hind coxa and usually first tergite yellow. Tergites 2–4 in side view continuously yellow along lower margins. 14

2(1) Mid and front femora with only fine black or pale vestiture below, apart from preapical setae. 3

– Mid femur, and sometimes front femur, with 1 or more distinct ventral setae, either isolated on basal half (aa) or as a posteroventral row (bb) 10

(aa)

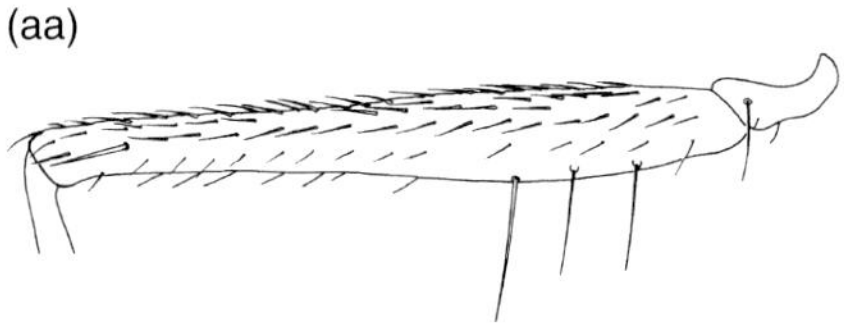

(bb)

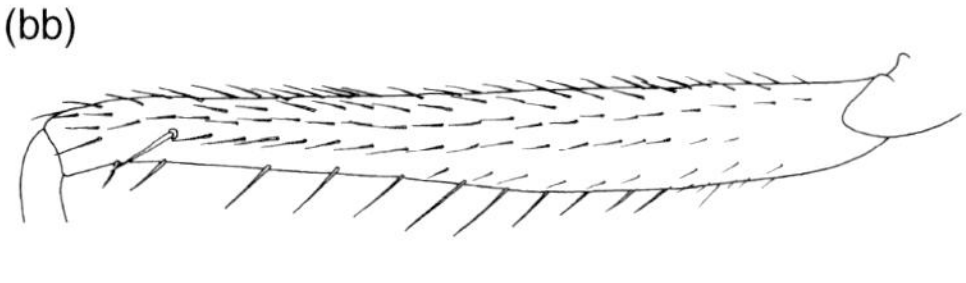

3(2) Hind basitarsus at base below without a long projection but often with other modifications including claws or flattened setae (see figs couplets 4–7)........ 4
– Hind basitarsus at base below with a long projection bearing setae (aa, two examples)........ 8

(aa)

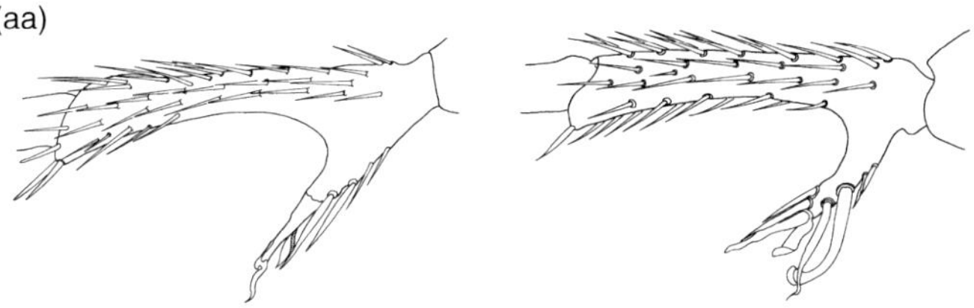

4(3) Hind basitarsus with a pair of claws ventrally near base (a). Front tibia with a well-developed anterodorsal comb of small setae, usually well over half width of shaft........ 5
– Hind basitarsus with different modification or none. Front tibia with comb scarcely evident, its width about one-third that of shaft........ 6

(a)

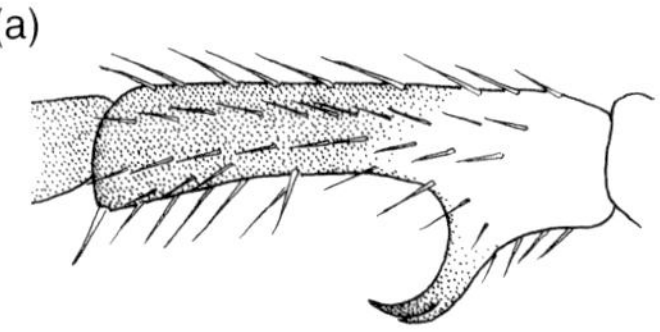

5(4) Hind femur and tibia with dark apical rings (a) which may extend to their bases so shaft is entirely black. Hind basitarsus entirely black. Hind tibia with an anterior fringe composed of longer thicker hairs, their average length 1.8 times width of shaft at penultimate anterodorsal seta, rarely as short as 1.4 times; hairs nearly as stout as anterodorsal setae (b, dorsal view). Hind basitarsus with the hook-bearing column usually more upright, its convex edge making an angle of 55–85° with tarsus axis (c, two examples)........ ***pallipes***
– Hind femora and tibia entirely pale, rarely slightly dusky beyond anteroventral preapical seta. Hind basitarsus pale on about basal third. Hind tibia with an anterior fringe composed of shorter thinner hairs, their average length 1.5 times width of shaft at penultimate anterodorsal seta, rarely as long as 1.9 times; hairs obviously thinner than anterodorsal setae, only two-thirds as thick (bb, dorsal view). Hind basitarsus with the hook-bearing column usually more slanting and curved, convex edge making an angle of 45–70° with tarsus axis (cc)........ ***pseudospicatum***

(a)

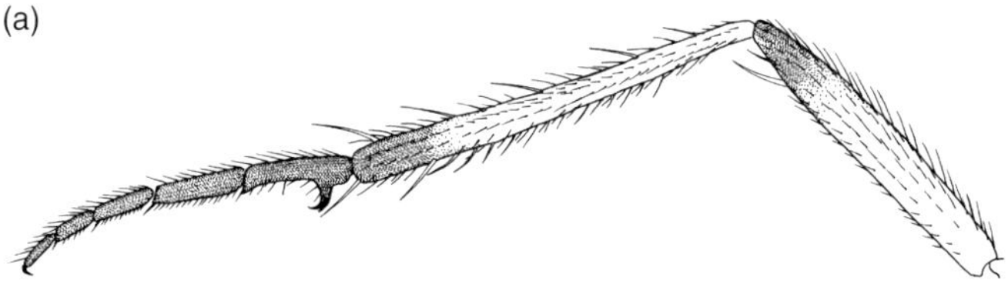

(c)

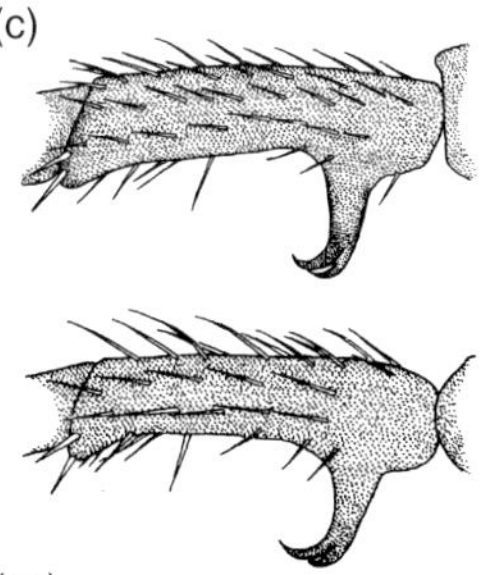

(b)

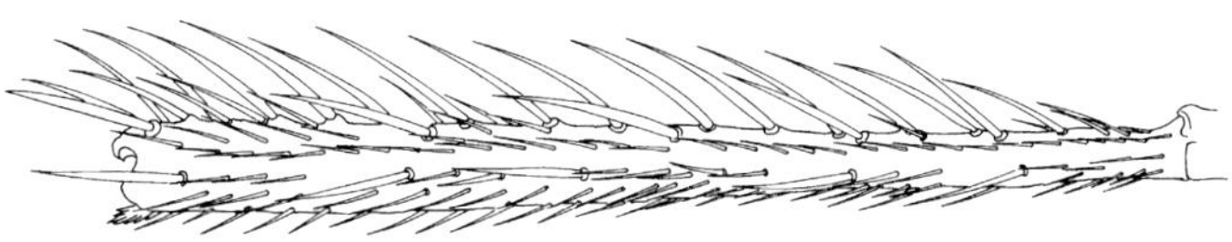

(cc)

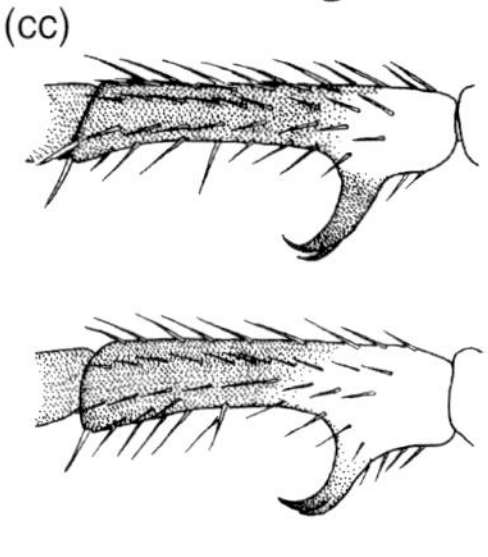

(bb)

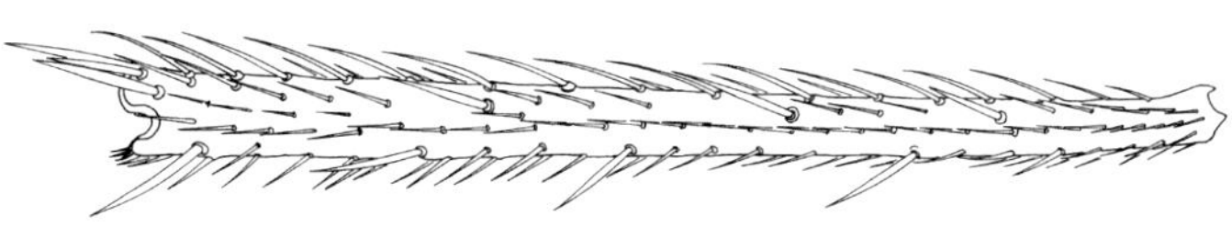

6(4) Tergites extensively yellow. Hind basitarsus simple. (See also couplet 13.) .... ***macula***

– Tergites entirely dark. Hind basitarsus with modified ventral setae. 7

7(6) All femora and front coxa mainly yellow. Hind tibia black, wide and flattened, with a shallow groove running along whole of anterior and posterior faces (a). Front basitarsus unmodified, cylindrical. Hind basitarsus with 1 longer ventral seta inclined towards tibia and 1 shorter nearly upright seta (b). .... ***sulcipes***

– All femora and front coxa mainly black. Hind tibia yellow on basal half, black on apical half, cylindrical, not markedly wide, without grooves. Front basitarsus with its tip projecting ventrally under tarsomere 2 (cc). Hind basitarsus broad at base bearing 2 white oval and 1 black sickle-shaped setae and with an anterior fringe of long hairs (bb). .... ***zelleri***

(a)

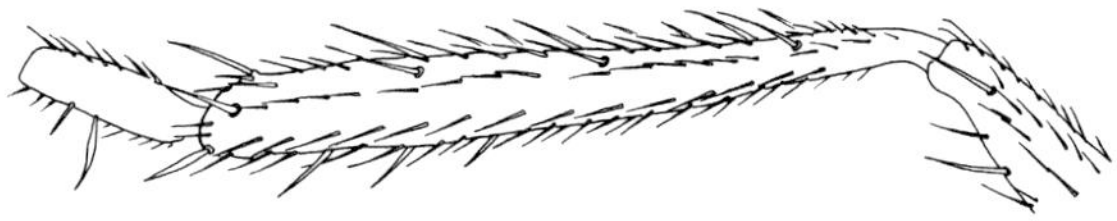

(b)

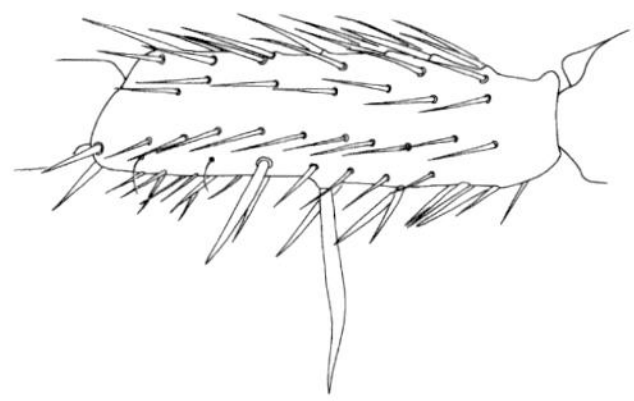

(cc)

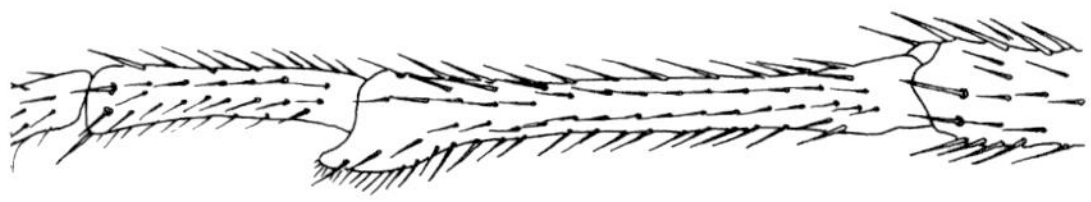

(bb)

8(3) Front basitarsus simple, cylindrical. Mid tarsomeres 4 and 5 cylindrical. Hind coxa yellow below lateral seta. .... ***fuscipes***

– Front basitarsus with swollen tip (figs a and aa of next couplet). Mid tarsomeres 4 and 5 flat and broad (aa). Hind coxa entirely or mostly black. 9

(aa)

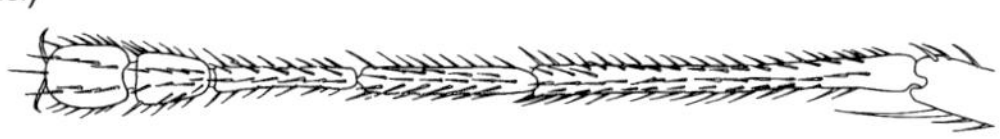

9(8) Front basitarsus with a posteroventral fringe along whole length of segment, hairs as long as width of shaft seen in dorsal view; anterior hairs at tip longer than width of shaft (a). Front tarsomere 2 quadrate with an expanded base in lateral view (b), with anteroventral fringe of long hairs (a, c). Front tibia with posterior fine hairs as long as width of shaft (d, anterodorsal view); anterodorsal comb of short weak setae about half width of shaft, width of their bases one-quarter of the gap separating them (e, posterodorsal view). Front femur with a dark dorsal streak along its entire length. Tergites always entirely green .... ***submonilis***

– Front basitarsus with lateral fringes restricted to swollen tip, hairs shorter than width of shaft (aa, dorsal view). Front tarsomere 2 almost cylindrical, slightly wider on basal half (bb), with only short vestiture (aa, cc). Front tibia with posterior hairs distinctly shorter than width of shaft; dorsal comb of stout setae as long as width of shaft, width of their bases about half the gap separating their bases (ee, posterodorsal view). Tergites 2 and 3 nearly always with yellow patches..***monile***

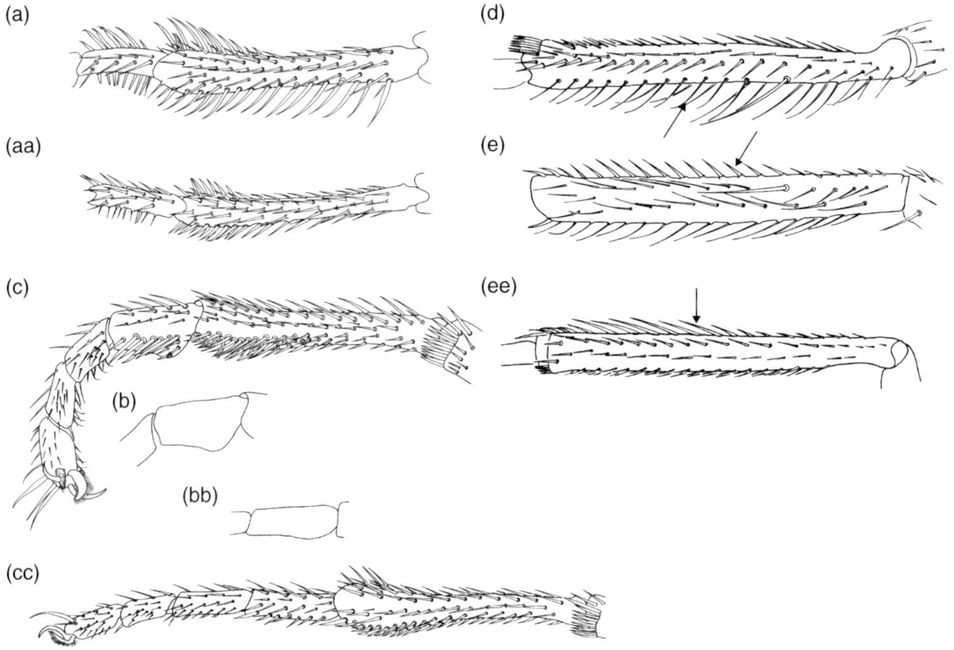

10(2) Mid and front femora with 1 or several long fine ventral setae confined to basal quarter....... 11

– Mid and sometimes front femur with a row of ventral setae or hairs occupying at least basal half, or a single fine and sometimes indistinct ventral seta arising well beyond basal quarter... 12

11(10) Front and mid femora with 3–5 setae near base (a). Arista about same length as postpedicel (b). Face narrow and tapered, its narrowest width equal to width of front basitarsus. Frons shining metallic purple-blue. Hind tarsus simple, at most with some ventral hairs as long as width of shaft (c).. ***pumilum***

– Front and mid femora with 1 ventral seta near base (aa). Arista very short, about half length of postpedicel (bb). Face broad and nearly parallel-sided, width equal to that of front femur. Frons black and dulled by dense brown dust. Hind tarsomere 2 with a long upright blunt-ended ventral seta, basitarsus with a shorter simple seta (cc).................. ***filiger***

(a)

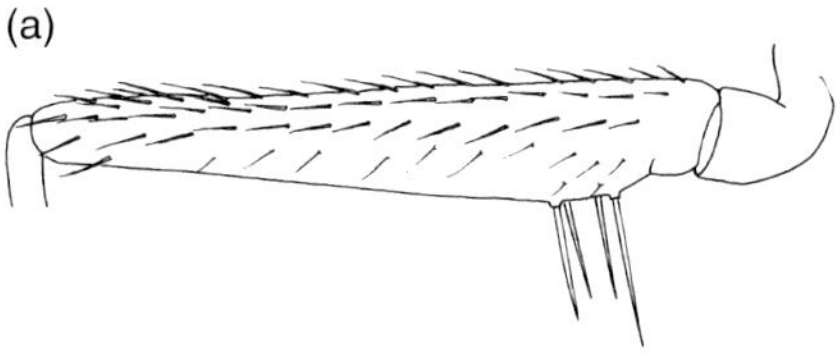

(b)

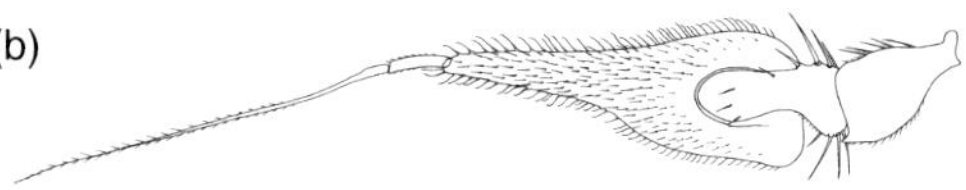

(bb)

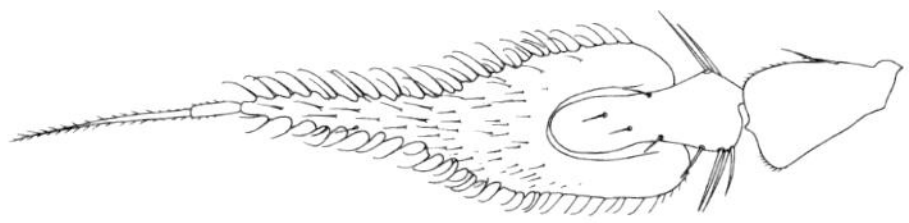

(aa)

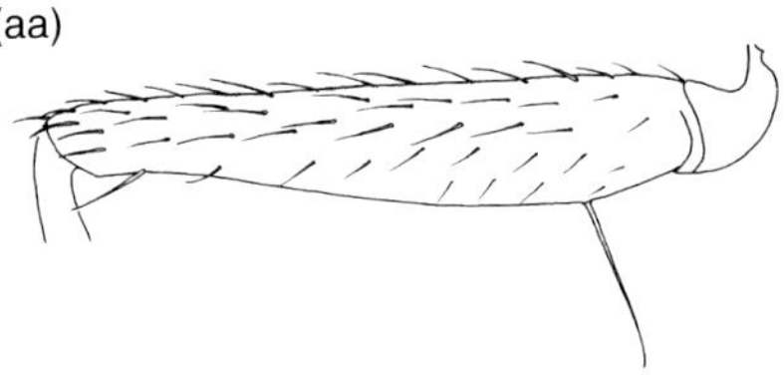

(c)

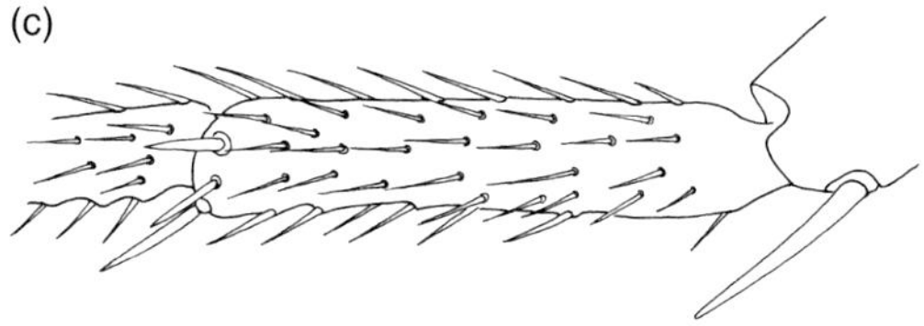

(cc)

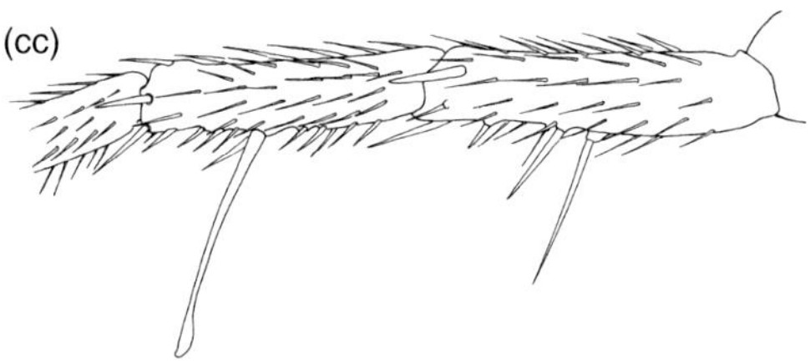

12(10) Hind basitarsus with a pair of small ventral claws on swollen basal half (a). Mid femur with a posteroventral row of setae from near base to tip, setae as long as width of mid tibia (not femur) (b). Front coxa with apical setae white. Postpedicel short triangular, less than 3 times as long as deep (c).. ***denticulatum***

– Hind basitarsus simple, unarmed ventrally. Mid femur with a different arrangement of ventral setae (see figs in following couplet). Front coxa with apical setae black in contrast to white hairs. Postpedicel as *denticulatum* or long and tapered..13

(a)

(b)

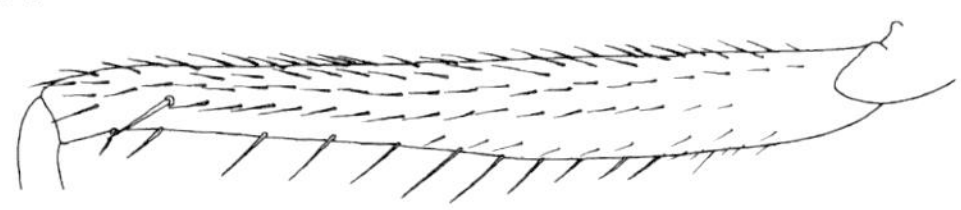

(c)

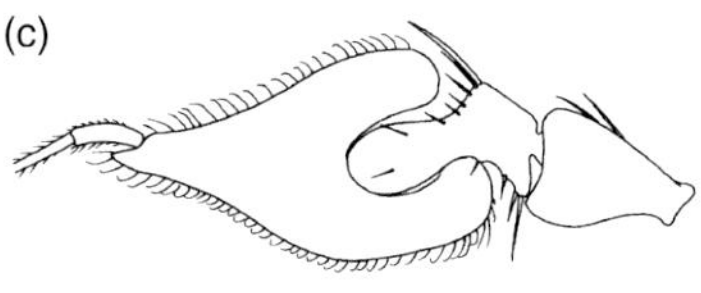

13(12) Tergites mainly yellow. Postpedicel with a narrow tapered apical half (a). Mid femur with a fine isolated ventral seta and row of shorter setae between this seta and base of femur (b). Front femur without ventral setae. Front tarsomeres long and slender; basitarsus about twice as long as tarsomere 2. Frons dull brown. M_1 with a small and sometimes indistinct brown spot on its apical section. .... ***macula***

– Tergites dark. Postpedicel with a triangular apical half (aa). Mid femur with a row of ventral setae on basal half, all similar in size and about as long as depth of femur (bb). Front femur with 4–5 small ventral setae on basal quarter (cc). Front tarsomeres 2–5 equally short; basitarsus about 4 times longer than tarsomere 2 (cc). Frons glossy violet. Vein M_1 without a spot. .... ***metathesis***

(a)

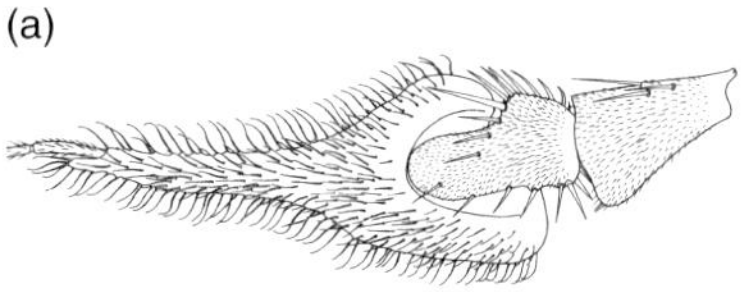

(b)

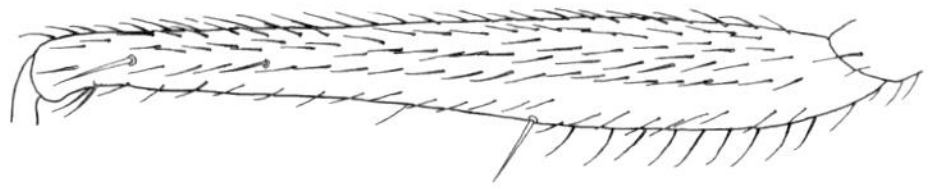

(aa)

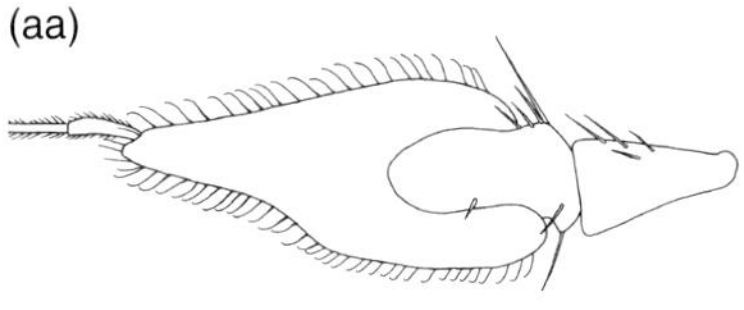

(bb)

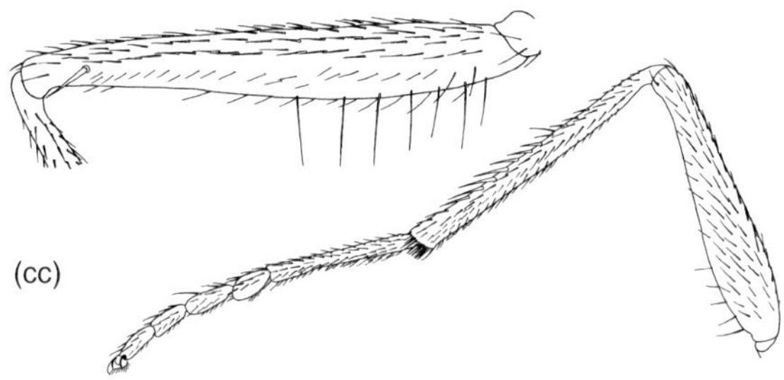

(cc)

14(1) Mid femur with modified ventral setae, either in a row or 2 very long. Postpedicel long (a, b). 15

– Mid femur without differentiated ventral setae, or at most a tiny seta at two-fifths, shorter than depth of femur (similar to *macula*, fig. b of couplet 13). Postpedicel short, triangular or rounded (aa, bb). 16

(a)

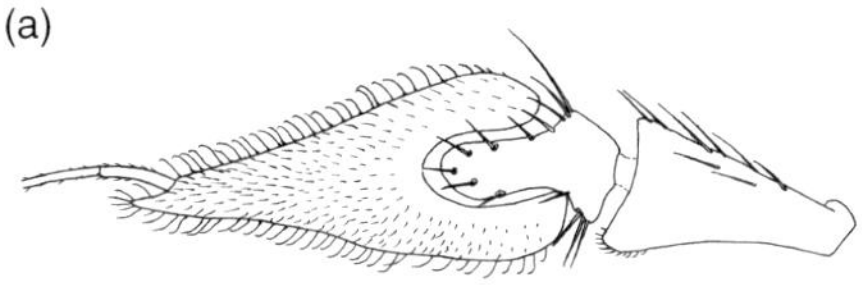

(b)

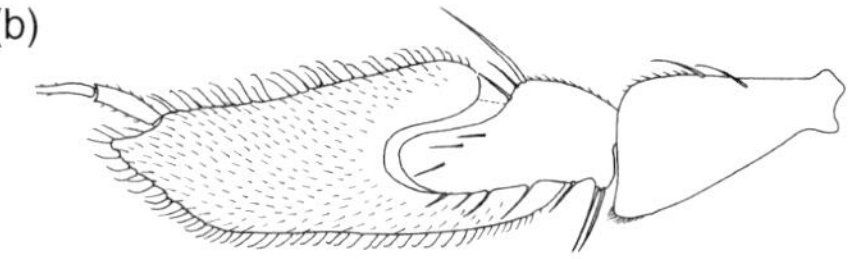

(aa)

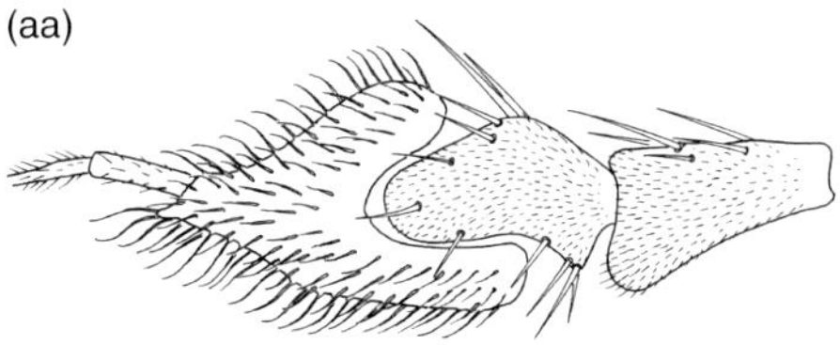

(bb)

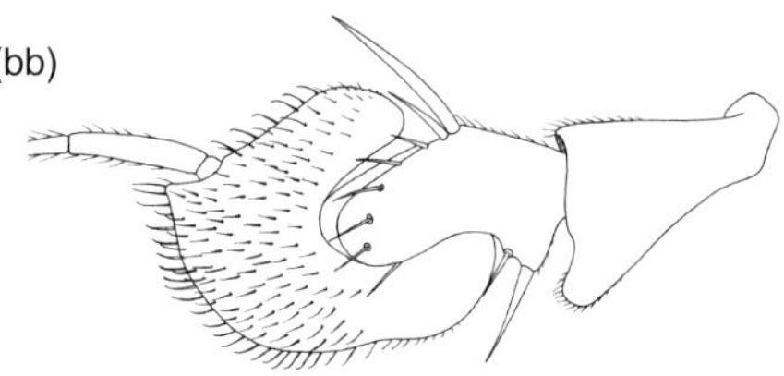

15(14) Mid tarsomeres 4 and 5 strongly deepened (a). Hind tibia conspicuously black at its slightly expanded tip (b). Mid femur with a row of ventral setae on basal two-thirds (c). Hind basitarsus ventrally with a long kinked seta and a short straight one beyond it (d). Antenna – fig. b of previous couplet. .... ***tarsatum***

– Mid tarsus cylindrical, simple. Hind tibia almost entirely yellow, shaft widening only slightly and gradually (bb). Mid femur at basal third with 2 long ventral setae nearly twice as long as depth of shaft (cc). Hind tarsus without modified setae but with dense short ventral hair (dd). Antenna – fig. a of previous couplet.. ***mikii***

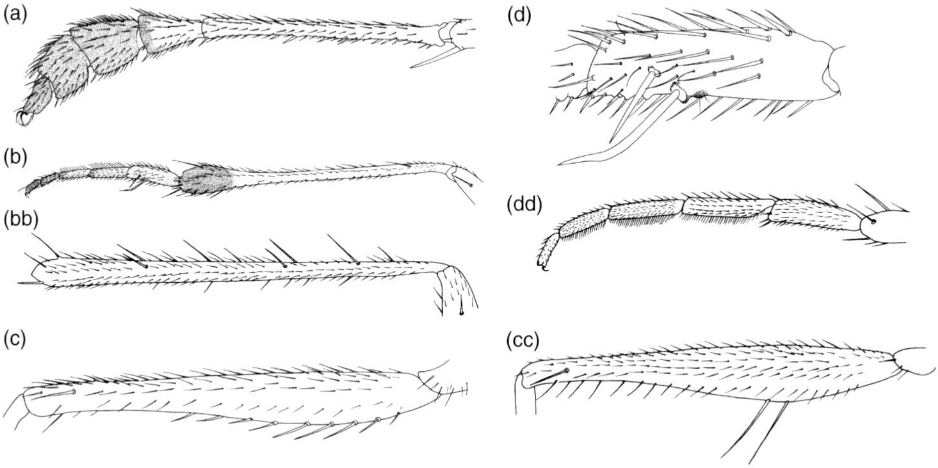

16(14) Large species, wing length 3.9–5.1 mm. Antenna black, postpedicel nearly round (a). Mid coxa black. Hind basitarsus slightly excavated ventrally with a cluster of 3–4 slightly flattened setae on a prominence before excavation (b). Eyes narrowly separated by about width of front basitarsus (c).. ***aulicum***

– Smaller species, wing length 2.5–3.2 mm. Antenna yellow with dark tip on triangular postpedicel (aa). Mid coxa yellow. Hind basitarsus simple, cylindrical (bb). Eyes nearly touching on face (cc).. ***bicolorellum***

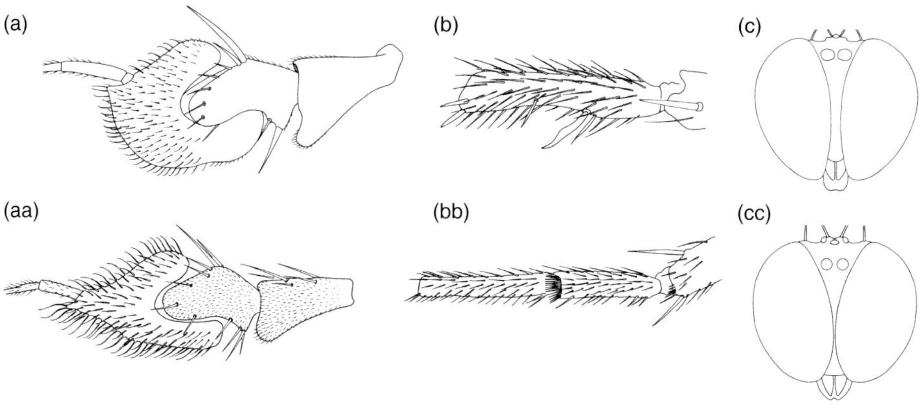

Key to species of *Syntormon* (females)

1 Metepimeron (above hind coxa) dark, same colour as pleura. Tergites partly yellow or entirely dark.. 2

– Metepimeron yellow (ignore pale reflections from pubescence). Tergites 1–4 mainly yellow laterally.. 14

2(1) Frons shining violet or blue, rarely greenish blue, sometimes shining only centrally with dusted margins. 3
– Frons brown, bronze or green, shiny or dull. 10

3(2) Antennal scape bare (figs for couplet 4). Front coxa with apical setae usually all pale. 4
– Antennal scape with tiny dorsal hairs, sometimes reduced to just 1 (fig. b of couplet 8). Front coxa usually with at least 1 black apical seta. 7

4(3,13) Postpedicel longer than deep, tip pointed, arista almost apical (a). Anal lobe of wing broad, as wide as depth of hind femur (b). Larger species, wing length 3.6–4.1 mm. ....***pallipes*** and ***pseudospicatum***
– Postpedicel almost round, blunt-tipped, arista distinctly dorsal (aa). Anal lobe narrow, as wide as width of hind tibia (not femur). Smaller species, wing length 2.8–3.7 mm.... 5

(a)

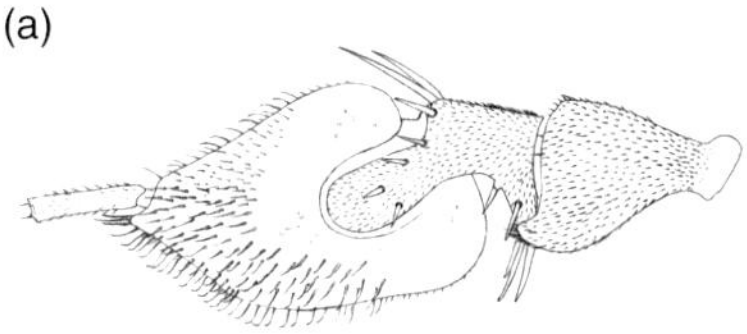

(b)

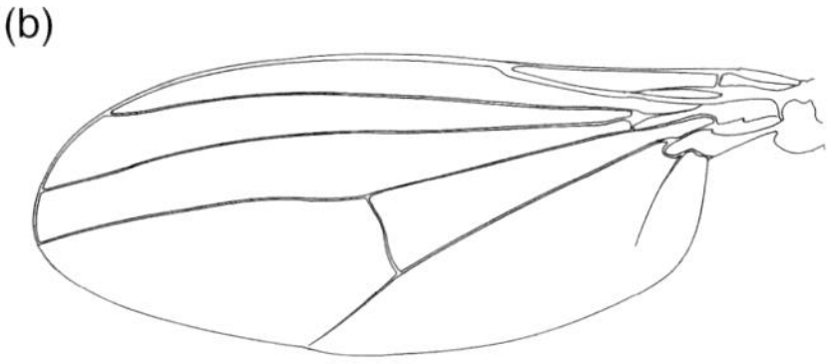

(aa)

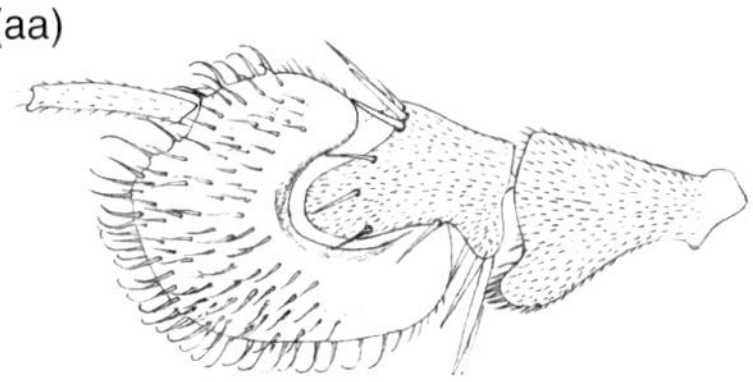

5(4) Hind coxa yellow below lateral seta and usually yellow to base on anterior face. Tergites 2 and 3 always broadly yellow laterally. Hind femur slightly darkened at extreme tip by knee. Mid coxa with a seta on anterolateral face....***fuscipes***
– Hind coxa dark almost to tip and on anterior face. Tergites 2 and 3 with at most small anterolateral yellow patches, rarely more extensive. Hind femur entirely yellow to extreme tip. Mid coxa lacking a seta on anterolateral face. 6

6(5) Front tibia with posterodorsal seta distinct and usually longer than width of shaft (a). Front coxa dark on basal fifth to quarter. Front femur usually with a dark shade along dorsal surface, visible in side view when faint, sometimes absent. Tergites 2 and 3 always dark. Sternites dark....***submonilis***
– Front tibia with posterodorsal seta absent or minute. Front coxa dark only at extreme upper outer corner. Front femur entirely yellow with no dorsal darkening. Tergites 2 and 3 usually with small anterolateral yellow marks. Sternites 1–3 pale. ....***monile***

(a)

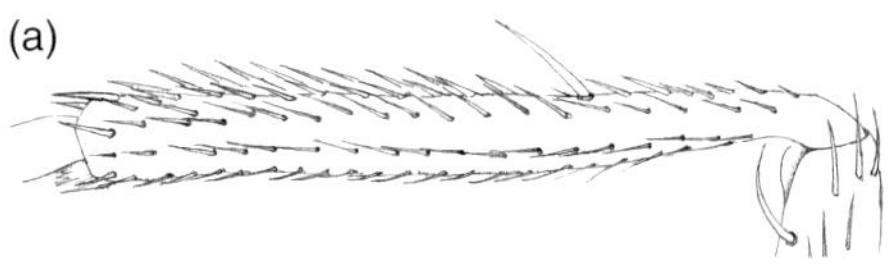

7(3) Face with a pair of hairs above protruding clypeus (a). Front tibia with a strong posterodorsal seta twice as long as width of shaft; anterodorsal comb scarcely differentiated from adjacent vestiture (b). Larger species, wing length about 4.0 mm......... ***setosum*** form of ***mikii***

– Face without hairs. Front tibia with a weak posterodorsal seta at most 1.5 times longer than width of shaft; setulae of anterodorsal comb more clearly differentiated from adjacent vestiture (bb). Smaller species, wing length about 3.0 mm... 8

(a)

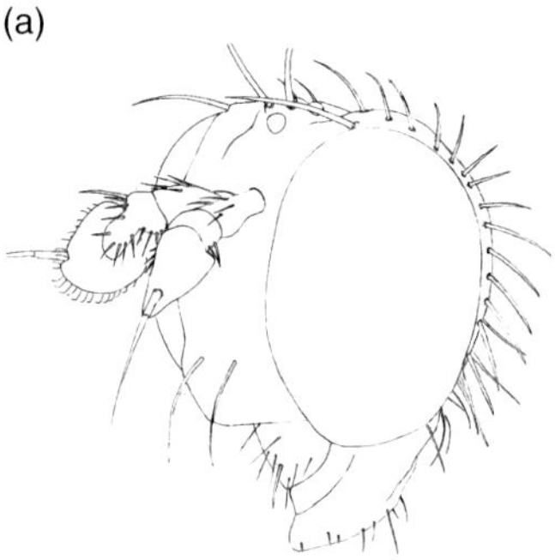

(b)

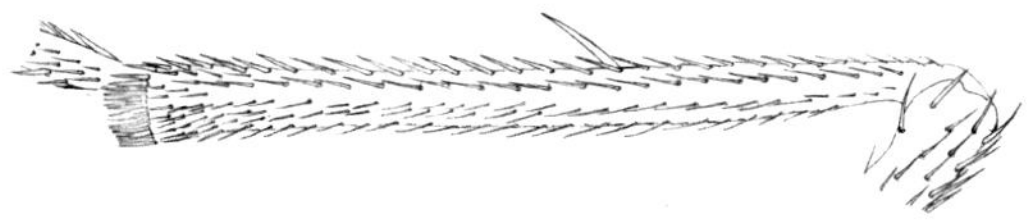

(bb)

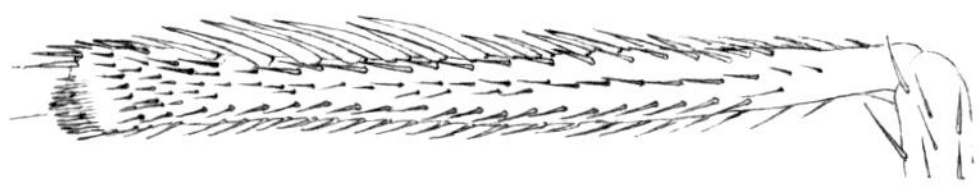

8(7) Front tibia with setulae of anterodorsal comb shorter than width of shaft. Hind femur with 2-3 anteroventral hairs before anterior preapical seta about twice as long as nearby vestiture (a). Hind tibia becoming gradually darker towards tip; hind tarsus entirely dark (a). Mid coxa with anterolateral seta shorter and thinner than lateral seta of hind coxa. Postpedicel rounded with a dorsal arista (b)... ***sulcipes***

– Front tibia with setulae of anterodorsal comb longer than width of shaft (fig. bb of couplet 7). Hind femur without any longer anteroventral hairs before anterior preapical seta (may have 1 longer hair beyond this seta). Hind tibia yellow, not or scarcely darker towards tip; hind basitarsus entirely or partly pale (may be dark late in season). Mid coxa and postpedicel as above or different. 9

(a)

(b)

9(8) Postpedicel tip conical, arista apical (a). Front coxa and femur ringed with black on basal quarter to half (b). Mid coxa with anterolateral seta weak, clearly shorter and thinner than lateral seta on hind coxa.. ***pumilum***

– Postpedicel tip rounded without a pronounced point, arista subapical (similar to fig. b of couplet 8). Front coxa and femur usually yellow but may be dark-ringed late in the year. Mid coxa with anterolateral seta almost as long and strong as lateral seta on hind coxa. .... ***metathesis***

(a)

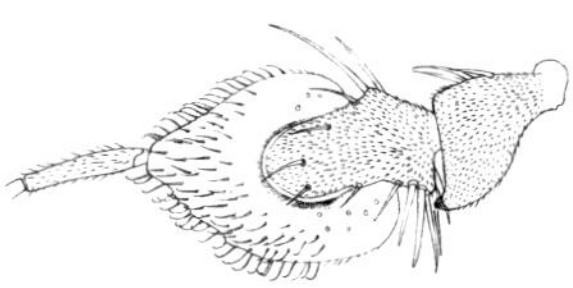

(b)

10(2) Front coxa entirely or mainly dark. Front femur dark on basal half or more. 11

– Front coxa entirely or mainly yellow, dark at most on basal third on outer face. Front femur entirely yellow. 12

11(10) Mid tibia with 2–3 anteroventral setae (a). Frons dulled by dust or microsculpture, usually brown, or if shining then not evenly across its whole width. Face dusting brownish. Facial protrusion large, its lower margin equal to length of postpedicel, and upper margin sloping away markedly from eye profile (b) although less pronounced in fresh or wet-preserved than in dry material. .... ***denticulatum***

– Mid tibia with 1 anteroventral seta. Frons shining green or brown across its whole width, including on wrinkled margin. Face dusting pale grey although green or coppery ground colour may show through. Facial protrusion less pronounced, its lower margin about two-thirds length of postpedicel, and upper margin usually sloping away only gently from eye profile (bb). .... ***zelleri***

(a)

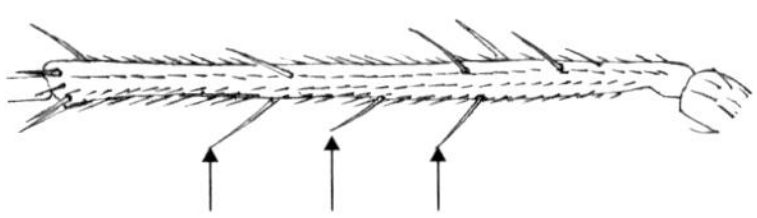

(b)

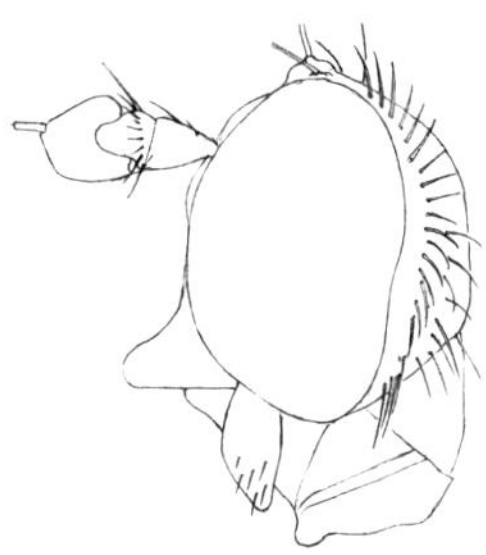

(bb)

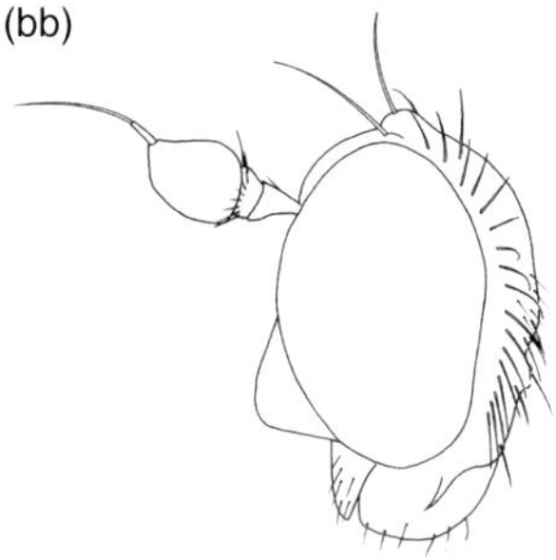

12(10) Wing with a distinct small brown spot on apical section of M_1 (a). Legs brownish yellow. Antennal scape with tiny dorsal hairs (b). Arista clearly subapical, distance from tip of postpedicel about equal to basal width of arista (b). Face with a pair of tiny hairs in middle. .... ***macula***

– Wing without a spot. Legs clear yellow. Antennal scape bare. Arista apical or almost so. Face bare. 13

(a)

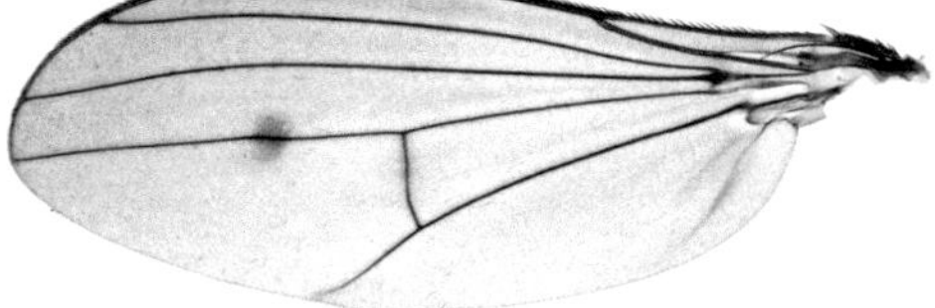

(b)

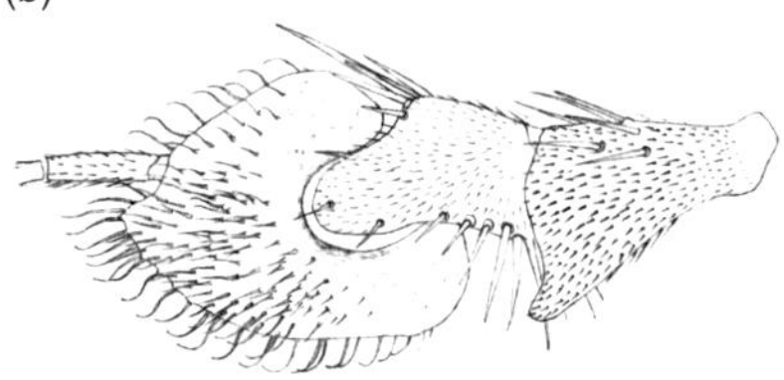

13(12) Large species, wing length more than 3.5 mm. Scutum green with poorly defined and shifting dusted areas. Front coxa with apical setae white. Apical section of M_4 about equal to dm-m. Front tibia with a small posterodorsal seta, shorter than width of shaft, at about two-fifths from base. .... ***pallipes*** and ***pseudospicatum***

– Small species, wing length less than 3.0 mm. Scutum dark brown with pale greyish brown dust stripes between acrostichal and dorsocentral rows. Front coxa with apical setae black. Apical section of M_4 almost twice as long as dm-m. Front tibia with a posterodorsal seta about 1.5 times width of shaft at midway. .... ***filiger***

14(1) Mid coxa dark on outer face and on part of anterior face, at least darker than front coxa. Slightly larger species, wing length 3.8–5.3 mm. 15

– Mid coxa yellow. Smaller species, wing length 2.7–3.6 mm. 16

15(14) Face without hairs above clypeal protrusion. Tergite 2 yellow with a narrow black hind margin about one-sixth length of tergite, the yellow rarely interrupted by a dark mid-dorsal stripe (a). Postpedicel round; scape without dorsal hairs (b). Front tibia with a posterodorsal seta placed at a quarter to third from base (sometimes absent). .... ***aulicum***

– Face with a pair of short black hairs above clypeal protrusion (fig. a of couplet 7). Tergite 2 yellow with a narrow black hind margin and mid-dorsally with a black stripe along its whole length (aa). Postpedicel longer than deep; scape with tiny dorsal hairs (bb). Front tibia with a posterodorsal seta placed shortly before midway along shaft... ***mikii***

(a)

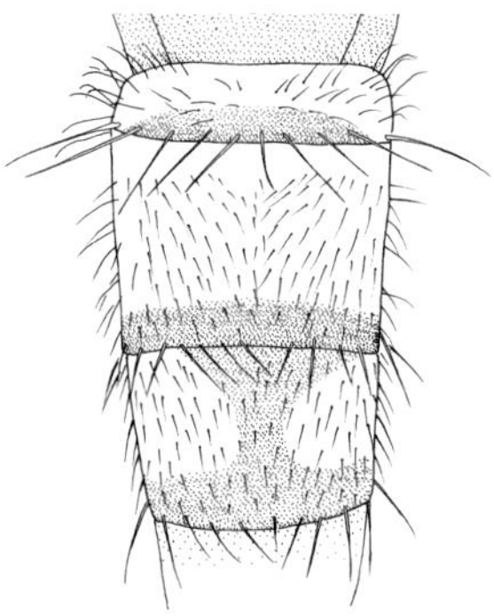

(aa)

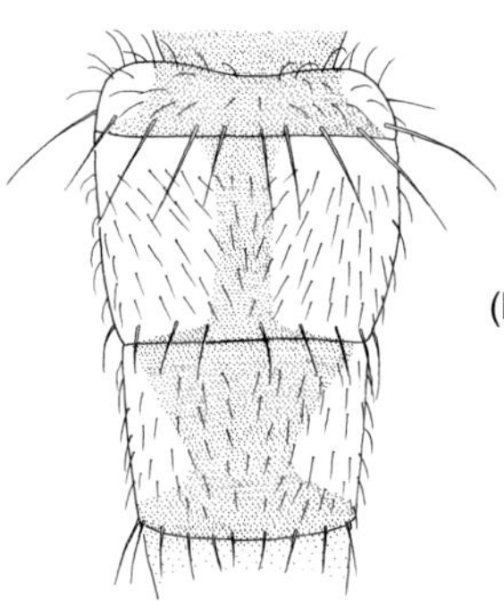

(b)

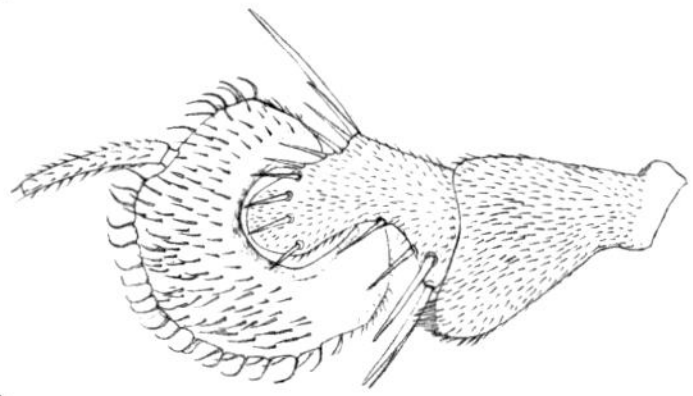

(bb)

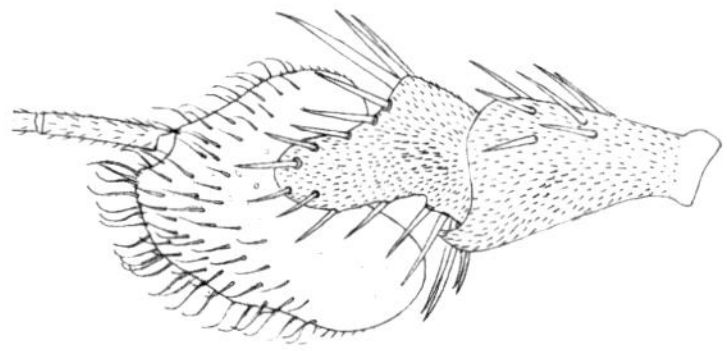

16(14) Scape, pedicel and most of postpedicel yellow. Postpedicel with a tiny pointed tip below arista (a). Face narrow, 3 times higher than width near mouth edge (b), scarcely protruding at mouth edge in profile (c). Frons green to blue.................................... ***bicolorellum***

– Antenna entirely black. Postpedicel with a rounded tip. Face broad, about 1.3 times higher than width near mouth (bb), strongly protruding at mouth edge in profile (cc). Frons violet. .... ***tarsatum***

(a)

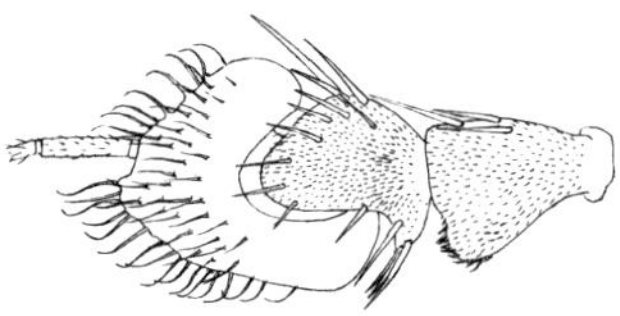

(b)

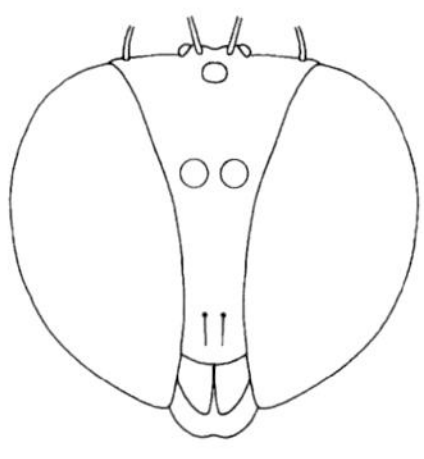

(c)

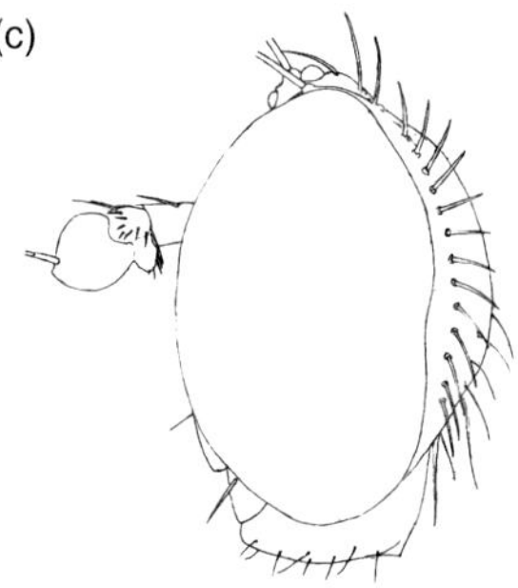

(bb)

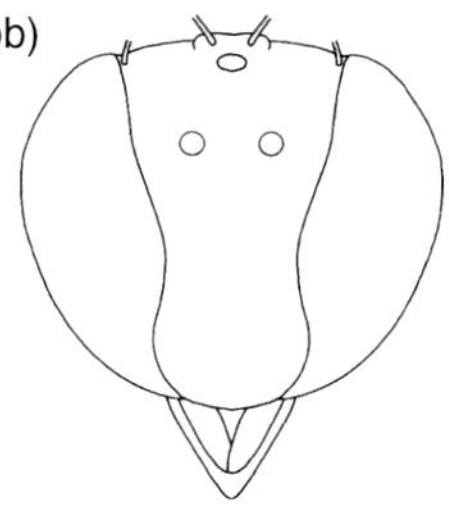

(cc)

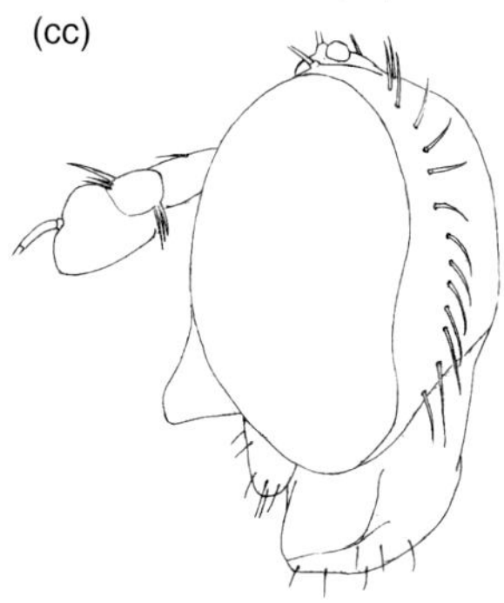

Syntormon aulicum (Meigen, 1824) (Plate 7)

This is the largest species of the genus. Among the species with a predominantly yellow abdomen, hind part of thorax, coxae and legs, *aulicum* is distinguished by its dark mid coxa and particularly short rounded postpedicel, as well as by its hind basitarsus modification. Females also have the coxal colour pattern of yellow: dark: yellow (from front to hind) but, unlike in males, *mikii* also has a dark mid coxa. Females of the east European species *punctatum* resemble *aulicum* in having a yellow metepimeron, extensively yellow tergites and no hairs on the face, but are distinguished by the arista being apical and having uniserial acrostichal setae (biserial in *aulicum*). Wing length ♂: 4.5 (3.9–5.1) mm, ♀: 4.9 (4.5–5.3) mm.

Syntormon aulicum has a sparse distribution in England and Wales, and may be more frequent in the west, but is rare in southern Scotland. It is found in well-vegetated and mainly unshaded wetlands, including the margins of streams, rivers, ponds and ditches, and at seepages, marshes, wet grasslands and at freshwater seepages at upper saltmarsh. It is occasionally found in carr woodland. Water at the ground surface appears to be a key requirement. Flight: v–x, peak in vii.

Syntormon bicolorellum (Zetterstedt, 1843)

As the only *Syntormon* with largely yellow antennae, this species should not be confused with others. It is a pale species with a mainly yellow abdomen, metepimeron, legs and coxae. The 'thumb' projection of the pedicel into the postpedicel is less obvious than in other species, being broader and rather variable in width, and this can cause problems in recognising the species as a *Syntormon*. There is a fine pale ventral seta on the mid femur, as in *macula* but even more inconspicuous. In females the largely yellow antennae and narrow and nearly flat face are unique among the British species. They have a pair of tiny hairs below the middle of the face, as in *macula* and *mikii* in which the hairs are much longer. The female looks very similar to those of *Hercostomus fulvicaudis* and *H. rothi* with the confusion made worse as these species have the strong thumb-like extension into the postpedicel characteristic of *Syntormon*; they are differentiated by *bicolorellum* having uniserial acrostichal setae and parallel veins R_{4+5} and M_1 (biserial and converging, respectively, in the two *Hercostomus* species).

Females of the non-British *luteicorne* Parent also have largely yellow antennae but differ in their entirely green tergites, black calypter hairs and having a more protruding face bearing longer hairs (Speight *et al.* 1995). The undescribed male of *S. luteicorne* presumably also has yellow antennae. Wing length ♂: 2.8 (2.5–3.2) mm, ♀: 3.1 (2.7–3.4) mm.

Syntormon bicolorellum is widespread and moderately frequent in England but very scarce in south Wales and southern Scotland. It is a species of shaded, usually damp, sites, so is most often found in woodland or scrub but sometimes in the shade cast by tall vegetation such as reeds. It shows no strong fidelity to saturated ground but can be found in a range of wetness from seepages and stream margins to almost dry woodland. Flight: ii–iii, v–xi, peak in vii.

Syntormon denticulatum (Zetterstedt, 1843) (Plate 8)

The legs of this large dark species are particularly variable in colour, ranging from almost entirely black to almost entirely yellow, although the front femur remains black on at least the basal half. The two small spines on the hind basitarsus resemble those of *aulicum* which is a largely pale species. A row of ventral setae on the mid femur is found in a few other species but they extend almost to the tip only in *denticulatum*. Females are among the few species with a dull brown frons, either dusted or with a lumpy surface showing only weak reflections. The dorsum is also usually dark brown with a pair of conspicuous pale dust spots outside the fourth and fifth dorsocentral setae, which are particularly obvious in slightly anterior view. It is the only species with two to three anteroventral setae on the mid tibia, all other species having only one such seta. Other differences from *zelleri*, which is most resembles, are given in that species account. Wing length ♂: 3.5 (2.9–3.9) mm, ♀: 4.2 (4.0–4.6) mm.

This is a common and widespread species in England and mainly coastal Wales but rare in Scotland, although recorded in Orkney and the Outer Hebrides. It is found in many wetlands including acid and base-rich mires, seepages, wet grasslands, the shores of rivers, ponds and ditches, and carr woodland. There is no apparent preference for open or shaded sites. The presence of water at the ground surface is a constant feature. Flight: i–xi, mainly vi–vii.

Syntormon filiger Verrall, 1912

The colouration of *filiger* is not found in other species, being sombre with scarcely a hint of green, a conspicuous pattern of grey-dusted lines and blotches on the scutum, and a dull dark frons. The postpedicel is particularly narrow on the apical half with dense and quite long pubescence, similar to that of *zelleri* and *macula*. The second tergite sometimes has small yellow side-patches and the leg colour varies, more typically being mainly yellow but with increasing amounts of black on the coxae and femora in specimens from more northern latitudes. Females are coloured similarly to the males, including the scutum dusting. The long apical section of vein M_4 relative to the crossvein dm-m is unique in British *Syntormon*. Wing length ♂: 2.1 (1.8–2.3) mm, ♀: 2.5 (2.2–2.6) mm.

Syntormon filiger is recorded sparsely on the coast of East Anglia and the Thames estuary, Solent in Hampshire, Morecombe Bay in Lancashire, several locations on the Welsh coast and the Solway and Firth of Forth in Scotland. It is a restricted to saltmarsh and possibly shows a preference for brackish pools with sea club-rush. It has also been recorded from an inland saltmarsh in Cheshire (Drake *et al.* 2019). Flight: v–ix.

Syntormon fuscipes (von Roser, 1840)

This medium-sized species is rather slender and moderately pale, with second and third tergites always largely yellow in side view and the fourth only obscurely yellow on the lower edge. Unusually for a species with a dark metepimeron, the hind coxa is yellow on its apical half. The femora and tibiae are usually entirely yellow although sometimes the hind femur is darkened at the tip. The hind basitarsus is very similar to those of *monile* and *submonilis* at low magnification but the simple front and mid tarsi separate it from this pair, as well as its simple, short and rather uniform hairing and poorly developed comb on the front tibia. Females differ from other species with yellow-marked tergites in having bicoloured mid and hind coxae. Additionally, the front coxa is entirely yellow at the outer corner at its base (slightly or obviously darkened here in *monile*, *submonilis* and *pallipes*). The weak comb and posterodorsal seta of the front tibia are similar to those of *monile* or the seta may be completely absent (stronger in *submonilis*). Wing length ♂: 3.0 (2.9–3.3) mm, ♀: 3.5 (3.1–3.7) mm.

Syntormon fuscipes is widespread although not frequent in England and Wales, and rare in Scotland northwards to the Outer Hebrides. It is a species of seepages and small streams, where shallow flowing water appears to be the common requirement. It is found in this habitat in wet pastures and woodlands, valley mires and on soft-rock cliffs. The sites may be shaded or sunlit. Flight: v–viii.

Syntormon macula Oldenberg, 1927

Males have conspicuously yellow-marked tergites and predominantly yellow legs, but poorly developed modification of the leg setae restricted to a fine short seta at the middle of the mid femur preceded by a row of even shorter fine hairs. The postpedicel has a long narrow apical half, resembling that of *filiger* and several other species which all have well-developed leg modifications. Females differ remarkably from males, being drab brown or dark green with a pale grey-dusted and entirely brown abdomen, and legs usually some shade of brown. They resemble *denticulatum* in size, colour and dull frons but are distinct in having a conspicuous wing spot (faint and much less obvious in males), a pair of small facial hairs and usually dark apical setae on the front coxa. The mismatch between males and females in colour and flight period, and the infrequency with which both occur at the same site may suggest that the sexes have not been correctly associated, and that the opposite sexes remain unrecognised. Wing length ♂: 3.4 (3.2–3.6) mm, ♀: 4.1 (3.7–4.4) mm.

Syntormon macula is an uncommon species found in southern England and the east of Wales, roughly south of a line from the Mersey to the Thames estuary, with a pronounced western bias, and in East Anglia. It is most often found by stony streams and rivers under trees although not necessarily in woodland, with occasional records from open sites. Females fly early in the year in cold conditions as well as in summer but males are known only in warmer months (d'Assis-Fonseca 1948, Drake 2021a, b). Records of females far outnumber those of males. Flight ♀: iii–x, peak in iv–v; ♂: v–x, mainly viii.

Syntormon metathesis (Loew, 1850)

Males superficially resemble those of *submonilis* and *monile* in body and leg colour, having mainly yellow femora and tibia, although darker towards the end of the year, and the second tergite being entirely dark or with a small yellow patch. A simple hind basitarsus is also found in *pumilum* which has several long setae at the base of the front and mid femora and in a few species which have extensive yellow on the tergites. The line of conspicuous ventral setae on the mid femur differs from the short setae of *denticulatum*, *macula* and *tarsatum* and form a longer line than in *filiger*, *mikii* and *pumilum*. Among females in the group of small species with a glossy violet frons, *metathesis* is distinguished by having both a long seta on the mid coxa and robust black apicals on the front coxa. This is helpful with individuals found late in the year which can have the same leg colour pattern as *pumilum*, although in midsummer the front coxa and femur are entirely yellow. The lateral margin of the second tergite and the sternites may sometimes be pale. Wing length ♂: 2.8 (2.7–2.9) mm, ♀: 3.0 (2.9–3.2) mm.

Syntormon metathesis has been recorded from fens in Norfolk and Kent (Drake 2023). Flight: iii, vii, xii; probably all year.

Syntormon mikii Strobl, 1899

This is a large species with a mainly yellow abdomen and legs that are completely yellow from coxa to basitarsus, except for a narrow dark streak on the mid coxa. It thus superficially resembles *aulicum* from which it is distinguished by the pair of long ventral setae on the mid femur which are unique to *mikii* and a simple hind tarsus. Females also resemble *aulicum* but have slightly more extensive dark marks on the basal tergites. The pair of face hairs are distinct in *mikii* and are also present but less obvious in *macula* and *bicolorellum*. *Syntormon setosum* Parent, 1938 is very similar and known only from females that differ from *mikii* in having darker metepimeron and hind coxa, and entirely dark abdomen. It is possible that this is merely a dark colour form of *mikii* (Speight and Meuffels 1989, Drake 2021a), and this view is followed here. Dyte (1990) suggests that *S. flexibilis* Becker, 1922 may be synonymous with *setosum*; this is a tramp species from North America and Taiwan. Wing length ♂: 3.4 (3.0–3.7) mm, ♀: 3.9 (3.8–4.1) mm.

Syntormon mikii is a scarce coastal species recorded in East Anglia and southwest England from Hampshire to north Devon. The form *setosum* has been recorded in Dorset and Devon. Although coastal, the records include none from the saline parts of saltmarsh but come from a variety of apparently freshwater habitats close to the shoreline, including soft-rock seepages, streams or seepages flowing onto the sea-shore or upper saltmarsh, dune pools, wet woodland, wet pasture and reedbeds. There are confirmed inland populations a few kilometres from the nearest coast in Dorset and Devon. Flight: i, iv–x, peak in vii; probably all year.

Syntormon monile (Haliday in Walker, 1851)

This small dark green species usually has small lateral yellow patches up to half the tergite's length but rarely entirely green. It closely resembles *submonilis* but is reliably distinguished by the simpler shape of the front basitarsus; additionally in dry specimens, the front tarsus remains extended whereas that of *submonilis* is usually strongly curled and twisted, similar to the figure in the key. Females resemble *submonilis* and some individuals may not be separable. Undoubted *monile* have no, or only a tiny, posterodorsal seta at the basal third of the front tibia, and pale sternites; other characters that appear to be constant are some yellow lateral marks on the tergites (rarely a complete ring), and the yellow colour of the mid and hind trochanters running continuously into the yellow base of the coxae. Further differences are given in the accounts for *submonilis* and *fuscipes*. The explanation for the confusingly different endings of the two species names is given by Drake (2021a). Wing length ♂: 2.4 (2.2–2.6) mm, ♀: 3.0 (2.9–3.2) mm.

As *monile* and *submonilis* have been confused, their distribution is uncertain, but the combined records are widespread in Britain from Cornwall to northern Scotland, but scarcer in Wales, Scotland and northern upland England. Records of confirmed *monile* suggest an equally wide distribution in Britain, reaching Orkney, but it is far less frequent than *submonilis*. Its habitat is mainly seepages, tiny streams and saturated fen and rushy grasslands, usually with a moderate cover of taller herbaceous vegetation; these may be acidic or base-rich. It is less often found by the margins of standing water and rarely in woodland, unlike *submonilis*. Flight: v–ix, mainly vi–vii.

Syntormon pallipes **(Fabricius, 1794) (Plate 8)**

This is a medium-sized moderately shining green species with a pair of claws on a column at the base of the hind basitarsus, as also found in *pseudospicatum*. The abdomen may be entirely green or the second and third tergites may have large yellow lateral patches, and both colour forms occur in the same population. Hind leg colour varies widely from typically yellow with an apical black ring on both the femur and tibia, to almost entirely black. Occasional pale individuals have almost no apical dark rings and if their hind tibial hairing is weak they cannot be assigned to either *pallipes* or *pseudospicatum*. Females of *pallipes* and *pseudospicatum* are not separable but this pair of species is readily recognised by their shining purple frons and relatively long postpedicel tapering to a strictly apical or just subapical arista. Some specimens have a green shiny frons which resembles that of *zelleri* but have almost entirely yellow legs with only the basal third of the front coxa dark, compared to the rather more extensively dark front coxa and femur of *zelleri*; they also differ from *zelleri* in having a seta on the mid coxa, absent in *zelleri*. Like males, there are two colour forms which occur together. Wing length ♂: 3.1 (2.8–3.3) mm, ♀: 3.8 (3.6–4.1) mm (females from inland populations).

Syntormon pallipes is a very common and widespread species in England and much of Wales, but in Scotland it is common only near the coast, although also found well inland. It occurs north to Shetland and westwards to the Outer Hebrides. It is a wetland species with no obvious habitat preferences although it is less frequent in shaded sites, can be particularly abundant on saltmarsh, and may prefer mesotrophic sites over oligotrophic and acidic wetlands. Flight: ii–xi, long season, mainly v–vii.

Syntormon pseudospicatum **Strobl, 1899**

Syntormon pseudospicatum resembles *pallipes* in having a pair of claws on a column at the base of the hind basitarsus. It differs most obviously in the entirely yellow hind tibia and basal half of the hind basitarsus, but it shows none of the wide variation in leg colour found in *pallipes*. Like *pallipes*, the tergites may be entirely dark green or the second and third may have large yellow lateral patches. See the account for *pallipes* for further characters. Females are not separable from those of *pallipes*, although specimens with entirely yellow mid and hind trochanters may represent *pseudospicatum*. Wing length ♂: 3.0 (2.6–3.4) mm. There remains disagreement whether *pseudospicatum* is a merely a form of *pallipes* (Chursina and Grichanov 2019, Drake 2020b) but they are treated as two species here as their habitat associations are sufficiently distinct to warrant recording them separately.

This is a widespread coastal species found around much of the British coast including tiny offshore Scottish islands. It can be frequent on saltmarshes and brackish marshes but is very rarely found inland. Flight: iv–x, mainly v–vii.

Syntormon pumilum **(Meigen, 1824)**

Syntormon pumilum is a small subshining green species with an entirely dark green abdomen, yellow legs with both the front femur and coxa dark on their basal halves and yellow on their apical halves, giving a characteristic alternating black-yellow pattern. The ventral setae on the front and mid femora are unique among British *Syntormon*. The front tarsus is unusual in having tarsomeres 2 to 5 dumpy and together equalling the length of the basitarsus, and the second tarsomere projecting ventrally under the third. It is one of the few species with a simple hind basitarsus. Females resemble the males in colour, including the front leg pattern. They differ from most other species in their strictly apical arista and in having a pair of pale dust spots on the occiput, conspicuous in dorsal view, although these are present in the similar *metathesis*; further differences are given in the account for *metathesis*. Wing length ♂: 2.3 (2.1–2.6) mm, ♀: 2.8 (2.6–2.9) mm.

Syntormon pumilum is moderately common and widespread with no obvious change in frequency throughout Britain, occurring north to Orkney. It is found in a wide range of wetlands that include both still and flowing water, fresh to brackish, and acid to base-rich. It is rarely found in the shade of trees so is absent from woodlands. Flight: i, iv–x, xii, peak in vi (probably all year).

Syntormon submonilis Negrobov, 1975

Syntormon submonilis is a small dark green species never having any yellow on the abdomen. It closely resembles *S. monile* but is distinguished by the distinctive modification of the front tarsus, which looks somewhat distorted in dry specimens as it is always twisted and curled to a right angle, and by the dark streak on the front femur. The hind basitarsus is almost identical to those of *S. monile* and *S. fuscipes*. Females can usually be distinguished from those of *monile* by having a small but distinct posterodorsal seta at the basal third of the front tibia, and dark sternites. Other characters of *monile* are the mid and hind trochanters being darker than the base of their femora, the mid and hind tarsi being dark from the tip of the basitarsus outwards, and the postpedicel being more conical so that its anterior face is slightly wider from the peg of the pedicel to the front margin than to ventral and dorsal margins (in *monile* the postpedicel is more rounded so that the anterior face is about equally wide from the peg to the margins in all directions). Wing length ♂: 2.6 (2.6–2.8) mm, ♀: 3.2 (2.8–3.4) mm.

Although recognised in Britain since 1993 as *silvianum* Pârvu, 1989 (Hodge 1993b, Drake 2021a), *submonilis* and *monile* continued to be confused. *S. submonilis* is widespread in England and Wales, and scarce in Scotland north to Sutherland. In well recorded areas *submonilis* is much more frequent than *monile* and this is likely to hold true for all of Britain. *S. submonilis* is a species of seepages, tiny streams and valley mire, and is sometimes found in saturated pastures. Most sites are shaded, within woodland or under trees, unlike those of *monile*. Flight: v–viii, mainly vi–vii.

Syntormon sulcipes (Meigen, 1824)

This small species has a dark green thorax and coppery green tergite discs. The outstanding feature is the strikingly flattened, grooved and wide hind tibia which is almost entirely black and contrasts strongly with the bright yellow of the rest of the legs, including the trochanters. Females are distinguished among the group with a violet or blue frons by the combination of a clearly dorsal arista and their hind tibia darkening gradually to the tip, where the dark colour runs on to the basitarsus, unlike the yellow of the tibia's tip and base of basitarsus of these other species. Wing length ♂: 2.6 (2.4–2.8) mm, ♀: 3.1 (2.9–3.3) mm.

S. bulgariensis Negrobov and Kechev, 2012 is very similar but the hind femur of the male is black rather than yellow in *sulcipes*.

Syntormon sulcipes is moderately frequent in western and northern Britain north to Sutherland, scarce in southern England and absent from most of lowland England. It is restricted to acid sites, mainly mires and seepages, and carr and wet oak or birch woods on acidic rocks. Flight: v–ix, mainly vi–vii.

Syntormon tarsatum (Fallén, 1823)

At low magnification this species is instantly recognisable by its flattened black 'flag' on the mid tarsus and the black and slightly broadened tip of the hind tibia, which both contrast strongly with the otherwise yellow legs. The abdomen is largely yellow when viewed from the side. Unlike other species with a yellow metepimeron, the first tergite is often dark. Females in lateral view are continuously pale on the lower half with yellow lateral patches on at least the second and third tergites, all sternites, all coxae, the metepimeron and all legs to at least the tip of the basitarsi. Wing length ♂: 3.1 (2.8–3.3) mm, ♀: 3.4 (3.2–3.6) mm.

This species has a clear northern and western distribution with a strong outlying population in the fens of East Anglia. It appears to be more frequent in Scotland than in most of England and Wales, although it can be frequent locally in southwest England. Its habitat is mainly acid wetlands, including mires, swamp at the margins of ponds and lakes, carr woodland and occasionally wooded stony streams. The East Anglian fen population is found mainly in reedbeds or swamp rather than typical herb-fen. Flight: vi–ix, mainly vi–vii.

Syntormon zelleri (Loew, 1850)

This is a medium-sized dark subshining green species with dark coxae and predominantly dark femora, giving the whole fly a sombre appearance, superficially resembling *denticulatum*. The antenna is particularly long and, together with the arista, is about 3 times as long as the head. No other species has a ventral projection at the tip of the front basitarsus. Females also resemble *denticulatum* in size and colour. Although the colour of the mid and hind legs varies widely, the typical pattern is yellow with the mid femur black on the basal quarter and the hind femur black on the apical quarter and at the base ventrally. Additional differences from *denticulatum* (in parentheses) are the peg of the pedicel being clearly wider than the sections of postpedicel above and below it (equal or narrower), a longer and more pointed postpedicel with the arista just slightly dorsal (blunter and arista practically apical), no conspicuous dust spots on the scutum outside of the dorsocentral row and behind the suture (conspicuous), and longer acrostichal setae about the length of the upper postocular setae (much shorter). The European species *punctatum* is found from Germany eastwards but could occur in Britain; it resembles *zelleri* in having dark femora (as do dark specimens of *denticulatum* and *filiger*) but is distinguished in the male by the hind basitarsus having two long basal setae. Wing length ♂: 3.1 (2.9–3.4) mm, ♀: 3.9 (3.7–4.2) mm.

The distribution is mainly northern and western but with concentrations of records in the northern half of Scotland and southwest England and Wales. It also occurs in northeast England but some records for lowland southeast England may be errors. It is found in acid wetlands, including pools, seepages and small streams on acid mires and grasslands, moorlands and montane heath and, at lower altitudes, valley mire, wet heaths and seepages on soft-rock cliffs. Sites are rarely wooded, except for occasional birch or sparse oak woodland. Flight: iii–x, mainly vi–vii.

SYSTENUS Loew, 1857 (Medeterinae)

Systenus is a genus of medium-sized species which in the male is recognised by the combination of an apical or subapical arista on an elongate postpedicel, no anterior preapical seta on the mid femur, and the hypopygium exposed on a short stalk formed by the last tergite (Plate 8). In females, the genus is recognised by the combination of an apical arista on an oval or almost round postpedicel, no anterior preapical seta on the mid femur, large palps and, to distinguish *Systenus* from *Medetera*, the apical sections of R_{2+3} and M_1 not obviously parallel. Both sexes have mainly pale legs, often including the coxae and most tarsomeres, and tibial vestiture arranged in tidy rows. Males may be confused with those of *Syntormon* or *Rhaphium* which often have similar large and strongly tapered antennae and an apical arista, the similarity with *Rhaphium* being more pronounced as their hypopygia are superficially similar, but species of *Systenus* differ from *Syntormon* in lacking the peg on the pedicel and from *Rhaphium* in having no anterior preapical seta on the mid femur. The slightly bulging clypeus of females is either glossy violet or green, or thinly dusted so that the shining colour shows through, as in some species of *Medetera*.

Systenus larvae develop in rot-holes, detritus-filled cavities, sappy decaying wood and sap-runs of many species of deciduous trees. Dyte (1959) suggests that oviposition in oozing sap wounds is the usual procedure for the genus. Adults are infrequently caught by sweep-netting but more often by trapping using Malaise traps and interception traps placed over rot-holes and similar likely breeding sites. Compared with most other dolichopodids, a far larger number of records derive from flies reared from decaying wood and rot-hole material (e.g. Crossley 2001). The records show that perhaps all British species have little preference for particular species of tree. With the exceptions of *mallochi* and *tenur*, records are unusually evenly spread over England, if somewhat thinly, suggesting that the species are probably more common than usually assumed.

Eleven species occur in the Palaearctic and six in Britain (Negrobov *et al.* 2020). *Systenus* was originally placed in the Rhaphiinae (e.g. d'Assis-Fonseca 1978), moved by Robinson (1970) to its own subfamily, Systeninae, then to Medeterinae (Bickel 1986). The inclusion of *Systenus* within the Medeterinae was confirmed using molecular data (Pollet *et al.* 2011).

Key to species of *Systenus* (males)

1 Veins R_{4+5} and M_1 converge strongly near wing margin, their tips separated by not more than one-third length of crossvein dm-m (a) ... 2

– Veins R_{4+5} and M_1 much less convergent, almost parallel beyond dm-m, their tips separated by little less than length of dm-m (aa) ... 4

(a)

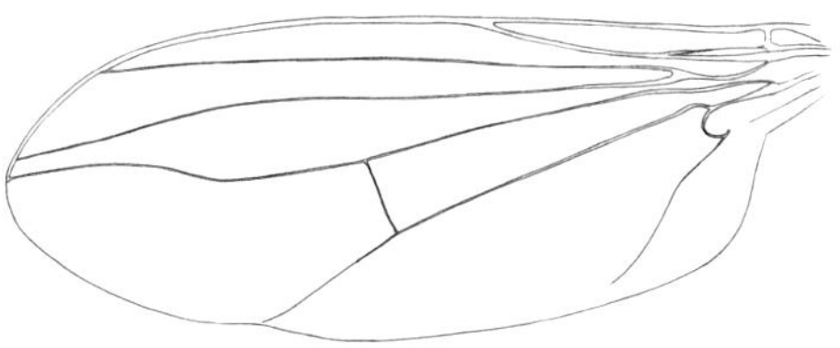

(aa)

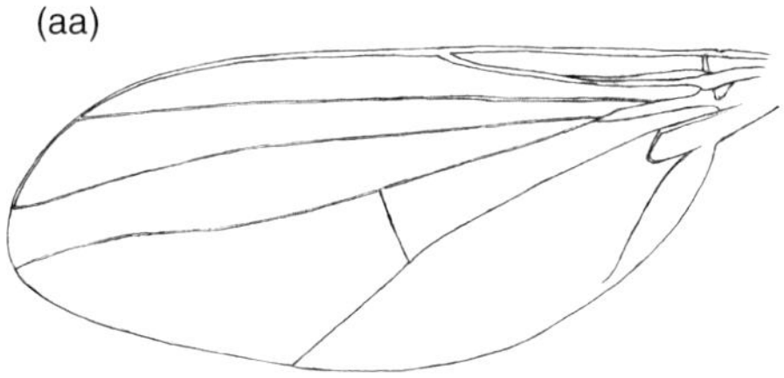

2(1) Wing not concave on hind margin behind wing tip, without an apical black spot (fig. a of couplet 1). Scape and pedicel black, postpedicel mainly black but yellow ventrally on basal third to half, at least 4 times longer than deep (a). Arista with long pubescence, more than twice the width of basal section. Cercus with a single lobe ... ***pallipes***

– Wing concave on hind margin behind wing tip, with an apical black spot (figs in couplet 3). Scape and pedicel yellow, postpedicel colour either as in *pallipes* or entirely black, at most 3.5 times as long as wide and usually less (aa). Arista with inconspicuous pubescence. Cercus with 2 lobes ... 3

(a)

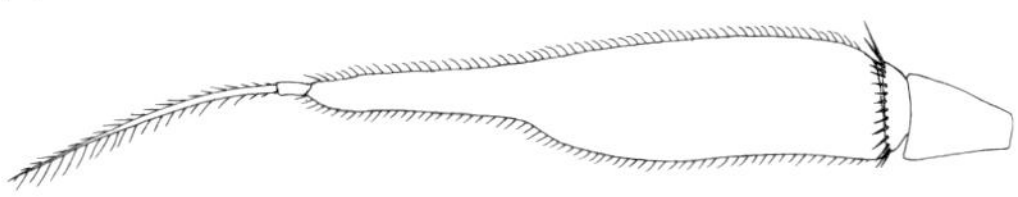

(aa)

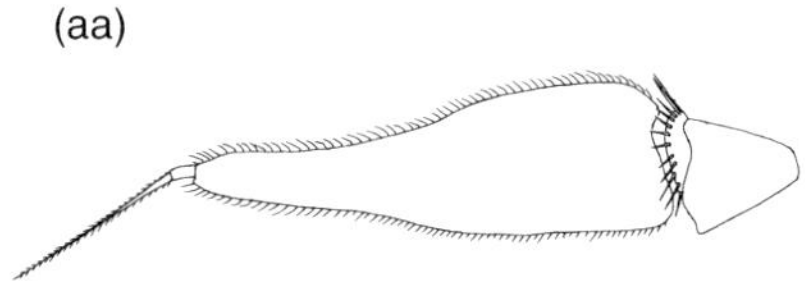

3(2) Postpedicel black dorsally and apically, yellow on ventral half of swollen section. Wing with a large spot entirely filling apical lobe (a). Mid tibia often with setulae on shaft in addition to a strong basal pair, but may be absent (as fig. a of female couplet 3). Cercus lateral lobe as long as median lobe, bearing a cluster of hairs at tip as well as along sides (b).......... ***scholtzii***

– Postpedicel almost entirely black, occasionally with a trace of yellow at extreme base. Wing with a small spot not filling all of apical lobe, separated from wing tip by a whitish area free of microtrichia (aa). Mid tibia with only a pair of long setae on basal half. Cercus lateral lobe not extending as far as median lobe, bearing hairs only along sides (bb)...................... ***mallochi***

(a)

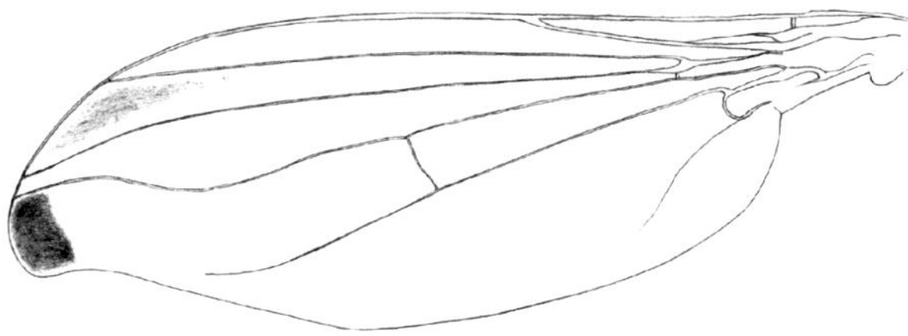

(b)

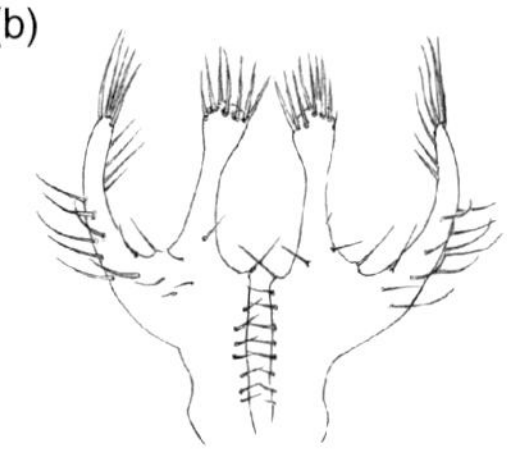

(aa)

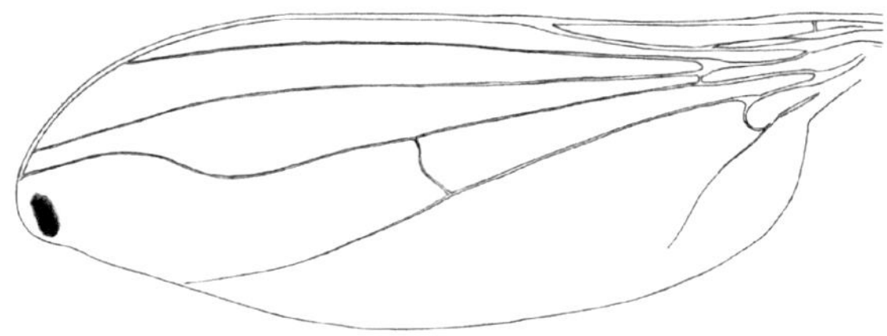

(bb)

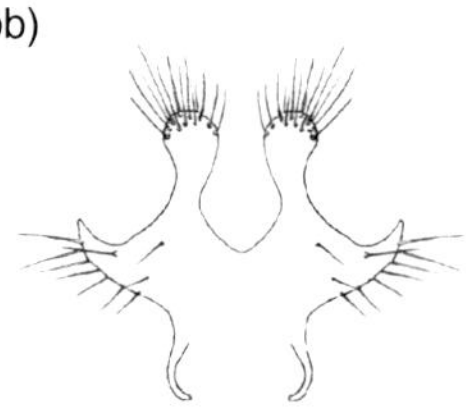

4(1) Hypopygium and sternite 8 entirely whitish yellow except for black tip of cercus (a). Cercus short with a broad truncated tip (a). Mid tibia dorsally with 4 strong setae, 2 at basal quarter and 2 on apical half (as fig. a of female couplet 5). Postpedicel more than 3 times longer than deep, and almost 2.5 times as long as arista (b). Apical section of M_4 about 2.5 times length of dm-m. .... ***leucurus***

– Hypopygium capsule and preceding segments dark, only cercus and surstylus pale. Cercus ribbon-shaped (aa). Mid tibia dorsally with only 2 long setae at base and very weak setae on apical half (as fig. a of female couplet 3). Postpedicel length less than 3 times its depth (bb, two species). Apical section of M_4 about twice length of dm-m. 5

(a)

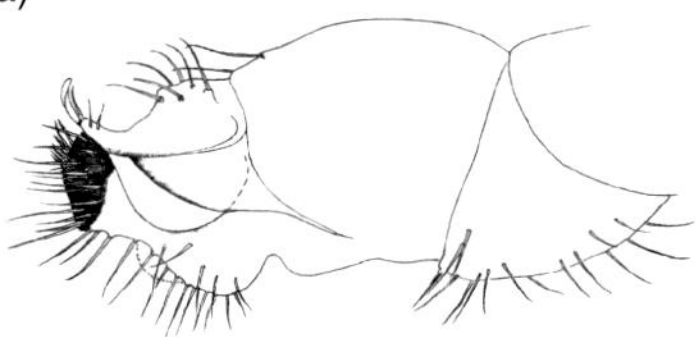

(b)

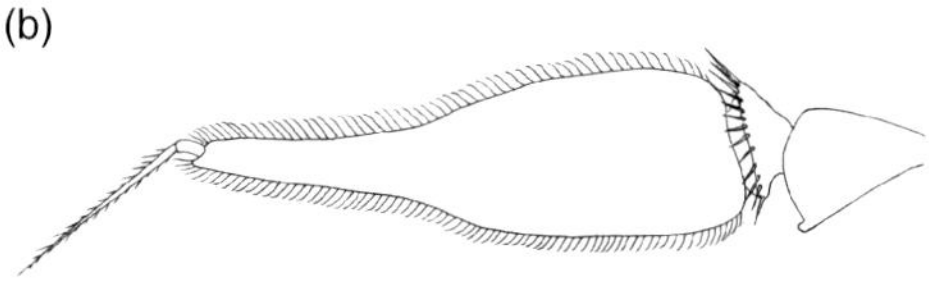

(aa)

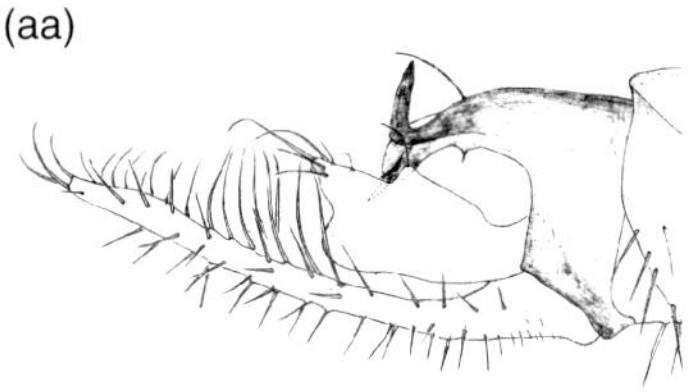

(bb)

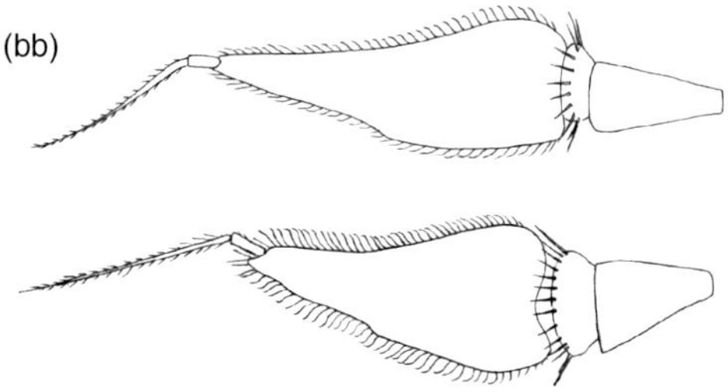

5(4) Antenna entirely black. Cercus twice as long as surstylus; epandrial lobe with a black hook-shaped tip (fig. aa of couplet 4)............***bipartitus***

– Scape and pedicel clear yellow. Cercus less than 1.5 times longer than surstylus; no dark hook on epandrial lobe (aa)............***tener***

(aa)

Key to species of *Systenus* (females)

1 Veins R_{4+5} and M_1 converge strongly near wing margin, their tips separated by not more than one-third length of crossvein dm-m (a)............2

– Veins R_{4+5} and M_1 much less convergent, almost parallel beyond dm-m, their tips separated by little less than length of dm-m (aa)............4

(a)

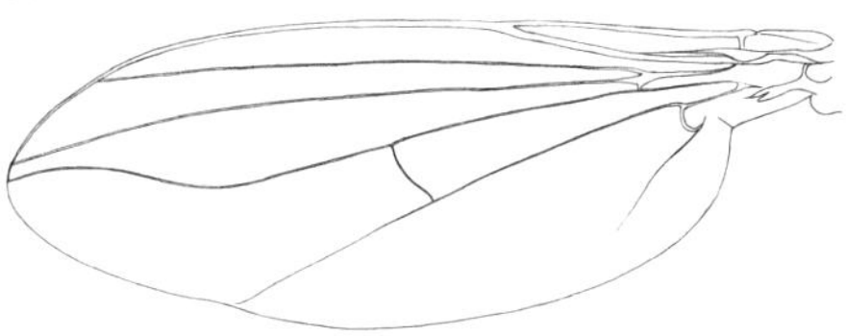

(aa)

2(1) Scape and pedicel black. Apical section of M_4 about twice length of dm-m; apical section of M_1 more gently sinuous (a). Hind femur entirely yellow............***pallipes***

– Scape and pedicel brownish yellow. Apical section of M_4 about 2.5–3 times length of dm-m; apical section of M_1 with more abrupt inflexions (fig. a of couplet 1). Hind femur slightly dusky at tip............3

(a)

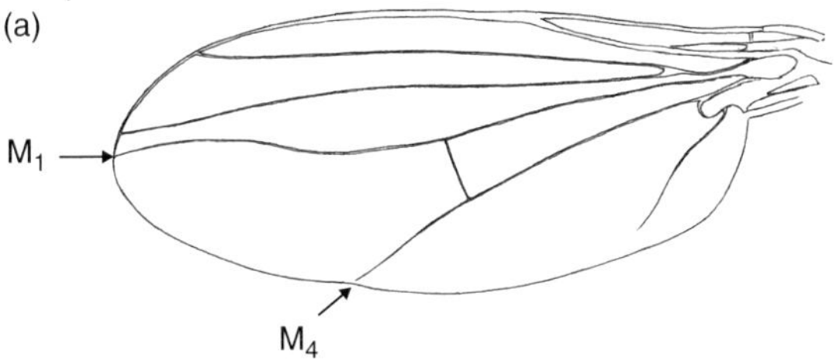

3(2) Postpedicel broadly yellow basally. Mid tibia with setulae on shaft beyond midway in addition to a strong pair on basal half (a)............***scholtzii***

– Postpedicel entirely black or at most slightly red at base. Mid tibia with only a pair of strong setae on basal half............***mallochi***

(a)

4(1) Scape and pedicel yellow............***tener***

– Antenna entirely black............5

5(4) Mid tibia dorsally with 4 strong setae, 2 at basal quarter and 2 on apical half (a). Hind femur at tip with only a dark dorsal streak, yellow ventrally. Arista black, 1.5 times as long as whole antenna (b). Apical section of M_4 at least 2.5 times longer than dm-m............................ ***leucurus***

– Mid tibia dorsally with only 2 long setae on basal half and very weak or no seta on apical half (as fig. a of couplet 3). Hind femur at tip broadly black in a complete ring. Arista pale, only slightly longer than whole antenna (about 1.25 times) (bb). Apical section of M_4 twice as long as dm-m...***bipartitus***

(a)

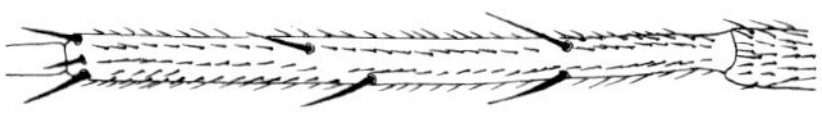

(b)

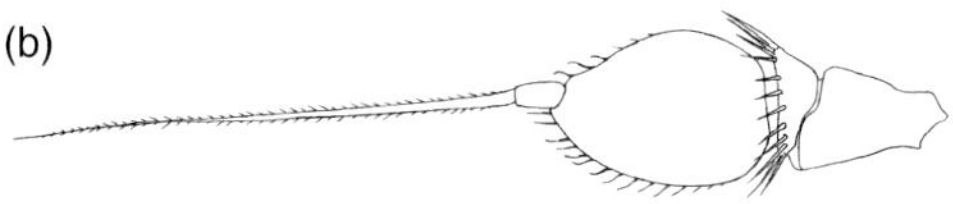

(bb)

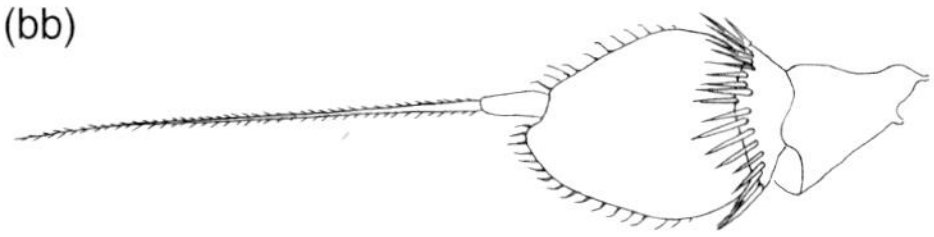

Systenus bipartitus (Loew, 1850) (Plate 8)

Males are distinguished by veins R_{4+5} and M_1 being almost parallel beyond dm-m, ribbon-shaped cerci that extend well beyond the other appendages, and the conspicuous black sharp hook on the epandrial lobe. Females resemble those of *leucurus* in having a glossy violet clypeus and similarly shaped entirely black antennae. The hind femur of *bipartitus* is strongly black on the apical quarter, the black colour extending around all surfaces rather than being pale on the ventral surface as in most other female *Systenus* whose femurs are slightly darkened at the tip; this may fade in specimens stored in alcohol. Wing length ♂: 2.9 (2.6–3.0) mm, ♀: 3.4 (2.9–3.7) mm.

Systenus bipartitus is widely distributed in England, reaching Angus in Scotland and apparently absent from Wales. It has been recorded mainly from deciduous woodlands and occasionally from wood-pasture or parks, and reared from sap-runs and rot-hole debris of ash, elm, horse chestnut and sycamore. Flight: vi–vii.

Systenus leucurus Loew, 1859

The very pale hypopygium and adjacent sternite 8 (appearing to be the last 'tergite') with entirely white hairs make this species unmistakable. Both sexes differ from most other species in having a pair of strong setae on the apical half of the mid tibia as well as a pair on the basal half, although female *pallipes* sometimes have this arrangement. Both sexes have entirely black antennae, as in *bipartitus*. Wing length ♂: 3.1 (2.9–3.3) mm, ♀: 3.6 (3.1–3.7) mm.

Systenus leucurus is widely distributed in England north to Yorkshire and Wales but apparently absent from Scotland. A large proportion of records relate to wood-pasture in old deer parks and ancient woodlands. It has been reared from rot-hole debris and sap-runs of beech, elm, horse chestnut, oak, poplar and sycamore. Flight: v–vii.

Systenus mallochi MacGowan, 1997

Males are readily separated from the similar *scholtzii* by the small wing spot and absence of shading at the tip of cell r_{2+3}. Females closely resemble those of *scholtzii*; whether the setulae on the mid tibia are always absent may need confirming as they are present in males. Wing length ♂: 3.8 (3.5–4.2) mm, ♀: 4.2 mm.

This is a sparsely recorded species with records from the eastern half of Britain from the Thames basin to Moray. Sites include deciduous woodlands and ancient parks, and the species has been reared from wet or water-filled rot-holes in ash, birch, oak and sycamore, and from elm damaged by goat moth *Cossus* (MacGowan 1997, Alexander *et al.* 2008). Flight: vi–viii.

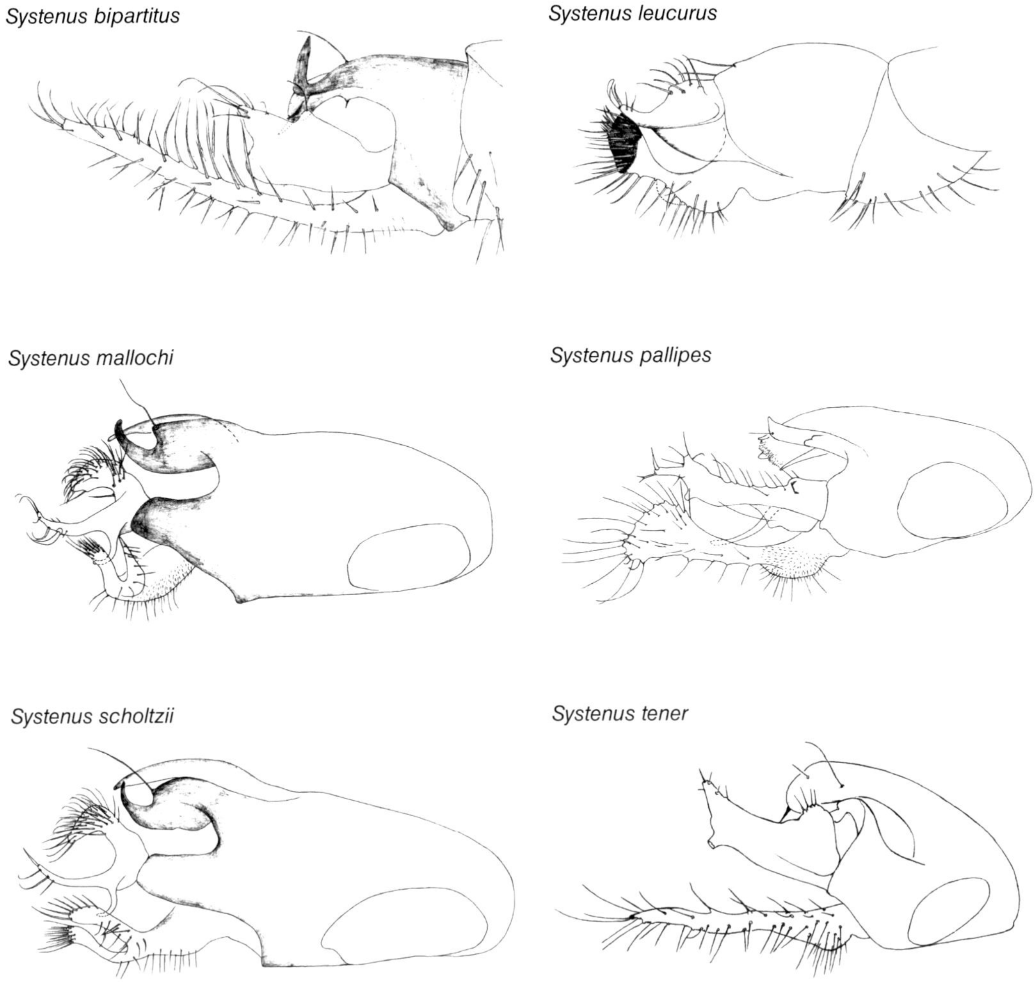

Fig. 38. *Systenus* hypopygia.

Systenus pallipes (von Roser, 1840)

Both sexes have venation similar to that of *scholtzii* and *mallochi* but males are distinguished by the lack of a spot and simple profile at the wing tip. Females are told by the shorter apical section of M_4 and by the more gently sinuous shape of M_1 compared to the more abrupt bends following the straight first half of this vein beyond dm-m in *scholtzii* and *mallochi*. Males have a particularly elongate postpedicel and a relatively long arista that has markedly longer pubescence than in other species. In females, the clypeus is often dusted although the shining violet ground colour shows through. Wing length ♂: 3.5 (3.2–4.0) mm, ♀: 3.6 (3.2–4.0) mm.

Systenus pallipes is widely distributed in England and present in Wales and southern Scotland, reaching northwards to Strathspey. Records are mainly from deciduous woodlands, wood pastures and parks but also willow and alder carr woodlands. It has been reared from rot-hole debris, sap-runs and under bark of beech, elm, horse chestnut, oak, poplar and sycamore; in Ireland, from birch (Speight 1987). Flight: v–ix.

Systenus scholtzii (Loew, 1850)

This is the largest British *Systenus* and is readily identified in the male by its large apical wing spot and grey shade at the tip of cell r_{2+3}, and by the continuously curved apical section of M_4 which becomes almost parallel to the wing margin at the vein's tip rather than ending at a distinct angle to the margin as in *mallochi*. Separating females from those of *mallochi* using the antennal colour characters used in the key works for material whose colours have not faded, but fading often occurs in specimens stored in alcohol. Wing length ♂: 4.1 (3.9–4.3) mm, ♀: 4.3 (3.9–4.7) mm.

Systenus scholtzii is widely distributed in England north to Cumbria and Wales but apparently absent from Scotland. Deciduous woodland is the main habitat, with fewer records from wood-pasture and parkland. It has been reared from ash, beech, elm, horse chestnut, oak and sycamore in Britain and from birch in Slovakia (Oboňa *et al.* 2012) and Finland (Kahanpää and Grichanov 2006). Flight: v–viii.

***Systenus tener* Loew, 1859**

This is the smallest British *Systenus*, characterised in both sexes by the combination of yellow basal antennal segments, a dark ring at the tip of the hind femur and veins R_{4+5} and M_1 being almost parallel beyond dm-m. The cercus is ribbon-shaped and extends beyond the other appendages, as in *bipartitus*, in contrast to the short cercus of other species. Females resemble *bipartitus*, differing in the colour of the antenna, the arista which is black and metepimeron which is yellow on its upper half compared to mostly black in *bipartitus*. Its postpedicel is short, resembling that of *bipartitus* rather than *leucurus*. Wing length ♂: 3.0 (2.8–3.2) mm, ♀: 3.1 (3.1–3.2) mm.

This is probably a genuinely uncommon species with a distribution limited to southern England between Kent and Herefordshire. It has been recorded from deciduous woodlands, parklands and an orchard, and has been reared from rot-holes in apple and beech in Britain, and from maple in Slovakia (Oboňa *et al.* 2012). Flight: vi–viii.

TACHYTRECHUS Haliday in Walker, 1851 (Dolichopodinae)

Tachytrechus is a distinctive genus of moderately large, robust and deep-bodied flies (Plate 8). The genus has the characters of the subfamily Dolichopodinae (scape hairy, arista dorsal, acrostichal setae biserial) and is distinguished by the shape of the face which is narrowest at its middle then widens to the mouth edge, and in having several strong anterodorsal setae on the hind femur rather than just one, as also found in a few *Dolichopus*. Veins R_{4+5} and M_1 end close together at their tips, crossvein dm-m is bordered by shaded membrane and often there is a small dark patch at the wing boss on M_1 so that at low magnification the wing appears to have one or two small spots.

There are 20 species of *Tachytrechus* in the Western Palaearctic, four in Britain. Several others may occur in Britain, all of which may be distinguished by the male genitalia. *T. ammobates* (Haliday, 1851) resembles *ripicola* in the shape of its front tarsus and triangular cercus but differs in having a white face (greenish yellow in *ripicola*) and the scape being yellow only ventrally (entirely yellow in *ripicola*). *T. ocior* Loew, 1869 superficially resembles *notatus* but the front tarsus is dorsoventrally flattened and has only a few very weak setae (strong setae on a cylindrical tibia in *notatus*). *T. genualis* Loew, 1857 also resembles *notatus* but has entirely black antennae, without even the tip of the scape being yellow as in *notatus* ssp. *obscuripes* Gerstäcker, 1864, and has a yellowish (not white) face.

Adults are often associated with bare mud, sand or peat by pools and seepage.

Key to species of *Tachytrechus* (males)

1 Front basitarsus only slightly thinner than tarsomeres 2–5, and widening gradually from middle to tip (a). Mid femur black with at most the apical quarter yellow (b). Scape black dorsally. Face silver. .... ***notatus***

– Front basitarsus far narrower than tarsomeres 2–5 (including hairs), and equally slender for most of its length. Mid femur yellow on at least the apical half. Scape yellow dorsally. Face silver, yellow or gold. 2

(a)

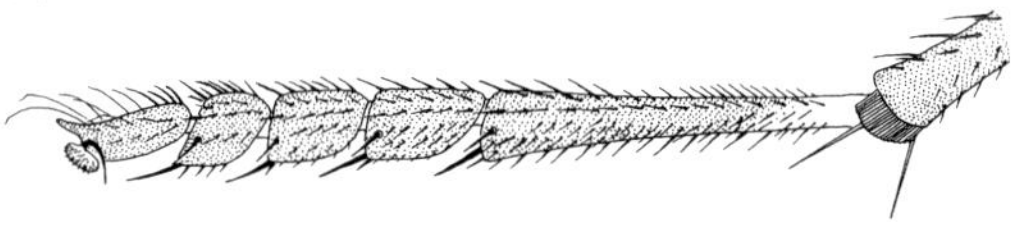

(b)

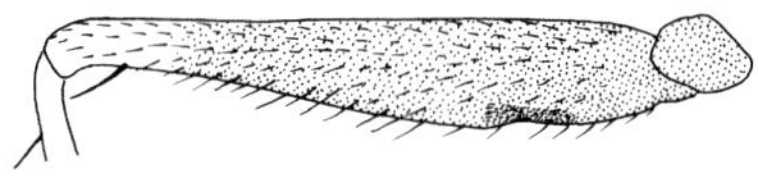

2 Front tarsomeres 3 and 4 deeper than long, their dorsal hairs far shorter than depth of segments (a). Mid femur black on basal half, yellow on apical half (b). Pedicel entirely yellow. Face silver and projecting well below lower eye margin (c). Crossvein dm-m conspicuously oblique. .... ***consobrinus***

– Front tarsomeres 3 and 4 not deeper than long, square to rectangular, their dorsal hairs at least as long as depth of segments (figs in next couplet). Mid femur mainly yellow, black only at base on dorsal surface but black to middle of ventral surface (bb). Pedicel dark dorsally. Face gold or pale yellow, not projecting beyond lower eye margin. Crossvein dm-m approximately at right angles to long veins..................... 3

(a)

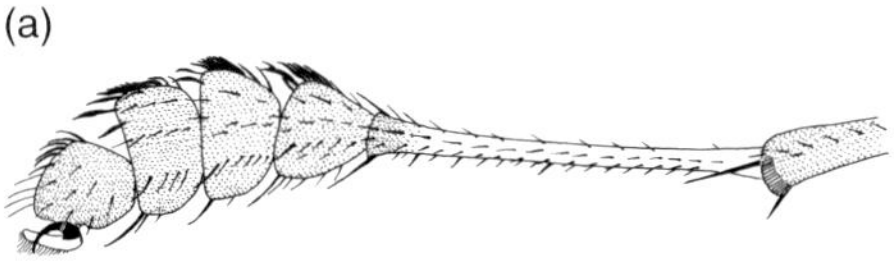

(b)

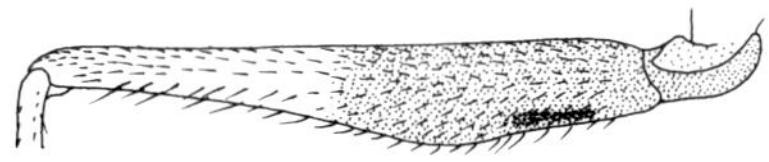

(c)

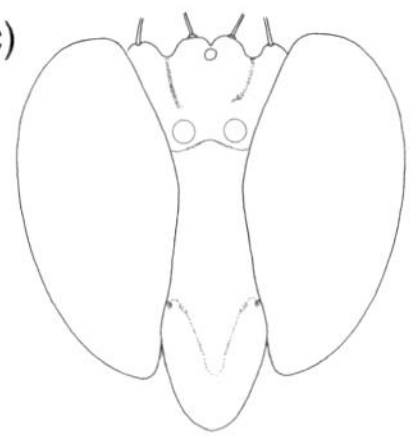

(bb)

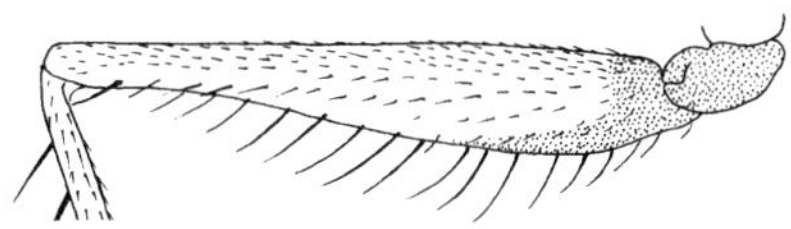

3 Front basitarsus almost twice as long as tarsomeres 2–5 combined, very slender (a). Cercus outer face with hairs shorter than those on margin (b)..................... ***insignis***

– Front basitarsus about 1.4 times as long as tarsomeres 2–5 combined (aa). Cercus outer face with hairs conspicuously longer than those on margin (bb)..................... ***ripicola***

(a)

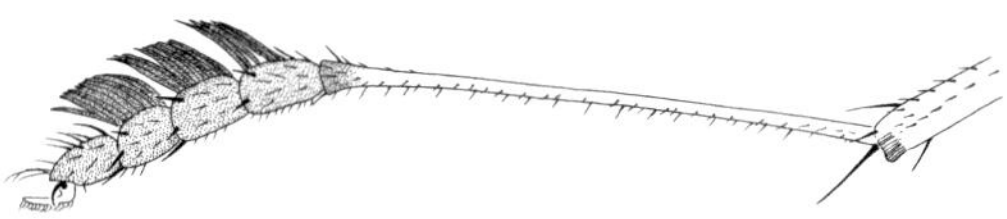

(b)

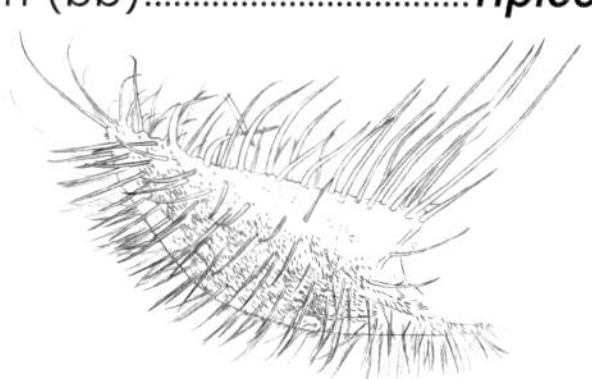

(aa)

(bb)

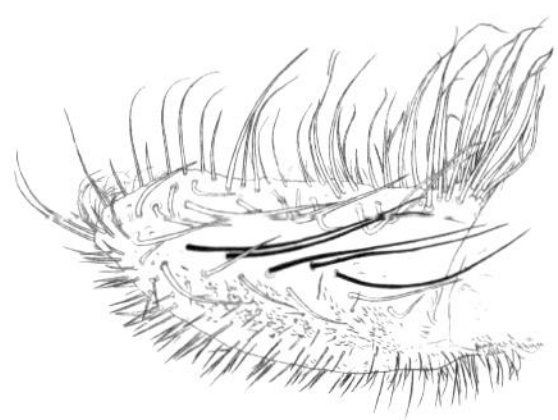

Key to species of *Tachytrechus* (females)

1 Mid femur with short ventral hairs, not longer than about half greatest depth of femur (a). Costa at base between h and R_1 conspicuously thickened, obviously wider than cell membrane between the bulge and vein Sc (b). Mid femur with 3–4 anterodorsal preapical setae. Front tibia darkened towards tip. Hind basitarsus dark.......2

– Mid femur with long ventral hairs, especially anteroventrally, about equal to or longer than greatest depth of femur (aa). Costa at base not markedly thick, not wider (usually narrower) than cell membrane behind thickest section and vein Sc (bb). Mid femur with 1–2 anterodorsal preapical setae. Front tibia almost entirely yellow. Hind basitarsus mainly or entirely yellow.......3

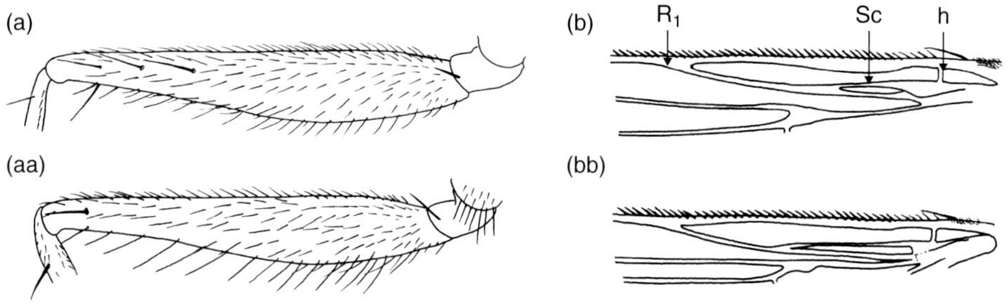

2 Face dusted yellowish brown. Clypeus ending below bottom of eye, usually with slightly pointed lower margin (a). Scape and pedicel entirely clear yellow. Front basitarsus slightly shorter than tarsomeres 2–5 combined, often hardly wider at tip than at base (b). ....***consobrinus***

– Face silver-dusted. Clypeus level with bottom of eye, with gently rounded margin (aa). Scape mainly black or sometimes dark only at base, pedicel partly darkish yellow. Front basitarsus as long as tarsomeres 2–5 combined, slightly wider at tip than at base (sometimes resembling that of males) (bb).......***notatus***

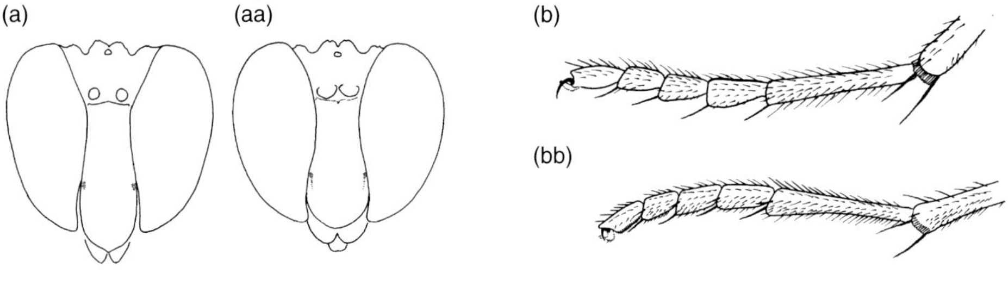

3 Postpronotal lobe with both setae about equal in length (range 0.85–1.2). Front tibia slightly shorter than tarsus (range 0.85–0.95) (needing measurement, not an estimate). ....***insignis***

– Postpronotal lobe with lower seta slightly shorter, about 0.7–0.9 times length of upper seta. Front tibia slightly longer than tarsus (range 1.0–1.1).......***ripicola***

Tachytrechus consobrinus (Haliday in Walker, 1851) (Plate 8)

The deep tarsal segments of the front leg contrast strongly with the narrow basitarsus but the dorsal fringe is short relative to these expanded tarsomeres. In both sexes, the scape and pedicel are entirely yellow whereas in other species one or both of these segments are partly or entirely black, the face is particularly long and, as in *notatus*, the costa is thickened in the section parallel to vein R_1. The male's face is silver-dusted, as in *notatus*, whereas the female's is yellowish brown rather than

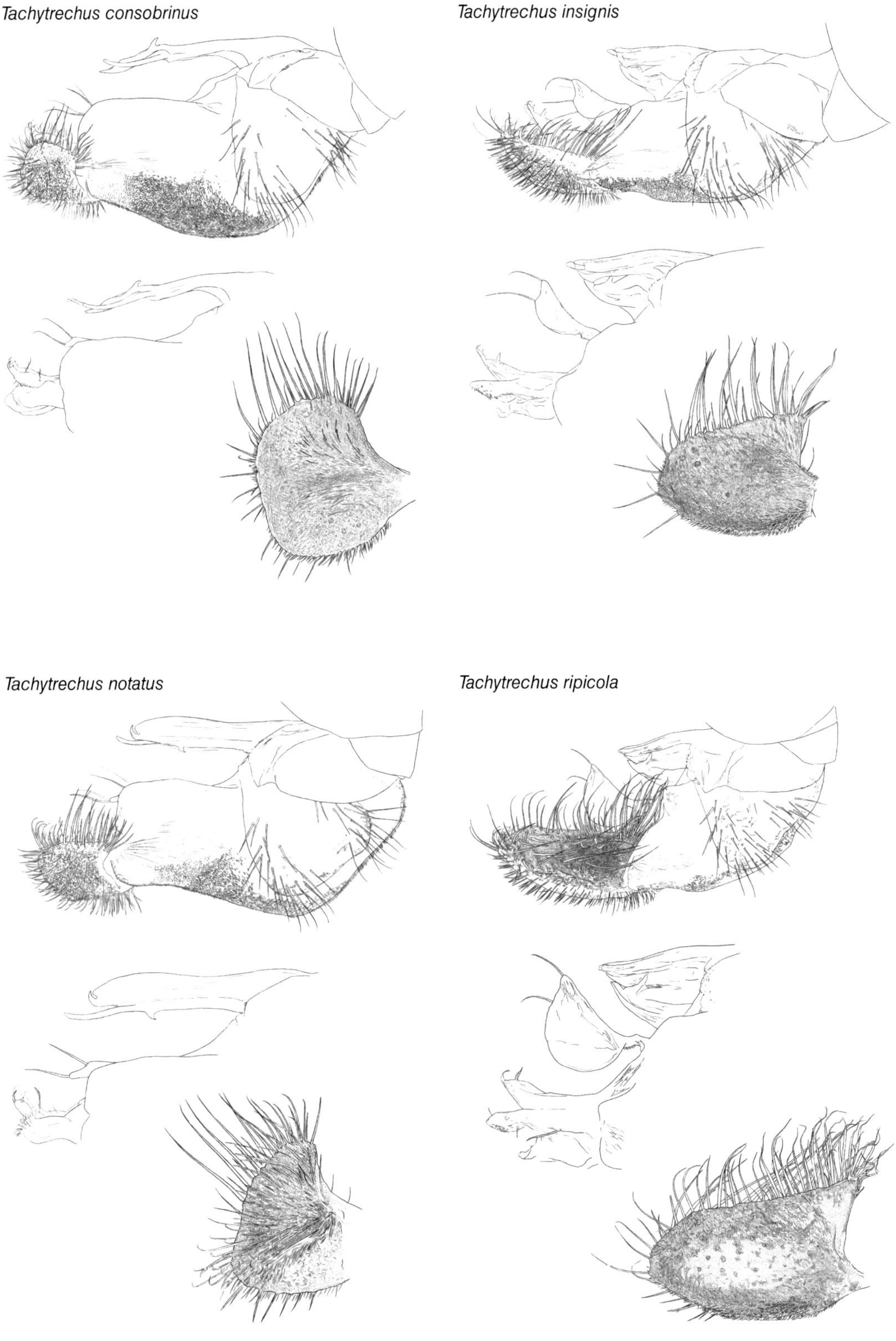

Fig. 39. *Tachytrechus* hypopygia, epandrial lobes and cercus.

white as in the other three species. The male hypandrium has a conspicuous thorn placed well before its tip. Wing length ♂: 4.6 (4.3–4.9) mm, ♀: 4.8 (4.3–5.6) mm.

Tachytrechus consobrinus is sparsely distributed in Britain except in the southwest from Hampshire to Devon. It lives almost exclusively by pools and seepages of acid mires. Flight: v–ix.

Tachytrechus insignis (Stannius, 1831)

This species and *ripicola* have a conspicuous dorsal fringe on the front tarsus but *insignis* is distinguished by its long slender front basitarsus, and by the cercus with its short hairs on the outer face. Females are easily separated from *consobrinus* and *notatus* but are almost inseparable from *ripicola* – see that account for details. The hypopygium is very similar to that of *ripicola* although, apart from the cercus, it differs in the curious apicoventral epandrial lobe being less expanded. Wing length ♂: 4.3 (3.9–4.6) mm, ♀: 4.9 (4.1–5.7) mm.

Tachytrechus insignis is sparsely distributed in Britain with most records from southeast Britain with a concentration in eastern counties. Most western records are coastal. Although it has been found in Sutherland and Moray, there are few records for Scotland, and Verrall (1904) noted that these late-season Scottish specimens differed from southern ones. It is closely associated with sand and gravel sites such as dunes, coastal gravels, gravel pits, quarries and heaths, where it often sits on wet sand by pools. Flight: v–x.

Tachytrechus notatus (Stannius, 1831)

The front tarsus is only moderately deepened and no segment is outstanding although the fifth tarsomere is white when viewed from above. The costa is thickened on the section parallel to vein R_1 and the face is silver-dusted, as in *consobrinus*. Females are distinguished by their scape being dark above compared to yellow in all other species, and the femora are usually more extensively black although this character is unreliable on its own. The male hypandrium has a tiny hook at its tip. Wing length ♂: 4.7 (4.5–4.9) mm, ♀: 5.2 (4.7–5.7) mm.

Tachytrechus notatus is the most frequently recorded of the genus with a distribution extending from Cornwall to Sutherland, and is as frequent in the north of Scotland as in the south of England. It occupies two distinct habitats: seepages on coastal cliffs, particularly those on soft rocks, and occasionally saltmarsh, and mires which are usually bogs but sometimes fen. In both types of habitat it is found mainly on pools, even if these are very small. Inland it also occurs at gravelly river shores and quarry pools which may have similar hydrological properties to those of soft-rock cliffs. Flight: iv–ix.

Tachytrechus ripicola Loew, 1857

This species closely resembles *insignis* from which it can be separated by the shape of the front tarsus and the exceptionally long hairs on the outer face of the cercus. It may sometimes be distinguished by having the long ventral hairs on the mid femora extending slightly further towards the tip than in *insignis*, some hairs being clearly beyond the yellow ventral area, although this character is not consistently reliable. Females closely resemble those of *insignis*. The mid and hind femora are more extensively black but not consistently. The measurements in the key are based on a small sample available to d'Assis-Fonseca (1978) whose key is used here. There is more variation and overlap than his key suggests and a reliable identification may not be possible; it fails with some continental specimens. It is essential to measure the length of leg segments as the tibia of *insignis* always looks longer even when it its measured length is shorter. Wing length ♂: 4.7 (4.0–4.9) mm, ♀: 5.2 (4.7–5.5) mm.

This is a rare species in Britain, recorded in the first half of the 20th century from dunes on the south and west Welsh coast but not subsequently, and from the Poole harbour area in Dorset where it was found again in 2014. It has been recorded only from dunes systems where is has been found beside pools, a lagoon and mud by a river. Flight: v–ix.

TELMATURGUS Mik, 1874 (Sympycninae)

There is only one Palaearctic species in the genus which is more speciose in the Afrotropical and Oriental realms (Grichanov 2011). In the British fauna the genus is recognised by the combination of uniserial acrostichal setae, three to four dorsocentral setae, no proepisternal seta and the arista placed close to the base of the postpedicel. More obviously, it is a tiny species with a violet-blue scutum and frons, the hind femur black on about the apical third to half, most of the hind tibia usually dark, and the postpedicel yellow basally (Plate 8). Among similarly small flies, a violet scutum is found in *Acropsilus*, *Chrysotus laesus* and *C. suavis* which differ from *tumidulus* in having almost entirely dark femora. Males have the unique combination of a swollen tip on the arista (found elsewhere only in *Rhaphium* in the British fauna), a single long ventral seta at the basal third of the hind femur (Fig. 40) and a highly modified front basitarsus. Females have a broad violet shiny face with a strongly protruding clypeus resembling that of *Syntormon*.

Telmaturgus tumidulus **(Raddatz, 1873) (Plate 8)**

Despite its small size there should be no problem recognising this distinctive species. Wing length ♂: 1.9 (1.8–2.0) mm, ♀: 2.2 (2.1–2.4) mm.

Telmaturgus tumidulus is a rare species recorded mainly in two clusters in Norfolk plus Suffolk, where the most extensive populations occur in the Norfolk fens, and Hampshire plus Dorset, with isolated occurrences in Devon and Gwynedd. It is found in wet mires, both acid bogs and base-rich fens, which are unshaded although may have tall reed. Flight: v–viii, mainly vi.

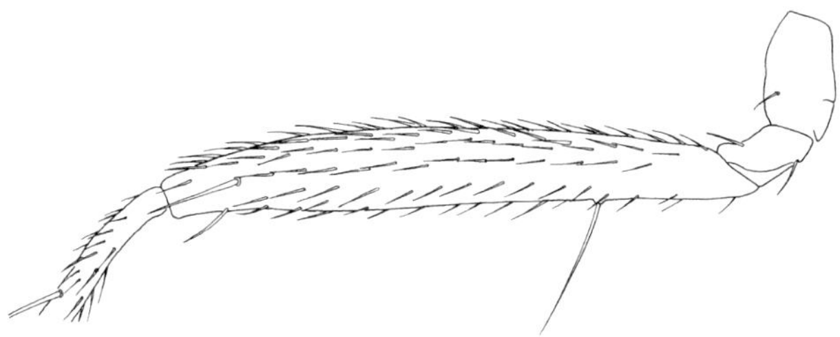

Fig. 40. *Telmaturgus tumidulus* male hind femur.

TEUCHOPHORUS Loew, 1857 (Sympycninae)

Teuchophorus are small to tiny dark metallic green flies with mainly yellow legs (Plate 8). The genus, along with *Sympycnus*, is recognised by the combination of small size, with a wing length less than 3 mm, a triangular postpedicel with a dorsal arista, the face tapering to the mouth edge, six (rarely five) dorsocentral setae, no proepisternal seta and the mid femur having an anterior preapical seta. Males are recognised by the thick black costa on the basal half of the wing and the ornamented hind tibia (Fig. 41), although these characters are less obvious in *simplex*. Both sexes have an oblique crossvein dm-m, immediately beyond which M_1 bows forward slightly for a short length.

Males can be identified using the ornamentation of the hind tibia alone (Fig. 41) but females are difficult to separate. The key to females requires careful measurement of some leg segments, and, as these measurements often overlap, two sets of values are presented to overcome some of the variation (Table 2). The key by d'Assis-Fonseca (1978) fails more often than the key given here, particularly as the first couplet is wrong. Additional clues to separate species are the habitat, as these wetland flies are fairly specific in their preferences, and in the flight period, which peaks later in *simplex*.

Five of the ten European species of *Teuchophorus* occur in Britain and none of the remaining species is likely to be found here. Grichanov *et al.* (2012) provide an illustrated key to the Palaearctic fauna.

Key to species of *Teuchophorus* (males)

1 Mid femur and tibia without conspicuous long ventral setae. Hind tibia simple, cylindrical, with a posteroventral fringe of equal-sized setae along most of its length (Fig. 41)..... ***simplex***

– Mid femur and tibia each with 2–4 long ventral setae (aa). Hind tibia with conspicuously modified shape and setae...2

(aa)

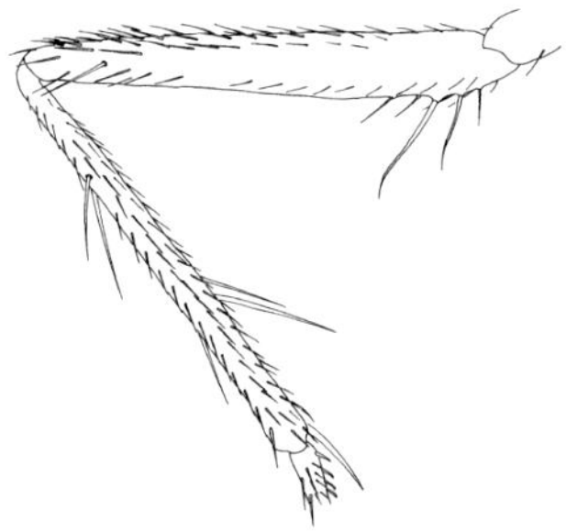

2(1) Hind tibia with modified setae arising at or just beyond halfway. Costal thickening large, about two-thirds length of cell r_1 and twice width of r_1 at base of thickening (a). Front basitarsus with several ventral setae as long as or longer than width of shaft (figs in next couplet)...................3

– Hind tibia with modified setae arising on thickened apical third. Costal thickening smaller, about half length of cell r_1 and only slightly wider than r_1 (aa). Front basitarsus with ventral setae not longer than width of shaft (bb)...4

(a)

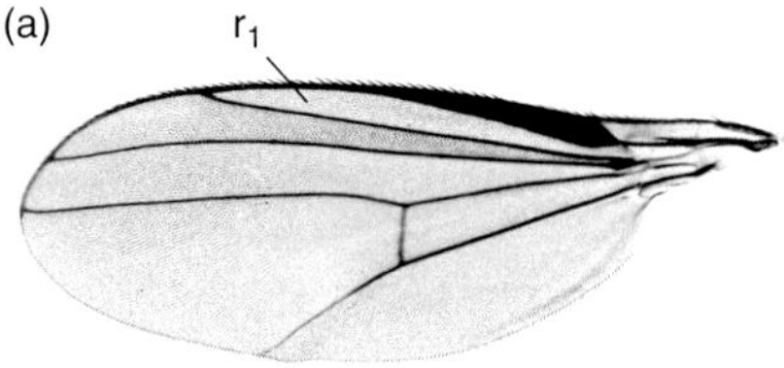

(aa)

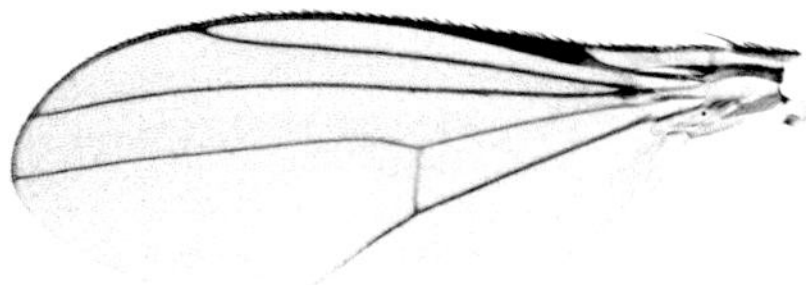

(bb)

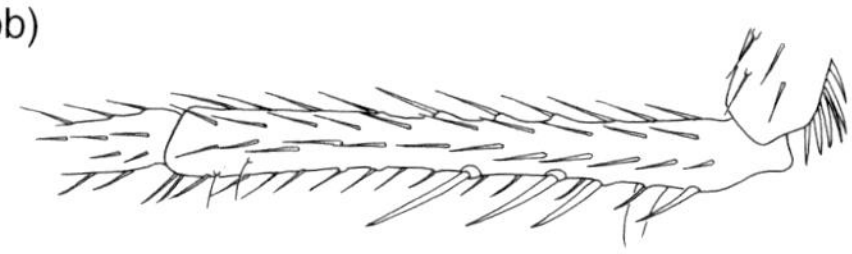

3(2) Hind tibia with ventral modification consisting of a fan of flat black setae on a short stem, and 4 long fine setae beyond it (Fig. 41). Front basitarsus, in addition to long basal hairs, with an apical ventral fringe of short curled hairs on the slightly widened tip (a)............ ***calcaratus***

– Hind tibia with ventral modification consisting of a dense cluster of blunt-ended black setae and a longer pale feathered seta at apical end of cluster; only 2 long fine setae between cluster and tip of tibia (Fig. 41). Front basitarsus with scarcely discernible apical fringe and tip not swollen (aa)... ***nigricosta***

(a)

(aa)

4(2) Hind tibia, on swollen section, with the strongest seta pointing apically (Fig. 41). Frons shining metallic violet.. ***spinigerellus***

– Hind tibia, on swollen section, with the strongest seta pointing basally and only a weak seta pointing apically (Fig. 41). Frons green.. ***monacanthus***

Key to species of *Teuchophorus* (females)

1 Frons vivid metallic violet, almost glossy, dusting restricted to a narrow anterior strip. Other characters as in first half of couplet 2; dorsal fringe of hind tibia with 2 rows of equally long hairs. Ratios of hind leg tibia : basitarsus and basitarsus : tarsomere 2 as *monacanthus*. .... ***spinigerellus***

– Frons green or greenish blue, but not steely purple-blue, either shining or extensively dusted. 2

2(1) Hind femur on basal half with a dorsal fringe of longer hairs, as long as width of tibia (not femur), grading to shorter apically (a). Hind basitarsus 0.7–1.1 times length of tarsomere 2 but rarely more than 1.0. Hind tibia 3.6-4.7 times longer than basitarsus (rarely less). Hind margin of wing behind vein CuA + CuP with no shadow vein parallel to wing margin, or one that is shorter than twice length of this vein........ 3

– Hind femur with a scarcely differentiated dorsal fringe, hairs much shorter than width of tibia (aa). Hind basitarsus 0.9–1.1 times length of tarsomere 2 but rarely less than 1.0. Hind tibia 3.0–3.7 times length of basitarsus (rarely more). Wing membrane behind CuA + CuP with a faint shadow vein running in a curve parallel and very close to margin........ 4

(a)

(aa)

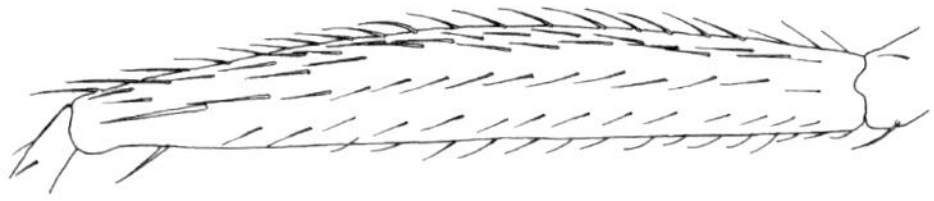

3(2) Slightly darker species: mid coxa and sometimes hind coxa darkened, at least basally, metepimeron dark, hind femur with an extensive dark smudge dorsally, hind tarsus mainly dark. Mid coxa with setae and hairs slightly browner. Hind tibia 3.6–4.1 times length of basitarsus; front basitarsus 2.1–2.9 times longer than tarsomere 2. Frons extensively dusted, leaving separated small shining lateral areas, as *nigricosta* below....... ***monacanthus***

– Paler species: mid and hind coxae, hind femur and metepimeron clear yellow, hind tarsus mainly yellow. Mid coxa with setae and hairs pale yellow. Hind tibia 4.1–4.7 times longer basitarsus; front basitarsus 1.8–2.3 times longer than tarsomere 2. Frons mainly shining, dusting restricted to narrow anterior strip, as *calcaratus* below................ ***simplex***

4(2) Hind femur with pale ventral hairs almost as long as dark dorsal hairs, easily discernible (a). Frons subshining and scarcely dusted either side of ocellar triangle and for half distance between front ocellus and antennae, dense dust restricted to a strip just behind antennae (b), better seen in oblique lateral view. Front basitarsus with ventral hairs usually as long as width of shaft, but sometimes shorter........ ***calcaratus***

– Hind femur with pale ventral hairs clearly shorter than dorsal dark hairs, and very fine (aa). Frons dusted from front ocellus to antennae, but subshining either side of ocellar triangle (bb). Front basitarsus with ventral hairs usually shorter than width of shaft................ ***nigricosta***

(a)

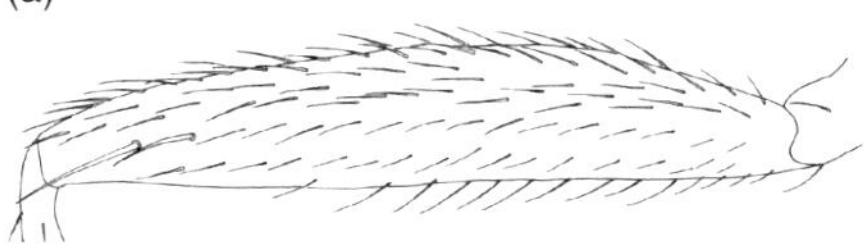

(b)

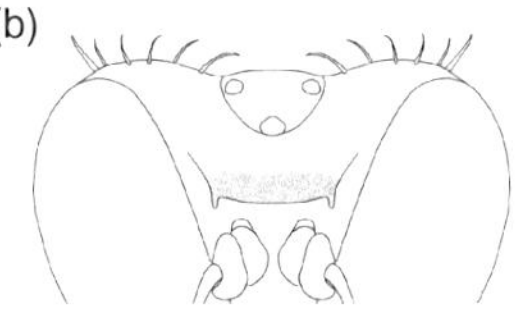

(aa)

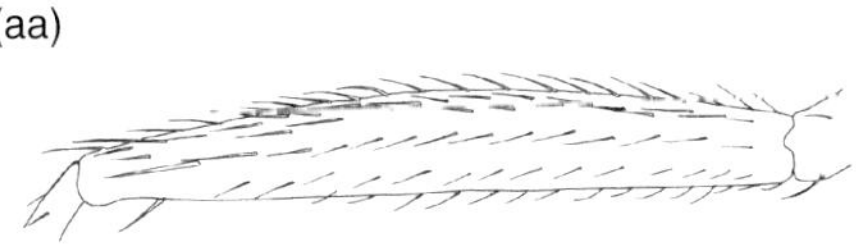

(bb)

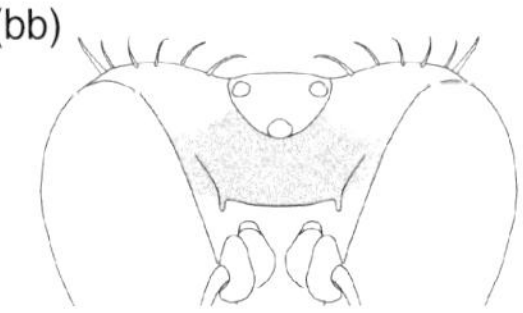

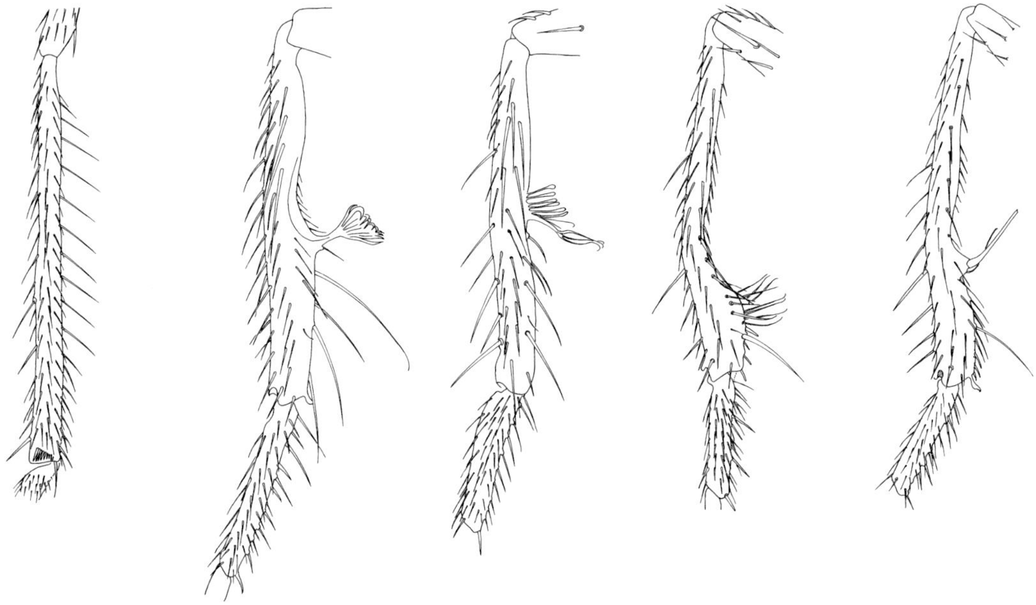

Teuchophorus simplex — *Teuchophorus calcaratus* — *Teuchophorus nigricosta* — *Teuchophorus spinigerellus* — *Teuchophorus monacanthus*

Fig. 41. *Teuchophorus* male hind tibiae, ordered as in the key.

Table 2. Ratios of leg segments of female *Teuchophorus* in the order they appear in the key.

	Front leg		Hind leg	
	tib/tar_1	tar_1/tar_2	tib/tar_1	tar_1/tar_2
spinigerellus	1.7–2.2	2.0–2.9	3.5–4.1	0.8–1.0
monacanthus	1.7–1.9	2.1–2.9	3.6–4.1	0.7–1.1
simplex	1.8–2.0	1.8–2.3	4.1–4.7	0.7–0.9
calcaratus	1.5–1.8	2.0–2.3	3.0–3.7	0.9–1.1
nigricosta	1.7–2.0	1.8–2.2	3.0–3.6	1.0–1.1

tar_1, tar_2 = tarsomeres 1 or 2; tib = tibia.

Teuchophorus calcaratus (Macquart, 1827) (Plate 8)

The ornamentation of the hind tibia and front basitarsus of males is particularly distinctive. It is a pale-legged species although the hind femora may be dark on the apical half except ventrally. Females closely resemble those of *nigricosta* in size, colour and ratios of leg segments, and there appear to be no characters other than those used in the key that help to separate them; even these overlap and mistakes will be inevitable. Wing length ♂: 2.3 (2.2–2.4) mm, ♀: 2.5 (2.4–2.6) mm.

Teuchophorus calcaratus is widely distributed in England and Wales, and reaches into southern Scotland. It has a strong association with rivers and streams, and is particularly frequent at finer exposed riverine sediments. The sites may be in shade or sunlit. Few other wetland types appear to support it. Flight: v–ix, mainly vi–vii.

Teuchophorus monacanthus **Loew, 1859**

The slightly swollen down-curved hind tibia superficially resembles that of *spinigerellus* but the basally pointing large thick seta is unmistakable. Unlike other British *Teuchophorus*, the legs are usually partly dark, particularly the hind femur and swollen part of the tibia; they have a slightly dirty appearance rather than clear yellow with contrasting dark areas. Females resemble those of *spinigerellus* in the ratios of leg segments but the frons is dusted green, not shiny purple-blue, and the mid coxa is almost always grey basally. Wing length ♂: 2.4 (2.3–2.5) mm, ♀: 2.3 (2.1–2.6) mm.

The species is widely distributed in England and Wales, and more sparsely in Scotland north to Sutherland. It is strongly associated with rivers and streams, in particular fine exposed riverine sediments, and seepages on soft-rock cliffs where the feature in common may be freely draining fine sediments; most sites are at least partly shaded. It is only rarely found in other wetlands, including fen and upper saltmarsh. Flight: v–ix, mainly vi–vii.

Teuchophorus nigricosta **(von Roser, 1840)**

This is a bright-legged species with the yellow of the coxae running onto the metepimeron in both sexes. The nearly cylindrical and thick hind male tibia distinguishes *nigricosta* from the other species with ornamentation on this segment. Females closely resemble those of *calcaratus* with which it may sometimes be confused using this key. Wing length ♂: 2.4 (2.2–2.6) mm, ♀: 2.5 (2.3–2.6) mm.

Teuchophorus nigricosta is widely distributed in England but scarce in Wales and with only a few records from central Scotland. It is found by shaded small streams and seepages, usually in deciduous woodlands but also on coastal soft-rock cliffs. It may be restricted to more base-rich geology. Flight: v–x, peak in vii.

Teuchophorus simplex **Mik, 1881**

This is a particularly bright-legged species with the yellow of the coxae usually extending onto the base of the metepimeron and meron above the mid coxa. Without looking closely, the legs appear unmodified but do have inconspicuous evenly spaced posteroventral setae on the hind tibia. Females are usually clearly distinct from the other British species by the combination of the fringe on the hind femur, bright yellow legs and the exceptionally short hind basitarsus which makes its tibia usually well over 4 times longer than it (up to about 4 times in *spinigerellus* and *monacanthus*), and the second tarsomere very clearly longer compared to usually shorter in the other two. Wing length ♂: 2.4 (2.2–2.6) mm, ♀: 1.9 (1.7–2.2) mm.

The distribution is mainly southern England and Wales with fewer records northward to Cumbria and Yorkshire. It is found by shaded small streams, seepages and springs in deciduous woodlands, sometimes by larger rivers with shores of fine sediment and occasionally in fens with springs. It is strongly associated with base-rich geology. Flight: v–x, mainly vii–viii.

Teuchophorus spinigerellus **(Zetterstedt, 1843)**

Both sexes have a vivid steely purple-blue almost glossy frons; some *simplex* may have a blue frons but lack the purple tinge. Usually the legs are entirely yellow although sometimes the hind tibia and femur may be darkened on the apical half or more. Females resemble those of *monacanthus* but are distinguished by the frons colour and nearly entirely yellow legs. Wing length ♂: 2.0 (1.9–2.1) mm, ♀: 2.1 (2.1–2.2) mm.

Teuchophorus spinigerellus is widely distributed in England and around the Welsh coast, and more sparsely on the Scottish coast north to the Moray Firth. It is found in stands of tall unshaded and ungrazed monocotyledons, mainly reed and less often sedges, sea club-rush and other tall species. The main habitats are reedbeds, fens, ditch systems in grasslands, and upper saltmarsh. Its apparent restriction to coastal sites in parts of the country is due to the association with reedbeds. Flight: v–ix, peak in vii.

THINOPHILUS Wahlberg, 1844 (Hydrophorinae)

Thinophilus may be distinguished by the large proboscis and large bright yellow palps that are wider than the small round postpedicel with its dorsal arista, dense whiskers on the postgena, and only one notopleural seta which is unusual but also found in *Scellus* and *Schoenophilus* in the Hydrophilinae (Plate 8). The tergites have conspicuous dusted patches at the sides. Acrostichal setae are absent, which is an uncommon condition in large species of dolichopodids.

Two of the nine Western Palaearctic species occur in Britain. The genus is more speciose around the Mediterranean and Baltic Sea but it is unlikely that other species will be found in Britain. Grichanov (1997) provides a key to the species of the west and central Palaearctic.

Species of *Thinophilus* are confined to the coast or saline sites inland. The British species are found on saltmarshes and some other brackish habitats.

Key to species of *Thinophilus* (males)

1 Coxae and femora mainly black. Front and mid legs strongly modified, front tibia at apex beneath with a close-set group of 3 long curved setae (a); front basitarsus excavated beneath, tarsomeres 3 and 4 strongly bristled dorsally. Mid leg tarsomere 4 and 5 pale, flattened and fringed with long black setae. Large species, wing length 5–6 mm. .... ***flavipalpis***

– Front coxa and all femora entirely yellow. Front and mid legs simple. Small species, wing length less than 4 mm. .... ***ruficornis***

(a)

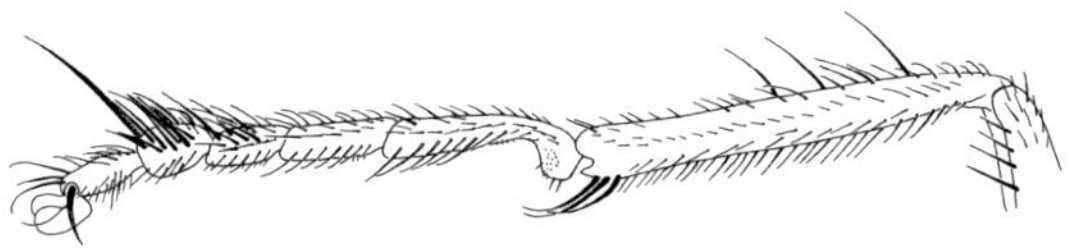

Key to species of *Thinophilus* (females)

1 Front coxa black, all hairs black with some stout and as long as greatest depth of front femur (a). Hind femur with a row of about 7–9 anterodorsal setae on apical two-thirds (b). Hind tibia with at least 5 strong ventral setae longer than width of tibia (c). Femora usually black but may be entirely brownish yellow. Large species, wing length 5–6.5 mm. .... ***flavipalpis***

– Front coxa yellow, hairs on upper and inner faces mainly short and pale (aa). Hind femur with 1–2 anterodorsal setae at apical fifth (bb). Hind tibia with only 2–3 anteroventral setae as long as width of tibia (cc). Femora always entirely yellow. Small species, wing length 3.5–4 mm. .... ***ruficornis***

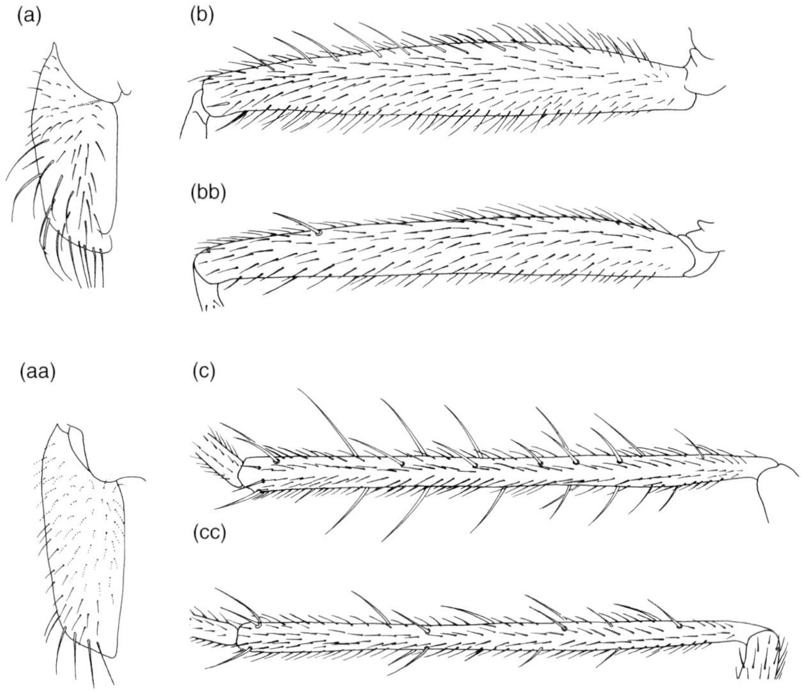

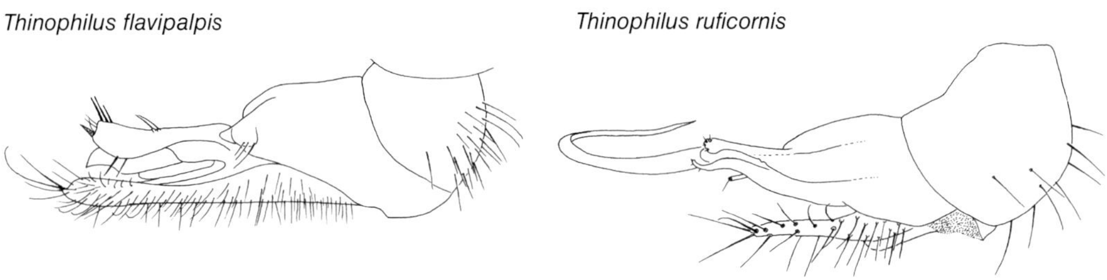

Fig. 42. *Thinophilus* hypopygia.

Thinophilus flavipalpis (Zetterstedt, 1843) (Plate 8)

This is a large robust species, dark brown with green and coppery tinges. Males are distinguished by their highly modified front and mid tarsi and front tibia. The cerci are long, slightly spatulate and end just beyond the tip of the other appendages. Females are distinguished from *ruficornis* by their strong leg setae. Their leg colour is variable so although any specimens with dark femora are *flavipalpis*, some have entirely yellow legs apart from the black front coxa. Wing length ♂: 5.5 (5.1–5.8) mm, ♀: 5.9 (5.4–6.3) mm.

Thinophilus flavipalpis is distributed around the English coast from Norfolk to the Bristol Channel, Morecombe Bay in Cumbria and Lancashire and on the Welsh coast, with long stretches with no records. It is confined to upper saltmarsh in moderately sheltered places, and also occurs on brackish lagoons and brackish dune slacks. Flight: v–ix, peak in vii.

Thinophilus ruficornis **(Haliday, 1838)**

This species is considerably smaller than *flavipalpis*. Both sexes have entirely yellow coxae and femora with rather weaker and sparser setae than in *flavipalpis*. The tergites in side view have a conspicuous pale grey-dusted triangle that obliterates the shiny cuticle at the lower margin, rather than being a parallel band to the margin in *flavipalpis*. Males have no secondary sexual characters; the hypopygium is particularly small and the cerci are narrow rods that end well before the other appendages. Wing length ♂: 3.4 (3.1–3.7) mm, ♀: 3.8 (3.4–4.1) mm.

Thinophilus ruficornis is distributed around the English and Welsh coasts from Norfolk to Morecambe Bay in Cumbria. It is found on upper saltmarsh and sometimes muddy estuaries, brackish lagoons and brackish ditches. It often occurs together with *T. flavipalpis* but it may have a slightly wider tolerance of brackish habitat above the upper saltmarsh where most are found. Flight: v–viii, peak in vii.

THRYPTICUS Gerstäcker, 1864 (Medeterinae)

Thrypticus are small to tiny bright green flies with only thin dusting on the body (Plate 8). They may be recognised by the combination of the concave occiput, a very short ocellar triangle where the ocelli lie at the corners of a very shallow isosceles triangle, distinct prescutellar depression, veins R_{4+5} and M_1 almost parallel, all the larger setae and hairs being pale, two lateral setae on the hind coxa, males with a large hypopygium slung under the abdomen and its cercus with only a short extension, and females with a blade-shaped ovipositor whose tip is usually visible. *Corindia* are very similar but differ in having a yellow scape and pedicel and different terminalia to those described here.

There are about 18 species in the Western Palaearctic of which 12 occur in Britain.

Many species of *Thrypticus* cannot be reliably identified using external characters alone. However, the hypopygia of British species are distinctive, usually in lateral view alone. The genitalia need to be exserted in pinned specimens and ideally mounted so they can be seen at a magnification of more than ×40 (the usual limit of a binocular microscope). The profile can be treated as a two-dimensional structure so that reference in the key to teeth and points are in fact artefacts of a foreshortened lateral view. Females can be difficult to identify but the tip of the ovipositor is often distinctive for most species and appears to show relatively little variation. In fresh material, the ovipositor can be pulled out to its full extent or eased out by squeezing the abdomen near its end so that at least the ovipositor's tip is visible. Drake *et al.* (2023) illustrate the ovipositors of British species.

Empidoidea in general are thought to have predatory larvae but *Thrypticus* is an exception, being herbivores mining within monocotyledons. The blade-shaped ovipositor is a defining feature of the genus and is used to insert eggs into the host plants (Bickel 1986). Reed is the host plant of *bellus* and *smaragdinus*, and rushes also of *bellus* (Dyte 1993b). However, in Britain sampling of stands of spike-rush often results in the collection of several species, including *cynicus*, *divisus* and *intercedens*, while *cuneatus* and *lichanus* have been collected where club-rush is abundant. Cordo *et al.* (2000) describe mines of South American *Thrypticus* in water hyacinth from water-level to about 10 cm above, and this may be true of *Thrypticus* in the much thinner host plants in Britain. Lübben (1908) describes the young larvae of *T. smaragdinus* moving from the rhizome tissue of reed into the young shoots where they eat their way up the stem but do not damage the growing tip; he implies that the female had oviposited near the rhizome.

Key to species of *Thrypticus* and *Corindia* (males)

1 Scutellum with several stout lateral setae in addition to a long apical pair (a). Legs almost entirely pale, including parts of coxae. Face narrow, its height more than 5 times its width at facial division (b). Proboscis and palp pale. Surstylus with numerous long hairs (c). Relatively large species, wing length 2.8–3.0 mm. .... ***smaragdinus***

– Scutellum with a pair of long apical setae but no or only 1 tiny lateral seta. Femora partly dark. Face broader, its height up to 2.5–4 times its width at facial division. Proboscis usually dark. Surstylus with a few inconspicuous hairs. Smaller species, wing length 1.4–2.5 mm.........2

(a) (b) (c)

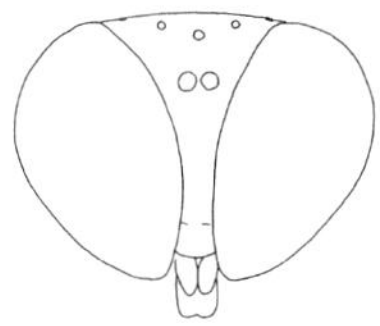

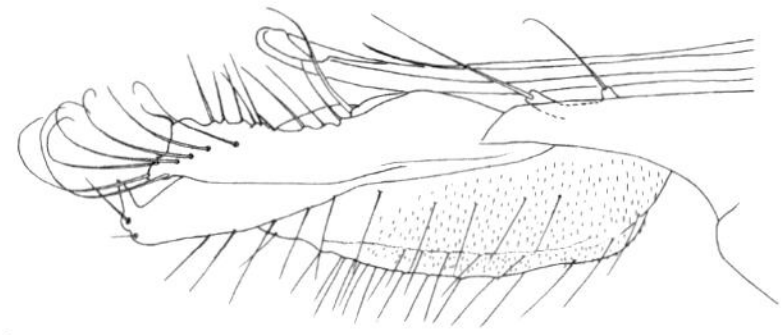

2(1) Wing widening markedly from base to well beyond midway, with a wide rounded tip and no anal lobe (a). Crossvein dm-m placed about halfway along M_1 measured from node at fork of R_{2+3} and R_{4+5}. Epistoma glossy. Front tarsus long and slender, 1.5 times longer than tibia (b). 6 dorsocentral setae, including a small one at front of row. Arista long, twice height of face. Cercus with a clearly differentiated triangular extension at tip..3

– Wing shape unremarkable, anal lobe developed. Crossvein dm-m well before middle of M_1 measured as above. Epistoma slightly dulled by micro-sculpturing. Front tarsus usually short, 1.1–1.4 times longer than tibia. 5 dorsocentral setae. Arista shorter, at most 1.6 times height of face. Cercus rarely (*divisus*) with a triangular projection at tip.....................................4

(a) (b)

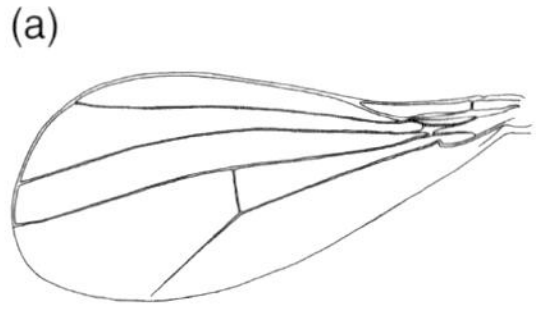

3(2) Triangular extension of cercus with its ventrally directed corner broad (uppermost in figure), terminal setae shorter than this projection (a). Surstylus tip projecting more on its dorsal corner (lower side in figure) (b)..***cuneatus***

– Triangular extension of cercus bearing a narrow finger-like ventrally directed projection, terminal setae much longer than projection (aa). Surstylus tip projecting more on its ventral corner (bb)...***lichanus***

(a) (b)

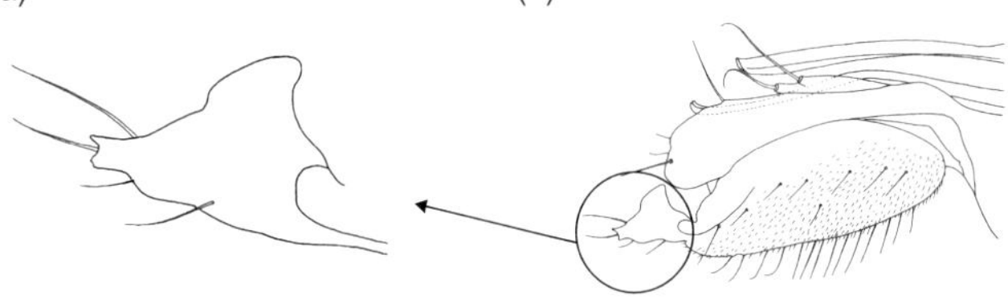

(aa) (bb)

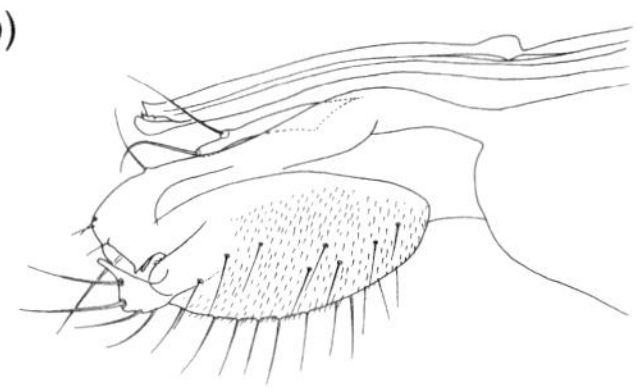

4(2) Cercus and often surstylus entirely dark, similar in shade to epandrium. Tibiae and tarsi dark. If in doubt, follow second half of couplet...5

– Cercus and surstylus mainly yellow, only tip of surstylus black. If these parts are brown but paler than epandrium, then tibiae and tarsi are pale...6

5(4) Mid tibia with a small anterodorsal seta at basal third (sometimes difficult to see) (a). Arista stout, scarcely tapering on apical half, ending abruptly as if broken (b), only slightly longer than face (about 1.25 times). Palp entirely black. Vein M_1 on apical third or more usually continuously and almost evenly curved posteriorly, R_{4+5} often curving anteriorly at tip so the two veins splay out at the costa (c) but sometimes parallel here. Hind trochanter usually obviously clearer yellow than front and often mid trochanters. Surstylus with an apical indentation forming two teeth (d). (Cerci and surstyli sometimes pale: see couplet 8.) .... ***laetus***

– Mid tibia without a seta at basal third. Arista tapering evening to fine tip, resembling the tapering of wing vein M_4, clearly longer than face (nearly 1.5 times). Palp pale around margin. Vein M_1 on apical third or more continuously straight and parallel to R_{4+5}. Hind trochanter colour usually similar in tone to front and mid trochanters. Surstylus tip gently rounded or slightly pointed (dd, variation shown).......... ***nigricauda***

(a)

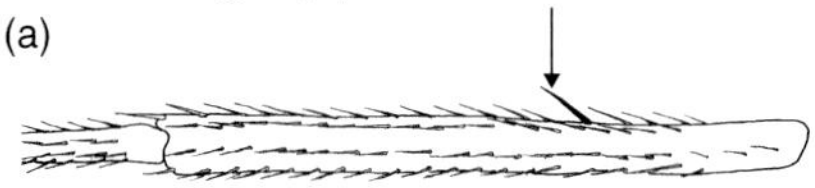

(b)

(c)

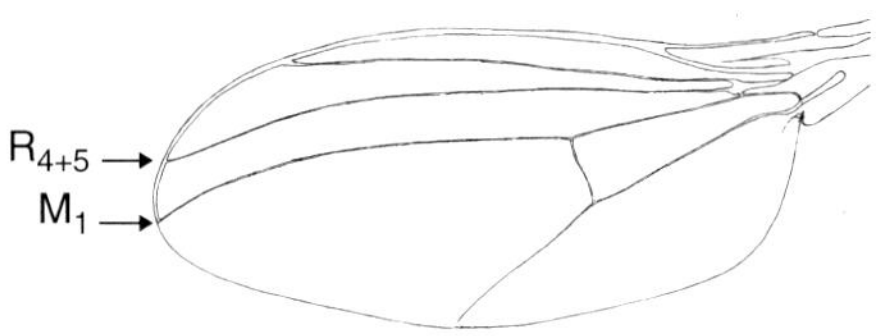

(d)

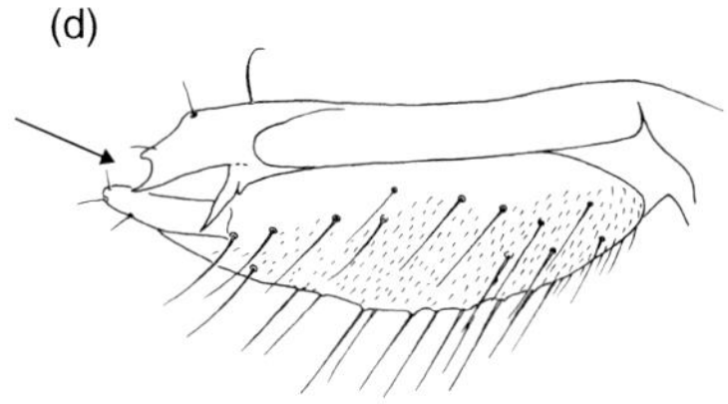

(dd)

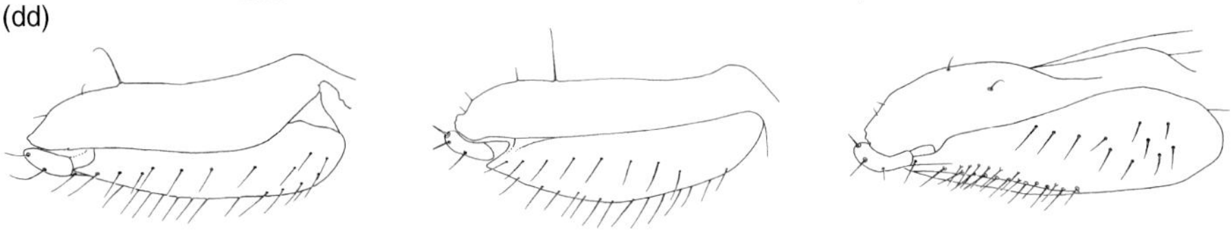

6(4) Scutellum without a tiny lateral seta. Front and mid tibiae usually mainly dark (but variable). 7

– Scutellum with a tiny lateral seta in front of apical seta (difficult to see and, if lost, the socket scar is very small) (aa). Front and mid tibiae usually mainly pale, at least ventrally (but variable).......... 10

Note: This couplet will cause trouble as the lateral setae are small so they can be difficult to see when present or to be certain that they are absent; leg colour is too variable to provide a reliable separation but is a good guide when obviously either dark or pale.

(aa)

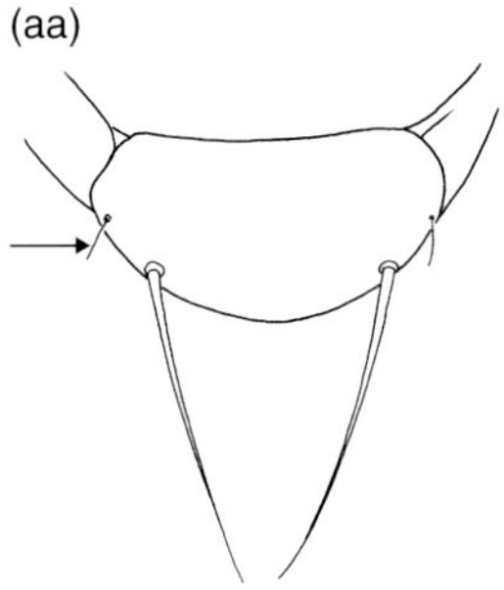

7(6) Arista stout, scarcely tapering on apical half, ending abruptly as if broken (a). See alternative couplet 7i below if arista is damaged. 8
– Arista tapering evenly to a fine tip, resembling the tapering of vein M_4 (aa). 9

(a)

(aa)

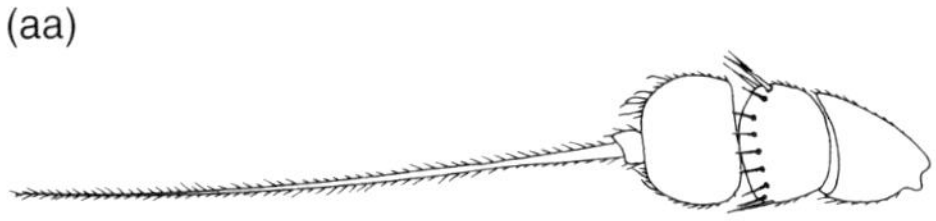

8(7) Vein M_1 on apical third or more continuously and almost evenly curved posteriorly, R_{4+5} often curving anteriorly at tip so the two veins splay out at the costa (fig. c of couplet 5). All femora with pale vestiture. Surstylus narrow in side view, not expanded at tip which has a small indentation above a pointed tooth (fig. d of couplet 5). Cercus tip simple and narrow with inconspicuous hairs (fig. d of couplet 5, *below* arrow). Arista stout and blunt-ended but pubescence scarcely visible (fig. a of couplet 7) .... ***laetus***
– Vein M_1 imperceptibly bowed on apical half (aa). All femora with dark vestiture except ventrally. Surstylus tip broad and truncated with several small teeth (bb). Cercus tip with a small appendage with a finger-like projection and hairs as long as appendage (bb). Arista markedly thick with discernible pubescence (cc). .... ***divisus***

(aa)

(bb)

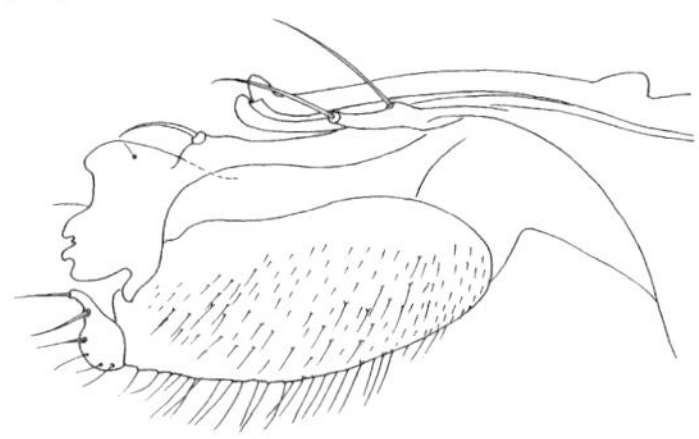

(cc)

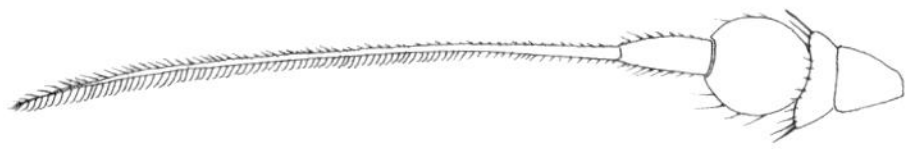

Alternative to couplets 7 and 8 if arista is damaged:

7i(6) All femora with dark vestiture except ventrally. Surstylus tip broad and truncated with several small teeth (fig. bb of couplet 8). Cercus tip with a small appendage with a finger-like projection and hairs as long as appendage (fig. bb of couplet 8) .... ***divisus***
– All femora with pale vestiture. Surstylus tip sloping to an acute or blunt tip. Cercus tip simple and narrow with inconspicuous hairs 8i

8i(7i) Vein M_1 on apical third continuously and almost evenly curved posteriorly, R_{4+5} often curving anteriorly at tip so the two veins splay out at the costa (fig. c of couplet 5). Surstylus narrow in side view, not expanded at tip which has a small indentation above a pointed tooth (fig. d of couplet 5) .... ***laetus***
– Vein M_1 on apical third or more continuously straight and parallel to R_{4+5} (aa). Surstylus slightly broadening towards tip, no distinct indentation just before tip 9

(aa)

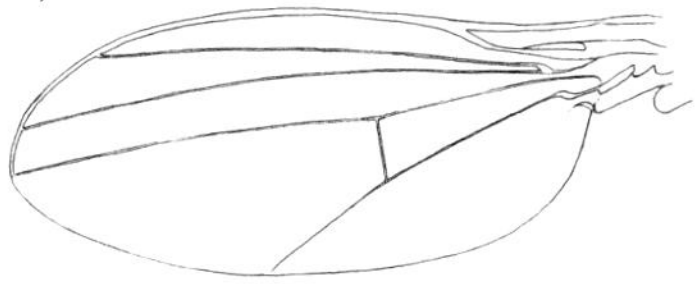

9(7,8i) Cercus and surstylus obviously longer than dorsal edge of epandrium (lower edge in figure) (a). Surstylus narrower, in ventral view elongate-oval, margins not so strongly flared outwards (b). Tip of surstylus with shallow indentation and a tiny point at extreme tip (c). Hypandrium with asymmetrical notch halfway along visible section, profile raised beyond notch (c). .... ***tarsalis***

– Cercus and surstylus about as long as dorsal edge of epandrium (aa). Surstylus broader, in ventral view broadly oval, margins strongly flared outwards (bb). Tip of surstylus with only small irregularities in its profile (cc). Hypandrium with a simple U-shaped notch, profile at same level on both sides of notch (cc, dd) .... ***bellus***

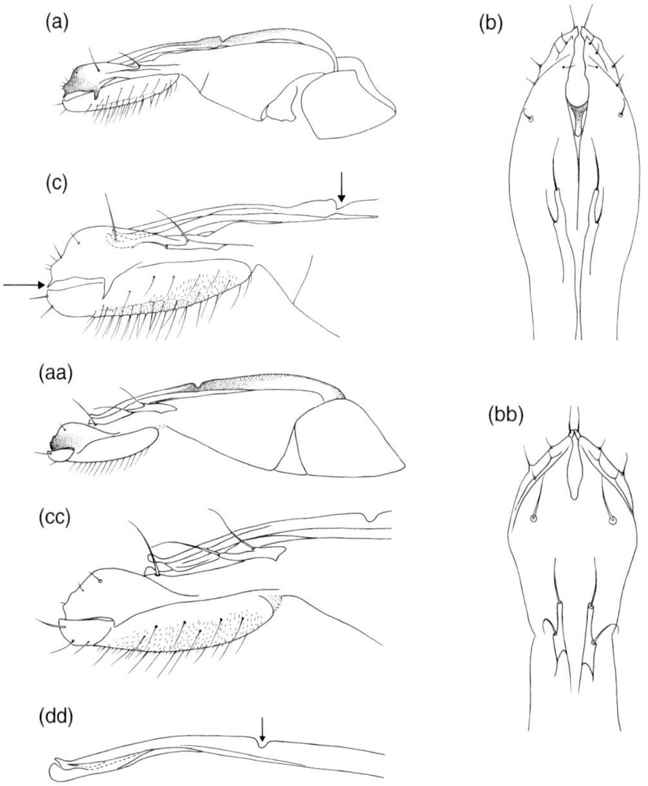

10(6) Antennal scape and pedicel yellow. .... ***Corindia viridis***

– Antenna entirely black. 11

11(10) Palp and proboscis pale. Arista relatively long, twice length of long head setae. Front coxa tip, trochanter and base of femur completely yellow on their anterior faces. Epistoma sides almost parallel on lower two-thirds (a). Surstylus with a marked square 'nose' at tip (b); surstylus dark at base, grading to yellow apically. [Hypandrium entirely black; phallus with a minute projection close to tip (c).] .... ***atomus***

– Palp and proboscis dark. Arista relatively short, about 1.7–1.8 times length of long head setae. Front coxa tip, trochanter and base of femur almost completely dark on their anterior faces (trochanter may be paler brown but not bright yellow). Epistoma more wedge-shaped, sides more convergent (aa). Surstylus tip straight or gently curved with no strong indentation; surstylus either entirely yellow or with dark tip. 12

(a)

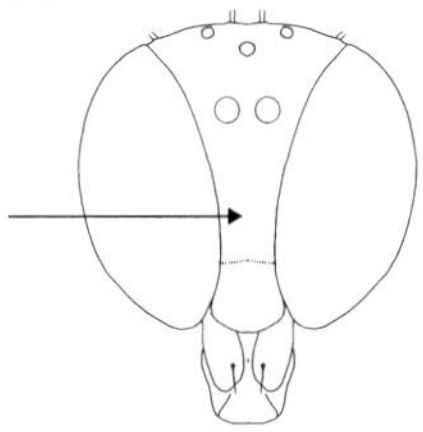

(b)

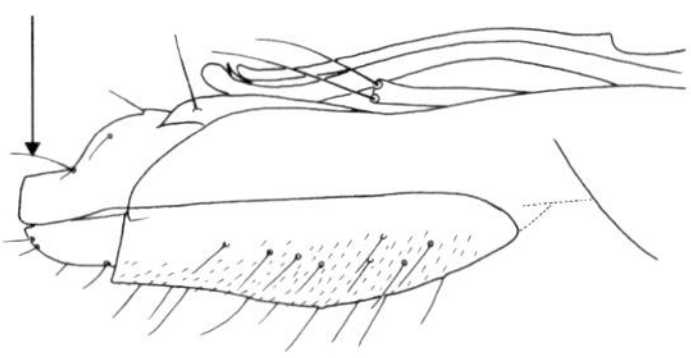

(c)

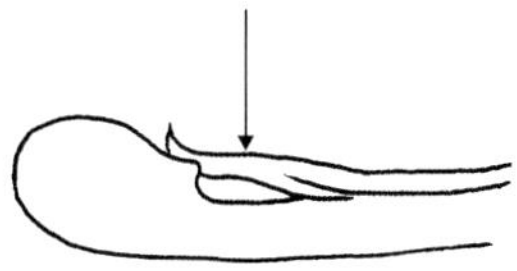

(aa)

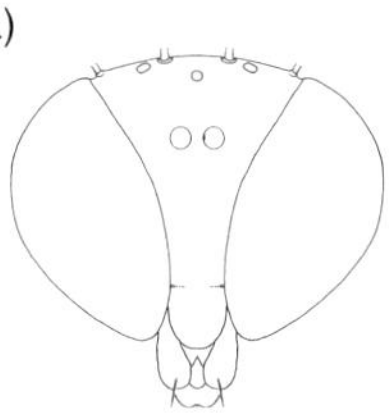

12(11) Surstylus ventral surface concave, rising to a triangular point before a straight slope to tip (a). Cercus narrow in profile. Hypandrium bent just before tip, without onion-shaped swelling halfway but a ventral projection only (b). Phallus tip with a barely perceptible split (c). .... ***cynicus***

– Surstylus ventral surface flat, tip gently rounded. Cercus broad and rounded in profile. Hypandrium straight beyond node at midpoint. Phallus tip distinctly forked or lobed. 13

(a)

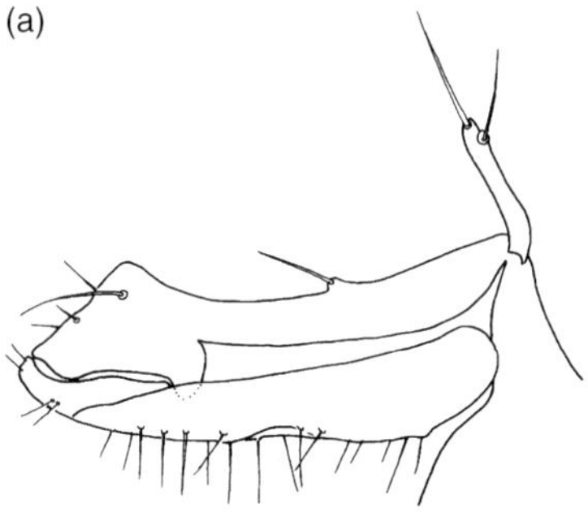

(b)

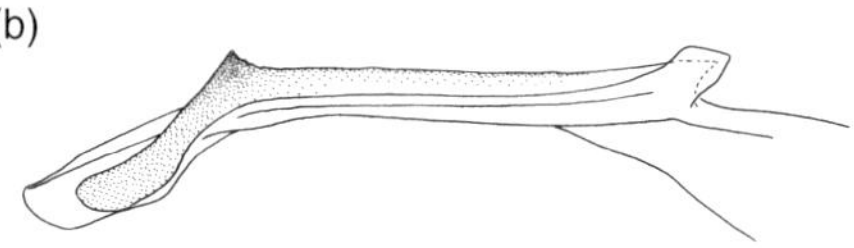

(c)

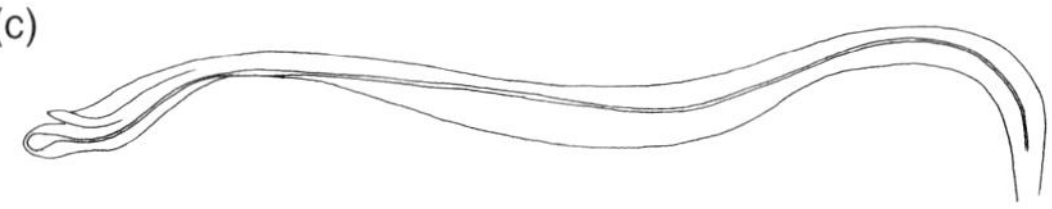

13(12) Surstylus tip at a shallow angle of about 45° to its long axis (a). Hypandrium swelling at about midway along visible section (b). Phallus tip appearing clearly forked, shorter arm longer than wide (with duct between the two arms of fork) (c). Epandrial lobe with seta on shaft separated from seta at tip (d). .... ***intercedens***

– Surstylus tip at a steep angle of about 80° to its long axis (aa). Hypandrium swelling beyond midway along visible section (bb). Phallus tip with long narrow apical arm and low broad flange (with duct between them) (cc). Epandrial lobe with both setae arising at tip or very close to tip (dd). .... ***pollinosus***

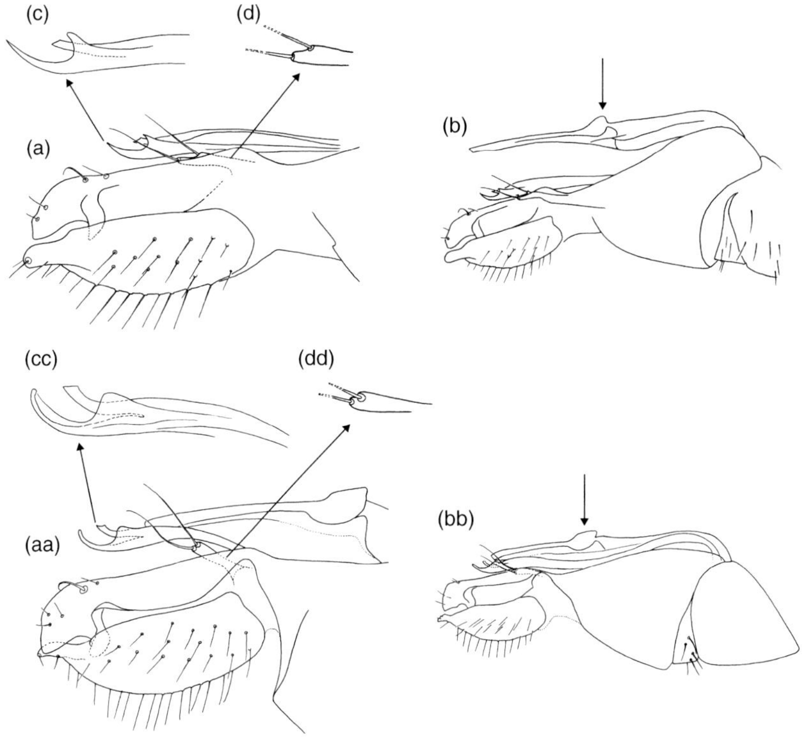

Key to species of *Thrypticus* and *Corindia* (females)

1 Legs including femora mainly or entirely yellow........ 2

– Legs with at least femora glossy green on all surfaces, may be yellow apically........ 4

2(1) Scape and pedicel yellow. Ovipositor in profile stocky with an almost circular toothed structure above paired lateral rods (a – lateral view, b – dorsal view). Mid tibia with a strong anterodorsal seta at basal third, twice as long as width of shaft (c). .... ***Corindia viridis***

– Antennae entirely black. Ovipositor blade-shaped in lateral view (aa). Mid tibia without or with a weak anterodorsal seta at basal third, at most 1.5 times width of shaft (cc). 3

(a) (b)

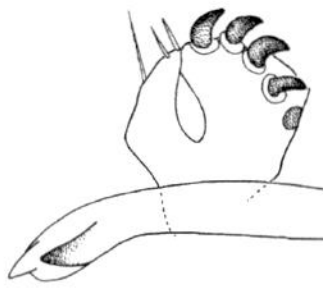

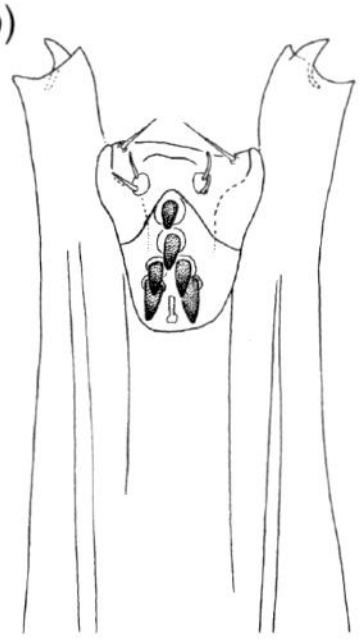

(c)

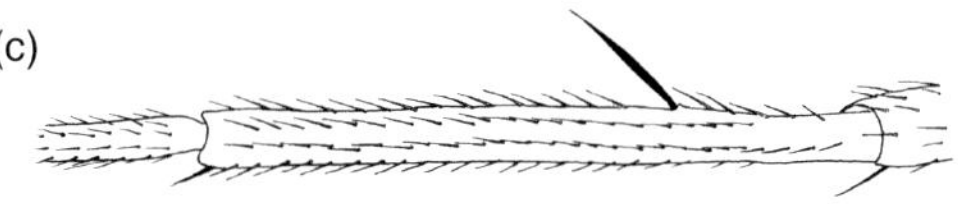

(aa)

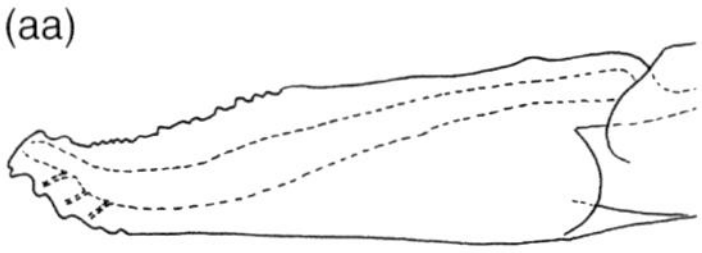

(cc)

3(2) Scutellum with several stout lateral setae in addition to long apical pair (a). Mid tibia without an anterodorsal setae. Claws thick, stout and short (b). Last section of ovipositor wedge-shaped, far shorter than central section (c). Larger species, wing length 2.4–3.8 mm........................***smaragdinus***

– Scutellum with 1 tiny marginal seta additional to the long apical pair (aa), which can be very small and difficult to see. Mid tibia with a small anterodorsal seta. Claws normal, slender. Last section of ovipositor elongate and similar in length to central section (cc). Tiny species, wing length 2.2 mm........................***atomus***

(a) (b) (c)

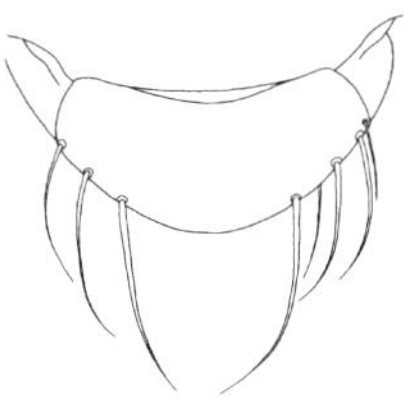

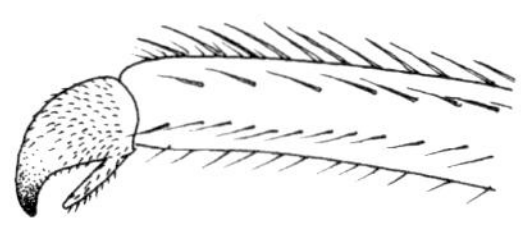

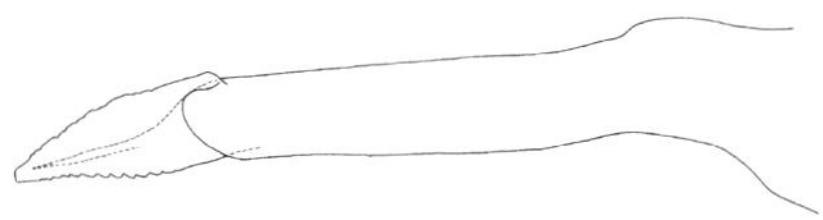

(aa)

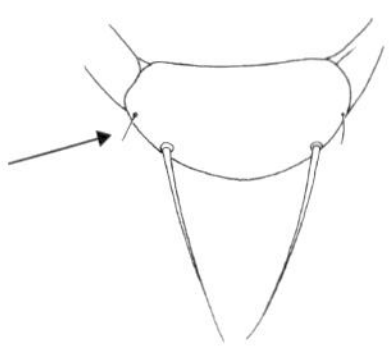

(cc)

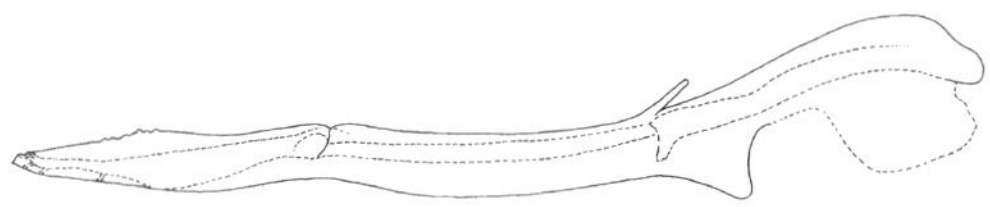

4(1) Mid tibia without an anterodorsal seta........................5

– Mid tibia with an anterodorsal seta at basal third (as fig. cc of couplet 2; sometimes small and difficult to see)........................6

5(4) Mid and hind trochanters and often base of femora usually yellow when viewed from below. Knees often widely yellow. Tarsi relatively short, ratio of tarsus to tibia less than 1.4 for front leg, and less than about 1.2 for hind leg. Palp usually pale around margin. Ovipositor tip slightly narrower and with a slight upturn (a)........................***nigricauda***

– Trochanters almost as dark as femora. Knees only very narrowly yellow. Tarsi relatively long, ratio of tarsus to tibia more than 1.4 for front leg and more than about 1.25 for hind leg. Palp entirely black. Ovipositor tip not upturned, dorsal edge straight (aa)......... ***cuneatus***, ***lichanus***

(a) (aa)

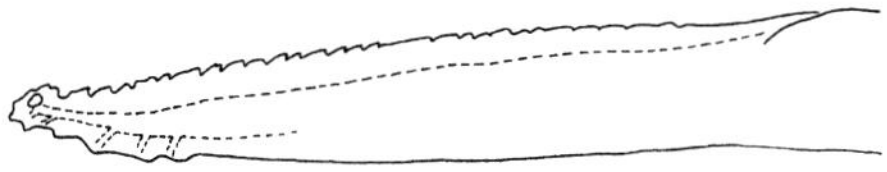

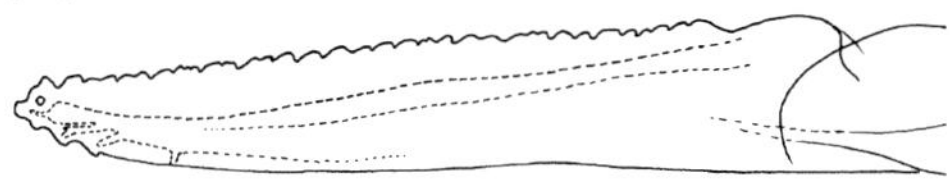

6(4) Scutellum with a tiny lateral seta additional to the long apical seta (as fig. aa of couplet 3; can be difficult to see; if lost, its socket needs high magnification to discern). Middle section of ovipositor with a pronounced curve in lateral view (a).. ***cynicus*, *intercedens*, *pollinosus***

– Scutellum lacking a lateral seta. Middle section of ovipositor nearly straight or only very gently curved (aa).. 7

(a)

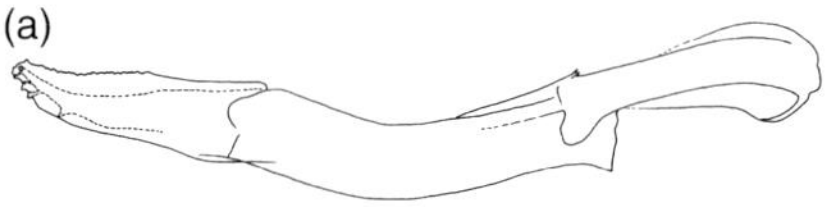

(aa) *laetus*

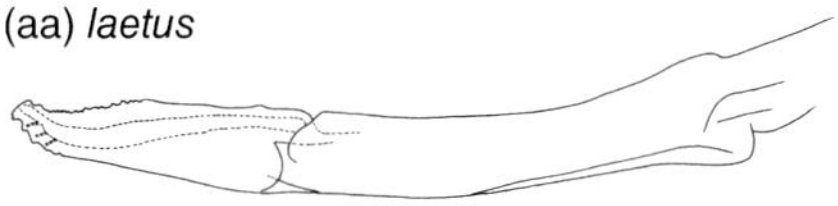

(aa) *bellus*

7(6) Ovipositor hardly upturned at tip (a); in cases of doubt, arista obviously pubescent and equally thick throughout its length.. 8

– Tip of ovipositor clearly upturned (aa). If arista not tapering to a fine tip, it is thin and lacks obvious pubescence.. 9

(a) *tarsalis*

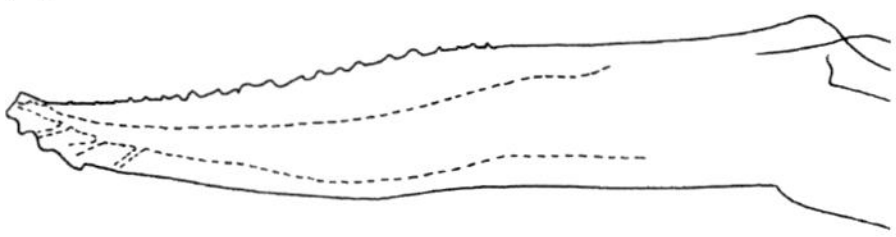

(aa) *laetus*

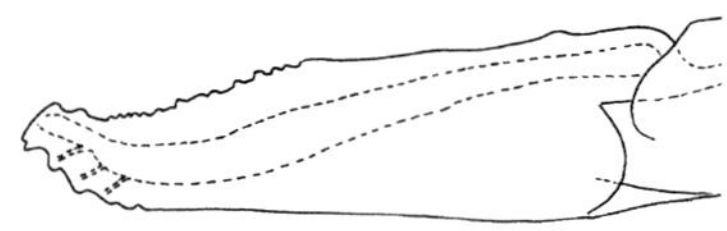

(a) *divisus*

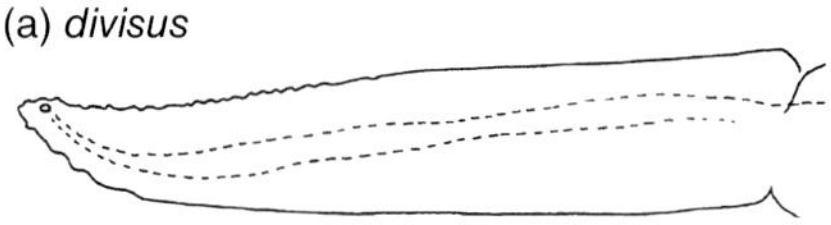

(aa) *bellus*

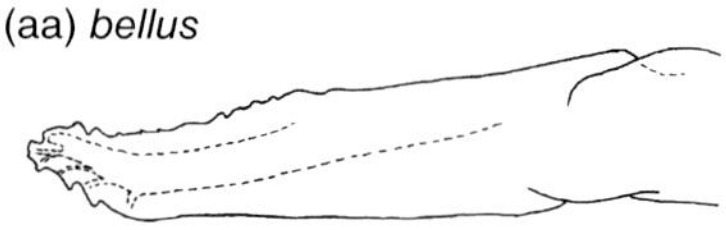

8(7) Arista obviously pubescent, equally broad throughout its length (a).. ***divisus***

– Arista fine and tapering to a point, pubescence barely visible (aa).. ***tarsalis***

(a)

(aa)

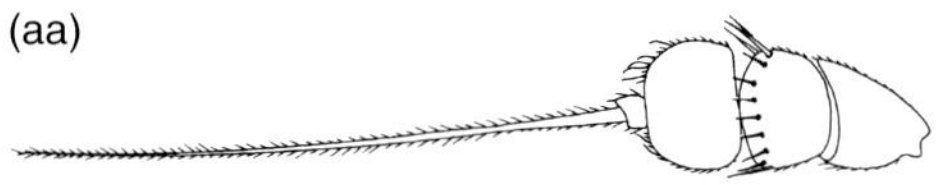

9(7) Arista usually stout, scarcely tapering on apical half, ending abruptly as if broken (a). .... ***laetus***

– Arista tapering evenly to a fine tip.. ***bellus***

(a)

***Thrypticus atomus* Frey, 1915**

Both sexes of this tiny species have markedly pale legs with only short dark section on the femora, similar only to *Corindia viridis* and the far larger *T. smaragdinus*. It is the only species with yellow palps and proboscis. The face, frons and sometimes parts of the thorax and tergites are violet or blue. The basal third of the cercus is dark, the rest is yellow. Wing length ♂: 1.7 (1.5–2.0) mm, ♀: 2.2 mm.

Records are from Norfolk, West Sussex and Yorkshire. The habitat includes fens, carr woodland in fens, and ponds and ditches on river-floodplain grasslands. Flight: v–vi.

Thrypticus atomus

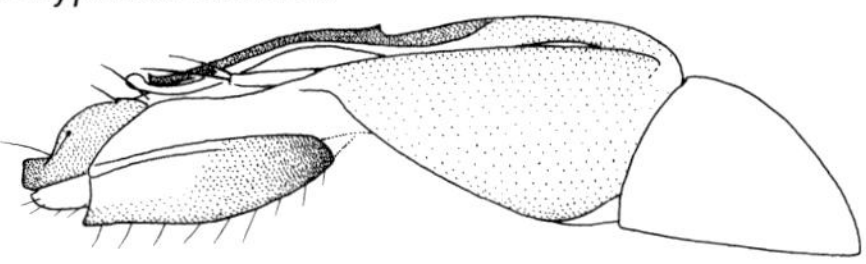

Thrypticus bellus

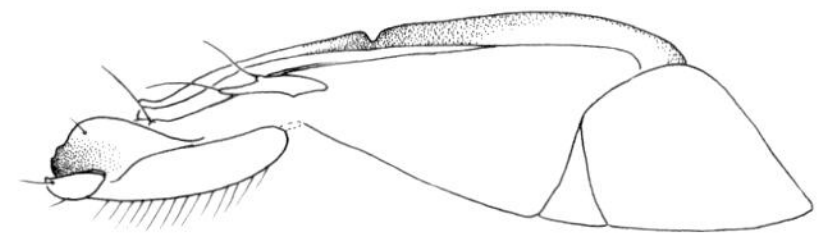

Thrypticus cuneatus

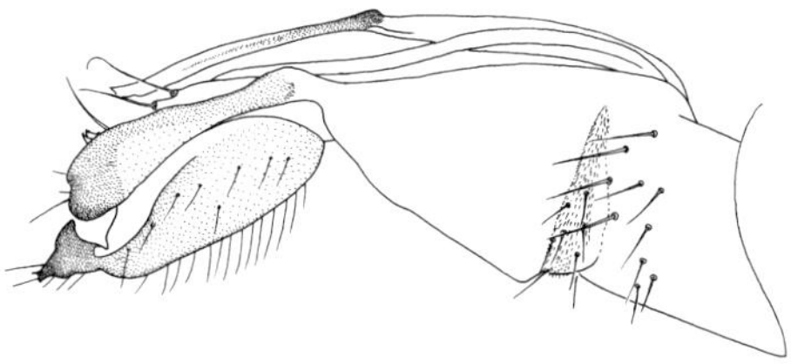

Thrypticus cynicus

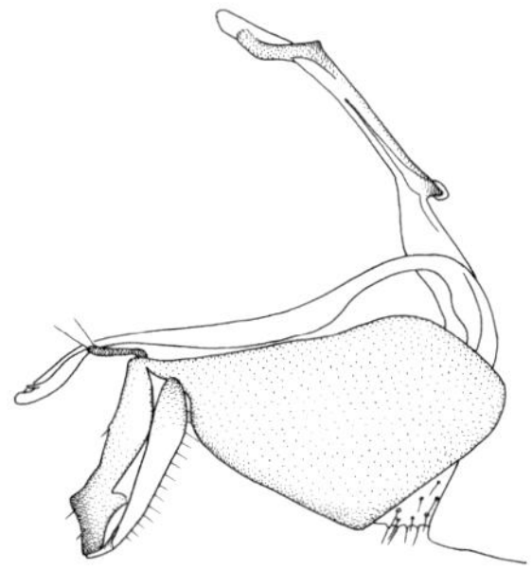

Thrypticus divisus

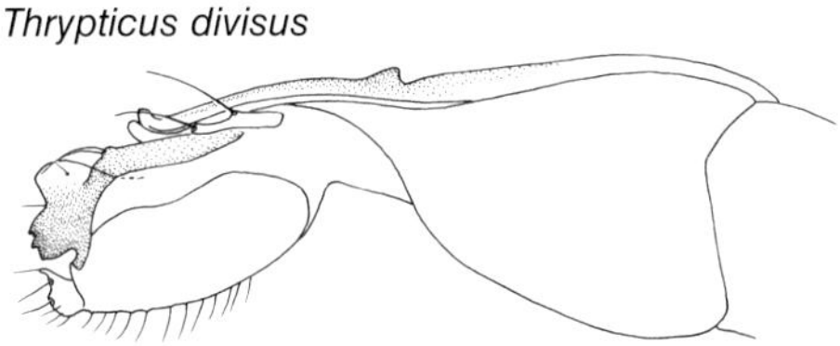

Thrypticus intercedens

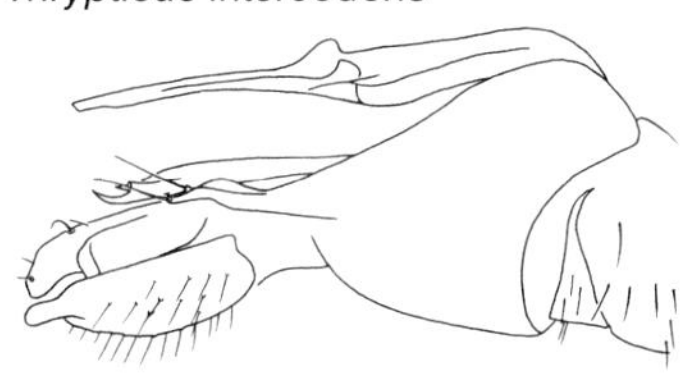

Thrypticus laetus

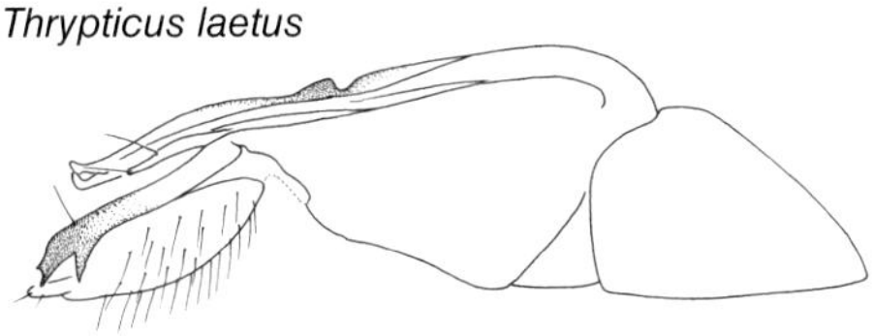

Thrypticus lichanus

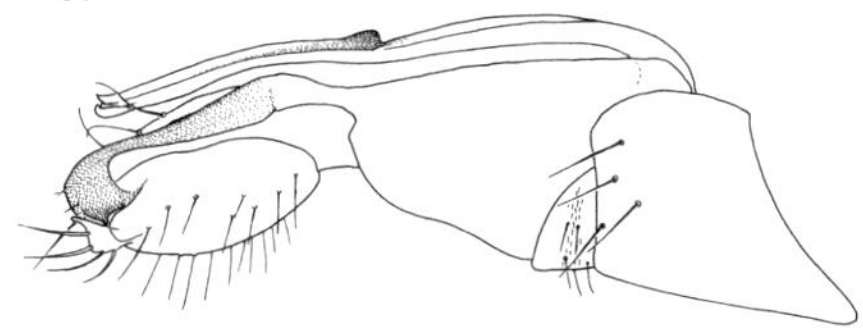

Thrypticus nigricauda

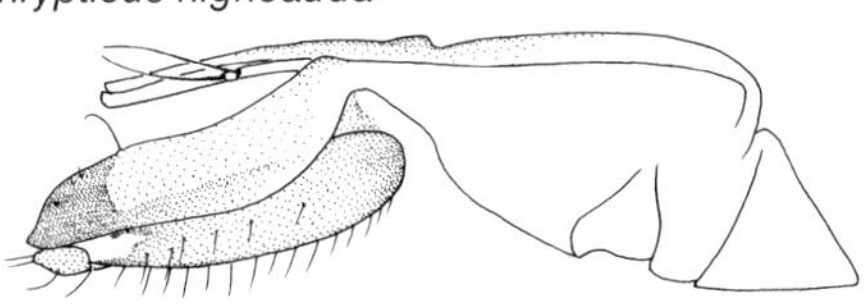

Thrypticus pollinosus

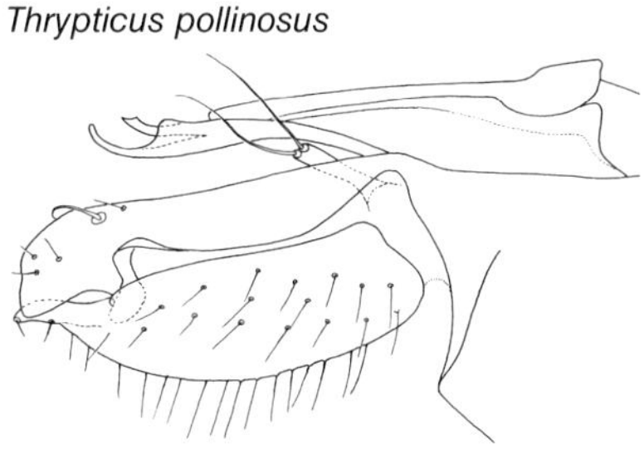

Thrypticus smaragdinus

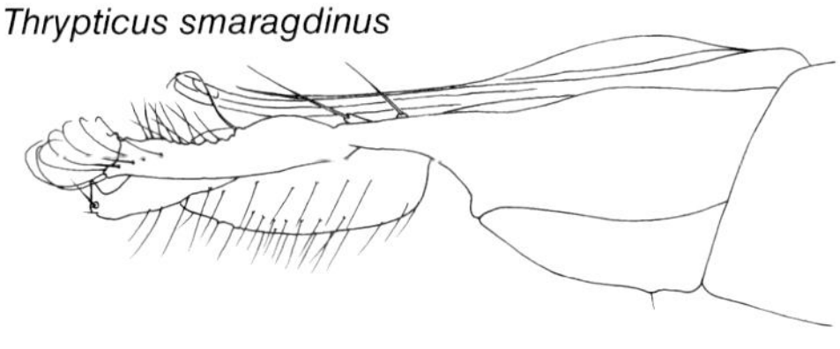

Thrypticus tarsalis

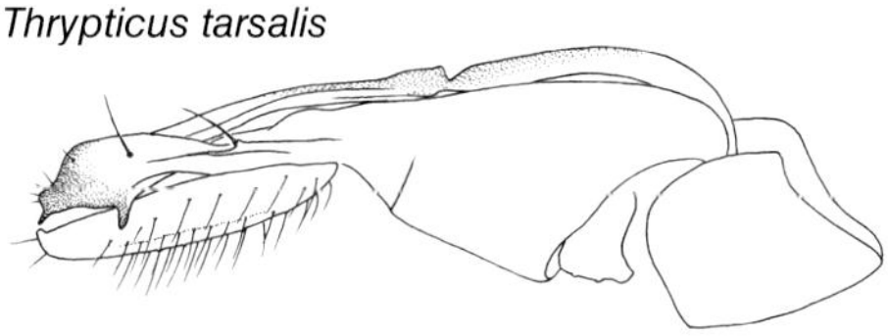

Fig. 43. *Thrypticus* hypopygia.

***Thrypticus bellus* Loew, 1869**

This species closely resembles *tarsalis* and they are reliably separated only using the male genitalia. They are both dark-legged, equally small and with a tapered arista. The yellow parts (knees, trochanters, genitalia) are brighter than in *tarsalis* where the yellow is often rather murky. Females closely resemble *laetus* in the extent of the bright yellow on the knees, the yellow dorsal surface of the mid tibia and the upturned tip to the ovipositor, but have a tapered arista (thick in *laetus*) and veins R_{4+5} and M_1 at their tips are either parallel or M_1 curves imperceptibly forward (splayed or parallel at their tips in *laetus*). *T. bellus* (and *laetus*) differ from *tarsalis* in having more extensively yellow areas on the legs, making those of *bellus* conspicuously bicoloured, and a strongly upturned tip to the ovipositor rather than almost straight. Wing length ♂: 1.8 (1.7–2.1) mm, ♀: 2.0 (1.8–2.2) mm.

Assuming that a reasonable number of the records are accurate, this species is widely distributed in Britain although apparently confined to the coast in Scotland. It appears to use a wide range of habitats from dune slacks and brackish marsh to upland acid moorland, and many types of water margins including ponds, ditches and washlands. Flight: v–ix, mainly vii–viii.

***Thrypticus cuneatus* (Becker, 1917)**

The distinctive wing shape is also found in *lichanus*. These two species have long, slender and entirely dark legs, apart from their narrowly pale knees; other species have relatively shorter legs and at least the trochanters are pale even if most the rest of the leg is dark. The two species are distinguished by their hypopygia and the detail can usually be seen without special preparation. There is small difference in the hind tibia of these species: *cuneatus* has slightly longer dorsal hairs on the apical half where they are conspicuously longer than those on the basal half, and about as long as the width of the hind basitarsus; in *lichanus* the hairs are slightly shorter than the width of the hind basitarsus. Females cannot be separated from those of *lichanus* and only with great care from *nigricauda*. Wing length ♂: 2.2 (2.1–2.3) mm, ♀: 2.3 mm (in association with male).

Records are from Cambridgeshire, Lancashire and Inverness-shire. One site had ponds with abundant common spike-rush and common club-rush. Flight: vii–viii.

***Thrypticus cynicus* Drake and Godfrey, 2023**

Externally, *cynicus* resembles *intercedens* and *pollinosus* in having pale tibiae, a short anterodorsal seta on the mid tibia and a tiny lateral seta on the scutellum. Its surstylus and hypandrium are unlike other British *Thrypticus*: the profile of the surstylus is particularly characteristic as the hypandrium has a small projection where, in other *Thrypticus*, there is only a small swelling, and it has a pronounced bend near the tip which can be seen without special preparation if the hypopygium is moved away from the abdomen. Females of these three species are indistinguishable. Wing length ♂: 1.8 (1.7–1.9) mm.

Thrypticus cynicus has been recorded from Kent, Berkshire, Northamptonshire and Norfolk. The sites were ponds, swamp-like wet grassland and wet wood-pasture (Drake *et al.* 2023). Flight: v–ix.

***Thrypticus divisus* (Strobl, 1880)**

Males have a thickly pubescent arista, unlike any other British species; *laetus* has a fairly stout arista but its pubescence is barely visible. *T. divisus* otherwise resembles other dark-legged species with pale trochanters and knees, such as *bellus* and *tarsalis*. The terminal extension of the surstylus resembles that of *lichanus*. Wing length ♂: 2.6 (2.4–2.9) mm, ♀: 3.0 mm.

The species distribution appears to be widespread in England, rare in Wales and confined to the coast in Scotland. The only information on habitat is from a Welsh lake shore dominated by common spike-rush. Flight: vi–viii.

***Thrypticus intercedens* Negrobov, 1967**

This tiny species closely resembles *cynicus* and *pollinosus* in having pale tibiae, a short anterodorsal seta on the mid tibia and a tiny lateral seta on the scutellum. It is closer to *pollinosus* and both have an entirely yellow dumpy surstylus and cercus which separates them from other species. The tiny lateral seta on the scutellum can be easily overlooked and, if lost, the socket scar is barely detectable. *T. intercedens* differs macroscopically from *pollinosus* in having less extensively dark legs, in particular the ventral surface of the front and mid tarsi is pale, rather than often dark in *pollinosus*, but these features vary too much to be definitive. The species can be reliably separated only by examining the hypopygium. Females are indistinguishable from those of *cynicus* and *pollinosus*. Wing length ♂: 1.8 (1.7–1.9) mm, ♀: 1.9 (1.8–1.9) mm (in association with males).

T. paludicola Negrobov, 1971 is almost identical to *intercedens*, differing in small details of the phallus tip and colour of the femora and surstylus (Negrobov and Naglis 2020). British specimens have the phallus shape of *intercedens* and the yellow-tipped femora and mainly yellow surstylus of *paludicola*. Given the uncertainty of the identification, the earlier of Negrobov's names is used for the British species, agreeing with the synonymy by Jonassen (1990).

This species is sparsely distributed in England from Devon to Southwest Yorkshire. The habitat is margins of ponds and ditches in herb-rich wet grasslands and fen. Flight: vi–vii.

Thrypticus laetus Verrall, 1912

Thrypticus laetus can be distinguished by the combination of dark genitalia, the mid tibia with a small anterodorsal seta, and a thick arista that scarcely tapers on the apical half and ends abruptly. Specimens whose veins R_{4+5} and M_1 diverge at the wing tip are unambiguously *laetus*, but often the two veins are parallel to the tip and M_1 does not always curve posteriorly. Usually the tibiae are pale but occasionally dark. Females strongly resemble *bellus* and *tarsalis*; see the account for *bellus*. Wing length ♂: 2.3 (1.9–2.4) mm, ♀: 2.3 (2.2–2.5) mm.

This species appears to be widespread in England and Wales but rare in Scotland. Habitats include the margins of lakes and river backwaters, and ponds and ditches in fens, wet grasslands and coastal dunes. Flight: v–ix.

Thrypticus lichanus Drake, 2023

This species closely resembles *cuneatus* – see the account for that species. Wing length ♂: 2.2 (1.9–2.4) mm, ♀: 2.3 (2.1–2.4) mm (in association with males).

It has been recorded from Cornwall, Devon, Dorset, Cambridgeshire and Norfolk, from ponds on clay and in dunes, and a ditch on coastal marsh, where potential host plants included common spike-rush and common club-rush (Drake *et al.* 2023). Flight: vi-vii.

Thrypticus nigricauda Wood, 1913

This species is distinguished by the combination of dark genitalia, pale-tipped palps, a long fine arista and lacking an anterodorsal seta on the mid tibia. Careful examination is needed to be sure that the tibial seta is genuinely absent since, in other species where it is present, it is short and often difficult to see. One specimen of *nigricauda* has been found with this seta. The surstylus of British material shows some variation in the shape of the tip; it is unclear whether this represents more than one species. Females resemble those of *cuneatus* and *lichanus* but specimens with paler knees and trochanters are *nigritarsis*; dark-legged specimens need considerable care to separate. Wing length ♂: 1.9 (1.6–2.2) mm, ♀: 2.1 (1.9–2.3) mm.

The species is widespread in lowland England with a concentration in East Anglia and the East Midlands, scarce in Wales and the northern half of England, and apparently absent from Scotland. Habitat include the margins of ponds, lakes and ditches in wet grasslands and fens. Flight: v–vii, peak in vii.

Thrypticus pollinosus Verrall, 1912

Darker-legged male *pollinosus* are no different from most other small British *Thrypticus*, whereas pale-legged specimens resemble *intercedens*. See the account for *intercedens* for separation of both sexes. Wing length ♂: 2.0 (1.8–2.2) mm, ♀: 2.3 (2.2–2.4) mm (with associated males).

Records are widespread in Britain north to Moray. They will probably include *atomus* and *intercedens* which run to *pollinosus* using d'Assis-Fonseca (1978). Habitat include margins of ponds, including those on a braided river shore, swampy grassland, and a freshwater seepage crossing a sandy sea-shore. Flight: vi–ix, peak in vii.

***Thrypticus smaragdinus* Gerstäcker, 1864**

There can be no mistaking this large species with almost entirely yellow legs, several scutellar setae and a narrow face. It differs in many other small features from other British *Thrypticus*. Females, in addition to the same marked characters as in males, have long thick tarsomeres that scarcely taper to the last one, which is longer than the fourth, and remarkably stout and finely haired claws unlike any other British dolichopodid. Wing length ♂: 2.9 (2.8–3.1) mm, ♀: 3.3 (2.4–3.8) mm.

The species is frequent in the Norfolk fens, with isolated records from Kent and Dorset. The habitat at nearly all sites is fen, almost invariably containing reed, or sometimes pure reedbeds. Flight: vi–viii, peak in vi.

***Thrypticus tarsalis* Parent, 1932 (Plate 8)**

This species closely resembles *bellus* and can be reliably separated only by using the male genitalia. See the account for *bellus* for a brief comparison. Females resemble those *bellus* and *laetus* but are particularly dark-legged in comparison; see the account for *bellus* for distinctions. The mid tibia's anterodorsal seta is often inconspicuous so easily overlooked, which would lead to misidentification as *nigricauda* but *tarsalis* is distinct in having a shorter arista which is about 0.75–0.85 the length of the front tibia compared to about equal in *nigricauda*, and the front basitarsus is slightly less than half the tibia's length compared to about half in *nigricauda*. Wing length ♂: 1.9 (1.4–2.1) mm, ♀: 1.9 (1.8–2.1) mm.

This species is widespread in lowland England and Wales, scarcer to the north and west, and absent from Scotland. Habitat include pond margins and fen ditches, and seepages in several situations including coastal soft-rock cliffs, moorland and wet hillside grasslands. Some sites are lightly shaded by scrub. Flight: v–vii, peak in vi.

XANTHOCHLORUS Loew, 1857 (Xanthochlorinae)

These small flies are recognised by being largely yellow, often including the thoracic setae, with rather slender legs and brilliantly green eyes in life (Plate 1). Dark specimens may be yellow only on the lower parts of the pleura, coxae and rest of legs, and the hypopygium. Other dolichopodids that are predominantly yellow are much larger (*Neurigona*) or have postpedicels that are long-triangular (*Achalcus*, *Anepsiomyia*) rather than scarcely longer than deep as in *Xanthochlorus*. In addition, the combination of pale upper postocular setae, no anterior preapical setae on the mid femur, a dorsal arista on a short but deep and almost rectangular postpedicel will identify the genus. Females of this genus are unique in having eight tergites, all of which are hairy. Males are almost devoid of secondary sexual characters. *Xanthochlorus* is the only genus in this subfamily.

The figures of the hypopygia show the morphologically dorsal (ventral-facing) view as this is often clearly visible in pinned specimens, rather than the ventral (internal) view, although the latter shows more useful characters in cleared preparations, particularly the well-developed parameres. The key given here is based on Chandler and Negrobov (2008).

Species of *Xanthochlorus* are found mainly in woodland and scrub, and show no pronounced preference for damp places. They may often be collected from foliage well above ground level.

Key to species of *Xanthochlorus* (males and females)

1 Scutum, scutellum, postnotum and, in males, pleura almost entirely dark, greenish or bronze, and grey-dusted. Epandrium almost spherical. Surstylus composed of two lobes which do not protrude conspicuously. Parameres in the form of a dark star. .... ***ornatus***

– Scutum and scutellum at most partly dark on their discs, more-or-less yellow at front and sides (rarely entirely dark), postnotum and pleura yellow. Epandrium ovoid. Surstylus composed of a single lobe that protrudes obviously. Parameres consist of rods or not visible. 2

2 Scutum with greenish or bronze colouration extending forward beyond the suture usually to level of second dorsocentral seta and outside of dorsocentral row, sometimes covering most of scutum or rarely extending only as far as penultimate (fourth) dorsocentral seta. Parameres forked near tip. Male hind femur with 2–3 slightly stouter hairs at base ventrally (a). (Surstylus curved in lateral view.).. ***galbanus***

– Scutum with greenish or bronze colouration restricted to a prescutellar patch between dorsocentral setae and usually not reaching penultimate (fourth) pair anteriorly, or dark patch sometimes absent. Parameres not forked. Male hind femur without or with only 1 slightly stouter hair at base ventrally...3

(a)

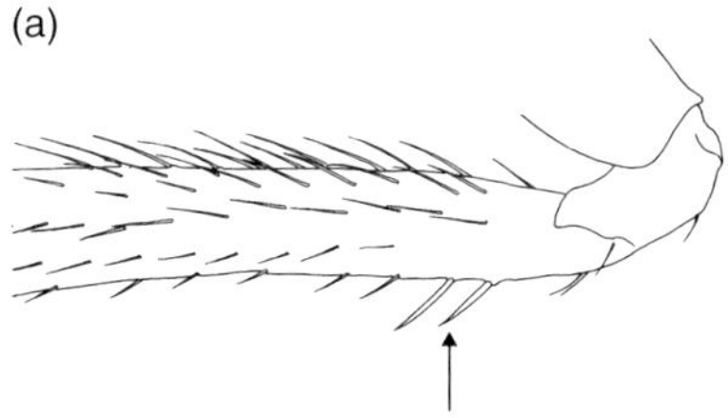

3 Surstyli curved in lateral view, flattened and slightly swollen at tip in dorsal view. Parameres consist of 1 long rod with tip bent at a right angle, and a shorter straight rod; phallus appearing as a third 'rod' between them. Prescutellar dark patch usually well-defined in both sexes. .... ***tenellus***

– Surstyli almost straight and cylindrical from all angles of view. Parameres scarcely visible but a flattened bilobed appendage arises at base of epandrium with an obvious broad hook projecting laterally. Prescutellar green patch often reduced or absent in females. .... ***silaceus***

Xanthochlorus galbanus Chandler and Negrobov, 2008 (Plate 8)

This species is usually distinctive in most of the scutum being green with the sides and front yellow, and it differs from *ornatus* in having a yellow postnotum. In dark females that may resemble pale *ornatus*, a greater part of the upper part of the pleura are yellow or pale brown rather than dark green or bronze. In pale females the green patch may stop just before the suture and in such cases they may be inseparable from dark *tenellus*, although in side view a dark narrow strip can usually be seen extending forward to the second dorsocentral seta. The male's curved surstylus superficially resembles that of *tenellus*, being curved in profile and flattened apically, but in *galbanus* it is scarcely broadened towards the tip when seen in ventral or dorsal view, and the parameres are conspicuously forked rather than single rods. Wing length ♂: 3.3 (3.0–3.5) mm, ♀: 3.6 (3.3–3.9) mm.

Xanthochlorus galbanus is widely distributed in England, although with few records in the centre of the country, Wales and northwards to central Scotland. It is a fairly common species of dry to wet deciduous woodlands or shaded places such as parklands and tall hedgerows. Flight: v–x, peak in vii.

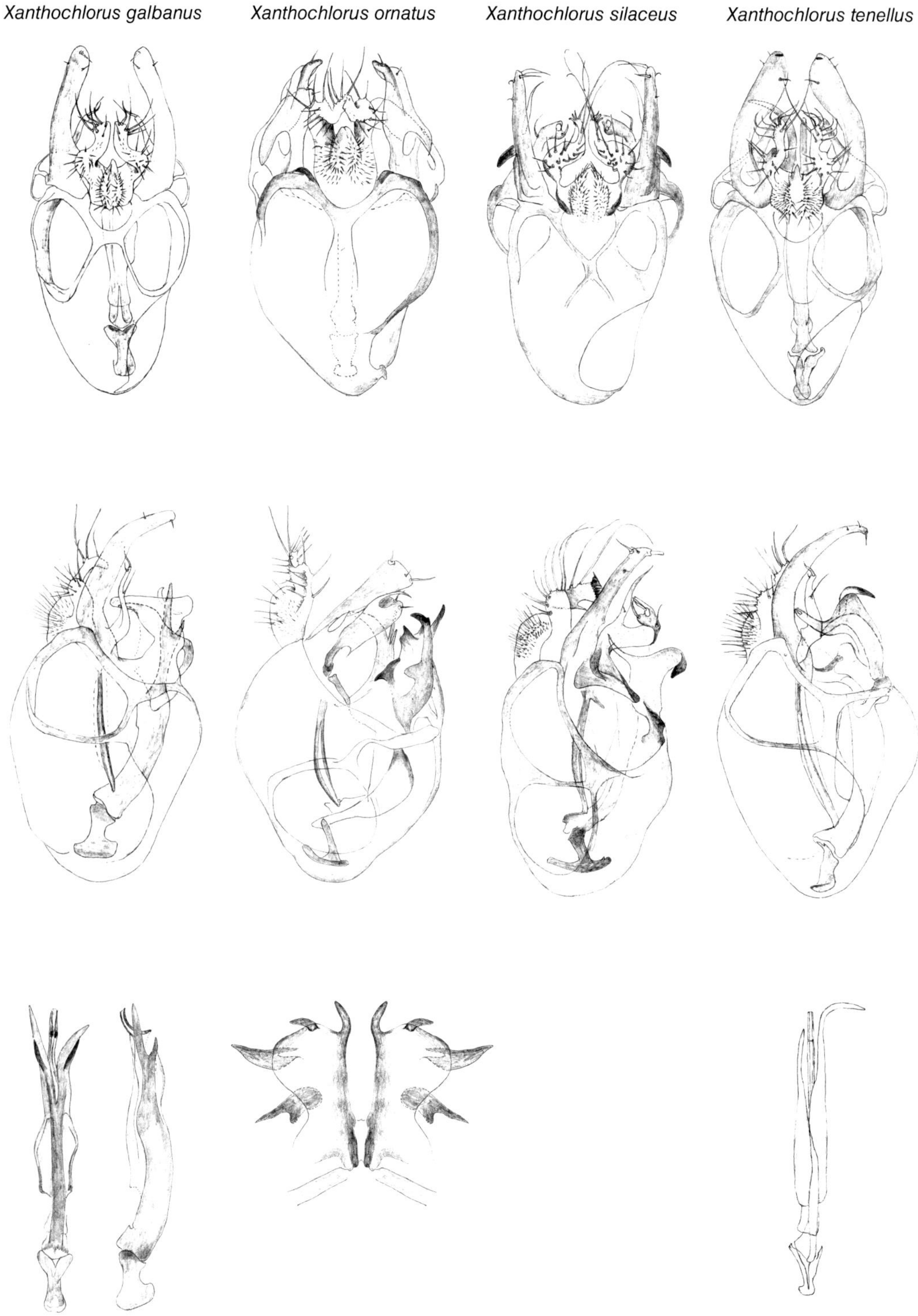

Fig. 44. *Xanthochlorus* hypopygia in dorsal and lateral views, and parameres.

Xanthochlorus ornatus **(Haliday, 1832)**

Xanthochlorus ornatus is the most distinct of the British species in having an entirely green scutum and postnotum. In paler females, in which the pleura may be yellow, the scutum remains green up to the neck in front and the postnotum is dark below the scutellum. Males are obvious in having a nearly round genital capsule, bilobed surstylus, and parameres in the form of a conspicuous black star which is often visible even when the capsule is not hinged out. Wing length ♂: 3.5 (3.2–3.7) mm, ♀: 3.7 (3.4–4.0) mm.

This is a common species found throughout Britain but is scarcer on the west side of Scotland. It is found in dry to wet deciduous woodlands, scrub and on isolated trees and bushes. Flight: v–x, peak in vi.

Xanthochlorus silaceus **Chandler and Negrobov, 2008**

Xanthochlorus silaceus usually has very little green on the scutum and may be entirely yellow so there is some overlap in the range of colour found in *tenellus*. Females may sometimes be inseparable. Males are distinct in having a straight cylindrical surstylus, whereas in *tenellus* it is curved in profile, and flattened and slightly spatulate when viewed from below. The parameres are indistinct but there is a flat bilobed appendage whose larger lobe points sideways and appears as a hook. The cercus, visible without special preparation, has particularly long hairs extending as far as the tip of the surstylus; in the other species these hairs are much shorter. Wing length ♂: 3.2 (3.1–3.3) mm, ♀: 3.2 (2.8–3.4) mm.

This appears to be an uncommon species recorded from the southern counties of England and Wales approximately south of a line from the Severn to the Thames. It has been found in damp and dry deciduous woodland, with a bias towards drier calcareous habitats, and has been reared from soil at the base of a willow. Flight: vi–viii.

Xanthochlorus tenellus **(Wiedemann, 1817)**

This species resembles *silaceus* in colour and the main differences are discussed under that species. The green patch of the scutum of occasional specimens reaches forward to the suture but scarcely extends laterally beyond the dorsocentral rows. Wing length ♂: 3.0 (2.8–3.2) mm, ♀: 3.5 (3.1–3.9) mm.

The distribution is uncertain owing to past confusion with *galbanus* and the non-British *luridus* Negrobov before the distinction was made between these species. It is probably widespread but uncommon in England with more records in the east, but sparse in Wales and Scotland. Flight: v–ix.

References

Alexander, K.N.A. 2020. Recent records of *Australachalcus melanotrichus* Mik (Diptera, Dolichopodidae) and observations on its habitat and range in Britain and Ireland. *Dipterists Digest (Second Series)* **27**, 181–184.

Alexander, K.N.A., Gibbs, D.J. and Speight, M.C.D. 2008 (2009). The Irish *Systenus alpinus* Vaillant reassessed as *Systenus mallochi* MacGowan (Diptera, Dolichopodidae). *Irish Naturalists' Journal* **29**, 95–96.

Allen, A.A. 1976. A new species of *Medetera* Fisch. (Dipt.: Dolichopodidae) in Britain. *Entomologist's Record* **88**, 77–79.

Andrews, H.W. 1939. Diptera from North Kent. *Entomologist's Record* **51**, 128–129.

Barták, M. and Kubík, S. 2022. Unexpected previously unknown diversity of the genus *Microphor* Macquart (Diptera, Dolichopodidae, Microphorinae) in the west Palaearctic. *Insects* **13**, 700.

Beaver, R.A. 1966. The biology and immature stages of two species of *Medetera* (Diptera, Dolichopodidae) associated with the bark beetle *Scolytus scolytus* (F.). *Proceedings of the Royal Entomological Society of London (A)* **41**, 145–154.

Becker, T. 1918. Dipterologische Studien, Dolichopodidae. Part 3. *Abhandlungen der Kaiserlichen Leopoldinisch-Carolinischen Deutschen Akademie der Naturforscher* **54**, 35–214.

Beling, E. 1882. Beitrag zur Metamorphose zweiflugeliger Insekten aus den Familien Tabanidae, Leptidae, Asilidae, Empidae, Dolichopidae und Syrphidae. *Archiv für Naturgeschichte* **48**, 186–240.

Bellstedt, R. 2007. Zur Limnofauna von Salzstellen in Thüringen (aquatische Coleoptera und Diptera, Dolichopodidae). In: Werres, W. (ed.) *Binnensalzstellen Mitteleuropas. Internationale Tagung Bad Frankenhausen 2005*. Hrsg. Thüringer Ministerium für Landwirtschaft, Naturschutz und Umwelt (TLMNU), Erfurt, pp. 102–116.

Bernasconi, M.V., Pollet, M., Varini-Ooijen, M. and Ward, P.I. 2007a. Phylogeny of European *Dolichopus* and *Gymnopternus* (Diptera, Dolichopodidae) and the significance of morphological characters inferred from molecular data. *European Journal of Entomology* **104**, 601–617.

Bernasconi, M.V, Pollet, M. and Ward, P.I. 2007b. Molecular systematics of Dolichopodidae (Diptera) inferred from COI and 12S rDNA gene sequences based on European exemplars. *Invertebrate Systematics* **21**, 453–470.

Bickel, D.J. 1985. A revision of the Nearctic *Medetera* (Diptera, Dolichopodidae). *United States Department of Agriculture Technical Bulletin* **1692**, 1–109.

Bickel, D.J. 1986. Australian species of *Systenus* (Diptera, Dolichopodidae). *Records of the Australian Museum* **38**, 263–270.

Bickel, D.J. 1987. A revision of the Oriental and Australasian *Medetera* (Diptera, Dolichopodidae). *Records of the Australian Museum* **39**, 195–259.

Bickel, D.J. 1994. The Australian Sciapodinae (Diptera, Dolichopodidae), with a review of the Oriental and Australasian faunas, and a world conspectus of the subfamily. *Records of the Australian Museum Supplement* **21**, 1–394.

Bickel, D.J. 1998. Australian, Melanesian, and Micronesian *Acropsilus* Mik (Diptera, Dolichopodidae). *Tijdschrift voor Entomologie* **141**, 1–17.

Bickel, D.J. 2022. Notes on the genus *Micromorphus* Mik (Diptera, Dolichopodidae), and the uncertain identity of its type-species, *M. albipes* (Zetterstedt). *Zootaxa* **5125**, 437–444.

Bowden, J. 1998. Observations on *Neurigona abdominalis* (Fallén) and other Dolichopodidae (Diptera). *Entomologist's Monthly Magazine* **134**, 61–64.

Brooks, S.E. 2005. Systematics and phylogeny of Dolichopodinae (Diptera, Dolichopodidae). *Zootaxa* **857**, 1–158.

Brooks, S.E. and Cumming, J.M. 2022. Revision of the Nearctic Microphorinae (Diptera, Dolichopodidae *sensu lato*). *Zootaxa* **5095**, 1–141.

Brooks, S.E. and Wheeler, T.A. 2005. *Ethiromyia*, a new genus of Holarctic Dolichopodinae (Diptera, Dolichopodidae). *Proceedings of the Entomological Society of Washington* **107**, 489–500.

Chandler, P. 1989. *Campsicnemus dasycnemus* (Loew) (Diptera, Dolichopodidae), first record for the British Isles from Ireland. *Irish Naturalists' Journal* **23**, 17–19.

Chandler, P.J. 1999. *Micropygus vagans* Parent (Diptera, Dolichopodidae), a New Zealand fly in the British Isles. *British Journal of Entomology and Natural History* **12**, 215–220.

Chandler, P.J. 1998. Checklists of Insects of the British Isles (New Series) Part 1, Diptera (incorporating a list of Irish Diptera). *Handbooks for the Identification of British Insects* **12**. Royal Entomological Society, London, xx + 234 pp.

Chandler, P.J. 2024. An Update of the 1998 Checklist of Diptera of the British Isles [updated 18 April 2024]. Available from: https://dipterists.org.uk/checklist (Accessed 18 December 2024).

Chandler, P.J. and Negrobov, O.P. 2008. The British species of *Xanthochlorus* Loew, 1857 (Diptera, Dolichopodidae), with description of two new species. *Dipterists Digest (Second Series)* **15**, 29–40.

Chandler, P.J. and Smith, J.E. 2005. *Micropygus vagans* (Parent) (Diptera, Dolichopodidae) new to England. *Dipterists Digest (Second Series)* **12**, 172.

Chursina, M.A. and Grichanov, I.Ya. 2019. Analysis of the differences between *Syntormon pallipes* and *S. pseudospicatus* (Diptera, Dolichopodidae), morphological and molecular data. *Zoosystematica Rossica* **28**, 305–316.

Chvála, M. 1983. The Empidoidea (Diptera) of Fennoscandia and Denmark. II. General Part. The families Hybotidae, Atelestidae and Microphoridae. *Fauna Entomologica Scandinavica* **12**, 1–279.

Chvála, M. 1986. Revision of Palaearctic Microphoridae (Diptera) 1. *Microphor* Macq. *Acta Entomologica Bohemoslovaca* **83**, 432–454.

Cicero, J.M., Adair, M.M, Adair, R.C., Hunter, W.B., Avery, P.B. and Mizell, R.F. 2017. Predatory behavior of long-legged flies (Diptera, Dolichopodidae) and their potential negative effects on the parasitoid biological control agent of the Asian Citrus Psyllid (Hemiptera, Liviidae). *Florida Entomologist* **100**, 485–487.

Clemons, L. 2003. A note on the apparent rarity of *Orthoceratium lacustre* (Scopoli) (Dipt.: Dolichopodidae) in Kent. *Entomologist's Record and Journal of Variation* **115**, 70–71.

Cole, J. 1987. Dolichopodidae difficulties. *Empid and Dolichopodid Study Group Newsheet* No. 3, 2–3.

Cole, J. 1990. The separation of some *Argyra* males, are *A. argyria* and *argentella* distinct species? *Empid and Dolichopodid Study Group Newsheet* No. 9, 3–4.

Cole, J. 1998. *Sciapus basilicus* Meuffels and Grootaert (Diptera, Dolichopodidae) new to Britain. *Dipterists Digest (Second Series)* **5**, 79.

Cole, J.H. 1989. Two species of *Medetera* Fischer (Diptera, Dolichopodidae) new to Britain. *British Journal of Entomology and Natural History* **2**, 115–118.

Cole, J.H. 1991. A new British species of *Neurigona* Rondani (Diptera: Dolichopodidae) from Anglesey. *British Journal of Entomology and Natural History* **4**, 66.

Collin, J.E. 1940. Critical notes on some recent synonymy affecting British species of Dolichopodidae (Diptera). *Entomologist's Monthly Magazine* **76**, 261–271.

Collin, J.E. 1941. The British species of the dolichopodid genus *Medeterus* Fisch. (Dipt.). *Entomologist's Monthly Magazine* **77**, 141–153.

Collin, J.E. 1943. A revised table of the British species of *Argyra* Mcq. (Dipt., Dolichopodidae). *Entomologist's Monthly Magazine* **79**, 114–117.

Collin, J.E. 1961. *British Flies. Volume VI: Empididae*. Cambridge University Press, Cambridge.

Cordo, H.A., Sosa, A.J. and Hernández, M.C. 2000. The petiole mining fly *Thrypticus* sp. (Diptera: Dolichopodidae), a new agent for the biological control of water hyacinth (*Eichhornia crassipes*). In: Spencer, N.R. (ed.) *Proceedings of the X International Symposium on Biological Control of Weeds, 4–14 July 1999*. United States Department of Agriculture, Agricultural Research Services, Sidney, MT and Montana State University, Bozeman, MT, pp. 315–323.

Couturier, G. 1974. Présence d'une glande exsertile chez les mâles de la sous-famille des Dolichopodinae [Dipt. Dolichopodidae]. *Bulletin de la Société entomologique de France* **79**, 240–248.

Crossley, R. 1988. Mating behaviour of *Liancalus virens* (Scop.) (Dolichopodidae). *Dipterists Digest (First Series)* **1**, 45–46.

Crossley, R. 1996. Using Diptera in assessing site quality, with particular reference to Empidoidea – a regional perspective. *The Naturalist* **122**, 155–157.

Crossley, R. 1998. *Sciapus maritimus* Becker (Diptera: Dolichopodidae) in Britain. *British Journal of Entomology and Natural History* **10**, 192.

Crossley, R. 2001. Entomological report, Diptera (Tipuloidea and Empidoidea). *The Naturalist* **126**, 193–196.

Crossley, R. 2009. Entomological report, Diptera (Tipuloidea and Empidoidea). *The Naturalist* **134**, 30–33.

Crossley, R. 2014. Notes on the distribution and habitat associations of dolichopodid flies in Yorkshire. *The Naturalist* **139**, 108–111.

Crossley, R. 2018. *Hercostomus* Loew, 1857 (Diptera: Dolichopodidae), a former 'dustbin' genus, with reference to the Yorkshire species-list. *The Naturalist* **143**, 145–149.

Crossley, R. 2020. Notes on the genus *Rhaphium* Meigen, 1803 (Diptera: Dolichopodidae) in Yorkshire. *The Naturalist* **145**, 67–73.

Cumming, J.M. and Wood, D.M. 2009. Adult morphology and terminology. In: Brown, B.V., Borkent, A., Cumming, J.M., Wood, D.M., Woodley, N.E. and Zumbado, M.A. (eds) *Manual of Central American Diptera. Volume 1*. NRC Research Press, Ottawa, pp. 9–50.

Cumming, J.M. and Wood, D.M. 2017. Adult morphology and terminology. In: Kirk-Spriggs, A.H. and Sinclair, B.J. (eds) *Manual of Afrotropical Diptera. Volume 1*. Introductory chapters and keys to Diptera families. *Suricata* 4. South African National Biodiversity Institute, Pretoria, pp. 89–133.

Cunningham, A. 2015. *Achalcus britannicus* Pollet (Diptera, Dolichopodidae) in Devon. *Dipterists Digest (Second Series)* **22**, 23.

d'Assis-Fonseca, E.C.M. 1948. *Syntormon macula*, Par. (Dipt., Dolichopodidae), an addition to the British list. *Entomologist's Record and Journal of Variation* **60**, 70–71.

d'Assis-Fonseca, E.C.M. 1955. Behaviour of *Neurigona pallida* Fall. (Dolichopodidae). *Entomologist's Record and Journal of Variation* **67**, 150–151.

d'Assis-Fonseca, E.C.M. 1978. Diptera Orthorrhapha Brachycera Dolichopodidae. *Handbooks for the Identification of British Insects* **9** (5), 1–90.

Denninger, E. 1950. Zur Synonymie einiger Dolichopodidenarten (Dipt.). *Jahreshefte des Vereins für vaterländische Naturkunde in Württemberg* **102–105**, 42–48.

Diestelhorst, O. and Lunau, K. 2001. Leben in der Krone – Farbschalenfaenge von Dolichopodiden im Kronenraum einer Buche. *Mitteilungen der Deutschen Gesellschaft für Allgemeine und Angewandte Entomologie* **13**, 543–546.

Doane, R.W. 1907. Notes on the habits of *Scellus virago* Ald. *Entomological News* **18**, 136–138.

Downes, J.A. and Smith, S.M. 1969. New or little known feeding habits of Empididae (Diptera). *Canadian Entomologist* **101**, 404–408.

Drake, C.M. 1999. Flower-feeding in *Ortochile nigrocoerulea* Latreille (Diptera, Dolichopodidae). *Dipterists Digest (Second Series)* **6**, 105.

Drake, C.M. 2002. Some wetland Diptera of a disused brick-pit. *British Journal of Entomology and Natural History* **15**, 9–23.

Drake, C.M. 2005. *Dolichophorus kerteszii* Lichtwardt (Diptera, Dolichopodidae) new to Britain. *Dipterists Digest (Second Series)* **12**, 1–2.

Drake, C.M. 2007a. *Cyrturella albosetosa* (Strobl) (Diptera, Dolichopodidae) in Norfolk, England. *Dipterists Digest (Second Series)* **14**, 165–167.

Drake, C.M. 2007b. *Rhaphium suave* (Loew) (Diptera, Dolichopodidae) new to Britain. *Dipterists Digest (Second Series)* **14**, 1–4.

Drake, C.M. 2008. *Achalcus nigropunctatus* Pollet and Brunhes, 1997 (Diptera, Dolichopodidae) new to Britain. *Dipterists Digest (Second Series)* **15**, 41–43.

Drake, C.M. 2013. The distribution and abundance of the rare flies *Dolichopus laticola* Verrall and *D. nigripes* Fallén in Norfolk fens (Diptera, Dolichopodidae). *Dipterists Digest (Second Series)* **20**, 191–199.

Drake, C.M. 2015. *Dolichopus 'pectinitarsis'* Stenhammar, a parasitised form of *D. plumipes* (Scopoli) (Diptera, Dolichopodidae), found in Britain. *Dipterists Digest (Second Series)* **22**, 173–178.

Drake, C.M. 2016. The relative importance to Diptera of pasture and ditch margins on an English grazing marsh. *Dipterists Digest (Second Series)* **23**, 1–22.

Drake, C.M. 2017. The British species of *Lamprochromus* Mik (Diptera, Dolichopodidae) including *L. kowarzi* Negrobov and Chalaja new to Britain. *Dipterists Digest (Second Series)* **24**, 115–128.

Drake, C.M. 2018a. A review of the status of the Dolichopodidae flies of Great Britain – Species Status No. 30. *Natural England Commissioned Report NECR 195*. Natural England, York.

Drake, C.M. 2018b. *Diaphorus winthemi* Meigen (Diptera, Dolichopodidae) is almost certainly not British. *Dipterists Digest (Second Series)* **25**, 21–24.

Drake, C.M. 2018c. Habitat associations of the rare flies *Dolichopus laticola* and *D. nigripes* (Diptera, Dolichopodidae) in the fens of Norfolk, England. *Dipterists Digest (Second Series)* **25**, 35–52.

Drake, C.M. 2019. *Nematoproctus praesectus* Loew (Diptera, Dolichopodidae) new to Britain, found together with *N. distendens* (Meigen), and notes on their habitat preferences. *Dipterists Digest (Second Series)* **26**, 151–160.

Drake, C.M. 2020a. Two species of *Chrysotus* (Diptera, Dolichopodidae) new to the British list. *Dipterists Digest (Second Series)* **27**, 127–137.

Drake, C.M. 2020b. The identity of *Syntormon pseudospicatum* Strobl (Diptera, Dolichopodidae). *Dipterists Digest (Second Series)* **27**, 61–82.

Drake, C.M. 2021a. Comments on the taxonomic status of some British species of *Syntormon* Loew, 1857 (Diptera, Dolichopodidae). *Dipterists Digest (Second Series)* **28**, 17–44.

Drake, C.M. 2023. *Syntormon metathesis* (Loew) (Diptera, Dolichopodidae) new to Britain. *Dipterists Digest (Second Series)* **30**, 55–59.

Drake, C.M. 2024. Four species of *Medetera* new to Britain with notes on unresolved identities (Diptera, Dolichopodidae). *Dipterists Digest (Second Series)* **31**, 74–79.

Drake, C.M. and Hunnisett, J. 2014. *Neurigona erichsoni* (Zetterstedt) (Diptera, Dolichopodidae) new to Britain, and a key to the British species. *Dipterists Digest (Second Series)* **21**, 175–183.

Drake, C.M. and Painter, D. 2023. *Hercostomus rusticus* (Meigen) (Diptera, Dolichopodidae) new to Britain in urban London. *Dipterists Digest (Second Series)* **30**, 70–79.

Drake, C.M. and Pollet, M.A.A. 2016 (2017). *Dolichopus calinotus* Loew (Diptera, Dolichopodidae) new to Britain and Belgium. *Dipterists Digest (Second Series)* **23**, 231–236.

Drake, C.M. and Welter-Schultes, F. 2025. Case 3852 – *Syntormon* Loew, 1857 (Diptera, Dolichopodidae), proposed confirmation of neuter gender. *Bulletin of Zoological Nomenclature* (in press)

Drake, C.M., Crellin, S.M., Jones, N.P., Spilling, C.R. and Wolton, R.J. 2019. Diptera at two inland saltmarshes in Cheshire and Staffordshire. *Dipterists Digest (Second Series)* **26**, 73–79.

Drake, C.M., Crossley, R., Smith, M.H. and Pollet, M. 2013. *Hercostomus rothi* (Zetterstedt) (Diptera, Dolichopodidae) *stat. nov.*, new to Britain. *Dipterists Digest (Second Series)* **20**, 121–130.

Drake, C.M., Godfrey, A. and Gibbs, D.J. 2023. *Thrypticus* and *Corindia* in Britain, with the description of two new species and the addition of two species to the British list (Diptera, Dolichopodidae, Medeterinae). *Dipterists Digest (Second Series)* **30**, 172–199.

Drake, C.M., Godfrey, A. and Perry, I. 2024. Two previously unillustrated species of *Medetera* described by British authors, *M. oscillans* and *M. unisetosa* (Diptera, Dolichopodidae). *Dipterists Digest (Second Series)* **31**, 166–1175.

Drake, C.M., Lott, D.A., Alexander, K.N.A. and Webb, J. 2007. Surveying terrestrial and freshwater invertebrates for conservation evaluation. *Natural England Research Report NERR005*. Natural England, York.

Drake, M. 2021b. *Syntormon macula* phenology. *Empidid and Dolichopodid Recording Scheme Newsletter* No. 26, 2.

Dyte, C.E. 1959. Some interesting habitats of larval Dolichopodidae (Diptera). *Entomologist's Monthly Magazine* **95**, 139–143.

Dyte, C.E. 1987. A third British species of *Xanthochlorus*. *Empid and Dolichopodid Study Group Newsheet* No. 3, 4.

Dyte, C.E. 1988a. Mate guarding and sex ratio in *Hydrophorus oceanus* (Macq.) (Diptera, Dolichopodidae). *The Entomologist* **107**, 122–126.

Dyte, C.E. 1988b. Some remarks on *Ortochile nigrocoerulea*. *Empid and Dolichopodid Study Group Newsheet* No. 5, 1–2.

Dyte, C.E. 1990. A putative swain for *Syntormon setosus*. *Empid and Dolichopodid Study Group Newsheet* No. 9, 6–7.

Dyte, C.E. 1993a. A new catalogue of Palaearctic Dolichopodidae. *Empid and Dolichopodid Study Group Newsheet* No. 12, 6–9.

Dyte, C.E. 1993b. The occurrence of *Thrypticus smaragdinus* Gerst. (Diptera: Dolichopodidae) in Britain, with remarks on plant hosts in the genus. *The Entomologist* **112**, 81–84.

Dyte, C.E. 1996. *Medetera bispinosa* Negrobov new to Britain. *Empid and Dolichopodid Study Group Newsheet* No. 14, 4–5.

Dyte, C.E. and Poulding, R.H. 1992. The distribution of *Aphrosylus mitis* (Diptera, Dolichopodidae). *Dipterists Digest (First Series)* **12**, 32–38.

Ebejer, M.J. 2021. Flies of the family Dolichopodidae (Diptera: Brachycera), excluding the Microphorinae and Parathalassinae, from the Maltese Islands with notes on their habitats and phenology. *Zoosystematica Rossica* **30**, 25–39.

Emeis, W. 1964. Untersuchungen über die ökologische Verbreitung der Dolichopodiden (Ins. Dipt.) in Schleswig-Holstein. *Schriften des Naturwissenschaftlichen Vereins für Schleswig-Holstein* **35**, 61–75.

Evenhuis, N.L. 1996. New records and synonymies of Hawaiian Diptera. *Bishop Museum Occasional Papers* **46**, 27–28.

Falk, S. and Crossley, R. 2005. *A review of the scarce and threatened flies of Great Britain. Part 3. Empidoidea. Species Status No. 3*. Joint Nature Conservation Committee, Peterborough.

Gelbič, I. and Olejníček, J. 2011. Ecology of Dolichopodidae (Diptera) in a wetland habitat and their potential role as bioindicators. *Central European Journal of Biology* **6**, 118–129.

Germann, C., Pollet, M. and Bernasconi, M.V. 2011b. Aspects of European *Argyra* systematics, molecular insights and morphology (Diptera: Dolichopodidae). *Entomologica Fennica* **22**, 5–14.

Germann, C., Pollet, M., Wimmer, C. and Bernasconi, M.V. 2011a. Molecular data sheds light on the classification of long-legged flies (Diptera: Dolichopodidae). *Invertebrate Systematics* **25**, 303–321.

Gibbs, D. 2006. *Dolichopus excisus* Loew, 1859 (Diptera, Dolichopodidae) new to Britain discovered in southern England. *Dipterists Digest (Second Series)* **13**, 5–10.

Gibbs, D. 2007 (2008). *Medetera insignis* Girschner, 1888 (Diptera, Dolichopodidae) new to Britain. *Dipterists Digest (Second Series)* **14**, 95–97.

Godfrey, A. and Drake, C.M. 2012. *Asyndetus latifrons* (Loew) (Diptera, Dolichopodidae) new to Britain. *Dipterists Digest (Second Series)* **19**, 39–44.

Godfrey, A. and Drake, C.M. 2024. *Hercostomus argentifrons* (Diptera, Dolichopodidae) in Britain. *Dipterists Digest (Second Series)* **31**, 133–138.

Grichanov, I. Ya. and Ahmadi, A. 2017. Palaearctic species of the genus *Lamprochromus* Mik, 1878 (Diptera, Dolichopodidae). *Far Eastern Entomologist* **336**, 1–12.

Grichanov, I.Ya. 1997. Notes on Afrotropical and Palaearctic species of the genus *Thinophilus* Wahlberg with descriptions of new species. *International Journal of Dipterological Research* **8**, 135–147.

Grichanov, I.Ya. 2006. A checklist and keys to North European genera and species of Dolichopodidae (Diptera). *VIZR RAAS (Plant Protection News Supplement)*, pp. 1–120. St Petersburg.

Grichanov, I.Ya. 2007. A checklist and keys to Dolichopodidae (Diptera) of the Caucasus and East Mediterranean. *VIZR RAAS (Plant Protection News Supplement)*, pp. 1–160. St Petersburg.

Grichanov, I.Ya. 2009. Review of the genus *Dolichophorus* Lichtwardt, 1902 (Diptera, Dolichopodidae, Medeterinae). *Far Eastern Entomologist* **201**, 1–16.

Grichanov, I.Ya. 2011. Species of the genus *Telmaturgus* Mik, 1874 (Diptera: Dolichopodidae). *Caucasian Entomological Bulletin* **7**, 229–232.

Grichanov, I.Ya. 2013. Systematic notes on West-Palaearctic species of the genus *Syntormon* Loew (Diptera, Dolichopodidae). *VIZR RAAS (Plant Protection News Supplement)*, pp. 3–26. St Petersburg.

Grichanov, I.Ya. 2017. Alphabetic list of generic and specific names of predatory flies of the epifamily Dolichopodoidae (Diptera). Second edition. *VIZR RAAS (Plant Protection News Supplement)* **23**, pp. 1–156. St Petersburg.

Grichanov, I.Ya. and Kazerani, F. 2014. A new species of *Sybistroma* Meigen (Diptera: Dolichopodidae) from the Middle East with a key to West-Palaearctic species of the genus. *Zootaxa* **3866**, 572–582.

Grichanov, I.Ya. and Negrobov, O.P. 2014. Palaearctic species of the genus *Sciapus* Zeller (Diptera, Dolichopodidae). *All-Russian Institute of Plant Protection RAAS* **13**, 1–85.

Grichanov, I.Ya. and Viklund, B. 2007. Dolichopodidae (Diptera) new to the fauna of Sweden. *International Journal of Dipterological Research* **18**, 113–118.

Grichanov, I.Ya., Negrobov, O.P. and Selivanova, O.V. 2011a. Keys to Palaearctic subfamilies and genera of the family Dolichopodidae (Diptera). *Cesa News* **62**, 13–46.

Grichanov, I.Ya., Negrobov, O.P. and Selivanova, O.V. 2012. Review of Palearctic *Teuchophorus* Loew (Diptera, Dolichopodidae) with an updated catalog and revised key to species. *Russian Entomological Journal* **21**, 89–96.

Grichanov, I.Ya., Selivanova, O.V. and Negrobov, O.P. 2011b. A brief synopsis of Palaearctic genera of the family Dolichopodidae (Diptera). *Ukrainska entomofaunistyka* **2**, 1–42.

Grootaert, P. and Meuffels, H.J.G. 1997. Dolichopodidae (Diptera) from Papua New Guinea. 15. *Scepastopyga* gen. nov. and the establishment of a new subfamily, the Achalcinae. *Journal of Natural History* **31**, 1587–1600.

Grootaert, P., Haghebaert, G. and Pollet, M. 1987. Some *Medetera* (Diptera, Dolichopodidae) associated with Scolytidae (Coleoptera) from elms. *Bulletin et Annales de la Société royale belge d'Entomologie* **123**, 380–382.

Halstead, A.J. 2003. Some scarce or local Diptera collected in 2002. 2002 Annual Exhibition. *British Journal of Entomology and Natural History* **16**, 179.

Hanson, M.W. 1986. *Liancalus virens* (Scop.) (Diptera, Dolichopodidae) on brickwork seepages in east London. *Journal of Natural HistoryProceedings and Transactions of the British Entomological and Natural History Society* **19**, 86.

Hodge, P.J. 1992. *Lamprochromus strobli* (Dolichopodidae) confirmed as a British species. *Dipterists Digest (First Series)* **12**, 28–29.

Hodge, P.J. 1993a. *Hercostomus verbekei* Pollet in Britain. *Empid and Dolichopodid Study Group Newsheet* No. 12, 2.

Hodge, P.J. 1993b. Two species of *Syntormon* confused under the name *monilis* (Haliday). *Empid and Dolichopodid Study Group Newsheet* No. 12, 3.

Howe, M. 1997. Abergavenny Field Meeting – June 1996. *Bulletin of the Dipterists Forum* **44**, 11–12.

Hulcr, J., Pollet, M., Ubik, K. and Vrkoč, J. 2005. Exploitation of kairomones and synomones by *Medetera* spp. (Diptera: Dolichopodidae), predators of spruce bark beetles. *European Journal of Entomology* **102**, 655–662.

Irwin, A.G. 2010a. Curating. In: Chandler, P.J. (ed.) *A Dipterist's Handbook*. The Amateur Entomologists' Society, London, pp. 81–95.

Irwin, A.G. 2010b. Equipment. In: Chandler, P.J. (ed.) *A Dipterist's Handbook*. The Amateur Entomologists' Society, London, pp. 75–81.

Jonassen, T. 1990. Notes on Norwegian Dolichopodidae (Diptera). *Fauna Norvegica (B)* **37**, 105–106.

Jonassen, T., Andersen, T. and Kvifte, G.M. 2013. Empidoidea (Diptera) from Finnmark, northern Norway. *Norwegian Journal of Entomology* **30**, 201–245.

Kahanpää, J. 2008. Nematode-induced demasulinisation of *Dolichopus* males (Diptera, Dolichopodidae) *Zootaxa* **1689**, 51–62.

Kahanpää, J. and Grichanov, I.Ya. 2006. Dolichopodidae (Diptera) new for the fauna of Finland. *Entomologica Fennica* **17**, 73–78.

Kassebeer, C.F. 1998. Die paläarktischen Arten der Gattung *Systenus* Loew, 1857 (Diptera, Dolichopodidae). 1. Nomenklatur. *Dipteron* **1**, 10–20.

Kejval, K. 2022. Novy nalez *Acropsilus niger* (Diptera, Dolichopodidae) v zapadnich Cechach. (New record of *Acropsilus niger* (Diptera, Dolichopodidae) from western Bohemia). *Zapadoceske entomologicke listy* **13**, 73–74.

Kejval, Z. and Pollet, M. 2023. The genus *Medetera* (Diptera: Dolichopodidae) in the Czech Republic with first records of twelve species. *Zootaxa* **5245**, 69–93.

Kenis, M., Wermelinger, B. and Grégoire, J.C. 2004. Research on parasitoids and predators of Scolytidae – a review. In: Lieutier, F., Day, K.R., Battisti, A., Grégoire, J.C. and Evans, H.F. (eds) *Bark and Wood Boring Insects in Living Trees in Europe, A Synthesis*. Kluwer Academic Publishers, Dordrecht, pp. 237–290.

Kowarz, F. 1869. Beitrag zur Dipteren-Fauna Ungarns. *Verhandlungen der Zoologisch-Botanischen Gesellschaft in Wien* **19**, 561–566.

Kowarz, F. 1874. Die Dipteren-Gattung *Chrysotus* Meig. *Verhandlungen der Zoologisch-Botanischen Gesellschaft in Wien* **24**, 453–478.

Land, M.F. 1993a. The visual control of courtship behaviour in the fly *Poecilobothrus nobilitatus*. *Journal of Comparative Physiology A* **173**, 595–603.

Land, M.F. 1993b. Chasing and pursuit in the dolichopodid fly *Poecilobothrus nobilitatus*. *Journal of Comparative Physiology A* **173**, 605–613.

Latreille, P.A. 1809. *Genera crustaceorum et insectorum secundum ordinen naturalem in familias disposita, iconibus exemplisque plurimis explicata*. Parisiis et Argentorati [= Paris and Strasbourg] **4**, 1–399, Figs I–XVI [1–16].

Laurence, B.R. 1948. Observations on *Microphorus crassipes* Macquart (Dipt., Empididae). *Entomologist's Monthly Magazine* **84**, 282–283.

Laurence, B.R. 1993. *Neurigona abdominalis* Fall. (Dipt., Dolichopodidae) in Norfolk. *Entomologist's Monthly Magazine* **129**, 166.

Lim, G.S., Hwang, W.S., Kutty, S.N., Meier, R. and Grootaert, P. 2010. Mitochondrial and nuclear markers support the monophyly of Dolichopodidae and suggest a rapid origin of the subfamilies (Diptera, Empidoidea). *Systematic Entomology* **35**, 59–70.

Lloyd, R.W. 1944. Dolichopodid fly associated with bark-beetles. *Entomologist's Monthly Magazine* **80**, 180.

Loew, H. 1859. Neue Beiträge zur Kenntniss der Dipteren. Sechster Beitrag. *Programm der Königlichen Realschule zu Meseritz*, 1–50.

Lübben, H. 1908. *Thrypticus smaragdinus* Gerst und seine Lebensgeschichte. *Zoologische Jahrbucher* **26**, 319–332.

Lunau, K. 1992. Mating behaviour in the long-legged fly *Poecilobothrus nobilitatus* L. (Diptera, Dolichopodidae), courtship behaviour, male signalling and mating success. *Zoologische Beiträge* **34**, 465–479.

Lunau, K. 1996. Das balzverhalten von langbeinfliegen (Diptera, Dolichopodidae). *Acta Albertina Ratisbonensia* **50**, 49–73.

Lunau, K., Middelmann, A. and Pianka, M. 2006. Density- and food-resource-dependent courtship behaviour in the fly *Poecilobothrus nobilitatus* L. (Diptera, Dolichopodidae). *Entomologie Heute* **18**, 123–132.

MacGowan, I. 1987. *A Preliminary Survey of Dolichopodidae (Diptera) in Scotland*. Nature Conservancy Council, Edinburgh.

MacGowan, I. 1997. *Systenus mallochi* sp. n. from Britain (Diptera, Dolichopodidae). *Dipterists Digest (Second Series)* **4**, 24–29.

MacGowan, I. 2001. *Medetera freyi* Thuneberg, *M. setiventris* Thuneberg and *M. fasciata* Frey (Diptera, Dolichopodidae) new to Britain with notes on the status of *Medetera striata* Parent. *Dipterists Digest (Second Series)* **8**, 85–90.

MacGowan, I. and Drake, C.M. 2023. The correct name for the montane *Hydrophorus* species (Diptera, Dolichopodidae) occurring in the British Isles. *Dipterists Digest (Second Series)* **30**, 51–54.

Maes, D. and Pollet, M. 1997. Dolichopodid communities (Diptera, Dolichopodidae) in "De Kempen" (eastern Belgium), biodiversity, faunistics and ecology. *Bulletin et Annales de la Société royale belge d'Entomologie* **133**, 419–438.

Maslova, O.O., Negrobov, O.P. and Selivanova, O.V. 2011. Russian fauna of the genus *Chrysotus* (Diptera, Dolichopodidae), Part 1. *Chrysotus cilipes* Meigen and *Ch. laesus* Wied, species-group. *Entomological Review* **91**, 969–972.

McAlpine, J.F. 1981. Morphology and terminology – adults. In: McAlpine, J.F. (ed.) *Manual of Nearctic Diptera. Volume 1. Canada Department of Agriculture Research Branch Monograph* No. 27, pp. 9–63.

Meigen, J.W. 1824. *Systematische Beschreibung der bekannten europäischen zweiflügeligen Insekten*. Hamm, Verlag der Schultz-Wundermann'schen Buchhandlung, **4**, vii + 428 pp. + plates 33–41.

Meuffels, H.J.G. and Grootaert, P. 1990. The identity of *Sciapus contristans* (Wiedemann, 1817) (Diptera, Dolichopodidae), and a revision of the species group of its relatives. *Bulletin de l'Institut Royal des Sciences Naturelles de Belgique* **60**, 161–178.

Moulton, J.K. and Wiegmann, B.M. 2007. The phylogenetic relationships of flies in the superfamily Empidoidea (Insecta, Diptera). *Molecular Phylogenetics and Evolution* **43**, 701–713.

Naglis, S. and Barták, M. 2015. Dolichopodidae (Diptera) from the Iberian Peninsula, with description of three new species. *Zootaxa* **3964**, 125–137.

Negrobov, O.P. 1977. 29. Dolichopodidae. In: Lindner, E. (ed.) *Die Fliegen der palaearktischen Region* **4** (5), issue 316, 347–386, plates 108–150.

Negrobov, O.P. 1978. 29. Dolichopodidae. In: Lindner, E. (ed.) *Die Fliegen der palaearktischen Region* **4** (5), issue 319, 387–418, plates 151–173.

Negrobov, O.P. 1979a. 29. Dolichopodidae. In: Lindner, E. (ed.) *Die Fliegen der palaearktischen Region* **4** (5), issue 321, 419–474, plates 174–188.

Negrobov, O.P. 1979b. 29. Dolichopodidae. In: Lindner, E. (ed.) *Die Fliegen der palaearktischen Region* **4** (5), issue 322, 475–530, plates 189–207.

Negrobov, O.P. 1986. On the system and phylogeny of flies of the family Dolichopodidae (Diptera) [in Russian + English summary]. *Entomologicheskoe Obozrenie* **65**, 182–186.

Negrobov, O.P. 1991. Family Dolichopodidae. In: Soós, Á. and Papp, L. (eds) *Catalogue of Palaearctic Diptera. Volume 7, Dolichopodidae – Platypezidae*. Elsevier, Amsterdam, pp. 11–139.

Negrobov, O.P. 2000. Revision of the Palaearctic species of the genus *Micromorphus* Mik (Diptera: Dolichopodidae). *International Journal of Dipterological Research* **11**, 19–26.

Negrobov, O.P. and Chandler, P.J. 2006 (2007). The status of *Chrysotus angulicornis* Kowarz (Diptera, Dolichopodidae) and its deletion from the British list. *Dipterists Digest (Second Series)* **13**, 103–109.

Negrobov, O.P. and Maslova, O.O. 1996. Revision of the Palearctic species of the genus *Chrysotus* Mg. (Diptera, Dolichopodidae) II. *Entomological Review* **75**, 110–121.

Negrobov, O.P. and Naglis, S. 2016. Palaearctic species of the genus *Medetera* (Diptera, Dolichopodidae). *Zoosystematica Rossica* **25**, 333–379.

Negrobov, O.P. and Naglis, S. 2020. Two new species of *Thrypticus* Gerstäcker from Turkey (Diptera, Dolichopodidae), with checklist and key to Palaearctic species. *Zootaxa* **4858**, 111–125.

Negrobov, O.P. and Stackelberg, A.A. 1971. 29. Dolichopodidae. In: Lindner, E. (ed.) *Die Fliegen der palaearktischen Region* **4** (5), issue 284, 225–256, plates 13–28.

Negrobov, O.P. and Stackelberg, A.A. 1972. 29. Dolichopodidae. In: Lindner, E. (ed.) *Die Fliegen der palaearktischen Region* **4** (5), issue 289, 257–302, plates 29–44.

Negrobov, O.P. and Stackelberg, A.A. 1974a. 29. Dolichopodidae. In: Lindner, E. (ed.) *Die Fliegen der palaearktischen Region* **4** (5), issue 302, 303–324, plates 45–76.

Negrobov, O.P. and Stackelberg, A.A. 1974b. 29. Dolichopodidae. In: Lindner, E. (ed.) *Die Fliegen der palaearktischen Region* **4** (5), issue 303, 325–346, plates 77–107.

Negrobov, O.P. Tsurikov, M.N. and Maslova, O.O. 2000. Revision of the Palaearctic species of the genus *Chrysotus* Mg. (Diptera, Dolichopodidae). III. *Entomological Review* **80**, 877–886.

Negrobov, O.P., Barkalov, A. and Selivanova, O.B. 2012. Species of the genus *Rhaphium* Meigen (Diptera, Dolichopodidae) from the Taimyr Peninsula (Russia), with description of a new species. *Zootaxa* **3548**, 75–87.

Negrobov, O.P., Barkalov, A.V. and Selivanova, O.V. 2014. Two new species of *Chrysotus* Meigen (Diptera, Dolichopodidae) from Siberia, with a key to the Siberian species. *Zootaxa* **3815**, 409–416.

Negrobov, O.P., Chursina, M.A. and Selivanova, O.V. 2018. Morphology and morphometry of the tibial organ of males of the genus *Dolichopus* Latreille, 1796 (Dolichopodidae, Diptera) [in Russian]. *Acta Biologica Sibirica* **4**, 46–51.

Negrobov, O.P., Grichanov, I.Ya and Selivanova, O.V. 2013. Palearctic species of the *Rhaphium albifrons* group (Diptera, Dolichopodidae). *Eurasian Entomological Journal* **12**, 601–606.

Negrobov, O.P., Manko, P. and Oboňa, J. 2020. A new species of *Systenus* Loew (Dolichopodidae, Diptera) from Slovakia. *Acta Musei Silesiae, Scientiae Naturales* **69**, 161–164.

Negrobov, O.P., Maslova, O.O. and Selivanova, O.V. 2007. New data on systematics of family Dolichopodidae (Diptera). *Russian Entomological Journal* **16**, 243–244.

Noble, M.J. 2016. New observations on the mating-related behaviours of the dolichopodid fly *Poecilobothrus nobilitatus* (Linnaeus). *Dipterists Digest (Second Series)* **23**, 67–74.

Oboňa, J., Pollet, M. and Naglis, S. 2012. First records of one genus and three species of long-legged flies (Diptera, Dolichopodidae) from Slovakia. *Folia faunistica Slovaca* **17**, 357–360.

Oosterbroek, P. 2006. *The European Families of the Diptera. Identification, Diagnosis, Biology*. KNNV Publishing, Utrecht.

Pape, T., Bickel, D. and Meier, R. 2009. *Diptera Diversity, Status, Challenges and Tools*. Koninklijke Brill NV, Leiden.

Parent, O. 1933. Étude monographique sur les diptères dolichopodides de Nouvelle-Zélande. *Mémoires de la Société Scientifique de Bruxelles (B)* **53**, 325–441.

Parent, O. 1938. Diptères Dolichopodidae. *Faune de France* **35**, 1–720. Lechevalier, Paris.

Parmenter, L. 1942. *Chrysotus pallipalpus* Van Duzee (= *elegans* Parent) (Dipt., Dolichopodidae) in Britain. *Entomologist's Monthly Magazine* **78**, 232–233.

Pârvu, C. 1984. Data about family Dolichopodidae (Diptera) in Romania (II). *Travaux du Muséum National d'Histoire Naturelle "Grigore Antipa"* **26**, 209–215.

Pârvu, C. 1997. Dolichopodidae (Diptera, Dolichopodidae) from Iza Valley, Maramures (Romania) (XII). *Travaux du Muséum National d'Histoire Naturelle "Grigore Antipa"* **37**, 113–124.

Pârvu, C. 2000. Dolichopodidae (Diptera) from Bucovina, Romania (XVII). *Travaux du Muséum National d'Histoire Naturelle "Grigore Antipa"* **42**, 143–155.

Pârvu, C. 2001. Dolichopodidae (Diptera) from Romania (XIX). Additional data from the Maramures depression, Romania. *Travaux du Muséum National d'Histoire Naturelle "Grigore Antipa"* **43**, 249–262.

Perry, I. 1999. *Campsicnemus umbripennis hispanicus* Strobl (Diptera, Dolichopodidae) new to Britain. *Dipterists Digest (Second Series)* **6**, 118–120.

Plant, A.R. 2006. Further records of *Sciapus basilicus* Meuffels and Grootaert, 1990 (Diptera, Dolichopodidae) from the River Usk in Wales. *Dipterists Digest (Second Series)* **13**, 42.

Plant, A.R. and Cole, J.H. 2005 (2006). *Microphor strobli* Chvála, 1986 (Diptera, Empidoidea, Microphoridae) new to Great Britain. *Dipterists Digest (Second Series)* **12**, 141–142.

Pollet, M. 1990. Phenetic and ecological relationships between species of the subgenus *Hercostomus* (*Gymnopternus*) in western Europe, with the description of two new species (Diptera: Dolichopodidae). *Systematic Entomology* **15**, 359–382.

Pollet, M. 1991. Contributions to the knowledge of dolichopodid flies in Belgium, 4. Ecology and faunistics of dolichopodid flies collected in some nature reserves (Diptera: Dolichopodidae). *Phegea* **19**, 75–85.

Pollet, M. 1993. Morphological and ecological characterization of *Hercostomus* (*Hercostomus*) *plagiatus* and a sibling species, *H. verbekei* sp. n. (Diptera, Dolichopodidae). *Zoologica Scripta* **22**, 101–109.

Pollet, M. 1996. Systematic revision and phylogeny of the Palaearctic species of the genus *Achalcus* Loew (Diptera: Dolichopodidae) with the description of four new species. *Systematic Entomology* **21**, 353–386.

Pollet, M. 2001. Dolichopodid biodiversity and site quality assessment of reed marshes and grasslands in Belgium (Diptera: Dolichopodidae). *Journal of Insect Conservation* **5**, 99–116.

Pollet, M. 2002. 4.6 Slankpootvliegen [Slender-legged flies]. In: Dekoninck, W., Versteirt, V. and Grootaert, P. (eds) Praktijkgericht onderzoek naar kansen en belangrijke stuurvariabelen voor natuurontwikkeling op gronden met voormalig intensief landbouwgebruik. Deel IV, Invertebraten. [Practice-oriented research into opportunities and important steering variables for nature development on land with former intensive agricultural use. Part IV, Invertebrates.] Study carried out on behalf of the Flemish Community within the framework of the Flemish Impulse Programme for Nature Development on behalf of the Flemish Minister responsible for Nature Conservation [in Dutch]. Report ENT.2002.01, pp. 167–195.

Pollet, M. 2003 (2004). A critical note on the systematic position of *Gymnopternus* (Diptera, Dolichopodidae). *Studia Dipterologica* **10**, 537–548.

Pollet, M. 2005. Systematic revision of Neotropical *Achalcus* and a related new genus (Diptera: Dolichopodidae, Achalcinae) with comments on their phylogeny, ecology and zoogeography. *Zoological Journal of the Linnaean Society* **143**, 27–73.

Pollet, M. 2009. Diptera as ecological indicators of habitat and habitat change. In: Pape, T., Bickel, D. and Meier, R. (eds) *Diptera Diversity, Status, Challenges and Tools*. Koninklijke Brill NV, Leiden, pp. 302–322.

Pollet, M. 2011. Fauna Europaea, Dolichopodidae. In: Pape, T. and Beuk, P. (eds) *Fauna Europaea, Diptera, Brachycera*. Fauna Europaea version 2.4. http://www.faunaeur.org (accessed 2003, no longer available).

Pollet, M. and Grootaert, P. 1987. Ecological data on Dolichopodidae (Diptera) from a woodland ecosystem, 1. Colour preference, detailed distribution and comparison of different sampling techniques. *Bulletin de l'Institut Royal des Sciences Naturelles de Belgique Entomologie* **57**, 173–186.

Pollet, M. and Grootaert, P. 1994. Optimizing the water trap technique to collect Empidoidea (Diptera). *Studia Dipterologica* **1**, 33–48.

Pollet, M. and Grootaert, P. 1996. An estimation of the natural value of dune habitats using Empidoidea (Diptera). *Biodiversity and Conservation* **5**, 859–880.

Pollet, M. and Grootaert, P. 1999. Dolichopodidae (Diptera), poorly known but excellent agents for site quality assessments and nature conservation. *Proceedings of the Section Experimental and Applied Entomology of the Netherlands Entomological Society (NEV), Amsterdam* **10**, 63–68.

Pollet, M. and Stark, A. 2018. The quest for the identity of *Orthoceratium lacustre* (Scopoli, 1763) reveals centuries of misidentifications (Diptera, Dolichopodidae). *ZooKeys* **782**, 49–79.

Pollet, M., Andrade, R., Goncalves, A., Andrade, P., Jacinto, V., Almeida, J., De Braekeleer, A., Calster, H. van and Brosens, D. 2019. Dipterological surveys in Portugal unveil 200 species of long-legged flies, with over 170 new to the country (Diptera: Dolichopodidae). *Zootaxa* **4649**, 1–69.

Pollet, M., Barták, M. and Roháček, J. 2009. Dolichopodidae. In: Roháček, J. and Ševčík, J. (eds), *Diptera of the Poľana Protected Landscape Area – Biosphere Reserve (Central Slovakia)*. SNC SR, Adminstration of the PLA – BR Poľana, Zvolen, pp. 141–151.

Pollet, M., De Braekeleer, A., Drake, C.M. and Van de Meutter, F. 2017. The rediscovery of *Orthoceratium lacustre* (Scopoli, 1763) (Diptera, Dolichopodidae) in Belgium, with data on its ecology and distribution in the Palaearctic region. *Biologia* **72**, 62–69.

Pollet, M., Dekonink, W. and Grootaert, P. 2005. Faunistics of the dolichopodid fauna (Diptera: Dolichopodidae) of the Voeren region (Belgium) with new records of species of special faunistic interest. *Bulletin de l'Institut Royal des Sciences Naturelles de Belgique Entomologie* **75**, 291–303.

Pollet, M., Drake, C.M., Jacobs, M. and Stark, A. 2025. Discovery of two new European tree trunk-dwelling *Medetera* species (Diptera, Dolichopodidae). *Taxonomy* **5** (1), 7.

Pollet, M., Germann, C. and Bernasconi, M.V. 2011. Phylogenetic analyses using molecular markers reveal ecological lineages in *Medetera* (Diptera: Dolichopodidae). *Canadian Entomologist* **143**, 662–673.

Pollet, M., Germann, C., Tanner, S. and Bernasconi, M.V. 2010. Hypotheses from mitochondrial DNA, congruence and conflict between DNA sequences and morphology in Dolichopodinae systematics (Diptera: Dolichopodidae). *Invertebrate Systematics* **24**, 32–50.

Pollet, M., Grootaert, P. and Meuffels, H. 1988. Dolichopodid species new to the Belgian fauna with notes on their habitat (Diptera: Dolichopodidae). *Entomologische Berichten, Amsterdam* **48**, 44–46.

Pollet, M., Meuffels, H. and Grootaert, P. 1992. Geographical distribution and habitat selection of species of *Hercostomus* subgenus *Gymnopternus* in the Benelux (Diptera: Dolichopodidae). *Faunal Inventories of Sites for Cartography and Nature Conservation: Proceedings of the 8th International Colloquium of the European Invertebrate Survey, Brussels, 9–10 September 1991.* Royal Belgian Institute of Natural Sciences, Brussels, pp. 101–113.

Pollet, M., Persson, M., Bøggild, E, and Crossley, R. 2015. A long-lasting taxonomic problem in European *Sympycnus* resolved, with the description of a new species and data on habitat preferences. *Zootaxa* **4032**, 81–102.

Pollet, M.A.A. and Hulcr, J. 2007. The updated check list of Dolichopodidae (Diptera) of the Czech Republic and Slovakia. Background information, data and considerations. *Biologia, Bratislava* **62**, 470–476.

Pollet, M.A.A., Brooks, S.E. and Cumming, J.M. 2004. Catalog of the Dolichopodidae (Diptera) of America north of Mexico. *Bulletin of the American Museum of Natural History* **283**, 1–114.

Pollet, M. 2003 (2004). A critical note on the systematic position of *Gymnopternus* (Diptera, Dolichopodidae). *Studia dipterologica* **10**, 537–548.

Poulding, R. 1992. The status of *Dolichopus andalusiacus* in West Cornwall. *Empid and Dolichopodid Study Group Newsheet* No. 11, 6–9.

Poulding, R.H. 1998. The larvae of *Aphrosylus celtiber* Haliday (Diptera, Dolichopodidae) as predators of the littoral barnacle *Chthamalus montagui* Southward (Cirripedia, Chthamalidae). *Dipterists Digest (Second Series)* **5**, 70–74.

Poulding, R.H. 2010. Plastron respiration in the larvae of the intertidal fly *Aphrosylus celtiber* Haliday (Diptera, Dolichopodidae) *Dipterists Digest (Second Series)* **17**, 117–122.

Poulding, R.H. and Tompsett, P.E. 2007. The marine intertidal life of the Helford River, Cornwall. *Newsletter of the Porcupine Marine Natural History Society* **22**, 15–18.

Ratnasingham, S. and Hebert, P.D.N. 2007. BOLD, The Barcode of Life Data System (http://www.barcodinglife.org). *Molecular Ecology Notes* **7**, 355–364.

Ringdahl, O. 1928. Förteckning över de i Sverige hittills iakttagna arterna av familjen Dolichopodidae (Diptera). *Entomologisk Tidskrift* **49**, 179–201.

Robinson, H. 1970. Family Dolichopodidae (Dolichopidae). In: Papavero, N. (ed.) *A Catalogue of the Diptera of the Americas South of the United States* **40**, 1–92. Universidade de São Paulo, Museu de Zoologia.

Roháček, J., Barták, M., Čelechovsky, A., Grootaert, P., Kanavalová, L., Mazánek, L. and Pollet, M. 2021. New and interesting records of Diptera on glacial sand deposits in Silesia (NE Czech Republic). Part 2 – Brachycera except for Schizophora. *Acta Musei Silesiae, Scientiae Naturales* **70**, 1–32.

Rotheray, G.E. and Shaw, M.R. 1989. *Medetera* cocoons under bark (Dipt., Dolichopodidae). *Entomologist's Monthly Magazine* **125**, 174.

Runyon, J.B. 2008. The cingulum, a unique structure of some Dolichopodidae. *Fly Times* **41**, 3.

Selivanova, O.V., Negrobov, O.P. and Maslova, O.O. 2019. New data on the systematics and fauna of *Dolichopus subpennatus* D'Assis Fonseca, 1976 and *Dolichopus pennatus* Meigen, 1824 (Dolichopodidae, Diptera). *Acta Biologica Sibirica* **5**, 111–114.

Sinclair, B.J. and Cumming, J.M. 2006. The morphology, higher-level phylogeny and classification of the Empidoidea (Diptera). *Zootaxa* **1180**, 1–172.

Sivell, D. 2016. *Liancalus virens* habitat preferences. *Empidid and Dolichopodid Recording Scheme Newsletter* No. 21, 1–2.

Sivell, D. 2019. *Sciapus pallens* (Wiedemann) (Diptera, Dolichopodidae) new to Britain. *Dipterists Digest (Second Series)* **26**, 197–202.

Skidmore, P. 1977. Diptera. Entomological reports for 1975–76. *Naturalist* **102**, 77–81.

Smith, K.G.V. 1959. A note on the courtship and predaceous behaviour of *Neurigona* species (Diptera, Dolichopodidae). *Entomologist's Monthly Magazine* **95**, 32–33.

Speight, M.C.D. 1987. *Achalcus melanotrichus*, *Systenus leucurus* and *S. pallipes* (Diptera, Dolichopodidae) new to Ireland, bred from rot-holes in *Betula*, *Fagus* and *Acer pseudoplatanus*, with a comment on the significance of these records. *Irish Naturalists' Journal* **22**, 250–252.

Speight, M.C.D. and Meuffels, H.J.G. 1989. *Campsicnemus compeditus*, *Melanostolus melancholicus*, *Syntormon setosus* and *Systenus pallidus* (Diptera, Dolichopodidae), insects new to Ireland. *Irish Naturalists' Journal* **23**, 92–97.

Speight, M.C.D., Blackith, R.M. and Blackith, R.E. 1995. *Bathycranium*, synonymised with *Syntormon*, distinction between *Parasyntormon* and *Syntormon* discussed and *S. bicolorellus* and *S. luteicornis* (Diptera: Dolichopodidae) redescribed. *Insecta Mundi* **9**, 351–362.

Speight, M.C.D., Blackith, R.M. and de Courcy Williams, M. 1992. *Calliopum elisae*, *Hercostomus nanus*, *Psacadina verbekei*, *Systenus alpinus* and *Zabrachia tenella*. Insects new to Ireland. *Irish Naturalists' Journal* **24**, 147–151.

Stackelberg, A. 1930. 29. Dolichopodidae. In: Lindner, E. (ed.) *Die Fliegen der Palaearktischen Region* **4** (5), issue 51, 1–64, plates 1–2.

Stackelberg, A. 1933. 29. Dolichopodidae. In: Lindner, E. (ed.) *Die Fliegen der Palaearktischen Region* **4** (5), issue 71, 65–128, plates 3–4.

Stackelberg, A. 1934. 29. Dolichopodidae. In: Lindner, E. (ed.) *Die Fliegen der Palaearktischen Region* **4** (5), issue 82, 129–176.

Stackelberg, A. 1941. 29. Dolichopodidae. In: Lindner, E. (ed.) *Die Fliegen der Palaearktischen Region* **4** (5), issue 138, 177–224, plates 5–12.

Steyskal, G. 1973. The North American Species of *Dolichopus* Latreille, Group B (Diptera, Dolichopodidae). *Journal of the Kansas Entomological Society* **46**, 347–359.

Ulrich, H. 1981. Zur systematischen Gliederung der Dolichopodiden (Diptera). *Bonner Zoologische Beitraege* **31**, 385–402.

Ulrich, H. 2004. Predation by adult Dolichopodidae (Diptera), a review of literature with an annotated prey–predator list. *Studia dipterologica* **11**, 369–403.

Verrall, G.H. 1875. Notes on some British Dolichopodidae, with descriptions of new species. *Entomologist's Monthly Magazine* **12**, 31–35.

Verrall, G.H. 1904. List of British Dolichopodidae, with tables and notes. *Entomologist's Monthly Magazine* **40**, 164–173, 194–199, 223–228, 241–245.

Verrall, G.H. 1912. Another hundred new British species of Diptera. *Entomologist's Monthly Magazine* **48**, 56–59.

Wolton, R. and Drake, C.M. 2015. *Rhaphium pectinatum* (Loew) (Diptera, Dolichopodidae) re-found in Britain. *Dipterists Digest (Second Series)* **22**, 127–130.

Woollatt, L.H. 1972. *Sciopus heteropygus* Par. (Dipt., Dolichopodidae) in Devon. *Entomologist's Monthly Magazine* **108**, 20.

Wootton, R.J. and Ennos, A.R. 1989. The implications of function on the origin and homologies of the dipterous wing. *Systematic Entomology* **14**, 507–520.

Yang, D., Zhu, Y., Wang, M. and Zhang, L. 2006. *World Catalog of Dolichopodidae (Insecta, Diptera)*. China Agricultural University Press, Beijing.

Zetterstedt, J.W. 1843. *Diptera Scandinaviae disposita et descripta*. Lundae **2**, 441–894.

Index

Index to species of Dolichopodidae covered in this guide. Names used by both d'Assis-Fonseca (1978) and Chandler 1998 but since regarded as synonyms or misidentifications are in italics (i.e. names embedded in the literature). Page numbers for species accounts are in **bold**, those for names appearing in keys are in *italics*, and the remaining entries are in plain text.